To our William

Other books by Adrian Bejan:

Entropy Generation through Heat and Fluid Flow, Wiley, 1982, 248 pages.
  (*Solutions Manual*, Wiley, 1984, 50 pages, available from the author.)
Convection Heat Transfer, Wiley, 1984, 477 pages.
  (*Solutions Manual*, Wiley, 1984, 218 pages, available from the publisher and
  from the author.)
Advanced Engineering Thermodynamics, Wiley, 1988, 759 pages.
  (*Solutions Manual*, Wiley, 1988, 153 pages, available from the publisher and
  from the author.)
Convection in Porous Media, with D. A. Nield, Springer-Verlag, 1992, 425 pages.

HEAT TRANSFER

Adrian Bejan

J.A. Jones Professor of Mechanical Engineering
Duke University

John Wiley & Sons, Inc.

New York   Singapore

Acquisitions Editor	Cliff Robichaud
Marketing Manager	Susan Elbe
Production Supervisor	Charlotte Hyland
Cover Designer	Madelyn Lesure
Manufacturing Manager	Andrea Price
Copy Editing Supervisor	Deborah Herbert
Illustration	Sigmund Malinowski

This book was set in Palatino by Digitype, Inc.

Library of Congress Cataloging in Publication Data:

Bejan, Adrian, 1948 –
    Heat transfer / Adrian Bejan.
        p.   cm.
    Includes bibliographical references and index.

    1. Heat—Transmission.
    QC320.B44   1993
    621.402'2—dc20   92-25535
                                    CIP

Printed in Singapore

10 9 8 7 6 5 4 3 2 1

PREFACE

The principal objective in writing this textbook for a first course in heat transfer is to present a rigorous and refreshing treatment of the subject, and to relate heat transfer to other disciplines, in particular, thermodynamics and fluid mechanics. This book is about "modern" heat transfer, that is, it is intended to reflect the changes currently taking place in the engineering profession and in engineering education.

Engineering Design

Engineering work and research are driven not only by human curiosity but also by the real needs of our society. For this reason, this book emphasizes design, or the synthesizing of two or more issues (i.e., engineering disciplines) into an answer with a practical meaning. The student is called on to recognize that there is more than one way in which to design a heat transfer apparatus — and to understand that knowing the worst design (the greatest pitfall, or trap) is often as important as knowing the optimal solution and all of the other options in between.

Design questions are drawn from many diverse areas. They include, for instance, insulations for walls with nonuniform temperature, insulations that must provide mechanical rigidity and support, the safe disposal of hot ash, the stabilization of superconductors, optimal packing of fibrous insulation, fins subjected to material constraint, and several applications covering the packaging and cooling of integrated electronic circuits. Each design question is presented in a fundamental way, so that (1) the answer can be used in more than one application, and (2) the work of arriving at the answer has the greater educational value.

Interdisciplinary Approach

This book teaches heat transfer as a discipline that works hand-in-hand with other disciplines in the evolution of a process, or in the successful performance of a device. Throughout its discussion emphasizes that real applications tend to be interdisciplinary, and that good work requires a solid foundation in all of the compartments of engineering education. The material presented in the chapter discussions and in the problems at the end of each chapter illustrates extensively the interfaces between heat transfer and other disciplines, such as metallurgy, structural mechanics, tribology, electrical engineering, superconductivity, cryogenics, and solar and geothermal energy production.

The emphasis of this book on the interdisciplinary character of real-life thermal engineering, however, does not detract from the teaching of heat transfer first as a self-standing *discipline*, that is, a field with its own scope, language, and rules. Interdisciplinary engineering can only be practiced by the engineer who is solidly grounded in the fundamentals of each discipline. To gloss over the fundamentals in the quick and currently popular pursuit of "interdisciplinary" topics would be to promote shallowness in the name of erudition and modernism.

Student Oriented

Early in my career, I discerned a need to present a course that treats each student as an intelligent human being, regardless of his or her background or aspirations. In this book, my goal is to reach simultaneously the curious, who are attuned to asking more questions, and the pragmatic, who prefer to get on with their calculations based on well-defined formulas.

Relationship to Fluid Mechanics, Thermodynamics, and History of Engineering

The book stresses the importance of fluid mechanics as a prerequisite for heat transfer analysis. To understand the flow configuration and its regime (laminar versus turbulent) is tremendously important not only in convection heat transfer calculations but also in *conjugate* situations in which convection is accompanied by conduction and radiation.

The role of thermodynamics in heat transfer is presented with great care. This begins with the calorimetric terminology, which is shared by both disciplines, and continues with the thermodynamic meaning of heat transfer engineering concepts, such as natural ("free") convection, heat exchanger pressure drop, and solar collector optimization.

Throughout the book I explain where each concept and formula comes from. In many instances, I provide the history of the concept and, sometimes, even the origin of the words that are used to name that concept.

Scale Analysis

This book makes a strong case for the use of simple and approximate analyses. Special emphasis is placed on order-of-magnitude calculations, or "scale analysis." I previously used this method in my graduate text on *Convection Heat Transfer* (Wiley, 1984). Since then, I have found that scale analysis can play an even greater role in a first course on heat transfer, where it can be used to solve conduction problems in addition to those of convection. Approximate methods, such as scale analysis and integral analysis, are emphasized because of their high rate of "return on investment." This has always been a central objective in engineering practice.

Problem-Solving Material

The emphasis on physical understanding and the freedom to choose between exact and approximate calculations is reflected also in the problem-solving material included in this course. The simplest problems are presented in the text as worked out *examples*, whereas the more challenging ones are proposed as *problems* at the end of each chapter. In later chapters, the *projects* that follow the problems require more time to solve. These projects usually have greater design content, and may be developed by the instructor into homework of longer duration.

It is always a good idea to begin solving a problem by executing a *good drawing*, and I have tried to demonstrate this to the student. Making a good drawing requires the student to explain to himself or herself the physics of the problem. Learning to take this preliminary step will be a benefit in the student's future career in engineering. The *Solutions Manual* that accompanies this book reflects the same care for problem solving (explanations, details, graphics) that is exhibited throughout the text. I wrote the solutions manual myself, as a "second book" in both content and appearance.

Looking Back at the Last Four Years

I am extremely fortunate to be and work at Duke University, which has provided me with a very friendly and supportive environment and, above all, freedom. It is a rare privilege to be allowed the freedom to think one's own thoughts.

Mary, my wife, was once again my chief collaborator and counselor. I owe every bit of my progress to her. Our family was also very supportive and a steady reminder of what is really important. I am deeply indebted to my parents-in-law, Terry Riordan and the late Bill Riordan, who actively supported my work, laboratory, and students.

Linda Hayes typed the entire manuscript while improving its English and organization. She also prepared the *Solutions Manual* on the word processor so that the manual, too, has a professional appearance. Her exquisite work is an important part of this finished project. I am very grateful to her for all of her help.

Several of my colleagues and students helped me during various stages of this project. I thank Katherine McKinney, Kathy Vickers, Dianne Himler, Eric Smith, Stoian Petrescu, Peter Jany, Dimos Poulikakos, John Georgiadis, Ira Katz, Michael Kazmierczak, P.V. Kadaba, Ab Hashemi, M. V. Vazquez, Ulrich Gösele, Sang W. Lee, Sung J. Kim, Jose Lage, Jong Lim, Alexandru Morega, and Alex Fowler. I treasure their support. I also appeal to the readers to communicate to me any errors that may persist in this final version.

Adrian Bejan
Durham, North Carolina
May 1992

ACKNOWLEDGMENTS

As I complete this book, it gives me great pleasure to look back and acknowledge those individuals who meant a lot to me, my work, and my morale during the past four years.

Dean A. E. Bergles, Rensselaer Polytechnic Institute

President H. K. H. Brodie, Duke University

Dean E. H. Dowell, Duke University

Professor A. Faghri, Wright State University

Professor R. M. Hochmuth, Duke University

Dr. J. H. Kim, Electric Power Research Institute

Professor J. H. Lienhard, University of Houston

Professor W. J. Minkowycz, University of Illinois, Chicago

Professor K. J. Renken, University of Wisconsin, Milwaukee

Cliff Robichaud, Engineering Editor, John Wiley & Sons

Chancellor C. L. Tien, University of California, Berkeley

Professor K. Vafai, Ohio State University

I am very grateful to the Lord Foundation of North Carolina for providing me with partial support during the summers of 1989 and 1990.

A. B.

ABOUT THE AUTHOR

Adrian Bejan received his B.S. (1972, Honors Course), M.S. (1972, Honors Course), and Ph.D. (1975) degrees in mechanical engineering, all from the Massachusetts Institute of Technology. From 1976 until 1978, he was a Fellow of the Miller Institute for Basic Research in Science, at the University of California, Berkeley. Before coming to Duke as a full professor in 1984, he taught at the University of Colorado, Boulder.

Professor Bejan is the author of three graduate-level textbooks, *Entropy Generation through Heat and Fluid Flow, Convection Heat Transfer,* and *Advanced Engineering Thermodynamics,* and is coauthor of a research monograph on *Convection in Porous Media.* In addition, he has written 190 technical articles on a variety of topics, including natural convection, combined heat and mass transfer, convection through porous media, transition to turbulence, second-law analysis and design, solar energy conversion, melting and solidification, condensation, cryogenics, applied superconductivity, and tribology.

Adrian Bejan is the 1990 recipient of the thermodynamics award of the American Society of Mechanical Engineers (ASME), the James Harry Potter Gold Medal. He also received the Gustus L. Larson Memorial Award of the ASME in 1988, and was elected Fellow of the ASME in 1987. He was awarded the J. A. Jones chair at Duke in 1989.

CONTENTS

LIST OF SYMBOLS

d	diameter (m)
D	diameter (m)
D	mass diffusivity, diffusion coefficient, or diffusion constant (m^2/s)
D_h	hydraulic diameter (m), eq. (6.28)
D_i	inner diameter (m)
D_o	outer diameter (m)
D_{12}	mass diffusivity of species 1 into species 2 (m^2/s)
e	specific internal energy (J/kg), eq. (5.12)
E	energy (J)
E	modulus of elasticity (N/m^2)
E	total hemispherical emissive power (W/m^2)
E_b	total hemispherical blackbody emissive power (W/m^2)
$E_{b,\lambda}$	monochromatic hemispherical blackbody emissive power ($W/m \cdot m^2$)
Ec	Eckert number, Appendix A
E_λ	monochromatic hemispherical emissive power ($W/m \cdot m^2$)
f	factor, Figs. 9.37–9.38
f	friction factor, eq. (6.24)
f_v	vortex shedding frequency (s^{-1}), eqs. (5.137) and (F.9)
f_v	fire pulsating frequency, eq. (F. 10)
F	correction factor, Figs. 9.16–9.20
F	force (N)
F	similarity streamfunction profile, eq. (7.45)
F_D	drag force (N)
F_n	normal force (N)
Fo	Fourier number (dimensionless time), eqs. (4.63)
F_r, F_θ, F_z	body forces per unit volume (N/m^3), Table 5.2
F_r, F_ϕ, F_θ	body forces per unit volume (N/m^3), Table 5.3
F_t	tangential force (N)
F_{12}	geometric view factor, eq. (10.33)
g	gravitational acceleration (m/s^2)
G	mass velocity ($kg/m^2 \cdot s$), eq. (9.58)
G	similarity vertical velocity profile, eq. (7.43)
G	total irradiation (W/m^2)
Gr_y	Grashof number based on temperature difference and height y, $Gr_y = g\beta y^3 \, \Delta T/v^2 = Ra_y/Pr$
Gz	Graetz number, eq. (6.53)
G_L	constant, Table 7.1
G_λ	monochromatic irradiation ($W/m^2 \cdot m$)
h	heat transfer coefficient for external flow ($W/m^2 \cdot K$), eq. (1.54)
h	heat transfer coefficient for internal flow ($W/m^2 \cdot K$), eq. (1.55)
h	Planck's constant (J·s), Appendix A
h	specific enthalpy (J/kg)
h_e	effective heat transfer coefficient ($W/m^2 \cdot K$), eq. (9.7)
h_f	specific enthalpy of saturated liquid (J/kg)
h_{fg}	latent heat of condensation (J/kg), $h_g - h_f$
h'_{fg}	augmented latent heat of condensation (J/kg), eqs. (8.10) and (8.17)
h''_{fg}	augmented latent heat of condensation (J/kg), eq. (8.41)
h_g	specific enthalpy of saturated vapor (J/kg)
h_m	local mass transfer coefficient (m/s), eq. (11.57)
h_s	specific enthalpy of saturated solid (J/kg)
h_{sf}	latent heat of melting, or of solidification (J/kg), $h_f - h_s$
h_x	local heat transfer coefficient ($W/m^2 \cdot K$) at position x
$\bar{h}_x$	average heat transfer coefficient ($W/m^2 \cdot K$) averaged over length x
$\bar{h}_D$	heat transfer coefficient ($W/m^2 \cdot K$) averaged over cylinder or sphere of diameter D
H	enthalpy (J)
H	enthalpy flowrate per unit length (W/m), eq. (8.5)
H	height (m)
H	Henry's constant (bar), Table 11.5
i	specific enthalpy (J/kg), eq. (5.16)
I_b	total intensity of blackbody radiation ($W/m^2 \cdot sr$)

$I_{b,\lambda}$	intensity of monochromatic blackbody radiation (W/m$^3 \cdot$sr)	L_0	Lorentz constant, 2.45×10^{-8} (V/K)^2
I_λ	intensity of monochromatic radiation (W/m$^3 \cdot$sr)	m	fin parameter (m^{-1}), eq. (2.92)
j_H	Colburn j_H factor, eq. (9.75)	m	integer
j_i	diffusion mass flux of species i (kg/m$^2 \cdot$s)	m	mass (kg)
$\hat{j}_i$	diffusion molar flux of species i (kmol/m$^2 \cdot$s)	$\dot{m}$	mass flowrate (kg/s)
J	electric current density (A/m^2)	$\dot{m}'$	mass flowrate per unit length (kg/s$\cdot$m)
J	radiosity (W/m^2)	$\dot{m}''$	mass flux (kg/m$^2 \cdot$s)
Ja	Jakob number, eq. (8.19)	$\dot{m}'''_i$	volumetric rate of species i generation (kg/m$^3 \cdot$s), eq. (11.2)
J_0	zeroth-order Bessel function of the first kind, Fig. 3.6 and Appendix E	M	dimensionless factor, eq. (6.84)
		M	molar mass of mixture (kg/kmol)
J_1	first-order Bessel function of the first kind, eq. (3.63)	M_i	molar mass of species i (kg/kmol)
k	Boltzmann's constant, Appendix A	n	direction normal to the boundary (m), Table 3.2
k	thermal conductivity (W/m$\cdot$K)	n	integer
k_{avg}	average thermal conductivity (W/m$\cdot$K), eq. (2.55)	N	number of moles in mixture sample (kmol)
k_e	thermal conductivity due to conduction electrons (W/m$\cdot$K)	N_i	number of moles of species i (kmol)
k_l	thermal conductivity due to lattice vibrations (W/m$\cdot$K)	NTU	number of heat transfer units, eq. (9.27)
k_i^{-1}	thermal resistivity due to impurity scattering (m$\cdot$K/W)	$\dot{N}$	molar flowrate (kmol/s)
k_p^{-1}	thermal resistivity due to phonon scattering (m$\cdot$K/W)	$\dot{N}'$	molar flowrate per unit length (kmol/m$\cdot$s)
k_s	sand roughness scale (mm), Fig. 6.14	Nu_x	Nusselt number based on the local heat transfer coefficient $h_x x/k$
K	constant coefficient	$\overline{Nu}_D$	overall Nusselt number based on the surface-averaged heat transfer coefficient $\overline{h}_D D/k$, where D is the diameter
K	permeability (cm^2), Appendix B		
K_c	contraction loss coefficient, Figs. 9.30 and 9.31	$\overline{Nu}_L^0$	constant, Table 7.1
K_e	enlargement loss coefficient, Figs. 9.30 and 9.31	$\overline{Nu}_x$	overall Nusselt number based on the x-averaged heat transfer coefficient $\overline{h}_x x/k$
l	equivalent length (m), eq. (7.84)	p	number of iterations, Chapter 3
l	length (m)	p	perimeter (m)
l	mixing length (m), eq. (5.109)	p	perimeter of contact with fluid (wetted perimeter) (m)
L	characteristic length (m), eq. (7.76)		
L	length (m)	P	dimensionless parameter, Figs. 9.16–9.20
$\mathcal{L}$	equivalent length (m), eqs. (5.140) and (7.85)	P	pressure (Pa or N/m^2)
		$\mathbf{P}$	mechanical power (W)
Le	Lewis number, Le = Sc/Pr = α/D, Table 11.7	Pe_D	Peclet number based on diameter, $U_\infty D/\alpha$, UD/α
L_c	corrected fin length (m), eq. (2.112)	Pe_x	Peclet number based on longitudinal length, $U_\infty x/\alpha$
L_e	equivalent length (m), Table 10.5	Pr	Prandtl number, Pr = ν/α

Pr_t	turbulent Prandtl number, $Pr_t = \varepsilon_M / \varepsilon_H$
P_i	partial pressure of component i (N/m^2)
q	heat transfer rate (W)
q_b	total heat transfer rate through the fin (W)
q_{tip}	heat transfer through the tip of the fin (W)
q'	heat transfer rate per unit length (W/m)
q''	heat flux (W/m^2)
$\dot{q}$	volumetric rate of internal heat generation (W/m^3)
$q''_{w,x}$	local wall heat flux (W/m^2)
$\bar{q}''_{w,x}$	x-averaged wall heat flux (W/m^2), eq. (5.80)
$q_{1 \to 2}$	one-way heat current (W) from 1 to 2
q_{1-2}	net heat current (W) from 1 to 2
Q	heat transfer (J)
Q'	heat transfer interaction per unit length (J/m)
Q''	heat transfer interaction per unit area (J/m^2)
r	radial coordinate (m), Figs. 1.8 and 1.9
r_i	inner radius (m)
r_o	outer radius (m)
$r_{o,c}$	critical outer radius (m), eqs. (2.44) and (2.47)
r_s	thermal resistance of the scale (m$^2 \cdot$K/W), eq. (9.5)
R	dimensionless parameter, Figs. 9.16–9.20
R	radius (m)
R	function of r only, Chapter 3
R	ideal gas constant (kJ/kg·K), Appendix D
$\bar{R}$	universal ideal gas constant, Appendix A
$Ra_{m,y}$	mass transfer Rayleigh number, eq. (11.102)
Ra_y	Rayleigh number based on temperature difference and height y, $Ra_y = g\beta y^3 \, \Delta T / \alpha v$
Ra_y^*	Rayleigh number based on heat flux and height y, $Ra_y^* = g\beta q''_w y^4 / \alpha v k$
Re	Reynolds number $V_{max}D_h/v$, eq. (9.70)
Re_D	Reynolds number based on diameter, $U_\infty D/v$, UD/v
Re_l	local Reynolds number, Appendix F
Re_x	Reynolds number based on longitudinal length, $U_\infty x/v$
Re_y	condensate film Reynolds number, $4\Gamma(y)/\mu_l$, eq. (8.22)
R_i	internal radiation resistance (m^{-2}), eq. (10.78)
R_r	radiation thermal resistance (m^{-2}), eq. (10.45)
R_t	thermal resistance (K/W), eq. (2.8)
s	empirical constant, Table 8.1
s_n	dimensionless characteristic values, Table 4.2
S	conduction shape factor, eq. (3.33) and the header of Table 3.3
S	entropy (J/K)
S	solubility coefficient (kmol/m$^3 \cdot$bar), Table 11.6
Sc	Schmidt number, $Sc = v/D$
Sh_x	local Sherwood number, eq. (11.65)
St	x-independent Stanton number $h/\rho c_p U$
Ste	Stefan number, eq. (4.119)
St_m	local mass transfer Stanton number, eq. (11.85)
St_x	local Stanton number $h_x/\rho c_p U_\infty$
t	thickness (m)
t	time (s)
t_c	transition time scale (s), eq. (4.9)
T	temperature (K or °C), eqs. (1.5) and Fig. 1.2
T_b	base temperature in fin analysis (K), Chapter 2
T_b	bulk, or mean temperature (K or °C)
T_c	center temperature (K), Chapter 4
T_i	initial temperature (K)
$T_{i,j}$	temperature of the control volume surrounding the node (i, j), Chapter 3
T_m	mean, or bulk temperature (K), eq. (6.33)
T_m	melting point temperature (K)

T_{sat}	saturation temperature (K)	x	Cartesian coordinate (m), Fig. 1.7
T_w	wall temperature (K or °C)	x_i	mole fraction
T_0	surface temperature (K), Chapter 4	x_{tr}	transition length (m)
T_0	reference temperature (K or °C)	X	flow entrance length (m), eqs. (6.4′) and (6.65)
T_∞	free-stream or reservoir temperature (K or °C)	X	function of x only, Chapter 3
u	specific internal energy (J/kg)	X_C	concentration entrance length (m), eq. (11.93)
u	velocity component in the x direction (m/s)	X_l	longitudinal pitch (m)
U	average longitudinal velocity (m/s)	X_t	transversal pitch (m)
U	internal energy (J)	X_T	thermal entrance length, eqs. (6.32) and (6.65)
U	mean velocity (m/s), eq. (6.1)	$X_l^{\bullet}$	dimensionless longitudinal pitch X_l/D
U	overall heat transfer coefficient (W/m²·K)	$X_t^{\bullet}$	dimensionless transversal pitch X_t/D
U_∞	free stream velocity (m/s)	X, Y, Z	body forces per unit volume (N/m³), Table 5.1
v	specific volume (m³/kg)	y	Cartesian coordinate (m), Fig. 1.7
v	velocity in the y direction (m/s)	Y	function of y only, Chapter 3
v_n	normal velocity (m/s)	Y_0	zeroth-order Bessel function of the second kind, Fig. 3.6
V	mean longitudinal velocity (m/s)		
V	volume (m³)	z	axial position in cylindrical coordinates (m), Fig. 1.8
$\mathcal{V}$	volume (m³), Chapter 9	z	Cartesian coordinate (m), Fig. 1.7
w	mechanical transfer rate, or power (W)	Z	function of z only, Chapter 3
W	width (m)		
W	work transfer (J)		
$\dot{W}$	work transfer rate, or power (W)		

Greek Letters

α	heat transfer area density (m²/m³), eq. (9.64)	δ	skin thickness, boundary layer thickness (m)
α	thermal diffusivity (m²/s), $\alpha = k/\rho c_P$	δ	velocity boundary layer thickness (m), eq. (5.25)
α	total absorptivity	$\delta^{\bullet}$	displacement thickness (m), eq. (5.58)
α	total hemispherical absorptivity	δ_s	thickness of shear layer (m), eq. (7.38)
α_0	temperature coefficient of electrical resistivity (°C⁻¹), Appendix B	δ_T	thermal boundary layer thickness (m), eq. (5.60)
α_λ	monocromatic hemispherical absorptivity	δ_{99}	velocity boundary layer thickness (m), eq. (5.57)
α_λ'	directional monochromatic absorptivity	ΔP	pressure drop (N/m²), eq. (6.27)
β	coefficient of volumetric thermal expansion (K⁻¹), eq. (5.18)	ΔT	temperature difference (K)
β_c	composition expansion coefficient (m³/kg), eq. (11.99)	ΔT_{lm}	log-mean temperature difference (K), eqs. (6.105) and (9.22)
Γ	condensate mass flowrate per unit length (kg/s·m), eq. (8.4)	$\Delta \epsilon$	correction term, Fig. 10.30
δ	film thickness (m), Chapter 8	ϵ	convergence criterion, eq. (3.99)
		ϵ	heat exchanger effectiveness, eq. (9.29)

ϵ overall surface efficiency, eq. (9.4)

ϵ total hemispherical emissivity

ϵ_f fin effectiveness, eq. (2.118)

ε_H thermal eddy diffusivity (m²/s), eq. (5.100)

ε_M momentum eddy diffusivity (m²/s), eq. (5.99)

ϵ_0 overall projected-surface effectiveness, eq. (2.79)

ϵ_λ monochromatic hemispherical emissivity

ϵ_λ' directional monochromatic emissivity

η fin efficiency, eq. (2.115)

η similarity variable

η_c compressor isentropic efficiency

η_p pump isentropic efficiency

θ angular coordinate (rad), Figs. 1.8 and 1.9

θ excess temperature (K), eq. (2.88)

θ momentum thickness (m), eq. (5.59)

θ similarity temperature profile, eq. (7.46)

θ thermal potential function (W/m), eq. (2.51)

θ_b excess temperature of fin base (K)

κ von Karman's constant, eq. (5.112)

κ_λ monochromatic extinction coefficient (m⁻¹)

λ characteristic value, Chapter 3

λ dimensionless parameter in the Stefan solution, eq. (4.118)

λ wavelength (m)

μ characteristic value, Chapter 3

μ viscosity (kg/s·m), eq. (5.10)

ν frequency (s⁻¹)

ν kinematic viscosity (m²/s), $\nu = \mu/\rho$

Π pressure drop number, $\Pi = \Delta P \cdot L^2/\mu\alpha$

ρ density (kg/m³)

ρ total reflectivity

ρ_e electrical resistivity (W·m/A²)

σ contraction ratio, eq. (9.53)

σ Stefan–Boltzmann constant, Appendix A

σ surface tension (N/m), Table 8.2

σ_{xx} normal stress (N/m²), eq. (5.8)

τ angle of enclosure inclination (rad), Fig. 7.23

τ total transmissivity

$\tau_{w,x}$ local wall shear stress (N/m²)

$\bar{\tau}_{w,x}$ x-averaged wall shear stress (N/m²), eq. (5.54)

τ_{xy} tangential stress (N/m²), eq. (5.8)

τ_λ monochromatic transmissivity

ϕ angle of wall inclination (rad), Fig. 7.10

ϕ angular coordinate (rad), Fig. 1.9

ϕ dimensionless temperature profile, eq. (6.45)

ϕ relative humidity, Appendix D

φ porosity, Appendix B

Φ viscous dissipation function (s⁻²), eq. (5.15)

Φ_i mass fraction ρ_i/ρ

χ correction factor, Figs. 9.37–9.38

ψ streamfunction (m²/s), eq. (5.45)

ω solid angle (sr)

ω specific humidity, or humidity ratio, Appendix D

Subscripts

$(\ )_a$ absorbed, Chapter 10

$(\ )_a$ air, Chapter 11

$(\ )_{acc}$ acceleration

$(\ )_{app}$ apparent

$(\ )_{avg}$ average

$(\ )_b$ base of the fin, Chapter 2

$(\ )_b$ black, Chapter 10

$(\ )_b$ bulk, mean

$(\ )_c$ carbon dioxide, Chapter 10

$(\ )_c$ centerline, center, midplane

$(\ )_c$ cold

$(\ )_c$ compressor

$(\ )_{eddy}$ eddy transport

$(\ )_f$ fluid

$(\ )_g$ gas, Chapter 10

$(\ )_h$ hot

$(\ )_i$ initial

$(\ )_i$ inner

( )$_i$ species i, Chapter 11

( )$_{in}$ initial

( )$_{in}$ inlet

( )$_l$ liquid

( )$_{max}$ maximum

( )$_{min}$ minimum

( )$_{mol}$ molecular diffusion

( )$_N$ north

( )$_E$ east

( )$_S$ south

( )$_W$ west

( )$_o$ outer

( )$_{opt}$ optimal

( )$_{out}$ outlet

( )$_p$ pump

( )$_r$ reflected, Chapter 10

( )$_{rad}$ radiation

( )$_{ref}$ reference

( )$_s$ shield, Chapter 10

( )$_s$ straight, Chapter 9

( )$_s$ surface, Chapter 10

( )$_{sat}$ saturation

( )$_v$ vapor

( )$_v$ water vapor

( )$_w$ wall

( )$_w$ water vapor, Chapter 10

( )$_0$ nozzle

( )$_0$ reference

( )$_0$ wall

( )$_\infty$ free stream

Superscripts

($^-$)   time averaged, or volume averaged

( )'   fluctuation, eq. (5.88)

( )$^+$ wall coordinates, eq. (5.117)

1

INTRODUCTION

1.1 FUNDAMENTAL CONCEPTS

The concepts of *heat transfer* and *temperature*, the key words of the discipline of heat transfer, are two of the most basic concepts of thermodynamics. In this section we review the thermodynamic meaning of these concepts so as to provide a rigorous foundation for the ideas that we explore in the present course. This review will show also what is different about the discipline of heat transfer, that is, how the territory covered by it differs from that of thermodynamics.

1.1.1 Heat Transfer

Heat transfer Q (joules) is one of the two types of energy interactions that are accounted for in any statement of the first law of thermodynamics. The other type of energy interaction is the work transfer W. For example, the first-law statement for an infinitesimal process executed by a *closed system* is

$$\delta Q - \delta W = dE \tag{1.1}$$

in which dE represents the change in the energy of the system. Energy is a thermodynamic property (a function of state, or a path-independent quantity), whereas the heat transfer and work transfer interactions are not. The per-unit-time equivalent of the first-law statement, eq., (1.1), is

$$q - w = \frac{dE}{dt} \tag{1.2}$$

where q (watts) is now the net *rate* of heat transfer experienced by the closed system. According to the usual engineering thermodynamics sign convention, the value of q is positive when the heat transfer enters the system, that is, when the system is being heated by the other system (e.g., the environment) with which it communicates thermally.

The work transfer rate w, on the other hand, is considered positive when it exits the system, that is, when the system does work on its environment. The work transfer can be of several origins, namely, mechanical, electrical, or magnetic. For example, the electrical work per unit time (i.e., the power) invested by an external battery in pushing a certain electric current through the system is

1

represented by a negative w value. This example will be encountered many times in this book, starting with the analysis of Fig. 1.4.

The terms q and w on the left side of the first law, eq. (1.2), are shorthand for the sums of all the heat transfer and work transfer interactions experienced by the closed system. In other words, an alternative way of writing eq. (1.2) is

$$\sum_i q_i - \sum_j w_j = \frac{dE}{dt} \tag{1.2'}$$

The first law of thermodynamics does not make any distinction between heat transfer and work transfer: to it, heat transfer and work transfer are both energy "interactions" (nonproperties) that must be distinguished from the energy change (property). The proper distinction between heat transfer and work transfer is made by the second law of thermodynamics, which for the same *closed system* and on a per-unit-time basis states that

$$\sum_i \frac{q_i}{T_i} \leq \frac{dS}{dt} \tag{1.3}$$

In this statement, S is the entropy inventory of the system (a property), and T_i is the Kelvin or Rankine temperature of the particular system boundary point i that is crossed by q_i. Each term of the type q_i/T_i represents the entropy transfer rate (watts/K) associated with the heat transfer rate q_i, as q_i passes through the boundary point of temperature T_i.

Comparing eqs. (1.2) and (1.3) we see the distinction between heat transfer and work transfer: only the former appears as a term in the statement of the second law. A work transfer interaction carries zero entropy, whereas superimposed on the heat transfer interaction q_i there is always an entropy transfer interaction q_i/T_i.

The preceding discussion of the concept of heat transfer is too abstract (and too recent) to have influenced the terminology of heat transfer engineering. The heat transfer definition that historically proved more appealing was one based on a phenomenological description of heat and work transfer interactions. Regarding the concept of work transfer, thermodynamics opted early for the mechanics definition: Work transfer is equal to the force experienced by the system times the displacement of the point of application of that force. The phenomenological description of heat transfer, on the other hand, amounts to the statement that heat transfer is the energy interaction driven by the temperature difference between the system and its environment.

1.1.2 Temperature

To review what is meant by temperature, consider two closed systems whose boundaries are such that both systems cannot experience work transfer (e.g., two arbitrary amounts of air sealed in rigid containers, where "arbitrary" means that the mass, volume, and pressure of each system are not specified, Fig. 1.1). If two systems of this kind are positioned sufficiently close to one another (in "contact"), it is generally observed that changes are induced in both systems. For the air-filled containers of the above example, these changes can be documented by recording the air pressure versus time. It is commonly observed that there exists a time interval beyond which the changes triggered by the proximity of the two systems cease. In general, the condition of the closed system is said to

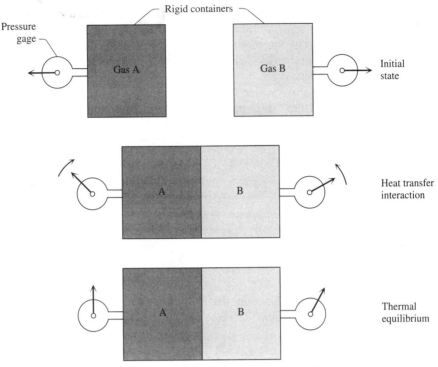

Figure 1.1 Heat transfer and the approach to thermal equilibrium, as a means of defining the concept of temperature.

be one of *equilibrium* when, after a sufficiently long period, changes cease to occur inside the system. And, since this particular closed system is incapable of experiencing work transfer interactions, the long-time condition illustrated in the above example is one of *thermal equilibrium.*

Let A and B be the closed systems that interact and reach thermal equilibrium in the preceding example. The same experiment can be repeated by using system A and a third system, C, which is also closed and unfit for work transfer. It is also a matter of common experience that if systems B and C are individually in thermal equilibrium with system A, then, when placed in direct communication, systems B and C do not undergo any changes as time passes. This second observation can be summarized as follows: If systems B and C are separately in thermal equilibrium with a third system, then they are in thermal equilibrium with each other. This statement is now recognized as the zeroth law of thermodynamics [1].*

Temperature T is the system property that determines whether the system is in thermal equilibrium with another system. Two systems, A and B, are in mutual thermal equilibrium when their temperatures are identical, that is, when

$$T_A = T_B \qquad (1.4)$$

The temperature of a system is measured by placing the system in thermal communication with a special system (a test system) called *thermometer.* The

---

*Numbers in square brackets indicate references at the end of each chapter.

thermometer has to be sufficiently smaller than the actual system so that the heat transfer interaction to the thermometer is negligible from the point of view of the system. The thermometer, on the other hand, is designed so that the same heat transfer interaction leads to measurable effects such as changes in volume or electrical resistance.

The temperature scales that are currently being employed are displayed in Fig. 1.2. Of critical importance in numerical engineering calculations are the relationships between the *absolute* temperatures recorded on the Kelvin and Rankine scales, and the popular Celsius and Fahrenheit scales:

$$T(^\circ C) = T(K) - 273.15$$

$$T(R) = \frac{9}{5}T(K) \tag{1.5}$$

$$T(^\circ F) = T(R) - 459.67$$

$$T(^\circ C) = \frac{5}{9}[T(^\circ F) - 32]$$

Figure 1.2 shows also that the Kelvin (or Celsius) degree is larger than the Rankine (or Fahrenheit) degree,

$$1 \text{ K, or } 1^\circ C = \frac{9}{5}(1R, \text{ or } 1^\circ F) \tag{1.6}$$

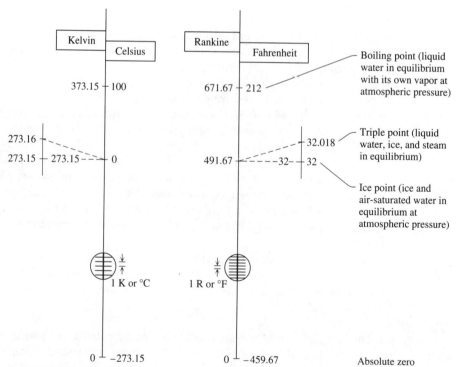

Figure 1.2 The relationships between the Kelvin, Celsius, Rankine, and Fahrenheit temperature scales [1].

1.1.3 Specific Heats

Two additional thermodynamic concepts that are essential in the present course are the specific heat at constant pressure c_P and the specific heat at constant volume c_v. In general, the two specific heats of a pure substance are different, the larger of the two being c_P.

The specific heat at constant pressure c_P can be measured during a process of isobaric heating in the apparatus shown in Fig. 1.3b. During such a process one can measure the heat input δQ, the temperature rise of the material sample dT, and the volume expansion dV. The definition of c_P is

$$c_P = \left(\frac{\delta Q}{m \, dT} \right)_P \qquad (1.7)$$

where m is the mass of the sample and, according to the first law of thermodynamics, $\delta Q = dU + \delta W$. Furthermore, since the work transfer during the same process is $\delta W = P \, dV$, and since P remains constant, the first law can be rewritten as $\delta Q = d(U + PV) = dH$, where dH is the enthalpy increase experienced by the sample. Finally, recalling the specific enthalpy notation h, the c_P definition (1.7) reduces to

$$c_P = \left(\frac{\partial h}{\partial T} \right)_P \qquad (1.8)$$

The specific heat at constant volume is defined based on the isochoric heating process described in Fig. 1.3a. Measuring both the heat transfer into the sample (δQ) and the temperature rise (dT), c_v is calculated by

$$c_v = \left(\frac{\delta Q}{m \, dT} \right)_v \qquad (1.9)$$

Invoking the first law for the heating process (and noting that the work transfer is zero), $\delta Q = dU$, we arrive at

$$c_v = \left(\frac{\partial u}{\partial T} \right)_v \qquad (1.10)$$

where u is the specific internal energy of the sample.

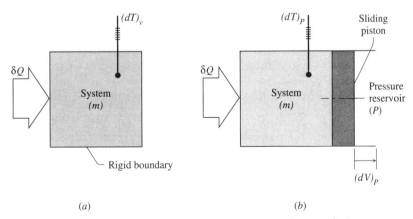

(a)   (b)

Figure 1.3 Two heating processes for measuring (a) the specific heat at constant volume and (b) the specific heat at constant pressure.

For any pure substance, c_P and c_v are, in general, functions of both temperature and pressure. Two special domains of thermodynamic behavior are particularly important in the present course, because in these domains the specific heats depend only on the temperature. First, in the *ideal gas* domain the functions $c_P(T)$ and $c_v(T)$ are related through Mayer's equation [1]

$$c_P(T) - c_v(T) = R \tag{1.11}$$

where R is the ideal gas constant of the substance that is being considered.

Second, in the *incompressible substance* domain the specific heats are equal to one another,

$$c_P(T) = c_v(T) = c(T) \tag{1.12}$$

and the better-known symbol for their common value is c. Examples of incompressible substances that occur frequently in this course are the solid bodies analyzed in the "conduction" chapters, and the moderately compressed (subcooled) liquids whose flow is studied in the "convection" chapters of the course.

1.2 THE OBJECTIVE OF HEAT TRANSFER

We can now look back at the phenomenological definition of heat transfer — the energy interaction triggered when $T_A \neq T_B$ — to recognize the objective of the discipline of heat transfer, which is to describe precisely the way in which the dissimilarity between T_A and T_B governs the magnitude of the heat transfer *rate* between the system (A) and its environment (B). The answer to this question is generally more complicated, because the heat transfer rate is influenced not only by the fact that T_A and T_B are different, but also by the physical configuration formed by the heat-exchanging entities,

$q =$ function (T_A, T_B, time, and, for both A and B, thermophysical properties,
size, geometric shape, relative movement, or flow) (1.13)

The definition of thermal equilibrium and temperature, Fig. 1.1, suggests that the heat transfer rate function, eq. (1.13), has the special property that

$$q = 0 \quad \text{when} \quad T_A = T_B \tag{1.14}$$

Simpler versions of the heat transfer function (1.13) will be adopted soon, by focusing on the three specific modes of heat transfer: conduction, convection, and radiation.

The discovery of the heat transfer rate function is only the starting point in the practice of heat transfer engineering. Even though the problems we encounter are extremely diverse, we can see better the importance of eq. (1.13) by focusing on three relatively large classes of engineering objectives:

1. **Thermal Insulations.** In this class the temperature extremes (T_A, T_B) experienced by the heat transfer medium are usually fixed. The key unknown is the heat transfer rate q, which is also named "heat loss" or, in cryogenics, "heat leak." If the problem is one of *thermal design*, then the ultimate objective is to minimize q, while keeping in mind certain economic and geometric constraints, such as the total cost of the heat transfer medium (the "insulation") or the total volume occupied by the system. The thermal design work consists of intelligently changing the constitu-

tion of the insulation (its size, material, shape, structure, flow pattern) so that q indeed decreases while T_A and T_B remain fixed.

2. **Heat Transfer Enhancement (Augmentation).** In heat exchanger design work, for example, the total heat transfer rate between the two streams is usually a prescribed quantity. Extremely desirable from a thermodynamic standpoint is the transfer of q across a minimum temperature difference $T_A - T_B$, because in this way the rate of entropy generation (or, proportionally, the destruction of useful mechanical power) is reduced (see Ref. [1], pp. 136–138). The key unknown in eq. (1.13) is the temperature difference: the design objective is to improve the thermal contact between the heat-exchanging entities, that is, to minimize the temperature difference $T_A - T_B$. This can be done by changing the flow patterns of the two streams and the shapes of the solid surfaces bathed by the fluid streams (by using fins, for example).

3. **Temperature Control.** There are several other applications in which the main concern is the overheating of the warm surface (T_A) that produces the heat transfer rate q. In a tightly packaged set of electronic circuits, q is generated by Joule heating, while the heat sink temperature T_B is provided by the ambient (e.g., a stream of atmospheric air). The temperature of the electrical conductors (T_A) cannot rise too much above the ambient temperature, because high temperatures threaten the error-free operation of the electrical circuitry. In this design problem, the heat transfer rate and flow configuration must vary in such a way that T_A does not exceed a certain ceiling value.

Examples of temperature control applications include also the cooling of nuclear reactor cores, and the cooling of the outer skin of space vehicles during reentry. In these examples the high temperature T_A must be kept below that critical domain where the mechanical strength characteristics of the hot surface deteriorate (e.g., the melting point).

1.3   CONDUCTION

1.3.1   The Fourier Law

Consider first the *solid* bar of material sketched in Fig. 1.4. The four long sides of this object are adiabatic (i.e., insulated perfectly) so that, if present, the transfer of heat will take place only in the longitudinal direction x. Any slice of thickness Δx in this bar constitutes a closed system in the thermodynamic sense used in the preceding section, because no mass is flowing through any of its six sides. Applied to the Δx slice system, then, the first law of thermodynamics, eq. (1.2), states that

$$q_x - q_{x+\Delta x} - w = \frac{\partial E}{\partial t} \tag{1.15}$$

in which the heat transfer rate q_x points in the positive x direction.

Equation (1.15) will be modified in three respects. First, the only relevant energy inventory component in the present stationary bar example is the internal energy; therefore, we write

$$E = (\rho A \, \Delta x) u \tag{1.16}$$

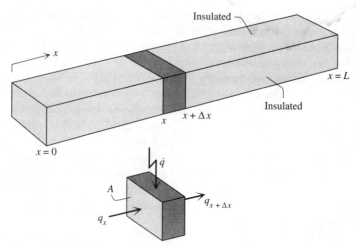

Figure 1.4 Unidirectional conduction through a solid body with internal heat generation.

where u is the specific internal energy, ρ is the density of the solid material, and $\rho A\,\Delta x$ is the mass of the system. We recall further that the internal energy change of an incompressible substance is proportional to its temperature change (Ref. [1], p. 126),

$$du = c\,dT \tag{1.17}$$

where the coefficient c is the lone specific heat of the solid, eq. (1.12). Combining eqs. (1.16) and (1.17), we can now replace the right side of eq. (1.15) with

$$\frac{\partial E}{\partial t} = \rho c A\,\Delta x \frac{\partial T}{\partial t} \tag{1.18}$$

In this statement it has been assumed that the temperature variation along the bar is sufficiently small so that the specific heat c may be treated as a constant (in general, c is a function of temperature). Note further that the partial derivative sign is used in this equation because the bar temperature is generally a function of both t and x.

The second modification of eq. (1.15) stems from the observation that in most cases the sign of the work transfer rate w is negative. For example, the rate of internal (volumetric) *heating caused by an electric current* that passes through the bar (i.e., the dissipation of electrical power) is represented by $-w$, not w. For this reason, we write

$$-w = (A\,\Delta x)\dot{q} \tag{1.19}$$

and recognize $\dot{q}$ (W/m^3) as the volumetric rate of internal heat generation in the solid. Equation (1.19) applies only to situations like the resistive heating example, where all the mechanical power is transformed into a volumetric heating effect. In general, the work transfer rates w_j of eq. (1.2′) can have additional effects such as the compression or stretching of the solid medium.

Third, guided by the observation that the functional form of the heat transfer rate must be such that q vanishes when the medium is isothermal, eq. (1.14), we *assume* that q_x is proportional to the local temperature difference in the x

direction, $q_x = C(T_x - T_{x + \Delta x})$. It is found experimentally that the proportionality factor C is itself proportional to $A/\Delta x$, that is, $C = kA/\Delta x$, where k is the *thermal conductivity* coefficient. In the limit $\Delta x \to 0$, the assumed expression for the local heat current in the x direction becomes

$$q_x = -kA\frac{\partial T}{\partial x} \tag{1.20}$$

This assumption is recognized in the field as the *Fourier law of heat conduction*,* or the Fourier law of thermal diffusion. It serves also as definition for the thermal conductivity coefficient k, which must be measured experimentally (see next section). Equation (1.20) and the thermal conductivity coefficient were first introduced by Biot [2].

The positive values that are exhibited by k (e.g., Fig. 1.6), coupled with the negative sign of the right side of eq. (1.20), are the fingerprint of the second law of thermodynamics, which demands that q_x must always flow in the direction of lower temperatures. Finally, the Fourier law can be used one more time in the rewriting of the second heat transfer current term of eq. (1.15):

$$q_{x + \Delta x} = q_x + \frac{\partial q_x}{\partial x}\Delta x \qquad \text{(Taylor series)}$$

$$= -A\left[k\frac{\partial T}{\partial x} + \frac{\partial}{\partial x}\left(k\frac{\partial T}{\partial x}\right)\Delta x\right] \tag{1.21}$$

The preceding modifications transform the first law, eq. (1.15), into a partial differential equation (the so-called "conduction" equation) for the temperature function $T(x,t)$:

$$\underbrace{\frac{\partial}{\partial x}\left(k\frac{\partial T}{\partial x}\right)}_{\substack{\text{Longitudinal} \\ \text{conduction}}} + \underbrace{\dot{q}}_{\substack{\text{Internal} \\ \text{heat} \\ \text{generation}}} = \underbrace{\rho c\frac{\partial T}{\partial t}}_{\substack{\text{Thermal} \\ \text{inertia}}} \tag{1.22}$$

This equation shows that *three* effects compete in the energy balance at any point inside the bar: the internal heat generation, the retarding effect of thermal inertia, and the *net* transfer of heat by longitudinal conduction. The word "net" must be stressed in connection with the longitudinal conduction term, because this term represents the *difference* between the conduction current that arrives at a given x and the conduction current that leaves. Note finally that the units of all the terms appearing in eq. (1.22) are W/m^3.

Thermal "inertia" means that a finite sample must be the recipient of net heat transfer if its temperature is to rise. When the net heat transfer input is fixed, the temperature rises faster when the group ρc is smaller. The group ρc represents the thermal inertia per unit of sample volume, or the specific heat capacity of the medium.

---

*Jean Baptiste Joseph Fourier (1768–1830) was a French mathematician and public servant (governor, prefect). He developed the general methodology (Fourier series, Chapters 3 and 4) for solving problems of heat conduction. Arguably the founder of the modern discipline of heat transfer, Fourier also had a great impact on the development of the field of applied mathematics.

If the variation of temperature along the bar is small enough so that the thermal conductivity may be treated as a constant, the one-dimensional conduction equation, eq. (1.22), assumes the simpler form

$$\frac{\partial^2 T}{\partial x^2} + \frac{\dot{q}}{k} = \frac{1}{\alpha}\frac{\partial T}{\partial t} \tag{1.23}$$

The new coefficient that is brought to light by this form is the *thermal diffusivity* of the conducting material,

$$\alpha = \frac{k}{\rho c} \tag{1.24}$$

In eq. (1.23) as well as in its three-dimensional equivalents, the thermal diffusivity α is to be treated as a constant, because ρ, c, and k have been assumed to be temperature independent. A compilation of thermal diffusivity data is presented in Appendixes B through D.

The conduction equation, eq. (1.22), was developed with reference to the solid material shaped as a bar in Fig. 1.4. It is not difficult to see that the same equation applies to the case where the bar is actually a column of *incompressible and motionless liquid or gas*.

A completely analogous derivation of the conduction equation can be constructed for a one-dimensional column of fluid that is not incompressible and whose pressure is constant and uniform. When the temperature variation along the bar is sufficiently small so that the local dilation of the fluid does not induce a significant movement (flow) in the x direction, the conduction equation that concludes this derivation is the same as eq. (1.22), except that c is now replaced by c_p. In this equation c_p is also treated as a constant. The reader can preview this equation by setting all the velocity components equal to zero in the more general, "convection" energy equation, eq. (5.20).

1.3.2 Thermal Conductivity

At least in principle, the thermal conductivity coefficient k can be determined by using the definition (1.20) and a conducting column apparatus such as that of Fig. 1.4, in which one can measure both the heat transfer rate and the temperature gradient. In the most general case, the measured k value will depend not only on the local thermodynamic state of the material sample (i.e., its temperature and pressure), but also on the orientation of the sample relative to the heat current q and on the point inside the sample where the k measurement is being performed (for example, the position x in Fig. 1.4). This general case is illustrated by means of Fig. 1.5a, in which the conducting material is anisotropic and nonhomogeneous.

The remainder of Fig. 1.5 shows the three special classes of materials for which the k function revealed by experiments is progressively simpler. In Fig. 1.5b the material is isotropic and nonhomogeneous. In this case the k value depends on the point where the measurement is made, but not on how the material is oriented relative to the heat current.

A homogeneous and anisotropic material is illustrated in Fig. 1.5c. Crystalline solids, meat, wood, and the windings of electrical machines can be described in this manner, provided the distance between adjacent fibers or laminae is much smaller than the size of the conducting sample. In such cases the measured k

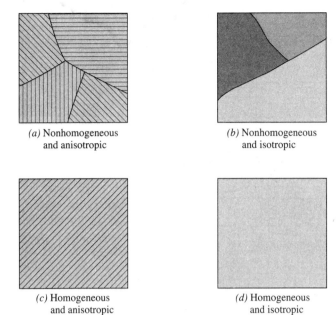

(a) Nonhomogeneous
and anisotropic

(b) Nonhomogeneous
and isotropic

(c) Homogeneous
and anisotropic

(d) Homogeneous
and isotropic

Figure 1.5 Classification of thermally conducting media in terms of their homogeneity and isotropy.

value depends on the orientation of the sample, and on the thermodynamic properties such as the temperature, but not on the point of measurement. Look at the first table of Appendix B and note the distinction made between the k values measured "perpendicular" and "parallel" to the fiber of the material (e.g., turkey breast, slate).

Most of the thermal conductivity data that are stored in the handbooks refer to homogeneous and isotropic materials (Fig. 1.5d). The types of relations that exist between k and T are illustrated in Fig. 1.6. Important to note is that the temperature can have a sizeable effect on k when the heat-conducting system occupies a wide range on the absolute temperature scale. The thermal conductivity differentiates between materials known as "good conductors" and "poor conductors": from the top to the bottom of Fig. 1.6, the k values decrease by six orders of magnitude.

Several theories aim at explaining the temperature trends exhibited in Fig. 1.6. In the case of *low pressure (ideal) gases*, the kinetic theory [6] argues that the energy transport that macroscopically is represented by the Fourier law, eq. (1.20), has its origin in the collisions between gas molecules. In the case of monatomic gases, for example, the thermal conductivity is expected to depend only on temperature,

$$k = k_0 \left(\frac{T}{T_0} \right)^n \tag{1.25}$$

in such a way that the theoretical exponent is $n = \frac{1}{2}$ (k_0 is the conductivity measured at the reference temperature T_0). In reality, the n exponent of a curve such as that of helium in Fig. 1.6 is somewhat larger than the theoretical value, $n \simeq 0.7$. The merit of the power law expression (1.25) is that, with an appropriate exponent n, it can be fitted to the conductivity data of any other gas, so as to

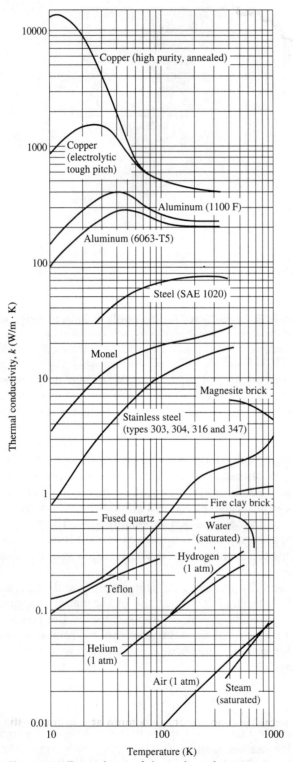

Figure 1.6 Dependence of thermal conductivity on temperature (the k data are from Refs. 3–5).

obtain an extremely compact $k(T)$ formula that holds over a wide temperature range.

In the same case of low-pressure monatomic gases, the gas density is proportional to P/T, while c_p is constant (see, for example, Ref. [1], p. 161). Consequently, the thermal diffusivity expression that corresponds to eq. (1.25) is

$$\alpha = \alpha_0 \left(\frac{T}{T_0}\right)^{n+1} \left(\frac{P}{P_0}\right)^{-1} \tag{1.26}$$

showing that α depends on both T and P and that it increases more rapidly than k as the temperature increases.

The thermal conductivity of *metallic solids* is attributed to the movement of conduction electrons (the "electron gas"), k_e, and the effect of lattice vibrations, k_l, the energy quanta of which are called phonons,

$$k = k_e + k_l \tag{1.27}$$

In metals, the electron movement plays the dominant role, such that, as a very good approximation,

$$k \cong k_e \tag{1.28}$$

The movement of the conduction electrons is impeded by scattering, which is the result of the interactions between electrons and phonons, as well as interactions between electrons and impurities and imperfections (e.g., fissures, boundaries) that may exist in the material. These two electron-scattering mechanisms are accounted for in an additive-type formula for the *thermal resistivity* (i.e., the inverse of thermal conductivity),

$$\frac{1}{k_e} = \frac{1}{k_p} + \frac{1}{k_i} \tag{1.29}$$

in which, according to the same electron conduction theory [7], the phonon-scattering resistivity (k_p^{-1}) and the impurity-scattering resistivity (k_i^{-1}) depend solely on the absolute temperature,

$$\frac{1}{k_p} = a_p T^2 \quad \text{and} \quad \frac{1}{k_i} = \frac{a_i}{T} \tag{1.30}$$

In these two relations the coefficients a_p and a_i are two characteristic constants of the metal.

Equations (1.30) show that at low temperatures the thermal resistivity is due primarily to impurity scattering, and that the effect of phonon scattering plays an important role at high temperatures. Putting eqs. (1.28), (1.29), and (1.30) together, one can see that the thermal conductivity of a metal obeys a temperature relation of the type

$$k = \frac{1}{a_p T^2 + (a_i/T)} \tag{1.31}$$

which, in general, shows a conductivity maximum at a characteristic absolute temperature:

$$k_{\max} = \frac{3}{2^{2/3}} a_p^{1/3} a_i^{2/3} \quad \text{at } T = \left(\frac{a_i}{2a_p}\right)^{1/3} \tag{1.32}$$

The k maximum shifts toward higher temperatures as the impurity-scattering effect a_i increases. These features are most evident in the shapes of the $k(T)$ curves of copper, Fig. 1.6, in which the impurity content increases as we shift from the "high purity" curve to the "electrolytic tough pitch" curve.

The maximum disappears entirely from the thermal conductivity curves of highly impure alloys such as the various types of stainless steel. In these cases the impurity-scattering resistivity overwhelms the phonon-scattering effect over a much wider temperature domain. Consequently, the simple formula

$$k \cong \frac{T}{a_i} \tag{1.33}$$

is a fairly good fit for the thermal conductivity data at temperatures below room temperature.

A more useful result contributed by the electron conduction theory stems from certain analogies that exist between the electron transport of energy (thermal diffusion) and the transport of electricity (electrical diffusion). This result is widely recognized as the *Wiedemann–Franz law* [7],

$$k\frac{\rho_e}{T} = L_0 \qquad \text{(constant)} \tag{1.34}$$

where ρ_e is the electrical resistivity of the metal.* The Lorentz constant, $L_0 = 2.45 \times 10^{-8}$ $(V/K)^2$, is a universal constant of all metals. Equation (1.34) holds particularly well in the two temperature extremes, that is, at cryogenic temperatures and at temperatures well above room temperature. At intermediate temperatures, the k value calculated with eq. (1.34) overestimates the measured thermal conductivity. The agreement between the Wiedemann–Franz law and k measurements at intermediate temperatures improves considerably as the impurity of the metal increases; therefore, eq. (1.34) is a good approximation for the thermal conductivity of an impure metal over the entire temperature range that characterizes most applications.

The practical value of eq. (1.34) is that it allows the engineer to estimate the thermal conductivity based on an electrical resistivity measurement, that is, based on a considerably simpler measurement. Furthermore, the Wiedemann–Franz law shows that when the thermal conductivity is nearly independent of temperature (e.g., copper above room temperature, Fig. 1.6), the electrical resistivity increases almost linearly with the temperature. This behavior has important consequences in the design of electrical cables that are thermally "stable," that is, safe against the threat of burn-up or thermal runaway instability (see Section 2.6.2).

The theoretical basis for the temperature effect on the thermal conductivity of dielectric solids and liquids can be explored further by reading Refs. [7–9]. One general conclusion to retain at this stage is that the thermal conductivity coefficient k is a *macroscopic*, or aggregate, fingerprint of phenomena that occur at the molecular level. In this sense, the thermal conductivity is similar to the properties (u, h, s, T, P, etc.) encountered in classical thermodynamics: the sample of conducting material whose k value is measured and later used in heat

---

*Recall that the electrical resistance of a conductor of length L and cross-sectional area A_c is $\rho_e L/A_c$, and that the units of ρ_e are $[W \cdot m/A^2]$.

transfer calculations contains a large enough number of molecules so that it (the sample) can be treated as a *continuum*.

The heat transfer theory that stands behind the conduction and convection parts of the present treatment is based, therefore, on the view that the conducting medium is a continuum. An entirely new brand of heat transfer theory will have to be developed for the microscale systems that contain no more than a few layers of molecules, as in the thin films envisioned for semiconductor integrated circuits.

1.3.3 Cartesian Coordinates

It is a simple matter to generalize the energy conservation statement, eq. (1.22), to the case where heat may be conducted in all directions. Switching, then, from Fig. 1.4 to Fig. 1.7, we recognize first the Cartesian system of coordinates (x,y,z) attached to the body that is the focus of this analysis. Note the good agreement—the good geometric fit—between the overall shape of the body and the coordinate system. The shape of the body almost "demands" that we use a Cartesian system, that is, not a cylindrical system (Fig. 1.8) or a spherical one (Fig. 1.9). The problem solver will soon learn to appreciate this observation, for the heat conduction analysis of a parallelepiped is the simplest (and, often, possible only) in a cartesian system.

The first law can be written immediately for the infinitesimal closed system of size $\Delta x\,\Delta y\,\Delta z$ enlarged on the right side of Fig. 1.7:

$$(q''_x - q''_{x+\Delta x})\Delta y\,\Delta z + (q''_y - q''_{y+\Delta y})\Delta x\,\Delta z +$$
$$(q''_z - q''_{z+\Delta z})\Delta x\,\Delta y + \dot{q}\Delta x\,\Delta y\,\Delta z = \rho c\Delta x\,\Delta y\,\Delta z\frac{\partial T}{\partial t} \quad (1.35)$$

In this statement each q'' term represents a *heat flux*, or heat transfer rate per unit area (W/m²). In eq. (1.20), for example, the corresponding heat flux would have been $q''_x = dq_x/dA$.

The Fourier law can be applied to each of the heat fluxes shown in Fig. 1.7:

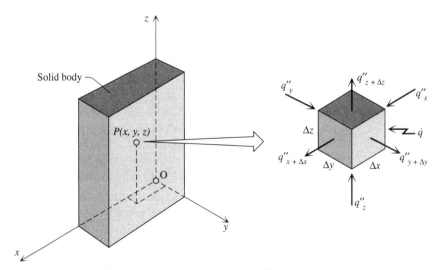

Figure 1.7 Three-dimensional Cartesian system of coordinates.

$$q_x'' = -k\frac{\partial T}{\partial x} \tag{1.36x}$$

$$q_y'' = -k\frac{\partial T}{\partial y} \tag{1.36y}$$

$$q_z'' = -k\frac{\partial T}{\partial z} \tag{1.36z}$$

and then substituted into eq. (1.35). Dividing the resulting equation by $\Delta x \, \Delta y \, \Delta z$ and invoking the limits $\Delta x \rightarrow 0$, $\Delta y \rightarrow 0$, and $\Delta z \rightarrow 0$, we obtain the general equation for energy conservation at a point in a Cartesian frame:

$$\frac{\partial}{\partial x}\left(k\frac{\partial T}{\partial x}\right) + \frac{\partial}{\partial y}\left(k\frac{\partial T}{\partial y}\right) + \frac{\partial}{\partial z}\left(k\frac{\partial T}{\partial z}\right) + \dot{q} = \rho c\frac{\partial T}{\partial t} \tag{1.37}$$

This form of the heat conduction equation is useful in applications where all of its terms have a meaning, namely, in a problem where (a) the temperature promises to be unsteady (time dependent), (b) the internal heating effect $\dot{q}$ is present, and (c) the thermal conductivity is not constant.

Several simpler versions of eq. (1.37) apply to the key heat conduction configurations covered by the present treatment. In the order of increasing simplicity, these configurations are as follows:

Unsteady, constant conductivity, with internal heat generation:

$$\frac{\partial^2 T}{\partial x^2} + \frac{\partial^2 T}{\partial y^2} + \frac{\partial^2 T}{\partial z^2} + \frac{\dot{q}}{k} = \frac{1}{\alpha}\frac{\partial T}{\partial t} \tag{1.38}$$

Unsteady, constant conductivity, without internal heat generation:

$$\frac{\partial^2 T}{\partial x^2} + \frac{\partial^2 T}{\partial y^2} + \frac{\partial^2 T}{\partial z^2} = \frac{1}{\alpha}\frac{\partial T}{\partial t} \tag{1.39}$$

Steady-state, constant conductivity, without internal heat generation:

$$\frac{\partial^2 T}{\partial x^2} + \frac{\partial^2 T}{\partial y^2} + \frac{\partial^2 T}{\partial z^2} = 0 \tag{1.40}$$

The sum of the first three terms appearing on the left side of eqs. (1.38) through (1.40) is often abbreviated as $\nabla^2 T$, where the $\nabla^2(\ \ )$ notation is shorthand for the Laplacian operator in three-dimensional Cartesian coordinates:

$$\nabla^2 = \frac{\partial^2}{\partial x^2} + \frac{\partial^2}{\partial y^2} + \frac{\partial^2}{\partial z^2} \tag{1.41}$$

1.3.4 Cylindrical Coordinates

The generalization of the heat conduction equation can be carried out in a completely analogous way in the cylindrical coordinate system (r, θ, z) defined in Fig. 1.8. Note again the good fit between the overall shape of the conducting body and the position of the coordinate system: the z axis coincides with the axis of symmetry of the body. The infinitesimal closed system that contains the

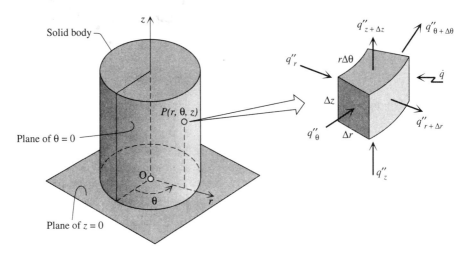

Figure 1.8 Cylindrical system of coordinates.

arbitrary point P now has the volume $\Delta r\, r\Delta\theta\, \Delta z$. In place of the Fourier law statements (1.36), we have the three heat fluxes in cylindrical coordinates:

$$q_r'' = -k\frac{\partial T}{\partial r} \qquad (1.42r)$$

$$q_\theta'' = -\frac{k}{r}\frac{\partial T}{\partial \theta} \qquad (1.42\,\theta)$$

$$q_z'' = -k\frac{\partial T}{\partial z} \qquad (1.42z)$$

In the end, the general energy conservation equation for unsteady conduction in a body with internal heat generation and variable conductivity becomes

$$\frac{1}{r}\frac{\partial}{\partial r}\left(kr\frac{\partial T}{\partial r}\right) + \frac{1}{r^2}\frac{\partial}{\partial \theta}\left(k\frac{\partial T}{\partial \theta}\right) + \frac{\partial}{\partial z}\left(k\frac{\partial T}{\partial z}\right) + \dot{q} = \rho c\frac{\partial T}{\partial t} \qquad (1.43)$$

The special forms of the conduction equation in cylindrical coordinates are as follows:

Unsteady, constant conductivity, with internal heat generation:

$$\frac{1}{r}\frac{\partial}{\partial r}\left(r\frac{\partial T}{\partial r}\right) + \frac{1}{r^2}\frac{\partial^2 T}{\partial \theta^2} + \frac{\partial^2 T}{\partial z^2} + \frac{\dot{q}}{k} = \frac{1}{\alpha}\frac{\partial T}{\partial t} \qquad (1.44)$$

Unsteady, constant conductivity, without internal heat generation:

$$\frac{1}{r}\frac{\partial}{\partial r}\left(r\frac{\partial T}{\partial r}\right) + \frac{1}{r^2}\frac{\partial^2 T}{\partial \theta^2} + \frac{\partial^2 T}{\partial z^2} = \frac{1}{\alpha}\frac{\partial T}{\partial t} \qquad (1.45)$$

Steady-state, constant conductivity, without internal heat generation:

$$\frac{1}{r}\frac{\partial}{\partial r}\left(r\frac{\partial T}{\partial r}\right) + \frac{1}{r^2}\frac{\partial^2 T}{\partial \theta^2} + \frac{\partial^2 T}{\partial z^2} = 0 \qquad (1.46)$$

Expanding the first term on the left side of eq. (1.46), we note the Laplacian operator in cylindrical coordinates:

$$\nabla^2 = \frac{\partial^2}{\partial r^2} + \frac{1}{r}\frac{\partial}{\partial r} + \frac{1}{r^2}\frac{\partial^2}{\partial \theta^2} + \frac{\partial^2}{\partial z^2} \tag{1.47}$$

The abbreviated version of eq. (1.46), for example, is $\nabla^2 T = 0$.

1.3.5   Spherical Coordinates

The conduction of heat in objects bounded by spherical surfaces is best handled in the spherical system of coordinates (r, ϕ, θ) defined in Fig. 1.9. The relationships between heat fluxes and temperature gradients at any point P in the conducting medium are

$$q_r'' = -k\frac{\partial T}{\partial r} \tag{1.48r}$$

$$q_\phi'' = -\frac{k}{r}\frac{\partial T}{\partial \phi} \tag{1.48ϕ}$$

$$q_\theta'' = -\frac{k}{r \sin \phi}\frac{\partial T}{\partial \theta} \tag{1.48θ}$$

The energy conservation equation for a body with time-dependent temperature, variable conductivity, and internal heat generation is

$$\frac{1}{r^2}\frac{\partial}{\partial r}\left(kr^2\frac{\partial T}{\partial r}\right) + \frac{1}{r^2 \sin \phi}\frac{\partial}{\partial \phi}\left(k \sin \phi \frac{\partial T}{\partial \phi}\right) +$$

$$\frac{1}{r^2 \sin^2 \phi}\frac{\partial}{\partial \theta}\left(k\frac{\partial T}{\partial \theta}\right) + \dot{q} = \rho c \frac{\partial T}{\partial t} \tag{1.49}$$

In the order of increasing simplicity, the special versions of the conduction equation in spherical coordinates are:

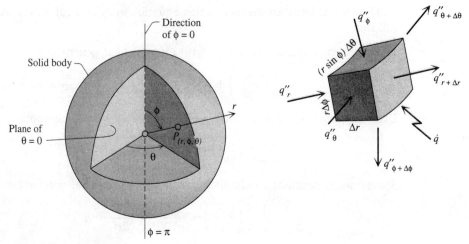

Figure 1.9 Spherical system of coordinates.

Unsteady, constant conductivity, with internal heat generation:

$$\frac{1}{r^2}\frac{\partial}{\partial r}\left(r^2\frac{\partial T}{\partial r}\right) + \frac{1}{r^2\sin\phi}\frac{\partial}{\partial\phi}\left(\sin\phi\frac{\partial T}{\partial\phi}\right) +$$

$$\frac{1}{r^2\sin^2\phi}\frac{\partial}{\partial\theta}\left(\frac{\partial T}{\partial\theta}\right) + \frac{\dot{q}}{k} = \frac{1}{\alpha}\frac{\partial T}{\partial t} \quad (1.50)$$

Unsteady, constant conductivity, without internal heat generation:

$$\frac{1}{r^2}\frac{\partial}{\partial r}\left(r^2\frac{\partial T}{\partial r}\right) + \frac{1}{r^2\sin\phi}\frac{\partial}{\partial\phi}\left(\sin\phi\frac{\partial T}{\partial\phi}\right) +$$

$$\frac{1}{r^2\sin^2\phi}\frac{\partial}{\partial\theta}\left(\frac{\partial T}{\partial\theta}\right) = \frac{1}{\alpha}\frac{\partial T}{\partial t} \quad (1.51)$$

Steady-state, constant conductivity, without internal heat generation:

$$\frac{1}{r^2}\frac{\partial}{\partial r}\left(r^2\frac{\partial T}{\partial r}\right) + \frac{1}{r^2\sin\phi}\frac{\partial}{\partial\phi}\left(\sin\phi\frac{\partial T}{\partial\phi}\right) + \frac{1}{r^2\sin^2\phi}\frac{\partial}{\partial\theta}\left(\frac{\partial T}{\partial\theta}\right) = 0 \quad (1.52)$$

The last equation can also be written as $\nabla^2 T = 0$, in which ∇^2 is the Laplacian operator in the spherical coordinates defined in Fig. 1.9:

$$\nabla^2 = \frac{\partial^2}{\partial r^2} + \frac{2}{r}\frac{\partial}{\partial r} + \frac{1}{r^2\sin\phi}\frac{\partial}{\partial\phi}\left(\sin\phi\frac{\partial}{\partial\phi}\right) + \frac{1}{r^2\sin^2\phi}\frac{\partial^2}{\partial\theta^2} \quad (1.53)$$

1.3.6 Initial and Boundary Conditions

The specific conduction problems that are addressed beginning with Chapter 2 will show that the heat conduction equation is not sufficient for determining the temperature distribution through the conducting body. In problems where the temperature field is time dependent, in addition to recognizing the proper form of the conduction equation, the problem solver must also be able to recognize the proper *initial condition* and the proper *boundary conditions* that characterize the given heat transfer configuration.

For example, if the problem is to determine the temperature history inside a metallic bar that is quenched by sudden immersion in a pool of oil, the initial condition is the statement that in the beginning (i.e., at the time $t = 0$) the temperature is uniform throughout the bar, and it is equal to a known high temperature. In the same problem the boundary conditions are statements that are made with regard to the temperature, the heat flux, or the relationship between temperature and heat flux at each point around the periphery of the conducting medium. The writing of boundary conditions becomes clearer after we learn the concept of heat transfer coefficient (see next section), and after we attack analytically a few specific problems.

In the simpler class of steady-state conduction problems, the sought temperature distribution depends only on the position inside the conducting body. The conduction equation does not contain the $\partial T/\partial t$ term [e.g., eq. (1.40)]; and since this equation will not be integrated in time, the specification of an initial condition is not required. Boundary conditions, however, are required over the entire surface that defines (surrounds) the conducting medium. The correct specification of these conditions is one of the objectives of Chapters 2 and 3.

Example 1.1

Thermal Conductivity, Conduction in a Slab

The figure shows an arrangement designed for measuring the thermal conductivity of polystyrene. Two polystyrene slabs sandwich between them a thin metallic plate heater. This arrangement is itself sandwiched between two copper plates. It is also sealed along the top, bottom, and the sides to prevent the flow of water into the polystyrene.

The electric plate heater is powered by a battery, and, acting as a resistance in the circuit, generates 0.1 W over each square centimeter of its extent. The temperature of the outer surfaces of the polystyrene layers is held at $T_s = 0\,°C$ by contact with an equilibrium mixture of ice and water. Note that the temperature drop across each copper plate is negligible. Thermocouples imbedded in the metallic plate heater indicate that the plate temperature T_p is uniform and equal to $62.5\,°C$. Calculate the polystyrene thermal conductivity k measured in this experiment.

Solution.   The geometry is one-dimensional, in the sense that the heating generated by the metallic strip can only escape laterally, that is, unidirectionally. Since there are two sides to the metallic strip, the heat transfer rate per unit area (heat flux) conducted through one of the polystyrene layers is $q_x'' = 0.05\ \text{W/cm}^2 = 500\ \text{W/m}^2$. Note that this heat flux is oriented in the positive x direction.

Since q_x'' is known, we can calculate the thermal conductivity k if we know the temperature gradient that drives q_x'' away from one face of the plate heater:

$$q_x'' = -k\left(\frac{dT}{dx}\right)_{x=0} \tag{1}$$

In other words,

$$k = \frac{q_x''}{-(dT/dx)_{x=0}} \tag{2}$$

The problem reduces to finding the temperature gradient $(dT/dx)_{x=0}$. For this we solve the conduction equation, eq. (1.23), in the polystyrene half-slab $0 \leq x \leq L$. Because in the present case the temperature is steady and $\dot{q} = 0$ in the polystyrene, eq. (1.23) reduces to

$$\frac{d^2T}{dx^2} = 0 \tag{3}$$

The general solution is

$$T = c_1 x + c_2 \tag{4}$$

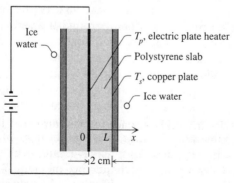

Figure E1.1

in which the constants of integration (c_1, c_2) are pinpointed by the temperature boundary conditions

$$T = T_p \quad \text{at} \quad x = 0 \tag{5}$$

$$T = T_s \quad \text{at} \quad x = L \tag{6}$$

The important observation at this stage is that the temperature distribution across the polystyrene is linear, eq. (4), which means that the temperature gradient dT/dx is the same at every x:

$$\frac{dT}{dx} = \left(\frac{dT}{dx}\right)_{x=0} = c_1 \tag{7}$$

In conclusion, we have to calculate only c_1, not c_1 and c_2. For this we use eqs. (5) and (6),

$$T_p = c_2 \tag{8}$$

$$T_s = c_1 L + c_2 \tag{9}$$

and obtain

$$c_1 = \frac{T_s - T_p}{L} \tag{10}$$

or

$$\frac{dT}{dx} = \frac{0°C - 62.5°C}{0.02 \text{ m}} = -3125 \text{ °C/m} \tag{11}$$

By substituting this temperature gradient and $q_x'' = 500$ W/m^2 in eq. (2), we arrive at the wanted result:

$$k = \frac{q_x''}{-dT/dx} = 500 \frac{W}{m^2} \frac{m}{3125°C}$$

$$= 0.16 \text{ W/m·K} \tag{12}$$

1.4   CONVECTION

Convective heat transfer, or, simply, convection, is the heat transfer process that is executed by the flow of a fluid. The fluid acts as carrier or conveyor belt for the energy that it draws from (or delivers to) a solid wall. The two most common entities that engage in convective-type heat transfer interactions — the entities that in eq. (1.13) were labeled A and B — are the solid wall and the fluid stream with which the wall comes in direct contact. Convection is that special heat transfer mechanism in which the characteristics of the flow (e.g., velocity distribution, turbulence) affect greatly the heat transfer rate between the wall and the stream.

The geometric relationship between wall and stream is illustrated by the two convective configurations shown in Figs. 1.10 and 1.11. In the *external flow* configuration, Fig. 1.10, the flow engulfs the body with which it interacts thermally. The transition from the body surface temperature (T_w) to the fluid temperature far into the stream (T_∞) is made by a special region of the flow (called "boundary layer" in Section 5.3) that coats the solid wall. Across the same flow region, the velocity of the fluid decreases from its free stream value to zero at the wall.

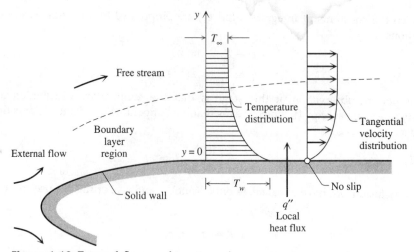

Figure 1.10 External flow configuration of convective heat transfer.

The key question is how the temperature extremes (T_w, T_∞) and the flow dictate the magnitude of the heat transfer rate between the body and the stream. The same question can be formulated on a per-unit-area basis, namely, "What is the relationship between the local heat flux q'' and the temperature extremes and the flow?" For if we know the local heat flux q'', we can presumably integrate it over the entire surface of the body to determine the total body-to-stream heat transfer rate q.

The traditional approach consists of defining first the external-flow *heat transfer coefficient h* by writing*

$$q'' = h(T_w - T_\infty) \tag{1.54}$$

so that [cf. eq. (1.14)] the heat flux vanishes when the stream is in thermal equilibrium with the wall. The question formulated in the preceding paragraph reduces, then, to finding the heat transfer coefficient and, in particular, the manner in which h is influenced by the characteristics of the flow. For this, the analyst must know not only the temperature distribution in the near-wall fluid, but also the flow (velocity) distribution. Out of necessity, then, convection analyses are based not only on fluid flow generalizations of the energy conservation statements of Section 1.3, but also on statements that account for the conservation of mass and momentum in the flow field. This more extensive mathematical apparatus is introduced in Section 5.2.

When the heat transfer surface surrounds and guides the stream, the convective configuration is said to be one of *internal flow*. Figure 1.11 shows the main features of the temperature and velocity distributions in the vicinity of the wall. The key heat transfer question can be reduced again to the problem of finding a "heat transfer coefficient"; however, this time there is no free-stream temperature (T_∞) that can be identified in the stream. The internal flow alternative to defining the heat transfer coefficient is

---

*The heat transfer coefficient symbol h should not be confused with that of specific enthalpy.

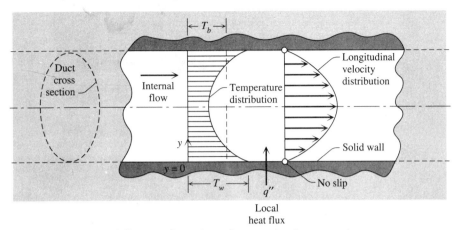

Figure 1.11 Internal flow configuration of convective heat transfer.

$$q'' = h(T_w - T_b) \tag{1.55}$$

in which the *bulk temperature* T_b is a certain cross section-averaged temperature of the stream, which is also called *mean temperature*. The proper definition of this average will be discussed in connection with eq. (6.33). The important thought to retain at this stage is that the definition of heat transfer coefficient in internal flow, eq. (1.55), differs from that used in external flow, eq. (1.54).

The path to evaluating h in either external or internal flow can be seen by looking at the fluid side of the solid wall. Figures 1.10 and 1.11 show that the infinitesimally thin fluid layer situated at $y = 0^+$ is stuck to the wall; in fluid mechanics, this feature is recognized as the *no slip condition* at a solid boundary. Since the $y = 0^+$ fluid layer is not moving, the wall heat flux that traverses it (q'') is ruled by pure conduction; the Fourier law, eq. (1.36y), applies to this layer as well:

$$q'' = -k\left(\frac{\partial T}{\partial y}\right)_{y=0} \tag{1.56}$$

where k is the thermal conductivity of the fluid and T is the temperature distribution in the fluid. Combining eqs. (1.54), (1.55), and (1.56) we discover that

$$h = -\frac{k}{T_w - T_\infty}\left(\frac{\partial T}{\partial y}\right)_{y=0^+} \quad \text{(external flow)} \tag{1.57}$$

$$h = -\frac{k}{T_w - T_b}\left(\frac{\partial T}{\partial y}\right)_{y=0^+} \quad \text{(internal flow)} \tag{1.58}$$

In conclusion, to calculate h we must first determine the temperature distribution *in the fluid situated close to the wall*. The temperature distribution, in turn, depends on the velocity distribution. Therefore, a prerequisite for sound calculations of convective heat transfer is an understanding of the flow that makes contact with the wall, that is, an understanding of the *fluid mechanics* of the convection configuration.

Figure 1.12 shows the dramatic effect that the type of fluid and the flow regime have on the order of magnitude of the heat transfer coefficient. These

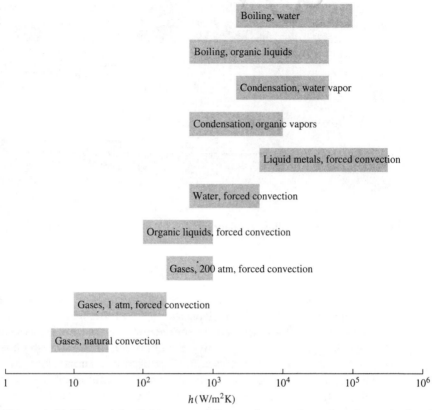

Figure 1.12 Effect of the fluid type and flow regime on the order of magnitude of the heat transfer coefficient.

convection regimes are analyzed systematically beginning with Chapter 5. First will be the external flow configurations, in which the emphasis will fall naturally on the flow region situated next to the wall — *the boundary layer.* The heat transfer to internal flows, or *duct flows,* will be discussed in Chapter 6. The flows of Chapters 5 and 6 together represent the larger class of *forced convection,* because in each of these configurations the fluid is forced (by a fan or a pump) to flow past the solid walls of interest. Flows that happen "naturally" are those driven by the buoyancy effect felt by the relatively warmer regions of a flow. These flows are examples of *natural, or free, convection* and form the subject of Chapter 7.

The forced and natural convection topics of Chapters 5 through 7 refer to *single-phase* flows, in which the fluid does not experience a change of phase. For example, the near-wall liquid remains as liquid (does not turn into vapor) while it is heated by the wall. Topics of convection with change of phase are the *condensation* and *boiling* problems that are analyzed in Chapter 8. The multitude of convection configurations treated in Chapters 5 through 8 will find application in Chapter 9, in which we study the *heat exchanger* as a device with the function of promoting heat transfer between two or more fluids at different temperatures.

Historically, eq. (1.54) was first written down by Fourier [10], who introduced in this way the concept of heat transfer coefficient ("external conductivity," in his terminology), to which he gave the symbol h. Fourier emphasized also the fundamental difference between h and the "proper" thermal conductivity k. More than 100 years earlier, Newton [11] had published an essay in which he reported that the rate of temperature decrease (dT/dt) of a body immersed in a fluid is at all times proportional to the body–fluid temperature difference ($T - T_\infty$). This is why beginning with Fourier's contemporaries (e.g., Peclet [12]), eq. (1.54) acquired the name "Newton's law of cooling." Worth noting, however, is that today Newton's statement can be written [13] as $dT/dt = b(T - T_\infty)$, in which the b coefficient (assumed constant) accounts for the ratio h/c, that is, the heat transfer coefficient divided by the specific heat. The exact form of the b coefficient will be derived in eq. (4.15). The concepts of heat transfer coefficient and specific heat were unknown in Newton's time.

Convection currents in liquids were first discovered experimentally by Count Rumford [14, 15], who also visualized the flow by suspending neutrally buoyant particles in the liquid. The heat transfer effect of such currents was named "convection" by Prout in 1834 [16].

Example 1.2

External Flow

An amount of fine coal powder has been stockpiled as the 2-m-deep layer shown in the figure. Due to a reaction between coal particles and atmospheric gases, the layer generates heat volumetrically at the uniform rate $\dot{q} = 20$ W/m³. The coal temperature is in the steady state. The ambient air temperature is $T_\infty = 25°C$, and the heat transfer coefficient at the upper surface of the coal stockpile is $h = 5$ W/m²·K. Calculate the temperature of the upper surface, T_w.

Solution. Expressed per unit of surface area covered by the coal pile, the heat generated by the entire layer of coal is

$$q'' = \dot{q}H = 20\,\frac{W}{m^3}2\text{ m} = 40\text{ W/m}^2 \tag{1}$$

This heat transfer rate must escape through the upper surface of temperature T_w,

$$q'' = h(T_w - T_\infty) \tag{2}$$

because the lower surface has been modeled as insulated (zero heat flux). Equation (2) delivers the unknown temperature of the upper surface of the coal stockpile,

$$T_w = T_\infty + \frac{q''}{h} = 25°C + 40\,\frac{W}{m^2}\frac{m^2\cdot K}{5\,W}$$

$$= 25°C + 8°C = 33°C$$

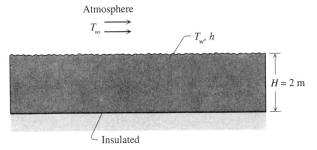

Figure E1.2

Example 1.3

Conduction in Series with Convection

Attached to a flat wall of temperature T_b is a plate of thickness b, length L, and width W (see figure). This plate is made of a highly conductive metal and, as a consequence, its temperature is practically uniform. The plate is bathed on all its exposed sides by a fluid of temperature T_∞. The heat transfer coefficient has the same value h on all the surfaces wetted by the fluid.

The plate described until now is attached to the wall by means of a layer of glue of thickness t and thermal conductivity k. Derive an expression for the heat transfer rate that passes from T_b to T_∞ through the glue–plate system. Under what conditions is this heat transfer rate controlled (impeded the most) by the layer of glue?

Solution. The temperature T of the isothermal plate is unknown. The heat transfer rate q is first conducted across the layer of glue:

$$q = k\frac{bW}{t}(T_b - T) \tag{1}$$

Later, the same q is convected away from the lateral surfaces of the plate:

$$q = h(2WL + 2Lb + Wb)(T - T_\infty) \tag{2}$$

By eliminating q between eqs. (1) and (2) we obtain an expression for the plate temperature:

$$T = \frac{1}{1+B}T_b + \frac{B}{1+B}T_\infty \tag{3}$$

where B denotes the dimensionless group

$$B = \frac{ht}{k}\left(2\frac{L}{b} + 2\frac{L}{W} + 1\right) \tag{4}$$

Finally, by substituting the T expression (3) in eq. (1) or eq. (2), we obtain the wanted relation for the heat transfer rate:

$$q = k\frac{bW}{t}(T_b - T_\infty)\frac{B}{1+B} \tag{5}$$

This expression shows that when h, t, k, L, W, and b are such that the dimensionless B number is considerably greater than 1, the heat transfer rate is "controlled" by the conduction across the glue layer:

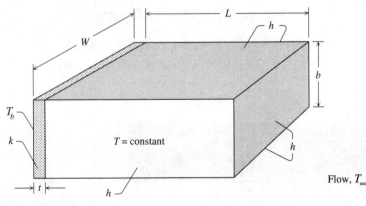

Figure E1.3

$$q \cong k\frac{bW}{t}(T_b - T_\infty) \qquad (B \gg 1) \tag{6}$$

Equation (3) indicates that in the same limit the plate temperature approaches the fluid temperature T_∞. This is why eqs. (6) and (1) look the same.

1.5   RADIATION

Next to conduction and convection, a third distinct mechanism for heat transfer is that of thermal radiation. This mechanism is covered in detail in Chapter 10. In this introductory segment we focus only on the most essential aspect that distinguishes radiation from both conduction and convection.

Consider the evacuated inner space that surrounds the spherical container B shown in Fig. 1.13. The "vacuum jacket" is an insulation feature employed in the design of many devices, for example, storage vessels for cryogenic liquids, nitrogen and helium in particular. In most applications, the evacuated space serves adequately as an insulation, precisely because the material (the continuum) that would have acted as conduit for conduction and convection is absent. Had the annular gap been filled with a solid, even with one of low conductivity, it would have been penetrated radially by conduction heat fluxes similar to the q_r'' arrow shown in Fig. 1.9. Had a fluid been present in the gap (e.g., air, or the vapor of the paint that coats the two facing surfaces), buoyancy would have driven a closed-loop flow (a "circulation") through the gap. Riding on this flow would have been a convective heat current of the kind that will be analyzed in the natural convection segments of this book.

Experience shows that even when the annular gap is completely evacuated, a finite heat transfer current still leaves the outer (warm) shell and lands on the inner (cold) shell. This net heat transfer rate is due to thermal radiation, that is, to the interaction of two bodies that can affect one another *from a distance.*

The thermal radiation effect can be explained on the basis of electromagnetic wave theory. The net radiation heat transfer rate (from A to B in Fig. 1.13) represents the difference between the energy stream with which the warm surface "bombards" the cold surface and the weaker energy stream emanating

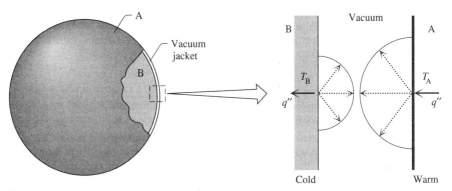

Figure 1.13 Thermal radiation across an evacuated space.

from the cold surface. Under certain conditions (see Section 10.4.4), the net radiation current q_{A-B} is described by

$$q_{A-B} = \beta A(T_A^4 - T_B^4) \tag{1.59}$$

in which A is the area of either shell, and T_A and T_B are the absolute (Kelvin, or Rankine) temperatures of the two surfaces.

The proportionality factor labeled β depends in general on T_A and T_B, the relative size of the two surfaces, on their degree of smoothness and cleanliness, and on the materials out of which they are made. Special forms of this proportionality factor can be seen by examining eqs. (10.84) through (10.86); however, the derivation and explanation of these forms is better postponed until Chapter 10. Until then, it is worth noting that eq. (1.59) is another special case of the heat transfer rate function presented in eqs. (1.13) and (1.14).

In conclusion, the feature that distinguishes the radiation heat transfer mode from both conduction and convection is that radiation can proceed even in the absence of a continuous medium. Of course, radiation heat transfer can also occur when a sufficiently transparent continuous medium (e.g., air) separates the heat-exchanging surfaces, as in the case of most of the domestic and solar heating arrangements that surround us. Indeed, in many engineering problems the challenge is to evaluate the total heat transfer rate when the radiation, convection, and conduction mechanisms participate simultaneously.

REFERENCES

1. A. Bejan, *Advanced Engineering Thermodynamics*, Wiley, New York, 1988.
2. J. B. Biot, *Traité de Physique*, Vol. 4, Paris, 1816, p. 669.
3. R. B. Scott, *Cryogenic Engineering*, Van Nostrand, Princeton, NJ, 1959, pp. 344–345.
4. W. M. Rohsenow and H. Y. Choi, *Heat, Mass and Momentum Transfer*, Prentice-Hall, Englewood Cliffs, NJ, 1961, p. 518.
5. A. Bejan, *Convection Heat Transfer*, Wiley, New York, 1984, pp. 462–465.
6. N. V. Tsederberg, *Thermal Conductivity of Gases and Liquids*, MIT Press, Cambridge, MA, 1965, Chapter II.
7. R. G. Scurlock, *Low Temperature Behavior of Solids: An Introduction*, Dover, New York, 1966, pp. 51–64.
8. P. G. Klemens, Theory of the thermal conductivity of solids, in R. P. Tye, Ed., *Thermal Conductivity*, Vol. 1, Academic Press, London, 1969, pp. 1–68.
9. E. McLaughlin, Theory of the thermal conductivity of fluids, in R. P. Tye, Ed., *Thermal Conductivity*, Vol. 2, Academic Press, London, 1969, pp. 1–64.
10. J. Fourier, *Analytical Theory of Heat*, translated, with notes, by A. Freeman, G. E. Stechert & Co., New York, 1878.
11. I. Newton, Scala Graduum Caloris, Calorum Descriptiones & Signa, *Phil. Trans. Roy. Soc. London*, Vol. 8, 1701, pp. 824–829; translated from Latin in *Phil. Trans. Roy. Soc. London*, Abridged, Vol. IV (1694–1702), 1809, pp. 572–575.
12. E. Péclet, *Traité de la chaleur considérée dans ses applications*, 3rd edition, Victor Masson, Paris, 1860, Vol. 1, p. 364.
13. A. E. Bergles, Enhancement of convective heat transfer: Newton's legacy pursued, in E. T. Layton, Jr., and J. H. Lienhard, Eds., *History of Heat Transfer*, American Society of Mechanical Engineers, New York, 1988, pp. 53–64.

14. Count of Rumford, *Essays, Political, Economical and Philosophical*, Vol. II, T. Cadell and W. Davis, London, 1798, Essay VII.

15. S. C. Brown, The discovery of convection currents by Benjamin Thompson, Count of Rumford, *Am. J. Physics*, Vol. 15, 1947, pp. 273–274.

16. W. Prout, *Bridgewater Treatises*, Vol. 8, Carey, Lea & Blanchard, Philadelphia, 1834, p. 65.

PROBLEMS

Conduction

1.1 The one-dimensional conductor shown in Fig. 1.4 stretches from $x = 0$ to $x = L$, and the respective end temperatures are maintained at T_0 and T_L. The four long sides of the conductor are insulated. The temperature distribution along the conductor is steady, that is, independent of time.

Let q_0 represent the heat current that enters the conductor through the $x = 0$ cross section. Assume further that the value of q_0 is positive, in other words, that the heat is conducted in the positive x direction. Invoke the second law to prove that q_0 flows toward lower temperatures, for example, by showing that T_L cannot be greater than T_0.

1.2 Derive the heat conduction equation corresponding to the cylindrical coordinates shown in Fig. 1.8. Consider the case where the conduction phenomenon is unsteady, the thermal conductivity is a function of temperature, and the effect of internal heat generation is present.

1.3 Consider the statement of conservation of energy in the infinitesimal chunk of material shown on the right side of Fig. 1.9. Derive from this statement the conduction equation in spherical coordinates, for the case of unsteady conduction with temperature-dependent thermal conductivity, and with internal heat generation.

1.4 The figure shows the distribution of temperature with depth in the rock formation of a geothermal field. The vertical direction x points downward, and $x = 0$ represents the ground level. The average thermal conductivity of the rock formation near the earth's surface is $k \cong 17.4$ W/m·K. Calculate the heat flux that crosses the $x = 0$ surface into the atmosphere. Estimate also the heat transfer rate

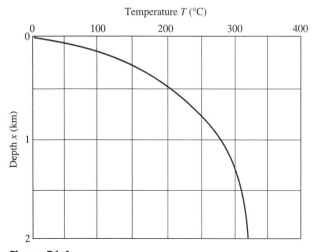

Figure P1.4

released by the rock formation into the atmosphere through a ground-level patch of area 1 km².

Convection

1.5   The flat heat exchanger surface shown in the figure is bathed by a stream of city water. In time, the surface becomes covered with a 0.1-mm-thin solid layer consisting of an accumulation of solids collected from the water stream. This process is called "fouling," and it is in many ways similar to the formation of plaque on teeth. The thermal conductivity of the accumulated solid is 0.6 W/m·K. The heat flux from the heat exchanger surface to the water stream is 0.1 W/cm². Calculate the temperature difference across the accumulated layer; in other words, how much does the temperature of the wetted surface drop because of fouling?

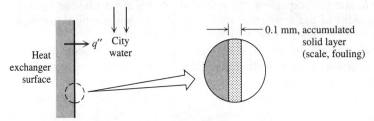

Figure P1.5

1.6   With reference to the illustration accompanying Example 1.2 in the text, consider the case where the layer of coal loses heat to the ground at the rate of 30 W/m². Chemical reactions heat the layer of coal at the rate of 50 W/m³, which is distributed uniformly throughout the coal volume. The coal layer is 2 m deep, the ambient air temperature is 30°C, and the heat transfer coefficient at the upper surface is $h = 15$ W/m²·K. Calculate the temperature of the upper surface of the layer of coal.

1.7   During a cold winter, the surface of a river develops a layer of ice of unknown thickness L. Known are the lake water temperature $T_w = 4°C$, the atmospheric air temperature $T_a = -30°C$, and the temperature of the underside of the ice layer $T_0 = 0°C$. The thermal conductivity of ice is $k = 2.25$ W/m·K. The convective heat transfer coefficients on the water and air sides of the ice layer are $h_w = 500$ W/m²·K and $h_a = 100$ W/m²·K, respectively. Calculate the temperature of the upper surface of the ice layer, T_s, and the ice thickness L.

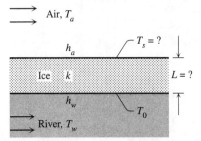

Figure P1.7

1.8   Under certain environmental conditions, the temperature of the human skin (30°C) is lower than the core temperature of the body (36.5°C). The transition between the two temperatures occurs across a subskin layer with an approximate thickness of 1 cm, which acts as an insulating coat. The thermal conductivity of the living tissue in this layer is approximately 0.42 W/m·K.

(a) Estimate the heat flux that escapes through the skin surface. Treat the subskin tissue as a motionless conducting medium.

(b) The ambient air temperature under the same conditions is 20°C. Calculate the heat transfer coefficient between skin and ambient air.

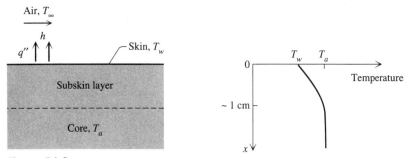

Figure P1.8

1.9 The electric reheating oven shown in the figure is used to heat a continuous sheet of metal, from 25°C to 1000°C. The oven and sheet are two-dimensional, that is, sufficiently wide in the direction perpendicular to the plane of the figure. The metal speed is such that each point on the sheet spends two hours inside the oven. The sheet thickness is 1 cm, the ambient temperature is 25°C, and the temperature of the outer surface of the oven (uniform all around) is 50°C. The heat transfer coefficient between the outer surface and the surrounding air is 20 W/m²·K. The metal is 304 stainless steel.

Calculate the electric power input to the oven, per unit oven width, q'. What percentage of this power input is lost by convection to the atmosphere?

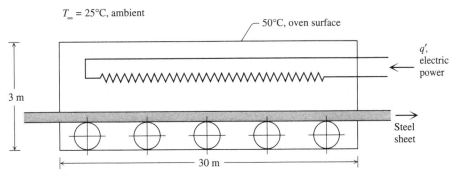

Figure P1.9

1.10 The oven described in the preceding problem has a constant-thickness wall made out of firebrick. The temperature of the internal surface is approximately 900°C, the temperature of the external surface is 50°C, and the ambient temperature is 25°C. The convection heat transfer coefficient at the external surface is 20 W/m²·K. Calculate the heat flux through the wall, and the wall thickness.

2
UNIDIRECTIONAL STEADY CONDUCTION

2.1 THIN WALLS

2.1.1 Thermal Resistance

Consider as a first and simplest conduction problem the insulation effect provided by a layer of a certain material whose thermal conductivity k is constant. The thickness of this layer is L. Across the layer, the temperature varies from T_0 (the temperature of the surface of the enveloped body, Fig. 2.1) to the temperature of the outer skin of the layer, T_L. The question addressed below is how this temperature difference drives the heat transfer that leaks through the layer.

The analytical treatment of this problem is particularly simple in the case when the layer looks "thin" compared with the dimensions of the covered body. For example, if the layer thickness L is much smaller than the radius of curvature of the surface, we can treat the layer as locally "plane," as is demonstrated on the right side of Fig. 2.1. Such a description is, of course, ideal in the case of a layer that is plane throughout its extent, like a sheet of glass, or the plasterboard covering of the wall in a house. This is why the present conduction configuration is sometimes referred to as the "plane wall," even though the associated analysis applies to the more general class of problems collected under the "thin wall" title.

In the Cartesian coordinate system attached to the inner surface of the layer, the temperature of the layer material can only vary in the transversal direction x. The equation that governs this temperature variation is the one for steady conduction in a material with constant conductivity and no internal heat generation, eq. (1.40):

$$\frac{d^2T}{dx^2} = 0 \tag{2.1}$$

The boundary conditions that apply to the sought $T(x)$ distribution are

$$T = T_0 \quad \text{at} \quad x = 0 \tag{2.2}$$

and

$$T = T_L \quad \text{at} \quad x = L \tag{2.3}$$

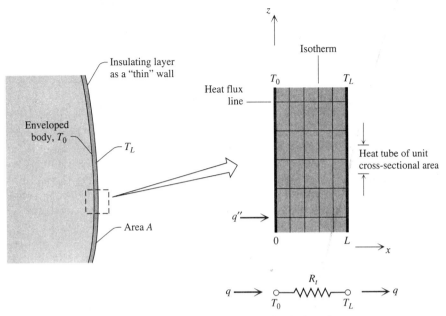

Figure 2.1 Thermal resistance posed by a sufficiently thin wall.

The solution to eq. (2.1) is $T = c_1x + c_2$, in which the two constants of integration are determined from the boundary conditions (2.2) and (2.3). In conclusion, the solution to eq. (2.1) is a *linear* temperature distribution that reaches T_0 and T_L at the two boundaries of the conducting medium:

$$T = T_0 + (T_L - T_0)\frac{x}{L} \tag{2.4}$$

This result is what is needed to calculate the heat flux q'' that locally* traverses the thin layer (or thin wall), eq. (1.36x):

$$q'' = -k\frac{dT}{dx} = \frac{k}{L}(T_0 - T_L) \tag{2.5}$$

Note that the heat flux q'' is defined positive when pointing in the positive x direction, and that it is conserved as it travels from $x = 0$ to $x = L$. This last feature is due to the linearity of the temperature distribution, which makes the gradient dT/dx independent of x.

Figure 2.1 further shows that the *isotherms* (the constant-T lines) are equidistant and parallel to the two faces of the wall. Perpendicular to the family of isotherms are the *heat flux lines*, which indicate the path followed by q'' through the conducting medium. The space contained between two adjacent flux lines is the *heat tube* aligned with the direction of q''. The heat tubes have the same thickness because the heat flux q'' is distributed uniformly over the face of the wall.

---

*That is, at the particular (y,z) location pinpointed in the detail drawing on the right side of Fig. 2.1, where the y direction is normal to the plane of the figure.

If the total surface covered by the thin layer is A, the total heat transfer rate that crosses the layer is simply

$$q = q''A = \frac{kA}{L}(T_0 - T_L) \tag{2.6}$$

We learn from this result that the steady leakage of heat across the layer is proportional to the temperature difference that drives it, $T_0 - T_L$, and the special group of physical quantities kA/L. The heat transfer rate q increases as the cross-sectional area A of its flow increases, as the layer is made out of increasingly more thermally conductive materials and as the layer becomes thinner. The kA/L group is sometimes referred to as the *thermal conductance* of the layer.

The inverse of kA/L, or the *thermal resistance* of the layer, is used just as frequently:

$$R_t = \frac{L}{kA} \tag{2.7}$$

Combining eqs. (2.6) and (2.7) we can see that the "fall" of q across the temperature "drop" $(T_0 - T_L)$, and "through" the resistance R_t,

$$q = \frac{T_0 - T_L}{R_t} \tag{2.8}$$

is completely analogous to that of an electrical current in a single-resistance circuit. This analogy, which also gives R_t its name, is stressed further by the electrical resistance symbol used on the right side of Fig. 2.1.

2.1.2   Composite Walls

The thermal resistance concept is particularly useful in the problem of estimating the heat transfer rate through a *composite wall* (Fig. 2.2). In such a wall two or more sheets are sandwiched and, together, bridge the temperature gap from T_0 to T_L. Each sheet has its own thermal resistance,

$$R_{t,i} = \frac{L_i}{k_i A} \qquad (i = 1, 2, 3) \tag{2.9}$$

where k_i is the thermal conductivity of the sheet material and L_i is the sheet thickness.

Assuming that two adjacent sheets have the same temperature at the interface (i.e., that the "thermal contact resistance" is zero), for the three sheets illustrated in Fig. 2.2 we can write [cf. eq. (2.8)]

$$T_0 - T_1 = qR_{t,1}$$
$$T_1 - T_2 = qR_{t,2} \tag{2.10}$$
$$T_2 - T_L = qR_{t,3}$$

In these equations T_1 and T_2 are the interface temperatures and q is the heat transfer rate. Note that q remains constant as it traverses the entire sandwich. Summing up the three equations listed above, we obtain the composite-wall

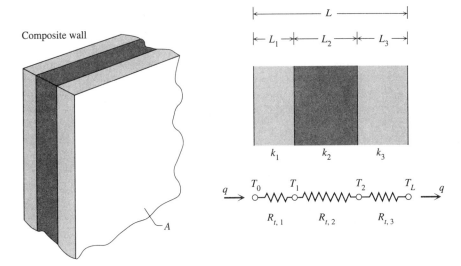

Figure 2.2 Composite wall and the structure of its thermal resistance.

equivalent of eq. (2.6), namely, a relationship between the heat transfer rate and the *overall* temperature difference that drives it:

$$q = \frac{T_0 - T_L}{\dfrac{L_1}{k_1 A} + \dfrac{L_2}{k_2 A} + \dfrac{L_3}{k_3 A}} \tag{2.11}$$

The denominator appearing on the right side of eq. (2.11) is the *overall thermal resistance* of the three-layer sandwich. It has as many terms as there are resistances in the string shown on the right side of Fig. 2.2.

2.1.3   Overall Heat Transfer Coefficient

In the preceding analysis it was assumed that the extreme temperatures (T_0, T_L) felt by the thin wall are known. That would be true in an application where both the inner body enveloped by the thin wall and the outer body are isothermal media (e.g., high-conductivity solids) at T_0 and T_L, respectively.

More frequent are the applications in which the two sides of the thin wall are exposed to fluids in motion. A common example is the metallic wall that separates two streams in a heat exchanger (more on this in Chapter 9) and the single-pane glass window. The latter is bathed by two jets driven in opposite directions by the effect of buoyancy. Both examples are represented well by the system illustrated in Fig. 2.3.

The distinguishing feature of this new configuration is that the known temperatures are the extreme fluid temperatures $(T_{\text{hot}}$ and $T_{\text{cold}})$, not the surface temperatures of the thin wall. Recall that these extreme fluid temperatures represent either bulk temperatures when the respective fluids flow through ducts or free-stream temperatures when the respective flows are of the "external" type (see Section 1.4). For each face of the thin wall we can invoke the heat flux relation (1.55):

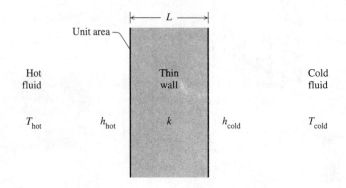

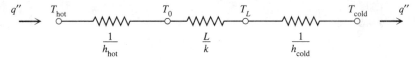

Figure 2.3 Thin wall sandwiched between two flows: the definition of overall heat transfer coefficient.

$$T_{hot} - T_0 = \frac{q''}{h_{hot}} \tag{2.12}$$

$$T_L - T_{cold} = \frac{q''}{h_{cold}} \tag{2.13}$$

where h_{hot} and h_{cold} are the respective heat transfer coefficients (these are assumed known from analyses of the convective heat transfer processes on the two fluid sides). The heat flux q'' is conserved as it passes from the hot fluid into the cold fluid. The same q'' overcomes the thermal resistance of the solid wall itself; therefore, according to eq. (2.5),

$$T_0 - T_L = \frac{L}{k} q'' \tag{2.14}$$

Adding eqs. (2.12) through (2.14) side by side, we obtain a compact proportionality between the heat flux and the overall (fluid-to-fluid) temperature difference:

$$T_{hot} - T_{cold} = \left(\frac{1}{h_{hot}} + \frac{L}{k} + \frac{1}{h_{cold}} \right) q'' \tag{2.15}$$

The form of this equation is similar to that of eqs. (2.12) and (2.13), in which the denominator of the right-hand side was occupied by a heat transfer coefficient. For this reason it is customary to rewrite eq. (2.15) as

$$T_{hot} - T_{cold} = \frac{q''}{U} \tag{2.16}$$

or as

$$q'' = U(T_{hot} - T_{cold}) \tag{2.16'}$$

where the *overall heat transfer coefficient U* is defined by

$$\frac{1}{U} = \frac{1}{h_{\text{hot}}} + \frac{L}{k} + \frac{1}{h_{\text{cold}}} \tag{2.17}$$

Equation (2.17) and the electrical resistance analog of this heat transfer configuration (Fig. 2.3) show that the inverse of U is equal to the sum of three thermal resistances. The first and third of these are of the "convection" type, while the second is the "conduction" thermal resistance of the solid wall itself. The right side of eq. (2.17) stresses also the conceptual difference between heat transfer coefficient and thermal conductivity: the units of h are $W/m^2 \cdot K$, whereas the units of k are $W/m \cdot K$. In other words, it is the ratio (thermal conductivity)/(thickness) that has the same units as the heat transfer coefficient.

Example 2.1

Composite Wall, Overall Heat Transfer Coefficient

The wall of a large incubator for eggs contains an 8-cm-thick layer of fiberglass sandwiched between two plywood sheets with a thickness of 1 cm. The outside temperature is $T_c = 10°C$, and the heat transfer coefficient at the outer plywood surface is $h_1 = 5\,W/m^2 \cdot K$. The corresponding conditions on the wall surface that faces the eggs are $T_h = 40°C$ and $h_3 = 20\,W/m^2 \cdot K$. The heat transfer coefficient is higher on the warm side of the wall because a fan recirculates the air that comes in contact with the eggs. Calculate the heat flux through the wall of the incubator.

Solution.   According to eq. (2.16′), to calculate the heat flux through the wall we must first determine the overall heat transfer coefficient U. An expression for $1/U$ can be constructed by looking at eq. (2.17) and keeping in mind the structure of the composite wall (see the right side of the figure):

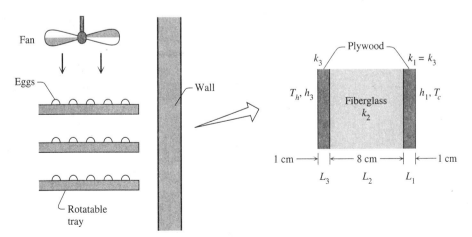

Figure E2.1

$$\frac{1}{U} = \frac{1}{h_3} + \frac{L_3}{k_3} + \frac{L_2}{k_2} + \frac{L_1}{k_1} + \frac{1}{h_1}$$

$$= \left(\frac{1}{20} + \frac{0.01}{0.11} + \frac{0.08}{0.035} + \frac{0.01}{0.11} + \frac{1}{5} \right) \frac{\mathrm{m}^2 \cdot \mathrm{K}}{\mathrm{W}}$$

$$= (0.05 + 0.09 + 2.29 + 0.09 + 0.2) \ \mathrm{m}^2 \cdot \mathrm{K}/\mathrm{W}$$

$$= 2.72 \ \mathrm{m}^2 \cdot \mathrm{K}/\mathrm{W} \tag{1}$$

The conductivities of plywood (k_1, k_3) and fiberglass (k_2) were collected from Appendix B. The third line in the calculation of $1/U$ shows that the overall thermal resistance of the composite wall is dominated by the middle compartment—the fiberglass layer. The heat flux can finally be calculated using eq. (2.16'):

$$q'' = U(T_h - T_c) = \frac{1}{2.72} \frac{\mathrm{W}}{\mathrm{m}^2 \cdot \mathrm{K}} (40 - 10)^\circ \mathrm{C}$$

$$= 11 \ \mathrm{W}/\mathrm{m}^2 \tag{2}$$

2.2   CYLINDRICAL SHELLS

A wall whose thickness is not negligible relative to the radius of curvature of the wall surface must be analyzed by a method that takes the curvature into account. In this section we focus on the class of cylindrical shells through which the heat transfer proceeds purely in the radial direction.

Figure 2.4 shows a cylindrical shell of length l, inner radius r_i, and outer radius r_o. The temperatures of the inner and outer cylindrical surfaces are T_i and

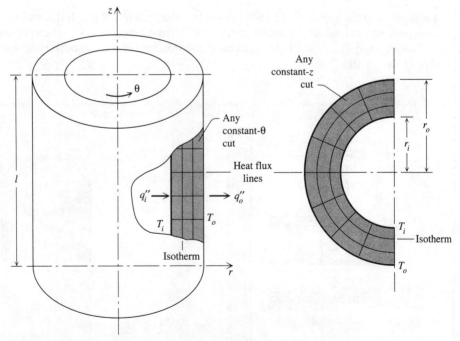

Figure 2.4 Radial conduction through a cylindrical shell.

T_o. Heat does not flow in the longitudinal direction (z), because the top and bottom cross sections are perfectly insulated. Or, we can think of the l-tall cylinder as simply a segment of an infinitely long pipe that extends above and below the system shown in the figure.

The challenge is to predict the total heat transfer rate q driven through the shell by the temperature difference $T_i - T_o$. The calculation would be simple if we knew the radial heat flux through the inner surface, q_i'', because the total heat transfer rate is the area integral of this flux:

$$q = (2\pi r_i l)q_i'' \tag{2.18}$$

We focus therefore on q_i'', which according to eq. (1.42r) is

$$q_i'' = -k\left(\frac{dT}{dr}\right)_{r=r_i} \tag{2.19}$$

This brings us immediately to the problem of determining the temperature distribution $T(r)$, which consists of solving the conduction equation in a medium with constant thermal conductivity and without volumetric heat generation, eq. (1.46):

$$\frac{1}{r}\frac{d}{dr}\left(r\frac{dT}{dr}\right) = 0 \tag{2.20}$$

subject to the temperature boundary conditions

$$T = T_i \quad \text{at} \quad r = r_i \tag{2.21}$$

and

$$T = T_o \quad \text{at} \quad r = r_o \tag{2.22}$$

The solution is obtained by integrating eq. (2.20) twice in r, in the following sequence:

$$\frac{d}{dr}\left(r\frac{dT}{dr}\right) = 0 \tag{2.23}$$

$$r\frac{dT}{dr} = C_1 \tag{2.24}$$

$$\frac{dT}{dr} = \frac{C_1}{r} \tag{2.25}$$

$$T = C_1 \ln r + C_2 \tag{2.26}$$

By subjecting the solution (2.26) to the boundary conditions (2.21) and (2.22) we obtain the system

$$T_i = C_1 \ln r_i + C_2 \tag{2.27}$$

$$T_o = C_1 \ln r_o + C_2 \tag{2.28}$$

and, after eliminating C_2,

$$C_1 = \frac{T_i - T_o}{\ln (r_i/r_o)} \tag{2.29}$$

Finally, we subtract eq. (2.27) from eq. (2.26),

$$T - T_i = C_1 \ln \frac{r}{r_i} \tag{2.30}$$

and use the C_1 value determined in eq. (2.29),

$$T = T_i - (T_i - T_o)\frac{\ln (r/r_i)}{\ln (r_o/r_i)} \tag{2.31}$$

In combination with eqs. (2.18) and (2.19), this solution yields

$$q = \frac{2\pi k l}{\ln (r_o/r_i)}(T_i - T_o) \tag{2.32}$$

In conclusion, the thermal resistance of the cylindrical shell increases as the logarithm of the radii ratio:

$$R_t = \frac{\ln (r_o/r_i)}{2\pi k l} \tag{2.33}$$

It is not difficult to show that this thermal resistance becomes identical to that of the thin wall, eq. (2.7), in the limit where r_i approaches r_o. The effect of the wall curvature is to increase the density of the heat flux lines toward smaller radii (Fig. 2.4), as a sign that the heat flux q'' in that region is larger. The same effect can be demonstrated analytically, by noting that it is the total heat transfer rate q (and not the heat flux q'') that is conserved as it flows radially outward through the shell:

$$q = (2\pi r_i l)q_i'' = (2\pi r l)q'' \tag{2.34}$$

From this we learn that at any radial distance inside the shell, r, the heat flux varies as $1/r$:

$$q'' = \frac{r_i}{r}q_i'' \tag{2.35}$$

The composite wall and fluid contact complications of this problem, which in the preceding section were illustrated by means of Figs. 2.2 and 2.3, can be analyzed by tracing the steps between eqs. (2.9) and (2.17). For the cylindrical shell configuration, we review only the end solution to the more general problem in which the shell—bathed by fluid streams on both sides—is also a composite wall. For the three-layer shell drawn in Fig. 2.5, the heat transfer rate is given by the now familiar formula

$$q = \frac{T_{\text{hot}} - T_{\text{cold}}}{R_t} \tag{2.36}$$

in which the overall thermal resistance of the shell, R_t, is defined by

$$R_t = \frac{1}{h_i A_i} + \frac{\ln (r_1/r_i)}{2\pi k_1 l} + \frac{\ln (r_2/r_1)}{2\pi k_2 l} + \frac{\ln (r_o/r_2)}{2\pi k_3 l} + \frac{1}{h_o A_o} \tag{2.37}$$

In this expression, A_i and A_o are the areas of the innermost and outermost cylindrical surfaces,

$$A_i = 2\pi r_i l \qquad A_o = 2\pi r_o l \tag{2.38}$$

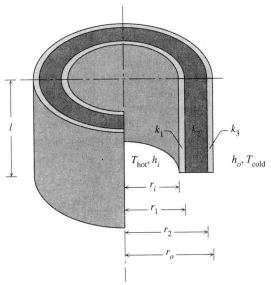

Figure 2.5 Composite cylindrical shell with convective heat transfer on both sides.

and h_i and h_o are the respective heat transfer coefficients. Note the presence of three conduction-type terms on the right side of eq. (2.37): the number of these terms matches the number of layers in the constitution of the composite shell. Each term of the sum on the right side of eq. (2.37) represents one thermal resistance in a "series" of the kind shown in Figs. 2.2 and 2.3.

2.3 SPHERICAL SHELLS

The spherical shell geometry can be analyzed similarly, by noting first that when the inner and outer surfaces are isothermal (T_i, T_o) the temperature inside the shell can only vary with the radial position. With reference to the notation defined in Fig. 2.6, the total heat current q (watts) that traverses the shell from T_i to T_o turns out to be (Problem 2.2)

$$q = 4\pi k \frac{r_i r_o}{r_o - r_i}(T_i - T_o) \tag{2.39}$$

The thermal resistance of a spherical shell is therefore

$$R_t = \frac{1}{4\pi k}\left(\frac{1}{r_i} - \frac{1}{r_o}\right) \tag{2.40}$$

This expression becomes identical to eq. (2.7) in the thin-wall limit $r_i \rightarrow r_o$.

The overall thermal resistance to a more complicated spherical shell design can be calculated in the manner demonstrated by eq. (2.37) for a cylindrical shell. For example, if the shell experiences convective heat transfer on both sides (h_i, h_o) and if it is composed of two layers (k_1, k_2), then the overall thermal resistance between the inner fluid and the outer fluid is

$$R_t = \frac{1}{h_i A_i} + \frac{1}{4\pi k_1}\left(\frac{1}{r_i} - \frac{1}{r_1}\right) + \frac{1}{4\pi k_2}\left(\frac{1}{r_1} - \frac{1}{r_o}\right) + \frac{1}{h_o A_o} \tag{2.41}$$

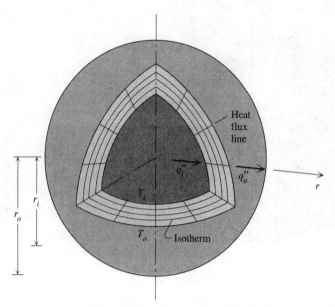

Figure 2.6 Radial conduction through a spherical shell.

The radius r_1 indicates the position of the interface between the k_1 and k_2 layers, while $A_i = 4\pi r_i^2$ and $A_o = 4\pi r_o^2$ are respectively the inner and outer areas of the composite spherical shell.

2.4 CRITICAL INSULATION RADIUS

A special application of the thermal resistance formulas developed thus far is the problem of determining the thickness of annular insulation that should be applied on the outer surface of a cylindrical wall of known temperature, T_i. The function of the insulation positioned between radii r_i and r_o, Fig. 2.7, is to reduce the total heat transfer rate between the inner body and the ambient fluid, T_∞. An example of such a design is the annular layer of foam insulation installed around the pipe carrying steam or hot water in a building.

The total heat transfer rate q behaves as the inverse of the overall thermal resistance R_t, because $q = (T_i - T_\infty)/R_t$. Assuming that the heat transfer coefficient between the insulating layer and the ambient fluid is a known constant h, the overall thermal resistance is the sum of the "cylindrical shell" resistance of the insulation, plus the external convective heat transfer resistance, eq. (2.37):

$$R_t = \frac{\ln(r_o/r_i)}{2\pi k l} + \frac{1}{(2\pi r_o l)h} \tag{2.42}$$

By solving the equation $\partial R_t/\partial r_o = 0$, it is easy to show that R_t reaches a *minimum* (or q reaches a maximum),

$$R_{t,\text{min}} = \frac{\ln(k/hr_i) + 1}{2\pi k l} \tag{2.43}$$

and that this minimum occurs when the outer surface of the insulation reaches the *critical radius*:

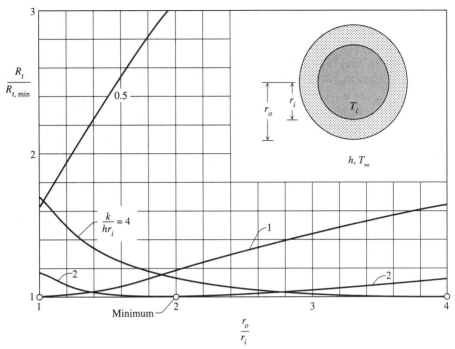

Figure 2.7 Effect of the outer radius on the overall thermal resistance of a cylindrical insulation layer.

$$r_{o,c} = \frac{k}{h} \quad \text{(cylinder)} \tag{2.44}$$

Figure 2.7 shows the behavior of the thermal resistance in terms of the dimensionless ratio $R_t/R_{t,\min}$ versus r_o/r_i. The R_t minimum shifts toward lower r_o values as the dimensionless group k/hr_i decreases, that is, as r_i increases and the inner cylinder becomes "thick."

The effect of building a thicker layer of insulation depends on how the radius of the bare cylinder, r_i, compares with the critical outer radius $r_{o,c}$, eq. (2.44). In the case of "thick" bare cylinders,

$$r_i > r_{o,c} \quad \text{or} \quad \frac{k}{hr_i} < 1 \tag{2.45}$$

the addition of insulation always translates into a higher R_t value, that is, into an "insulation effect." In the opposite extreme, where the bare cylinder is so "thin" that

$$r_i < r_{o,c} \quad \text{or} \quad \frac{k}{hr_i} > 1 \tag{2.46}$$

the wrapping of the very first layer of insulation induces a decrease in the overall thermal resistance. This initial effect is one of heat transfer enhancement, not insulation. Only when enough material has been added so that r_o exceeds $r_{o,c}$, the thickening of the insulation layer increases the R_t value and reduces q.

The behavior of R_t in the thin bare cylinder limit may seem paradoxical, because an insulation effect is intuitively expected when a finite-thickness layer of finite-k material is wrapped on anything. The explanation for the initial drop in R_t as r_o increases lies in the fact that, when the bare cylinder is sufficiently thin, the thickening of the insulation layer leads to an *increase* in the external area ($2\pi r_o l$) that is bathed by fluid, that is, by the ultimate heat sink (T_∞).

This effect will become clearer after reading the segment on extended surfaces later in this chapter (Section 2.7): in that sense, the wrapping of a finite-k layer around a thin cylinder "extends" the surface through which the system rejects heat to the ambient flow. For example, the thermal resistance between a thin electrical wire and the surrounding air can be reduced by coating the wire with a layer of dielectric material. In this case the thermal function of the dielectric coating is to facilitate the Joule heat rejection to the ambient.

The same conclusions can be drawn with regard to the insulation built around a spherical object of radius r_i. The critical outer radius of the insulating layer is in this case

$$r_{o,c} = 2\frac{k}{h} \qquad \text{(sphere)} \tag{2.47}$$

where k and h are again the thermal conductivity of the layer material and the heat transfer coefficient at the layer–ambient interface, respectively.

There is no "critical thickness" of insulation in an application where both the bare surface and the insulating layer are sufficiently plane, as in the thin-wall systems treated in Section 2.1. In this geometry the thickening of the insulation layer always leads to a higher R_t value. This conclusion can be drawn without actually performing an analysis, because the "plane" bare surface (or "thin" insulation layer) is the most extreme case of the class of thick bare cylinders discussed in connection with eq. (2.45).

In summary, there exists a critical insulation thickness to worry about in the design of cylindrical and spherical layers of insulation, but not in the sizing of plane or nearly plane layers. The critical thickness feature is therefore a reflection of the finite curvature of the bare surface on which the insulation is mounted.

Example 2.2

Cylindrical Shell, Critical Radius

An uninsulated wire suspended in air generates Joule heating at the rate $q' = 1 \, \text{W/m}$. The wire is a bare cylinder of radius $r_i = 0.5$ mm, and the temperature difference between it and the atmosphere is $30°C$.

It is proposed to cover this wire with a plastic sleeve of electrical insulation, the outer radius of which will be $r_o = 1$ mm. The thermal conductivity of the plastic material is $k = 0.35 \, \text{W/m·K}$. Will the plastic sleeve improve the wire–ambient thermal contact, or will it provide a thermal insulation effect? To verify your answer, calculate the new wire–ambient temperature difference when the wire is encased in plastic.

Solution.   The plastic sleeve of electrical insulation promises to have a heat transfer augmentation effect (i.e., it promises to reduce the overall thermal resistance) if the radius of the bare wire is smaller than the critical radius of insulation. To calculate $r_{o,c}$, eq. (2.44), we must first calculate the heat transfer coefficient:

$$q' = 2\pi r_i h (T_i - T_\infty)$$

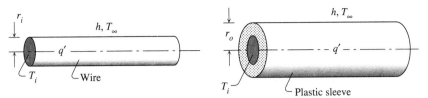

Figure E2.2

$$h = \frac{q'}{2\pi r_i(T_i - T_\infty)} = 1\,\frac{\text{W}}{\text{m}}\,\frac{1}{2\pi}\,\frac{1}{0.5 \times 10^{-3}\text{m}}\,\frac{1}{30}\,\frac{1}{\text{K}}$$

$$= 10.6\ \text{W/m}^2 \cdot \text{K} \tag{1}$$

The critical radius

$$r_{o,c} = \frac{k}{h} = \frac{0.35\ \text{W}}{\text{m} \cdot \text{K}}\,\frac{\text{m}^2 \cdot \text{K}}{10.6\ \text{W}} = 3.3\ \text{cm} \tag{2}$$

is much greater than the radius of the bare wire and greater than the outer radius of the plastic sleeve. We can expect, then, a heat transfer augmentation effect (a decrease in $T_i - T_\infty$) from the presence of the plastic sleeve.

The new wire–ambient temperature difference $T_i - T_\infty$ follows from the R_t definition

$$T_i - T_\infty = R_t q$$

$$= R_t l q' \tag{3}$$

where, according to eq. (2.42), the $R_t l$ value of the wire encased in plastic is

$$R_t l = \frac{\ln(r_o/r_i)}{2\pi k} + \frac{1}{2\pi r_o h}$$

$$= \frac{\ln(1/0.5)}{2\pi}\,\frac{\text{m} \cdot \text{K}}{0.35\ \text{W}} + \frac{1}{2\pi \times 10^{-3}\ \text{m}}\,\frac{\text{m}^2 \cdot \text{K}}{10.6\ \text{W}}$$

$$= 15.3\ \text{m} \cdot \text{K/W} \tag{4}$$

In conclusion, the temperature difference is almost half of what it was before the installation of the plastic coating:

$$T_i - T_\infty = R_t l q' = 15.3\,\frac{\text{m} \cdot \text{K}}{\text{W}}\,1\,\frac{\text{W}}{\text{m}}$$

$$= 15.3\ \text{K} \tag{5}$$

2.5   VARIABLE THERMAL CONDUCTIVITY

In all the problems treated so far in this chapter, the thermal conductivity of the material was assumed constant. In this section we learn that the preceding thermal resistance formulas are simple cases of a more general result that holds true for an arbitrary thermal conductivity function $k(T)$.

The most basic geometric feature of the thin wall, the cylindrical shell, and the spherical shell is the unidirectionality of the conduction phenomenon. That unique direction was transversal (x) in Fig. 2.1, radial (r) in Fig. 2.4, and again

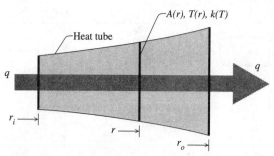

Figure 2.8 Unidirectional conduction through a solid with temperature-dependent thermal conductivity.

radial (r) in Fig. 2.6. This common feature is shown schematically in Fig. 2.8, in which $A(r)$ is the cross-sectional area penetrated by q at the position r along its path. The total heat transfer rate q is conserved as it flows through the heat tube; however, the cross-sectional area A may vary with r. In a cylindrical shell, for example, A is proportional to r, whereas in the thin-wall limit A is a constant.

At any position r, the material is characterized by a unique temperature $T(r)$ and, since $k = k(T)$, a unique thermal conductivity. The derivation of the thermal resistance formula consists of rearranging the Fourier law,

$$q = A(r)q''(r) = -A(r)k(T)\frac{dT}{dr} \tag{2.48}$$

so that the T and r variables are separated:

$$q\frac{dr}{A(r)} = -k(T)\,dT \tag{2.49}$$

This can be readily integrated because q is constant:

$$q\int_{r_i}^{r_o} \frac{dr}{A(r)} = -\int_{T_i}^{T_o} k(T)\,dT \tag{2.50}$$

The integral on the right side can be rewritten in terms of the *thermal potential* function [1],

$$\theta(T) = \int_{T_{\text{ref}}}^{T} k(T')\,dT' \tag{2.51}$$

in which T_{ref} is a reference temperature, for example, absolute zero in low-temperature applications, or the standard environmental temperature (25°C, or 298.15 K) in applications near room temperature. In this way, the entire $k(T)$ information of a chart like Fig. 1.6 can be projected on a corresponding $\theta(T)$ chart.

The total heat transfer rate formula (2.50) assumes now the new form

$$q = \frac{\theta(T_i) - \theta(T_o)}{\displaystyle\int_{r_i}^{r_o} \frac{dr}{A(r)}} \tag{2.52}$$

This result is due to Kirchhoff [1]. It is equivalent to writing that the thermal resistance R_t of the unidirectional heat path

$$q = \frac{T_i - T_o}{R_t} \tag{2.53}$$

is now given by the general formula

$$R_t = \frac{1}{k_{\text{avg}}} \int_{r_i}^{r_o} \frac{dr}{A(r)} \tag{2.54}$$

where k_{avg} is the *average thermal conductivity* defined by

$$k_{\text{avg}} = \frac{\theta(T_i) - \theta(T_o)}{T_i - T_o} \tag{2.55}$$

To see that the above thermal resistance formula is indeed general, the reader may wish to verify that eqs. (2.7), (2.33), and (2.40) are special cases of eq. (2.54) when k is constant. The value of eq. (2.54) is that it extends the applicability of the thermal resistance concept to a unidirectionally conducting layer whose thermal conductivity is a strong function of temperature.

2.6   INTERNAL HEAT GENERATION

2.6.1   Uniform Heating

It is already evident that the present coverage of conduction heat transfer processes is one that starts with the simplest configurations and proceeds toward the more complicated and general. In the realm of unidirectional conduction (the simplest geometric configuration possible), we treated first the class of constant-k problems and, only as a summary, the variable-k generalization of the same problems. In this section we execute the next step in the direction of more complicated models, and focus on the interplay between steady conduction and internal heat generation. The effect of internal heat generation was tacitly assumed absent in all the cases analyzed until now.

The essential feature of the steady internal heating problem can be illustrated by considering the thin-wall system shown in Fig. 2.9a. The conducting material experiences a *uniform* volumetric heating rate $\dot{q}$ (W/m³): its temperature distribution $T(x)$ reaches a steady state because the wall is bathed by a fluid of fixed temperature T_∞, which plays the role of heat sink. The same fluid flow washes both sides of the wall; therefore, the heat transfer coefficient h (assumed constant) has the same value on both sides.

The key unknown here is not the total heat transfer rate, because this is already known as the product between $\dot{q}$ and the total volume of the conducting slab. Note that in the steady state *all* the heat that is being generated inside the slab must be transferred to the fluid reservoir. The real question is how much warmer the innards of the slab must become to be able to drive this heat transfer rate to the sides.

Since the unknown is $T(x)$, we turn our attention to eq. (1.38), in which the right-hand side is zero:

$$\frac{d^2T}{dx^2} + \frac{\dot{q}}{k} = 0 \tag{2.56}$$

The *two* boundary conditions required by the *second order* of this ordinary differential equation are of the "convective" type shown in eq. (1.54):

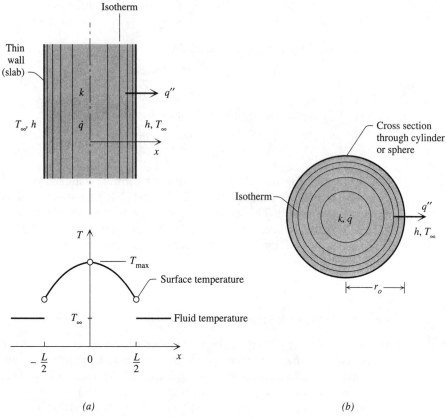

Figure 2.9 Steady temperature distribution due to uniform internal heat generation in a slab (*a*) and in a cylinder or sphere (*b*).

$$-q'' = h(T - T_\infty) \qquad \text{at } x = -\frac{L}{2} \tag{2.57}$$

$$q'' = h(T - T_\infty) \qquad \text{at } x = \frac{L}{2} \tag{2.58}$$

The negative sign is needed on the left side of eq. (2.57) because (a) q'' is considered positive when pointing in the x direction, and (b) in the h definition (1.54), q'' was assumed positive when pointing toward the fluid. After invoking the Fourier law on the left side of both eqs. (2.57) and (2.58), the boundary conditions become finally

$$k\frac{dT}{dx} = h(T - T_\infty) \qquad \text{at } x = -\frac{L}{2} \tag{2.59}$$

$$-k\frac{dT}{dx} = h(T - T_\infty) \qquad \text{at } x = \frac{L}{2} \tag{2.60}$$

We note in passing that the geometry and boundary conditions in Fig. 2.9*a* are symmetric about the midplane. Consequently, we can replace either one of the above conditions by the statement that the sought temperature distribution must also be symmetric: analytically, this condition is written as $dT/dx = 0$ at $x = 0$.

The solution to eq. (2.56) has the general form $T = (-\dot{q}/k)(x^2/2) + c_1 x + c_2$. The integration constants are determined by invoking the boundary conditions (2.59) and (2.60):

$$T(x) - T_\infty = \frac{\dot{q}L^2}{8k}\left[1 - \left(\frac{x}{L/2}\right)^2\right] + \frac{\dot{q}L}{2h} \qquad (2.61)$$

This parabolic temperature distribution is illustrated in the lower part of Fig. 2.9a. The isotherms are closer together near the $x = -L/2$ and $x = L/2$ boundaries, indicating that the heat flux (or the slope $|dT/dx|$) is greatest on the boundaries. The maximum temperature occurs in the midplane of the slab, $T_{max} = T(0)$, and its value increases as $\dot{q}$ increases:

$$T_{max} - T_\infty = \frac{\dot{q}L^2}{8k}\left(1 + \frac{4}{Bi}\right) \qquad (2.62)$$

Related results are the surface temperature $T(L/2) = T_\infty + (\dot{q}L/2h)$ and the temperature difference between the midplane and the slab surface, $T(0) - T(L/2) = \dot{q}L^2/8k$.

The dimensionless quantity Bi is recognized as the *Biot number*[*] based on the overall thickness L,

$$Bi = \frac{hL}{k} \qquad (2.63)$$

The Biot number can be viewed as a "dimensionless heat transfer coefficient." The order of magnitude of this group divides the possible applications of Fig. 2.9a into two important categories:

Bi >> 1. In this range the thermal contact between solid boundary and fluid flow is "good," so that the boundary temperature $T(\pm L/2)$ approaches the temperature of the fluid T_∞.

Bi << 1. In cases where the solid–fluid thermal contact is "poor," or when the solid is an extremely good thermal conductor, the boundary temperature is nearly the same as the maximum (midplane) temperature of the slab.

In summary, the temperature profile drawn in Fig. 2.9a becomes flatter as Bi decreases.

Completely analogous conclusions are reached in the study of a *cylindrical body* that is being heated volumetrically, Fig. 2.9b. The radial distribution of temperature inside the cylinder is given by

$$T(r) - T_\infty = \frac{\dot{q}r_o^2}{4k}\left[1 - \left(\frac{r}{r_o}\right)^2\right] + \frac{\dot{q}r_o}{2h} \qquad (2.64)$$

Similarly, the temperature distribution through a volumetrically heated *spherical body* is

---

[*]Jean Baptiste Biot (1774–1862) was a French physicist who worked for the first time on the problem of solid-body conduction with convection at the surface. He is known also for his work on electromagnetism (e.g., the Biot–Savart law), electrostatics, the properties of water at saturation, and the first balloon flight for scientific research.

$$T(r) - T_\infty = \frac{\dot{q} r_o^2}{6k} \left[1 - \left(\frac{r}{r_o} \right)^2 \right] + \frac{\dot{q} r_o}{3h} \tag{2.65}$$

where r_o is the outer radius of the sphere and h is the convective heat transfer coefficient.

2.6.2 Temperature-Dependent Heating: The Integral Method

Suppose now that the volumetric heating rate $\dot{q}$ is not constant and that it increases with the local temperature. In this case the conductor runs the risk of becoming thermally *unstable*, because higher temperatures lead to higher volumetric heating rates, which in turn lead again to higher temperatures.

An engineering application in which the potential thermal runaway instability must be avoided is the design of electrical conductors. At every point inside a cylindrical conductor (wire) the volumetric heating effect is proportional to the local electrical resistivity ρ_e:

$$\dot{q} = \rho_e J^2 \tag{2.66}$$

where J is the current density (amperes/m^2). The Wiedemann–Franz law, eq. (1.34), showed already that in most electrical conductors ρ_e increases with the temperature; a linear approximation of this relationship is [2]

$$\rho_e \cong \rho_{e,o} + \rho'_e (T - T_o) \tag{2.67}$$

where $\rho_{e,o} = \rho_e(T_o)$ is the electrical resistivity value at the outer radius of the conductor (r_o) and $\rho'_e = (d\rho_e / dT)_{T = T_o}$. The temperature of the outer surface T_o is assumed known and fixed; in other words, the heat transfer coefficient between the wire and the external flow is sufficiently large (recall the Bi $\gg 1$ range noted in the preceding section).

The equation for steady conduction in a cylindrical body with constant thermal conductivity and finite internal heat generation rate is [cf. eq. (1.44)]

$$\frac{1}{r} \frac{d}{dr} \left(r \frac{dT}{dr} \right) + \frac{\dot{q}}{k} = 0 \tag{2.68}$$

In view of eqs. (2.66) and (2.67), the $\dot{q}/k$ term is a linear function of T; therefore, the conduction equation reads

$$\frac{1}{r} \frac{d}{dr} \left(r \frac{dT}{dr} \right) + C_1 + C_2 T = 0 \tag{2.69}$$

where C_1 and C_2 are two empirical constants of the conductor. In particular, the second constant is proportional to J^2, which is a key electrical design parameter:

$$C_2 = \frac{J^2}{k} \left(\frac{d\rho_e}{dT} \right)_{T = T_o} \tag{2.70}$$

The relevant boundary conditions are two:

$$\frac{dT}{dr} = 0 \quad \text{at } r = 0 \quad \text{(symmetry about the centerline)} \tag{2.71}$$

and

$$T = T_o \quad \text{at } r = r_o \quad \text{(very good contact with an external heat sink)} \tag{2.72}$$

An exact solution to the problem (2.69) through (2.72) can be developed in terms of Bessel functions. Since the teaching of Bessel function analysis is beyond the scope of the present treatment, we use the above problem as an opportunity to outline an approximate method of solution — *the integral method.*

The motivation for developing a solution of the "approximate" kind stems from the observation that not all the features of the exact $T(r)$ solution are of critical importance. For example, in the exact temperature distributions illustrated in Fig. 2.9, the key engineering result is the maximum temperature that occurs at the innermost point, fiber, or plane of the solid body (T_{max} is important because it may approach the temperature range in which the mechanical characteristics of the body deteriorate). Relative to T_{max}, the *shape* of the $T(r)$ profile is of less importance — after all, the main features of the shape can be anticipated by simply reading the boundary conditions (2.71) and (2.72)!

The integral method consists of two decisions concerning the unknown of the problem, in this case, the temperature profile $T(r)$:

Step 1. The shape of the profile is considered known. This step is a "guess" educated by the reading of the boundary conditions (2.71) and (2.72) and the review of exact solutions that have been developed for related problems (e.g., Fig. 2.9). In the integral analysis presented below, it is assumed that the shape is parabolic; however, the *amplitude* of the profile $T_{max} - T_o$ remains unknown.

Step 2. The conduction equation (2.69) is integrated over the entire domain over which the assumed shape is supposed to hold. In the present problem that domain is the circular cross section of radius r_o. The resulting equation — the *integral conduction equation* — is used finally, to determine the unknown amplitude of the temperature profile.

This two-step structure of the integral method becomes clearer as we perform the actual analysis.

First, we assume a parabolic shape for the temperature distribution:

$$T = T_o + (T_{max} - T_o)\left[1 - \left(\frac{r}{r_o}\right)^2\right] \tag{2.73}$$

Note that this $T(r)$ function satisfies the boundary conditions (2.71) and (2.72). Second, we perform term-by-term the area integral

$$\int_0^{2\pi}\int_0^{r_o} [\text{eq. (2.69)}]\, r\, dr\, d\theta \tag{2.74}$$

where θ is the angular coordinate (Fig. 1.8). This operation leads to the integral conduction equation

$$\left(r\frac{dT}{dr}\right)_{r=r_o} + C_1\frac{r_o^2}{2} + C_2\int_0^{r_o} Tr\,dr = 0 \tag{2.75}$$

which can be solved for $T_{max} - T_o$ after using eq. (2.73). The final result,

$$T_{max} - T_o = \frac{2(C_1 + C_2 T_o)}{(8/r_o^2) - C_2} \tag{2.76}$$

shows that the maximum temperature remains finite only if $C_2 < 8/r_o^2$. This inequality must be satisfied if a steady-state temperature distribution $T(r)$ is to exist.

The above conclusion translates into the following design criterion for the avoidance of thermal runaway instability:

$$J < \frac{2^{3/2}}{r_0}\left(\frac{k}{\rho_e'}\right)^{1/2} \tag{2.77}$$

In the corresponding criterion generated by the exact (Bessel function) solution, the $2^{3/2} = 2.828$ factor is replaced by 2.405, that is, by a value that is only 15 percent smaller. This comparison illustrates the degree of accuracy that can be expected in an integral solution.

2.7 EXTENDED SURFACES (FINS)

2.7.1 The Enhancement of Heat Transfer

The augmentation of heat transfer, or the enhancement of the thermal contact between a solid surface and the fluid that bathes it, is one of the more common propositions in thermal design. This objective defined the special class of applications that was labeled (2) in Section 1.2. In the present section we have a first opportunity to see how a change in the geometry of the surface leads to a better thermal contact with the fluid. The technique consists of shaping certain portions of the surface so that they protrude into the fluid: the external surface of the solid protrusions constitutes the *extended surface* of the wall–fluid contact area, and the protrusions themselves are the *fins*.

The enhancement of heat transfer that is brought about by the use of fins is illustrated in Fig. 2.10. In the absence of fins, the heat transfer rate between the solid wall and the external flow is

$$q_0 = hA_0(T_b - T_\infty) \tag{2.78}$$

where A_0 is the extent of the original (bare) surface and h is the wall–fluid heat transfer coefficient. In this case the heat flux q_0/A_0 is distributed uniformly over the bare surface, because both h and the wall–fluid temperature difference $T_b - T_\infty$ are assumed constant.

The right side of Fig. 2.10 shows what happens to the distribution of wall heat flux when fins are present. For clarity in the figure and the following

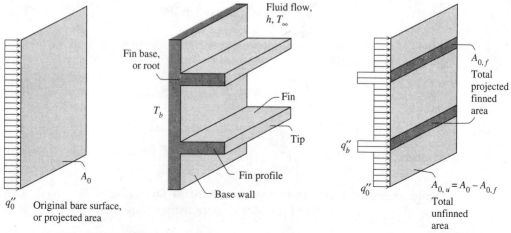

Figure 2.10 Increase in wall heat flux over the area covered by fins.

argument, it was assumed that the h value on the fin surfaces is the same as on the bare portions of the base surface. In practice, the heat transfer coefficient would be lower on the base surface than on the fin.

To begin with, the use of fins leads to an increase in the total area of contact between solid surfaces and fluid, because the sum of the total external area of all the fins plus the original area without any fins is greater than the bare surface A_0. This increase in the total area of contact does not guarantee an increase in the total heat transfer rate q between the solid body and the fluid. A net increase (i.e., $q > q_0$) is registered when the fin material has a sufficiently high thermal conductivity so that the temperature of the fin surface is comparable with the temperature of the base surface, T_b. In such a case the heat flux that passes into the fluid through the lateral surface of the fin (i.e., vertically in Fig. 2.10) is comparable in magnitude with the bare surface heat flux q_0/A_0. And since the total heat transfer rate that is discharged into the fluid through the wetted (exposed) surface of the fin must be equal to the heat transfer rate that—through its root—the fin "pulls" out of the wall, it is clear that the presence of the fins augments the distribution of heat flux over the projected area A_0.

The new heat flux distribution q/A_0 has the original bare surface heat flux q_0/A_0 as background level for a series of heat flux "spikes." Each spike occurs over that spot of the projected area that is now occupied by the base (root) of one fin. The degree to which the addition of fins augments the total heat transfer rate between solid wall and fluid is expressed by the *overall projected-surface effectiveness* ratio

$$\epsilon_0 = \frac{q}{q_0} = \frac{q}{hA_0(T_b - T_\infty)} \tag{2.79}$$

Considering the difficulty of manufacturing a fin-covered surface, meaningful augmentation of heat transfer occurs when ϵ_0 is significantly greater than 1.

The basic heat transfer problem in the design of finned surfaces reduces to being able to predict the heat transfer rate q, or the overall effectiveness ϵ_0, that is associated with a certain fin surface geometry, temperature difference $T_b - T_\infty$, convective heat transfer coefficient h, and thermal conductivity of fin material. The road that we are about to follow becomes evident if we regard the original bare surface A_0 as the sum of the total projected area covered by the roots of fins $A_{0,f}$ and the remaining unfinned portion $A_{0,u}$:

$$A_0 = A_{0,f} + A_{0,u} \tag{2.80}$$

Then, by looking at the ragged shape of the heat flux distribution when fins are present, q/A_0, we see that the total heat transfer rate q is simply the sum of two contributions:

$$q = q_b'' A_{0,f} + hA_{0,u}(T_b - T_\infty) \tag{2.81}$$

In the first term on the right side, q_b'' is the average heat flux through the base of one fin. Therefore, the calculation of the fin–base heat flux is the focus of the analysis described next.

2.7.2   Constant Cross-Sectional Area

The simplest fin geometry to consider—the geometry that goes back to the founding fathers of conduction heat transfer, Biot and Fourier [3]—is that of

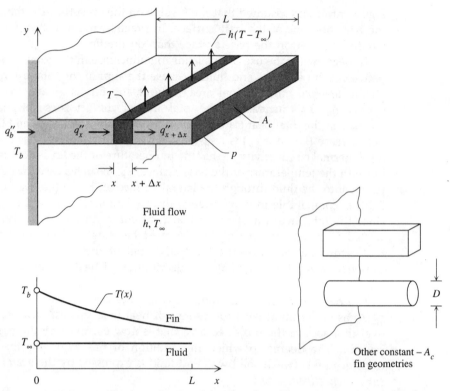

Figure 2.11 Longitudinal conduction through a fin with constant cross-sectional area.

the fin whose cross-sectional area A_c is independent of the longitudinal position x along the fin. Figure 2.11 features one example of such a fin, namely, the plate–fin geometry that appeared earlier in Fig. 2.10. Other examples are the cylindrical spine with constant circular cross section (the "pin fin") and the prismatic fin inserted in the lower-right portion of Fig. 2.11.

The Longitudinal Conduction Model. The key assumption on which the analysis rests is that the temperature T inside the fin is only a function of x. Accordingly, heat is conducted through the fin longitudinally, even though in reality most of this heat current escapes into the fluid through the lateral (exposed) surface. The $T = T(x)$ assumption is a good approximation only under special conditions [e.g., eqs. (2.127–2.130)], on which we focus soon. The immediate benefit of making this assumption is that the chief unknown (the fin–base heat flux) is the y-independent quantity.*

$$q_b'' = -k\left(\frac{dT}{dx}\right)_{x=0} \tag{2.82}$$

---

*If the $T = T(x)$ assumption does not apply, then T is a function of both x and y, where y is the vertical coordinate in Fig. 2.11. The base heat flux in this case is still given by eq. (2.82); however, q_b'' is a function of y.

The next necessary step is therefore the derivation of the fin temperature distribution $T(x)$. For this, we cannot invoke blindly the conduction equation for one-dimensional conduction in Cartesian coordinates, eq. (1.23), because unlike the solid bar of Fig. 1.4, the lateral surface of the plate fin of Fig. 2.11 is *not insulated*. Indeed, the only reason the plate "works" as a fin is that it makes thermal contact with the fluid (h, T_∞) all along its lateral surface.

The steady-state conservation of energy in the fin slice of thickness Δx pictured in Fig. 2.11 requires that

$$q_x'' A_c - q_{x+\Delta x}'' A_c - (p\Delta x)h(T - T_\infty) = 0 \qquad (2.83)$$

The last term in this balance represents the heat transfer rate from the Δx-thin system to the surrounding fluid. This term is written in the usual way, as the product between the local fin–fluid heat flux $h(T - T_\infty)$, and the area touched by fluid, $p\Delta x$. The length labeled p is the *wetted perimeter* of the fin cross section, that is, the line of contact with the fluid. The Fourier law and the assumption that the thermal conductivity of the fin material, k, is constant allow us to rewrite the first two terms of eq. (2.83) as

$$A_c(q_x'' - q_{x+\Delta x}'') = -A_c \frac{dq_x''}{dx}\Delta x = -A_c \frac{d}{dx}\left(-k\frac{dT}{dx}\right)\Delta x$$
$$= kA_c \frac{d^2T}{dx^2}\Delta x \qquad (2.84)$$

Combining eqs. (2.83) and (2.84) and dividing by Δx we obtain the conduction equation for a fin with constant cross section:

$$kA_c \frac{d^2T}{dx^2} \quad - hp(T - T_\infty) = 0 \qquad (2.85)$$

$$\underbrace{\qquad\qquad}_{\substack{\text{Longitudinal} \\ \text{conduction}}} \quad \underbrace{\qquad\qquad}_{\substack{\text{Lateral} \\ \text{convection}}}$$

This equation expresses a perfect balance between the net amount of conduction heat transfer that arrives longitudinally at the location x and the convective heat transfer that leaves the fin through the lateral line of contact with the fluid.

Long Fin. Consider first the case of a fin that is so long that its tip region is in thermal equilibrium with the surrounding fluid:

$$T \rightarrow T_\infty \quad \text{as} \quad x \rightarrow \infty \qquad (2.86)$$

The other boundary condition on the fin temperature $T(x)$ is the statement that the temperature at the root of the fin is the same as that of the base wall:

$$T = T_b \quad \text{at} \quad x = 0 \qquad (2.87)$$

The $T(x)$ problem consists of solving eq. (2.85) subject to eqs. (2.86) and (2.87); however, it is customary to first restate the entire problem in terms of the *excess temperature* function

$$\theta(x) = T(x) - T_\infty \qquad (2.88)$$

The new problem statement is therefore

$$\frac{d^2\theta}{dx^2} - m^2\theta = 0 \tag{2.89}$$

$$\theta = \theta_b \quad \text{at } x = 0 \quad \text{(note: } \theta_b = T_b - T_\infty) \tag{2.90}$$

$$\theta \to 0 \quad \text{as } x \to \infty \tag{2.91}$$

where m is a crucial parameter of the fin–fluid arrangement,

$$m = \left(\frac{hp}{kA_c}\right)^{1/2} \tag{2.92}$$

In what follows we assume that h has the same value along the fin surface; therefore, we treat m as a constant. Under this condition the general solution to eq. (2.89) is

$$\theta(x) = C_1 \exp(-mx) + C_2 \exp(mx) \tag{2.93}$$

for which the constants C_2 and C_1 are determined sequentially by invoking the boundary conditions (2.91) and (2.90):

$$C_2 = 0 \quad C_1 = \theta_b \tag{2.94}$$

In conclusion, the temperature of the fin material decays exponentially to the fluid temperature sufficiently far from the base,

$$\theta = \theta_b \exp(-mx) \tag{2.95}$$

that is, when the dimensionless product mx is significantly greater than 1. The lateral convective heat flux, $h(T - T_\infty) = h\theta$, decays to zero in the same manner as the temperature excess function $\theta(x)$. If L is the actual length of the fin, Fig. 2.11, then the fin can be regarded as being "long enough" [as in the boundary condition (2.91)] when

$$mL \gg 1 \tag{2.96}$$

The heat flux through the base is calculated next by using eq. (2.82), which yields $q_b'' = k\theta_b m$. Therefore, the total heat current that the fin pulls from the base wall is

$$q_b = q_b'' A_c = \theta_b (kA_c hp)^{1/2} \tag{2.97}$$

In this remarkably simple result, we see how all the physical parameters of the fin–fluid configuration affect the ultimate size of the important quantity q_b. The total heat transfer rate drawn from the base wall increases if any of the parameters θ_b, k, A_c, h, or p increases. In particular, q_b is equally sensitive to changes in the fin conductivity and the fin–fluid heat transfer coefficient.

Finite-Length Fin with Insulated Tip.   Most fin designs do not meet the long-fin criterion (2.96); therefore, their "finite" length L must be taken into consideration. The physical consequence of the finite length is that the temperature of the tip $T(L)$ is higher than the ambient temperature T_∞. In general, this temperature difference drives a convective heat transfer rate directly through the tip of the fin,

$$q_{\text{tip}} = hA_c[T(L) - T_\infty] \tag{2.98}$$

For simplicity, it is assumed that the h value on the tip is the same as that on the lateral surfaces of the fin.

An intermediate step toward the more general result developed in the next subsection is the $T(x)$ solution for a fin with insulated tip:

$$\frac{dT}{dx} = 0 \quad \text{or} \quad \frac{d\theta}{dx} = 0 \quad \text{at } x = L \tag{2.99}$$

This limiting case is a good approximation when the actual heat current that passes through the tip is negligible relative to the total heat current drawn from the base wall,

$$q_b >> q_{\text{tip}} \tag{2.100}$$

The general solution to the fin conduction equation is still the expression listed in eq. (2.93) or, alternatively,*

$$\theta(x) = C_1' \sinh(mx) + C_2' \cosh(mx) \tag{2.101}$$

The boundary conditions (2.90) and (2.99) produce, in order,

$$\theta_b = C_2' \tag{2.102}$$

$$0 = C_1' \cosh(mL) + C_2' \sinh(mL) \tag{2.103}$$

and, after substituting C_1' and C_2' into eq. (2.101),

$$\theta = -\theta_b \tanh(mL) \sinh(mx) + \theta_b \cosh(mx)$$

$$= \theta_b \frac{\cosh\,[m\,(L-x)]}{\cosh(mL)} \tag{2.104}$$

The main features of this solution are illustrated in Fig. 2.12 (left side). The finite excess temperature that occurs at the tip can be deduced by setting $x = L$ in eq. (2.104):

$$\theta(L) = \frac{\theta_b}{\cosh(mL)} \tag{2.105}$$

Finally, the total heat transfer rate flowing from the base wall into the root of the fin is

$$q_b = A_c \left(-k\frac{dT}{dx}\right)_{x=0}$$

$$= \theta_b (kA_c hp)^{1/2} \tanh(mL) \tag{2.106}$$

---

*As a reminder, the connection between exponentials and hyperbolic sine and cosine is expressed by

$$\exp(u) = \cosh(u) + \sinh(u)$$

$$\exp(-u) = \cosh(u) - \sinh(u)$$

Other useful relations and numerical values are listed in Appendix E. If we replace the exponentials in this manner in eq. (2.93), we obtain

$$\theta(x) = (C_2 - C_1) \sinh(mx) + (C_1 + C_2) \cosh(mx)$$

This shows that the new constants C_1' and C_2' employed in eq. (2.101) are shorthand for the constants $C_2 - C_1$ and, respectively, $C_1 + C_2$.

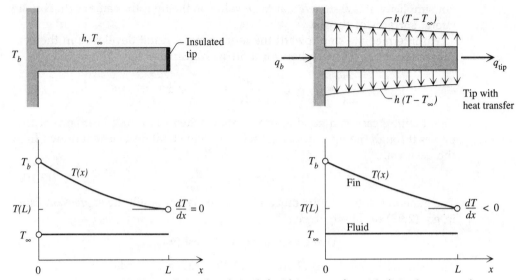

Figure 2.12 Fin with insulated tip (left side) versus fin with finite heat transfer rate through the tip (right side).

It pays to take another look at the long-fin results of the preceding subsection to see that they are the special limit $mL \rightarrow \infty$ of the solution that was just completed. In that limit the tip excess temperature $\theta(L)$ of eq. (2.105) approaches zero, as in eq. (2.91), and the total heat transfer rate through the fin, eq. (2.106), approaches the simpler result listed in eq. (2.97).

The "insulated tip" assumption is an adequate approximation when the tip heat transfer condition (2.100) is satisfied. Combining eqs. (2.98) and (2.105) into a formula for q_{tip}, we can use also eq. (2.106) to rewrite the (2.100) inequality as

$$\frac{q_{tip}}{q_b} = \frac{1}{\sinh(mL)} \left(\frac{hA_c}{kp} \right)^{1/2} \ll 1 \qquad (2.107)$$

Note first that in the case of finite-length fins, mL and $\sinh(mL)$ are finite quantities. This leaves hA_c/kp as the principal dimensionless group that governs the relative size of the heat transfer rate through the tip. A fin with a heat transfer coefficient and a cross-sectional area so small that the tip thermal conductance hA_c is much smaller than the value of kp is represented adequately by the insulated tip model. Of course, eq. (2.107) shows that the same model is applicable also in the limit $mL \rightarrow \infty$ (recall that in the long fin case the temperature gradient at the tip is zero, because the temperature near the tip is constant and equal to T_∞).

The Effect of Heat Transfer Through the Tip. The insulated tip assumption of the preceding subsection can be relaxed by replacing the boundary condition (2.99) with

$$-kA_c\frac{d\theta}{dx} = hA_c\theta \qquad \text{at} \qquad x = L \qquad (2.108)$$

in which we continue to assume that the h value on the tip surface is the same as on the lateral surfaces. In this boundary condition the left side represents the conduction heat transfer rate that arrives at the solid side of the tip surface, while the right side is the convection heat transfer rate that is removed by the fluid flow. Both sides of the equation are equal to q_{tip}, eq. (2.98). The temperature distribution in this general case [Fig. 2.12 (right side)] can be determined by combining the solution (2.101) with the boundary conditions (2.90) and (2.108):

$$\theta = \theta_b \frac{\cosh[m(L-x)] + (h/mk)\sinh[m(L-x)]}{\cosh(mL) + (h/mk)\sinh(mL)} \qquad (2.109)$$

The total heat transfer rate could be calculated next, by using the top line of eq. (2.106). However, a much more compact q_b formula is recommended by an approximate geometric argument advanced by Harper and Brown [4]. The upper part of Fig. 2.13 shows the actual fin with heat transfer through the tip. The length of this constant cross section fin is L. The lower part of the figure shows an alternative exit for the tip heat transfer rate, namely, a fin length $(L_c - L)$ that is attached to the end of the original fin. The right side of this additional chunk is insulated. The heat current q_{tip} enters it from the left (through the $x = L$ plane) and leaves it by crossing the top and bottom surfaces.

This geometric construction suggests that the total heat transfer rate of the original fin, q_b, is the same as that of a longer fin with insulated tip. According to eq. (2.106), then,

$$q_b = \theta_b (kA_c hp)^{1/2} \tanh(mL_c) \qquad (2.110)$$

in which L_c is the so-called *corrected length* that defines the longer fin. The problem of calculating q_b reduces to that of estimating the proper L_c value for the original fin. This last step consists of invoking the conservation of energy in the added chunk of length $L_c - L$,

$$hA_c[T(L) - T_\infty] = hp(L_c - L)[T(L) - T_\infty] \qquad (2.111)$$

where the left side is the heat current through the tip of the actual fin and the right side is the convective heat transfer through the lateral area of the added

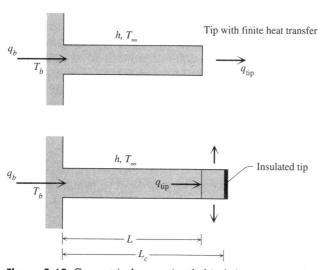

Figure 2.13 Geometrical reasoning behind the concept of corrected length, L_c.

segment, $p(L_c - L)$. In writing eq. (2.111), we are assuming that the $L_c - L$ segment is short enough so that its temperature is approximately equal to the tip temperature of the original fin, $T(L)$. The corrected length formula dictated by eq. (2.111) is

$$L_c = L + \frac{A_c}{p} \tag{2.112}$$

For example, in the case of a plate fin of thickness t and width W, we have $A_c = tW$ and $p = 2(W + t) \cong 2W$. Therefore,

$$L_c = L + \frac{t}{2} \quad \text{(plate fin)} \tag{2.113}$$

Similarly, the geometry of a pin fin of constant diameter D is characterized by $A_c = \pi D^2/4$ and $p = \pi D$. Therefore,

$$L_c = L + \frac{D}{4} \quad \text{(pin fin, or cylindrical spine)} \tag{2.114}$$

The corrected length formula (2.110) is always an accurate substitute for the exact q_b formula based on eq. (2.109) when the fin is "long" ($mL \gg 1$), because in this limit the heat transfer through the tip is negligible, eq. (2.107). In the opposite extreme, $mL \ll 1$, the error associated with using the approximate formula (2.110) instead of the exact q_b expression based on eq. (2.109) is approximately equal to $hA_c/3kp$, provided the Biot number based on fin thickness is small.

Fin Efficiency Versus Fin Effectiveness. A dimensionless parameter that describes how well the fin functions as an extension of the base surface is the *fin efficiency* η:

$$\eta = \frac{\text{actual heat transfer rate}}{\substack{\text{maximum heat transfer rate when} \\ \text{the entire fin is at } T_b}} = \frac{q_b}{hpL_c\theta_b} \tag{2.115}$$

The denominator in this definition is greater than q_b because the true fin temperature decreases along the fin. Consequently, the actual convective heat flux that crosses the wetted surface, $h\theta$, is consistently smaller than the ideal value $h\theta_b$ listed in the denominator (see the graphic definition of η in Fig. 2.14). The range covered by the fin efficiency of any fin is therefore $0 < \eta < 1$.

In view of the corrected q_b formula (2.110), the efficiency of a fin with constant cross section is

$$\eta = \frac{\tanh(mL_c)}{mL_c} \tag{2.116}$$

for which m is defined in eq. (2.92). This function is displayed in Fig. 2.14, next to the efficiency curves of other fin geometries. Note that in the case of the wide plate fin, the abscissa parameter used in Fig. 2.14,

$$L_c \left(\frac{2h}{kt} \right)^{1/2} \tag{2.117}$$

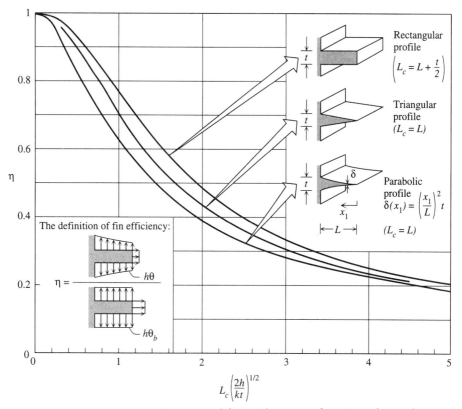

Figure 2.14 Efficiency of two-dimensional fins with rectangular, triangular, and parabolic profiles (drawn after Ref. 5).

is identical to the group mL_c. The figure shows that the efficiency drops significantly when the abscissa parameter is considerably greater than 1: long fins are "poor" because their tip temperature approaches the fluid temperature, and, as a consequence, the convective heat fluxes $h\theta$ that cover the wetted surface of the fin are considerably smaller than their ceiling value $h\theta_b$ (see the lower left corner of Fig. 2.14).

An alternative figure of merit is the *fin effectiveness*(ϵ_f):

$$\epsilon_f = \frac{\text{total fin heat transfer}}{\begin{array}{c}\text{the heat transfer that would have}\\ \text{occurred through the base area}\\ \text{in the absence of the fin}\end{array}} = \frac{q_b}{hA_c\theta_b} \qquad (2.118)$$

If the fin is to perform its heat transfer augmentation function properly, then ϵ_f must be definitely (i.e., sizably) greater than 1. It follows that in the case of a good fin the effectiveness value is greater than the efficiency value, the relation between the two being

$$\frac{\epsilon_f}{\eta} = \frac{pL_c}{A_c} = \frac{\text{total fluid contact area}}{\text{cross-sectional area}} \qquad (2.119)$$

The fin effectiveness ϵ_f is also greater than the overall projected-surface effectiveness ϵ_0, which was defined in eq. (2.79). The relationship between ϵ_0 and ϵ_f is obtained by combining eqs. (2.79), (2.81), and (2.118):

$$\epsilon_0 = \epsilon_f \frac{A_{0,f}}{A_0} + \frac{A_{0,u}}{A_0} \tag{2.120}$$

To summarize the contribution of the analysis presented in this section, the total heat transfer rate q_b can be calculated based on formulas such as eq. (2.110) or on fin efficiency charts of the type exhibited in Fig. 2.14. Many more q_b results have been developed for other fin geometries; these can be found in heat transfer handbooks (e.g., Ref. [6]).

2.7.3 Variable Cross-Sectional Area

The heat transfer enhancement capability of a fin with variable cross-sectional area can be analyzed similarly. Crucial in this more general analysis is again the assumption that the conduction heat transfer is oriented longitudinally through the fin.

Figure 2.15 shows a fin geometry in which the cross-sectional area A_c and wetted perimeter p are two unspecified functions of the longitudinal position x. The conduction equation for this configuration can be deduced by recognizing first the steady-state energy conservation requirement for the slice of thickness Δx,

$$q_x - q_{x+\Delta x} - (p\Delta x)h(T - T_\infty) = 0 \tag{2.121}$$

in which the longitudinal heat current is proportional to the local temperature gradient, $q_x = -kA_c(dT/dx)$. When A_c varies along the fin, the most compact form to which eq. (2.121) can be brought is

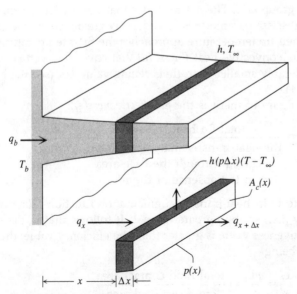

Figure 2.15 Longitudinal conduction through a fin with variable cross-sectional area and wetted perimeter.

$$\frac{d}{dx}\left(kA_c\frac{dT}{dx}\right) - hp(T - T_\infty) = 0 \qquad (2.122)$$

In this statement, as in eq. (2.85) earlier, the wetted perimeter p can be a function of x.

The objective in the analysis of a specified variable geometry, $A_c(x)$ and $p(x)$, is to calculate the total heat transfer rate between the fin and the surrounding fluid. This heat transfer rate is equal to the total heat current that passes through the base of the fin:

$$q_b = -\left(kA_c(x)\frac{dT}{dx}\right)_{x=0} \qquad (2.123)$$

The end result is usually catalogued in terms of the respective fin efficiency η, which is defined in the same manner as in the first part of eq. (2.115):

$$\eta = \frac{q_b}{hA_{exp}(T_b - T_\infty)} \qquad (2.124)$$

where A_{exp} is the exposed surface of the fin, that is, the area bathed by fluid.

Three examples of fin efficiencies of variable-A_c fin geometries are presented in Figs. 2.14 and 2.16. Note that in the two-dimensional fins with triangular and parabolic profiles shown in Fig. 2.14, only the cross-sectional area varies with the longitudinal position (the wetted perimeter is twice the width of the fin, i.e., constant). In a disc-shaped fin, Fig. 2.16, both A_c and p vary with the local radial

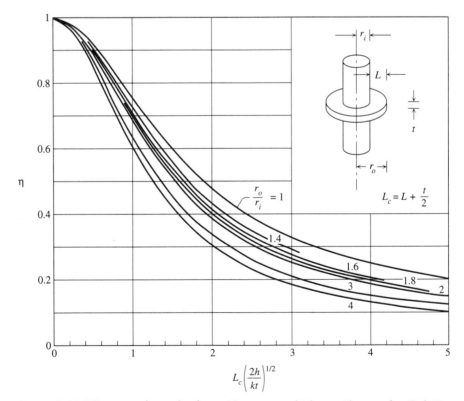

Figure 2.16 Efficiency of annular fins with constant thickness (drawn after Ref. 5).

position r, which here plays the role of longitudinal coordinate for the unidirectional heat current, as it did in Fig. 2.8.

2.7.4 When the Unidirectional Conduction Model Is Valid

The fin analyses discussed until now are all based on the assumption that the conduction process is unidirectional. This is why "extended surfaces" have been included as a special topic in this chapter. In fins, however, the unidirectional conduction model can only be valid approximately. Figure 2.17 shows that the heat flux lines inside the fin are not truly longitudinal and parallel, because most of these lines are destined to cross the lateral fin surface that is bathed by fluid. Even in the base plane $x = 0$, the heat flux lines are not truly equidistant and parallel.

A key question, then, is: "Under what conditions is the longitudinal conduction model valid?" Consider for this purpose the two-dimensional fin geometry with rectangular profile (the plate fin, Fig. 2.17). A heat flux line that intersects the exposed surface of the fin is oriented longitudinally "sufficiently" if the longitudinal heat flux q_x'' is considerably greater than the transversal conduction heat flux q_y'' that crosses the exposed area,

$$q_x'' >> q_y'' \tag{2.125}$$

As the "scale," or representative order of magnitude of q_x'', we can use the heat flux through the base of the fin, $q_b'' = q_b/A_c$, which follws from eq. (2.110). Next, the transversal heat flux q_y'' is the same as the convective heat flux on the fluid side of the exposed surface, $h\theta$: the order of magnitude of this heat flux is $h\theta_b$. Putting these conclusions together, in place of the inequality (2.125) we have

$$\frac{q_b}{A_c} >> h\theta_b \tag{2.126}$$

In view of eq. (2.118), this validity condition is the same as

$$\epsilon_f >> 1 \tag{2.127}$$

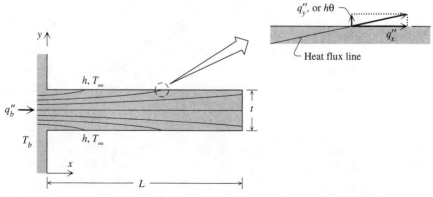

Figure 2.17 Pattern of heat flux lines through a two-dimensional fin with rectangular profile (plate fin).

The conclusion we have just reached is that the validity of the unidirectional conduction model improves as the single-fin effectiveness ϵ_f becomes considerably greater than 1. Conversely, the unidirectional conduction model is a poor description for a fin whose calculated ϵ_f value is only marginally greater than 1. Such a low ϵ_f value is first and foremost a warning that the formulas that were employed in the calculation [e.g., eq. (2.110)] *do not apply*. In other words, a calculated value of $\epsilon_f \cong 1$ using a unidirectional conduction model does not necessarily mean that the fin is ineffective as a heat transfer augmentation device. It means instead that a two-dimensional conduction theory (e.g., Chapter 3) should be used for the purpose of calculating the true heat transfer rate through the base of the fin and, eventually, the correct ϵ_f value.

The validity criterion (2.127) acquires new meaning if we rewrite it using eqs. (2.106) and (2.118):

$$\left(\frac{kp}{hA_c} \right)^{1/2} \tanh(mL) \gg 1 \tag{2.128}$$

The order of magnitude of $\tanh(mL)$ is 1, especially in long fins (Section 2.7.2). What remains is the front factor, $(kp/hA_c)^{1/2}$, which assumes special forms for specific fin geometries. For example, for a plate fin of thickness t and width W, the order-of-magnitude inequality (2.128) reduces to

$$\left(\frac{ht}{k} \right)^{1/2} \ll 1 \quad \text{(plate fin)} \tag{2.129}$$

Similarly, the validity criterion for unidirectional conduction in a pin fin of length L and diameter D becomes*

$$\left(\frac{hD}{k} \right)^{1/2} \ll 1 \quad \text{(pin fin)} \tag{2.130}$$

On the left-hand side of both eqs. (2.129) and (2.130) we see the *Biot number* based on the transversal length scale of the fin. In conclusion, the use of the unidirectional conduction model is particularly appropriate in the analysis of fins where the thermal conductivity is so great (or the convective heat transfer coefficient is so small) that the thickness-based Biot number is, in an order-of-magnitude sense, smaller than 1.

2.7.5 Optimization Subject to Volume Constraint

The sizing procedure for a fin generally requires the selection of more than one physical dimension (length). In a plate fin, for example, there are two such dimensions, the length of the fin, L, and the thickness, t. One of these dimensions can be selected optimally when the total volume of fin material is fixed [7]. Such a constraint is often justified by the high cost of the high-thermal-conductivity metals that are employed in the manufacture of finned surfaces (e.g., copper, aluminum) and by the cost associated with the weight of the fin (e.g., the finned cylinders of air-cooled motors for model airplanes and motorcycles).

---

*Numerical factors of the order of 1 (e.g., the factor 2) have been left out of the order-of-magnitude statements (2.129) and (2.130).

To select one of the two dimensions of the fin "optimally" means to maximize the total heat transfer rate q_b when the fin volume V is known and fixed:

$$V = LtW \tag{2.131}$$

For simplicity, let us assume that the fin profile is sufficiently slender ($t<<L$) so that $L_c \cong L$. This means that q_b can be calculated by using the insulated tip formula (2.106), in which $A_c = tW$ and $p = 2W$, where W is the width of the plate fin. Eliminating L using eq. (2.131), we obtain

$$q_b = at^{1/2}\tanh(bt^{-3/2}) \tag{2.132}$$

in which a and b are two constants,

$$a = \theta_b W(2kh)^{1/2} \qquad b = \frac{V}{W}\left(\frac{2h}{k}\right)^{1/2} \tag{2.133}$$

The maximum of the total heat current q_b with respect to the plate fin thickness t is found next by solving the equation $dq_b/dt = 0$, which is the same as the transcendental equation

$$\sinh(2bt^{-3/2}) = 6bt^{-3/2} \tag{2.134}$$

The lone nontrivial solution of this equation is $bt^{-3/2} = 1.4192$; rearranging this result, we obtain the optimum plate fin thickness

$$t = 0.998\left(\frac{V}{W}\right)^{2/3}\left(\frac{h}{k}\right)^{1/3} \tag{2.135}$$

and, in view of the volume constraint (2.131), the corresponding length of the plate fin

$$L = 1.002\left(\frac{V}{W}\right)^{1/3}\left(\frac{h}{k}\right)^{-1/3} \tag{2.136}$$

Equations (2.135) and (2.136) can be used for the purpose of sizing the plate fin. Instructive is the corresponding slenderness ratio of the rectangular profile of the plate fin, t/L, which, after some manipulation can be arranged as follows:

$$\frac{t}{L} = 0.996\left(\frac{ht}{k}\right)^{1/2} \tag{2.137}$$

The conclusion drawn at the end of the preceding section was that the thickness-based Biot number (ht/k) must be small if the longitudinal conduction model is to be valid. Consequently, the slenderness ratio recommended by this optimum design must be smaller than 1 also. This discovery, incidentally, justifies the assumption adopted immediately under eq. (2.131), at the start of this section.

Figure 2.18 shows geometrically a sequence of slenderness ratios calculated with eq. (2.137): the optimum rectangular profile becomes more slender as the Biot number (ht/k) decreases. Note that throughout this sequence the volume of the plate fin (in Fig. 2.18, the profile area $t \times L$) remains constant.

It can be shown that the q_b extremum pinpointed by the solution to eq. (2.134) is indeed a maximum [7]. That maximum value is

$$q_{b,\text{max}} = 1.258kW\theta_b\left(\frac{ht}{k}\right)^{1/2} \tag{2.138}$$

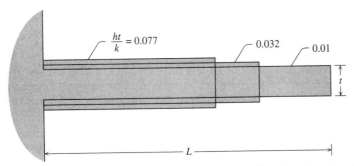

Figure 2.18 Scale drawing of the optimum profile of a plate fin of fixed volume.

where we note again the presence of the smaller-than-1 Biot number ht/k.

The maximization of q_b subject to constant fin volume—illustrated here by means of the plate fin geometry—has attracted considerable attention. Schmidt [7] argued intuitively that there exists not only an optimum fin size when the profile shape is specified, but also an optimum profile shape that maximizes the total heat transfer rate q_b. According to him, the optimum shape (a parabolic profile, it turns out) must be such that the temperature varies linearly along the fin. Schmidt's predictions were confirmed later by Duffin [8], who subjected the q_b maximization problem to variational calculus. The optimum profile shape of fins with temperature-dependent conductivity was determined by Jany and Bejan [9], who also showed that the fin shape optimization idea has an important analog in the design of long ducts for fluid flow. The optimum dimensions for a plate fin with fixed volume and transversal laminar boundary layers were determined by Bejan [10]. The same problem forms the subject of Project 5.2 at the end of Chapter 5.

Worth noting is that the classical conclusions regarding the optimum shaping of fins have become somewhat controversial, beginning with a 1974 critique published by Maday [11]. An overview of the debate that followed was presented by Snider and Kraus [12]. These authors point out that Schmidt neglected the so-called "length of arc," that is, the additional contribution made to the lateral (exposed) area by the slope of the edge of the fin profile.

Poulikakos and Bejan [13] have shown that optimum fin shapes and dimensions can be determined also based on purely thermodynamic grounds. For example, there exists an optimum pin fin diameter and an optimum length for which the thermodynamic irreversibility (entropy generation rate) of the fin–fluid arrangement is minimum. During this optimization procedure, the heat transfer rate through the base (the "duty" of the fin, q_b) is held constant. A sketch of this method is available also in Reference [14].

The present treatment of conduction heat transfer through fins was based on the simplest fin model possible. Key in this model are the assumptions that the conduction process is unidirectional [15, 16] (Section 2.7.4) and that the base temperature T_b is a known quantity. When these assumptions break down, one must rely on a multidirectional conduction model, as in the opening section to Chapter 3. Another important simplification resulted from the assumption that the heat transfer coefficient is equal to the same constant h all along the lateral surface of the fin [17, 18].

Example 2.3

Long Fin, Tip Temperature

During an investigation of the effect of temperature on the life of a battery, the battery is positioned on a plate heater of temperature $T_b = 50°C$. The battery is cylindrical, with a diameter $D = 3$ cm and a height $L = 6$ cm. The ambient temperature is $T_\infty = 20°C$, and the heat transfer coefficient at the outer cylindrical surface is $h = 5$ W/m²·K.

The outer skin of the battery is made out of stainless-steel sheet with a thickness $t = 0.5$ mm. The interior of the battery is such a poor conductor that the heat transfer between it and the stainless-steel shell can be neglected.

(a) Determine whether the stainless steel shell can be treated as a long fin.

(b) Calculate the temperature at the top edge of the shell, that is, at $x = L$ in the figure.

Solution.   To determine whether the stainless-steel skin is a "long fin," eq. (2.96), we first calculate the m parameter, eq. (2.92):

$$m = \left(\frac{hp}{kA_c}\right)^{1/2} = \left(\frac{hp}{kpt}\right)^{1/2} = \left(\frac{h}{kt}\right)^{1/2}$$

$$= \left(\frac{5\,\text{W}}{\text{m}^2\cdot\text{K}}\frac{\text{m}\cdot\text{K}}{15\,\text{W}}\frac{1}{0.5\times10^{-3}\text{m}}\right)^{1/2}$$

$$= 25.8 \text{ m}^{-1} \tag{1}$$

The long-fin criterion (2.96) is not satisfied, because in the present case mL is not much greater than 1:

$$mL = (25.8 \text{ m}^{-1})(0.06 \text{ m}) = 1.55 \tag{2}$$

The top-edge temperature can be estimated by modeling the edge as a tip with heat transfer. Substituting $x = L$ in eq. (2.109), we write

$$\frac{\theta}{\theta_b} = \frac{1}{\cosh(mL) + \dfrac{h}{mk}\sinh(mL)}$$

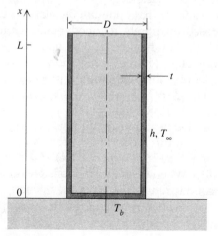

Figure E2.3

in which $mL = 1.55$ and

$$\frac{h}{mk} = \frac{5\,\text{W}}{\text{m}^2 \cdot \text{K}} \frac{\text{m}}{25.8} \frac{\text{m} \cdot \text{K}}{15\,\text{W}} = 0.013$$

We obtain in this way

$$\frac{\theta}{\theta_b} = 0.402 = \frac{T - T_\infty}{T_b - T_\infty} \tag{3}$$

which means that the top-edge temperature is $T = 32.1°\text{C}$.

The top edge could have been modeled more simply as an insulated tip [cf. eq. (2.105)]:

$$\frac{\theta}{\theta_b} = \frac{1}{\cosh(mL)} = 0.407 \tag{4}$$

and the top-edge temperature would have been $T \cong 32.2°\text{C}$. Since this estimate is practically equal to the one based on the more rigorous formula (2.109), we conclude that effect of heat transfer through the top edge (as a fin tip) is negligible.

It is worth going back to the long-fin criterion (2) to see the error that would have been made if the steel skin were treated as a long fin. Under the long-fin assumption, the top-edge temperature is given by eq. (2.95) with $x = L$,

$$\frac{\theta}{\theta_b} = \exp(-mL) = 0.212 \tag{5}$$

which means that $T = 26.4°\text{C}$. This estimate turns out to be lower than the correct value of $32.1°\text{C}$, because the long-fin assumption means that the fin continues beyond $x = L$ and that the cooling effect provided by this extension "depresses" the $x = L$ temperature.

2.8   EXTENDED SURFACES WITH RELATIVE MOTION AND INTERNAL HEAT GENERATION

2.8.1   The General Conduction Equation

The unidirectional conduction model of the classical fin finds application also in the analysis of slender bodies with considerably more complicated functions. The main complications that form the subject of this section are the presence of solid motion (flow) through the envelope that defines the slender system, and the effect of internal heat generation. Configurations that may be modeled as long fins with longitudinal solid motion occur in the extrusion of plastics, and the drawing of wires and artificial fibers. Suspended electrical conductors of various shapes may be treated as fins with internal heat generation.

The longitudinal conduction equation that holds in this general case can be written with reference to the control volume of length Δx shown in Fig. 2.19:

$$q_x - q_{x+\Delta x} - hp\Delta x(T - T_\infty) + \dot{m}h_x - \dot{m}h_{x+\Delta x} + \dot{q}A_c\Delta x = 0 \tag{2.139}$$

in which h_x is the specific enthalpy of the solid at the location x. Treating the solid as an incompressible substance, we recall that [14]

$$dh_x = c\,dT + \frac{1}{\rho}dP \tag{2.140}$$

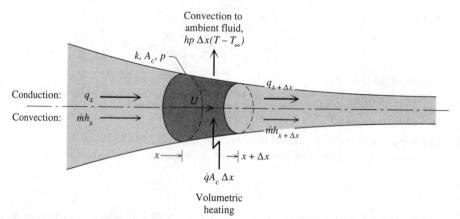

Figure 2.19 Conservation of energy in a slender body with solid movement (U) and internal heat generation ($\dot{q}$).

Therefore, at constant pressure, $dh_x = c\,dT$. The terms multiplied by the solid flowrate $\dot{m}$ in eq. (2.139) yield

$$\dot{m}(h_x - h_{x+\Delta x}) = -\dot{m}\frac{dh_x}{dx}\Delta x = -\dot{m}c\frac{dT}{dx}\Delta x \qquad (2.141)$$

Implicit in this derivation is the statement that the flowrate $\dot{m}$ is conserved from one cross section to the next:

$$\dot{m} = \rho A_c U = \text{constant} \qquad (2.142)$$

The longitudinal velocity U, the cross-sectional area A_c, and the wetted perimeter p are, in general, functions of x. Substituting eqs. (2.141) and (2.142) and the Fourier law $q_x = -kA_c(dT/dx)$ into the conduction equation (2.139) and dividing the result by Δx, we obtain finally

$$\frac{d}{dx}\left(kA_c\frac{dT}{dx}\right) - hp(T - T_\infty) - \rho c A_c U\frac{dT}{dx} + \dot{q}A_c = 0 \qquad (2.143)$$

Relative to the most recent derivation of this sort, eq. (2.122), this new conduction equation accounts for two additional effects, namely, the longitudinal flow of the solid material (the third term) and the effect of volumetric heating (the fourth term). The longitudinal variations of the cross-sectional area A_c and the solid velocity U are related through the mass conservation statement (2.142). It is instructive to look back at eqs. (2.85) and (2.122) and to recognize that they are simpler cases of the general steady-conduction equation for $T(x)$, eq. (2.143).

2.8.2 Plastics Extrusion and Wire Drawing

As a first application of eq. (2.143), consider determining the temperature distribution along a hot plastic fiber that is being extruded [19] through an opening of cross-sectional area A_c (Fig. 2.20). The supply of plastic material upstream of the opening is at a high temperature (T_b). The fiber material travels longitudinally with the velocity U, and is cooled gradually by the contact with the ambient fluid (h, T_∞). The simplest model of this manufacturing process is to

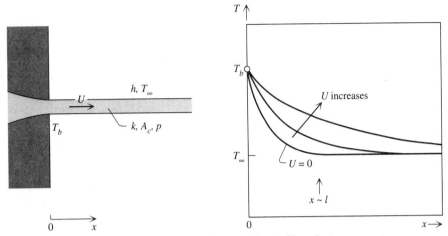

Figure 2.20 Temperature distribution along a plastic fiber during extrusion.

assume that both A_c and U are constant (i.e., independent of x) and that the fiber is sufficiently long so that it reaches thermal equilibrium with the ambient before it is rolled up on a spool. The same model applies to the analysis of the temperature distribution along a metallic wire that, after being heated, is drawn down to a smaller diameter by being pulled through a dye.

Without the volumetric heating term, eq. (2.143) reduces to

$$\frac{d^2\theta}{dx^2} - \frac{U}{\alpha}\frac{d\theta}{dx} - m^2\theta = 0 \tag{2.144}$$

where $\theta = T - T_\infty$, and m is defined in eq. (2.92). The necessary boundary conditions are

$$\theta = \theta_b \quad \text{at} \quad x = 0 \tag{2.145}$$

$$\theta \to 0 \quad \text{as} \quad x \to \infty \tag{2.146}$$

The solution to the problem (2.144)–(2.146) is straightforward:

$$\theta(x) = \theta_b \exp\left(-\frac{x}{l}\right) \tag{2.147}$$

where l is the characteristic length within which the fiber temperature approaches the temperature of the surrounding fluid:

$$l = \left\{\left[\left(\frac{U}{2\alpha}\right)^2 + m^2\right]^{1/2} - \frac{U}{2\alpha}\right\}^{-1} \tag{2.148}$$

Two limiting cases of this result are particularly interesting. First, in the limit of high speeds, $U/2\alpha \gg m$, the "cooling" length l is proportional to the speed of the fiber:

$$l \cong \frac{U}{\alpha m^2} \qquad \left(\frac{U}{2\alpha m} \gg 1\right) \tag{2.149}$$

In the opposite limit, $U/2\alpha \ll m$, the cooling length approaches a constant dictated by the "fin" characteristics of the fiber:

$$l \cong \frac{1}{m} \qquad \left(\frac{U}{2\alpha m} \ll 1\right) \tag{2.150}$$

In this second limit, the fiber approaches the "long" fin with constant cross section, eqs. (2.86) through (2.97). The general behavior of the temperature distribution along the fiber is illustrated in Fig. 2.20.

2.8.3 Electrical Cables

Figure 2.21 shows one way in which a current-carrying cable is attached to a supporting wall ($x = 0$). The other end of the cable is not shown, for it is assumed that the cable is so long that the two supports do not communicate thermally through the cable. Of interest in such applications is the temperature distribution along the cable, and to what extent the Joule heating generated in the cable is dumped via conduction into the support (T_b).

The temperature distribution along the cable can be determined by solving the following problem [cf. eq. (2.143)]:

$$\frac{d^2\theta}{dx^2} - m^2\theta + \frac{\dot{q}}{k} = 0 \tag{2.151}$$

$$\theta = \theta_b \quad \text{at} \quad x = 0 \tag{2.152}$$

$$\theta \to \text{finite value} \quad \text{as} \quad x \to \infty \tag{2.153}$$

where θ is the excess temperature function $T(x) - T_\infty$. The solution to the above problem is

$$\theta(x) = \theta_b \exp(-mx) + \frac{\dot{q}}{m^2 k}\left[1 - \exp(-mx)\right] \tag{2.154}$$

for which the value of the fin parameter m is listed in eq. (2.92). The key features of this solution are presented in Fig. 2.21. The interaction by longitudinal conduction with the $x = 0$ support is always felt over the long-fin length

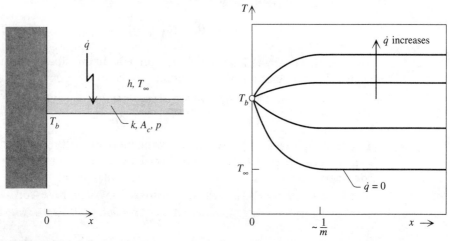

Figure 2.21 Temperature distribution along an electrical cable with volumetric heating effect.

scale $1/m$. Beyond this length, the cable temperature becomes independent of x, namely, $\theta \cong \dot{q}/(m^2 k)$; this shows that the distant section of the cable becomes proportionally warmer as $\dot{q}$ increases. Whether the support will be heated or cooled by the cable depends on how strong the $\dot{q}$ effect is. By calculating the total heat transfer rate through the root of the cable (out of the support), $q_b = -kA_c(d\theta/dx)_{x=0}$, it is easy to show that the support will be heated by the cable ($q_b < 0$) if

$$\frac{\dot{q}A_c}{hp\theta_b} > 1 \tag{2.155}$$

The group on the left side of this inequality is therefore an important dimensionless parameter of the cable. When this group is exactly equal to 1, the entire cable is isothermal, that is, it is at the same temperature as the support.

REFERENCES

1. G. Kirchhoff, *Vorlesungen über die Theorie der Wärme*, Teubner, Leipzig, 1894, p. 13.

2. H. S. Carslaw and J. C. Jaeger, *Conduction of Heat in Solids*, 2nd edition, Oxford University Press, Oxford, 1959, p. 150.

3. I. Grattan-Guiness, *Joseph Fourier 1768–1830*, with J. R. Ravetz, M.I.T. Press, Cambridge, MA, 1972.

4. W. B. Harper and D. R. Brown, Mathematical equations for heat conduction in the fins of air-cooled engines, NACA Report No. 158, 1922.

5. K. A. Gardner, Efficiency of extended surfaces, *Trans. ASME*, Vol. 67, pp. 621–631, 1945.

6. W. M. Rohsenow, J. P. Hartnett and E. N. Ganic, *Handbook of Heat Transfer Applications*, 2nd edition, McGraw-Hill, New York, 1985, Chapter 4.

7. E. Schmidt, Die Wärmeübertragung durch Rippen, *Z. Ver. Dt. Ing.*, Vol. 70, pp. 885–889 and 947–951, 1926.

8. R. J. Duffin, A variational problem relating to cooling fins, *J. Math. Mech.*, Vol. 8, pp. 47–56, 1959.

9. P. Jany and A. Bejan, Ernst Schmidt's approach to fin optimization: an extension to fins with variable conductivity and the design of ducts for fluid flow, *Int. J. Heat Mass Transfer*, Vol. 31, pp. 1635–1644, 1988.

10. A. Bejan, *Convection Heat Transfer*, Wiley, New York, 1984, pp. 62–63.

11. C. J. Maday, The minimum weight one-dimensional straight cooling fin, *J. Engng. Ind.*, Vol. 96, pp. 161–165, 1974.

12. A. D. Snider and A. D. Kraus, The quest for the optimum longitudinal fin profile, ASME HTD, Vol. 64, pp. 43–48, 1986.

13. D. Poulikakos and A. Bejan, Fin geometry for minimum entropy generation in forced convection, *J. Heat Transfer*, Vol. 104, 1982, pp. 616–623.

14. A. Bejan, *Advanced Engineering Thermodynamics*, Wiley, New York, 1988.

15. R. K. Irey, Errors in the one-dimensional fin solution, *J. Heat Transfer*, Vol. 90, pp. 175–176, 1968.

16. W. Lau and C. W. Tan, Errors in one-dimensional heat transfer analyses in straight and annular fins, *J. Heat Transfer*, Vol. 95, pp. 549–551, 1973.

17. J. W. Stachiewicz, Effect of variation of local film coefficients on fin performance, *J. Heat Transfer*, Vol. 91, pp. 21–26, 1969.

18. D. C. Look, Jr., Two-dimensional fin performance: Bi (top surface) ≥ Bi (bottom surface), *J. Heat Transfer*, Vol. 110, pp. 780–782, 1988.

19. E. G. Fisher, *Extrusion of Plastics*, Wiley, New York, 1976.

20. A. Bejan, Theory of heat transfer from a surface covered with hair, *J. Heat Transfer*, Vol. 112, pp. 662–667, 1990.

PROBLEMS

Steady Conduction without Internal Heating

2.1 Consider the thin-wall limit of the thermal resistance of a cylindrical shell with isothermal inner and outer surfaces, eq. (2.33). Writing L for the shell thickness and A for the size of the cylindrical surface in this limit, show that eq. (2.33) reduces to the thin-wall formula (2.7).

2.2 Derive the expression for the thermal resistance of a spherical shell with constant thermal conductivity, eq. (2.40). Show that when the shell is sufficiently thin, this expression becomes identical to eq. (2.7). *Hint:* Follow the analysis exhibited for the cylindrical shell in Section 2.2.

2.3 An isothermal spherical object of radius r_i is coated with a constant-k material of outer radius r_o and exposed to an external flow. The heat transfer coefficient h is a known constant. Determine analytically the critical outer radius of this spherical shell insulation. Is the overall thermal resistance minimum or maximum when r_o is equal to the critical radius?

2.4 Derive the steady-state temperature distribution through a cylindrical body that is being heated internally such that the volumetric rate $\dot{q}$ is a constant (Fig. 2.9b). The thermal conductivity of the solid material (k) and the cylinder–fluid heat transfer coefficient (h) are also constant.

2.5 The pipe shown in the figure carries a water stream of temperature $T_w = 4°C$. It functions in a cold climate, the effect of which is to lower the pipe wall temperature to $T_s = -10°C$. A cylindrical sheet of ice builds over the pipe wall. The inner diameter of the pipe wall is 8 cm, and the convective heat transfer coefficient at the inner surface of the ice layer is $h = 1000 \ W/m^2 \cdot K$. Calculate the thickness of the ice layer.

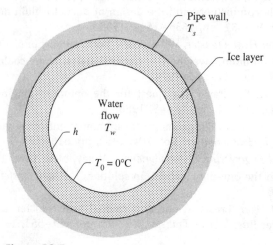

Figure P2.5

2.6   The spherical probe shown in the figure is immersed in a bath of paraffin (*n*-octadecane) of temperature $T_\infty = 35°C$. The wall of the spherical probe is maintained at a lower temperature, $T_s = 10°C$. The solidification point of this paraffin falls between these two temperatures, $T_m = 27.5°C$, and this leads to the formation of a layer of solid paraffin all around the spherical object. Calculate the thickness of the solidified paraffin layer, considering that the radius of the spherical object is $r_i = 1$ cm, the conductivity of the solid paraffin is $k = 0.36$ W/m·K, and the heat transfer coefficient at the outer surface of the solidified layer is assumed constant, $h = 100$ W/m²·K.

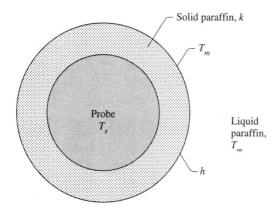

Figure P2.6

2.7   The insulated box shown in the figure is designed to keep the air trapped in it at a high temperature, $T_h = 50°C$. The outside temperature is $T_c = 10°C$. To keep the temperature of the trapped air constant, the heat transfer that leaks through the insulation, q, is made up by an electrical resistance heater placed in the center of the box. Calculate the electrical power dissipated in the heater.

The dimensions of the internal (air) space are $x = 1$ m, $y = 0.4$ m, and $z = 0.3$ m. The insulating wall consists of a 10-cm-thick layer of fiberglass sandwiched between two plywood sheets. The plywood sheet is 1 cm thick. The heat transfer coefficients on the internal and external surfaces of the wall are, respectively, $h_h = 5$ W/m²·K and $h_c = 15$ W/m²·K.

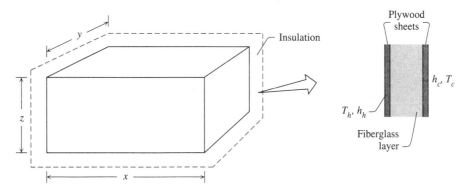

Figure P2.7

2.8 The homogeneous and isotropic porous medium shown on the left side of the figure is saturated with a stagnant fluid of thermal conductivity k_f. The thermal conductivity of the solid material (the matrix) is k_s. The porosity (void fraction) of the medium, or the ratio of the fluid space divided by the total space, is equal to φ.

This L-thick layer of porous material is placed between two parallel walls at different temperatures, T_h and T_c. To determine the conduction heat flux from T_h to T_c across the layer, model the porous medium as a succession of parallel columns of solid and fluid arranged in such a way that the void fraction is equal to the same φ. Show that the heat flux is

$$q'' = k_{\text{eff}} \frac{T_h - T_c}{L}$$

in which k_{eff} is the "effective" thermal conductivity of the porous medium:

$$k_{\text{eff}} = (1 - \varphi)k_s + \varphi k_f$$

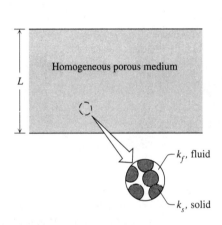

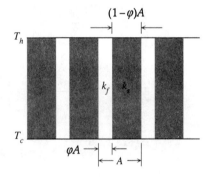

Figure P2.8

2.9 The temperature of the human body drops from the core (or arterial blood) level $T_a = 36.5°C$ to the skin level T_w across a subskin layer of thickness $\delta \sim 1$ cm and conductivity $k \cong 0.42$ W/m·K.

(a) Calculate the skin temperature when the body is swept by the flow of 20°C air with a heat transfer coefficient of 30 W/m²·K.

(b) Calculate the skin temperature when the external fluid is 10°C water and the skin–water heat transfer coefficient is 500 W/m²·K.

(c) How much greater is the rate of body heat loss to water (b) than to air (a)?

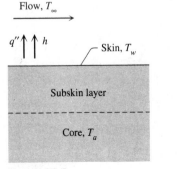

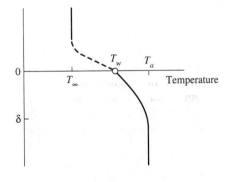

Figure P2.9

2.10 The wall of a steam pipe has a temperature $T_i = 100°C$ and an outer radius $r_i = 4$ cm. The ambient air temperature is $T_\infty = 15°C$, and the heat transfer coefficient between the cylindrical surface and ambient air is $h = 10$ W/m²·K.

You would like to build a 1-cm-thick layer of polystyrene insulation around this bare steam pipe. Will this design change have a "thermal insulation" effect? In other words, will the heat transfer rate from the steam to the atmosphere decrease if you install the polystyrene shell? To verify your answer, calculate the steam–air heat transfer rate when the pipe wall is bare and when it is covered by the polystyrene layer. Assume the same h value in both cases.

2.11 The mechanical support for a low-temperature apparatus (a cryostat) has the geometry shown in the figure. It consists of two pieces joined end-to-end, their lengths and cross-sectional areas being L_1, A_1 and L_2, A_2, respectively. The two pieces are made out of the same material. The thermal conductivity of the material varies strongly with the temperature, $k(T)$.

The lateral surface of the entire support is insulated perfectly. Both pieces are sufficiently slender, so that the conduction heat transfer is practically unidirectional through the entire support (i.e., in the x direction in the figure). As a designer, you have a choice between design "a" and design "b". A "good" design is one in which the conduction heat transfer rate through the support (from T_{high} to T_{low}) is small. Which of the two designs is better?

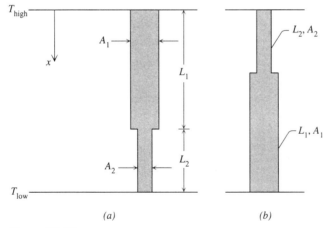

(a)   (b)

Figure P2.11

Internal Heat Generation

2.12 Consider a solid sphere immersed in a fluid flow, with which it communicates thermally through a constant heat transfer coefficient h (Fig. 2.9b). The rate of internal heat generation in the sphere, $\dot{q}$, is constant. Determine the steady-state temperature distribution inside the spherical body, by assuming that its thermal conductivity k is a known constant.

2.13 The resistivity of a certain cylindrical electrical conductor increases linearly with the temperature, as shown in eq. (2.67). The outer surface (radius r_o) is exposed to a fluid of temperature T_∞ through a constant heat transfer coefficient h. The thermal conductivity of the conductor, k, is also constant.

Repeating the integral analysis steps outlined in Section 2.6.2, show that the conductor temperature is steady and finite if

$$J < \frac{2^{3/2}}{r_o} \left(\frac{k}{\rho'_e} \right)^{1/2} \left(\frac{\text{Bi}}{\text{Bi} + 4} \right)^{1/2}$$

where $\rho'_e = d\rho_e/dT$, $\text{Bi} = hr_o/k$, and J is the electric current density. Show that this criterion is more stringent than eq. (2.77), namely, that the finiteness of h (or Bi) reduces the domain of stable operation of the electrical conductor.

2.14 The coal powder shown in the figure is stockpiled in a layer of constant thickness $H = 2$ m. This layer is heated volumetrically at the rate $\dot{q} = 50$ W/m³, which is due to the chemical reaction between coal particles and the interstitial air. The effective thermal conductivity of the coal layer is $k = 0.2$ W/m·K, and the heat transfer coefficient at the upper surface is $h = 2$ W/m²·K. The temperature of the lower surface of the coal layer equals the temperature of the ground, $T_0 = 25°$C. The atmospheric air temperature is also 25°C.

(a) Show that the maximum temperature registered inside the coal layer is

$$T_{\text{max}} = T_0 + \frac{\dot{q}H^2}{2k} \left(\frac{1 + (\text{Bi}/2)}{1 + \text{Bi}} \right)^2$$

where Bi is the Biot number hH/k. Show also that, in general, this temperature maximum is located in a horizontal plane situated between $y = H/2$ and $y = H$.

(b) Calculate numerically the maximum temperature and the position of this hottest plane relative to ground level.

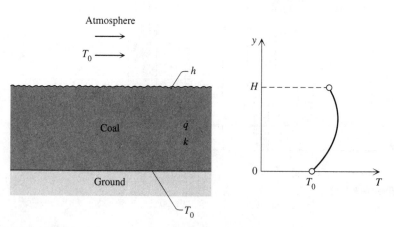

Figure P2.14

2.15 Inserted in the middle of the conducting slab shown in the figure is a plane heat source, that is, an infinitesimally thin blade that generates heat at the rate q'' per unit of blade area. The generated heat, q'', proceeds to the right and to the left, away from the midplane. The temperature of the two lateral surfaces is fixed at T_o. Determine analytically the steady temperature distribution through the entire slab of thickness L. Derive also an expression for the temperature of the plane heat source.

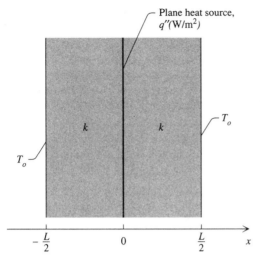

Figure P2.15

2.16 The long cylindrical rod shown in the figure is being heated steadily by a line-shaped heat source situated along its axis. The source generates heat at the rate q' per unit axial length. The temperature of the external surface $r = r_o$ is fixed at T_o. Determine analytically the steady temperature distribution in the rod cross section. Show that the temperature at points close to the line heat source blows up as $\ln(r_o/r)$ when $r \to 0$.

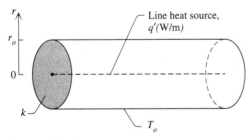

Figure P2.16

2.17 Embedded in the geometric center of a conducting spherical body is a steady, point-size source of heat q. The temperature of the spherical surface is maintained at T_o by contact with an external fluid. Determine analytically the temperature distribution through the spherical body, and show that the core temperature blows up as r_o/r when $r \to 0$.

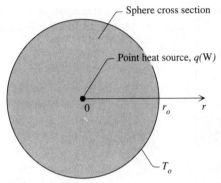

Figure P2.17

Extended Surfaces (Fins)

2.18 Show that the heat current through the base of a fin is equal to the integral of the heat flux over the total area bathed by fluid (A_{total}):

$$q_b = \int_{A_{total}} h(T - T_\infty) \, dA_{total}$$

This can be demonstrated in two ways:

(a) Integrate the fin conduction equation, eq. (2.85), from $x = 0$ to $x = L$, while looking at Fig. 2.11, or

(b) Apply the first law of thermodynamics to the entire fin as a control volume. Along this second route, you will find that the relation listed above holds for any fin geometry.

2.19 The temperature distribution along the two-dimensional fin with sharp tip shown in the figure is *linear*:

$$\theta(x) = \frac{x}{L}\theta_b$$

The width of this fin, W, is not shown (it would be normal to the plane of the figure). Note also that x points toward the base of the fin and that the tip is in thermal equilibrium with the surrounding fluid, $\theta(0) = 0$. Recognizing that this fin

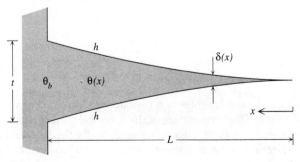

Figure P2.19

must have a *variable* cross-sectional area (because a constant-A_c fin does not have a linear temperature distribution), show that the profile thickness δ is parabolic in x:

$$\delta = \frac{h}{k}x^2$$

Derive an expression for the total heat transfer rate through the base of this fin, q_b. Show that if the total volume of the parabolic profile fin, V, is the same as the volume of the plate fin optimized in Section 2.7.5, then q_b is 14.8 percent larger than the maximum heat transfer rate that can be accommodated by the plate fin, eq. (2.138).

2.20 Consider the process of longitudinal conduction through the two-dimensional fin of triangular profile shown in the figure. The width of the fin (not shown) is W. Demonstrate that the temperature distribution along this fin, $\theta(x)$, must satisfy the differential equation

$$x\frac{d^2\theta}{dx^2} + \frac{d\theta}{dx} - a\theta = 0$$

where $a = 2hL/kt$. Solve this equation by assuming the series expression

$$\theta = c_0 + c_1x + c_2x^2 + \cdots + c_nx^n + \cdots$$

where $c_0, c_1, c_2 \cdots$ are constant coefficients. Show that the $\theta(x)$ solution is

$$\theta(x) = \theta_b\frac{F(ax)}{F(aL)} \qquad \text{where } F(ax) = \sum_{n=0}^{\infty}\frac{(ax)^n}{(n!)^2}$$

The $F(ax)$ series is better known as the modified Bessel function of the first kind, of order zero, $I_0[2(ax)^{1/2}]$.

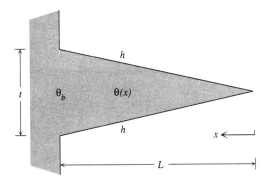

Figure P2.20

2.21 Derive an expression for the total heat transfer rate q_b corresponding to the exact temperature distribution (2.109). Compare this result with the approximate corrected-length formula (2.110), and show that:

(a) When the fin is "long," $mL \gg 1$, the approximate formula is accurate.

(b) When the fin is "short," $mL \ll 1$, the error associated with using eq. (2.110) instead of eq. (2.109) is approximately equal to

$$\frac{hA_c}{3kp}$$

Show further that this error is small when the Biot number based on fin transversal dimension (e.g., thickness, if plate fin) is small when compared with 1.

2.22 The idea of a constant fluid temperature (T_∞) all around a fin is only an approximation of the fluid temperature that might occur in an actual application. In laminar flow parallel to the base wall, for example, the fluid temperature varies in the direction normal to the wall, $T_\infty(x)$. Consider the case where the variation of T_∞ mimicks that of the fin temperature, $T(x)$, such that at every point along the fin the fin–fluid temperature difference is a known constant,

$$T(x) - T_\infty(x) = \Delta T \quad \text{(constant)}$$

Assume also that the remaining parameters of the fin (k, A_c, h, p) are known constants, and that there is heat transfer through the tip. Determine the temperature distribution along the fin, $T(x)$, and the total heat transfer rate through the base, q_b.

2.23 The ring shown in the figure is made out of a thin wire of length L, cross-sectional area A_c, cross-sectional perimeter p, and thermal conductivity k. The heat transfer coefficient between the wire and the ambient air is h. One point ($x = 0$) on the ring is heated by internal means so that its temperature T_b is maintained above the ambient temperature T_∞. Determine analytically the following:

(a) The temperature distribution along the ring.
(b) The temperature at the point that is diametrically opposed to $x = 0$.
(c) The total heat transfer rate q_b that must be applied at $x = 0$.

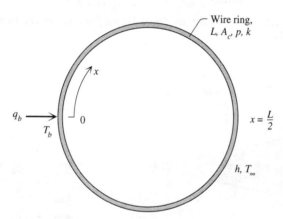

Figure P2.23

2.24 The Japanese teapot *kyusu* has a hollow handle that feels cool to the touch even when the pot contains boiling water. This handle can be modeled as a hollow-cylinder fin, as shown on the right side of the figure, where

$$\begin{aligned}
D_o &= 2 \text{ cm} & h_o &= 10 \text{ W/m}^2 \cdot \text{K} & T_b &= 50°\text{C} \\
D_i &= 1.5 \text{ cm} & h_i &= 2 \text{ W/m}^2 \cdot \text{K} & T_\infty &= 15°\text{C} \\
L &= 7 \text{ cm}
\end{aligned}$$

The handle is made out of porcelain. The heat transfer coefficient on the outer surface of the handle is considerably larger than on the inner surface, because the buoyancy-driven air flow is more intense around the handle.

(a) Calculate the appropriate parameter m of the hollow-cylinder fin and show that this fin is a "long fin."

(b) Calculate the distance from the base of the handle where the handle temperature is 30°C.

(c) Calculate the instantaneous heat transfer rate through the base of the handle.

Figure P2.24

2.25 The copper fin (k) with the shape and dimensions shown in the figure is soldered to a wall of temperature T_w. The solder film (k_s) has the thickness t, which is exaggerated in the figure. The heat transfer coefficient between the fin and the surrounding fluid has the same value h on all the exposed surfaces. Derive an expression for the total heat transfer rate through the fin, as a function of only the dimensions and properties identified on the drawing.

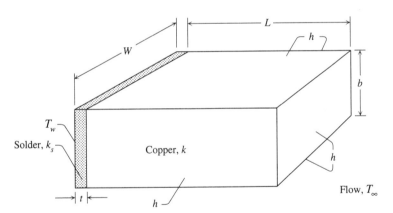

Figure P2.25

2.26 Like everything else that we see from the first moments of conscious life, hair and fur are things that we take for granted. Their purpose seems obvious not only with respect to our own needs, but also for the survival of numerous other species. The insulation effect of hair can be explained through the argument that when the hair strands are sufficiently dense, they trap a blanket of air in the tight spaces created between the strands. Well known for its low thermal conductivity, this air blanket insulates the warm skin against the colder atmosphere.

The hair-covered surface is considerably more interesting when approached from the point of view—the function—of the individual hair strand. The thermal conductivity of hair material (roughly that of human skin, 0.37 W/m·K) is 14 times greater than that of ambient air. It seems that one effect of the hair strands is to "extend" the warm surface into the cold atmosphere, in the same way that a population of cylindrical spines (pin fins) extends the heat transfer surface of a heat exchanger. This finning effect is, most certainly, an unwanted by-product of the true function of the hair strands, which is to slow down the breeze that would otherwise blow by. This effect is examined in Ref. [20].

The fur of a lynx, for example, has a density of 9000 strands/cm² and a cylindrical hair strand with a diameter of 27×10^{-6} m. The heat transfer coefficient between the hair strand and the surrounding air is 100 W/m²·K, and the heat transfer coefficient between the bare portions of the skin and the same air flow is 10 W/m²·K. The temperature difference between the skin surface and the air flow happens to be 10 K. The fur-covered area under consideration is a square with side 10 cm.

(a) Calculate the heat transfer rate that leaves the skin through the roots of all the hair strands. Treat each hair strand as a long fin.

(b) Calculate the heat transfer rate released by the bare portions of the skin. Compare this contribution with the heat transfer current calculated in part (a).

(c) Estimate the order of magnitude of the distance of conduction penetration along the hair strand. In other words, for what length near its root is the hair strand significantly warmer than the surrounding air?

(d) Verify that the heat transfer through one hair strand can be indeed modeled as one of unidirectional conduction (cf. Section 2.7.4).

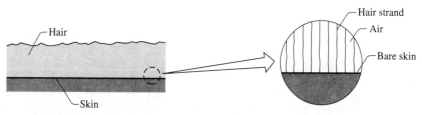

Figure P2.26

2.27   The junction between two electrical leads suspended in air is defective (i.e., resistive) and generates heat at the known rate q. On both sides of the plane of the junction, the area and wetted perimeter of the lead cross section are A_c and p, respectively. The thermal conductivities and heat transfer coefficients are different, k_1, h_1 and k_2, h_2.

(a) Develop a formula for calculating the rise of the junction temperature T_0 above the ambient temperature T_∞.

(b) Show that the fraction of the junction heating rate that escapes to one side of the junction (q_1/q or q_2/q) is a function of only the dimensionless "partition parameter"

$$R_{1,2} = \left(\frac{k_1 h_1}{k_2 h_2}\right)^{1/2}$$

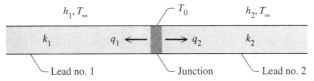

h_1, T_∞ T_0 h_2, T_∞

k_1 $q_1 \leftarrow$ $\rightarrow q_2$ k_2

Lead no. 1   Junction   Lead no. 2

Figure P2.27

2.28 A semi-infinite electrical cable with constant parameters (k, A_c, h, p) is attached to a base wall of temperature T_b. The electrical resistivity of the cable material is such that the Joule heating effect $\dot{q}$ increases linearly with the local temperature of the cable, $T(x)$:

$$\frac{\dot{q}}{k} = C_1 + C_2(T - T_\infty)$$

In this expression, C_1 and C_2 are two known constants, whereas T_∞ (also constant) is the temperature of the surrounding fluid. Determine the temperature distribution $T(x)$ and the condition that must be met to avoid the thermal runaway (burnup) of the cable.

2.29 Determine the optimal diameter D for which the total heat transfer rate q_b through a pin fin is maximum, when the total volume V of the fin is fixed. Assume that the pin fin is sufficiently slender, so that $L_c \cong L$, where L is the length of the fin.

Determine also the maximum heat transfer rate that corresponds to this optimum design, $q_{b,max}$. Show that under the same conditions the pin slenderness ratio D/L is proportional to the square root of the Biot number based on diameter, hD/k.

PROJECTS

2.1 How to Distribute a Finite Amount of Insulation on a Wall with Nonuniform Temperature

The temperature of the wall of a long oven increases linearly in the longitudinal direction, $T(x)$. The cold end ($x = 0$) is at the ambient temperature T_0, and the other end ($x = L$) reaches the high temperature T_L. The width of the wall in the direction perpendicular to the plane of the figure is W. The wall must be insulated with respect to the T_0 surroundings, by using a layer of insulation of thermal conductivity k, and variable thickness t. The insulation temperature at the wall surface is $T(x)$, and at the exposed surface (in contact with room air) is T_0. The amount (volume V) of this insulation material is fixed by economic considerations,

$$V = \int_0^L W\, t(x)\, dx$$

As a designer, you have several options for distributing this limited amount of insulation over the length L, that is, for choosing the insulation thickness $t(x)$. Your objective is to minimize the total rate of heat transfer through the wall surface $W \times L$.

(a) Assume that the thickness is uniform, and derive an expression for the total rate of heat transfer.

(b) It occurs to you later that no insulation is actually needed at $x = 0$, because there the wall-ambient temperature difference is zero. Perhaps a better way to distribute the insulation would be to use a thickness that increases proportionally with x. Derive an expression for the total heat transfer rate, and compare it with the

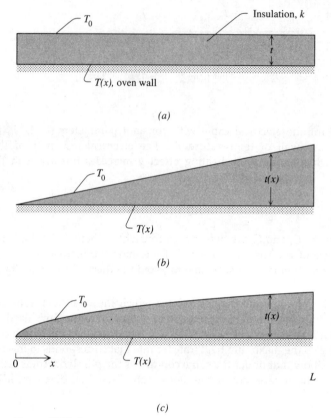

Figure PR2.1

result obtained in part (a). What do you do next? Do you give up on the idea of using a nonuniform insulation thickness for minimum heat loss?

(c) Before you give up, try a design that falls between (a) and (b), namely an insulation thickness that increases as $x^{1/2}$ in the longitudinal direction. Once again, determine the total heat transfer rate through the insulation, and compare it to the previous results.

Note: It can be shown that this third alternative is in fact the optimal design, that is, the best way of distributing the fixed amount of insulation material. In the even more general case where the wall temperature $T(x)$ does not vary linearly, the optimal thickness function is given by

$$t_{opt}(x) = \frac{t_{avg}L[T(x) - T_0]^{1/2}}{\displaystyle\int_0^L (T - T_0)^{1/2}\, dx}$$

where t_{avg} is the L-averaged thickness of the insulating layer, $t_{avg} = V/WL$. Note that t_{avg} is fixed by the volume constraint recognized early in the problem statement. The complete solution is beyond the objective of the present treatment, and is given in the Solutions Manual.

2.2 The Best Shape of a Unidirectional Thermal Conductance with Variable Thermal Conductivity

The purpose of a long piece of thermally conducting material is to maximize the thermal contact between two bodies of different temperatures (T_0, T_L) situated a

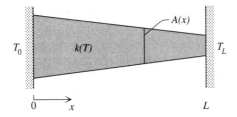

Figure PR2.2

distance L apart. The heat transfer through this strut is unidirectional. The amount (volume V) of conducting material is fixed,

$$V = \int_0^L A(x)\, dx$$

while the thermal conductivity increases linearly with the temperature,

$$k(T) = \bar{k}[1 + b(2\theta - 1)]$$

In this relation, $\bar{k}$ is the thermal conductivity averaged over the temperature interval spanned by the material, b is a positive dimensionless constant proportional to dk/dT, and

$$\theta = \frac{T - T_0}{T_L - T_0}$$

The challenge is to come up with the strut shape $A(x)$ that maximizes the heat transfer between the two bodies (assume that T_L is high, and T_0 is low). If you select the strut with x-independent geometry, $A =$ constant, you may doubt that it is the best choice because near the cold end ($x = 0$) the thermal conductivity is lower than near the warm end ($x = L$). Perhaps, if we increase the cross-sectional area near the cold end we will compensate for the lower thermal conductivity, and increase the overall thermal conductance of the strut. Investigate the goodness of this idea by assuming a strut shape that is tapered in the x direction,

$$A(x) = \bar{A}\left[1 - a\left(2\frac{x}{L} - 1\right)\right]$$

where $\bar{A}$ is the L-averaged cross-sectional area, $\bar{A} = V/L$, and a is a positive dimensionless number proportional to the taper dA/dx. Determine the best geometry $A(x)$ [that is, the best taper parameter a] that maximizes the overall thermal conductance of the strut. Does the best geometry depend on the thermal conductivity parameter b?

Note: The solution to the general design problem in which A can be any function of x, and k any function of T, is beyond the level of this course and is given in the Solutions Manual.

2.3 How to Shape a Bar That Must Be Stiff in Tension and a Good Thermal Insulator

A low temperature apparatus (T_0) is supported from room temperature (T_L) through a vertical bar in tension. This bar must be strong enough to carry the weight of the apparatus (F), slender enough to allow the smallest heat transfer rate q to flow from T_L to T_0, and rigid enough to fix the position of the apparatus on the vertical line. The rigidity requirement means that the total elastic elongation ΔL caused by the longitudinal force F must be as small as possible.

At certain temperatures below room temperature, the thermal conductivity and modulus of elasticity of structural materials vary roughly as

$$k(T) \cong \bar{k}[1 + b(2\theta - 1)]$$

$$E(T) \cong \bar{E}[1 - c(2\theta - 1)]$$

where $\bar{k}$ and $\bar{E}$ are the temperature averaged values

$$\bar{k} = \frac{1}{T_L - T_0} \int_{T_0}^{T_L} k\, dT \qquad \bar{E} = \frac{1}{T_L - T_0} \int_{T_0}^{T_L} E\, dT$$

b and c are two positive numbers less than 1, proportional to dk/dT and $-dE/dT$, and

$$\theta = \frac{T - T_0}{T_L - T_0}$$

In other words, k decreases and E increases toward lower temperatures. These characteristics of structural materials give the designer the idea that a tapered bar $(0 < a < 1)$

$$A(x) = \bar{A}\left[1 + a\left(2\frac{x}{L} - 1\right)\right], \qquad (0 < a < 1)$$

$$\bar{A} = \frac{1}{L} \int_0^L A(x)\, dx, \qquad \text{(constant)}$$

might exhibit a smaller elongation in tension than the bar with x-independent geometry $(a = 0)$. It can be argued that the overall elongation is reduced when more material is positioned near the warm end, where E is smaller than near the cold end.

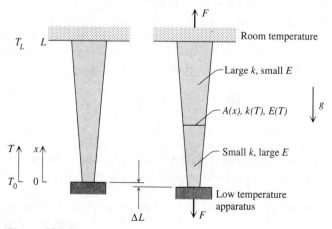

Figure PR2.3

Note further that by holding the L-averaged cross-sectional area fixed, the total amount (weight) of the structural material used is fixed, regardless of the shape chosen for the bar.

Evaluate the merit of this proposal by determining the effect of taper (a) not only on the total elongation ΔL but also on the heat leak q. It is advisable to conduct the analysis using the dimensionless heat transfer rate

$$\tilde{q} = \frac{q}{\overline{kA}(T_L - T_0)/L}$$

and the dimensionless total elongation

$$\delta = \frac{\Delta L}{L} \frac{\overline{AE}}{F}$$

Determine δ and $\tilde{q}$ as functions of geometry (a) and choice of structural material (b,c), and comment on how you might design a rigid support in tension so that it can serve as thermal insulation at the same time.

2.4 How to Shape a Bar That Must Be Stiff in Bending and a Good Thermal Insulator

The horizontal beam with variable thickness $H(x)$ shown in the figure connects a body of weight F and temperature T_0 to a support of temperature T_L. The beam geometry is slender and two-dimensional, with the width B measured in the direction perpendicular to the figure. The distance L, and the amount of beam material are fixed. The thermal conductivity (k) and modulus of elasticity (E) of this material are known constants.

The function of the beam is to support the weight F as rigidly as possible, while impeding the transfer of heat from T_L to T_0. It is, simultaneously, a mechanical support and a thermal insulation. The designer is interested in arranging the beam material [the thickness, $H(x)$] in such a way that the tip deflection y_0 and the end-to-end heat transfer rate q are minimized.

One possibility is to shape the beam cross section such that the thickness H increases as x^n as x increases, where the exponent n is a number between 0 and 1. Determine analytically y_0 and q as functions of n, and comment on how the shape of the beam cross section (n) affects the success of the design.

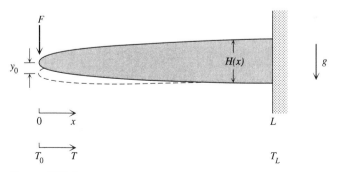

Figure PR2.4

2.5 How to Taper a Beam That Must Be Stiff in Bending and a Good Thermal Insulator: The Effect of Temperature-Dependent Properties

The function of the horizontal beam shown in the figure is to support the weight F most rigidly, that is, with minimum tip displacement downward, y_0. The beam is slender and two-dimensional, with the width B measured in the direction perpendicular to the plane of the figure. The thickness varies linearly with x,

$$H(x) = \overline{H}\left[1 + a\left(2\frac{x}{L} - 1\right)\right], \quad (-1 < a < 1)$$

where the dimensionless taper parameter a is a design variable, and $\overline{H}$ is the L-averaged thickness. The dimensions (L, B, $\overline{H}$) are fixed, and this means that the amount of structural material ($LB\overline{H}$) is also fixed.

This mechanical support bridges the temperature gap $T_0 - T_L$, which causes the temperature variation $T(x)$ and the flow of heat in the longitudinal direction. More critical to this design is the way in which the local temperature $T(x)$ affects the properties of the structural material. Assume that the modulus of elasticity and the thermal conductivity vary linearly with the temperature,

$$E(T) = \overline{E}[1 - c(2\theta - 1)], \quad (-1 < c < 1)$$
$$k(T) = \overline{k}[1 + b(2\theta - 1)], \quad (-1 < b < 1)$$

where $\overline{E}$ and $\overline{k}$ are the averages of $E(T)$ and $k(T)$ over the interval $T_0 - T_L$, and in which

$$\theta = \frac{T - T_0}{T_L - T_0}$$

Derive the relations necessary for evaluating numerically the tip displacement y_0, as a function of taper (a) and choice of structural material (b,c). Choose the material (for example, set $b = -0.5$, $c = 0.5$), and minimize y_0 numerically with respect to a. Other students may do the same work while using other (b,c) pairs. In this way one can determine the optimal taper (a_{opt}) for minimum tip deflection, as a function of the temperature dependent properties of the chosen structural material (b,c).

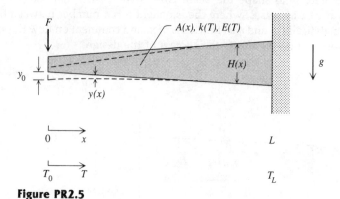

Figure PR2.5

3

MULTIDIRECTIONAL STEADY CONDUCTION

3.1 ANALYTICAL SOLUTIONS

3.1.1 Two-Dimensional Conduction in Cartesian Coordinates

In this chapter we focus on a more complicated class of conduction problems, the distinguishing feature of which is the multidirectionality of the heat transfer process. Unlike in the problems of the preceding chapter, where the heat flux lines pointed always in only one direction (longitudinally, or radially), in the present chapter we consider those situations where heat is conducted in more than one direction. A situation of this kind was identified already in the discussion of when the unidirectional conduction model is no longer appropriate for description of the heat transfer through a fin (Section 2.7.4). We learned then that the heat flux lines flare out in two or more directions when the order of magnitude of the Biot number based on thickness (ht/k) is larger than 1.

As a first example of steady conduction in two dimensions, consider the rectangular-profile plate fin precisely in the limit where the unidirectional conduction assumption breaks down, $Bi \to \infty$. An infinite Biot number ht/k means that the heat transfer coefficient h is infinite. Furthermore, if the heat flux through the fluid-bathed surface is finite, the "infinite h" limit is one where the temperature difference between surface and fluid is zero [cf. eq. (1.54)]. Figure 3.1 shows the temperature boundary conditions that surround the rectangular profile in this limit. The thickness of the profile is now labeled H instead of t, because soon t will be used as symbol for time. Note further the constant temperature T_b maintained along the base $(x = 0)$, and the fact that the remaining three sides of the profile are in thermal equilibrium with the ambient fluid (T_∞).

Homogeneous Boundary Conditions. The problem consists of determining the temperature field inside the two-dimensional rectangular domain, $T(x,y)$, because only when $T(x,y)$ is known can we evaluate the total heat transfer rate through the base:

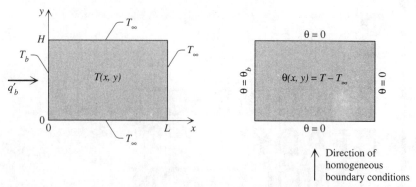

Figure 3.1 Two-dimensional conducting medium with isothermal boundaries (left side), and the emergence of a direction of homogeneous boundary conditions (right side).

$$q'_b = \int_0^H q''_b \, dy = -k \int_0^H \left(\frac{\partial T}{\partial x}\right)_{x=0} dy \tag{3.1}$$

The total heat current q'_b (W/m) is expressed per unit width, that is, per unit length in the direction normal to the plane of Fig. 3.1. The unknown temperature field $T(x,y)$ must satisfy the equation for two-dimensional steady conduction in a medium with constant thermal conductivity and zero internal heat generation rate, eq. (1.40):

$$\frac{\partial^2 T}{\partial x^2} + \frac{\partial^2 T}{\partial y^2} = 0 \tag{3.2}$$

This equation is often referred to as *Laplace's equation*. The boundary conditions indicated in Fig. 3.1 are written finally as

$$T = T_b \qquad \text{at} \qquad x = 0 \tag{3.3a}$$

$$T = T_\infty \qquad \text{at} \qquad x = L \tag{3.3b}$$

$$T = T_\infty \qquad \text{at} \qquad y = 0 \tag{3.3c}$$

$$T = T_\infty \qquad \text{at} \qquad y = H \tag{3.3d}$$

The most important aspect of the following analysis is the analytical *method*, that is, the structure, or sequence of steps that must be recognized en route to the final answer. The same structure holds together in the analysis of any other multidirectional steady conduction problem that might demand a solution. The first step in the present problem is the adoption of the excess temperature function as the unknown,

$$\theta = T - T_\infty \tag{3.4}$$

In terms of θ, the two-dimensional conduction equation is now written as

$$\frac{\partial^2 \theta}{\partial x^2} + \frac{\partial^2 \theta}{\partial y^2} = 0 \tag{3.5}$$

This equation must be solved subject to four conditions:

$$\theta = \theta_b \quad \text{at} \quad x = 0 \qquad (3.6a)$$

$$\theta = 0 \quad \text{at} \quad x = L \qquad (3.6b)$$

$$\theta = 0 \quad \text{at} \quad y = 0 \qquad (3.6c)$$

$$\theta = 0 \quad \text{at} \quad y = H \qquad (3.6d)$$

The effect of using θ (instead of T) as the unknown of the problem is that the last three boundary conditions are now *homogeneous*. Note the zeros appearing on the right side of eqs. (3.6b)–(3.6d): in the original formulation, eqs. (3.3b)–(3.3d), the right side of each equation was occupied by a constant.

Special among the homogeneous conditions engineered above are the last two, because *both* eqs. (3.6c) and (3.6d) refer to special positions marked along the same direction (y). In the x direction, on the other hand, only the $x = L$ condition is homogeneous. We conclude that in the present problem the y direction is *the direction of homogeneous boundary conditions*. The identification of such a direction is critical; this is why the original problem statement, eqs. (3.2)–(3.3), which contained no homogeneous boundary conditions, had to be restated in terms of the excess temperature θ.

Separation of Variables. The second step is the search for a product-type solution,

$$\theta(x,y) = X(x)Y(y) \qquad (3.7)$$

where $X(x)$ and $Y(y)$ are two unknown functions. Substituting this expression into eq. (3.5) and dividing the result by the product XY yields

$$\underbrace{\frac{X''}{X}}_{\substack{\text{Not a} \\ \text{function} \\ \text{of } y}} + \underbrace{\frac{Y''}{Y}}_{\substack{\text{Not a} \\ \text{function} \\ \text{of } x}} = 0 \qquad (3.8)$$

What the product expression (3.7) accomplishes is the separation of the x and y variables in the conduction equation. The first term in eq. (3.8) is, at best, a function of x only. Similarly, the second term can be either a constant or a function of y. The only way in which both terms can coexist in the same equation for *any* values of x and y is if both terms are constant.

Let the positive constant λ^2 be the value of the first term, X''/X. Equation (3.8) splits then into two separate equations for $X(x)$ and $Y(y)$:

$$X'' - \lambda^2 X = 0 \qquad (3.9)$$

$$Y'' + \lambda^2 Y = 0 \qquad (3.10)$$

The general solutions to these linear ordinary differential equations are well known:

$$X = C_1 \sinh(\lambda x) + C_2 \cosh(\lambda x) \qquad (3.11)$$

$$Y = C_3 \sin(\lambda y) + C_4 \cos(\lambda y) \qquad (3.12)$$

Substituting them into eq. (3.7) yields the general solution for θ,

$$\theta = K[\sinh(\lambda x) + A\cosh(\lambda x)][\sin(\lambda y) + B\cos(\lambda y)] \tag{3.13}$$

where the new constants K, A, and B are shorthand for $C_1 C_3$, C_2/C_1, and C_4/C_3, respectively. We note also that in the special case when $\lambda = 0$ the general solution (3.11)–(3.12) degenerates into the linear forms $X = C_1 x + C_2$ and $Y = C_3 y + C_4$. In the same case, eq. (3.13) becomes $\theta_0 = K(x + A)(y + B)$. Soon we shall find that the $\lambda = 0$ case (i.e., θ_0) contributes zero to the ultimate solution that we seek. For this reason we continue the analysis by writing the general solution as in eq. (3.13).

As an aside, note that in the statement above eq. (3.9), we had the choice to assign the positive constant λ^2 to the second term of eq. (3.8), writing instead of eqs. (3.9)–(3.10)

$$X'' + \lambda^2 X = 0 \tag{3.9$'$}$$

$$Y'' - \lambda^2 Y = 0 \tag{3.10$'$}$$

The end result of this alternative route is the general solution

$$\theta = K'[\sin(\lambda x) + A'\cos(\lambda x)][\sinh(\lambda y) + B'\cosh(\lambda y)] \tag{3.13$'$}$$

in which K', A', and B' are three new constants. That the general solution (3.13$'$) is equivalent to the earlier form (3.13) can be seen by replacing λ by $i\lambda$ in eq. (3.13).

To choose between a θ solution written as in eq. (3.13) and one arranged as in eq. (3.13$'$) is to choose between y and, respectively, x as the direction in which the temperature variation is described by *multivalued functions* (sine, cosine). This choice is dictated by the boundary conditions of the problem. In the next section we learn that in the present problem the correct choice is to orient the multivalued functions (called *characteristic functions*) in the y direction, eq. (3.13), because in Fig. 3.1 the y direction is the direction of homogeneous boundary conditions.

Orthogonality.   Continuing then with the θ expression listed in eq. (3.13), we note that the general solution depends on four unknowns, namely, K, A, B, and λ. These unknowns will be pinpointed one by one by the four boundary conditions (3.6a)–(3.6d), of which the lone nonhomogeneous condition (3.6a) will be invoked last.

We begin with the tip condition (3.6b), which, combined with eq. (3.13), requires that

$$\sinh(\lambda L) + A\cosh(\lambda L) = 0 \tag{3.14}$$

This equation pinpoints the value of the A constant; substituting this value in the first pair of brackets of the θ expression (3.13) we obtain

$$\sinh(\lambda x) + A\cosh(\lambda x) = \frac{\sinh[\lambda(x - L)]}{\cosh(\lambda L)} \tag{3.15}$$

According to the observation made under eq. (3.13), in the special case of $\lambda = 0$ we obtain $A = -L$ in place of eq. (3.14), so that the θ solution in this case reduces to $\theta_0 = K(x - L)(y + B)$.

Next, the statement that the lower side of the rectangular profile is isothermal, eq. (3.6c), is obeyed by the θ expression (3.13) only if

$$B = 0 \tag{3.16}$$

In summary, the two boundary conditions that have been invoked thus far have the effect of trimming the θ expression down to the new form:

$$\theta = K \frac{\sinh[\lambda(x - L)]}{\cosh(\lambda L)} \sin(\lambda y) \tag{3.17}$$

Again, in the case of $\lambda = 0$ the boundary condition (3.6c) requires $B = 0$, and the "condensed" solution that replaces eq. (3.17) is $\theta_0 = K(x - L)y$.

The remaining two constants, λ and K, require special care. The isothermal surface condition at $y = H$, eq. (3.6d), is satisfied by eq. (3.17) only if

$$\sin(\lambda H) = 0 \tag{3.18}$$

This equation has an infinity of solutions* λH, which are called *characteristic values, proper values,* or *eigenvalues†*

$$\lambda H = n\pi \qquad (n = 0, \pm 1, \pm 2, \ldots) \tag{3.19}$$

Among these, $n = 0$ represents the $\lambda = 0$ case that we have been commenting on. In that special case the (3.6d) boundary condition is satisfied by the θ_0 solution listed at the end of the preceding paragraph only if $K = 0$. In conclusion, the solution that accounts for the case $\lambda = 0$ is the banal solution $\theta_0 = 0$. This contributes zero to the complete θ solution that we seek, and, for this reason, we drop the case $n = 0$ from the summation that will be made in eq. (3.21).

If we now substitute $\lambda = n\pi/H$ into the θ expression (3.17), we find that we can write down a θ_n solution (led by its own constant K_n) for each value of the integer n,

$$\theta_n = K_n \frac{\sinh[n\pi(x - L)/H]}{\cosh(n\pi L/H)} \sin\left(\frac{n\pi y}{H}\right) \tag{3.20}$$

All these θ_n solutions are part of the θ solution: we have no right to discard any of the θ_ns other than θ_0; therefore, we retain all of them in the sum

$$\theta = \sum_{n=1}^{\infty} \theta_n = \sum_{n=1}^{\infty} K_n \frac{\sinh[n\pi(x - L)/H]}{\cosh(n\pi L/H)} \sin\left(\frac{n\pi y}{H}\right) \tag{3.21}$$

Since each θ_n satisfies the conduction equation (3.5) and the boundary conditions (3.6a)–(3.6d), their sum θ satisfies also the same problem statement. Note further that it is sufficient to extend the sum over only the nonnegative values of the integer n, because the θ_n contribution to the sum accounts also for the contribution that would have been made by the θ_{-n} term (both θ_n and θ_{-n} depend in the same manner on y and x).

---

*Note that had we continued the analysis with eq. (3.13′) instead of eq. (3.13), we would have been unable to satisfy the boundary condition at $y = H$, that is, unable to determine λ. In conclusion, the choice between eq. (3.13) and (3.13′) must be made in such a way that the multivalued functions (sine, cosine) are aligned in the same direction as the pair of homogeneous boundary conditions.

†The term "eigenvalue" originates from the German words *eigen*, which means "proper," and *Eigenwerte*, which refers to a hidden, intrinsic feature, or capacity (private communication by Dr. Ruud Henkes).

The values of the leading coefficients K_n are determined finally by invoking the remaining boundary condition, eq. (3.6a). In view of eq. (3.21), this condition requires that

$$\theta_b = -\sum_{n=1}^{\infty} K_n \tanh\left(\frac{n\pi L}{H}\right) \sin\left(\frac{n\pi y}{H}\right) \tag{3.22}$$

The key is the *orthogonality* property of the sine function, which is expressed by the well-known integral

$$\int_0^H \sin\left(n\frac{\pi y}{H}\right) \sin\left(m\frac{\pi y}{H}\right) dy = \begin{cases} 0 & \text{if } m \neq n \\ H/2 & \text{if } m = n \end{cases} \tag{3.23}$$

This integral has a finite value only when the new integer m is equal to n. Therefore, if we multiply both sides of eq. (3.22) by $\sin(m\pi y/H)$ and then integrate both sides from $y = 0$ to $y = H$, the only term that yields a non-zero value on the right side of eq. (3.22) is the $n = m$ term. This operation consists of writing, in order,

$$\int_0^H \theta_b \sin\left(m\frac{\pi y}{H}\right) dy = -\int_0^H \sum_{n=1}^{\infty} K_n \tanh\left(n\frac{\pi L}{H}\right) \sin\left(n\frac{\pi y}{H}\right) \sin\left(m\frac{\pi y}{H}\right) dy$$

$$= -K_m \tanh\left(m\frac{\pi L}{H}\right)\frac{H}{2} \tag{3.24}$$

The integral on the left side reduces to

$$\int_0^H \theta_b \sin\left(m\frac{\pi y}{H}\right) dy = \theta_b \frac{H}{m\pi}[1 - \cos(m\pi)] \tag{3.25}$$

where

$$1 - \cos(m\pi) = \begin{cases} 0 & \text{if } m = \text{even} \\ 2 & \text{if } m = \text{odd} \end{cases} \tag{3.26}$$

Combining eqs. (3.24), (3.25), and (3.26), we conclude that the only nonzero K_m coefficients correspond to the odd values of the integer m;

$$K_m = -\frac{4\theta_b}{m\pi \tanh(m\pi L/H)} \qquad (m = 1, 3, 5, \dots) \tag{3.27}$$

Looking back at the analysis started with eq. (3.23), we see that the orthogonality property has the power to filter or sift out of the infinite series (3.22) the value of a particular coefficient K_n, namely, the coefficient that corresponds to $n = m$. The complete θ solution is now easy to write, by substituting eq. (3.27) into the sum (3.21) and by choosing the group $2n + 1$ as the new counter of odd integers (note that now n starts from zero, $n = 0, 1, 2, \dots$):

$$\theta = \frac{4\theta_b}{\pi} \sum_{n=0}^{\infty} \frac{\sinh[(2n+1)\pi(L-x)/H]}{\sinh[(2n+1)(\pi L/H)]} \frac{\sin[(2n+1)(\pi y/H)]}{2n+1} \tag{3.28}$$

The network of isotherms and flux lines (or heat tubes) that corresponds to this solution is presented in Fig. 3.2. Each heat tube carries one tenth of the total heat transfer rate q_b. The heat tubes are denser near the lower-left and upper-left corners because in those corner regions, two isothermal boundaries intersect

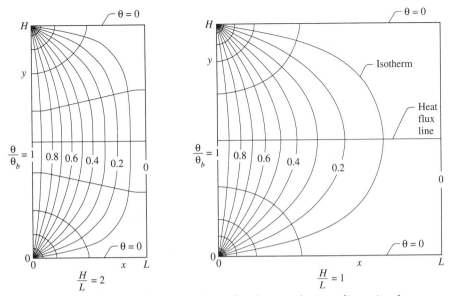

Figure 3.2 Patterns of isotherms and heat flux lines in the two-dimensional rectangular domain defined in Fig. 3.1.

(note the finite temperature difference θ_b between the two intersecting boundaries). Figure 3.2 shows also that when the rectangular domain becomes taller (i.e., when H/L increases), the isotherms become vertical and equidistant. The $H/L \to \infty$ limit of this trend was recognized already in the thin wall (or plane wall) illustrated in Fig. 2.1.

The total heat transfer rate through the $x = 0$ base is obtained by integrating the heat flux from $y = 0$ to $y = H$:

$$q'_b = -k \int_0^H \left(\frac{\partial \theta}{\partial x} \right)_{x=0} dy$$

$$= \frac{8}{\pi} k \theta_b \sum_{n=0}^\infty \frac{1}{(2n+1) \tanh[(2n+1)(\pi L/H)]} \tag{3.29}$$

This quantity is expressed in watts per meter, that is, per unit length, in the direction normal to the plane of Fig. 3.1.

At the end of any stretch of analysis, especially after a lengthy one, it is useful to check whether the end result agrees with much simpler solutions that are already known. In the limit $L/H \to 0$, for example, the rectangular profile of Fig. 3.1 reduces to that of a "thin wall" (Fig. 2.1), through which the conduction is oriented now purely in the x direction. Therefore, the value of q'_b in this limit should be simply $kH\theta_b/L$. It turns out that the q'_b formula (3.29) agrees with this expected result,

$$\lim_{L/H \to 0} q'_b = \frac{8}{\pi^2} k \theta_b \frac{H}{L} \left(1 + \frac{1}{3^2} + \frac{1}{5^2} + \cdots \right) = \frac{kH\theta_b}{L} \tag{3.30}$$

because the infinite sum listed in parentheses has the finite value of $\pi^2/8$. In the opposite limit, $L/H \to \infty$, the total heat transfer rate is infinite,

$$\lim_{L/H \to \infty} q_b' = \frac{8}{\pi} k\theta_b \left(1 + \frac{1}{3} + \frac{1}{5} + \cdots \right) = \infty \qquad (3.31)$$

because this time the series inside the parentheses does not converge.

The key engineering result of the analysis is the formula for the total heat transfer rate driven by the temperature difference between the base and the remaining three sides of the rectangular profile ($\theta_b = T_b - T_\infty$). If W is the width measured in the direction normal to the plane of Fig. 3.1 (note that $W \!>\!> H$ if the present problem is to be two-dimensional), then the total number of watts that pass through the $W \times H$ base area is

$$q_b = q_b' W \qquad (3.32)$$

In view of the final answer, eq. (3.29), this total heat transfer rate can be expressed alternatively as

$$q_b = S k \theta_b \qquad (3.33)$$

where S is a *shape factor* (in units of m), the magnitude of which depends only on the geometry (i.e., shape and dimensions) of the conducting body:

$$S = \frac{8}{\pi} W \sum_{n=0}^{\infty} \frac{1}{(2n+1)\tanh[(2n+1)(\pi L/H)]} \qquad (3.34)$$

In this alternative formulation the challenge of calculating the total heat transfer rate q_b reduces to figuring out the shape factor: it is the formula for S, eq. (3.34), that the *Fourier analysis* of Sections 3.1.2 through 3.1.4 puts in the hands of the engineer. The solutions to steady conduction problems in a diversity of multidimensional configurations have been catalogued in terms of the respective S formulas, as will be seen shortly in the discussion of Table 3.3.

Example 3.1

Temperature Series Solution

To get a feel for how the temperature series solution (3.28) converges, evaluate the temperature in the center of the rectangular domain of Fig. 3.1 at $x = L/2$ and $y = H/2$. Consider three cases—$L/H = 1$, 2, and 3. In each case, calculate the dimensionless center temperature θ_c/θ_b by retaining the first meaningful terms of the series expression. In this way, show that the series converges faster as L/H increases.

Solution. Substituting $x = L/2$ and $y = H/2$ on the right side of eq. (3.28) we obtain the center temperature θ_c:

$$\frac{\theta_c}{\theta_b} = \frac{4}{\pi} \sum_{n=0}^{\infty} \frac{\sinh[(2n+1)(\pi/2)(L/H)]}{\sinh[(2n+1)\pi L/H]} \frac{\sin[(2n+1)\pi/2]}{2n+1}$$

$$= \frac{2}{\pi} \sum_{n=0}^{\infty} \frac{\sin[(2n+1)\pi/2]}{(2n+1)\cosh[(2n+1)(\pi/2)(L/H)]}$$

Note that to arrive at this final expression, we had to invoke the identity $\sinh(2u) = 2\sinh u \cosh u$.

The dimensionless center temperature θ_c/θ_b is clearly a function of the slenderness ratio L/H. Note further that θ_c decreases monotonically as L/H increases. To illustrate the effect of the slenderness ratio on the convergence of the infinite series solution for the center temperature, we evaluate the first meaningful terms of the series, for specific values of L/H:

$$\frac{L}{H} = 1: \quad \frac{\theta_c}{\theta_b} = 0.2537 - 0.0038 + 0.0001 - \cdots = 0.250$$

$$\frac{L}{H} = 2: \quad \frac{\theta_c}{\theta_b} = 0.05492 - 0.00003 + \cdots \qquad = 0.0549$$

$$\frac{L}{H} = 3: \quad \frac{\theta_c}{\theta_b} = 0.01144 - 3.08 \times 10^{-7} + \cdots \qquad = 0.01144$$

These calculations show that the convergence of the series improves as the ratio L/H increases; in fact, when $L/H \geq 2$, the first term is practically equal to the sum of the series.

3.1.2 Heat Flux Boundary Conditions

The analytical solutions for steady conduction in two or more dimensions are as diverse as the shapes and the boundary conditions of the conduction domains. A very large volume of such solutions has been generated in the almost 200 years since Fourier [1], that is, since the discovery of the analytical method. A small representative sample of this volume is found in the advanced treatises on conduction heat transfer and on mass diffusion, for example, in Refs. [2–7]. What unites these analytical solutions is the method, which — by reviewing the headings of the preceding analysis — consists of three important steps:

1. The identification of the direction of homogeneous boundary conditions
2. The separation of variables, and
3. The use of the orthogonality property of the characteristic functions, to determine the coefficients of the series solution.

To illustrate the generality of the method and the diversity of conduction problem statements that may confront the engineer, consider the phenomenon of steady conduction in the two-dimensional domain shown in Fig. 3.3. The conduction pattern is two-dimensional (i.e., only in the plane of Fig. 3.3) because the actual solid body has a width W much larger than the thickness H, the width W being measured in the direction normal to the plane of Fig. 3.3.

This new geometry is, of course, related to the one already treated in connection with Fig. 3.1. There are, however, two important differences: first, the *infinite* length of the rectangular cross section in Fig. 3.3; and, second, the thermal contact with a fluid flow across a *finite* heat transfer coefficient h, or a finite Biot number hH/k. It is worth noting that this new geometry is the same as that of an infinitely long plate fin, Fig. 2.11, except that this time we treat the temperature field as two-dimensional (i.e., as illustrated in Fig. 2.17).

In terms of the actual temperature $T(x,y)$, the problem statement consists of writing the conduction equation (3.2) and the following four boundary conditions:

$$T = T_b \qquad\qquad \text{at} \qquad x = 0 \qquad\qquad (3.35a)$$

$$T \rightarrow T_\infty \qquad\qquad \text{as} \qquad x \rightarrow \infty \qquad\qquad (3.35b)$$

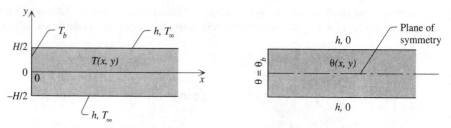

Figure 3.3 Infinitely long plate fin with finite heat transfer coefficient on both sides.

$$k\frac{\partial T}{\partial y} = h(T - T_\infty) \qquad \text{at} \qquad y = -\frac{H}{2} \qquad (3.35c)$$

$$-k\frac{\partial T}{\partial y} = h(T - T_\infty) \qquad \text{at} \qquad y = \frac{H}{2} \qquad (3.35d)$$

For example, the last boundary condition states that the conduction heat flux that arrives (from below) at the $y = H/2$ surface, $-k(\partial T/\partial y)$, is equal to the convective heat flux that enters the fluid realm, $h(T - T_\infty)$. The analogous boundary condition along the lower surface, eq. (3.35c), does not have a minus sign in front of the $k(\partial T/\partial y)$ term, because on the lower surface the heat flux $h(T - T_\infty)$ points in the negative y direction.

Before plunging blindly into any kind of laborious problem solving, it pays to examine the geometry of the heat transfer configuration, to see whether some of its features can be exploited for the purpose of simplifying the analysis. In this case Fig. 3.3 shows that the y boundary conditions are *symmetric* about the horizontal midplane of the rectangular profile: this is why the x axis was positioned already in that plane. The symmetry about the $y = 0$ plane allows us to replace either one of eqs. (3.35c) and (3.35d) with the simpler adiabatic (zero heat flux) midplane condition

$$\frac{\partial T}{\partial y} = 0 \qquad \text{at} \qquad y = 0 \qquad (3.36)$$

In many cases the identification of a zero heat flux condition is useful because eq. (3.36) is a *homogeneous* boundary condition. The search for a direction of homogeneous boundary conditions is not over, however, because the remaining condition in y [say, eq. (3.35d)] is nonhomogeneous (note the presence of the $-hT_\infty$ constant on the right side). The y direction does emerge as the direction of homogeneous boundary conditions only after restating the problem in terms of the excess temperature function $\theta = T - T_\infty$. The new problem statement consists of the conduction equation (3.5) and, in order, the θ counterparts of the boundary conditions (3.35a, b), (3.36), and (3.35d):

$$\theta = \theta_b \qquad \text{at} \qquad x = 0 \qquad (3.37a)$$

$$\theta \to 0 \qquad \text{as} \qquad x \to \infty \qquad (3.37b)$$

$$\frac{\partial \theta}{\partial y} = 0 \qquad \text{at} \qquad y = 0 \qquad (3.37c)$$

$$\frac{h}{k}\theta + \frac{\partial \theta}{\partial y} = 0 \qquad \text{at} \qquad y = \frac{H}{2} \qquad (3.37d)$$

The separation of variables and the selection of y as the direction of characteristic functions leads to the θ expression (3.13). That expression cannot be used directly in the present analysis, because at first glance, at least, it does not satisfy the boundary condition (3.37b). More useful is an equivalent of eq. (3.13), which is arrived at by noting the relation

$$C_1 \sinh(\lambda x) + C_2 \cosh(\lambda x) = C_3 e^{\lambda x} + C_4 e^{-\lambda x} \tag{3.38}$$

where $C_3 = (C_1 + C_2)/2$ and $C_4 = (C_2 - C_1)/2$. It is permissible, therefore, to replace eq. (3.13) with

$$\theta = (A_1 e^{\lambda x} + A_2 e^{-\lambda x}) [B_1 \sin(\lambda y) + B_2 \cos(\lambda y)] \tag{3.39}$$

in which A_1, A_2, B_1, and B_2 are all constant coefficients. Two of these coefficients can be determined by simply inspecting the boundary conditions (3.37b) and (3.37c):

$$A_1 = 0 \qquad B_1 = 0 \tag{3.40}$$

In this way the θ expression reduces to

$$\theta = K e^{-\lambda x} \cos(\lambda y) \tag{3.41}$$

where K replaces now the constant product $A_2 B_2$. Note that in the special case when $\lambda = 0$, eq. (3.13) reduces to $\theta_0 = K(x + A)(y + B)$. This expression satisfies the boundary condition (3.37b) only if $K = 0$. In conclusion, θ_0 contributes zero to the ultimate solution, and for this reason we continue the analysis with eq. (3.41).

The expression (3.41) contains only two constants, K and λ, because there are only two boundary conditions that have not been invoked yet, namely, eqs. (3.37a) and (3.37d). Substituting the θ expression (3.41) first into the boundary condition (3.37d) leads to the transcendental equation

$$a \tan(a) = \frac{hH}{2k} \tag{3.42}$$

where a is the dimensionless group

$$a = \lambda \frac{H}{2} \tag{3.43}$$

The constant appearing on the right side of eq. (3.42) is the Biot number based on the half-thickness of the plate. Since $\tan a$ is periodic in a, eq. (3.42) has an infinity of solutions a_n $(n = 1, 2, 3, \ldots)$ for any fixed value of $hH/2k$. Furthermore, if a_n is a solution of eq. (3.42), then $-a_n$ is also a solution.

The first six of the positive characteristic values a_n are listed in Table 3.1. A look at this table teaches us that the form of the characteristic values of a certain conduction configuration is rarely as simple as in eq. (3.19). More often, the characteristic values must be determined numerically by solving multiple-solutions equations such as eq. (3.42). Furthermore, each characteristic value is, in general, a function of a dimensionless parameter, namely, the Biot number based on the solid–fluid heat transfer coefficient and a linear dimension of the conducting body.

The series expression for θ is constructed in the same manner as in eq. (3.21),

$$\theta = \sum_{n=1}^{\infty} K_n e^{-\lambda_n x} \cos(\lambda_n y) \tag{3.44}$$

Table 3.1 The First Six Roots of Eq. (3.42)[a]

$hH/2k$	a_1	a_2	a_3	a_4	a_5	a_6
0	0	π	2π	3π	4π	5π
0.01	0.0998	3.1448	6.2848	9.4258	12.5672	15.7086
0.1	0.3111	3.1731	6.2991	9.4354	12.5743	15.7143
1	0.8603	3.4256	6.4373	9.5293	12.6453	15.7713
3	1.1925	3.8088	6.7040	9.7240	12.7966	15.8945
10	1.4289	4.3058	7.2281	10.2003	13.2142	16.2594
30	1.5202	4.5615	7.6057	10.6543	13.7085	16.7691
100	1.5552	4.6658	7.7764	10.8871	13.9981	17.1093
∞	$\dfrac{\pi}{2}$	$\dfrac{3\pi}{2}$	$\dfrac{5\pi}{2}$	$\dfrac{7\pi}{2}$	$\dfrac{9\pi}{2}$	$\dfrac{11\pi}{2}$

[a]Adapted from Ref. [2].

where $\lambda_n = 2a_n/H$. Only the positive λ_n values are counted in this series, because the corresponding negative values do not contribute any new terms to the series [note that $\cos(-\lambda_n y) = \cos(\lambda_n y)$].

The final step of the analytical method—the use of the orthogonality property of $\cos(\lambda_n y)$—delivers the values of the coefficients K_n. This step consists of multiplying both sides of eq. (3.44) by $\cos(\lambda_m y)$ and then integrating both sides from $y = 0$ to $y = H/2$. The final result can be written as

$$K_n = 2\theta_b \frac{\sin(a_n)}{a_n + \sin(a_n)\cos(a_n)} \tag{3.45}$$

In combination with the series (3.44), this result completes the solution to the steady two-dimensional conduction problem. This series solution is noted for its rapid convergence. Even though today such series are evaluated routinely on the computer, it is worth trying out a few numerical examples manually, to develop a feel for how the series behaves.

The corresponding solution for the total heat transfer rate (watts) through the $x = 0$ surface of the solid body is

$$q_b = W \int_{-H/2}^{H/2} -k\left(\frac{\partial \theta}{\partial x}\right)_{x=0} dy$$

$$= 4k\theta_b W \sum_{n=1}^{\infty} \frac{\sin^2(a_n)}{a_n + \sin(a_n)\cos(a_n)} \tag{3.46}$$

It is instructive to verify numerically that when the Biot number $hH/2k$ is smaller than 1, the q_b series (3.46) converges on values that agree within a few percentage points with the one-dimensional conduction estimate made possible by eq. (2.97). The one-dimensional model overestimates the q_b value produced by the more exact, two-dimensional model (3.46).* The comparison of eq. (3.46) with eq. (2.97) forms the subject of Problem 3.1.

---

*With regard to plate fins, even the two-dimensional model of Fig. 3.3 is approximate, because the actual fin root temperature is depressed below T_b by the heat current q_b that is pulled out of the base wall. To the rest of the base wall (as a semi-infinite solid of temperature T_b), the root of the fin looks like a heat sink; therefore, the fin temperature at $x = 0$ must be somewhat lower than T_b.

3.1.3 Superposition of Solutions

Table 3.2 summarizes the three types of boundary conditions we have encountered thus far. Their homogeneous counterparts are listed in the right-hand column. Worth noting is that the last condition—specified heat transfer coefficient and fluid temperature (h, T_∞)—is somewhat more general than the first two. For example, if the heat transfer coefficient is infinite, then the boundary temperature becomes "specified," because in this limit it is equal to the specified fluid temperature T_∞. On the other hand, if the boundary temperature T_w is specified *in addition* to h and T_∞, then the boundary heat flux is also specified, because $q'' = h(T_w - T_\infty)$.

A step-by-step comparison of the two analyses of the problems posed in Figs. 3.1 and 3.3 shows that the boundary conditions can involve the boundary temperature, the heat flux, or a relationship between the boundary temperature and heat flux. Figures 3.1 and 3.3 represent two of the simplest cases, because

Table 3.2 Three Types of Boundary Conditions and the Corresponding Homogeneous Forms

Type of Boundary Condition[a]		Nonhomogeneous Form	Homogeneous Form
1. Specified boundary temperature (T_w)		$T = T_w$	$T = 0$
2. Specified boundary heat flux (q'')		$-k\dfrac{\partial T}{\partial n} = q''$	$\dfrac{\partial T}{\partial n} = 0$
3. Specified heat transfer coefficient and fluid temperature (h, T_∞). Note that the sign in front of $\partial T/\partial n$ depends on the direction of n (see sketch)		$-k\dfrac{\partial T}{\partial n} = h(T - T_\infty)$	$\dfrac{h}{k}T + \dfrac{\partial T}{\partial n} = 0$

[a]These three boundary condition types are recognized (respectively) by the alternative names:
1. Boundary condition of the *first kind*, or of *Dirichlet* type,
2. Boundary condition of the *second kind*, or of *Neumann* type, and
3. Boundary condition of the *third kind*, or of *Robin* type.
The conditions specified at infinity (e.g., in infinite or semi-infinite solids, Chapter 4) are said to be of the *zeroth kind*.

the specified boundary temperature [e.g., eq. (3.3a)], boundary heat flux [e.g., eq. (3.37c)], heat transfer coefficient h, and fluid temperature T_∞ are all constant.

In these simplest cases it was possible to identify the direction of homogeneous boundary conditions by just using the transformation $\theta = T - T_\infty$. There are other heat transfer configurations that require additional ingenuity. In Fig. 3.4, for example, the use of the excess temperature function $\theta = T - T_\infty$ is not sufficient, because the temperatures of *two* boundaries (T_a, T_b) differ from the background temperature (T_∞). Note that with a finite θ_b on the left surface and a finite θ_a on the bottom surface, neither x nor y is a direction of homogeneous boundary conditions.

The $\theta(x,y)$ problem can still be solved analytically by noting the relationship

$$\theta(x,y) = \theta_1(x,y) + \theta_2(x,y) \tag{3.47}$$

where θ_1 and θ_2 are the solutions of two simpler conduction problems (Fig. 3.4):

Conduction Equations

$$\frac{\partial^2 \theta_1}{\partial x^2} + \frac{\partial^2 \theta_1}{\partial y^2} = 0 \qquad \frac{\partial^2 \theta_2}{\partial x^2} + \frac{\partial^2 \theta_2}{\partial y^2} = 0 \tag{3.48a}$$

Boundary Conditions

$$x = 0: \qquad \theta_1 = \theta_b \qquad \theta_2 = 0 \tag{3.48b}$$

$$x = L: \qquad \theta_1 = 0 \qquad \theta_2 = 0 \tag{3.48c}$$

$$y = 0: \qquad \theta_1 = 0 \qquad \theta_2 = \theta_a \tag{3.48d}$$

$$y = H: \qquad \theta_1 = 0 \qquad \theta_2 = 0 \tag{3.48e}$$

Both $\theta_1(x,y)$ and $\theta_2(x,y)$ have analytical solutions because they both have one direction of homogeneous boundary conditions (y and x, respectively). The θ_1 solution is the same as in eq. (3.28), whereas the θ_2 solution is obtained from eq. (3.28) by replacing θ_b with θ_a, (x,L) with (y,H), and (y,H) with (x,L). It is easy to

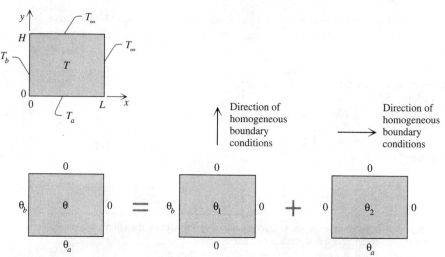

Figure 3.4 The superposition of two known solutions ($\theta_1 + \theta_2$) as a way of constructing the solution for a problem without a direction of homogeneous boundary conditions (θ).

verify that superimposing θ_1 and θ_2, the conduction equations *and* their respective boundary conditions [i.e., by writing $\theta_1 + \theta_2$, eq. (3.47)] leads back to the original problem statement for the $\theta(x,y)$ problem (see Fig. 3.4, left side):

$$\frac{\partial^2 \theta}{\partial x^2} + \frac{\partial^2 \theta}{\partial y^2} = 0 \qquad (3.49a)$$

$$\theta = \theta_b \qquad\qquad \text{at} \qquad x = 0 \qquad\qquad (3.49b)$$

$$\theta = 0 \qquad\qquad \text{at} \qquad x = L \qquad\qquad (3.49c)$$

$$\theta = \theta_a \qquad\qquad \text{at} \qquad y = 0 \qquad\qquad (3.49d)$$

$$\theta = 0 \qquad\qquad \text{at} \qquad y = H \qquad\qquad (3.49e)$$

Heat transfer configurations with more complicated boundary conditions than in Fig. 3.4 may require superposition rules more complicated than eq. (3.47). For example, it may be necessary to use more than two building block solutions (θ_1, θ_2, θ_3, . . .) to construct the needed temperature field (θ). The efficient use of the method depends on the problem-solver's understanding of the geometry of the problem, and, certainly, on experience with similar problems. In the present course, this experience can be gained by trying to solve the problems proposed at the end of this chapter.

3.1.4   Cylindrical Coordinates

The analytical method outlined so far consists of the three basic steps reviewed at the start of Section 3.1.2. The method was illustrated solely by means of two-dimensional problems stated in Cartesian coordinates (Figs. 3.1–3.4). This section and the next will show that the analytical method is general, that is, that it can be used to determine temperature fields in three dimensions and in coordinate systems other than the Cartesian system.

Consider the cylindrical object shown in Fig. 3.5. The lateral surface and the left end of this object are isothermal (T_∞), while the right end-surface is maintained at a different temperature (T_b). It is easy to see that this conduction heat transfer configuration is the "cylindrical" counterpart of the two-dimensional Cartesian problem solved in connection with Fig. 3.1. The present temperature field is also two-dimensional, $T(r,z)$, because the boundary conditions are such that there can be no temperature variation in the circumferential direction (θ in Fig. 1.8).

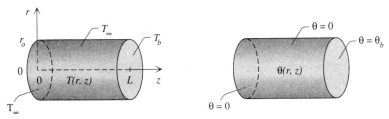

Figure 3.5 Cylindrical object with a temperature difference between one end face and the rest of its surface.

The problem statement begins with the equation for steady conduction in a constant-k medium without internal heat generation,

$$\frac{\partial^2 T}{\partial r^2} + \frac{1}{r}\frac{\partial T}{\partial r} + \frac{\partial^2 T}{\partial z^2} = 0 \tag{3.50}$$

and ends with the four boundary conditions

$$T = T_\infty \quad \text{at} \quad z = 0 \tag{3.51a}$$

$$T = T_b \quad \text{at} \quad z = L \tag{3.51b}$$

$$\frac{\partial T}{\partial r} = 0 \quad \text{at} \quad r = 0 \tag{3.51c}$$

$$T = T_\infty \quad \text{at} \quad r = r_o \tag{3.51d}$$

Novel among these conditions is the third, the zero-heat flux condition (3.51c), which states that the sought temperature distribution $T(r,z)$ is symmetric about the centerline.

The radial direction r emerges as the direction of homogeneous boundary conditions, after introducing $\theta = T - T_\infty$ as the new unknown:

$$\frac{\partial^2 \theta}{\partial r^2} + \frac{1}{r}\frac{\partial \theta}{\partial r} + \frac{\partial^2 \theta}{\partial z^2} = 0 \tag{3.52}$$

$$\theta = 0 \quad \text{at} \quad z = 0 \tag{3.53a}$$

$$\theta = \theta_b \quad \text{at} \quad z = L \tag{3.53b}$$

$$\frac{\partial \theta}{\partial r} = 0 \quad \text{at} \quad r = 0 \tag{3.53c}$$

$$\theta = 0 \quad \text{at} \quad r = r_o \tag{3.53d}$$

Next, the separation of variables is achieved by setting $\theta(r,z) = R(r)Z(z)$ in eq. (3.52):

$$\frac{R''}{R} + \frac{1}{r}\frac{R'}{R} + \frac{Z''}{Z} = 0 \tag{3.54}$$

This result generates two separate equations for $R(r)$ and $Z(z)$,

$$\frac{R''}{R} + \frac{1}{r}\frac{R'}{R} = -\lambda^2 \tag{3.55}$$

$$\frac{Z''}{Z} = \lambda^2 \tag{3.56}$$

where λ^2 is a positive constant. The general solution to eq. (3.56) is the same as that of eq. (3.9):

$$Z = C_3 \sinh(\lambda z) + C_4 \cosh(\lambda z) \tag{3.57}$$

The $R(r)$ equation (3.55) can be solved by expanding R in a power series in r, by following the method of solution described in Problem 2.20. The general solution assumes the form

$$R = C_1 J_0(\lambda r) + C_2 Y_0(\lambda r) \tag{3.58}$$

where J_0 is the zeroth-order Bessel function of the first kind and Y_0 is the zeroth-order Bessel function of the second kind. The respective infinite-series expressions for J_0 and Y_0 can be found in more advanced treatises on conduction [2–7] and on Bessel functions [8]. Figure 3.6 illustrates the behavior of these two functions, which play the role of characteristic functions in the present problem. Representative values of the function J_0 are tabulated in Appendix E.

Figure 3.6 shows that Y_0 becomes infinite on the cylinder centerline and that $dJ_0/dr = 0$ at $r = 0$. Consequently, the centerline boundary condition (3.53c) requires that $C_2 = 0$ in eq. (3.58). The left-end boundary condition (3.53a), on the other hand, requires that $C_4 = 0$ in eq. (3.57). Combining the resulting (shorter) versions of eqs. (3.57) and (3.58) into the product expression $\theta = RZ$ yields

$$\theta = K\sinh(\lambda z)J_0(\lambda r) \tag{3.59}$$

It is worth noting that in the special case of $\lambda = 0$, the general solution $\theta = RZ$ degenerates into $\theta_0 = K(\ln r + A)(z + B)$, where K, A, and B are constants. The centerline boundary condition (3.53c) requires in this case $K = 0$, meaning that $\theta_0 = 0$. In conclusion, θ_0 contributes zero to the final θ solution, and it is correct to continue the analysis with eq. (3.59), which holds for $\lambda \neq 0$.

The $r = r_o$ boundary condition, eq. (3.53d), pinpoints the characteristic values of the θ problem:

$$J_0(\lambda r_o) = 0 \tag{3.60}$$

According to Fig. 3.6, there is an infinity of dimensionless characteristic values $\lambda_n r_o (n = 1, 2, \ldots)$ that satisfy eq. (3.60); the first of these values is $\lambda_1 r_o = 2.405$. The θ solution can therefore be constructed as an infinite series of J_0 characteristic functions:

$$\theta = \sum_{n=1}^{\infty} K_n\sinh(\lambda_n z)J_0(\lambda_n r) \tag{3.61}$$

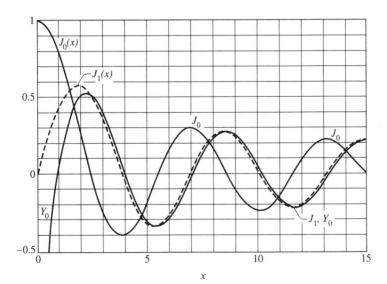

Figure 3.6 The behavior of the Bessel functions J_0, Y_0, and J_1 (adapted from Ref. 9).

The final step amounts to combining the last of the boundary conditions, eq. (3.53b),

$$\theta_b = \sum_{n=1}^{\infty} K_n \sinh(\lambda_n L)\, J_0(\lambda_n r) \qquad (3.61')$$

with the orthogonality property of the J_0 function, which in cylindrical coordinates is expressed by the integral*

$$\int_0^{r_o} r J_0(\lambda_n r)\, J_0(\lambda_m r)\, dr = 0 \qquad \text{if } m \neq n \qquad (3.62)$$

Therefore, the K_n coefficients of the θ series can be identified by multiplying both sides of eq. (3.61) by $r J_0(\lambda_m r)$ and integrating from $r = 0$ to $r = r_o$. This final step is omitted, because it requires the use of certain identities that exist between Bessel functions [see, for example, Ref. [4], pp. 132–143]. The end result for the steady-state temperature distribution $\theta(r,z)$ is

$$\theta = 2\theta_b \sum_{n=1}^{\infty} \frac{J_0(\lambda_n r)\, \sinh(\lambda_n z)}{\lambda_n r_o\, J_1(\lambda_n r_o)\, \sinh(\lambda_n L)} \qquad (3.63)$$

where J_1 is the first-order Bessel function of the first kind (Fig. 3.6). This series solution is obviously similar to eq. (3.28): the main difference is that instead of the sine and cosine functions encountered in Cartesian coordinates, in cylindrical coordinates the role of characteristic functions is played by Bessel functions. Dexterity in the manipulation of Bessel functions and the construction of solutions such as eq. (3.63) can be achieved by studying a specialized treatise on conduction [2–7].

3.1.5   Three-Dimensional Conduction in Cartesian Coordinates

Consider, finally, the more general case in which the temperature field is three-dimensional. In Fig. 3.1, for example, T will depend not only on x and y, but also on z when the width W of the plate fin is comparable with the thickness H. This general case is illustrated in Fig. 3.7, which shows that five faces of the prism are cooled down to the temperature T_{∞} while the sixth face ($x = 0$) is heated to a different temperature (T_b). The right side of Fig. 3.7 shows the equivalent problem statement in terms of the excess temperature function $\theta = T - T_{\infty}$, namely,

Conduction Equation

$$\frac{\partial^2 \theta}{\partial x^2} + \frac{\partial^2 \theta}{\partial y^2} + \frac{\partial^2 \theta}{\partial z^2} = 0 \qquad (3.64)$$

Boundary Conditions

$$\theta = \theta_b \qquad \text{at} \quad x = 0 \qquad\qquad\qquad\qquad (3.65a)$$

$$\theta = 0 \qquad \text{at} \quad x = L \qquad\qquad\qquad\qquad (3.65b)$$

$$\theta = 0 \qquad \text{at} \quad y = 0 \quad \text{and} \quad y = H \qquad (3.65c, d)$$

$$\theta = 0 \qquad \text{at} \quad z = 0 \quad \text{and} \quad z = W \qquad (3.65e, f)$$

---

*Note the presence of r as a factor in the integrand, that is, compare eqs. (3.62) and (3.23).

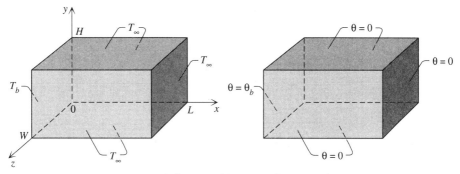

Figure 3.7 Three-dimensional (finite width) generalization of the heat transfer configuration treated in Fig. 3.1.

The variables in eq. (3.64) can be separated by adopting the product solution $\theta = X(x)Y(y)Z(z)$,

$$\underbrace{\frac{X''}{X}}_{\lambda^2 + \mu^2} + \underbrace{\frac{Y''}{Y}}_{-\lambda^2} + \underbrace{\frac{Z''}{Z}}_{-\mu^2} = 0 \tag{3.66}$$

where λ^2 and μ^2 are *two* positive numbers. Equation (3.66) breaks up into three ordinary differential equations, the general solutions to which contribute as factors in the assumed product solution for θ:

$$\theta = \{C_1 \sinh[(\lambda^2 + \mu^2)^{1/2}x] + C_2 \cosh[(\lambda^2 + \mu^2)^{1/2}x]\} \times \\ [C_3 \sin(\lambda y) + C_4 \cos(\lambda y)][C_5 \sin(\mu z) + C_6 \cos(\mu z)] \tag{3.67}$$

Beyond this point the solution proceeds along the same path as in Sections 3.1.1 through 3.1.4. One important difference is that in the present problem *two* directions of homogeneous boundary conditions are needed (y and z) to determine two infinite series of characteristic functions, λ_n and μ_n. In the special case when $\lambda = 0$ and $\mu = 0$, the solution has the form $\theta_{0,0} = (C_1 x + C_2)(C_3 y + C_4)(C_5 z + C_6)$ in place of eq. (3.67).

3.2 APPROXIMATE METHODS

3.2.1 The Integral Method

The geometrical layout and boundary conditions are in many cases more complicated than the configurations analyzed in the preceding sections. When an exact analytical solution does not exist, or even when one exists but it is hampered by a lengthy analysis and the numerical work of calculating the characteristic values, progress can still be made based on approximate methods.

As a first example, consider again the integral method of Section 2.6.2, in which the object was a problem of unidirectional conduction with internal heat generation. In this section we focus on Fig. 3.8, which shows the two-dimensional Cartesian counterpart of the problem of Fig. 2.9a. The rectangular area emphasized in the figure represents a cross section through an electrical conductor with uniform rate of internal heat generation $\dot{q}$.

When the external heat transfer coefficient (or the Biot number hL/k, or

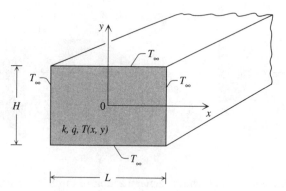

Figure 3.8 Bar with rectangular cross section and uniform rate of internal heat generation.

hH/k) is high enough, the temperature of the outer edge of the cross section is equal to the ambient temperature T_∞. The temperature everywhere inside the conductor, $T(x,y)$, must be greater than T_∞ to drive the locally generated $\dot{q}$ toward the periphery of the cross section. The maximum temperature occurs, therefore, in the geometrical center of the cross section, because the point ($x = 0, y = 0$) is located the farthest from all the boundaries. Worth keeping in mind is that in problems such as this (finite-size body with internal heat generation), a key challenge is to predict the maximum temperature that occurs inside the body.

Assuming that the thermal conductivity of the solid is also constant, the steady conduction equation in $x-y$ coordinates is

$$\frac{\partial^2 T}{\partial x^2} + \frac{\partial^2 T}{\partial y^2} = -\frac{\dot{q}}{k} \tag{3.68}$$

In this equation and throughout the remainder of this section, $\dot{q}$ is a constant, that is, it has the same value throughout the conducting medium. The isothermal boundary conditions specified around the periphery,

$$T = T_\infty \quad \text{at} \quad x = \pm\frac{L}{2} \tag{3.69a, b}$$

$$T = T_\infty \quad \text{at} \quad y = \pm\frac{H}{2} \tag{3.69c, d}$$

can all be made homogeneous by adopting the excess temperature $\theta = T - T_\infty$ as the new unknown. Unfortunately, the next step of the exact analysis (the separation of variables) is not nearly as straightforward, because it requires the imaginative use of the superposition of simpler solutions (see Problem 3.2). The source of this difficulty is the additional term $\dot{q}/k$ that appears now on the right side of eq. (3.68): because of this constant, eq. (3.68) is nonhomogeneous, in contrast to the equation for steady conduction without internal heat generation, eq. (3.2), which is homogeneous.*

A much more direct approach to estimating $T_{\max}$ is based on the integral

---

*Equations (3.68) and (3.2) are respectively known as Poisson's and Laplace's equations.

method, which consists of two steps (review Section 2.6.2). The first step is the assumption of a reasonable shape for the $T(x,y)$ surface, for example,

$$T(x,y) = T_\infty + \theta_{\max} \left[1 - \left(\frac{x}{L/2} \right)^2 \right] \left[1 - \left(\frac{y}{H/2} \right)^2 \right] \qquad (3.70)$$

This expression satisfies the boundary conditions (3.69a–d). The constant $\theta_{\max}$ is unknown and equal to the maximum temperature difference, $\theta_{\max} = T(0,0) - T_\infty$.

The second step consists of substituting the assumed $T(x,y)$ expression into eq. (3.68) and then integrating both sides of eq. (3.68) over the entire rectangular domain:

$$\int_{-L/2}^{L/2} \int_{-H/2}^{H/2} \left(\frac{\partial^2 T}{\partial x^2} + \frac{\partial^2 T}{\partial y^2} \right) dy\, dx = -\frac{\dot{q}}{k} HL \qquad (3.71)$$

After some algebra, this equation delivers the wanted result:

$$\theta_{\max} = \frac{3\dot{q}}{16k} \frac{H^2 L^2}{H^2 + L^2} \qquad (3.72)$$

The maximum temperature difference increases proportionally with the ratio $\dot{q}/k$ and with the square of the smaller of the two sides of the rectangular cross section. The formula (3.72) approaches the exact solution when the cross section is flat ($H \gg L$, or $H \ll L$). It is the least accurate in the case of a square cross section, where it overestimates the exact maximum temperature difference by 27 percent.

3.2.2 The Method of Scale Analysis

By focusing again on the problem of Fig. 3.8, we have the opportunity to become acquainted with an even simpler and more direct method, namely, *scale analysis*, or *order-of-magnitude analysis*. The unknown is still the maximum temperature difference between the center of the cross section and the periphery, $\theta_{\max}$. This time, however, we estimate only the order of magnitude of each of the three terms of the conduction equation (3.68). From the approximate relationship that emerges, we deduce the value of $\theta_{\max}$.

The first term, $\partial^2 T/\partial x^2$, represents the curvature of the temperature distribution in the x direction. The curvature represents the change in the slope $\partial T/\partial x$; therefore, with reference to Fig. 3.8,

$$\frac{\partial^2 T}{\partial x^2} \sim \frac{\left(\frac{\partial T}{\partial x} \right)_{x=L/2} - \left(\frac{\partial T}{\partial x} \right)_{x=0}}{(L/2) - 0} \qquad (3.73)$$

in which the sign $\sim$ means *is of the same order of magnitude as*. Symmetry dictates that $(\partial T/\partial x)_{x=0} = 0$. About the temperature gradient at the surface, the most we can say is that it is proportional to the maximum temperature difference and inversely proportional to the distance between the point of temperature maximum and the edge of the cross section,

$$\left(\frac{\partial T}{\partial x} \right)_{x=L/2} \sim -\frac{\theta_{\max}}{L/2} \qquad (3.74)$$

Combining eqs. (3.74) and (3.73) leads to the conclusion that the x curvature must be negative and of order

$$\frac{\partial^2 T}{\partial x^2} \sim -\frac{\theta_{max}}{(L/2)^2} \tag{3.75}$$

The same argument can be used in conjunction with the second term of eq. (3.68) to conclude that its scale is

$$\frac{\partial^2 T}{\partial y^2} \sim -\frac{\theta_{max}}{(H/2)^2} \tag{3.76}$$

The order of magnitude of the right side of eq. (3.68) is indicated already, $\dot{q}/k$. Substituting these three scales in place of the respective terms in eq. (3.68) yields

$$\frac{\theta_{max}}{(L/2)^2} + \frac{\theta_{max}}{(H/2)^2} \sim \frac{\dot{q}}{k} \tag{3.77}$$

In other words,

$$\theta_{max} \sim \frac{\dot{q}}{4k} \frac{H^2 L^2}{H^2 + L^2} \tag{3.78}$$

This scale analysis result is only 33 percent larger than the integral analysis estimate (3.72). Its deviation from the exact result (Problem 3.2) is as much as 70 percent in the case of the square cross section. In conclusion, the scale analysis produces a compact and extremely inexpensive result that matches within a factor of order 1 the exact solution to the same problem. The method of scale analysis is used extensively in the convection heat transfer field [10] and in the convection chapters of the present course.

3.2.3   The Graphic Method

The more complicated conduction heat transfer geometries are nowadays handled routinely on the computer. An introduction to numerical methods is presented in the next section. Before the age of computers, an effective way of producing an estimate for the total heat transfer rate through a complicated (curved, crooked) geometry was based on the graphic method that we are about to see [11]. This graphic method has a place in heat transfer even today, because it gives the problem solver a chance to develop a better feel for "how heat flows" through a complicated body.

The idea that stands behind the method becomes clearer if we reexamine a conduction heat transfer pattern that we know very well. The left side of Fig. 3.9 shows the rectangular cross section of a slab that is very long in the direction normal to the plane of the figure. A finite temperature difference $(T_{hot} - T_{cold})$ is maintained between the top and bottom surfaces, while the two lateral surfaces are perfectly insulated. The conduction pattern is unidirectional (downward, from T_{hot} to T_{cold}); in other words, it consists of a family of parallel heat flux lines as on the right side of Fig. 2.1. Normal to the heat flux lines are the isotherms, which are also normal to the insulated lateral surfaces.

The network of flux lines and isotherms of Fig. 3.9a was constructed by first dividing the total heat current q (from T_{hot} to T_{cold}) into n minicurrents of equal size:

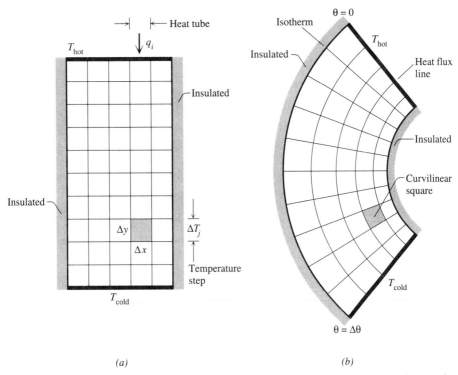

Figure 3.9 Networks of orthogonal heat flux lines and isotherms: (*a*) square loops of equal size; (*b*) curvilinear square loops of unequal size.

$$q_i = \frac{q}{n} \qquad (i = 1, 2, \ldots n) \tag{3.79}$$

Each minicurrent flows through a heat tube, that is, through the space between two adjacent flux lines. That the minicurrents are equal is indicated by the fact that in Fig. 3.9*a* the heat flux lines are equidistant.

The second move in the construction of the network was the drawing of *enough* isotherms so that each loop is a *square*. It turned out that $m = 10$ temperature steps were needed to complete the network. Each square experiences the same temperature difference in the vertical direction, namely,

$$\Delta T_j = \frac{T_{\text{hot}} - T_{\text{cold}}}{m} \qquad (j = 1, 2, \ldots, m) \tag{3.80}$$

According to the Fourier law, the minicurrent that passes through the (i,j) square is

$$q_i = k \Delta x W \frac{\Delta T_j}{\Delta y} = kW \Delta T_j \tag{3.81}$$

where Δx and Δy are the *equal* sides of the square loop and W is the width measured in the direction normal to the plane of Fig. 3.9*a*. Eliminating q_i and ΔT_j between eqs. (3.79)–(3.81), we find that the total heat transfer rate can be calculated by simply counting the heat tubes (n) and the temperature steps (m):

$$q = \frac{n}{m} Wk(T_{hot} - T_{cold}) \qquad (3.82)$$

This result can be compared with the definition of the shape factor S, eq. (3.33), to conclude that

$$S = \frac{n}{m}W \qquad (3.83)$$

Equations (3.82) and (3.83) apply not only to Fig. 3.9a but also to two-dimensional problems in which the flux lines and isotherms are curved. The chief requirement is that each curvilinear loop in the network must look "square." When this requirement is met, the analysis contained between eqs. (3.79) and (3.82) applies unchanged.

An example of a network with curved heat tubes and wedge-shaped temperature steps is presented in Fig. 3.9b. This new conduction pattern is obviously the result of "bending" the body of Fig. 3.9a. While trying to construct a network of curvilinear squares in Fig. 3.9b, the "wedge" shape of the temperature steps forced me to draw narrower heat tubes near the inner part of the bend.

Thinner heat tubes mean a higher heat flux; therefore, a larger share of the total heat current q prefers to flow through the region of smaller radius. Another way of looking at Fig. 3.9b is to recognize that the length of the inner heat tubes is shorter than the length of the outer heat tubes, and that the heat transfer from T_{hot} to T_{cold} will naturally favor the shorter path.

The graphics-based shape factor formula (3.83) is particularly useful when dealing with a configuration like the one sketched in Fig. 3.10a, for which an

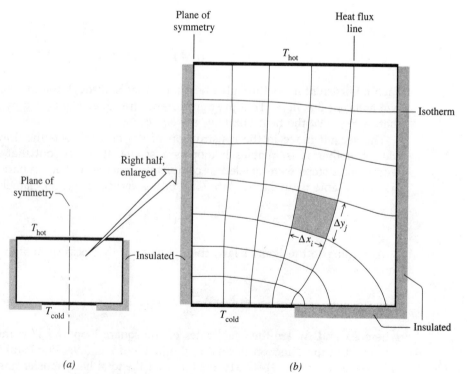

(a)   *(b)*

Figure 3.10 The manual construction of the heat flux lines and isotherms for a complicated two-dimensional conduction configuration.

exact analytical solution does not exist. This particular configuration is important in the work of predicting the "contact resistance" between two bodies the
surfaces of which are rough (i.e., bodies that make contact through the asperities of their surfaces). The state of the art in this area is presented by Yovanovich [12] and Madhusudana and Fletcher [13].

The network of flux lines and isotherms of Fig. 3.10b was constructed by
hand, by first noting the vertical plane of symmetry of the problem (Fig. 3.10a).
Therefore, it was sufficient to draw the network over only one-half of the
domain; in it, the vertical plane of symmetry appears as an adiabatic wall (a flux
line). All the isotherms are perpendicular to the adiabatic boundaries, and all
the flux lines are perpendicular to the specified isothermal boundaries (T_{hot},
T_{cold}).

The most difficult phase of the drawing is the arranging (and rearranging!) of
the flux lines and isotherms so that (a) they intersect at right angles, and (b) each
curvilinear loop appears square ($\Delta x_i \approx \Delta y_j$). The lower right portion of the
network of Fig. 3.10b is clearly in violation of this second requirement. A better
drawing can be made by constructing a finer mesh. One could, for example,
start out with 10 instead of only 4 heat flux lines. A finer mesh will improve the
accuracy that the ratio n/m brings to the calculation of the shape factor, eq.
(3.83).

The heat transfer literature contains a large volume of shape factor results for
the more common geometrical configurations that are encountered in actual
applications. Table 3.3 shows a representative sample of what is presently
available. Complementary tabulations can be found in Refs. [14–16], which
served also as sources for the information compiled in Table 3.3. Most of these
results have been determined based on advanced analytical methods (e.g.,
conformal mapping, the superposition of heat sources and sinks) and experiments that exploit the analogy between the conduction of heat and that of
electricity. They were not developed based on the graphic method. Table 3.3 is
exhibited at this stage only because the graphic method made it easier to explain
the meaning of the shape factor S.

An approximate formula for calculating the conduction shape factor for a

Table 3.3 The Shape Factors (S) of Several Geometrical Configurations Involving Isothermal
Surfaces, $q_{1 \to 2} = Sk(T_1 - T_2)$

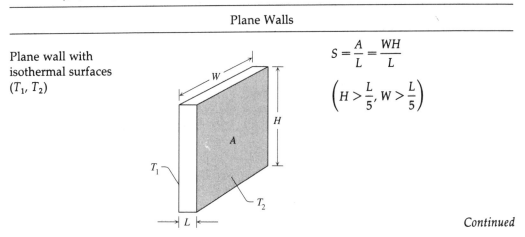

Plane Walls		
Plane wall with isothermal surfaces (T_1, T_2)	W, A, H, T_1, T_2, L	$S = \dfrac{A}{L} = \dfrac{WH}{L}$ $\left(H > \dfrac{L}{5}, W > \dfrac{L}{5} \right)$

Continued

Table 3.3, (*continued*)

Plane Walls

Edge prism, joining
two plane walls
with isothermal
surfaces:

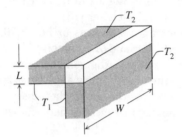

$$S = 0.54W$$
$$\left(W > \frac{L}{5}\right)$$

T_1 = temperature of internal surfaces of plane walls

T_2 = temperature of external surfaces of plane walls

T_2 = temperature of two exposed faces of the prism

Corner cube, joining
three plane walls.
The temperature
difference $T_1 - T_2$ is
maintained between
the inner and outer
surfaces of the
three-wall structure.

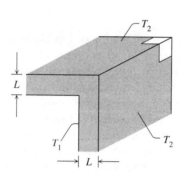

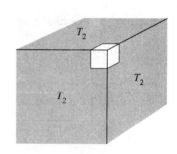

$$S = 0.15L$$

Cylindrical Surfaces

Bar with square
cross section,
isothermal outer
surface, and
cylindrical hole
through the center

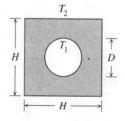

$$S = \frac{2\pi L}{\ln(1.08H/D)}$$

(L = length normal to the plane of
the figure)

Infinite slab with
isothermal surfaces
(T_2) and thickness
$2H$, with an
isothermal
cylindrical hole (T_1)
positioned midway
between the surfaces

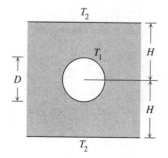

$$S = \frac{2\pi L}{\ln(8H/\pi D)}$$

($L >> D$, $H > D/2$, L = length of
cylindrical hole)

Continued

Table 3.3, (*continued*)

Cylinder of length L and diameter D, with cylindrical hole positioned eccentrically	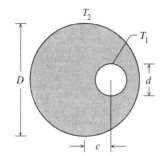	$$S = \frac{2\pi L}{\cosh^{-1}\left(\dfrac{D^2 + d^2 - 4c^2}{2Dd}\right)}$$ $(L \gg D)$
Semi-infinite medium with isothermal surface (T_2) and isothermal cylindrical hole (T_1) of length L, parallel to the surface (i.e., normal to the plane of the figure)	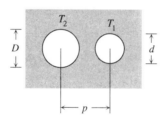	$$S = \frac{2\pi L}{\cosh^{-1}(2H/D)}$$ $(L \gg D)$ $$S \cong \frac{2\pi L}{\ln(4H/D)} \quad \text{if } H/D > 3/2$$
Infinite medium with two parallel isothermal cylindrical holes of length L	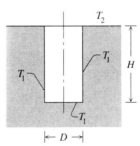	$$S = \frac{2\pi L}{\cosh^{-1}\left(\dfrac{4p^2 - D^2 - d^2}{2Dd}\right)}$$ $(L \gg D, d, p)$
Semi-infinite medium with isothermal surface (T_2) and with a cylindrical hole (T_1) drilled to a depth H normal to the surface. The hole is open. The medium temperature far from the hole is T_2		$$S = \frac{2\pi H}{\ln(4H/D)}$$ $(H \gg D)$

Continued

Table 3.3, (*continued*)

Spherical Surfaces

Semi-infinite
medium with
isothermal surface
(T_2) and isothermal
spherical cavity

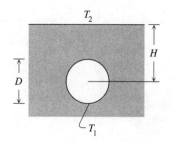

$$S = \frac{2\pi D}{1 - (D/4H)}$$

$$(H > D/2)$$

Spherical cavity in an infinite
medium:
$$S \cong 2\pi D \quad \text{if } H/D > 3$$

Semi-infinite
medium with
insulated surface
and far-field
temperature T_2
containing an
isothermal spherical
cavity (T_1)

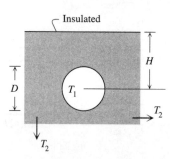

$$S = \frac{2\pi D}{1 + (D/4H)}$$

$$(H > D/2)$$

Infinite medium
with two isothermal
spherical cavities

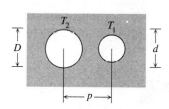

$$S = \frac{4\pi}{\dfrac{d}{D}\left[1 - \dfrac{(D/2p)^4}{1 - (d/2p)^2}\right] - \dfrac{d}{p}}$$

$$(p/D > 3)$$

Hemispherical
isothermal dimple
(T_1) into the
insulated surface of
a semi-infinite
medium with far-
field temperature T_2

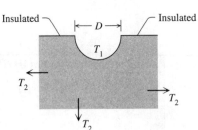

$S = \pi D$

Thin Discs and Plates

Semi-infinite
medium with
isothermal surface
(T_2) and isothermal
disc (T_1) parallel to
the surface

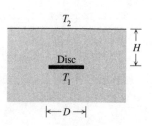

$$S = \frac{2\pi D}{(\pi/D) - \tan^{-1}(D/4H)}$$

$$(H/D > 1)$$

Disc attached to the surface:
$$S = 2D \quad (H = 0)$$

Continued

Table 3.3, (continued)

Semi-infinite medium with insulated surface and far-field temperature T_2, containing an isothermal disc (T_1) parallel to the surface	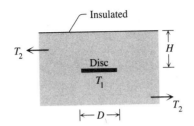	$S = \dfrac{2\pi D}{(\pi/2) + \tan^{-1}(D/4H)}$ $(H/D > 1)$
Infinite medium with two parallel, coaxial and isothermal discs	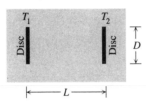	$S = \dfrac{2\pi D}{(\pi/2) - \tan^{-1}(D/2L)}$ $(L/D > 2)$
Semi-infinite medium with isothermal surface (T_2) and isothermal plate (T_1) parallel to the surface (W is the width of the plate, in the direction normal to the plane of the figure)	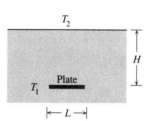	$S = \dfrac{2\pi L}{\ln(4L/W)} \quad (H \gg L)$ Plate attached to the surface: $S = \dfrac{\pi L}{\ln(4L/W)} \quad (H = 0)$

layer of uniform thickness surrounding a body of arbitrary shape has been developed by Hassani and Hollands [17]. The heat transfer concept of "shape factor" was introduced in 1913 by Langmuir et al. [18].

Example 3.2

Shape Factors

The cubical experimental chamber shown on the left side of the figure is surrounded by a layer of fiberglass insulation of thickness $L = 0.15$ m. The side of the cubical chamber is $H = 1$ m, and the chamber wall temperature is $T_h = 50°C$. The temperature of the external surface of the insulating layer is $T_c = 20°C$. Calculate the total heat transfer rate that leaks through the insulating layer.

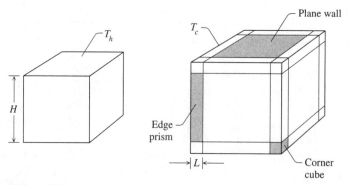

Figure E3.2

Solution. The heat transfer rate through the entire insulation can be broken up into the following currents:

$$q = 6q_{\substack{plane \\ wall}} + 12q_{\substack{edge \\ prism}} + 8q_{\substack{corner \\ cube}}$$

$$= 6S_{\substack{plane \\ wall}}k\Delta T + 12S_{\substack{edge \\ prism}}k\Delta T + 8S_{\substack{corner \\ cube}}k\,\Delta T$$

$$= Sk\,\Delta T \qquad (1)$$

where $\Delta T = T_h - T_c$. The overall shape factor S can be evaluated with the help of Table 3.3:

$$S = 6S_{\substack{plane \\ wall}} + 12S_{\substack{edge \\ prism}} + 8S_{\substack{corner \\ cube}}$$

$$= \left[6\frac{1}{0.15} + (12 \times 0.54 \times 1) + (8 \times 0.15 \times 0.15)\right]m$$

$$= (40 + 6.48 + 0.18)\,m = 46.66\,m \qquad (2)$$

The total heat transfer rate is therefore equal to

$$q = Sk\Delta T = 46.66\ m \times 0.035\ W/m\cdot K \times (50-20)°C$$

$$= 49\ W$$

The decomposition of the overall shape factor S into three contributions, eq. (2), shows that the six "plane walls" of the insulating wrapping account for $40/46.66$, or 86 percent of the entire heat transfer rate q.

3.3 NUMERICAL METHODS

3.3.1 Finite-Difference Conduction Equations

The most commonly used method today is the numerical approach, in which the continuous conducting medium is replaced by an array of discrete points (nodes). As their name suggests,* these nodes are the "knots" or points of intersection of the lines of the network (grid) that covers the conducting medium. There are no rigid rules for how one should spread the net over the

---

*The origin of the word "node" is the Latin noun *nodus* (knot).

conducting medium; however, the approximate conduction equations that we are about to derive assume their simplest form when the grid lines are parallel to the boundaries (i.e., when the boundaries themselves are lines of the grid).

Consider the two-dimensional conduction problem illustrated in Fig. 3.11. Its complicated shape rules out the possibility of deriving infinite series solutions of the type exhibited in Section 3.1. The crooked shape of the domain hampers also the analytical work and accuracy associated with an integral solution (Section 3.2.1). It is advisable to proceed in this case along the numerical path and to cover the conducting region with the two families of orthogonal lines shown in the figure. One very simple choice is to make both the horizontal and vertical lines equidistant, in which the respective spacings $(\Delta x, \Delta y)$ are two known constants.

The temperature distribution over the conducting medium, $T(x,y)$, is represented now by the finite number of node temperatures $T_{i,j}$. The indexes i and j count the grid lines in the x and y directions, respectively. In a physical sense, each node temperature accounts not only for the temperature at the point (i,j), but also for the temperature of the small (finite-size) region that surrounds the point (i,j). The small "isothermal" region of temperature $T_{i,j}$ is the *control volume* drawn around the node (i,j). If this node is *internal* (i.e., not a point on the boundary), then the control volume is represented by the shaded rectangle $\Delta x \, \Delta y$ that has the point (i,j) as its center (Fig. 3.11, right). The control volume boundaries are halfway between the grid points. When the node (i,j) is situated on one of the boundaries or in a corner of the conducting domain, the respective control volume is somewhat smaller than $\Delta x \, \Delta y$: these special cases are illustrated by the shaded patches in the left column of Table 3.4.

The equation for steady conduction in two directions, eq. (3.68), can be replaced with an analogous statement of energy conservation in each of the control volumes that have been identified. Consider first the internal control volume that has been isolated to the right in Fig. 3.11. The volume of this

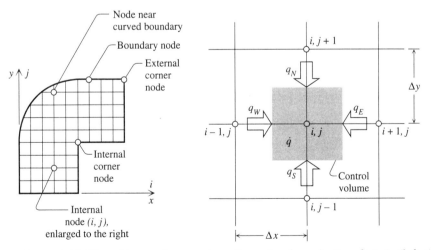

Figure 3.11 The grid spread over a two-dimensional conduction domain (left side), and the steady-state conservation of energy in the control volume associated with an internal node (right side).

Table 3.4 Finite-Difference Conduction Equations for the Nodes of a Uniform Square Grid ($\Delta x = \Delta y = $ Constant) Placed Over a Two-Dimensional Conduction Domain Without Internal Heat Generation ($\dot{q} = 0$)

Internal Node

$$T_{i-1,j} + T_{i+1,j} + T_{i,j-1} + T_{i,j+1} - 4T_{i,j} = 0$$

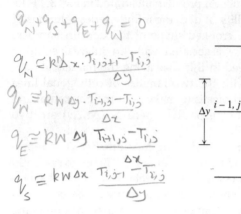

$$q_N + q_S + q_E + q_W = 0$$

$$q_N \cong k\Delta x \cdot \frac{T_{i,j+1} - T_{i,j}}{\Delta y}$$

$$q_W \cong k w \Delta y \cdot \frac{T_{i-1,j} - T_{i,j}}{\Delta x}$$

$$q_E \cong k w \Delta y \frac{T_{i+1,j} - T_{i,j}}{\Delta x}$$

$$q_S \cong k w \Delta x \frac{T_{i,j-1} - T_{i,j}}{\Delta y}$$

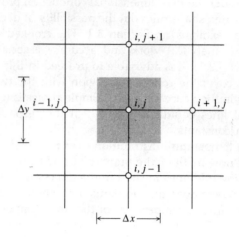

Surface Node[a]

$$2T_{i-1,j} + T_{i,j-1} + T_{i,j+1} + 2\frac{h\Delta x}{k}T_\infty - 2\left(\frac{h\Delta x}{k} + 2\right)T_{i,j} = 0$$

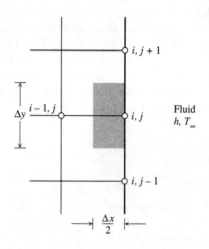

External Corner Node[a]

$$T_{i,j-1} + T_{i-1,j} + 2\frac{h\Delta x}{k}T_\infty - 2\left(1 + \frac{h\Delta x}{k}\right)T_{i,j} = 0$$

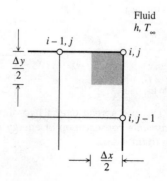

Continued

Table 3.4, (continued)

Internal Corner Node[a]

$$2(T_{i-1,j} + T_{i,j+1}) + T_{i+1,j} + T_{i,j-1} + 2\frac{h\,\Delta x}{k}T_\infty - 2\left(3 + \frac{h\,\Delta x}{k}\right)T_{i,j} = 0$$

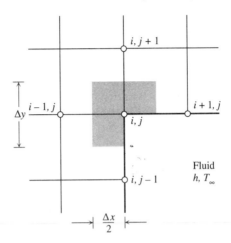

Node Near Curved Boundary with Specified Temperature

$$\frac{2}{1+a}T_{i+1,j} + \frac{2}{1+b}T_{i,j-1} + \frac{2T_1}{a(1+a)} + \frac{2T_2}{b(1+b)} - 2\left(\frac{1}{a} + \frac{1}{b}\right)T_{i,j} = 0$$

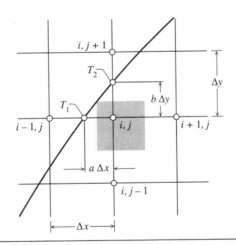

[a]Two special limits of this equation are:
adiabatic boundary, $h\,\Delta x/k = 0$;
boundary with specified temperature, $T_{i,j} = T_\infty$, that is, $h\,\Delta x/k \to \infty$.

internal element is $(\Delta x) \times (\Delta y) \times W$, where W is the width of the two-dimensional body, in the direction normal to the plane of Fig. 3.11. In the steady state this closed thermodynamic system receives heat transfer from four directions: north (q_N), east (q_E), south (q_S), and west (q_W). As in the derivation of eq. (1.37), these four heat transfer rates can be evaluated by invoking the proportionality between the heat flux and the respective temperature gradient. When the control volumes are sufficiently small, each temperature gradient is approximately equal to the temperature difference between two adjacent nodes, divided by the distance that separates these nodes:

$$q_N \cong kW\Delta x \frac{T_{i,j+1} - T_{i,j}}{\Delta y}$$

$$q_W \cong kW\Delta y \frac{T_{i-1,j} - T_{i,j}}{\Delta x} \qquad\qquad q_E \cong kW\Delta y \frac{T_{i+1,j} - T_{i,j}}{\Delta x} \quad (3.84)$$

$$q_S \cong kW\Delta x \frac{T_{i,j-1} - T_{i,j}}{\Delta y}$$

Applying now the first law of thermodynamics to the internal control volume, we write

$$q_N + q_E + q_S + q_W + \dot{q}\,\Delta x\,\Delta y\,W = 0 \qquad\qquad (3.85)$$

Substituting the heat current approximations (3.84) into the first law (3.85), we obtain the *finite-difference* conduction equation for the internal node (i,j):

$$2\left(\frac{\Delta x}{\Delta y} + \frac{\Delta y}{\Delta x}\right)T_{i,j} \cong \frac{\Delta y}{\Delta x}(T_{i-1,j} + T_{i+1,j}) +$$

$$\frac{\Delta x}{\Delta y}(T_{i,j-1} + T_{i,j+1}) + \frac{\dot{q}}{k}\Delta x\,\Delta y \quad (3.86)$$

An even simpler expression holds in the case when the loops of the grid are square ($\Delta x = \Delta y$):

$$T_{i,j} \cong \frac{1}{4}(T_{i-1,j} + T_{i+1,j} + T_{i,j-1} + T_{i,j+1}) + \frac{\dot{q}}{4k}(\Delta x)^2 \qquad (3.87)$$

This approximate equation allows us to calculate the internal node temperature $T_{i,j}$ by using the temperature values at the four closest nodes. One equation of this type can be written for every internal node of the uniform and square grid.

The finite-difference conduction equation assumes a *special* form in each of the cases when the node is *not* internal. These special forms are summarized in Table 3.4 for uniform square grid problems without internal heat generation ($\dot{q} = 0$). For example, in the case of an *external corner* (the third entry in Table 3.4), the corresponding control volume has the smaller size $(\Delta x/2) \times (\Delta y/2) \times W$. The corresponding finite-difference approximations for the heat currents that enter this control volume from the four directions associated with the cardinal points are

$$q_N \cong hW\frac{\Delta x}{2}(T_\infty - T_{i,j})$$

$$q_W \cong kW\frac{\Delta y}{2}\frac{(T_{i-1,j} - T_{i,j})}{\Delta x} \qquad\qquad q_E \cong hW\frac{\Delta y}{2}(T_\infty - T_{i,j}) \quad (3.88)$$

$$q_S \cong kW\frac{\Delta x}{2}\frac{(T_{i,j-1} - T_{i,j})}{\Delta y}$$

Substituted into the first law of thermodynamics (3.85), and setting $\dot{q} = 0$ and $\Delta x = \Delta y$, these approximations lead to the "external corner node" equation listed in Table 3.4. In that equation, T_∞ and h are the *local* values* of the external fluid temperature and the heat transfer coefficient, respectively. Unlike in the analytical solutions reported until now, T_∞ and h are not necessarily constant. The dimensionless heat transfer coefficient that emerges in the finite-difference equation, $h \, \Delta x/k$, is the *grid-size Biot number*.

3.3.2 The Matrix Inversion Method

In order to learn how the preceding relationships can be used to generate a sufficiently accurate solution for a temperature field $T(x,y)$, consider the bar with square cross section drawn on the right side of Fig. 3.12. This problem stems from questioning the second entry listed in Table 3.3, which shows that the conduction shape factor of an edge prism is $S = 0.54 \, W$. The edge prism joins two perpendicular walls with the same thickness L, and with a constant temperature difference, ΔT, maintained between the outer and inner surfaces of the enclosure formed by the two walls (Fig. 3.12, left side).

The square cross section bar shown on the right side of the figure is the enlarged version of the edge prism, after it was cut and removed. To calculate its shape factor, we must first determine the temperature field $T(x,y)$ and, from it, the total heat transfer rate through the entire surface of temperature ΔT. The conduction problem is not complete unless we specify boundary conditions along the left side and bottom of the square cross section (i.e., along the cuts). One idea would be to assume that along these two surfaces the temperature

---

*The values associated with the boundary point (i,j) or with the fluid region that bathes the point (i,j).

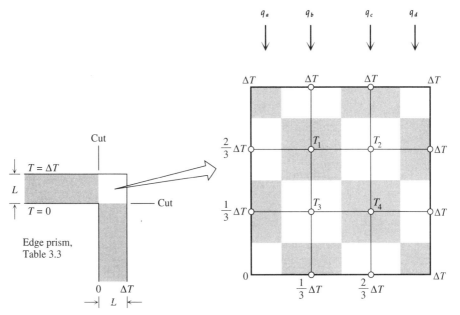

Figure 3.12 Bar with square cross section, as a model of the edge prism listed as the second entry in Table 3.3.

varies linearly, so that the temperature of the lower left corner is $T = 0$. This assumption is recommended by the thought that (sufficiently far from edge prism region) the temperature varies linearly across each of the walls that are joined on the left side of Fig. 3.12.

For the sake of brevity in presentation, we select a coarse grid consisting of only four horizontal lines and four vertical lines. The grid has a total of 16 nodes, out of which only four are internal. Consequently, there are only four unknowns in this problem, namely, the internal node temperatures labeled T_1, T_2, T_3, and T_4. We obtain the needed four equations by writing the equivalent of eq. (3.87) for each of the internal nodes:

$$
\begin{aligned}
4T_1 - T_2 \ - T_3 \quad\quad\ &= \tfrac{5}{3}\Delta T \\
-T_1 + 4T_2 \quad\quad\ - T_4 &= 2\,\Delta T \\
-T_1 \quad\quad + 4T_3 - T_4 &= \tfrac{2}{3}\Delta T \\
- T_2 \ - T_3 \ + 4T_4 &= \tfrac{5}{3}\Delta T
\end{aligned}
\tag{3.89}
$$

In general, when as many as n node temperatures are unknown, this system of equations reads

$$[A][T] = [C] \tag{3.90}$$

or, more explicitly,

$$
\begin{bmatrix}
a_{11} & a_{12} \dots a_{1n} \\
a_{21} & a_{22} \quad\ \vdots \\
\vdots & \quad\quad \vdots \\
 & \quad\quad \vdots \\
a_{n1} & \dots\dots a_{nn}
\end{bmatrix}
\begin{bmatrix}
T_1 \\
T_2 \\
\vdots \\
\vdots \\
T_n
\end{bmatrix}
=
\begin{bmatrix}
C_1 \\
C_2 \\
\vdots \\
\vdots \\
C_n
\end{bmatrix}
\tag{3.91}
$$

In the present problem, the elements of the $[A]$ matrix are

$$
[A] =
\begin{bmatrix}
4 & -1 & -1 & 0 \\
-1 & 4 & 0 & -1 \\
-1 & 0 & 4 & -1 \\
0 & -1 & -1 & 4
\end{bmatrix}
\tag{3.92}
$$

while the elements of the $[C]$ vector are the quantities listed on the right sides of eqs. (3.89).

Assuming that the determinant of $[A]$ is nonzero, the unknown node temperatures (the elements of the vector $[T]$) are obtained by solving eq. (3.90),

$$[T] = [B][C] \tag{3.93}$$

in which the new matrix $[B]$ is the inverse of $[A]$,

$$
[B] = [A]^{-1} =
\begin{bmatrix}
b_{11} & b_{12} & \dots & b_{1n} \\
b_{21} & b_{22} & & \vdots \\
\vdots & & & \vdots \\
\vdots & & & \vdots \\
b_{n1} & \dots & \dots & b_{nn}
\end{bmatrix}
\tag{3.94}
$$

Equation (3.93) is shorthand for the system

$$T_1 = b_{11}C_1 + b_{12}C_2 + \cdots + b_{1n}C_n$$
$$\vdots$$
$$T_n = b_{n1}C_1 + b_{n2}C_2 + \cdots + b_{nn}C_{nn}$$

(3.95)

In conclusion, the key step in the calculation of the node temperatures is the inversion of the [A] matrix, namely, the evaluation of the elements of the [B] matrix [19]. This step is aided by the software packages that accompany many of the computers that are in use today, large and small:

$$[B] = \begin{bmatrix} 0.2916 & 0.0833 & 0.0833 & 0.0416 \\ 0.0833 & 0.2916 & 0.0416 & 0.0833 \\ 0.0833 & 0.0416 & 0.2916 & 0.0833 \\ 0.0416 & 0.0833 & 0.0833 & 0.2916 \end{bmatrix}$$

(3.96)

The temperatures at the four internal nodes of Fig. 3.12 (right) are obtained finally by writing four equations of type (3.95); the results are

$$T_1 = 0.7778\,\Delta T \qquad T_2 = 0.8889\,\Delta T$$
$$T_3 = 0.5556\,\Delta T \qquad T_4 = 0.7778\,\Delta T$$

(3.97)

This solution is symmetric about the rising diagonal of the square cross section (note that $T_1 = T_4$), because the boundary temperatures specified around the square are symmetric about the same diagonal.

The matrix inversion method outlined in this section is not recommended because it is numerically inefficient, especially in cases where the grid is fine and the [A] matrix contains a large number of elements [20]. In such cases a much more efficient path is provided by the iterative procedure that is described next.

3.3.3   The Gauss–Seidel Iteration Method

Continuing with the example of Fig. 3.12, the four-unknown system (3.89) can be rearranged so that each equation shows on the left side only the unknown temperature of the node (control volume) for which that equation was written:

$$T_1 = \tfrac{1}{4}\,(T_2 + T_3 + \tfrac{5}{3}\,\Delta T)$$
$$T_2 = \tfrac{1}{4}\,(T_1 + T_4 + 2\,\Delta T)$$
$$T_3 = \tfrac{1}{4}\,(T_1 + T_4 + \tfrac{2}{3}\,\Delta T)$$
$$T_4 = \tfrac{1}{4}\,(T_2 + T_3 + \tfrac{5}{3}\,\Delta T)$$

(3.98)

The first step in this new procedure is to assign an initial set of values to the unknown node temperatures $T_1, \ldots, T_4$. In the present problem a reasonable starting value for these four temperatures is $0.5\,\Delta T$, because this value happens to fall between the temperature extremes that have been specified around the square boundary.

The second step consists of using each of eqs. (3.98) to update (i.e., to obtain a new estimate of) the wanted node temperatures. For example, the first of these equations provides a new estimate of T_1. Next, this updated T_1 value and the still not updated value of T_4 are used in the second of eqs. (3.98), to update the value of T_2. Said another way, the values that are used on the right side of one of the eqs. (3.98) represent the *latest* temperature information that is available. In this manner one complete sweep through the stack of four equations (i.e., through the interior of the square, Fig. 3.12 right) produces an updated set of values $T_1, \ldots, T_4$. This once-through sweep constitutes the *first iteration*.

In the third and subsequent steps of this procedure, the sweep through eqs. (3.98) is repeated until the calculated node temperatures converge on a set of values that no longer vary from one iteration to the next. In the numerical example of Fig. 3.12, the updated values found after each additional iteration are

p	$T_1/\Delta T$	$T_2/\Delta T$	$T_3/\Delta T$	$T_4/\Delta T$
0	0.5	0.5	0.5	0.5
1	0.6458	0.8229	0.4167	0.6458
2	0.7448	0.8724	0.4896	0.7448
3	0.7695	0.8848	0.5391	0.7695
4	0.7757	0.8879	0.5514	0.7757
5	0.7773	0.8886	0.5545	0.7773

where p indicates the number of iterations. The values listed in this table were obtained by sweeping through system (3.98) in the sequence $T_3 \rightarrow T_4 \rightarrow T_1 \rightarrow T_2$. Comparing the bottom row of the table with the exact solution to system (3.98) [i.e., with eqs. (3.97)],* we see that just five iterations generate a solution that reproduces very well the exact solution.

The calculation is terminated when the change in each node temperature between two consecutive iterations is sufficiently small,

$$\left| \frac{T_{i,j}^{(p+1)} - T_{i,j}^{(p)}}{T_{i,j}^{(p)}} \right| < \epsilon \tag{3.99}$$

where $T_{i,j}$ is the more general (Fig. 3.11) notation for the unknown node temperatures $T_1, \ldots, T_4$ of Fig. 3.12. The number ϵ has a dimensionless value much smaller than 1, for example, 10^{-4} or 10^{-5}. The ϵ value is selected before the start of the entire iterative procedure, for it is a trade-off between the accuracy of the eventual solution and the cost of the computation (number of iterations, p).

A trade-off between accuracy and cost is also made in choosing the fineness of the grid, which determines the total number of equations of type (3.87) that will appear in system (3.98). The accuracy increases with the number of these equations, n, that is, with the only geometrical parameter of the uniform square grid.

In summary, the node temperatures recorded at the end of the Gauss–Seidel iteration procedure are functions of two parameters chosen by the designer of the numerical solution,

$$T_{i,j} = T_{i,j}(\epsilon, n) \tag{3.100}$$

---

*Equations (3.97) are the "exact" solution of system (3.98), not of the problem of finding the temperature distribution over the square domain of Fig. 3.12. Recall that the four equations of system (3.98) are all of type (3.87), that is, finite-difference *approximate* equations.

The smallness of ϵ and the greatness of n are both limited by the computational time needed for generating the solution.

Returning to the edge prism problem that suggested the numerical exercise of Fig. 3.12., it is instructive to calculate the conduction shape factor of the square cross section. For this, we must calculate the total heat transfer rate through the region, q, and divide it by $k\Delta T$. The heat current q enters through the top and right-side surfaces of the conducting body, that is, through the surfaces of temperature ΔT. The symmetry about the diagonal $0 - T_3 - T_2 - \Delta T$ means that $q/2$ enters through the top surface and another $q/2$ enters through the right-side surface. Therefore, looking only along the top surface, we recognize the four minicurrents associated with the four control volumes that touch the boundary:

$$\frac{q}{2} = q_a + q_b + q_c + q_d$$

$$= kW\frac{L}{6}\frac{\Delta T - \frac{2}{3}\Delta T}{L/3} + kW\frac{L}{3}\frac{\Delta T - T_1}{L/3} +$$

$$kW\frac{L}{3}\frac{\Delta T - T_2}{L/3} + kW\frac{L}{6}\frac{\Delta T - \Delta T}{L/3} \quad (3.101)$$

The T_1 and T_2 values needed in this calculation are listed in eqs. (3.97); in the end, the right side of eq. (3.101) reduces to $0.5\,kW\,\Delta T$, meaning that the conduction shape factor is

$$S = \frac{q}{k\,\Delta T} = W \quad (3.102)$$

This S value is 85 percent larger than the shape factor listed in the second case of Table 3.3. The discrepancy is due to the linear temperature distributions that have been assumed along the left side and bottom of the square cross section. In the actual edge prism region (Fig. 3.12, left), the temperature in the two "cuts" is dominated by the temperature of the outer surface, ΔT. In other words, the temperatures that have been assumed for the left side and bottom of the square (Fig. 3.12, right) are lower than the real temperatures in the two cuts of the edge prism configuration.

A final observation concerns the expressions listed for the minicurrents q_a, q_b, q_c, and q_d in eq. (3.101). These expressions represent the heat currents across the bottom faces of the four control volumes that line the top boundary. Each bottom face current happens to be equal to the respective current that enters through the top face of the control volume, because the direction of heat flow is vertical (recall that the top boundary of the square is isothermal). For example, if the top boundary is not isothermal, the calculation of the heat currents q_a, q_b, q_c, and q_d must account also for the heat flow parallel to the boundary, that is, across the vertical faces of the four control volumes that touch the top boundary.

REFERENCES

1. J. Fourier, *Analytical Theory of Heat*, translated, with notes, by A. Freeman, G. E. Stechert & Co., New York, 1878. (The French edition appeared in 1822, as

Théorie Analytique de la Chaleur; this theory was described first in a paper submitted by Fourier to the *Institut de France* in 1807.)

2. H. S. Carslaw and J. C. Jaeger, *Conduction of Heat in Solids*, 2nd edition, Oxford University Press, Oxford, 1959.

3. P. J. Schneider, *Conduction Heat Transfer*, Addison Wesley, Reading, MA, 1955.

4. V. S. Arpaci, *Conduction Heat Transfer*, Addison Wesley, Reading, MA, 1966.

5. G. E. Myers, *Analytical Methods in Conduction Heat Transfer*, McGraw-Hill, New York, 1971.

6. M. N. Ozisik, *Heat Conduction*, Wiley, New York, 1980.

7. S. Kakac and Y. Yener, *Heat Conduction*, Hemisphere, Washington, DC, 1985.

8. G. N. Watson, *Theory of Bessel Functions*, Cambridge University Press, Cambridge, 1966.

9. M. Abramowitz and I. Stegun, Eds., *Handbook of Mathematical Functions*, NBS AMS 55, Washington, DC, 1964.

10. A. Bejan, *Convection Heat Transfer*, Wiley, New York, 1984.

11. T. Lehmann, Graphische Methode zur Bestimmung des Kraftlinienverlaufes in der Luft, *Elektrotechnische Zeitschrift*, Vol. 42, 1909, pp. 995–998, and 1019–1022.

12. M. M. Yovanovich, Recent developments in thermal contact, gap and joint conductance theories and Experiment, in C. L. Tien, V. P. Carey, and J. K. Ferrel, Eds., *Heat Transfer—1986*, Vol. 1, Hemisphere, New York, pp. 35–45, 1986.

13. C. V. Madhusudana and L. S. Fletcher, Contact heat transfer—the last decade, *AIAA J*, Vol. 24, pp. 510–523, 1986.

14. R. V. Andrews, Solving conductive heat transfer problems with electrical-analogue shape factors, *Chem. Eng. Progress*, Vol. 51, No. 2, pp. 67–71, 1955.

15. J. E. Sunderland and K. R. Johnson, Shape factors for heat conduction through bodies with isothermal or convective boundary conditions, *Trans. ASHRAE*, Vol. 70, pp. 237–241, 1964.

16. E. Hahne and U. Grigull, Shape factor and shape resistance for steady multidimensional heat conduction (in German), *Int. J. Heat Mass Transfer*, Vol. 18, pp. 751–767, 1975.

17. A. V. Hassani and K. G. T. Hollands, Conduction shape factor for a region of uniform thickness surrounding a three-dimensional body of arbitrary shape, *J. Heat Transfer*, Vol. 112, pp. 492–495, 1990.

18. I. Langmuir, E. Q. Adams, and G. S. Meikle, Flow of heat through furnace walls: the shape factor, *Trans. Am. Electrochem. Soc.*, Vol. 24, pp. 53–81, 1913.

19. W. J. Minkowycz, E. M. Sparrow, G. E. Schneider, and R. H. Pletcher, *Handbook of Numerical Heat Transfer*, Wiley, New York, 1988, Chapter 1.

20. J. R. Rice, *Matrix Computations and Mathematical Software*, McGraw-Hill, New York, 1981, p. 23.

PROBLEMS

Analytical Methods

3.1 Compare the two-dimensional conduction estimate (3.46) with the simpler, one-dimensional formula (2.97), and show that the ratio $q_{b,(3.46)}/q_{b,(2.97)}$ depends only on the Biot number $hH/2k$. Using the first six characteristic values listed in Table 3.1, show that the q_b ratio decreases steadily as the Biot number increases. In other words, show that the one-dimensional model (2.97) overestimates the value of q_b and that it is accurate only when the Biot number is small.

3.2 The exact analytical solution for the temperature distribution inside the cross section with uniform heat generation shown in Fig. 3.8 in the text can be constructed by adding together the solutions to two simpler problems (see sketch):

$$\theta(x,y) = \theta_1(y) + \theta_2(x,y)$$

The volumetric rate of internal heat generation is uniform, $\dot{q}$ = constant. The first contribution (θ_1) is the solution to a unidirectional conduction problem with internal heat generation, whereas the second contribution (θ_2) is the solution to a two-dimensional conduction problem without internal heat generation.

(a) Determine the complete analytical solution for $\theta(x,y)$. You may wish to position the origin of the (x,y) system in the center of the cross section.

(b) Show that when the cross section is square, the maximum temperature at the center of the cross-section is

$$\theta_{max} = 0.07367 \, \frac{\dot{q} \, H^2}{k}$$

Compare this result with the approximate result based on integral analysis, eq. (3.72).

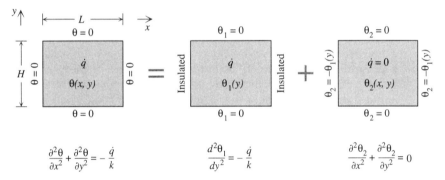

Figure P3.2

3.3 Consider the steady conduction problem in the sector of the annulus of Fig. 3.9b and explain why the temperature distribution cannot be a function of radius (i.e., why the isotherms must be oriented radially). Rely on a geometrical argument by noting the symmetric shape of the conducting domain. Write down the exact analytical solution for the temperature distribution $T(\theta)$ and for the distribution of circumferential heat flux, q''_θ. Show that q''_θ is indeed more intense at small radii, that is, in the "inside lanes" of the curved track.

3.4 Reconsider the two-dimensional plate fin problem of Fig. 3.1 by focusing on the case when L may be regarded as infinitely greater than H. Step by step, perform the equivalent of the analysis contained between eqs. (3.2) and (3.28), and show that the temperature distribution is

$$\theta = \frac{4\theta_b}{\pi} \sum_{n=0}^{\infty} \frac{1}{2n+1} \exp[-(2n+1)(\pi \, x/H)]\sin[(2n+1)(\pi \, y/H)]$$

3.5 The semi-infinite plate of thickness H shown in the figure has a linear temperature distribution imposed on the left boundary,

$$\theta = by \quad \text{at} \quad x = 0$$

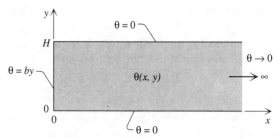

Figure P3.5

The remaining boundaries and the distant end to the right ($x \to \infty$) are all at the same temperature ($\theta = 0$). By following the steps outlined in Sections 3.1.2 through 3.1.4 in the text, show that the temperature distribution inside the plate is

$$\theta = bH\frac{2}{\pi} \sum_{n=0}^{\infty} \frac{(-1)^{n+1}}{n} \exp\left(-n\pi\frac{x}{H}\right) \sin\left(n\pi\frac{y}{H}\right)$$

where the constant b is the temperature gradient along the left boundary.

3.6   A temperature difference θ_a is maintained between the top and bottom surfaces of the semi-infinite plate shown on the left side of the figure. The temperature distribution inside the plate, $\theta(x,y)$, cannot be determined directly: note that y is not a direction of homogeneous boundary conditions and x is not such a direction either because there is no distinct "boundary" to the right of the conducting medium. Instead, $\theta(x,y)$ can be obtained by adding the solutions to the simpler problems outlined on the right side of the figure. The first of these, $\theta_1(y)$, is the solution to one-dimensional conduction in the y direction. The $\theta_2(x,y)$ problem is the same as Problem 3.5. Determine the temperature distribution $\theta(x,y)$.

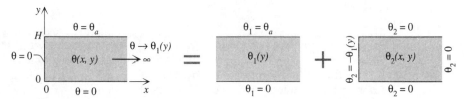

Figure P3.6

3.7   The figure shows the actual distribution of temperature in the cross section through the semi-infinite plate shown on the left of the figure in Problem 3.6. This plot was made using the series solution derived in Problem 3.6. Note that the temperature field $\theta(x,y)$ has two distinct regions—a distant region (far to the right) in which θ is practically independent of x, and an "end" region of approximate length δ, in which θ depends on both x and y.

Write down the conduction equation satisfied by θ in the end region, and replace each term in this equation with its respective scale. From the approximate algebraic equation that results, deduce the proper scale of the end-region length δ. Compare your conclusion with the length scale revealed by the exact drawing.

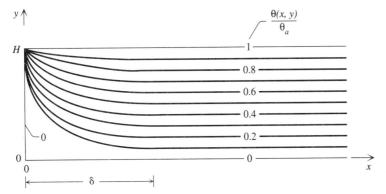

Figure P3.7

3.8 The figure shows the triangular cross section through a long bar. A finite temperature difference (θ_b) is maintained between the two sides that are mutually perpendicular. The hypotenuse is perfectly insulated. Determine analytically the temperature distribution for steady conduction in the triangular area, $\theta(x,y)$. (*Hint*: Exploit the geometrical relationship that might exist between the given triangle and a square cross section of side L.)

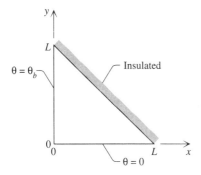

Figure P3.8

3.9 The figure shows the square cross section of a long bar. Two adjoining sides are at the same temperature, T_c, while the remaining two sides are at different temperatures, T_a and T_b. Determine the steady-state temperature distribution inside the square cross section by superimposing two solutions of the type developed in the text for the problem of Fig. 3.1. In other words, rely on eq. (3.28) and the principle of superposition in order to deduce the series solution for the present problem.

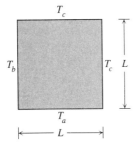

Figure P3.9

3.10 Derive an expression for the heat transfer rate through the $x = L$ tip of the two-dimensional fin shown in Fig. 3.1. Use the temperature distribution (3.28) as a starting point in this derivation. Show that the heat transfer rate through the tip is a function of the slenderness ratio L/H. Determine the tip heat transfer rate formula for the case $L/H = 2$.

3.11 (a) Determine the temperature distribution $T(x,y)$ inside semi-infinite slab of thickness H shown in the figure. Conduct your analysis by following the steps outlined in Section 3.3.1.

(b) Note the geometrical relationship between the present problem and the problem solved in the text based on Fig. 3.3. Write down the solution to the present problem by appropriately interpreting the solution listed in eqs. (3.44) and (3.45). Compare your solution with the one developed step by step in part (a).

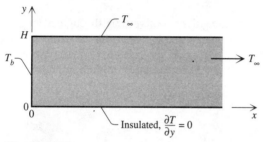

Figure P3.11

Conduction Shape Factor

3.12 The pipe that supplies drinking water to a community in a cold region of the globe is buried at a depth $H = 3$ m below the ground surface. The external surface of the pipe can be modeled as an isothermal cylinder of diameter $D = 0.5$ m and temperature 4°C. The ground surface is at 0°C, and the thermal conductivity of the soil is $k = 1\,\text{W/m} \cdot \text{K}$. Calculate the heat transfer rate from the pipe to the surrounding soil, q' (W/m).

During a very harsh winter, the top layer of the soil freezes to a depth of 1 m. This new position of the freezing front can be modeled as an isothermal plane of temperature 0°C situated at $H = 2$ m above the pipe centerline. Calculate again the per-unit-pipe-length heat transfer rate q', and the increase in q' that is due to the 1-m advancement of the freezing front.

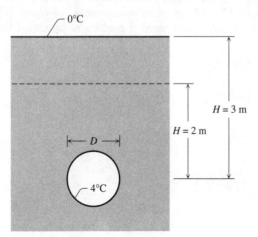

Figure P3.12

3.13 The sphere of radius r_i is surrounded by a constant-thickness shell of conductivity k and outer radius r_o. This assembly is buried in an infinite stationary medium of conductivity k_∞.

Derive an expression for the overall thermal resistance R_t between the r_i sphere and the infinite medium. Show that R_t varies monotonically with the shell outer radius; in other words, show that the "critical radius" concept of Section 2.4 does not apply. Under what circumstances does the spherical shell provide a thermal insulation effect for the r_i sphere? Conversely, what relationship must exist between k and k_∞ if the addition of the shell is to reduce the thermal resistance between the r_i sphere and the infinite medium?

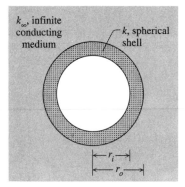

Figure P3.13

3.14 The figure shows a cross-sectional view through a horizontal pipe buried at a depth $H = 2$ m beneath the ground surface. The pipe of outer diameter $D_i = 26$ cm carries pressurized steam of temperature $T_s = 120°C$. A 7-cm-thick layer of insulation of conductivity $k_{ins} = 0.2\,W/m \cdot K$ is wrapped around the pipe wall. The ground may be modeled as dry soil with the thermal conductivity $k_\infty = 1\,W/m \cdot K$. The ground surface is isothermal and at the average seasonal temperature $T_\infty = 10°C$.

Calculate the heat transfer rate from the steam pipe, expressed per unit pipe length. Model the conduction heat transfer through the insulation as purely radial, and use eq. (2.42) for the overall thermal resistance between T_s and T_∞. The apparent heat transfer coefficient h at the outer surface of the insulation can be evaluated based on the appropriate shape factor S listed in Table 3.3.

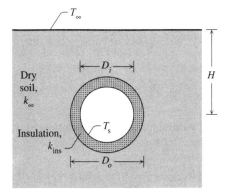

Figure P3.14

3.15 The working space of an experimental apparatus is shaped as a parallelepiped with the dimensions $x = 1$ m, $y = 0.5$ m, and $z = 0.3$ m. This space is covered all around by a layer of fiberglass insulation, with a uniform thickness $L = 0.1$ m. The temperature difference everywhere across this insulating layer is $\Delta T = 10°C$. Calculate the total heat transfer rate through the insulation.

Numerical Methods

3.16 The node (i,j) shown in the figure is situated on a boundary along which the heat flux q'' is specified. The grid is square, $\Delta x = \Delta y$. Invoke the steady-state conservation of energy in the control volume associated with the (i,j) node, to demonstrate that the node temperature is given by

$$T_{i,j} \cong \frac{1}{4} (T_{i,j-1} + T_{i,j+1} + 2T_{i-1,j}) + \frac{q''}{2k} \Delta x$$

Verify that the adiabatic boundary limit of this result agrees with the corresponding formula listed in Table 3.4.

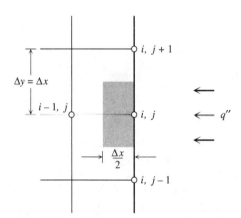

Figure P3.16

3.17 The conduction shape factor of the square domain of Fig. 3.12 was evaluated in the text by means of the total heat transfer rate that *enters* the domain, eq. (3.101). Perform the alternative calculation in which you evaluate the total heat transfer rate that leaves through the left and bottom sides of the square, and compare the resulting shape factor S with the estimate listed in eq. (3.102).

3.18 The plate fin with constant cross section shown in the figure is covered with a string of $n + 1$ equidistant nodes ($\Delta x = L/n$ = constant). The heat transfer coefficient h is known. Show that the temperature of an "internal" node, T_i, is related to the adjacent node temperatures via

$$T_{i-1} - [2 + (m\Delta x)^2] T_i + T_{i+1} + (m\Delta x)^2 T_\infty = 0$$

where $m^2 = hp/kA_c$ and p is the total wetted perimeter of the cross section A_c. Show further that the corresponding relation for the tip node temperature T_{n+1} is

$$T_n - \left[\frac{h\Delta x}{k} + 1 + \tfrac{1}{2} (m\Delta x)^2\right] T_{n+1} + \left[\frac{h\Delta x}{k} + \tfrac{1}{2} (m\Delta x)^2\right] T_\infty = 0$$

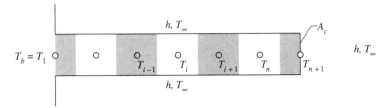

Figure P3.18

3.19 To develop an understanding for the trade-off between accuracy and computational cost in the implementation of the Gauss–Seidel iteration, consider the square cross section case ($H = L$) of the bar with uniform internal heat generation ($\dot{q}$, k) shown in Fig. 3.8. The exact solution for the maximum temperature in the center of this cross section is listed in Problem 3.2, part (b).

Develop a numerical solution for the steady temperature distribution over the square cross section. Follow these steps:

1. Nondimensionalize all the temperatures, that is, instead of the unknown $\theta = T - T_\infty$ use the *dimensionless excess temperature*

$$\hat{\theta} = \frac{\theta}{\dot{q}L^2/k}$$

2. Cover the square domain with a uniform grid with square loops of size $\Delta x = L/N$, where N is an even integer that will be specified later.
3. Begin a numerical program by assigning the value $\hat{\theta} = 0$ to all the boundary and corner nodes.
4. Assign the initial value $\hat{\theta} = 0.04$ to all the internal nodes.
5. Write the dimensionless $\hat{\theta}$ equivalent of eq. (3.87) to update the internal node temperatures during each iteration.
6. Select $\epsilon = 10^{-4}$ as convergence criterion in eq. (3.99).
7. Select $N = 4$ as a first case (a very coarse grid), run the program, and record the following:

The number of iterations needed (p).
The calculated temperature for the center of the square cross section, $\hat{\theta}_c$ (note that according to Problem 3.2, the *exact* value of $\hat{\theta}_c$ should be 0.07367).
A measure of the "error" (inaccuracy) of the calculated $\hat{\theta}_c$:

$$\text{Error} = \left| \frac{0.07367 - \hat{\theta}_c}{0.07367} \right|$$

A measure of the "cost" (e.g., internal node calculations) associated with producing one solution:

$$\text{Cost} = p(n-1)^2$$

8. Repeat step 7 for several finer grids, for example, $N = 8, 12, 16$, and so on.
9. Select a more stringent convergence criterion, $\epsilon = 10^{-5}$, and execute again the steps 7 and 8.
10. Plot the resulting information as two functions of ϵ and N:

$$\text{Error} = f_1(\epsilon, N) \qquad \text{Cost} = f_2(\epsilon, N)$$

and discuss the relationship (trade-offs) between the trends revealed by these functions.

3.20 The right side of the two-dimensional conducting body shown in the figure is plated with a thin layer of high-conductivity metal (k_p, δ). The exterior of this surface layer is exposed to a convective flow (h, T_∞). The surface layer is so thin that its temperature is a function of y only. This temperature is equal to the temperature of the boundary of the inner body of conductivity k.

Derive the expression for the boundary condition that applies along the right side of the two-dimensional conducting body.

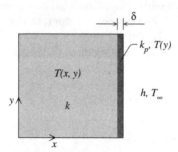

Figure P3.20

3.21 The finite-difference boundary conditions listed in Table 3.4 are for steady two-dimensional conduction without internal heat generation. Show that when the local volumetric heat generation rate is finite, $\dot{q}$, the boundary conditions assume the following forms:

(a) Surface node:

$$2T_{i-1,j} + T_{i,j-1} + T_{i,j+1} + 2\frac{h\Delta x}{k}T_\infty - 2\left(\frac{h\Delta x}{k} + 2\right)T_{i,j} + \frac{\dot{q}}{k}(\Delta x)^2 = 0$$

(b) External corner node:

$$T_{i,j-1} + T_{i-1,j} + 2\frac{h\Delta x}{k}T_\infty - 2\left(1 + \frac{h\Delta x}{k}\right)T_{i,j} + \frac{\dot{q}}{2k}(\Delta x)^2 = 0$$

The finite-difference conduction equation for an internal control volume with heat generation is listed in eq. (3.87).

3.22 The conducting bar with square cross section shown in the figure has the side $L = 10$ cm and the thermal conductivity 0.1 W/m·K. Only half of the base is cooled to $T_c = 0°$C, while the upper surface is heated entirely to $T_h = 10°$C. The remaining surfaces are perfectly insulated.

(a) Determine numerically the temperature distribution in the square cross section.

(b) Calculate the total heat transfer rate from T_h to T_c. Compare this result with the heat transfer rate that would occur if the entire base surface is cooled to $0°$C. Comment on the heat transfer reduction associated with "strangling" the flow of heat by forcing it to pass through only one half of the base surface.

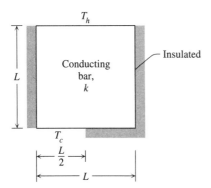

Figure P3.22

3.23 The elbow-shaped conducting medium shown in the figure is the basic component of the insulation built around a square duct. The upper-left and lower-right ends of the elbow are perfectly insulated, because of the symmetry of the four-elbow assembly around the square duct. The inner surface is isothermal at $T_h = 20°C$, and the outer surface is isothermal at $T_c = 0°C$. The design is characterized also by $k = 0.5$ W/m·K, and $L = 20$ cm.

(a) Calculate the total heat transfer rate from T_h to T_c, and the temperature distribution in the elbow-shaped region.

(b) Compare the heat transfer rate calculated in part (a) with the more approximate estimate based on the shape factors S listed in Table 3.3. If you were to solve a similar problem in the future, which method would you choose, the complete finite-difference method or the shape-factor method?

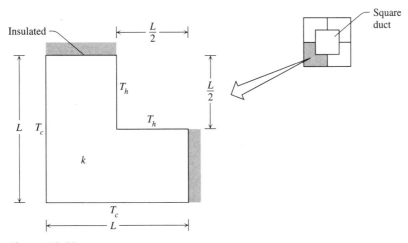

Figure P3.23

3.24 The structural wall of temperature T_b is reinforced with two-dimensional ribs with rectangular profile $L \times 2L$, of the type shown in the figure. Assume that the base surface (rib root) is at $T_b = 40°C$, the surrounding fluid is at $T_\infty = 20°C$, the heat transfer coefficient is uniform along the wetted surfaces, $h = 500$ W/m²K, the rib thickness is $L = 1$ cm, and the rib thermal conductivity is $k = 10$W/m·K.

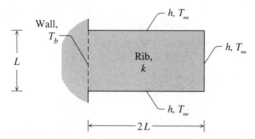

Figure P3.24

Determine numerically the temperature distribution in the rib cross section, and the rate of heat transfer through the rib. Can this rib be treated as a fin with unidirectional heat transfer, as in Section 2.7?

3.25 The two-dimensional flange shown in the figure is bathed by a fluid of temperature $T_\infty = 20°C$, through a uniform heat transfer coefficient $h = 2W/m^2K$. The flange thickness is $L = 5$ cm, and its thermal conductivity is $k = 100W/m \cdot K$. The temperature in the plane of the neck (the dash line) is approximately uniform at $T_b = 100°C$.

(a) Note the symmetry about the vertical midplane, and determine numerically the temperature distribution in the flange cross section, and the rate of heat transfer from T_b to T_∞.

(b) Does the calculated temperature vary appreciably over the flange cross section? If the answer is "no," can you think of a much simpler way of calculating the heat transfer rate through the flange, from T_b to T_∞?

(c) Calculate the Biot number based on the flange length scale (L, or $2L$), and compare it with the threshold value recognized earlier in the fin criterion (2.129). Should you have expected the conclusion reached in part (b)? If the objective was to calculate only the heat transfer rate through the flange, was it really necessary to perform the finite difference calculations?

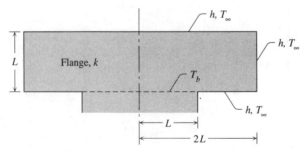

Figure P3.25

PROJECTS

3.1 Numerical Solution for Steady Conduction with Volumetric Heat Generation

Coal powder has been stockpiled in a long bed that has the vertical cross section shown in the figure. The bed is infinitely long in the direction normal to the plane of the figure. The coal powder generates heat volumetrically at the uniform rate $\dot{q} = 50$ W/m^3. Other relevant numerical data are:

$$h = 5\,\text{W/m}^2\cdot\text{K} \qquad H = 5\text{ m} \qquad T_\infty = 25°\text{C}$$
$$k = 0.2\,\text{W/m}\cdot\text{K} \qquad L = 30\text{ m}$$

(a) Develop a numerical solution for the maximum steady-state temperature $T_{\max}$ and its location inside the bed cross section. Apply the Gauss–Seidel iteration method only to the right half of the bed cross section. Determine the temperature distribution over the $(L/2) \times H$ space in terms of the dimensionless temperature

$$\tilde{\theta} = \frac{T - T_\infty}{\dot{q}H^2/k}$$

Use a uniform square grid and set $\epsilon = 10^{-4}$ in eq. (3.99). Demonstrate also that your grid is sufficiently fine to no longer influence the accuracy of your final result.

(b) Compare the numerical $T_{\max}$ estimate with the theoretical result listed in the statement of Problem 2.14. Explain why the numerical $T_{\max}$ value is smaller than the theoretical value.

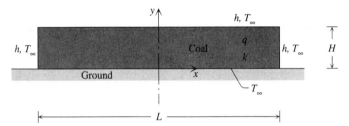

Figure PR3.1

3.2 Numerical Solution for Steady Conduction in a Medium with Temperature-Dependent Thermal Conductivity

The figure shows the cross section through a bar that is long in the direction perpendicular to the paper. The bar acts as mechanical support between an isothermal body of temperature T_0 and one of temperature T_L. Its other surfaces are surrounded by vacuum and may be assumed adiabatic. The bar material has thermal conductivity that increases linearly with the temperature:

$$k(T) = k_0 + a(T - T_0)$$

The a constant and the parameters listed in the figure (T_0, T_L, H, L) are known. You are asked to demonstrate numerically the manner in which the variation of thermal conductivity and the slenderness ratio L/H influence the heat transfer from T_0 to T_L. Note that in this case the equation for steady conduction is

$$\frac{\partial}{\partial x}\left(k\frac{\partial T}{\partial x}\right) + \frac{\partial}{\partial y}\left(k\frac{\partial T}{\partial y}\right) = 0$$

and nondimensionalize the entire problem statement by using the new variables

$$\xi = \frac{x}{H} \qquad\qquad \eta = \frac{y}{H}$$

$$\theta = \frac{T - T_0}{T_L - T_0} \qquad A = \frac{a}{k_0}(T_L - T_0)$$

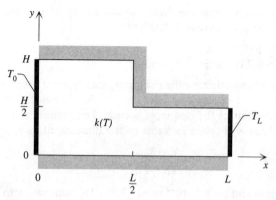

Figure PR3.2

Show that the overall thermal resistance between the two ends of the bar is a function of L/H and A. Cover the domain with a square grid. Determine the steady-state temperature distribution inside the bar for the following combinations $(L/H, A)$:

$$A = 0, 0.25, 0.5, 0.75$$

$$L/H = 2, 4, 10$$

and report the corresponding thermal resistance. Comment on the effect of L/H and A on the thermal resistance. Construct an accuracy test to show that the chosen grid is fine enough.

4

TIME-DEPENDENT CONDUCTION

4.1 THE IMMERSION COOLING OR HEATING OF A CONDUCTING BODY

The most basic problem of time-dependent conduction is the calculation of the temperature history inside a conducting body that is immersed suddenly in a bath of fluid at a different temperature. This problem finds applications in many areas, for example, in the heat treating (e.g., quenching) of special alloys and in the cooldown of large-scale superconducting magnets. It is represented most succinctly by the model shown in Fig. 4.1, in which a body of initial temperature T_i, volume V, density ρ, specific heat c, thermal conductivity k, and external (wetted) area A is surrounded by a fluid of temperature T_∞. The coefficient h for convective heat transfer across the exposed area of the body is assumed constant.

The first solid layers that feel the effect of thermal contact with the surrounding fluid are the peripheral ones, that is, the "skin" region situated immediately under the wetted area A. This region assumes temperatures that continuously bridge the gap between the core temperature of the body (still at the initial level T_i) and the surface temperature T_0. The latter assumes an intermediate value between T_i and T_∞ and, as the time increases, approaches the bath temperature T_∞.

This discussion becomes more meaningful if we think of a metal-quenching application such that the initial temperature of the body is greater than the fluid temperature. This assumption is evident in the drawing on the right side of Fig. 4.1, which is an enlargement of the temperature distribution across the skin layer. The surviving hot core is surrounded by a colder region (the skin layer), and the size of the hot core shrinks one layer at a time as the thickness of the skin layer, δ, increases. There comes a time t_c when δ has grown all the way to the center of the body, that is, a time when the solid no longer has a *distinct core* and a *distinct skin layer*. When the time greatly exceeds this time of complete "thermal penetration," the body temperature becomes essentially equal to the surface temperature T_0. In other words, at sufficiently long times, the tempera-

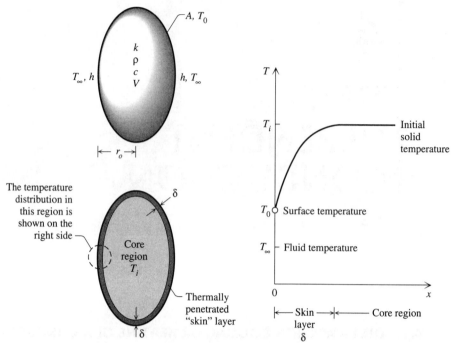

Figure 4.1 Formation of a thermally penetrated "skin" layer under the surface of a body immersed suddenly in a fluid.

ture distribution through the body is for all practical purposes a function of only the time. This temperature, $T(t)$, approaches the bath temperature level as t continues to increase.

The preceding description is summarized by the three-frame scenario presented in Fig. 4.2. The purpose of this figure is to show that only early in the life of the process is the body temperature a function of both time and spatial position. The long-time behavior is quite different (i.e., much simpler) in that the entire body is represented by a single instantaneous temperature $T(t)$.

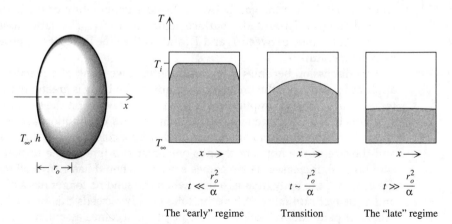

Figure 4.2 Growth of the skin layer into the core region, and the transition from the early regime to the late regime.

Critically important, then, is the special time scale t_c that marks the transition between the two regimes. This scale can be determined based on a very simple argument.

Examine one more time the enlarged view of the instantaneous temperature distribution across the skin layer, Fig. 4.1. If the skin thickness δ is much smaller than the distance to the center of the body (labeled L or r_o in the charts of Figs. 4.7 through 4.15), then the conduction through the skin region may be treated as *one-dimensional*, in the same manner as the curved thin wall seen earlier in Fig. 2.1. According to eq. (1.23), the equation for time-dependent conduction in the direction x without internal heat generation is

$$\frac{\partial^2 T}{\partial x^2} = \frac{1}{\alpha}\frac{\partial T}{\partial t} \tag{4.1}$$

In the skin region of thickness δ, the order of magnitude of the curvature of the temperature profile is the same as the change in the slope $\partial T/\partial x$ across the distance δ:

$$\frac{\partial^2 T}{\partial x^2} \sim \frac{\left(\dfrac{\partial T}{\partial x}\right)_{x \sim \delta} - \left(\dfrac{\partial T}{\partial x}\right)_{x=0}}{\delta - 0} \tag{4.2}$$

Figure 4.1 suggests the following temperature gradient scales:

$$\left(\frac{\partial T}{\partial x}\right)_{x \sim \delta} \sim 0 \qquad \left(\frac{\partial T}{\partial x}\right)_{x=0} \sim \frac{T_i - T_0}{\delta} \tag{4.3}$$

Therefore, after substituting these into eq. (4.2), the scale of the x-curvature becomes

$$\frac{\partial^2 T}{\partial x^2} \sim -\frac{T_i - T_0}{\delta^2} \tag{4.4}$$

Next, we focus on the right side of the conduction equation (4.1), which refers to the thermal inertia effect of the conducting material. The correct order of magnitude of that term can be deduced by arguing that the average temperature of the δ-thick region drops from the initial level T_i to a value comparable* with T_0 during the time interval of length t:

$$\frac{\partial T}{\partial t} \sim \frac{T_0 - T_i}{t - 0} \tag{4.5}$$

Combining this result with eq. (4.4) into the equality of scales that is required by the conduction equation (4.1),

$$-\frac{T_i - T_0}{\delta^2} \sim \frac{1}{\alpha}\frac{T_0 - T_i}{t} \tag{4.6}$$

we draw the important conclusion that the skin thickness δ increases as the square root of the elapsed time,

$$\delta \sim (\alpha t)^{1/2} \tag{4.7}$$

---

*Note that the lower the surface temperature T_0, the lower is the average temperature of the skin layer.

The transition time t_c is reached when δ has grown to the point that it is comparable with the transversal dimension of the whole body:

$$\delta \sim r_o \qquad \text{at} \qquad t \sim t_c \qquad (4.8)$$

In view of eq. (4.7), the transition time scale is

$$t_c \sim \frac{r_o^2}{\alpha} \qquad (4.9)$$

This time scale distinguishes between the *early regime*, when the skin layer and the untouched core are distinct,

$$t << \frac{r_o^2}{\alpha} \qquad T = T(x,t) \qquad (4.10)$$

and the *late regime*, when the instantaneous temperature has practically the same value throughout the body,

$$t >> \frac{r_o^2}{\alpha} \qquad T \cong T(t) \qquad (4.11)$$

The time criterion and the two regimes that have just been identified are the skeleton of an *approximate* description of the temperature history at a point inside a conducting body of arbitrary shape. In the next two sections we analyze these regimes separately.

4.2   THE LUMPED CAPACITANCE MODEL (THE "LATE" REGIME)

Consider first the late regime, when the temperature gradients inside the body have all decayed and the body temperature is well approximated by the single value $T(t)$. We consider this regime first because it is analytically the simplest. The analysis consists of writing the first law of thermodynamics for the body of Fig. 4.1, by treating it as a closed system that experiences zero work transfer, eq. (1.2):

$$q = \frac{dE}{dt} \qquad (4.12)$$

The net heat transfer rate *into* the body, q, is proportional to the fluid–body temperature difference and the wetted area:

$$q = hA(T_\infty - T) \qquad (4.13)$$

The heat transfer coefficient h is assumed constant, although this is only an approximation of the real situation (in natural convection, for example, h will be larger at the bottom of the immersed hot object, Chapter 7). The time rate of change in the energy inventory of the system can be rewritten by invoking the incompressible substance model (1.17):

$$\frac{dE}{dt} = \rho c V \frac{dT}{dt} \qquad (4.14)$$

Substituting eqs. (4.13) and (4.14) together into eq. (4.12) yields

$$-\frac{hA}{\rho cV}(T - T_\infty) = \frac{dT}{dt} \tag{4.15}$$

This equation for $T(t)$ can be integrated easily by noting, first, the "starting" condition

$$T = T_1 \quad \text{at} \quad t = t_c \tag{4.16}$$

The resulting solution shows that the body–fluid temperature difference decays exponentially,

$$\frac{T - T_\infty}{T_1 - T_\infty} = \exp\left[-\frac{hA}{\rho cV}(t - t_c)\right] \tag{4.17}$$

and that the characteristic time of this decay is $\rho cV/hA$. As might have been expected, it takes longer for the body to reach equilibrium with the surrounding fluid when its *lumped capacitance* ρcV is large and/or its product hA is small.

Under what circumstances does the exponential decay (4.17) *alone* characterize the temperature history inside the body of Fig. 4.1? It does so when the starting temperature T_1 postulated in eq. (4.16) is nearly the same as the initial temperature T_i specified in the original problem statement. At the start of the exponential decay, $t = t_c$, the skin layer had just reached the center of the body. The temperature gradient across the body at this moment is of order $(T_i - T_0)/r_o$; therefore, the conduction heat flux that escapes through the exposed area A is

$$q'' \sim k\frac{T_i - T_0}{r_o} \tag{4.18}$$

This heat flux must be equal to the convective heat flux that enters the fluid through the wetted side of the same surface:

$$k\frac{T_i - T_0}{r_o} \sim h(T_0 - T_\infty) \tag{4.19}$$

This equation can be rearranged as follows,

$$T_i - T_0 \sim \frac{\text{Bi}}{1 + \text{Bi}}(T_i - T_\infty) \quad \text{where} \quad \text{Bi} = \frac{hr_o}{k} \tag{4.20}$$

to show that the temperature variation across the body, $T_i - T_0$, is negligible relative to the overall difference $T_i - T_\infty$ only when the Biot number hr_o/k is small:

$$\frac{hr_o}{k} \ll 1 \tag{4.21}$$

Therefore, it is in this limit that the starting temperature T_1 used in eq. (4.16) is nearly the same as the true initial temperature of the body, T_i. The exponential decay (4.17) alone is an adequate description of the body temperature history when the Biot number is small, eq. (4.21), and when the time exceeds the critical time t_c, eq. (4.11). The lumped capacitance regime described by eq. (4.17) covers most of the life of the time-dependent conduction process when the decay time constant $\rho c V / h A$ greatly exceeds the duration of the early regime, t_c.

4.3   THE SEMI-INFINITE SOLID MODEL (THE "EARLY" REGIME)

4.3.1   Constant Surface Temperature

The analysis of the early regime is more complicated, because when the skin layer is distinct the temperature inside the solid is a function of both x and t. Relative to the thin layer of thickness δ, which steadily expands according to eq. (4.7), the isothermal core looks like a *semi-infinite solid* of temperature T_i. The unsteady conduction in the region close to the surface can be studied in the one-dimensional system of Fig. 4.3, in which the x axis points toward the core of the solid.

Consider first the limiting case in which the heat transfer coefficient between the fluid and the $x = 0$ surface is so large that, once exposed, the surface assumes the same temperature as the fluid, $T_0 = T_\infty$. The complete problem statement for determining $T(x,t)$ near the surface consists of eq. (4.1) and the initial and boundary conditions that are evident in the figure:

Conduction equation

$$\frac{\partial^2 T}{\partial x^2} = \frac{1}{\alpha} \frac{\partial T}{\partial t} \tag{4.22}$$

Initial condition

$$T = T_i \quad \text{at} \quad t = 0 \tag{4.23}$$

Boundary conditions

$$T = T_\infty \quad \text{at} \quad x = 0 \tag{4.24}$$

$$T \to T_i \quad \text{as} \quad x \to \infty \tag{4.25}$$

By reading these conditions and the statement that δ increases as $(\alpha t)^{1/2}$, eq. (4.7), we can already sketch the general outlook of the family of curves $T(x,t)$ (see the left side of Fig. 4.3). The temperature gradient along each T-vs.-x curve becomes shallower as the time increases, that is, as the effect of having dropped the surface temperature from T_i to T_∞ "diffuses" into the semi-infinite solid.

The key to solving the problem (4.22)–(4.25) is the observation that all the T-vs.-x curves are *similar*. Each curve starts out from $T = T_\infty$ and aims asymptotically for T_i as x becomes sufficiently large. Furthermore, each curve has only one "knee." The scale analysis completed in the preceding section showed that the skin layer (i.e., the curved portion of the T-vs.-x curve) expands to the right as $(\alpha t)^{1/2}$; therefore, it is possible to replot the entire $T(x,t)$ family as a single curve $T(\eta)$, where the dimensionless *similarity variable* η is defined as the ratio between the actual x and the scale of the skin layer thickness:

$$\eta = \frac{x}{(\alpha t)^{1/2}} \tag{4.26}$$

This graphic alternative is shown on the right side of Fig. 4.3. What remains is to convert the $T(x,t)$ problem (4.22)–(4.25) into a new statement for the

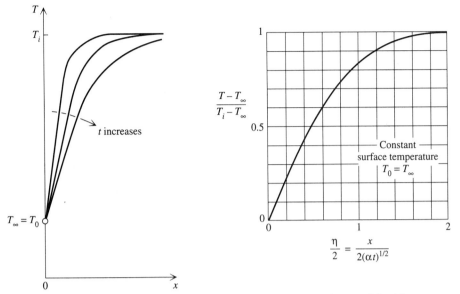

Figure 4.3 Penetration of the conduction effect into a semi-infinite solid with isothermal surface (left side), and the condensation of all the T-vs.-x curves into a single similarity profile (right side).

similarity temperature profile $T(\eta)$. Term after term, the conduction equation (4.22) is transformed as follows:

$$\frac{\partial T}{\partial x} = \frac{dT}{d\eta}\frac{\partial \eta}{\partial x} = \frac{dT}{d\eta}\frac{1}{(\alpha t)^{1/2}} \tag{4.27}$$

$$\frac{\partial^2 T}{\partial x^2} = \frac{d}{d\eta}\left(\frac{\partial T}{\partial x}\right)\frac{\partial \eta}{\partial x} = \frac{d^2 T}{d\eta^2}\frac{1}{\alpha t} \tag{4.28}$$

$$\frac{\partial T}{\partial t} = \frac{dT}{d\eta}\frac{\partial \eta}{\partial t} = \frac{dT}{d\eta}\left(-\frac{x}{2\alpha^{1/2}\,t^{3/2}}\right) \tag{4.29}$$

Substituting eqs. (4.28) and (4.29) into the original equation (4.22), we obtain an ordinary differential equation for $T(\eta)$:

$$\frac{d^2 T}{d\eta^2} + \frac{\eta}{2}\frac{dT}{d\eta} = 0 \tag{4.30}$$

The initial and boundary conditions (4.23)–(4.25) are represented by

$$T = T_\infty \quad \text{at} \quad \eta = 0 \tag{4.31}$$

$$T \to T_i \quad \text{as} \quad \eta \to \infty \tag{4.32}$$

in which eq. (4.32) accounts for both eqs. (4.25) and (4.23) [review the η definition, eq. (4.26)].

The $T(\eta)$ problem (4.30)–(4.32) can be solved by first separating the variables in eq. (4.30),

$$\frac{d(T')}{T'} = -\frac{\eta}{2}\,d\eta \quad \text{where} \quad T' = \frac{dT}{d\eta} \tag{4.33}$$

Integrated twice in η, this equation yields sequentially

$$\ln T' = -\frac{\eta^2}{4} + \ln C_1 \tag{4.34}$$

$$\frac{dT}{d\eta} = C_1 \exp\left(-\frac{\eta^2}{4}\right) \tag{4.35}$$

$$T = C_1 \int_0^\eta \exp\left(-\frac{\beta^2}{4}\right) d\beta + C_2 \tag{4.36}$$

where β is a dummy variable and, according to eq. (4.31), $C_2 = T_\infty$:

$$T - T_\infty = C_1 \int_0^\eta \exp\left[-\left(\frac{\beta}{2}\right)^2\right] d\beta \tag{4.37}$$

On the right side, we see the emergence of the *error function* described in Appendix E,

$$\operatorname{erf}(x) = \frac{2}{\pi^{1/2}} \int_0^x \exp(-m^2) dm \tag{4.38}$$

which has the noteworthy properties

$$\operatorname{erf}(0) = 0 \qquad \operatorname{erf}(\infty) = 1 \tag{4.39a, b}$$

$$\frac{d}{dx}[\operatorname{erf}(x)]_{x=0} = \frac{2}{\pi^{1/2}} = 1.1284 \tag{4.40}$$

The right side of eq. (4.37) can be reshaped by introducing the notation $m = \beta/2$,

$$\begin{aligned}
T - T_\infty &= 2C_1 \int_0^\eta \exp\left[-\left(\frac{\beta}{2}\right)^2\right] d\left(\frac{\beta}{2}\right) \\
&= 2C_1 \int_0^{\eta/2} \exp(-m^2) dm \\
&= \underbrace{2C_1 \frac{\pi^{1/2}}{2}}_{C_3} \frac{2}{\pi^{1/2}} \int_0^{\eta/2} \exp(-m^2) dm \\
&= C_3 \operatorname{erf}\left(\frac{\eta}{2}\right) \tag{4.41}
\end{aligned}$$

The constant C_3 is finally pinpointed by invoking the remaining boundary condition, eq. (4.32), while keeping in mind eq. (4.39b). This operation yields $C_3 = T_i - T_\infty$. Replacing η with $x/(\alpha t)^{1/2}$ in eq. (4.41), the analytical solution for $T(x,t)$ becomes

$$\frac{T - T_\infty}{T_i - T_\infty} = \operatorname{erf}\left[\frac{x}{2(\alpha t)^{1/2}}\right] \tag{4.42}$$

This solution is displayed in two forms, $T(x,t)$ and $T(\eta)$, in Fig. 4.3. The corresponding instantaneous surface heat flux formula is

$$q''(t) = -k\left(\frac{\partial T}{\partial x}\right)_{x=0} = -k\frac{T_i - T_\infty}{(\pi\alpha t)^{1/2}} \tag{4.43}$$

where the sign of q'' is positive when q'' points in the positive x direction.

This solution is valid in the early regime, eq. (4.10), *and* when the surface temperature can indeed be modeled as fixed and equal to the fluid temperature, $T_0 = T_\infty$. [Review the large-h assumption made before adopting the isothermal surface condition, eq. (4.24).] We shall learn soon that the $T_0 = T_\infty$ assumption holds as soon as the Biot number based on the thermal penetration depth δ, namely, the group $(\alpha t)^{1/2}h/k$, is considerably greater than 1 (see the right side of Fig. 4.4).

4.3.2   Constant Heat Flux Surface

There are several other analytical solutions that, although unrelated to the metal quenching problem of Fig. 4.1, are quite useful in the calculation of transient temperatures and heat transfer in the superficial layers of objects heated by thermal radiation. One such result is temperature field near the surface of a semi-infinite solid that, starting with the time $t = 0$, is exposed to a constant (i.e., time-independent) heat flux q'',

$$T(x,t) - T_i = 2\frac{q''}{k}\left(\frac{\alpha t}{\pi}\right)^{1/2} \exp\left(-\frac{x^2}{4\alpha t}\right) - \frac{q''}{k}x \operatorname{erfc}\left[\frac{x}{2(\alpha t)^{1/2}}\right] \qquad (4.44)$$

where T_i is the initial temperature of the solid. The new function erfc( ) is the *complementary error function* which, in association with eq. (4.38), is defined by

$$\operatorname{erfc}(x) = 1 - \operatorname{erf}(x) \qquad (4.45)$$

The surface temperature history $T_0(t)$ that corresponds to the constant flux configuration is

$$T_0(t) - T_i = T(0,t) - T_i = 2\frac{q''}{k}\left(\frac{\alpha t}{\pi}\right)^{1/2} \qquad (q'' = \text{constant}) \qquad (4.46)$$

This formula shows that the temperature of the surface exposed to the flux q'' increases monotonically as $t^{1/2}$. Worth noting is the similarity between this formula and isothermal surface formula (4.43), which can be rearranged as

$$T_\infty - T_i = \frac{q''(t)}{k}(\pi\alpha t)^{1/2} \qquad (T_0 = T_\infty) \qquad (4.43')$$

In both eqs. (4.43′) and (4.46), the ratio $(T_0 - T_i)/q''$ is proportional to the group $(\alpha t)^{1/2}/k$, which is the same as δ/k. This general conclusion can also be drawn based on scale analysis, by stating that the conduction heat flux (on the solid side of the exposed surface) is proportional to the local temperature gradient, $q'' \sim k(T_0 - T_i)/\delta$.

4.3.3   Surface in Contact with Fluid

The more general version of the isothermal surface problem treated in Section 4.3.1 is the problem of time-dependent conduction in a semi-infinite solid in contact with a fluid of a different temperature. The statement of this problem is the same as in eqs. (4.22) through (4.25), except that the surface condition (4.24) is now replaced by

$$h(T_\infty - T) = -k\left(\frac{\partial T}{\partial x}\right) \quad \text{at} \quad x = 0 \qquad (4.47)$$

Convection heat flux arriving from the fluid

Conduction heat flux entering the solid

The solution for the instantaneous temperature distribution through the solid is [1]

$$\frac{T(x,t) - T_\infty}{T_i - T_\infty} = \text{erf}\left[\frac{x}{2(\alpha t)^{1/2}}\right] +$$
$$\exp\left(\frac{hx}{k} + \frac{h^2\alpha t}{k^2}\right)\text{erfc}\left[\frac{x}{2(\alpha t)^{1/2}} + \frac{h}{k}(\alpha t)^{1/2}\right] \quad (4.48)$$

This temperature distribution depends on only two dimensionless groups, the dimensionless distance from the surface, x/δ, and the Biot number based on the thermal penetration depth, $h\delta/k$, or, as plotted on Fig. 4.4,

$$\frac{x}{2(\alpha t)^{1/2}} \quad \text{and} \quad \frac{h}{k}(\alpha t)^{1/2} \qquad (4.49)$$

The left side of Fig. 4.4 shows the effect of the Biot number $h(\alpha t)^{1/2}/k$ on the temperature profile. The lowest curve shown on this graph is the same as the curve plotted on the right side of Fig. 4.3, because the constant surface temperature model of Fig. 4.3 is valid when the δ-based Biot number is much greater than 1.

The right side of Fig. 4.4 shows the evolution of the surface temperature as

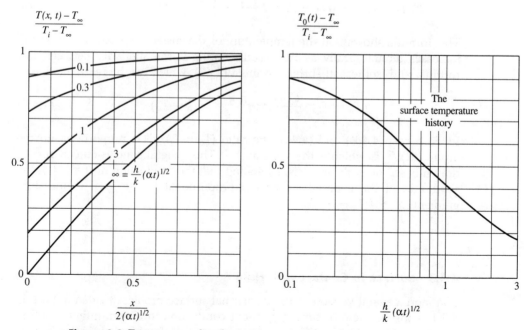

Figure 4.4 Temperature distribution in an isothermal semi-infinite solid (T_i) that is placed suddenly in contact with a fluid of a different temperature (T_∞).

the time increases. If the heat transfer coefficient is finite, the earliest temperatures of the surface resemble the initial temperature of the solid, T_i, no matter how large the value of h. The surface temperature $T_0(t)$ approaches the temperature of the fluid bath, T_∞, only after a time interval long enough such that

$$\frac{h}{k}(\alpha t)^{1/2} \gg 1 \tag{4.50}$$

This time criterion can be rewritten in terms of the dimensionless time group brought to light by eqs. (4.10) and (4.11),

$$\frac{hr_o}{k}\left(\frac{\alpha t}{r_o^2}\right)^{1/2} \gg 1 \tag{4.51}$$

where hr_o/k is now the Biot number based on the transversal dimension of the body. The combined message of the "transitions" identified by the time inequalities (4.10), (4.11), and (4.51) is presented in Fig. 4.5. In this figure the time increases from left to right. The right half of the plane represents the late regime, when the lumped system model of Section 4.2 is valid. The left half of the plane accounts for the shorter times $\alpha t/r_o^2$, that is, for the early regime. The line of slope $-\frac{1}{2}$ (note the log–log grid) represents eq. (4.51): only above this line, namely, when the Biot number hr_o/k is sufficiently large, does the constant surface temperature solution of Section 4.3.1 hold true. Below the inclined line, the appropriate solution to use is eq. (4.48).

The structure of the transient conduction phenomenon inside a body immersed suddenly in a fluid is now clear. Regardless of the geometric shape of the body (slab, cylinder, sphere), the conduction phenomenon is described

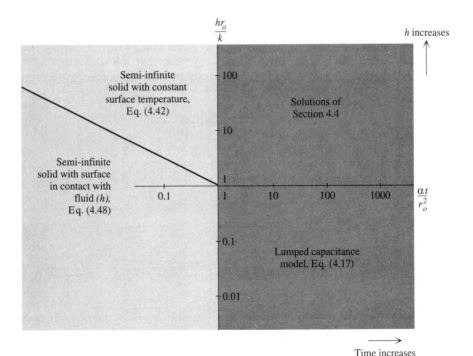

Figure 4.5 The domains of applicability of the lumped capacitance and semi-infinite solid models (note the logarithmic scale).

adequately by an appropriate sequence of much simpler regimes, for which the temperature distribution and heat flux solutions are analytically simple. In particular, for applications in which the Biot number hr_o/k is smaller than 1, it is permissible to use the finite-h semi-infinite solid conduction result (4.48) when $\alpha t/r_o^2 \ll 1$, followed in time by the lumped capacitance model (4.17), which holds when $\alpha t/r_o^2 \gg 1$.

Figure 4.5 also identifies the only segment of the domain in which the simple solutions developed until now are not adequate. That segment is the upper-right quadrant (large hr_o/k and large $\alpha t/r_o^2$), that is, applications in which (a) the surface temperature has already been pulled down to a level comparable with the temperature of the surrounding fluid, and (b) the conduction effect has penetrated to the center of the body. For this reason we must consider the exact solutions that are described next, one exact solution for each body shape that is specified. These exact solutions *are valid for all values* of hr_o/k and $\alpha t/r_o^2$, not just in the upper-right quadrant of Fig. 4.5. They constitute, therefore, the means by which one can assess the accuracy associated with using the lumped system and semi-infinite solid models discussed until now.

Example 4.1

Semi-infinite Medium, Fixed Surface Temperature

Hot tea of temperature 70°C is poured into a porcelain cup whose wall is initially at the temperature $T_i = 25°C$. Assume that the surface of the porcelain wall instantly assumes the tea temperature $T_\infty = 70°C$. The thickness of the porcelain wall is 6 mm. Estimate the time that passes until the wall temperature rises to 30°C at a point situated at 2 mm under the wetted surface. Show that during this short time the conduction process may be modeled according to the semi-infinite medium with fixed side temperature treated in Fig. 4.3.

Solution.   The temperature ($T = 30°C$) at the location $x = 2$ mm under the heated surface is nearly the same as the initial temperature at every point inside the porcelain wall. This means that most of the heating that has been experienced by the wall is located to the left of $x = 2$ mm, near the $T_\infty = 70°C$ surface. The model we can use is the unidirectional conduction through a semi-infinite solid, eq. (4.42):

$$\frac{T - T_\infty}{T_i - T_\infty} = \text{erf}\left[\frac{x}{2(\alpha t)^{1/2}}\right]$$

$$\frac{30 - 70}{25 - 70} = 0.889 = \text{erf}\left[\frac{x}{2(\alpha t)^{1/2}}\right]$$

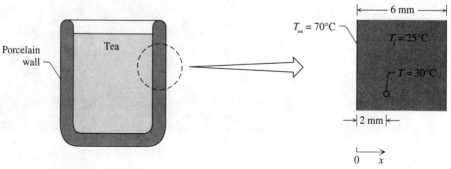

Figure E4.1

Interpolating between the error function values listed in Appendix E, we find

$$\frac{x}{2(\alpha t)^{1/2}} \cong 1.14$$

$$t \cong \frac{x^2}{(2.28)^2 \alpha} = \frac{4 \text{ mm}^2}{5.2} \frac{\text{s}}{0.004 \text{ cm}^2} = 1.92 \text{ s}$$

Now that we have an estimate for the time needed to witness the temperature 30°C at the depth $x = 2$ mm, we can evaluate the goodness of the semi-infinite medium model adopted when we used eq. (4.42). For this, we invoke eq. (4.7) and calculate the scale of thermal penetration:

$$\delta \sim (\alpha t)^{1/2} = \left(0.004 \frac{\text{cm}^2}{\text{s}} 1.92 \text{ s}\right)^{1/2}$$

$$= 0.88 \text{ mm}$$

This length scale is much smaller than the overall thickness of the porcelain wall; therefore, the time t is short enough so that, to the time-dependent heating that proceeds from the left, the wall "looks" semi-infinite.

Example 4.2

Lumped Capacitance, Varying Ambient Temperature

Consider a conducting body of density ρ, specific heat c, volume V, and surface area A that can be treated as a lumped thermal capacitance of temperature $T(t)$. Beginning with the time $t = 0$, this body makes thermal contact with a fluid flow through a constant heat transfer coefficient h. The ambient fluid temperature increases linearly in time:

$$T_\infty = at \qquad (a = \text{constant})$$

Initially, the body and the surrounding fluid are in thermal equilibrium, $T(0) = 0$. Determine the temperature history of the body, $T(t)$.

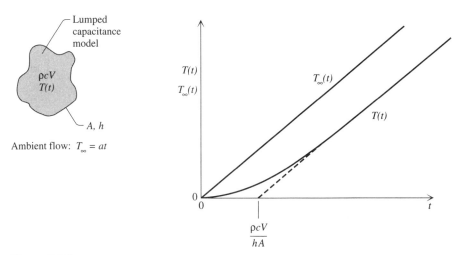

Figure E4.2

Solution.   When the temperature of the surrounding fluid is a function of time, $T_\infty(t)$, the lumped capacitance energy conservation equation (4.15) becomes

$$\frac{dT}{dt} + \frac{hA}{\rho cV}T - \frac{hA}{\rho cV}T_\infty(t) = 0 \tag{1}$$

Worth noting is that this is a linear differential equation of type

$$\frac{dy}{dx} + P(x)y + Q(x) = 0 \tag{2}$$

the general solution of which

$$y(x) = \exp\left[-\int P(x)\,dx\right] \cdot \left\{C - \int Q(x)\exp\left[\int P(x)\,dx\right]dx\right\} \tag{3}$$

contains the constant of integration C. In the present example the ambient temperature increases linearly in time:

$$T_\infty = at \qquad (a = \text{constant}) \tag{4}$$

and the solution (3) reduces to

$$T(t) = C\exp\left(-\frac{hA}{\rho cV}t\right) + a\left(t - \frac{\rho cV}{hA}\right) \tag{5}$$

From the initial condition that at $t = 0$ the body and the ambient are in thermal equilibrium,

$$T(0) = 0 \tag{6}$$

we learn that $C = a\rho cV/hA$, so that the $T(t)$ solution becomes

$$T(t) = a\frac{\rho cV}{hA}\left[\exp\left(-\frac{hA}{\rho cV}t\right) - 1\right] + at \tag{7}$$

The behavior of this temperature function is illustrated in the figure. In particular, the body temperature remains unchanged in the early stages of the heat transfer process,

$$\frac{dT}{dt} = 0 \qquad \text{at} \qquad t = 0 \tag{8}$$

while in the late stages it increases linearly in time:

$$T \cong a\left(t - \frac{\rho cV}{hA}\right) \tag{9}$$

The time represented by the group $\rho cV/hA$ is the interval (the time lag) by which the linear rise of the body temperature is delayed with respect to the linear rise of the ambient temperature.

4.4 UNIDIRECTIONAL CONDUCTION

4.4.1 The Constant-Thickness Plate

Consider first the case of the plane wall geometry,* that is, a slab of thickness $2L$ and initial temperature T_i, both sides of which are exposed suddenly to a

---

*Or, more generally, the thin wall geometry discussed in connection with Fig. 2.1.

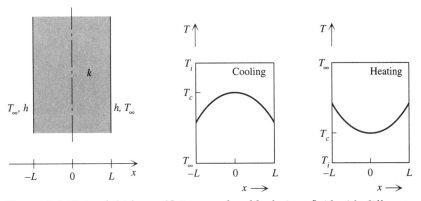

Figure 4.6 Plate of thickness $2L$ immersed suddenly in a fluid with different temperature.

convective medium of distant temperature T_∞. The heat transfer coefficient is equal to the same constant h on both surfaces of the slab (Fig. 4.6). In view of the large height and width of the slab, we expect the temperature distribution inside the solid to depend only on the transversal position x and time t. The only difference between the present problem and the semi-infinite solid placed in contact with fluid (Section 4.3.3) is that now the extent of the solid is finite in the x direction.

In terms of the excess temperature function $\theta(x,t) = T(x,t) - T_\infty$, the complete mathematical statement of the problem of Fig. 4.6 is as follows:

Conduction Equation

$$\frac{\partial^2 \theta}{\partial x^2} = \frac{1}{\alpha}\frac{\partial \theta}{\partial t} \tag{4.52}$$

Initial Condition

$$\theta = \theta_i \quad \text{at} \quad t = 0 \tag{4.53}$$

Boundary Conditions

$$\frac{\partial \theta}{\partial x} = 0 \quad \text{at} \quad x = 0 \tag{4.54}$$

$$-k\frac{\partial \theta}{\partial x} = h\theta \quad \text{at} \quad x = L \tag{4.55}$$

where $\theta_i = T_i - T_\infty$. Noteworthy in this statement are the boundary conditions (4.54) and (4.55), because they are both homogeneous. This means that x is a direction of homogeneous boundary conditions (review Section 3.1.1), which is why we have chosen $\theta(x,t)$ as the unknown instead of $T(x,t)$.

The separation of variables in the conduction equation (4.52) is achieved by assuming (i.e., trying out) the product solution $\theta(x,t) = X(x)\tau(t)$,

$$\underbrace{\frac{X''}{X}}_{-\lambda^2} = \underbrace{\frac{1}{\alpha}\frac{\tau'}{\tau}}_{-\lambda^2} \tag{4.56}$$

In this way eq. (4.56) splits into two separate equations for $X(x)$ and $\tau(t)$, in which λ^2 is an undetermined positive number. The general solutions to these two equations can then be recombined into the product solution for the temperature excess function

$$\theta(x,t) = [C_1 \sin(\lambda x) + C_2 \cos(\lambda x)] \exp(-\alpha\lambda^2 t) \tag{4.57}$$

Worth noting is that in the special case when $\lambda = 0$, this solution degenerates into $\theta_0 = C_1 x + C_2$. It is easy to see that the boundary conditions (4.54) and (4.55) require $C_1 = 0$ and $C_2 = 0$, so that $\theta_0 = 0$. In this way the $\lambda = 0$ solution contributes zero to the $\theta(x,t)$ solution that we are trying to construct, and this means that we can proceed based on the general solution (4.57), which holds for $\lambda \neq 0$.

Invoking first the symmetry condition (4.54), we learn that the constant C_1 must be zero. Next, the surface condition (4.55) and what remains of the solution (4.57) require that

$$a_n \tan(a_n) = \frac{hL}{k} \tag{4.58}$$

where $a_n = \lambda_n L$ are the characteristic values. The first six roots of eq. (4.58) are listed in Table 3.1, in which the left column represents now the values of the present Biot number, hL/k. In summary, the $\theta(x,t)$ solution can be constructed as an infinite series with yet unknown coefficients:

$$\theta(x,t) = \sum_{n=1}^{\infty} K_n \cos(\lambda_n x) \exp(-\alpha\lambda_n^2 t) \tag{4.59}$$

The last step consists of applying the initial condition (4.53),

$$\theta_i = \sum_{n=1}^{\infty} K_n \cos(\lambda_n x) \tag{4.60}$$

and using the orthogonality property of the characteristic functions $\cos(\lambda_n x)$,

$$\theta_i \int_0^L \cos(\lambda_n x) dx = K_n \int_0^L \cos^2(\lambda_n x) dx \tag{4.61}$$

Following the evaluation of the integrals, this step identifies the coefficients K_n, which substituted back into the series (4.59) complete the solution:

$$\frac{\theta(x,t)}{\theta_i} = \frac{T(x,t) - T_\infty}{T_i - T_\infty}$$

$$= 2 \sum_{n=1}^{\infty} \frac{\sin(a_n)}{a_n + \sin(a_n)\cos(a_n)} \cos\left(a_n \frac{x}{L}\right) \exp\left(-a_n^2 \frac{\alpha t}{L^2}\right) \tag{4.62}$$

Expressed in dimensionless form as $(T - T_\infty)/(T_i - T_\infty)$, the temperature distribution determined above depends on three dimensionless groups:

$$\frac{x}{L} \qquad \frac{\alpha t}{L^2} = \text{Fo}, \qquad \frac{hL}{k} = \text{Bi} \tag{4.63}$$

The Biot number hL/k influences the solution (4.62) through the characteristic values a_n, as shown in eq. (4.58). The characteristic values a_n have been listed in Table 3.1, in which the left most column contains the values that now represent the group hL/k. The temperature in a certain plane (x/L = constant) depends only on the Biot number hL/k and the so-called *Fourier number*, Fo = $\alpha t/L^2$. Note that "Fourier number" is another way of saying "dimensionless time."

Figure 4.7 reproduces Heisler's [2] chart for the history of the temperature in the midplane of the plate, $T_c(t) = T(0,t)$. The lines drawn on the figure correspond to fixed values of the *inverse* Biot number, k/hL. These lines appear almost straight because of the time exponentials of the solution (4.62) and the semilogarithmic scale of Fig. 4.7. The sharp breaks (corners) in the lines are due to the changes in the size of the divisions marked on the abscissa: there are four such division sizes. This method of plotting the temperature history had been used earlier by Hottel [3]. The great merit of the method is that it expands greatly the Fourier number range that is covered by the abscissa.

The temperature in a plane other than the midplane of the plate can be calculated by multiplying the readings furnished by Figs. 4.7 and 4.8. The latter shows how the excess temperature in an arbitrary plane, $T(x,t) - T_\infty$, compares with the corresponding (simultaneous) value in the midplane, $T_c(t) - T_\infty$. Therefore, the temperature in any plane can be calculated as the product of two readings:

$$\left(\frac{T(x,t) - T_\infty}{T_i - T_\infty} \right)_{\text{plate}} = \left(\frac{T(x,t) - T_\infty}{T_c(t) - T_\infty} \right)_{\text{Fig. 4.8}} \times \left(\frac{T_c(t) - T_\infty}{T_i - T_\infty} \right)_{\text{Fig. 4.7}} \qquad (4.64)$$

Another quantity of interest is the total heat transfer interaction that occurs between the solid and the ambient fluid during the finite time interval $0-t$. Considering only the right half of the plate ($0 < x < L$, Fig. 4.6), the *maximum* heat transfer that can take place is

$$Q_i = \rho W H L c (T_i - T_\infty) \qquad (4.65)$$

where H and W are the large height and large width of the plate, respectively.

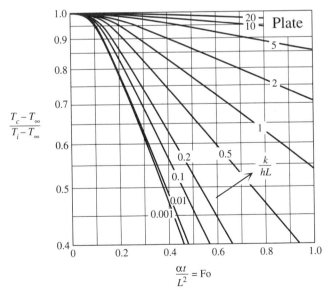

Figure 4.7 Temperature history in the midplane of a plate immersed suddenly in a fluid of a different temperature (L = plate half-thickness). (Drawn after Heisler [2].)

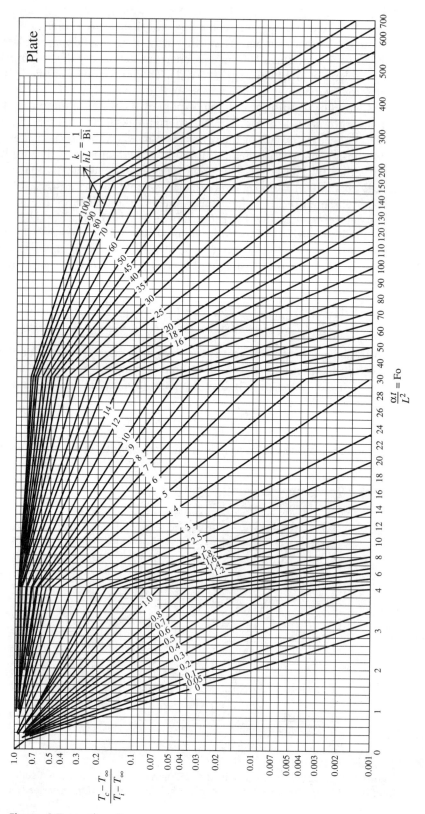

Figure 4.7 (*continued*)

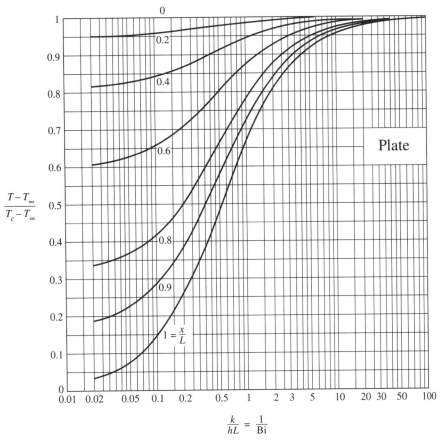

Figure 4.8 Relationship between the temperature in any plane (x) and the temperature in the midplane ($x = 0$, Fig. 4.7) of a plate immersed suddenly in a fluid of a different temperature (L = plate half-thickness). (Drawn after Heisler [2].)

Note that the area $W \times H$ is normal to the x direction. The maximum heat transfer Q_i (joules) is equal to the drop in the half-plate internal energy, from the initial state (T_i) to a state of thermal equilibrium with the ambient (T_∞). The actual heat transfer during the time interval $0-t$ is always smaller than Q_i: its value is given by the integral

$$Q(t) = WH \int_0^t q'' \, dt \qquad (4.66)$$

where q'' is the conduction heat flux that leaves the half-plate through the $x = L$ surface.

Figure 4.9 shows Gröber's classical chart [4, 5] for the ratio $Q(t)/Q_i$ obtained by dividing eqs. (4.65) and (4.66). This ratio always starts from zero and, at sufficiently long times, approaches 1. The dimensionless time parameter used on the abscissa of Fig. 4.9 is the same as the Fourier number times the Biot number squared,

$$\left(\frac{h}{k}\right)^2 \alpha t = \left(\frac{hL}{k}\right)^2 \frac{\alpha t}{L^2} \qquad (4.67)$$

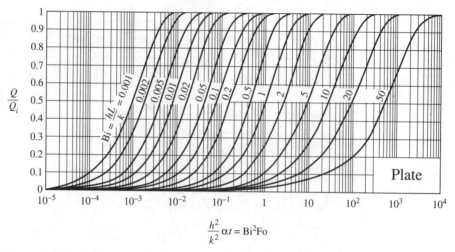

$$\frac{h^2}{k^2}\alpha t = \mathrm{Bi}^2\mathrm{Fo}$$

Figure 4.9 Total heat transfer between a plate and the surrounding fluid, as a function of the total time of exposure t. (Drawn after H. Gröber, S. Erk, and U. Grigull, *Fundamentals of Heat Transfer*, McGraw-Hill, New York, 1961; used with permission.)

Each constant-hL/k curve drawn on the figure makes the transition from the $Q/Q_i = 0$ level to the $Q/Q_i = 1$ level in a relatively narrow, "characteristic" time range. This characteristic time depends on the Biot number.

4.4.2 The Long Cylinder

Consider next the time-dependent temperature field in a long cylindrical rod of radius r_o and initial temperature T_i, which is placed suddenly in contact with a fluid flow of temperature T_∞ across a constant heat transfer coefficient h. The only feature that differentiates between this problem and the constant-thickness slab of Fig. 4.6 is the outer radius r_o of the cylinder, which now becomes the transversal dimension of the problem (i.e., the length to be used in defining the Biot and Fourier numbers). Since the temperature distribution inside the rod depends only on radial position and time, by reading eqs. (4.62) and (4.63) we expect from the start a dimensionless temperature $[T(r,t) - T_\infty]/(T_i - T_\infty)$ that depends on only three dimensionless groups, namely,

$$\frac{r}{r_o} \qquad \frac{\alpha t}{r_o^2} = \mathrm{Fo} \qquad \frac{hr_o}{k} = \mathrm{Bi} \qquad (4.68)$$

The exact solution for the temperature field is available in the form of an infinite series (see, for example, Ref. [6]),

$$\frac{T(r,t) - T_\infty}{T_i - T_\infty} = \sum_{n=1}^{\infty} K_n J_0\left(b_n \frac{r}{r_o}\right)\exp(-b_n^2\,\mathrm{Fo}) \qquad (4.69)$$

in which the K_n coefficients are given by

$$K_n = \frac{2\mathrm{Bi}}{(b_n^2 + \mathrm{Bi}^2)J_0(b_n)} \qquad (4.70)$$

and where the characteristic values b_n are the roots of the equation

$$b_n J_1(b_n) - \text{Bi} J_0(b_n) = 0 \qquad (4.71)$$

The Bessel functions of the first kind, J_0 and J_1, have been illustrated in Fig. 3.6. A representative set of J_0 and J_1 values is also tabulated in Appendix E.

The first six roots of eq. (4.71) are reported as functions of the Biot number in Ref. [1]. Table 4.1 shows only the first two characteristic values and the values of the corresponding K_n coefficients. In many cases the series (4.69) converges so rapidly that it is approximated adequately by only the first term, or the first two terms. The approximation is particularly good in the late regime, that is, when the Fourier number $\alpha t / r_o^2$ is of order 1, or greater (Fig. 4.5).

The same solution is displayed in Fig. 4.10, which shows the temperature history on the centerline of the cylinder, $T_c(t) = T(0,t)$. The temperature at other radial positions can be calculated by multiplying the readings provided by Figs. 4.10 and 4.11:

$$\left(\frac{T(r,t) - T_\infty}{T_i - T_\infty} \right)_{\text{cylinder}} = \left(\frac{T(r,t) - T_\infty}{T_c(t) - T_\infty} \right)_{\text{Fig. 4.11}} \times \left(\frac{T_c(t) - T_\infty}{T_i - T_\infty} \right)_{\text{Fig. 4.10}} \qquad (4.72)$$

Finally, Fig. 4.12 shows the evolution of the ratio $Q(t)/Q_i$, where $Q(t)$ is the total heat transfer between solid and fluid during the time interval $0-t$, and Q_i is the maximum heat transfer that occurs when the cylinder reaches equilibrium with the fluid, $Q_i = \rho V c (T_i - T_\infty)$, where V is the cylinder volume. The behavior of the Q/Q_i ratio of a cylinder, Fig. 4.12, is very similar to that of the Q/Q_i ratio of a constant-thickness plate, Fig. 4.9.

4.4.3 The Sphere

Similar means exist for calculating the temperature distribution inside a sphere that is placed in contact with a fluid flow of a different temperature. The transversal dimension in this case is the sphere radius r_o. The dimensionless temperature difference $[T(r,t) - T_\infty]/(T_i - T_\infty)$ depends on the same three dimensionless parameters as the solution for the cylinder, eq. (4.68), except that the numbers Fo and Bi are based now on the radius of a sphere, not a cylinder.

Table 4.1 The Constants Needed for Evaluating the First Two Terms in the Series Solution for the Time-Dependent Temperature Inside a Long Cylinder, Eq. (4.69)

Bi = hr_o/k	b_1	K_1	b_2	K_2
0.01	0.1412	1.0025	3.8343	−0.00338
0.03	0.2439	1.0075	3.8395	−0.01012
0.1	0.4417	1.0246	3.8577	−0.0334
0.3	0.7465	1.0712	3.9091	−0.0983
1	1.2558	1.2071	4.0795	−0.2904
3	1.7887	1.4191	4.4634	−0.6315
10	2.1795	1.5677	5.0332	−0.9581
30	2.3261	1.5973	5.3410	−1.0487
100	2.3809	1.6015	5.4652	−1.0605

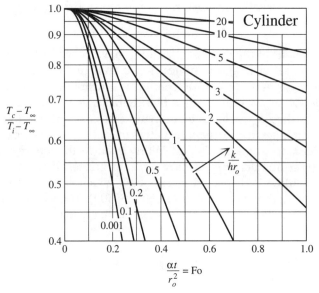

Figure 4.10 Temperature history on the centerline of a long cylinder immersed suddenly in a fluid of a different temperature (r_o = cylinder radius). (Drawn after Heisler [2].)

The series solution for the instantaneous temperature distribution in a sphere is (see, for example, Ref. [6])

$$\frac{T(r,t) - T_\infty}{T_i - T_\infty} = \sum_{n=1}^{\infty} K_n \frac{\sin(s_n r/r_o)}{s_n r/r_o} \exp(-s_n^2 \text{Fo}) \qquad (4.73)$$

where

$$K_n = 2 \frac{\sin(s_n) - s_n \cos(s_n)}{s_n - \sin(s_n)\cos(s_n)} \qquad (4.74)$$

The characteristic values s_n are the roots of the equation

$$s_n \cot(s_n) = 1 - \text{Bi} \qquad (4.75)$$

The first two roots (s_1, s_2) and the corresponding coefficients (K_1, K_2) are presented in Table 4.2. The first term of the series (4.73) is a very good approximation for the entire series when the Fourier number is not much smaller than 1. Note the presence of $-s_n^2$ in the argument of the exponential: When Fo is not negligibly small, the argument becomes very large (and negative), rendering the second and the subsequent terms in the series negligible relative to the first term.

The sphere temperature distribution is presented graphically in Figs. 4.13 through 4.15. The first of these graphs shows the history of the temperature right in the center of the sphere, $T(0,t) = T_c(t)$. Figure 4.14 provides the "correction" factor needed for calculating the simultaneous temperature at any other radius, $T(r,t)$:

$$\left(\frac{T(r,t) - T_\infty}{T_i - T_\infty}\right)_{\text{sphere}} = \left(\frac{T(r,t) - T_\infty}{T_c(t) - T_\infty}\right)_{\text{Fig. 4.14}} \times \left(\frac{T_c(t) - T_\infty}{T_i - T_\infty}\right)_{\text{Fig. 4.13}} \qquad (4.76)$$

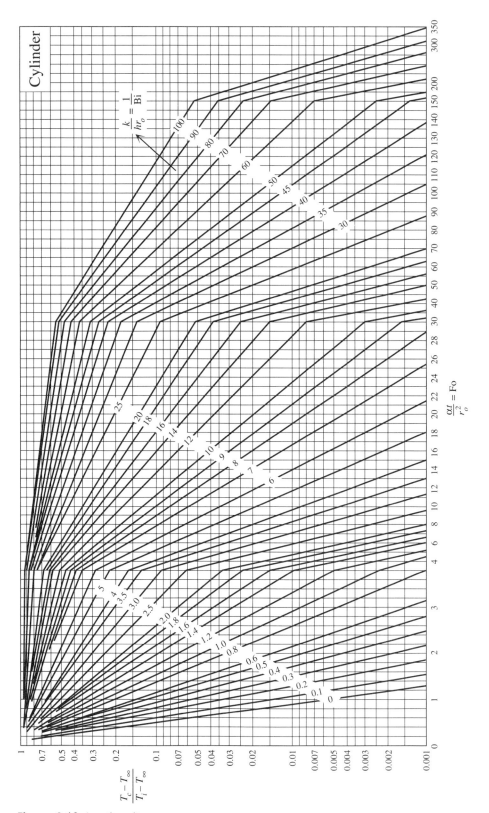

Figure 4.10 (*continued*)

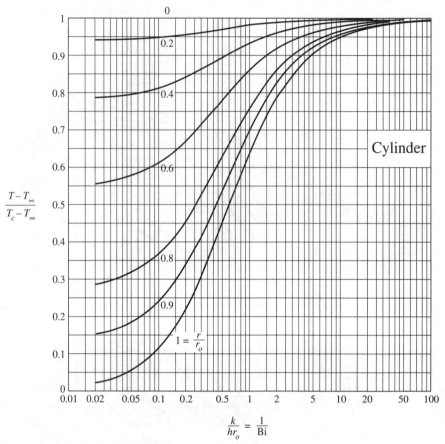

Figure 4.11 Relationship between the temperature at any radius (r) and the temperature on the centerline ($r = 0$, Fig. 4.10) of a long cylinder immersed suddenly in a fluid of a different temperature. (Drawn after Heisler [2].)

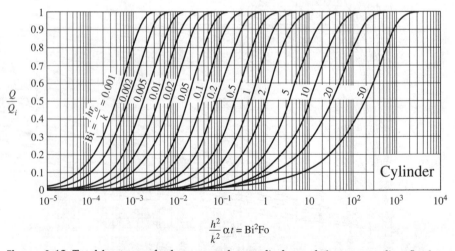

Figure 4.12 Total heat transfer between a long cylinder and the surrounding fluid, as a function of the total time of exposure t. (Drawn after H. Gröber, S. Erk, and U. Grigull, *Fundamentals of Heat Transfer*, McGraw-Hill, New York, 1961; used with permission.)

Table 4.2 The Constants Needed for Evaluating the First Two Terms in the Series Solution for the Time-Dependent Temperature Inside a Sphere, Eq. (4.73)

Bi = hr_o/k	s_1	K_1	s_2	K_2
0.01	0.1730	1.0030	4.4956	−0.00449
0.03	0.2991	1.0090	4.5001	−0.01369
0.1	0.5423	1.0298	4.5157	−0.04544
0.3	0.9208	1.0880	4.5601	−0.1345
1	1.5708	1.2732	4.7124	−0.4244
3	2.2889	1.6227	5.0870	−1.0288
10	2.8363	1.9249	5.7172	−1.7381
30	3.0372	1.9898	6.0740	−1.9593
100	3.1102	1.9990	6.2204	−1.9961

The total heat transfer ratio $Q(t)/Q_i$ is shown as a function of Fourier number and Biot number in Fig. 4.15. The denominator of this ratio is the ceiling value to which $Q(t)$ aspires as t increases, namely, $Q_i = \rho V c(T_i - T_\infty)$, where V is the volume of the sphere $4\pi r_o^3/3$.

4.4.4 Plate, Cylinder, and Sphere with Fixed Surface Temperature

Figures 4.9, 4.12, and 4.15 show that, in general, the actual/maximum heat transfer ratio Q/Q_i is a function of the Biot number and the Fourier number.

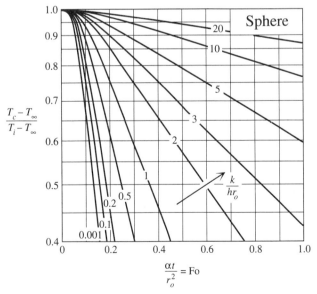

Figure 4.13 Temperature history in the center of a sphere immersed suddenly in a fluid of a different temperature (r_o = sphere radius). (Drawn after Heisler [2].)

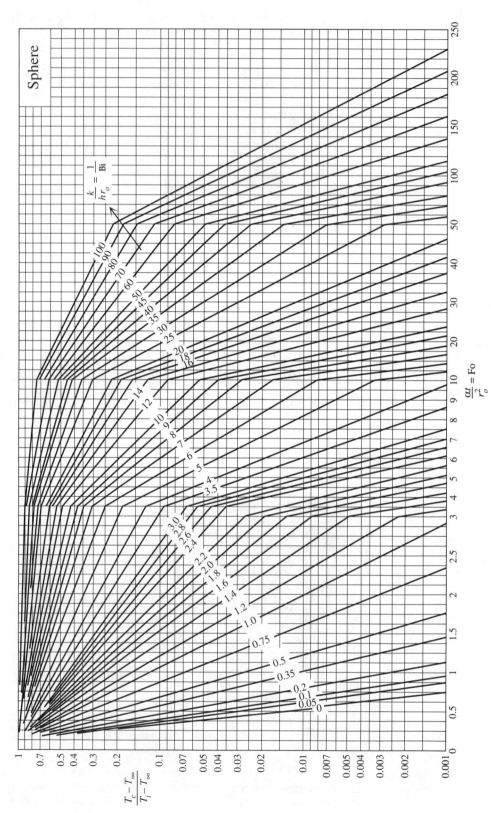

Figure 4.13 (*continued*)

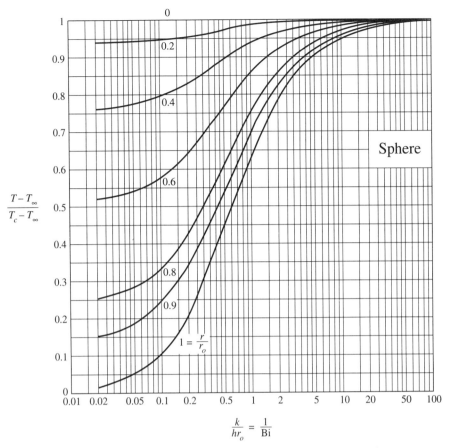

Figure 4.14 Relationship between the temperature at any radius (r) and the temperature in the center ($r = 0$, Fig. 4.13) of a sphere immersed suddenly in a fluid of a different temperature. (Drawn after Heisler [2].)

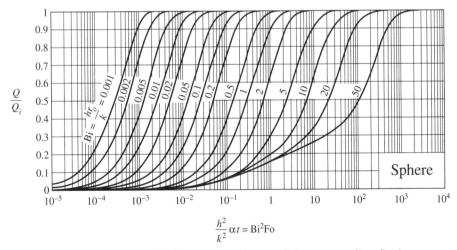

Figure 4.15 Total heat transfer between a sphere and the surrounding fluid, as a function of the total time of exposure t. (Drawn after H. Gröber, S. Erk, and U. Grigull, *Fundamentals of Heat Transfer*, McGraw-Hill, New York, 1961; used with permission.)

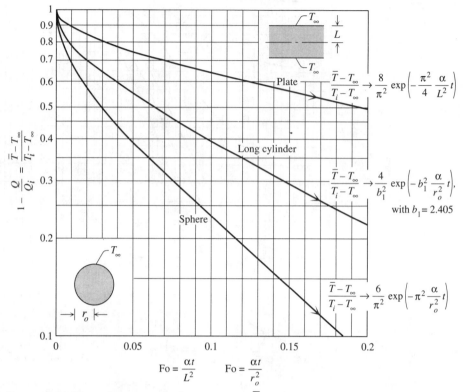

Figure 4.16 The volume-averaged temperature $\overline{T}$, and the total heat transfer from a body with fixed surface temperature. The three curves in this figure represent the Bi $= \infty$ limit of the charts presented in Figs. 4.9, 4.12, and 4.15.

This relationship is considerably simpler in the Bi $\rightarrow \infty$ limit, where the heat transfer coefficient is sufficiently large so that the surface of the immersed body assumes the temperature of the surrounding fluid, T_{∞}. In this limit the Q/Q_i ratio is a function of only the dimensionless time represented by the Fourier number.

Figure 4.16 shows these limiting relations for the three basic geometries considered in this section — the plate, the long cylinder, and the sphere. Plotted on the ordinate is the group $1 - (Q/Q_i)$, whose value at $t = 0$ is 1. This group is the same as the temperature difference ratio $(\overline{T} - T_{\infty})/(T_i - T_{\infty})$, in which $\overline{T}$ is the instantaneous temperature *averaged* over the entire volume V of the body, $\overline{T} = \int_V T \, dV$. Written in terms of $\overline{T}$, the total heat transfer from the body to the surrounding fluid during the time interval $0-t$ is $Q = \rho V c (T_i - \overline{T})$.

The exponential expressions listed on the right side of Fig. 4.16 are the asymptotes that are approached by the curves as the time increases. Note that the exponentials appear as straight lines on a semilogarithmic graph such as Fig. 4.16. Since the plotted curves look straight when the Fourier number is greater than approximately 0.1, in the large Fo range we may use the exponential expressions as a more accurate alternative to the reading of the graph.

Example 4.3

The Immersion Cooling of a Steel Plate

The purpose of this exercise is to illustrate the use of the charts exhibited in this section, and to show that alternative estimates can be made using the analytical solutions on which the charts are based. Consider a 1.6-cm-thick plate of carbon steel at the initial temperature $T_i = 600°C$. This plate is plunged at $t = 0$ in a bath of water at the temperature $T_\infty = 15°C$. The heat transfer coefficient is assumed constant and equal to $h = 10^4$ W/m²·K. The properties of carbon steel are approximated by the constants $k = 40$ W/m·K, and $\alpha = 0.1$ cm²/s. Calculate the time t when the temperature in the midplane of the steel plate drops to $T_c = 100°C$. Determine also the corresponding temperature in a plane situated 0.2 cm under one of the cooled surfaces. For the same time interval t, calculate the heat transfer released by the plate, as a fraction of total heat transfer that would be released in the limit $t \to \infty$.

Solution. To deduce from Fig. 4.7 the time when $T_c = 100°C$ (namely, the corresponding Fourier number), we first calculate

$$\text{Bi} = \frac{hL}{k} = \frac{10^4 \text{ W}}{\text{m}^2 \cdot \text{K}} \frac{0.008 \text{ m}}{40 \text{ W/m} \cdot \text{K}} = 2 \tag{1}$$

$$\frac{T_c - T_\infty}{T_i - T_\infty} = \frac{(100 - 15)°C}{(600 - 15)°C} = 0.145 \tag{2}$$

Next, we identify on Fig. 4.7 the curve labeled $1/\text{Bi} = 0.5$; and, squinting hard enough, we read the corresponding abscissa value

$$\frac{\alpha t}{L^2} \cong 1.8 \tag{3}$$

which corresponds to the time interval

$$t = 1.8 \frac{0.8^2 \text{ cm}^2}{0.1 \text{ cm}^2/\text{s}} = 11.5 \text{ s} \tag{4}$$

A more exact alternative is to rely on eq. (4.62) instead of Fig. 4.7 and to retain only the first term in the series

$$\frac{T_c - T_\infty}{T_i - T_\infty} \cong 2 \frac{\sin a_1}{a_1 + \sin a_1 \cos a_1} \exp(-a_1^2 \text{ Fo}) \tag{5}$$

We know that the first term (i.e., a single exponential) is an adequate approximation because, in the Fo range of interest, the $(T_c - T_\infty)/(T_i - T_\infty)$ curve appears as a straight line on the semilogarithmic grid of Fig. 4.7. The first eigenvalue a_1 follows from tables such as Table 3.1, in which the half-thickness Biot number listed in the leftmost column is equal to 2:

$$a_1 = 1.0769 \tag{6}$$

After we substitute the numerical a_1, T_c, T_∞, and T_i values into eq. (5), we obtain the new Fourier number

$$\text{Fo} = 1.807 \tag{7}$$

which corresponds to $t = 11.6$ s. This time estimate is nearly the same as the one obtained graphically, eq. (4).

Consider now the temperature that we might expect in the off-center plane located at $x = L - 0.2$ cm $= 0.6$ cm, at the time $t = 11.6$ s. For this we use Fig. 4.8, on which we know $x/L = 6/8 = 0.75$ and $1/\text{Bi} = 0.5$. The correct point is located between the curves labeled $x/L = 0.6$ and $x/L = 0.8$, and closer to the $x/L = 0.8$ curve. Therefore, we conclude *approximately* that

$$\frac{T - T_\infty}{T_c - T_\infty} \cong 0.7 \tag{8}$$

which means that the sought temperature is

$$T \cong 15°C + 0.7(100 - 15)°C = 74.5°C \tag{9}$$

Again, a more accurate alternative is to use the first term of the series solution (4.62),

$$\frac{T - T_\infty}{T_i - T_\infty} = 2\frac{\sin a_1}{a_1 + \sin a_1 \cos a_1}\cos\left(a_1\frac{x}{L}\right)\exp(-a_1^2 \text{ Fo}) \tag{10}$$

for which we know a_1, x/L, and Fo. The numerical value of the entire right side of eq. (10) turns out to be 0.1001; therefore, the temperature in the $x = 0.6$ cm plane is

$$T = 15°C + 0.1001(600 - 15)°C = 73.6°C \tag{11}$$

Finally, we can read Fig. 4.9 and estimate the total heat transfer that left the slab during the interval $0-t$, as a fraction of the ultimate (long time) heat transfer total Q_i. For this reading we need two numbers, Bi $= 2$ and

$$\text{Bi}^2\text{Fo} = 4 \times 1.807 = 7.23 \tag{12}$$

and these lead to the ordinate value of

$$\frac{Q}{Q_i} \cong 0.84 \tag{13}$$

4.5 MULTIDIRECTIONAL CONDUCTION

The preceding three sections outlined both analytical and graphic methods for calculating temperature distributions and total heat transfer interactions in *unidirectional* time-dependent conduction configurations. The reason so much space was dedicated to unidirectional problems is that their solutions can be "combined" in order to construct the needed solutions for time-dependent conduction in more complex geometries.

The simplest way of illustrating this combination procedure is to consider the immersion cooling or heating of a long bar with rectangular cross section. This problem is stated analytically in the left column of Fig. 4.17. It consists of

The rectangular domain of half-length L and half-height H.

The equation for simultaneous conduction in the x and y directions, eq. (a) in Fig. 4.17.

The boundary conditions in the x direction, eqs.(b) and (c) in Fig. 4.17.

The boundary conditions in the y direction, eqs. (d) and (e) in Fig. 4.17.

The initial condition, eq. (f) in Fig. 4.17.

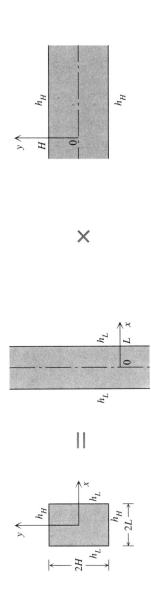

	Long bar with rectangular cross section, $2L \times 2H$		Plate, L-half-thickness		Plate, H-half-thickness	
Conduction equation:	$\dfrac{\partial^2\theta}{\partial x^2} + \dfrac{\partial^2\theta}{\partial y^2} = \dfrac{1}{\alpha}\dfrac{\partial\theta}{\partial t}$	(a)	$\dfrac{\partial^2\theta_L}{\partial x^2} = \dfrac{1}{\alpha}\dfrac{\partial\theta_L}{\partial t}$	(a)$_L$	$\dfrac{\partial^2\theta_H}{\partial y^2} = \dfrac{1}{\alpha}\dfrac{\partial\theta_H}{\partial t}$	(a)$_H$
$x = 0$:	$\dfrac{\partial\theta}{\partial x} = 0$	(b)	$\dfrac{\partial\theta_L}{\partial x} = 0$	(b)$_L$	$\dfrac{\partial\theta_H}{\partial x} = 0$	(b)$_H$
$x = L$:	$\dfrac{h_L}{k}\theta + \dfrac{\partial\theta}{\partial x} = 0$	(c)	$\dfrac{h_L}{k}\theta_L + \dfrac{\partial\theta_L}{\partial x} = 0$	(c)$_L$	$\theta_H = \theta_H(y,t)$	(c)$_H$
$y = 0$:	$\dfrac{\partial\theta}{\partial y} = 0$	(d)	$\dfrac{\partial\theta_L}{\partial y} = 0$	(d)$_L$	$\dfrac{\partial\theta_H}{\partial y} = 0$	(d)$_H$
$y = H$:	$\dfrac{h_H}{k}\theta + \dfrac{\partial\theta}{\partial y} = 0$	(e)	$\theta_L = \theta_L(x,t)$	(e)$_L$	$\dfrac{h_H}{k}\theta_H + \dfrac{\partial\theta_H}{\partial y} = 0$	(e)$_H$
$t = 0$:	$\theta = \theta_i$	(f)	$\theta_L = \theta_{i,L}$	(f)$_L$	$\theta_H = \theta_{i,H}$	(f)$_H$

Figure 4.17 The time-dependent temperature in a bar immersed in fluid, as the product of the temperature distributions in two perpendicular plates.

This problem is formulated already in terms of the excess temperature function $\theta(x,y,t) = T(x,y,t) - T_\infty$, in which T_∞ is the actual temperature of the surrounding fluid.

The second and third columns of the figure outline two simpler problems. The $\theta_L(x,t)$ problem shown in the second column is identical to that of Fig. 4.6, namely, the unidirectional time-dependent distribution of temperature in a long and wide plate of half-thickness L. The solution to this problem is amply documented in Section 4.4.1, in which x represented the distance to the midplane of the plate. Geometrically, the bar with rectangular cross section is the "intersection" in space of the two plates sketched at the top of Fig. 4.17.

The $\theta_H(y,t)$ problem listed in the right column of Fig. 4.17 is the same as the unidirectional problem of Section 4.4.1, with the only difference being that now the distance to the midplane is called y and the half-thickness of the plate is H. Therefore, the $\theta_H(y,t)$ distribution can be calculated based on Figs. 4.7 and 4.8, by first replacing x with y, and L with H.

It remains to show that the wanted solution $\theta(x,y,t)$, is equal to the *product* of the one-dimensional solutions $\theta_L(x,t)$ and $\theta_H(y,t)$:

$$\theta(x,y,t) = \theta_L(x,t) \cdot \theta_H(y,t) \tag{4.77}$$

as soon as θ_L and θ_H obey all the equations aligned in the middle and right columns of Fig. 4.17. To begin with, substituting the product formula (4.77) into the conduction equation of the original problem, eq. (a) in Fig. 4.17, yields

$$\underbrace{\left(\frac{\partial^2\theta_L}{\partial x^2} - \frac{1}{\alpha}\frac{\partial\theta_L}{\partial t}\right)}_{=\,0}\theta_H + \underbrace{\left(\frac{\partial^2\theta_H}{\partial y^2} - \frac{1}{\alpha}\frac{\partial\theta_H}{\partial t}\right)}_{=\,0}\theta_L = 0 \tag{4.78}$$

Eq. (a)_L, Fig. 4.17   Eq. (a)_H, Fig. 4.17

This result shows that, indeed, the original equation (a) in Fig. 4.17 is satisfied if the unidirectional temperature distributions satisfy their respective conduction equations, eqs. (a)$_L$ and (a)$_H$.

Next, any of the four original boundary conditions (b) through (e) in Fig. 4.17 is satisfied as soon as the corresponding boundary conditions are satisfied by θ_L and θ_H. For example, by combining the boundary conditions (b)$_L$ and (b)$_H$ through the product formula (4.77), we obtain the original boundary condition for the vertical midplane, eq. (b). Note that the h value need not be the same all around the rectangular cross section: the h value must be the same only on a pair of parallel faces.

The last observation in this argument concerns the initial condition of the original problem, eq. (f) in Fig. 4.17. According to the product solution (4.77), this condition is respected only if the initial temperatures of the long plates ($\theta_{i,L}$, $\theta_{i,H}$) are such that

$$\theta_i = \theta_{i,L} \cdot \theta_{i,H} \tag{4.79}$$

Dividing eqs. (4.77) and (4.79) side by side, we reach the important conclusion that the *dimensionless* temperature distribution of the original (two-dimen-

sional) problem, θ/θ_i, is the same as the product of the *dimensionless* temperature distributions in the two "plate" geometries:

$$\left[\frac{\theta(x,y,t)}{\theta_i}\right]_{\substack{\text{bar,} \\ 2L \times 2H}} = \left[\frac{\theta(x,t)}{\theta_i}\right]_{\substack{\text{plate,} \\ L = \text{half-thickness}}} \times \left[\frac{\theta(y,t)}{\theta_i}\right]_{\substack{\text{plate,} \\ H = \text{half-thickness}}} \tag{4.80}$$

In conclusion, the value of $\theta(x,y,t)/\theta_i$ for the long bar of rectangular cross
section can be calculated by reading Figs. 4.7 and 4.8 *twice*, first for the plate of
half-thickness L and later for the plate of half-thickness H.

As a more recent extension to the theorem outlined here in Fig. 4.17, it has
been shown [7] that a special formula can be used to calculate the total heat
transfer Q (joules) that crosses the boundary of the rectangular domain during
the time interval $0-t$:

$$\frac{Q(t)}{Q_i} = \left(\frac{Q}{Q_i}\right)_L + \left(\frac{Q}{Q_i}\right)_H - \left(\frac{Q}{Q_i}\right)_L \left(\frac{Q}{Q_i}\right)_H \tag{4.81}$$

In this expression the subscripts L and H denote the plates of half-thickness L
and H, respectively. Each of the Q_i denominators represents the maximum
value of the respective $Q(t)$ function, that is, $Q_i = Q(t \to \infty)$.

Solutions analogous to eqs. (4.80) and (4.81) hold in the case of other
multidirectional conduction geometries. The rule in constructing the appropriate factors that participate in the product is to recognize the two simpler bodies
the intersection of which is the body of interest. Figure 4.18 shows that a *short
cylinder* of length $2L$ and radius r_o can be viewed as the intersection of a long
cylinder and a perpendicular plate, each with its own h value at the exposed
surfaces. Therefore, the dimensionless temperature distribution inside a short
cylinder immersed suddenly in a fluid can be calculated with the formula

$$\left[\frac{\theta(r,x,t)}{\theta_i}\right]_{\substack{\text{short} \\ \text{cylinder,} \\ L = \text{half-length} \\ r_o = \text{radius}}} = \left[\frac{\theta(r,t)}{\theta_i}\right]_{\substack{\text{long} \\ \text{cylinder,} \\ r_o = \text{radius} \\ \text{Figs. 4.10} \\ \text{and 4.11}}} \times \left[\frac{\theta(x,t)}{\theta_i}\right]_{\substack{\text{plate,} \\ L = \text{half-thickness} \\ \text{Figs. 4.7} \\ \text{and 4.8}}} \tag{4.82}$$

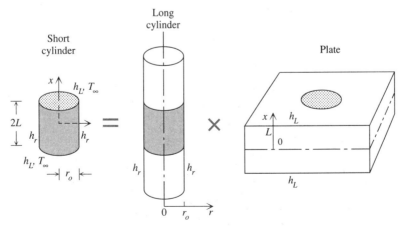

Figure 4.18 The time-dependent temperature of a short cylinder immersed in fluid,
as the product of the temperatures in a long cylinder and in a perpendicular plate.

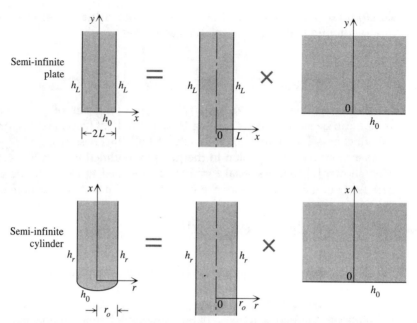

Figure 4.19 Multiplication rules for the temperature distribution in a semi-infinite plate (top), and the temperature distribution in a semi-infinite cylinder (bottom).

Additional examples of two-body "intersections" that can be used to reconstruct solutions for multidirectional time-dependent conduction are the *semi-infinite* counterparts of the rectangular cross section shown on the left side of Fig. 4.17 and the cylinder drawn on the left side of Fig. 4.18. The temperature distribution in a *semi-infinite plate* of half-thickness L, immersed suddenly in a fluid at a different temperature, is therefore (Fig. 4.19, top)

$$\left[\frac{\theta(x,y,t)}{\theta_i}\right]_{\substack{\text{semi-infinite} \\ \text{plate,} \\ L = \text{half-thickness}}} = \left[\frac{\theta(x,t)}{\theta_i}\right]_{\substack{\text{infinite} \\ \text{plate,} \\ L = \text{half-thickness} \\ \text{Figs. 4.7} \\ \text{and 4.8}}} \times \left[\frac{\theta(y,t)}{\theta_i}\right]_{\substack{\text{semi-infinite} \\ \text{medium,} \\ y = \text{normal to} \\ \text{the surface} \\ \text{eq. (4.48)}}} \quad (4.83)$$

Likewise, the temperature distribution in a *semi-infinite cylinder* exposed to a fluid along the base and the cylindrical surface is (Fig. 4.19, bottom)

$$\left[\frac{\theta(r,x,t)}{\theta_i}\right]_{\substack{\text{semi-infinite} \\ \text{cylinder,} \\ r_0 = \text{radius}}} = \left[\frac{\theta(r,t)}{\theta_i}\right]_{\substack{\text{infinite} \\ \text{cylinder,} \\ r_0 = \text{radius} \\ \text{Figs. 4.10} \\ \text{and 4.11}}} \times \left[\frac{\theta(x,t)}{\theta_i}\right]_{\substack{\text{semi-infinite} \\ \text{medium,} \\ x = \text{normal to} \\ \text{the surface} \\ \text{eq. (4.48)}}} \quad (4.84)$$

In these examples as well as in the case of the short cylinder (Fig. 4.18), the total heat transfer interaction can be calculated using eq. (4.81), in which a new set of subscripts may be used to indicate the *two* simpler bodies identified on the right side of each of eqs. (4.82), (4.83), and (4.84).

A more complicated body shape that can be handled in the same manner is the *parallelepiped* shown on the left side of Fig. 4.20. This body can be viewed as the intersection of three plates that are mutually perpendicular. Therefore,

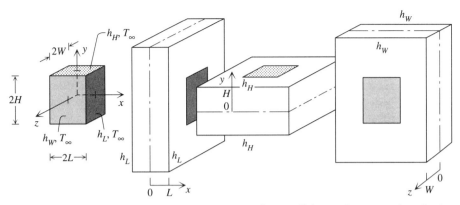

Figure 4.20 The time-dependent temperature of a parallelepiped immersed in fluid, as the product of the temperatures in three mutually perpendicular plates.

$$\left[\frac{\theta(x,y,z,t)}{\theta_i}\right]_{\substack{\text{parallelepiped,} \\ 2L \times 2H \times 2W}} = \left[\frac{\theta(x,t)}{\theta_i}\right]_{\substack{\text{plate,} \\ L = \text{half-thickness}}}$$

$$\times \left[\frac{\theta(y,t)}{\theta_i}\right]_{\substack{\text{plate,} \\ H = \text{half-} \\ \text{thickness}}} \times \left[\frac{\theta(z,t)}{\theta_i}\right]_{\substack{\text{plate,} \\ W = \text{half-} \\ \text{thickness}}} \quad (4.85)$$

The figure shows that the values x, y, and z mark the position relative to a Cartesian system with the origin in the center of the body. This means that on the right side of eq. (4.85), the x, y, and z values represent the distances to the midplanes of the respective plates.

The total heat transfer interaction between the three-dimensional body and the surrounding fluid during the interval $0-t$ is [7]

$$\frac{Q(t)}{Q_i} = \left(\frac{Q}{Q_i}\right)_L + \left(\frac{Q}{Q_i}\right)_H \left[1 - \left(\frac{Q}{Q_i}\right)_L\right] + \left(\frac{Q}{Q_i}\right)_W \left[1 - \left(\frac{Q}{Q_i}\right)_L\right]\left[1 - \left(\frac{Q}{Q_i}\right)_H\right]$$
$$(4.86)$$

The subscripts L, H, and W are shorthand for the three plates identified in Fig. 4.20 and on the right side of eq. (4.85).

4.6 CONCENTRATED SOURCES AND SINKS

4.6.1 Instantaneous (One-Shot) Sources and Sinks

In this section and the next we consider the class of time-dependent conduction problems in which the key feature is the generation (or absorption) of heat transfer in a very small region—a "concentrated" region—of the conducting medium. The small region will be modeled by means of a plane, a straight line, or a point. When heat transfer is *released* into the medium from this small region, the process will be one of time-dependent conduction in the vicinity of a *heat source*. Common phenomena that can be described in terms of concentrated heat sources are underground fissures filled with geothermal steam, underground explosions, canisters of nuclear and chemical waste, and electrical cables buried underground.

When the small region described above *receives* heat transfer from the surrounding (infinite) medium, the region functions as a concentrated *heat sink*. A good example of a heat sink is the buried heat exchanger (duct) through which a heat pump receives heat transfer from the ambient (the soil) in order to augment it and deposit it into a building.

Consider first the direction x through an infinite medium with constant properties (k, α, ρ, c), Fig. 4.21. The equation for time-dependent conduction in the x direction is

$$\frac{\partial^2 \theta}{\partial x^2} = \frac{1}{\alpha} \frac{\partial \theta}{\partial t} \tag{4.87}$$

where $\theta(x,t) = T(x,t) - T_\infty$ is the excess temperature relative to the background (far-field) temperature of the medium (T_∞). We note that eq. (4.87) is satisfied at all times by a function of the form

$$\theta(x,t) = \frac{K}{(\alpha t)^{1/2}} \exp\left(-\frac{x^2}{4\alpha t}\right) \tag{4.88}$$

where K is a constant.

This excess temperature distribution exhibits the behavior illustrated in Fig. 4.21. At times close to $t = 0$, the θ-vs.-x curve has a very sharp spike in the plane $x = 0$. The height and sharpness of this spike diminish as the time increases; in fact, as $t \to \infty$, the excess temperature θ drops to zero.

Another interesting feature of the θ-vs.-x curve is that the area trapped under it is independent of time. The area can be evaluated by recognizing the error function definition (4.38) and eq. (4.39b),

$$\int_{-\infty}^{\infty} \theta \, dx = 2\pi^{1/2} K \qquad \text{(constant)} \tag{4.89}$$

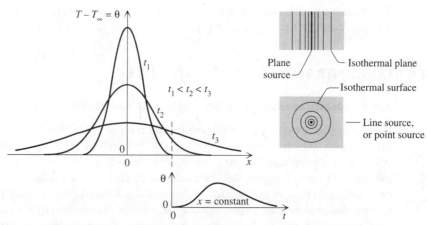

Figure 4.21 General features of the temperature distribution in the vicinity of an instantaneous (one-shot) heat source.

The integral listed on the left side of eq. (4.89) is proportional to the internal energy* inventory of the entire medium:

$$\int_{-\infty}^{\infty} \rho(u - u_\infty)A\,dx = \int_{-\infty}^{\infty} \rho c(T - T_\infty)A\,dx$$

$$= \rho cA \int_{-\infty}^{\infty} \theta\,dx \tag{4.90}$$

where A is the presumably large area of the plane normal to the x direction. Taken together, eqs. (4.89) and (4.90) indicate that the internal energy inventory of the medium is conserved in time. This inventory could have been deposited in the very beginning of the process (at $t = 0$) as a "one-shot" deposit of heat transfer Q (joules) in the $x = 0$ plane:

$$\int_{-\infty}^{\infty} \rho(u - u_\infty)A\,dx = Q \tag{4.91}$$

Combining eqs. (4.89), (4.90), and (4.91), we find that the constant K is proportional to the heat transfer release,

$$K = \frac{Q''}{2\pi^{1/2}\rho c} \tag{4.92}$$

where $Q'' = Q/A$ is the "strength" of the instantaneous plane heat source. In summary, the excess temperature in the vicinity of the $x = 0$ plane in which Q'' (J/m^2) is released once at $t = 0$ is [cf. eq. (4.88)]

$$\theta(x,t) = \frac{Q''}{2\rho c(\pi\alpha t)^{1/2}}\exp\left(-\frac{x^2}{4\alpha t}\right) \quad \text{(instantaneous plane source)} \tag{4.93}$$

From this, we learn that the temperature of the heat source (the plane $x = 0$) decreases as $t^{-1/2}$. Figure 4.21 shows that in any other constant-x plane, the temperature reaches a maximum level at a particular time following the one-shot release of Q''.

Formulas similar to eq. (4.93) can be derived for the temperature variation in the vicinity of sources of other shapes. For example, the excess temperature near a one-shot line heat source is

$$\theta(r,t) = \frac{Q'}{4\rho c\pi\alpha t}\exp\left(-\frac{r^2}{4\alpha t}\right) \quad \text{(instantaneous line source)} \tag{4.94}$$

In this expression, Q' is the strength of the instantaneous line source, that is, the number of "joules per meter" that are released only once at $t = 0$. The radial distance r is measured away from the line source. Equation (4.94) shows that the temperature along the line in which Q' was deposited ($r = 0$) decreases as t^{-1} as the time increases. This temperature decrease is steeper than the temperature in the plane of the Q'' source, eq. (4.93), because the line source is exposed to the conducting medium in all directions (over an angle of 2π, or 360°) around the $r = 0$ axis.

---

*The specific internal energy is measured relative to the reference state provided by the distant temperature of the medium, $u_\infty = u(T_\infty)$.

The temperature history near an instantaneous point source is described by

$$\theta(r,t) = \frac{Q}{8\rho c(\pi \alpha t)^{3/2}} \exp\left(-\frac{r^2}{4\alpha t}\right) \qquad \text{(instantaneous point source)} \qquad (4.95)$$

The strength Q (J) represents the heat transfer released into the medium at the time $t = 0$, and the radial distance r is measured away from the point source ($r = 0$). The temperature at $r = 0$ decreases as $t^{-3/2}$ [i.e., faster than in the plane of the source (4.93) and in the line of the source (4.94)] because now the source is "touched" by a cold medium from all the directions of the sphere circumscribed to the point source.

4.6.2 Persistent (Continuous) Sources and Sinks

All the concentrated sources that have been covered in the preceding section (plane, line, point) are of the instantaneous (one-shot) kind. The time-dependent temperature distributions and conduction processes that are induced by sources that persist in time can be determined analytically by superposing the effects of a large number of instantaneous sources.

The basic idea behind this superposition procedure can be illustrated by considering again the plane source geometry, Fig. 4.21. Assume that at the time $t = 0$ the plane $x = 0$ receives the per-unit-area heat input Q_0''; the temperature distribution that results is indicated in eq. (4.93), namely,

$$\theta_0(x,t) = \frac{Q_0''}{2\rho c(\pi \alpha t)^{1/2}} \exp\left(-\frac{x^2}{4\alpha t}\right) \qquad (4.96)$$

Assume also that at a subsequent moment, $t = t_1$, the plane $x = 0$ receives a new "shot," that is, a new deposit of heat transfer per unit of plane area, Q_1''. If this instantaneous heat source were to occur alone, that is, in the absence of the original shot Q_0'', then the temperature variation (decaying hump) triggered by Q_1'' could also be written by just reading eq. (4.93),

$$\theta_1(x,t) = \frac{Q_1''}{2\rho c[\pi \alpha(t - t_1)]^{1/2}} \exp\left[-\frac{x^2}{4\alpha(t - t_1)}\right] \qquad (4.97)$$

In this $\theta_1(x,t)$ expression the value of $t - t_1$ counts the time that elapses after the release of Q_1''.

If the Q_1'' source takes place in the presence (on the background) of the temperature hump created by Q_0'' at $t = 0$, then the temperature distribution after the time $t = t_1$ is simply the sum of $\theta_0(x,t)$ and $\theta_1(x,t)$. To summarize, the temperature distribution $\theta(x,t)$ during the entire time interval $t > 0$ is described by

$$\theta(x,t) = \begin{cases} \theta_0(x,t) & 0 < t < t_1 \\ \\ \theta_0(x,t) + \theta_1(x,t) & t_1 < t \end{cases} \qquad (4.98)$$

where θ_0 and θ_1 are the expressions listed in eqs. (4.96) and (4.97). It can be shown quite easily that the sum $\theta = \theta_0 + \theta_1$ satisfies the conduction equation (4.87), the far-field conditions ($\theta \rightarrow 0$ as $x \rightarrow \pm\infty$), and the energy conservation statement that the internal energy inventory* in the entire medium is equal to $Q_0'' + Q_1''$ when $t > t_1$.

More entries (lines) can be added to the two-line sequence started in eq. (4.98) if additional shots of size Q_i'' are deposited at the times t_i in the source plane $x = 0$. For example, after the time $t = t_n$ (i.e., after $n + 1$ shots), the temperature distribution is given by

$$\theta(x,t) = \theta_0 + \theta_1 + \theta_2 + \cdots + \theta_n \tag{4.99}$$

A *continuous* heat source in the plane $x = 0$ has the same effect as a sequence of a very large number of small instantaneous plane sources of equal size:

$$\Delta Q'' = q'' \, \Delta t \tag{4.100}$$

where q'' (W/m^2) is the heat transfer deposited per unit area and unit time, and Δt is the short duration of each shot. As Δt becomes infinitesimally small, the number of terms in the finite-time interval sum, eq. (4.99), becomes infinite, and the sum is replaced by the time integral

$$\theta(x,t) = \int_0^t \frac{q''}{2\rho c[\pi\alpha(t-\tau)]^{1/2}} \exp\left[-\frac{x^2}{4\alpha(t-\tau)}\right] d\tau \tag{4.101}$$

Under the integral sign, the dummy variable τ marks the time when each additional instantaneous source $q'' \, d\tau$ springs into action. The integral (4.101) can be evaluated [1], and the resulting expression is the temperature distribution near the $x = 0$ plane in which the continuous source q'' is turned on at $t = 0$:

$$\theta(x,t) = \frac{q''}{\rho c}\left(\frac{t}{\pi\alpha}\right)^{1/2}\exp\left(-\frac{x^2}{4\alpha t}\right) - \frac{q''|x|}{2k}\text{erfc}\left[\frac{|x|}{2(\alpha t)^{1/2}}\right]$$

(continuous plane source)   (4.102)

Placing $x = 0$ in this expression, we learn that the temperature in the plane of the source increases as $t^{1/2}$ forever:

$$\theta(0,t) = \frac{q''}{\rho c}\left(\frac{t}{\pi\alpha}\right)^{1/2} \tag{4.103}$$

In other words, even though the plane source persists at the constant level q'', the temperature in the source plane (and, for that matter, everywhere else in the medium) increases as the elapsed time t increases.

The temperature distribution near a continuous line source can be determined similarly by superposing the effects of a large number of small instantaneous line sources of equal size. The result is

$$\theta(r,t) = \frac{q'}{4\pi k}\int_{r^2/4\alpha t}^{\infty} \frac{e^{-u}}{u} du \qquad \text{(continuous line source)} \tag{4.104}$$

where q' (W/m) is the source strength. The integral listed on the right side of eq. (4.104) is the *exponential integral* function, the values of which are tabulated in Appendix E. At sufficiently long times and/or small radial distances, where the group $r^2/4\alpha t$ is smaller than 1, the temperature distribution approaches [1]

$$\theta(r,t) \cong \frac{q'}{4\pi k}\left[\ln\left(\frac{4\alpha t}{r^2}\right) - 0.5772\right] \qquad \left(\frac{r^2}{4\alpha t}\right) << 1 \tag{4.105}$$

---

*Per unit area of the source plane $x = 0$.

The effect of a continuous point source q (W) can also be determined by superposing the effects of a large number of instantaneous point sources of equal strength:

$$\theta(r,t) = \frac{q}{4\pi k r}\,\text{erfc}\left[\frac{r}{2(\alpha t)^{1/2}}\right] \qquad \text{(continuous point source)} \qquad (4.106)$$

Noting that erfc(0) = 1, we conclude that as the time increases and the argument $r/[2(\alpha t)^{1/2}]$ becomes considerably smaller than 1, the temperature distribution stabilizes at the level

$$\theta(r,\infty) = \frac{q}{4\pi k r} \qquad (4.107)$$

In the steady state the temperature decreases as r^{-1} away from the point heat source. The existence of a steady temperature distribution at large times distinguishes the continuous point source from the continuous line and plane sources, for which steady state solutions do not exist. The steady state can exist around a continuous point source because the spherical-radial geometry allows the released heat to escape in more directions than in the cylindrical-radial and one-dimensional Cartesian geometries. In other words, the infinite conducting medium provides a better (more effective) cooling effect to a point source than to a line source or a plane source.

Throughout these last two sections we referred to the concentrated heat transfer effect as that of a heat *source*. Exactly the same discussion and formulas apply to concentrated heat *sinks*: in these cases the numerical values of the source strengths (Q'', Q', Q, q'', q', q) are negative, and so are the corresponding excess temperatures θ.

4.6.3 Moving Heat Sources

One common feature of the concentrated sources and sinks treated until now is the symmetrical shape of the isotherms that develop around the location of the source. In this subsection we consider briefly the deformation of this temperature field when the source moves relative to the conductive medium in which it is imbedded. Specifically, we consider the *long-time* behavior of *steady* (continuous) sources that move with an *unchanging* velocity through the medium. We shall see that the temperature field that develops is a slender "thermal wake" that trails the source.

In Fig. 4.22 a continuous *line source* of strength q' (W/m) moves with the constant speed U through the conductive medium (k,α). The two-dimensional system (x,y) is perpendicular and attached to the line source: in the (x,y) plane, the medium moves in the positive x direction. The thermal wake left behind the source is two-dimensional and is represented by the temperature distribution $T(x,y)$. The latter is the unknown in this heat transfer problem. The temperature distribution $T(x,y)$ is steady in the eyes of an observer riding on the line source.

The model constructed on the left side of Fig. 4.22 is a simplified representation of what happens in the vicinity of the line of contact between two semi-infinite plates that are being welded together. Seen from above, the electric arc plays the role of the line heat source $q' = q/\Delta z$, where q (W) is the steady rate of heating provided by the arc to the plate-to-plate contact, and Δz is the thickness of the plates in the direction normal to the (x,y) plane. When the top and bottom

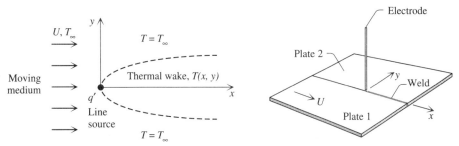

Figure 4.22 The thermal wake behind a continuous line source moving through a conductive medium (left), and the edge-to-edge arc welding of two semi-infinite plates (right).

surfaces of the plates are insulated enough (e.g., bathed in slowly moving air), the temperature distribution in the wake of the electric arc is nearly two-dimensional, $T(x,y)$. This thermal wake has the freshly formed weld as its line of symmetry.

The temperature distribution $T(x,y)$ can be determined analytically by solving the problem

$$U\frac{\partial T}{\partial x} = \alpha\frac{\partial^2 T}{\partial y^2} \tag{4.108}$$

$$T = T_\infty \quad \text{at} \quad y = \pm\infty \tag{4.109}$$

$$q' = \int_{-\infty}^{\infty} \rho c U(T - T_\infty)\,dy \tag{4.110}$$

in which the energy equation (4.108) acquires one "convective" term on the left side, $U(\partial T/\partial x)$. The origin of this term will become clear in Section 5.2.3. The integral condition (4.110) states that the energy that flows through any constant-x plane is conserved and equal to the energy rate deposited by the line source in the moving medium. The solution to eqs. (4.108), (4.109), and (4.110) forms the subject of Problem 4.23 and is analogous to the solution for the thermal wake behind a line source immersed in a turbulent stream with "grid-generated" (uniform) turbulence [8]:

$$T(x,y) - T_\infty = \frac{q'/\rho c}{(4\pi U\alpha x)^{1/2}}\exp\left(-\frac{Uy^2}{4\alpha x}\right) \tag{4.111}$$

This solution shows that the excess temperature in the centerplane of the wake, $T(x,0) - T_\infty$, decreases as $x^{-1/2}$ in the downstream direction. Note further that the argument of the exponential is constant when y^2/x is constant; this means that the approximate "width" of the thermal wake increases as $x^{1/2}$ in the downstream direction. This type of growth has been illustrated qualitatively on the left side of Fig. 4.22.

The thermal wake created behind a continuous *point source q* (W) in a moving medium can be analyzed based on a similar model (Fig. 4.23). In the cylindrical coordinate system (x,r) attached to the point source, the solution reads [8]

$$T(x,r) - T_\infty = \frac{q/\rho c}{4\pi\alpha x}\exp\left(-\frac{Ur^2}{4\alpha x}\right) \tag{4.112}$$

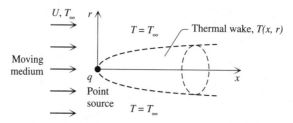

Figure 4.23 The thermal wake left behind a continuous point source in a moving medium.

The thermal wake has the shape of a body of revolution, the radius of which increases as $x^{1/2}$ in the downstream direction. The distribution of excess temperature in each constant-x plane has the same shape (Gaussian) as in the two-dimensional wake, eq. (4.111). The centerline excess temperature, $T(x,0) - T_\infty$, decreases as x^{-1} in the downstream direction, that is, faster than in the two-dimensional case.

An important aspect of any moving source problem is that the relative speed U must be sufficiently large for thermal wake solutions of types (4.111) and (4.112) to be valid. For example, in the extreme where the source moves so slowly that it is practically stationary, the temperature field is described by the persistent source solutions (4.104) and (4.106). Note that in the $U \to 0$ limit the isotherms around the source are almost circular (cylindrical or spherical).

The thermal wake solutions presented in this subsection hold when each thermal wake is slender (elongated in the U direction), not round. According to the argument of the exponentials in eqs. (4.111) and (4.112), the width of each thermal wake scales as $(\alpha x/U)^{1/2}$. Writing the wake slenderness condition as $(\alpha x/U)^{1/2} \ll x$, we conclude that the source speed is "large enough" when $Ux/\alpha \gg 1$. This condition also means that solutions of types (4.111) and (4.112) do not hold right next to the moving source, that is, at x values so small that the dimensionless number Ux/α is of order 1 or smaller.*

4.7   MELTING AND SOLIDIFICATION

In this section we revisit the time-dependent conduction in a semi-infinite solid exposed suddenly to a surface temperature T_0 that differs from the initial temperature T_i of the material (Section 4.3). For clarity, consider first the *heating* effect due to raising the surface temperature above the background temperature of the rest of the solid, $T_0 > T_i$.

If the solid is a crystalline substance (e.g., ice, high-purity metal), a known property of the solid is also the melting point T_m. When the temperature of the heated surface is less than the melting point, the semi-infinite medium remains solid forever and the formulas developed in Section 4.3 apply. On the other hand, if T_0 is greater than the melting point, a near-surface layer of finite

---

*This dimensionless group is the Peclet number, which is used more frequently in convection heat transfer, eq. (5.70).

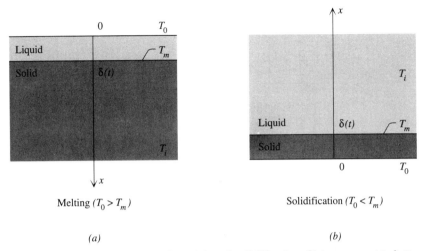

Figure 4.24 Unidirectional melting (*a*) and solidification (*b*) into a semi-infinite isothermal medium (T_i) with sudden temperature change at the boundary (T_0).

thickness will melt. The conduction process in this case continues to be time dependent; however, sandwiched between the semi-infinite solid and the heated boundary (T_0) is a liquid layer of finite thickness. This thickness and the temperatures of the still solid regions continue to increase as the time increases.

The boundary heating and melting phenomenon is illustrated in Fig. 4.24. Unlike in the all-solid case of Fig. 4.3, this time the figure is oriented so that the conduction process proceeds in the vertical direction and the liquid layer floats on top of the solid. It is being assumed that the conducting medium is a substance that expands upon melting, and that its liquid phase expands upon heating. The downward orientation of conduction in Fig. 4.24*a* is therefore a reminder that the entire medium (liquid + solid) is perfectly *motionless* in a gravitational field that points in the downward direction. In particular, the liquid layer is "stably stratified" because its uppermost layers are also the lightest (for more on this, see Section 7.4.3 on Bénard convection).

Figure 4.24*b* shows the analogous problem of unidirectional solidification in a semi-infinite liquid pool that is cooled suddenly along the bottom surface, $T_0 < T_m$. If, as in most substances, the solid phase is somewhat denser than the liquid phase, and if the liquid contracts upon cooling, the solid layer and the pool of liquid remain motionless.

To emphasize, the pure conduction analysis that is presented below applies to a perfectly motionless medium. In the case of water near the ice point (0°C)—that is, in an anomalous case where the solid is lighter than the liquid and the liquid expands upon cooling—both sides of Fig. 4.24, (*a*) and (*b*), would have to be rotated by 180°, so that motion is ruled out in a gravitational field that points downward.

The key engineering unknown in melting and solidification processes is the movement of the *melting front* $x = \delta(t)$, or the melting/solidification rate $d\delta/dt$. Consider first the melting process (Fig. 4.24*a*). The most basic features of the mechanism of conduction melting become visible if we assume first that the initial solid temperature is equal to the melting point, $T_i = T_m$. Since the inter-

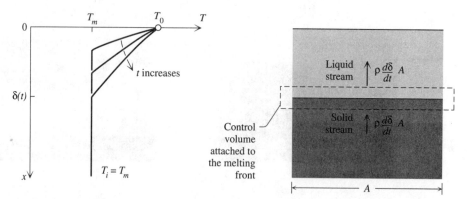

Figure 4.25 Melting of a semi-infinite solid of temperature equal to the melting point ($T_i = T_m$), and the flow through the infinitesimally thin control volume attached to the solid–liquid interface.

face between the solid and liquid regions (the melting front) is always at the melting point, the $T_i = T_m$ assumption means that the solid remains isothermal (at T_m, i.e., as a saturated solid) as the time increases. This feature is illustrated in Fig. 4.25, which also shows that the temperature difference between the heated boundary (T_0) and the remaining solid (T_m) is bridged by a δ-thick layer of liquid.

The movement of the melting front is governed by the conservation of energy in the plane $x = \delta(t)$. Consider a very wide and thin control volume that contains the melting front and moves downward with exactly the same speed as the melting front. The right side of Fig. 4.25 shows that a "stream" of solid flowrate $\rho A(d\delta/dt)$ enters the control volume from below, while another $\rho A(d\delta/dt)$ stream of liquid exits through the upper surface. The density ρ refers to the *liquid,** that is, to the region whose thickness is δ.

The first law of thermodynamics states that the net enthalpy flow out of the control volume must be balanced by the net heat transfer rate received by the same control volume:

$$\left(\rho A \frac{d\delta}{dt}\right)h_f - \left(\rho A \frac{d\delta}{dt}\right)h_s = -kA\left(\frac{\partial T}{\partial x}\right)_{x = \delta,\text{ liquid side}} \tag{4.113}$$

In this equation, A, h_f, and h_s are the frontal area of the control volume (Fig. 4.25), the specific enthalpy of the liquid, and the specific enthalpy of the solid, respectively. The term on the right side of eq. (4.113) represents the heat transfer rate that arrives from above, that is, on the liquid side of the melting front. There is no heat transfer term for the solid side of the melting front, because the solid region is isothermal. The coefficient k is therefore the thermal conductivity of the liquid.

---

*It is important to note that in this analysis the density of the solid (ρ_s) does not have to be exactly equal to the density of the liquid (ρ). The reason is the principle of mass conservation, according to which $\rho(d\delta/dt) = \rho_s\, v_s$, where v_s is the vertical velocity of the solid entering the control volume shown on the right side of Fig. 4.25.

The exact analysis of this melting phenomenon would continue with the derivation of the $T(x,t)$ distribution inside the liquid layer, because that result would permit the evaluation of the right side of eq. (4.113). The end result of the exact analysis is reported a little bit later, eq. (4.117). A much simpler approach is based on the observation that sufficiently early in the life of the melting process, when the melt layer is "thin," the temperature distribution is linear:

$$\frac{T(x,t) - T_m}{T_0 - T_m} \cong 1 - \frac{x}{\delta(t)} \tag{4.114}$$

Combining this with eq. (4.113), and using the *latent heat of melting* notation $h_{sf} = h_f - h_s$, we obtain

$$\delta\frac{d\delta}{dt} \cong \frac{k}{\rho h_{sf}}(T_0 - T_m) \tag{4.115}$$

Equation (4.115) can be integrated from the starting condition of "no liquid," $\delta(0) = 0$, to show that the melt thickness increases as $t^{1/2}$:

$$\delta(t) \cong \left[2\frac{kt}{\rho h_{sf}}(T_0 - T_m)\right]^{1/2} \tag{4.116}$$

The exact solution to the same problem (the melting of a solid at the melting point) was reported by Stefan [9]:

$$\pi^{1/2}\lambda\exp(\lambda^2)\operatorname{erf}(\lambda) = \frac{c(T_0 - T_m)}{h_{sf}} \tag{4.117}$$

where c is the specific heat of the liquid and the dimensionless number λ is shorthand for

$$\lambda = \frac{\delta}{2(\alpha t)^{1/2}} \tag{4.118}$$

The group appearing on the right side of eq. (4.117) is recognized today as the *Stefan number*:

$$\text{Ste} = \frac{c(T_0 - T_m)}{h_{sf}} \tag{4.119}$$

The number λ is small relative to 1 in the case of water freezing by conduction to an environment a few degrees below $0°C$. In the case of materials with high melting points, such as metals and rocks cooled by contact with room temperature surfaces, the value of λ is of the order of 1.

The Stefan number is a dimensionless measure of the degree of superheating $(T_0 - T_m)$ that is being experienced by the liquid. The exact solution (4.117) is presented in Fig. 4.26, next to the approximate solution (4.116). This figure shows that the approximate solution is quite accurate when the Stefan number is smaller than 1. In general, however, the approximate solution (4.116) overestimates the value of the melt thickness $\delta(t)$.

When the solid is initially at a temperature lower than the melting point, $T_i < T_0$, the melting phenomenon analyzed above is accompanied also by a process of time-dependent conduction into the solid, that is, on the solid side of

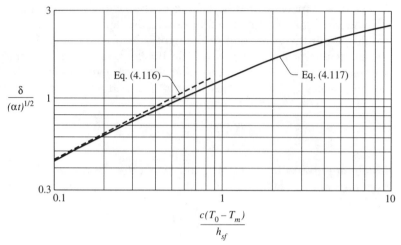

Figure 4.26 History of the melting (or solidification) front position in a semi-infinite medium of temperature equal to the melting point.

the melting front. The analytical solution for $\delta(t)$ in this more general case can be found in Ref. [1]: it was reported first by Franz Neumann in 1860.

Consider now briefly the *solidification* process illustrated on the right side of Fig. 4.24. In this case the bottom surface is colder than the melting front, $T_0 < T_m$. If the liquid is saturated ($T_i = T_m$), the thickness of the solidified layer $\delta(t)$ can be calculated with eqs. (4.116) and (4.117), in which the temperature difference $T_0 - T_m$ must now be replaced with $T_m - T_0$. In the resulting formulas, k, c, and α are now properties of the *solid* layer. The first analysis of solidification by time-dependent conduction was published in 1831 by Lamé and Clapeyron [10]. Other solutions for solidification and melting with plane or cylindrical liquid–solid interfaces can be found in Ref. [1]. Simple relationships for phase change processes in the spherical geometry have been developed by Milanez [11].

The present treatment of melting and solidification was based on the important assumption that the conduction process is unidirectional. Review the plane liquid–solid interfaces assumed in Figs. 4.24 and 4.25. Real melting and solidification processes are often accompanied by additional effects that can deform severely the liquid–solid interface. In such cases the conduction process is no longer unidirectional.

One example is shown in Fig. 4.27, where aluminum was solidified on the inside of a cold cylindrical surface (only a segment of the solidified shell is shown) [12]. The waviness of the solidification front appears to be caused by the nonuniform pressure (and thermal contact) that develops between the cold cylindrical surface and the newly solidified aluminum shell. The solidification rate is higher (i.e., "peaks" form) above areas of relatively high pressure between the outer shell and the aluminum layer.

Another example is the effect of natural convection (buoyancy-driven flow) in the liquid if the melting front is rotated away from the stable (no flow) position chosen in Fig. 4.24. A classical case of melting in the presence of

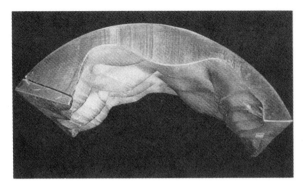

Figure 4.27 The waviness of the solidification front of a layer of 3003 aluminum solidified on the inside of a cold cylindrical surface with a diameter of 20 cm. (Courtesy of Dr. Louis G. Hector, Jr., Aluminum Company of America [12].)

natural convection is outlined in Fig. 4.28, which results from rotating Fig. 4.24*a* by 90 degrees. The left side is heated to the temperature T_0, which is above the melting point of the solid, T_m. The entire solid is at the melting point. The top and bottom sides of the two-dimensional system are impermeable and insulated. The liquid is driven clockwise by the heating imposed from the left side. The liquid rises along T_0 because it is warmer and *lighter* than the liquid near T_m. The latter descends along the melting front, which is colder.

As time passes, the melting front of Fig. 4.28 deviates considerably from the original (plane, vertical) shape. This deformation is due to the natural circulation, which brings the stream of T_0-heated liquid in direct contact with only the upper end of the liquid–solid interface. This is why the melting process is considerably faster near the top of the system. The current progress on calculating the melting rate in the natural-convection configuration of Fig. 4.28 has been reviewed in Ref. [13].

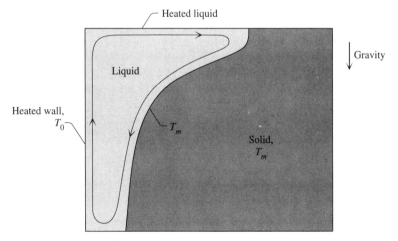

Figure 4.28 The deformed shape of the melting front when heating is from the side and the effect of natural convection is strong.

4.8   NUMERICAL METHODS

4.8.1   Discretization in Time and Space

Solutions to a wide variety of time-dependent conduction problems are now possible based on numerical methods. In Section 3.3 we saw the application of the method of finite differences to problems of *steady-state* conduction, in which the temperature and heat transfer patterns of interest were independent of time. In the present section we extend the finite-difference method to problems of time-dependent conduction.

The conceptual leap from finite differences in steady-state conduction to finite differences in time-dependent conduction can be appreciated by comparing the two-dimensional problems of Fig. 3.11 (right side) and Fig. 4.29. In both problems the space (conducting medium) is covered with a grid whose nodes serve as centers for a number of small control volumes. The uniform grid in these figures is generated by two families of equidistant parallel lines, one horizontal and the other vertical. The position of one node or control volume is pinpointed by two numbers, i for the horizontal position and j for the vertical position.

In the case of time-dependent conduction, Fig. 4.29, in addition to the discretization of the spatial domain (x,y), we see the discretization of the time domain $t \geq 0$. The latter is represented by a number of discrete "snapshots." The time between two consecutive snapshots is Δt, and for simplicity we assume that Δt does not vary from one frame to the next. These time frames are ordered on the time axis in the sequence indicated by the number m, so that m represents the current time frame, while $m - 1$ and $m + 1$ represent the frames immediately before and after the current frame. The very first frame in this sequence ($m = 0$) represents the initial temperature distribution, that is, the initial condition specified at $t = 0$.

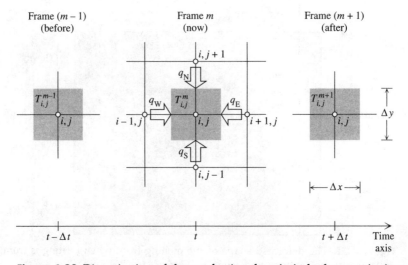

Figure 4.29 Discretization of the conducting domain in both space (x,y) and time (t).

The division of the conducting space into discrete control volumes (i, j) and time frames (m) means that the time-dependent temperature field $T(x, y, t,)$ is replaced by a three-dimensional array of temperature values $T_{i,j}^m$. These numerical values can be determined by applying the first law of thermodynamics to each control volume, and by properly accounting for the initial and boundary conditions that are an integral part of the problem statement.

Consider the example of Fig. 4.29, in which the control volume (i, j) is drawn around an "interior" node (for a classification of nodes, review the left side of Fig. 3.11). With reference to the m time frame pictured in the middle of the figure, we write the first law of thermodynamics by viewing the control volume as a closed system that behaves in a time-dependent manner [cf. eqs. (1.15)–(1.18)]:

$$\rho \Delta x \, \Delta y \, W c \frac{\partial T}{\partial t} = q_{\mathrm{N}} + q_{\mathrm{E}} + q_{\mathrm{S}} + q_{\mathrm{W}} \qquad (4.120)$$

The finite-difference expressions for the conduction heat currents arriving from the four cardinal points $(q_{\mathrm{N}}, q_{\mathrm{E}}, q_{\mathrm{S}}, q_{\mathrm{W}})$ are similar to those listed in eqs. (3.84), except that each temperature now carries the superscript m. The length W represents the long dimension of the conducting system, namely, the length measured in the direction perpendicular to the plane of Fig. 4.29. In the end, eq. (4.120) becomes

$$\frac{1}{\alpha} \frac{\partial T}{\partial t} = \frac{T_{i+1,j}^m + T_{i-1,j}^m - 2T_{i,j}^m}{(\Delta x)^2} + \frac{T_{i,j+1}^m + T_{i,j-1}^m - 2T_{i,j}^m}{(\Delta y)^2} \qquad (4.121)$$

The next step consists of approximating by means of finite differences the time derivative $\partial T / \partial t$. For this we have at least two options. We can look *forward* (into the future) and compare the present-time frame (m) with the next one $(m + 1)$,

$$\frac{\partial T}{\partial t} \cong \frac{T_{i,j}^{m+1} - T_{i,j}^m}{\Delta t} \qquad \text{(forward difference)} \qquad (4.122)$$

or we can look *backward* (into the past) by comparing the current frame with the one positioned in time immediately ahead of it,

$$\frac{\partial T}{\partial t} \cong \frac{T_{i,j}^m - T_{i,j}^{m-1}}{\Delta t} \qquad \text{(backward difference)} \qquad (4.123)$$

These alternatives lead to two distinct numerical algorithms, respectively, the so-called "explicit" and "implicit" methods. We focus on each of these methods in detail in the next two subsections.

4.8.2 The Explicit Method

If we combine eq. (4.121) with the forward-difference time derivative (4.122), and if we assume for the sake of simplicity that the uniform grid has a square loop ($\Delta x = \Delta y$), we obtain an expression for the *new* (next) temperature at the interior node:

$$T_{i,j}^{m+1} = \mathrm{Fo}(T_{i+1,j}^m + T_{i-1,j}^m + T_{i,j+1}^m + T_{i,j-1}^m) + (1 - 4\mathrm{Fo})T_{i,j}^m \qquad (4.124)$$

Table 4.3 Two-Dimensional Time-Dependent Conduction: The Finite-Difference Equations ($\Delta x = \Delta y$)

Internal node

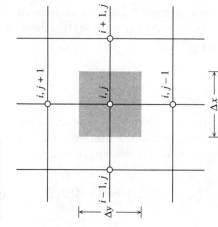

Explicit:

$$T_{i,j}^{m+1} = \text{Fo}(T_{i+1,j}^m + T_{i-1,j}^m + T_{i,j+1}^m + T_{i,j-1}^m) + (1 - 4\text{Fo})T_{i,j}^m$$

$$\text{(Stability condition: Fo} \leq \tfrac{1}{4})$$

Implicit:

$$(1 + 4\text{Fo})T_{i,j}^{m+1} - \text{Fo}\left(T_{i+1,j}^{m+1} + T_{i-1,j}^{m+1} + T_{i,j+1}^{m+1} + T_{i,j-1}^{m+1}\right) = T_{i,j}^m$$

Surface node[a,b]

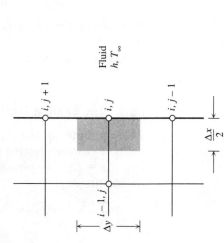

Explicit:

$$T_{i,j}^{m+1} = \text{Fo}(2T_{i-1,j}^m + T_{i,j+1}^m + T_{i,j-1}^m + 2\text{Bi}T_\infty) + (1 - 4\text{Fo} - 2\text{Bi Fo})T_{i,j}^m$$

$$\text{(Stability condition: Fo}(2 + \text{Bi}) \leq \tfrac{1}{2})$$

Implicit:

$$[1 + 2\text{Fo}(2 + \text{Bi})]T_{i,j}^{m+1} - \text{Fo}(2T_{i-1,j}^{m+1} + T_{i,j+1}^{m+1} + T_{i,j-1}^{m+1} + 2\text{Bi}T_\infty) = T_{i,j}^m$$

External corner node[a]

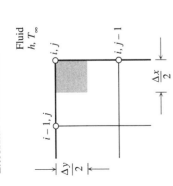

Explicit:

$$T_{i,j}^{m+1} = 2Fo(T_{i+1,j}^m + T_{i,j-1}^m + 2BiT_\infty) + (1 - 4Fo - 4Bi\,Fo)T_{i,j}^m$$

(Stability condition: $Fo(1 + Bi) \le \frac{1}{4}$)

Implicit:

$$[1 + 4Fo(1 + Bi)]T_{i,j}^{m+1} - 2Fo(T_{i-1,j}^{m+1} - T_{i,j-1}^{m+1} + 2BiT_\infty) = T_{i,j}^m$$

Internal corner node[a]

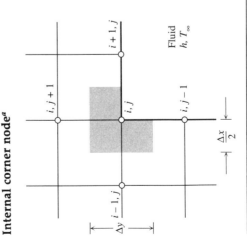

Explicit:

$$T_{i,j}^{m+1} = \tfrac{2}{3}Fo(T_{i+1,j}^m + 2T_{i-1,j}^m + 2T_{i,j+1}^m + T_{i,j-1}^m + 2BiT_\infty) + (1 - 4Fo - \tfrac{4}{3}Bi\,Fo)T_{i,j}^m$$

(Stability condition: $Fo(3 + Bi) \le \frac{3}{4}$)

Implicit:

$$\left[1 + 4Fo\left(1 + \frac{Bi}{3}\right)\right]T_{i,j}^{m+1} - \tfrac{2}{3}Fo(T_{i+1,j}^{m+1} + 2T_{i-1,j}^{m+1} + 2T_{i,j+1}^{m+1} + T_{i,j-1}^{m+1} + 2BiT_\infty) = T_{i,j}^m$$

[a]The adiabatic boundary limit of this case is obtained by setting Bi = 0.
[b]The equations for a surface with a specified distribution of heat flux are listed in Problem 4.30.

The *grid Fourier number* Fo in this expression is a way of nondimensionalizing the time step between two consecutive frames,[*]

$$\text{Fo} = \frac{\alpha \Delta t}{(\Delta x)^2} \tag{4.125}$$

On the right side of eq. (4.124) we see nothing but temperature information that is available at the present time (m). This information consists of the current temperature in the center of the control volume, $T_{i,j}^m$, and the current temperatures in the centers of the four neighboring control volumes. The computational method is called "explicit" because the new temperature at the node of interest, $T_{i,j}^{m+1}$, appears explicitly on the left side of eq. (4.124). This equation has to be used only once in order to determine the new temperature at the node (i,j). Invoked in connection with each of the remaining interior nodes that cover the conducting medium, eq. (4.124) delivers all the interior temperature values that would populate the new time frame labeled $m + 1$ in Fig. 4.29.

It is worth repeating that eq. (4.124) accounts for the conservation of energy only in an *interior* control volume, in a spatially *two-dimensional* conducting domain. The corresponding explicit forms of the energy conservation statement for other control volume types (surface, corner, etc.) are listed in Table 4.3. Relatively simpler expressions apply to control volumes in conductive domains that are spatially *one-dimensional*: these expressions are shown in Table 4.4.

The potential drawback of the explicit method is that, unless certain precautions are taken, the numerical solution can develop oscillations whose amplitude increases from one time frame to the next. This numerical "instability" can be avoided by selecting a time step (Fo) that is sufficiently small. It has been found that the numerical solution is stable whenever the coefficient that multiplies $T_{i,j}^m$ on the right side of equations such as eq. (4.124) is not negative. Therefore, in the case of an interior node in a two-dimensional conducting domain, the criterion for numerical stability is $(1 - 4\text{Fo}) \geq 0$, which translates into a small enough Δt so that

$$\text{Fo} \leq \frac{1}{4} \tag{4.126}$$

Reading the explicit finite-difference equations assembled in Table 4.3, we note that a different criterion for numerical stability applies to each type of control volume. For example, the stability condition for a control volume at a boundary with convective heat transfer is $(1 - 4\text{Fo} - 2\text{Bi}\,\text{Fo}) \geq 0$, in other words,

$$\text{Fo} \leq \frac{1}{2(2 + \text{Bi})} \tag{4.127}$$

where Bi is *the Biot number based on grid spacing,*[†]

$$\text{Bi} = \frac{h \Delta x}{k} \tag{4.128}$$

---

[*]The "grid" Fourier number should not be confused with the "macroscopic" Fourier number used earlier, for example, in eq. (4.63).

[†]This "grid" Biot number should not be confused with the "macroscopic" Biot number used previously, for example, in eq. (4.63).

Table 4.4 One-Dimensional Time-Dependent Conduction: The Finite-Difference Equations

Internal plane 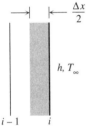	Explicit: $$T_i^{m+1} = \mathrm{Fo}(T_{i-1}^m + T_{i+1}^m) + (1 - 2\mathrm{Fo})T_i^m$$ (Stability condition: $\mathrm{Fo} \leq \tfrac{1}{2}$) Implicit: $$(1 + 2\mathrm{Fo})T_i^{m+1} - \mathrm{Fo}(T_{i+1}^{m+1} + T_{i+1}^{m+1}) = T_i^m$$

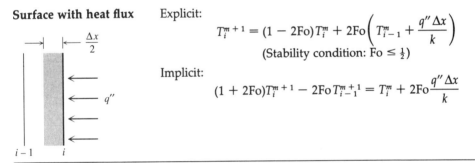

Surface with convection[a]	Explicit: $$T_i^{m+1} = 2\mathrm{Fo}(T_{i-1}^m + \mathrm{Bi}\,T_\infty) + (1 - 2\mathrm{Fo} - 2\mathrm{Bi}\,\mathrm{Fo})T_i^m$$ (Stability condition: $\mathrm{Fo}(1 + \mathrm{Bi}) \leq \tfrac{1}{2}$) Implicit: $$(1 + 2\mathrm{Fo} + 2\mathrm{Fo}\,\mathrm{Bi})T_i^{m+1} - 2\mathrm{Fo}\,T_{i-1}^{m+1} = 2\mathrm{Fo}\,\mathrm{Bi}\,T_\infty + T_i^m$$

Surface with heat flux	Explicit: $$T_i^{m+1} = (1 - 2\mathrm{Fo})T_i^m + 2\mathrm{Fo}\left(T_{i-1}^m + \frac{q''\,\Delta x}{k}\right)$$ (Stability condition: $\mathrm{Fo} \leq \tfrac{1}{2}$) Implicit: $$(1 + 2\mathrm{Fo})T_i^{m+1} - 2\mathrm{Fo}\,T_{i-1}^{m+1} = T_i^m + 2\mathrm{Fo}\frac{q''\,\Delta x}{k}$$

[a] The adiabatic boundary limit of this case is obtained by setting $\mathrm{Bi} = 0$.

As the Biot number is generally finite, the right side of eq. (4.127) is smaller than the right side of eq. (4.126). The stability criterion (4.127) is therefore more stringent (restrictive) than the interior node criterion (4.126). To guarantee the stability of the entire numerical solution, the time step (Fo) must be chosen small enough so that it satisfies the most stringent of the stability conditions recommended by the various explicit finite-difference equations that account for all the nodes.

In conclusion, the explicit method is a direct one, as each new time frame of numerical values $T_{i,j}^{m+1}$ is calculated in single-blow fashion by sweeping through the domain and using equations like eq. (4.124). The explicit method is only *conditionally* stable, the condition being that Fo must not exceed a certain number. With the Fo value chosen based on stability considerations, the actual space and time steps Δx and Δt, which together make up the Fourier number Fo, must be selected as trade-offs between numerical accuracy and computational cost. This give-and-take process is illustrated in the first part of Example 4.4.

4.8.3 The Implicit Method

The alternative path consists of relying on the backward difference (4.123) while estimating the left side of eq. (4.121). The resulting finite-difference equation for an interior control volume in two dimensions can be arranged as follows:

$$(1 + 4Fo)T_{i,j}^m - Fo(T_{i+1,j}^m + T_{i-1,j}^m + T_{i,j+1}^m + T_{i,j-1}^m) = T_{i,j}^{m-1} \quad (4.129)$$

This finite-difference equation is called "implicit" because the unknown node temperature in the current time frame, $T_{i,j}^m$, is related to four other unknowns — the temperatures at the four neighboring nodes *in the same (current) time frame*. Known in this equation is only the right side, because that side represents the node temperature that was calculated in the preceding time step. For these reasons the current node temperature $T_{i,j}^m$ can be calculated only after writing one equation of type (4.129) for every node, and solving the resulting system of equations *simultaneously* for all the node temperatures that would constitute the current time frame.

Tables 4.3 and 4.4 reinforce the observation that the exact form of the implicit finite-difference equation depends on the type of control volume (interior, surface, corner), the type of boundary condition, and whether the conducting domain is two-dimensional or one-dimensional. Note that, unlike in eq. (4.129), in the "implicit" equations listed in Tables 4.3 and 4.4 the unknown temperatures carry the superscript $m + 1$, while the most recent temperature is indicated by m.

The complete system of implicit finite-difference equations for the temperatures in the current time frame can be solved based on the matrix inversion method (Section 3.3.2) or on the Gauss–Seidel iteration method (Section 3.3.3). The use of the Gauss–Seidel iteration method is illustrated in the second part of Example 4.4.

One system of implicit finite-difference equations must be solved for each time step. The real-time interval between two consecutive time steps is proportional to the numerical value selected for the Fourier number. Unlike the explicit method, the implicit method is unconditionally stable: there is no upper limit to the numerical value that can be assigned to Fo. As long as the grid spacing Δx is fixed, a smaller Fo value will yield a denser (thicker stack) of time frame solutions and a lengthier and more expensive computation. The actual values of the selected steps Δx and Δt will be trade-offs between numerical accuracy, computational cost, and the detail (number of frames per unit time) that we want to see in the time evolution of the solution.

Example 4.4

The Effect of the Steps Δx and Δt on the Accuracy of The Explicit and Implicit Methods

The 4-cm-thick slab shown in Fig. E4.4a is initially at $T = 0°C$ throughout its volume. Its thermal diffusivity is $\alpha = 0.05$ cm^2/s. Beginning with the time $t = 0$ s, the temperature of the two faces of the slab are raised and maintained at $100°C$. Calculate the time when the

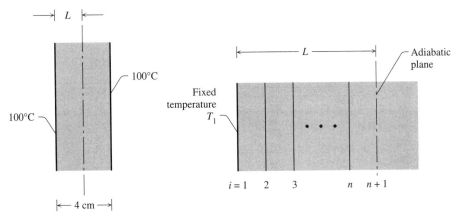

Figure E4.4(a)

temperature in the slab midplane reaches $10\,°C$. Use first (a) the explicit method and later (b) the implicit method, and investigate the effect of the space and time steps on the accuracy of your numerical results.

Solution. The symmetry of the boundary conditions about the vertical midplane makes it possible to focus on only the left half of the conducting slab. The left boundary of this L-thin half-slab is always at constant temperature, while its right boundary (the original midplane) is adiabatic.

Next, we divide the L thickness into n blades of thickness L/n, Fig. E4.4a. The temperature in each of the planes that define the blades is T_i^m, where $i = 1, 2, \ldots, n + 1$, and, in particular,

$$T_1^m = 100\,°C \qquad \text{for all } m \tag{1}$$

The initial temperature distribution is represented by the $m = 0$ values

$$T_i^0 = 0\,°C \qquad \text{for } i = 2, 3, \ldots, n+1 \tag{2}$$

a. *The explicit method*

There are $n - 1$ internal planes for which we write the explicit finite-difference equations (see Table 4.4)

$$T_i^{m+1} = \text{Fo}(T_{i-1}^m + T_{i+1}^m) + (1 - 2\text{Fo})T_i^m \qquad (i = 2, 3, \ldots, n) \tag{3}$$

The explicit equation for the adiabatic plane $i = n + 1$ is obtained by setting $\text{Bi} = 0$ in the second explicit case of Table 4.4:

$$T_{n+1}^{m+1} = 2\text{Fo}\,T_n^m + (1 - 2\text{Fo})T_{n+1}^m \tag{4}$$

In both eqs. (3) and (4), the stability constraint is $\text{Fo} \leq \frac{1}{2}$; therefore, it is safe to choose the value $\text{Fo} = 0.4$. The calculation of the temperatures for one complete time frame consists of applying eqs. (3) and then eq. (4), which is the same as sweeping the slab from $i = 2$ to $i = n + 1$. The sweep is then repeated to calculate the set of temperatures that belong to the next time frame.

In the case of a relatively coarse slicing of the slab, $n = 4$, the calculated temperatures turn out as follows:

Time Frame m	T_1 (°C)	T_2 (°C)	T_3 (°C)	T_4 (°C)	T_5 (°C)	t (s)
0	100	0	0	0	0	0
1	100	40	0	0	0	2
2	100	48	16	0	0	4
3	100	56	22.4	6.4	0	6
4	100	60.2	29.4	10.2	5.1	8
5	100	63.8	34.0	15.9	9.2	10
6	100	66.4	38.7	20.5	14.5	12

We terminated the calculation after the sixth frame because the midplane temperature (T_5) exceeded 10°C. The temperatures calculated in the first five time frames have been plotted on the left side of Fig. E4.4b.

The position of these time frames on the real time axis, $t(s)$, is determined by first calculating the time step Δt:

$$\Delta t = \text{Fo}\frac{(\Delta x)^2}{\alpha} = \text{Fo}\frac{L^2}{\alpha}\frac{1}{n^2}$$

$$= 0.4\frac{(2 \text{ cm})^2}{0.05 \text{ cm}^2/\text{s}}\frac{1}{n^2} = \frac{32 \text{ s}}{n^2} \tag{5}$$

In the case of the tabulated values ($n = 4$), the time step is $\Delta t = 32 \text{ s}/16 = 2$ s, hence the sequence of real times t listed in the rightmost column of the table. Interpolating linearly between the bottom two lines of the table, we conclude that $T_5 = 10$°C occurs at approximately $t \cong 10.29$ s.

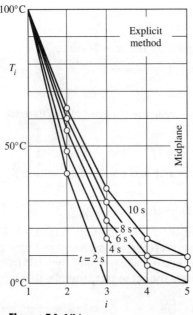

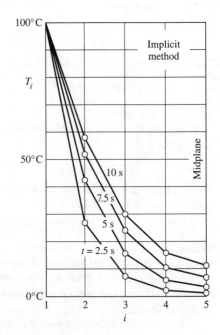

Figure E4.4(b)

A more refined solution can be obtained by dividing the slab into a larger number of thinner slices and by using a shorter time step. Equation (5) shows that both objectives (larger n, smaller Δt) can be achieved while holding the Fourier number fixed at Fo = 0.4. For example, when $n = 7$, the time step is $\Delta t = 0.65$ s. The next table illustrates the effect of a larger n (or smaller Δt) on the calculated time necessary for reaching a midplane temperature of 10°C:

n	Δt (s)	Time Steps Needed	t (s)
4	2	6	10.29
7	0.65	16	10.36
12	0.22	47	10.39

This second table was constructed by performing two more times the algorithm that generated the first table. In conclusion, despite the coarseness of the $n = 4$ grid, the first calculations provided a sufficiently accurate time answer.

b. *The implicit method*

With reference to the same division of the L-thin domain, we write the implicit equations for all the internal planes by pulling on the left side the unknown temperature of the plane of interest:

$$T_i^{m+1} = \frac{T_i^m}{1 + 2Fo} + \frac{Fo}{1 + 2Fo}(T_{i-1}^{m+1} + T_{i+1}^{m+1}) \qquad (i = 2,3, \ldots ,n) \tag{6}$$

The equation for the midplane ($i = n + 1$) is obtained by setting Bi = 0 in the implicit equation for a surface with convection in Table 4.4:

$$T_{n+1}^{m+1} = \frac{T_{n+1}^m}{1 + 2Fo} + \frac{2Fo}{1 + 2Fo} T_n^{m+1} \tag{7}$$

The Gauss–Seidel procedure consists of using eqs. (6) and (7) in the sequence $i = 2,3, \ldots ,n + 1$, while using the *most recent* temperature information in all the terms on the right side of these equations. The sweep through this stack of equations—the $(p + 1)^{th}$ iteration—is repeated until the temperature set of this latest iteration, $(T_i^{m+1})_{p+1}$, is almost the same as the set of values produced by the preceding iteration, $(T_i^{m+1})_p$. The iterations can be stopped on the basis of a convergence criterion of type (3.99); a more compact alternative is to use the "global" criterion

$$\left| \frac{S_{m+1}^{p+1} - S_{m+1}^p}{S_{m+1}^p} \right| < \epsilon' \tag{8}$$

where

$$S_{m+1}^p = \sum_{i=1}^{n+1} (T_i^{m+1})_p \tag{9}$$

The global criterion (8) is generally less stringent than the local convergence criterion that would have been written in the style of eq. (3.99). After one iteration has converged, the computation moves on to the next time frame by increasing the time index m by 1.

In the implicit method there is no upper bound on the value chosen for Fo. In the case of Fo = 0.5 and $n = 4$, which according to eq. (5) means $\Delta t = 2.5$ s, we obtain the following table of converged time frame solutions. The same results have been plotted on the right side of Fig. E4.4b. They were obtained by setting $\epsilon' = 10^{-4}$ in eq. (8). Interpolating between the last two lines of the table, we find that $T_5 = 10$°C occurs at $t = 9.25$ s.

Time Frame m	Iterations Needed	T_1 (°C)	T_2 (°C)	T_3 (°C)	T_4 (°C)	T_5 (°C)	t (s)
0		100	0	0	0	0	0
1	6	100	26.8	7.2	2.1	1.0	2.5
2	6	100	42.3	15.6	5.8	3.4	5.0
3	7	100	52.0	23.4	10.5	7.0	7.5
4	7	100	58.6	30.3	15.6	11.3	10.0

Now begins the *necessary* phase of convincing yourself (and others!) that the solution is sufficiently accurate. The solution accuracy depends on three numbers that can be chosen independently, n, Δt (or Fo), and ϵ'. Repeating solutions analogous to the one exhibited above, we find that the time t when $T_5 = 10°C$ is still sensitive to the step size Δt (all these calculations are based on $n = 4$):

ϵ'	$\Delta t = 2.5$ s (Fo = 0.5)	$\Delta t = 1$ s (Fo = 0.2)	$\Delta t = 0.5$ s (Fo = 0.1)
10^{-4}	$t = 9.25$ s	9.57 s	9.69 s
10^{-6}	$t = 9.25$ s	9.58 s	9.69 s

The accuracy improves and the solution becomes less sensitive as Δt (or Fo) becomes smaller. These tests show that $\epsilon' = 10^{-4}$ is already a small enough number to use in the convergence criterion (8).

Everything that we discussed until now must be repeated while using successively denser grids, to make sure that the final time estimate is also sufficiently insensitive to n. That n still has an influence on the $n = 4$ results exhibited until now can be expected, in view of the 7 percent discrepancy between the time estimates produced by the two methods (explicit vs. implicit) of this example. Indeed, if we set $n = 12$, Fo = 0.5 ($\Delta t = 0.28$ s), and $\epsilon' = 10^{-6}$, the implicit method yields the time answer $t = 10.29$ s, which is much closer to the value produced by the explicit method.

REFERENCES

1. H. S. Carslaw and J. C. Jaeger, *Conduction of Heat in Solids*, Oxford University Press, Oxford, 1959, p. 71.

2. M. P. Heisler, Temperature charts for induction and constant-temperature heating, *Trans. ASME*, Vol. 69, 1947, pp. 227–236.

3. W. H. McAdams, *Heat Transmission*, 2nd edition, McGraw-Hill, New York, 1942, Hottel charts, p. 36.

4. H. Gröber, *Einführung in die Lehre von der Wärmeübertragung*, Springer, Berlin, 1926.

5. H. Gröber, S. Erk, and U. Grigull, *Fundamentals of Heat Transfer*, McGraw-Hill, New York, 1961.

6. V. S. Arpaci, *Conduction Heat Transfer*, Addison-Wesley, Reading, MA, 1966, pp. 284, 288.

7. L. S. Langston, Heat transfer from multidimensional objects using one-dimensional solutions for heat loss, *Int. J. Heat Mass Transfer*, Vol. 25, 1982, pp. 149–150.

8. A. Bejan, *Convection Heat Transfer*, Wiley, New York, pp. 301–303.

9. J. Stefan, Über die Theorie der Eisbildung, insbesondere über die Eisbildung im Polarmeere, *Annalen Phys. Chem.*, Vol. 42, 1891, pp. 269–286.

10. G. Lamé and E. Clapeyron, Memoire sur la solidification par refroidissement d'un globe liquide, *Annales de Chimie et de Physique*, Vol. 47, 1831, pp. 250–256.

11. L. F. Milanez, Simplified relations for the phase change process in spherical geometry, *Int. J. Heat Mass Transfer*, Vol. 28, 1985, pp. 884–885.

12. L. G. Hector, Jr., A theory of growth instability during the early stages of metal casting, *Mechanics Research Communications*, Vol. 18, 1991, pp. 51–60.

13. A. Bejan, Z. Zhang, and J. H. Kim, Analytical advances on melting by natural convection, Keynote paper at the ICHMT International Symposium on Manufacturing and Materials Processing, Dubrovnik, Yugoslavia, August 27–31, 1990.

14. A. J. Fowler and A. Bejan, The effect of shrinkage on the cooking of meat, *Int. J. Heat Fluid Flow*, Vol. 12, 1991, pp. 375–383.

15. A. Bejan and C. L. Tien, Effect of axial conduction and metal-helium heat transfer on the local stability of superconducting composite media, *Cryogenics*, Vol. 18, 1978, pp. 433–441.

PROBLEMS

Lumped Capacitance Model

4.1 A sphere of carbon steel (0.5%C) with a diameter of 1 cm and an initial temperature of 100°C is exposed for 2 minutes to the atmosphere of temperature 10°C. The heat transfer coefficient between the spherical surface and the surrounding air flow is 20 W/m²·K.

   (a) Calculate the Biot number hr_o/k and the Fourier number $\alpha t/r_o^2$, and using Fig. 4.5 demonstrate that the cooling of the steel ball is described adequately by the lumped capacitance model.

   (b) Calculate the temperature of the steel ball at the end of 2 minutes of exposure to the atmosphere.

   (c) For the same point in time, estimate the order of magnitude of the temperature difference between the center and the surface of the steel ball. (*Hint*: Write that the scale of the convection heat flux at the surface is the same as the scale of the radial conduction heat flux through the ball.)

4.2 A solid conducting body of initial temperature $T_{1,0}$ is immersed suddenly in an amount of incompressible liquid of initial temperature $T_{2,0}$. The respective heat capacities of the immersed body and the liquid are $(mc)_1$ and $(mc)_2$, where m and c denote the respective masses and specific heats of the two entities. The external (wetted) area of the immersed body is A, and the heat transfer coefficient between the body and the liquid is h (constant).

Treating the immersed body and the surrounding liquid as two lumped capacitance systems, show that their respective temperatures vary according to the relations.

$$T_1(t) = T_{1,0} - \frac{T_{1,0} - T_{2,0}}{1 + [(mc)_1/(mc)_2]}(1 - e^{-nt})$$

$$T_2(t) = T_{2,0} + \frac{T_{1,0} + T_{2,0}}{1 + [(mc)_2/(mc)_1]}(1 - e^{-nt})$$

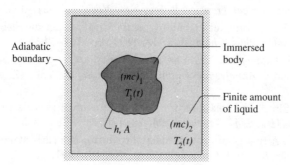

Figure P4.2

where

$$n = hA\frac{(mc)_1 + (mc)_2}{(mc)_1(mc)_2}$$

Note that what distinguishes this problem from the system treated in Section 4.2 is the finiteness of the liquid pool: because the amount of liquid is finite, the lumped liquid temperature T_2 varies under the influence of the heat transfer across the wetted surface A.

4.3   A small bottle containing 60°C-temperature milk is cooled by sudden immersion in a vessel containing 200 cm³ tap water at 10°C. The milk bottle and the amount of tap water can be modeled as two lumped capacitances, in accordance with the temperature history solutions listed in the preceding problem statement. The milk bottle is a cylinder of diameter $D = 4$ cm and height $H = 6$ cm. Its total mass and average specific heat are respectively $m_1 = 0.125$ kg and $c_1 = 4$ kJ/kg·K. The heat transfer coefficient between the bottle and the cold water is constant, $h = 350$ W/m²·K. The heat transfer between this ensemble and the atmosphere is negligible.

(a) Using the $T_1(t)$ solution listed in the preceding problem, calculate the temperature of the milk bottle after 1 minute of continuous immersion in cold water. Calculate also the corresponding water temperature T_2.

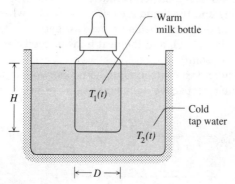

Figure P4.3

(b) As a simpler alternative, calculate the bottle temperature after 1 minute by making the (incorrect) assumption that the temperature of the surrounding water is fixed at $10°C$. Compare this new estimate with the T_1 value calculated in part (a).

4.4 A body of volume V, surface area A, density ρ, and specific heat c is initially at the temperature T_0. The body is immersed at $t = 0$ in a fluid reservoir of a higher temperature, $T_0 + \Delta T$. At the time $t = t_1$, it is removed from the hot fluid and plunged into a cold bath of temperature $T_0 - \Delta T$. Derive an expression for the time $t = t_2$ when the body temperature returns to its initial level T_0. Show that the time spent in the cold bath can never be greater than the time spent in the hot fluid.

4.5 During the daily cycle, the temperature of ambient air T_∞ varies in a way that is approximated by the sinusoid $T_\infty = T_0 + a \sin bt$. The average temperature T_0, and the oscillation amplitude a are known constants of the geographic region and annual season. The constant b is shorthand for $2\pi/\text{day}$.

Consider now the temperature history $T(t)$ of a body that is permanently exposed to $T_\infty(t)$. The body (e.g., a rock) is small enough to behave as a thermal capacitance (density ρ, specific heat c, volume V, surface area A). The heat transfer coefficient between A and the T_∞ air is the constant h.

Determine analytically the body temperature $T(t)$. Show that the body temperature oscillates sinusoidally, and that this oscillation lags behind the atmospheric temperature oscillation. Determine the time lag between the two sinusoids, and the amplitude of the body temperature oscillation.

Unidirectional Time-Dependent Conduction

4.6 The surface temperature history curve plotted on the right side of Fig. 4.4 can be approximated by using the simpler expression

$$\frac{T_0(t) - T_\infty}{T_i - T_\infty} = \frac{1}{1 + C(h/k)(\alpha t)^{1/2}}$$

where C is a constant approximately equal to 1. Show that this expression can be derived from the fluid contact boundary condition (4.47), by restating this condition as an equality between the respective scales of the two sides of the equal sign.

4.7 Consider the porcelain cup illustrated in Example 4.1 in the text, where the initial temperature of the porcelain wall is $T_i = 25°C$ and the liquid temperature is $T_\infty = 70°C$. The time when the hot liquid is poured into the cup is $t = 0$, and the location of the wetted surface of the porcelain wall is $x = 0$. Calculate the porcelain temperature at the time $t = 3$ s and at the distance $x = 2$ mm away from the wetted surface, using each of the following models:

(a) The porcelain wall is a semi-infinite conducting medium whose $x = 0$ surface assumes the temperature of the liquid, $T_0 = T_\infty = 70°C$.

(b) The wall is a semi-infinite medium whose $x = 0$ surface is in contact with the T_∞ liquid across a heat transfer coefficient $h = 1200$ W/m²·K, with the solution presented graphically in Fig. 4.4.

(c) The same model as in part (b), with the analytical solution given in eq. (4.48).

4.8 The surface of a brick wall of uniform initial temperature $T_i = 10°C$ absorbs solar thermal radiation at the rate $q'' = 100$ W/m², beginning with the time $t = 0$. The thermal conductivity and diffusivity of brickwork are $k = 0.5$ W/m·K and $\alpha = 0.005$ cm²/s. Calculate

(a) The surface temperature after 1 hour of exposure to the heat flux·q'', and

(b) The temperature after 1 hour at an interior point in the brick wall situated at $x = 10$ cm beneath the exposed surface.

4.9 Hot coffee of temperature $T_\infty = 90°C$ is poured into a cup whose wall is initially at the temperature $T_i = 10°C$. The porcelain wall of the cup is 0.5 cm thick. The heat transfer coefficient between the coffee and the wetted surface is $h = 700$ W/m$^2 \cdot$K. The external surface of the cup may be modeled as perfectly insulated.

Calculate the time that passes until the external surface of the cup becomes too hot to hold, for example, when it reaches the temperature of 50°C.

4.10 An isothermal icicle at the initial temperature $T_i = 0°C$ is placed inside a domestic freezer in which the air temperature is maintained at $T_\infty = -5°C$. The heat transfer coefficient between the ice surface and the surrounding air flow is $h = 10$ W/m$^2 \cdot$K. The icicle can be modeled as a long cylinder with a diameter of 2 cm.

Calculate the temperature reached by the icicle centerline after 10 minutes in the freezer. Perform this calculation in two ways: first, by reading Fig. 4.10; and later, by using the series solution for the time-dependent temperature in a long cylinder. Numerically, show that the first term of the series is a very good substitute for the sum of the entire series.

4.11 It is proposed to manufacture glass beads of diameter 0.5 mm by spraying them in 20°C air and allowing them to harden as they fall to the ground. Assume that the initial temperature of the glass bead is $T_i = 500°C$, the bead–air heat transfer coefficient is $h = 324$ W/m$^2 \cdot$K, and the constant ("terminal") downward velocity is $U = 3.8$ m/s. The glass properties are similar to those listed for window glass in Appendix B.

The design calls for the production of beads whose temperature in the center does not exceed 40°C. From what height should the beads be allowed to fall? Calculate the temperature of the bead surface when its center temperature reaches 40°C. Compare these two temperatures, and comment on whether the lumped capacitance model is applicable in the late stages of this cooling process.

4.12 The shape in which a potato is cut has an important effect on how fast each piece becomes cooked. Study this effect by considering three different shapes, each containing 5 grams of potato matter:

(a) Sphere.
(b) Cylinder with a length of 6 cm.
(c) Thin disc with a diameter of 4 cm.

Each piece is initially at the temperature $T_i = 30°C$. At the time $t = 0$, each piece is placed in a pot with water boiling at the temperature $T_\infty = 100°C$. The heat transfer coefficient is constant, $h = 2 \times 10^4$ W/m$^2 \cdot$K, and the properties of potato matter are approximately $\rho = 0.9$ g/cm^3, $k = 0.6$ W/m$\cdot$K, and $\alpha = 0.0017$ cm^2/s.

For each shape, calculate the time t until the volume-averaged temperature of the potato rises to 65°C. Compare the three times calculated in this manner, and comment on why only rarely potato is shaped as balls before cooking. (*Hint:* First, show that the Biot number is generally much greater than 1, and then rely exclusively on Fig. 4.16 to calculate the respective Fo values.)

4.13 The cooking of a beef steak can be modeled as time-dependent conduction in a slab. The thickness of the steak is 4 cm, its initial temperature is 25°C, and the properties of beef are $k = 0.4$ W/m$\cdot$K and $\alpha = 1.25 \times 10^{-7}$ m^2/s. The steak is cooked (with both sides exposed) in a convection oven of temperature 150°C. The heat transfer coefficient between meat and oven air is 60 W/m$^2 \cdot$K.

(a) The steak is cooked "well done" when its centerplane temperature reaches 80°C. Calculate the required cooking time by noting that the right side of eq. (4.62) is approximated well by the first term of the series.

(b) It is well known that meat shrinks during cooking. For example, the thickness of a steak cooked well done drops to about 75 percent of the original thickness. To see the extent to which the shrinking of meat shortens the cooking time [14], repeat part (a) by assuming that the steak thickness is equal to 3 cm throughout the cooking process.

(c) Is the real cooking time longer or shorter than the estimate obtained in part (b)?

4.14 Two solid blocks with different properties (see figure) and with the same initial temperature T_∞ are suddenly placed in contact and rubbed against each other. The friction between the two blocks generates the heat flux q'' in the interface. A fraction of this flux, q_1'', is absorbed by the upper block, while the remainder, q_2'', flows into the lower block. Determine the manner in which the properties of the two materials affect the ratio q_2''/q_1''. Derive the formulas needed for calculating q_1'', q_2'' and the interface temperature T_0.

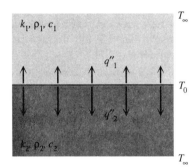

Figure P4.14

4.15 A 1-cm-thick slab of ice is pulled out of the freezer of a domestic refrigerator and exposed to room-temperature air. Its initial temperature is $-10°C$. The immediate heating felt by the ice slab causes melting at the surface. The superficial melting effect can be modeled by setting the ice surface temperature equal to $0°C$, which is equivalent to assuming an infinite Biot number in eq. (4.62).

Calculate the time that passes until the temperature in the centerplane of the slab rises from $-10°C$ to $-0.1°C$. For this, it is sufficient to use only the first term in the series of eq. (4.62).

4.16 A hot dog with a diameter of roughly 2 cm and initial temperature of $20°C$ is dropped into a pool of $95°C$ water. The heat transfer coefficient may be assumed to be constant and equal to $200 \ W/m^2 \cdot K$. The meat thermal conductivity and thermal diffusivity are approximately $0.4 \ W/m \cdot K$ and $0.0014 \ cm^2/s$.

(a) How long do we have to wait until the center of the hot dog warms up to $65°C$?

(b) Estimate the hot dog surface temperature at the time calculated in part (a).

4.17 The regenerative heat exchanger of a Stirling-cycle heat engine consists of a thick stack of stainless steel wire screens (gauzes). During each half-cycle, hot gas with a temperature of $600°C$ is forced to flow through the wire screens. The duration of this flow, or the exposure of the individual wire to hot gas, is 0.01 s. The heat transfer coefficient h between the wire surface and the hot gas varies inversely with the wire diameter D,

$$hD = 1\frac{W}{m \cdot K}$$

The thermodynamic mission of the wire screens is to store a large fraction of the energy carried by the hot gas, so that during the second half of the cycle that energy can be used to preheat a cold stream. The design is considered successful when during each blow the wire stores 90 percent or more of the maximum energy that it can store. How thin must the wire be to meet this requirement?

Concentrated Sources and Sinks

4.18   The purpose of this exercise is to demonstrate that a function of the exponential type shown in eq. (4.88) can be "expected" to be a solution of the conduction equation (4.87).

   (a) Assume that the unspecified function $f(x,t)$ satisfies the conduction equation (4.87). Show that if this is true, then the new function $g = \partial f/\partial x$ satisfies the conduction equation also.

   (b) Recall the solution for time-dependent conduction in a semi-infinite solid, eq. (4.41). Regard that solution as one example of an $f(x,t)$ function. Use the conclusion drawn in part (a), and report the form of the corresponding function $g(x,t)$. In this way you will arrive at the exponential-type expression (4.88), that is, you will expect that expression to be a solution of your equation (4.87).

4.19   Consider the temperature history in a constant-x plane parallel to the $x = 0$ plane of the instantaneous plane heat source Q''. Determine the time t when the temperature reaches a maximum in the constant-x plane. Derive also an expression for the maximum temperature at that location.

4.20   Consider the temperature history at a fixed distance r away from an instantaneous line heat source Q'. Determine the time t when the temperature becomes maximum at this location. Show that the maximum temperature at the radius r decreases as r^{-2} as r increases.

4.21   Examine the temperature history at a point P situated at a distance r away from the instantaneous point heat source Q. Determine the time t when the temperature reaches a maximum at point P. Report also the maximum temperature that is registered at this special moment.

4.22   At sufficiently long times, the temperature distribution around a continuous point source reaches the steady state described by eq. (4.107). Use this result to determine the shape factor S for steady conduction between an isothermal spherical object (radius r_1, temperature T_1) and the infinite conducting medium (k, T_∞) in which the object is buried. Compare your result with the appropriate S value deduced from Table 3.3.

4.23   The two-dimensional wake problem of Fig. 4.22 and eqs. (4.108) through (4.110) can be solved by replacing $T(x,y)$ with the new unknown $\theta(\eta)$, where

$$T(x,y) - T_\infty = \frac{q'/\rho c}{(U\alpha x)^{1/2}} \theta(\eta)$$

$$\eta = y\left(\frac{U}{\alpha x}\right)^{1/2}$$

Show that in this new notation the problem statement (4.108)–(4.110) reads

$$-\tfrac{1}{2}(\eta\theta)' = \theta''$$

$$\theta = 0 \quad \text{at} \quad \eta = \pm\infty$$

$$\int_{-\infty}^{\infty} \theta\, d\eta = 1$$

Solve this new problem for $\theta(\eta)$, and in this way prove the validity of eq. (4.111).

4.24   One of the proposed methods for the extraction of geothermal energy is the "hot dry rock" scheme illustrated in the figure. The rock is first fractured hydraulically to generate a crack through which cold high-pressure water will later be circulated. The crack space is approximated fairly well by a disc of diameter D which has a thickness much smaller than D. The two surfaces of this crack become the heat transfer area A between the hot rock and the cold water that circulates through the crack. The cold stream is pumped down through one well and returned to ground level through a parallel well.

Assume that the crack surface always has the same temperature as the cold water, namely, $T_c = 25\,^\circ C$. The temperature of the rock sufficiently far from the crack is $T_\infty = 200\,^\circ C$. The rock thermal properties are approximately the same as those of granite, $k = 2.9$ W/m·K and $\alpha = 0.012$ cm^2/s. The cold water circulates through the crack beginning with the time $t = 0$. The crack diameter is $D = 10$m.

(a) It is 10 days later. Calculate in an order-of-magnitude sense the thickness of the rock layer that has been affected (cooled) by the water flow. Verify that this thickness is considerably smaller than D. Calculate the total heat transfer rate from the rock medium to the water-cooled crack.

(b) After how many years will the cooled rock layer acquire a thickness of the same order as D? Assume that enough years have passed so that the cooled rock region can be modeled as a sphere of diameter D and temperature 25°C. Note that at long times the rock temperature distribution around this sphere becomes steady (time independent) [cf. eq. (4.107)]. Calculate the total heat transfer rate that "sinks" into the cooled region, and is removed steadily by the water stream.

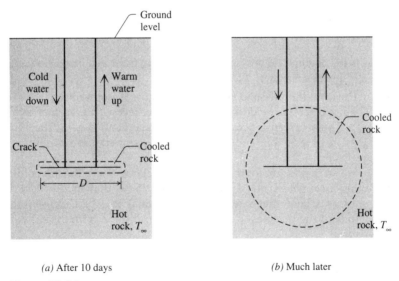

(a) After 10 days   (b) Much later

Figure P4.24

Melting and Solidification

4.25   The solid layer shown in the figure is initially at the melting point T_m. At a certain point in time ($t = 0$), the upper surface is heated to the temperature T_0, while the lower surface is cooled to the temperature T_L; in other words, $T_0 > T_m > T_L$. As the time increases, liquid is generated in a sublayer of thickness δ near the heated boundary.

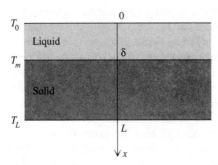

Figure P4.25

Determine the relationship between δ and time t, by assuming that the liquid and solid sublayers are sufficiently thin and the melting rate $d\delta/dt$ is sufficiently small so that temperature distributions in the two sublayers are nearly the same as in the case of steady unidirectional conduction. The liquid and solid phases have different thermal conductivities (k_f and k_s, respectively) and nearly the same density (ρ). Carry out this analysis in terms of the dimensionless front position and time,

$$X = \frac{\delta}{L} \qquad \tau = \frac{t}{\rho h_{sf} L^2 / [k_f(T_0 - T_m)]}$$

and the dimensionless degree of solid subcooling

$$B = \frac{k_s(T_m - T_L)}{k_f(T_0 - T_m)}$$

What is the asymptotic position of the melting front as t increases (i.e., what is the true steady-state value of X)?

4.26   The semi-infinite solid shown in the figure is isothermal at the melting point, T_m. The $x = 0$ surface is suddenly placed in contact with a warmer convective fluid, $T_\infty > T_m$. The heat transfer coefficient h is constant and known. The liquid layer that develops in time is thin enough so that the temperature distribution across it is always linear.

Derive an expression for the liquid film thickness δ as a function of the time t and the other parameters noted in the figure. Show that in the very beginning δ increases as t, and that later δ is proportional to $t^{1/2}$. Determine the order of magnitude of the thickness δ that marks the transition from the early regime to the late regime.

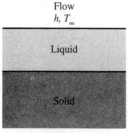
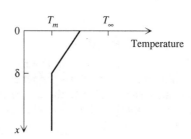

Figure P4.26

4.27 The chief question in the making of popsicles in the home freezer is "Dad, is it ready yet?" The finished popsicle is a bar of flavored ice frozen around a stick. The liquid is poured into a cavity that gives the frozen bar the dimensions shown in the figure. The bar is roughly two-dimensional, that is, sufficiently wide in the direction perpendicular to the figure. It has a slight taper to allow for the expansion during freezing, and to be pulled out of the cavity by means of the flat stick (tongue depressor) frozen in the middle.

Calculate the freezing time by assuming that all the liquid is originally at the freezing point. The properties of the frozen material are approximately equal to those of ice. The temperature of the cavity wall is maintained at $-5°C$.

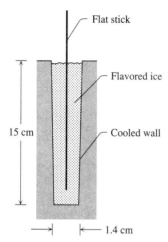

Figure P4.27

4.28 Fish is frozen by exposure to a strong flow of $-20°C$ air. Each fish rests on a 30-m-long perforated conveyor belt, which exposes the fish to cold air and dumps it frozen for packing and storage. Calculate the upper limit of the speed of the conveyor belt, that is, the speed that ensures the complete freezing of each fish. Model the fish as a 2-cm-thick slab with both sides maintained at $-20°C$. The original temperature of the fish is equal to the freezing point, which is $-2.2°C$. The approximate properties of frozen fish are $h_{sf} = 235$ kJ/kg, $c = 1.7$ kJ/kg·K, $k = 1$ W/m·K, and $\rho = 1000$ kg/m³.

Numerical Methods

4.29 With reference to the $\Delta x/2$-wide control volume shown in the last drawing of Table 4.4, derive the finite-difference equations for one-dimensional time-dependent conduction at a boundary with uniform heat flux q''.

4.30 The plane boundary of a two-dimensional conducting solid is heated by a certain (prescribed) distribution of heat flux $q''(y)$. Show that the appropriate finite-difference equations for the boundary control volume shaded in the figure are

Explicit:

$$T_{i,j}^{m+1} = (1 - 4\text{Fo})T_{i,j}^m + \text{Fo}(2T_{i-1,j}^m + T_{i,j+1}^m + T_{i,j-1}^m) + 2\text{Fo}\frac{q_j''\Delta x}{k}$$

Implicit:

$$(1 + 4\text{Fo})T_{i,j}^{m+1} + \text{Fo}(2T_{i-1,j}^{m+1} + T_{i,j+1}^{m+1} + T_{i,j-1}^{m+1}) = T_{i,j}^m + 2\text{Fo}\frac{q_j''\Delta x}{k}$$

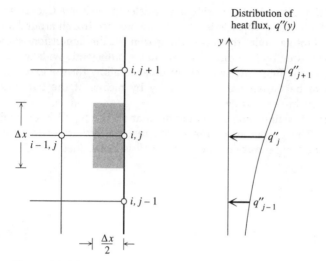

Figure P4.30

4.31 The initial temperature distribution in an infinite one-dimensional homogeneous and isotropic medium is illustrated in the figure. The material contained between the planes $x = -L$ and $x = L$ is at an initial temperature T_i higher than the rest of the medium, T_∞. Develop an implicit Gauss-Seidel iterative scheme for calculating the medium temperature history $T(x,t)$ for $t > 0$. Focus only on one half of the medium ($x > 0$), and note that $x = 0$ is a plane of symmetry.

(a) Express the conduction equation and the initial and boundary conditions in terms of the dimensionless variables

$$\theta = \frac{T - T_\infty}{T_i - T_\infty} \qquad \xi = \frac{x}{L} \qquad \tau = \frac{\alpha t}{L^2}$$

(b) The numerical $\theta(\xi,\tau)$ solution will depend on the fineness of your grid, which is represented by $\Delta\xi$ and $\Delta\tau$. As an accuracy test, estimate the time when θ reaches the value $\frac{2}{3}$ in the center plane ($\xi = 0$), and show how this time estimate depends on $\Delta\xi$ and $\Delta\tau$. Select appropriate values for $\Delta\xi$ and $\Delta\tau$.

(c) Report the time variation of temperature in the center plane, $\theta(0,\tau)$. Compare this result with the approximate estimate that can be derived from eq. (4.93) at $x = 0$, by modeling the 2L-thick region as an instantaneous plane heat source of strength $Q'' = 2L \, \rho c \, (T_i - T_\infty)$. How great must the time τ be for the approximate estimate to be adequate?

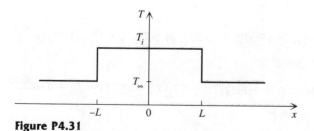

Figure P4.31

PROJECTS

4.1 Immersion Heating in a Flow with Time-Dependent Temperature

A slab with a thickness of 4 cm and initially at $0\,°C$ is immersed at the time $t = 0$ s in a flow whose free-stream temperature varies in accordance with the $T_\infty(t)$ curve shown in the figure. The heat transfer coefficient on the two faces of the slab is constant, $h = 10^3\,W/m^2 \cdot K$. The properties of the slab material are also constant, $\alpha = 0.05\ cm^2/s$ and $k = 1\,W/m \cdot K$.

Determine numerically the temperatures of the wetted surfaces and the mid-plane of the slab. Plot these temperatures as functions of time in the interval $0\ s \le t \le 90$ s. Comment on the way in which the maxima exhibited by these temperatures lag behind the peak of the $T_\infty(t)$ curve. Use Example 4.4 as a guide in setting up the numerical procedure.

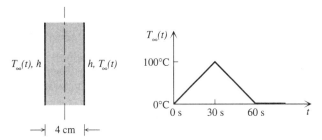

Figure PR4.1

4.2 Slab Immersed in a Flow: Dimensionless Temperature Distribution and History

The finite-difference equations developed in Section 4.8 have been expressed in terms of temperature (i.e., a certain number of degrees K or $°C$). The purpose of this

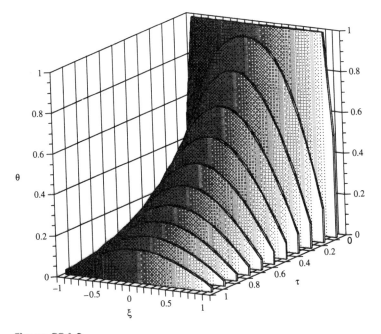

Figure PR4.2

project is to show how a numerical solution can be conducted in dimensionless form, to guarantee the *generality* of the calculated results.

Consider again the problem shown in Fig. 4.6 in the text, in which a slab of thickness $2L$ and initial temperature $T = T_i$ is exposed suddenly to a flowing fluid of temperature T_∞. The heat transfer coefficient on both surfaces of the slab is h. The value of this coefficient is such that the macroscopic Biot number hL/k is equal to 1.

Determine numerically the temperature inside the slab $\theta(\xi, \tau)$, where θ, ξ, and τ are all dimensionless:

$$\theta = \frac{T - T_\infty}{T_i - T_\infty} \qquad \xi = \frac{x}{L} \qquad \tau = \frac{\alpha t}{L^2}$$

The behavior of the dimensionless temperature is illustrated in the figure for a different macroscopic Biot number, $hL/k = 10$ (courtesy of Prof. R. M. Hochmuth, Duke University). Study the effect of grid fineness on the midplane and surface temperatures, to decide which grid is fine enough.

4.3 The Safe Discharge of Hot Ash from a Fluidized Bed Combustor

Hot ash of temperature $T_i = 900\,°C$ accumulates at the bottom of a fluidized bed combustor at the rate $\dot{m} = 500$ kg/h. The ash must be discharged and stored temporarily in a hopper for the purpose of lowering its temperature below $550\,°C$ before it can be shipped out in sacks. The cooling effect is due to combined convection and radiation at the upper surface of the ash pile through an "effective" heat transfer coefficient $h = 30\,W/m^2 \cdot K$. The atmospheric temperature is $T_\infty = 30\,°C$. The thermal properties of this particular ash are $k = 1\,W/m \cdot K$, $c = 1$ kJ/kg·K, and $\rho = 1900$ kg/m³.

The hopper has a depth $H = 1$ m and a horizontal cross-sectional area $A = 6$ m². Its bottom and side walls are very well insulated. The hopper is to be filled

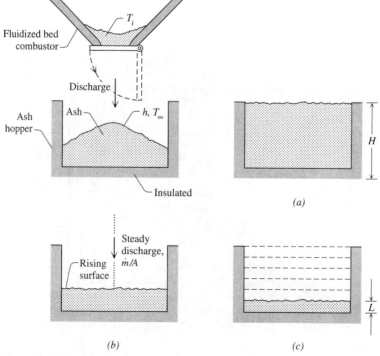

Figure PR4.3

intermittently, in a sequence of equal discharges that are spaced equally in time. As a designer of the safe operation of the ash hopper installation, you must determine how many (or how frequent) the discharges must be so that the temperature at the bottom of the ash pile settles to a level below the 550°C threshold.

Procedure. Assume that the top surface of the ash pile is always flat and that conduction proceeds in the vertical direction only. Before simulating the conduction in the ash numerically, it is wise to develop a feel for what goes on, especially for the required number of discharges. This can be accomplished in a few preliminary steps:

(a) Calculate the time t_1 needed for accumulating enough ash to fill the hopper once. Assume that the hopper is filled in one large discharge at $t = 0$, and calculate the temperature at the bottom of the ash pile at $t = t_1$. Is this a safe temperature, or do you recommend smaller and more frequent discharges?

(b) Consider the opposite limit in which the discharges are so many and so small that the filling of the hopper is steady, with a flowrate $\dot{m}$. Calculate the temperature of the rising top surface of the ash pile, and recognize that this temperature prevails all the way to the bottom of the pile. Is this a safe temperature?

(c) Let n be the number of discharges, which are spaced $t_n = t_1/n$ apart. Each of these adds a fresh layer of thickness $L = H/n$ to the pile. The first layer falls at $t = 0$. Calculate the bottom temperature of this first layer at $t = t_n$, and think of how this estimate might compare with the final bottom temperature of the full hopper (i.e., the temperature at $t = t_1$). Perform this calculation for several values of n, and obtain a rough estimate for the smallest number of discharges ($n_{\min}$) that guarantees temperatures lower than 550°C at the bottom of the ash pile.

(d) Finally, to verify the correctness of your $n_{\min}$ estimate, set $n = n_{\min}$ and simulate the temperature history in the ash pile numerically. Begin with the unidirectional time-dependent conduction in the very first layer (Fig. PR 4.3c), which lasts from $t = 0$ until $t = t_n$. Divide this L-thick layer into enough sublayers so that the numerical solution is sufficiently accurate. Rely on an appropriate set of finite-difference formulas from Table 4.4 (explicit or implicit), and through an accuracy test demonstrate that the space and time grid is sufficiently fine.

Immediately after $t = t_n$, the second discharge adds a new isothermal (T_i) layer on top of the first layer. Continue the numerical simulation (now in a $2L$-deep layer) until $t = 2t_n$, and report the evolution of the ash temperature at the bottom of the pile. If the bottom temperature exhibits significant variations, extend your numerical simulation into the next time interval, $2t_n < t < 3t_n$, when the depth of the ash pile is $3L$. Compare the asymptotic value of the bottom temperature with the quick estimate obtained in part (c) of the project.

4.4 Superconductor Local Stability Due to Longitudinal Conduction

The remarkable property of a *superconductor* is that its electrical resistivity vanishes below a critical temperature T_c. At temperatures immediately above T_c, the resistivity jumps to a finite level as the conductor returns to its *normal* (i.e., resistive) state. A very large current that at supercritical temperatures ($T > T_c$) would cause thermal runaway and burnup (Section 2.6.2) is carried without any resistive heating at subcritical temperatures ($T < T_c$).

Unfortunately, small disturbances (mechanical, magnetic) can locally raise the temperature to supercritical levels, $T_i > T_c$. This possibility leads to the important safety problem of making sure that, in time, the local temperature drops below T_c. The challenge then is to design a superconductor that is locally *stable*, so that the postulated normal (disturbed) zone disappears, and the superconducting state is reestablished.

One method of forcing the normal zone to collapse is by cooling it by axial

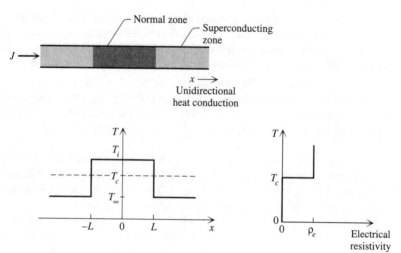

Figure PR4.4

conduction, using the surrounding cold (still superconducting) zones as heat sinks [15]. To greatly increase the effective thermal conductivity for longitudinal conduction, k, the superconducting material is drawn as a composite of pure superconducting strands embedded in a high purity copper matrix. The steady-state operating temperature of the composite as a superconductor is T_∞ (note that $T_\infty < T_c$).

Assume that the disturbance (hot spot) has the instantaneous temperature T_i and length $2L$. The operating current density of the composite conductor is J [amperes/m²], and the effective electric resistivity of the composite when $T > T_c$ is the constant ρ_e.

The study of instantaneous plane heat sources (Section 4.6.1.) shows that in the absence of steady volumetric heating ($J = 0$ in this case) any temperature bump T_i will decay eventually to the background level, T_∞. As J increases, however, one crosses a critical threshold J_c that distinguishes between stable disturbances (collapsing normal zones, $J < J_c$), and unstable disturbances (expanding normal zones, $J > J_c$). The objective of this project is to determine the critical current density (J_c) for a specified instantaneous normal zone (T_i, L). The behavior of the normal zone can be monitored by looking at the time history of the temperature in the plane of symmetry ($x = 0$, also the hottest plane). The normal zone disappears when the center plane temperature drops below T_c.

It is advisable to formulate the problem in terms of the following dimensionless variables,

$$\xi = \frac{x}{L} \qquad \tau = \frac{\alpha t}{L^2} \qquad \theta = \frac{T - T_\infty}{T_c - T_\infty}$$

$$B = \frac{T_i - T_\infty}{T_c - T_\infty} \qquad Q = \frac{\rho_e J^2 L^2}{k(T_c - T_\infty)}$$

and to note that the finite-difference equations of Table 4.4 must be expanded to include the effect of volumetric Joule heating ($\dot{q} = \rho_e J^2$) at those locations x where $T > T_c$. Develop an algorithm based on the implicit Gauss-Seidel iterative scheme. Confine your calculations to the right half of the unidirectional conductor, from $x = 0$ to a large enough x where $T \cong T_\infty$ at every time step.

Perform an accuracy test for selecting the proper steps $\Delta\xi$ and $\Delta\tau$. For the severity of the initial temperature disturbance select a B value between 1 and 2. Report the critical dimensionless volumetric heating rate Q_c that corresponds to the chosen B. Other students may solve the same problem for several other B values, in such a way that the class generates the curve $Q_c = Q_c(B)$. This curve serves as boundary between stable and unstable operating conditions.

5

EXTERNAL FORCED CONVECTION

5.1 CLASSIFICATION OF CONVECTION CONFIGURATIONS

Beginning with this chapter we turn our attention to the heat transfer mechanism called "convection," that is, the transport mechanism made possible by the motion of a fluid. Convection is the second of the three basic heat transfer mechanisms reviewed in Sections 1.3 through 1.5. What distinguishes a convection configuration from the conduction configurations studied in Chapters 2 through 4 is the fact that the "convective" heat transfer medium (e.g., fluid) is in motion.* Or, looking back at the mechanism of conduction, we can say that conduction is the heat transfer mechanism that survives when the motion ceases throughout what once was a convective medium.

In Section 1.4 we learned also that the fundamental problem in convection heat transfer consists in determining the relationship between the heat flux through the solid wall (q'') and the temperature difference ($T_w - T_\infty$) between the wall and the fluid flow that bathes it. Alternatively, this basic problem amounts to determining the convective heat transfer coefficient h (W/m²·K), the external-flow definition of which is worth repeating here:

$$h = \frac{q''}{T_w - T_\infty} \tag{5.1}$$

In the infinitesimally thin fluid layer that sticks to the solid wall, the transversal heat flux q'' is due to pure conduction because the fluid in that layer is practically motionless. In Section 1.4 this observation led to the Fourier law, eq. (1.56), in which y is the transversal distance measured away from the solid surface (toward the fluid) and k is the thermal conductivity of the fluid. The

---

*This important feature is stressed by the term "convection," which originates from the Latin verbs *convecto-are* and *convĕho-vĕhĕre* (to bring together, to carry into one place).

alternative definition of h that follows from eqs. (5.1) and (1.56) is eq. (1.57), namely,

$$h = -\frac{k}{T_w - T_\infty}\left(\frac{\partial T}{\partial y}\right)_{y = 0^+} \tag{5.2}$$

This expression shows that to determine h, we must first determine the temperature distribution in the thin fluid layer that coats the wall.

Equation (5.2) serves as introduction and objective for the convection heat transfer analyses that we are about to develop. One idea to recognize at this stage is that the flow and heat transfer configurations that can be categorized as "convective" are extremely diverse. It pays, then, to organize these configurations into a structure, or a small number of important classes of convection "problems." Figure 5.1 outlines one classification that distinguishes first between *external-flow* and *internal-flow* convection problems. These two classes have been defined already with the help of Figs. 1.10 and 1.11.

Figure 5.1 differentiates also between *forced-convection* and *free-convection* (*natural-convection*) configurations, that is, between the origins of the motion

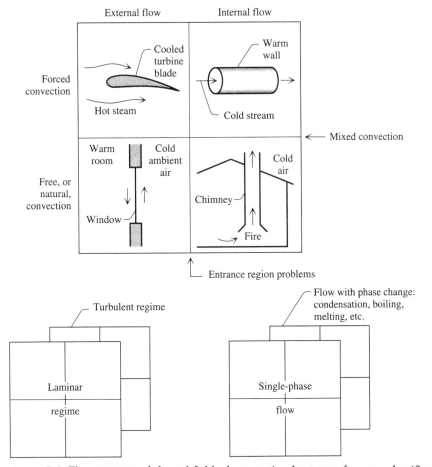

Figure 5.1 The structure of the subfield of convection heat transfer, or a classification of the main convection configurations.

experienced by the convective medium. The flow is "forced" when another (a different) entity pushes the fluid past the solid body against which the fluid rubs. In the case of internal forced convection (e.g., pipe flow), the external device that forces the flow is the pump or compressor. In an external forced-convection configuration such as the flow parallel to a plate fin in a compact heat exchanger, the different entity is again the pump or compressor that pushes the entire fluid stream through the heat exchanger.

In "free" or natural convection, the fluid motion occurs "by itself," without the assistance of an external mechanism. In Chapter 7, for example, all the flows are due to the effect of relative buoyancy (lightness, or heaviness) of the various regions of the flow field. It turns out that this motion-driving mechanism is intimately coupled to the heating or cooling effect experienced by the fluid because, generally speaking, fluids expand upon heating at constant pressure.

The ordering of convection heat transfer configurations is even more complicated than what is listed in the four-option menu shown in the upper part of Fig. 5.1. At the interface between forced convection and natural convection, we encounter a diversity of *mixed-convection* configurations. In a mixed-convection problem, the flow is driven simultaneously by an external device (e.g., pump) and by a built-in effect such as that of buoyancy.

Furthermore, at the interface between external-flow and internal-flow configurations, there is also a group of in-between problems the configurations of which are neither external flow nor internal flow. One example of this kind is in the entrance (mouth) region of the flow into a duct (see Section 6.1.1).

The convection classification scheme of Fig. 5.1 continues on at least two additional planes. One of these is represented by the transition from the laminar flow regime to the turbulent flow regime, that is, from a smooth flow to one with complicated fluctuations that are dominated by eddies of several sizes (Section 5.4). The heat transfer characteristics of the flow change dramatically from the laminar regime to the turbulent regime, and vice versa. Some problems are complicated further by the simultaneous presence of laminar and turbulent flow regions in the same flow field [e.g., eq. (5.132)].

Another direction in which convection problems exhibit still another degree of diversification is illustrated in the lower right-hand corner of Fig. 5.1. The eight flow categories identified until now (external vs. internal, forced vs. free, laminar vs. turbulent) were discussed in the context of a single-phase fluid. For example, a stream of nitrogen gas in a counterflow heat exchanger and the buoyancy-driven flow of air on the room side of a single-pane window remain single-phase (i.e., gaseous). The same flow categories are found also when the convective medium undergoes a change of phase. In the film condensation of steam on a vertical cold surface, for example, the flow consists of two distinct regions: a liquid (water) film attached to the wall and a buoyancy-driven steam flow on the outer side of the water layer (Section 8.1.1).

With these diverse configurations in mind, we begin a review of the three principles that govern the flow of fluid and the transport of energy through a convective medium. The purpose of the equations recommended by these principles is to allow us to determine sequentially:

(a) The flow field,
(b) The temperature field—in particular, the temperature distribution near the solid wall—and, finally,

(c) The heat transfer coefficient, or the relationship between wall heat flux and wall–fluid temperature difference.

In *forced-convection* problems, an auxiliary result made possible by the same analysis is the *friction force* between the flow and the solid wall. This additional result is needed to estimate the total power required by the external mechanism (e.g., pump) that forces the fluid to flow.

5.2 BASIC PRINCIPLES OF CONVECTION

5.2.1 The Mass Conservation Equation

Consider the two-dimensional flow geometry illustrated in Fig. 5.2, where the fluid velocity at any point in the flow field is described by the local velocity components u and v. These components are, in general, functions of location (x,y) and time (t). An important relation between the spatial changes in u and v is dictated by the principle of mass conservation. From engineering thermodynamics, we recall that the conservation of mass in a control volume of instantaneous mass inventory m requires (see, for example, Ref. [1], p. 28) that

$$\frac{\partial m}{\partial t} = \sum_{\text{in}} \dot{m} - \sum_{\text{out}} \dot{m} \tag{5.3}$$

The mass conservation statement that applies to the (u,v) flow of Fig. 5.2 can be obtained by analyzing the infinitesimal control volume $\Delta x\, \Delta y$ enlarged to the right of the figure. The left and bottom faces of this control volume are inlet ports in the sense of eq. (5.3), whereas the top and right faces are outlet ports. Applied to the $\Delta x\, \Delta y$ system, eq. (5.3) states that

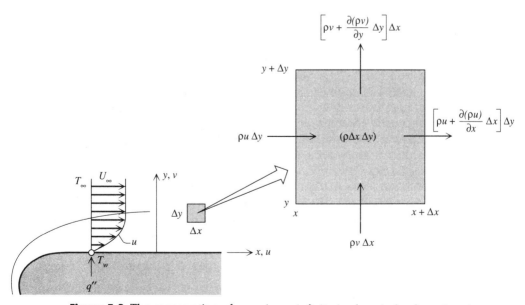

Figure 5.2 The conservation of mass in an infinitesimal control volume in a two-dimensional flow field.

$$\frac{\partial}{\partial t}(\rho \Delta x \Delta y) = \rho u\,\Delta y + \rho v\,\Delta x - \left[\rho u + \frac{\partial(\rho u)}{\partial x}\Delta x\right]\Delta y - \left[\rho v + \frac{\partial(\rho v)}{\partial y}\Delta y\right]\Delta x \quad (5.4)$$

where ρ is the local density of the fluid. Dividing eq. (5.4) by the constant-size product $\Delta x\,\Delta y$, we obtain

$$\frac{\partial \rho}{\partial t} + \frac{\partial(\rho u)}{\partial x} + \frac{\partial(\rho v)}{\partial y} = 0 \quad (5.5)$$

Of particular interest in the present course are the convection problems in which the fluid density may be regarded as (*approximated* by) a constant (ρ). According to the *nearly constant-density model*, then, the spatial and temporal variations in density are neglected relative to those of u and v, and eq. (5.5) reduces to

$$\frac{\partial u}{\partial x} + \frac{\partial v}{\partial y} = 0 \quad (5.6)$$

This is the *mass conservation* equation for two-dimensional constant-density flow in Cartesian coordinates. The same result is recognized also as the *continuity equation*, because of the continuity* of mass during its flow through the control volume $\Delta x\,\Delta y$. The three-dimensional counterpart of this equation is listed in Table 5.1. The corresponding mass conservation equations for constant-density flow in cylindrical and spherical coordinates are listed in Tables 5.2 and 5.3.

5.2.2 The Momentum Equations

The next principle we invoke is the *momentum principle*, or the *momentum theorem* encountered in the analysis of propulsion systems in engineering thermodynamics [2]:

$$\frac{\partial}{\partial t}(mv_n) = \sum F_n + \sum_{in} \dot{m}v_n - \sum_{out} \dot{m}v_n \quad (5.7)$$

In this equation, n is the particular direction for which eq. (5.7) has been written, and v_n and F_n are the projections of the fluid velocities and forces on the n direction. The product mv_n is the instantaneous inventory of n-direction momentum of the control volume. The terms labeled F_n account for the forces that act on the control volume. The terms of type $\dot{m}v_n$ represent the flow of n-direction momentum into and out of the control volume. In summary, eq. (5.7) is the control volume restatement of *Newton's second law of motion*.

Consider now the implications of eq. (5.7) with respect to the flow through the control volume $\Delta x\,\Delta y$ of Fig. 5.2. This control volume has been enlarged twice more in Fig. 5.3. There are two directions, x and y, in which we can invoke eq. (5.7). We begin with the x direction and list on Fig. 5.3 all the momentum flows and forces that act on the control volume $\Delta x\,\Delta y$.

The upper drawing shows the impact and reaction effects due to the flow of momentum into and out of the control volume. For example, the x momentum

---

*In this context, continuity means the absence of mass sources or sinks in the flow field.

Table 5.1 The Governing Equations for Nearly Constant-Property (ρ, μ, k) Fluid Flow and Convection in Cartesian Coordinates

Mass Conservation

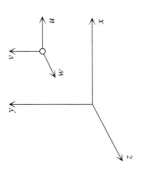

$$(m) \qquad \frac{\partial u}{\partial x} + \frac{\partial v}{\partial y} + \frac{\partial w}{\partial z} = 0$$

Momentum Equations

$$(M_x) \qquad \rho\left(\frac{\partial u}{\partial t} + u\frac{\partial u}{\partial x} + v\frac{\partial u}{\partial y} + w\frac{\partial u}{\partial z}\right) = -\frac{\partial P}{\partial x} + \mu\left(\frac{\partial^2 u}{\partial x^2} + \frac{\partial^2 u}{\partial y^2} + \frac{\partial^2 u}{\partial z^2}\right) + X$$

$$(M_y) \qquad \rho\left(\frac{\partial v}{\partial t} + u\frac{\partial v}{\partial x} + v\frac{\partial v}{\partial y} + w\frac{\partial v}{\partial z}\right) = -\frac{\partial P}{\partial y} + \mu\left(\frac{\partial^2 v}{\partial x^2} + \frac{\partial^2 v}{\partial y^2} + \frac{\partial^2 v}{\partial z^2}\right) + Y$$

$$(M_z) \qquad \rho\left(\frac{\partial w}{\partial t} + u\frac{\partial w}{\partial x} + v\frac{\partial w}{\partial y} + w\frac{\partial w}{\partial z}\right) = -\frac{\partial P}{\partial z} + \mu\left(\frac{\partial^2 w}{\partial x^2} + \frac{\partial^2 w}{\partial y^2} + \frac{\partial^2 w}{\partial z^2}\right) + Z$$

Energy Equation

$$(E) \qquad \rho c_P\left(\frac{\partial T}{\partial t} + u\frac{\partial T}{\partial x} + v\frac{\partial T}{\partial y} + w\frac{\partial T}{\partial z}\right) = k\left(\frac{\partial^2 T}{\partial x^2} + \frac{\partial^2 T}{\partial y^2} + \frac{\partial^2 T}{\partial z^2}\right) + \dot{q} + \mu\Phi$$

where

$$\Phi = 2\left[\left(\frac{\partial u}{\partial x}\right)^2 + \left(\frac{\partial v}{\partial y}\right)^2 + \left(\frac{\partial w}{\partial z}\right)^2\right] + \left(\frac{\partial u}{\partial y} + \frac{\partial v}{\partial x}\right)^2 + \left(\frac{\partial v}{\partial z} + \frac{\partial w}{\partial y}\right)^2 + \left(\frac{\partial w}{\partial x} + \frac{\partial u}{\partial z}\right)^2$$

Table 5.2 The Governing Equations for Nearly Constant-Property (ρ, μ, k) Flow and Convection in Cylindrical Coordinates

Mass Conservation

(m)
$$\frac{\partial v_r}{\partial r} + \frac{v_r}{r} + \frac{1}{r}\frac{\partial v_\theta}{\partial \theta} + \frac{\partial v_z}{\partial z} = 0$$

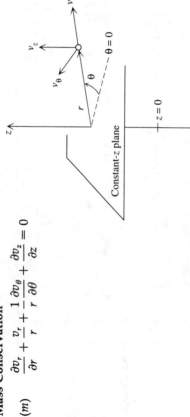

Constant-z plane

$\theta = 0$

$z = 0$

Momentum Equations

(M_r)
$$\rho\left(\frac{\partial v_r}{\partial t} + v_r\frac{\partial v_r}{\partial r} + \frac{v_\theta}{r}\frac{\partial v_r}{\partial \theta} - \frac{v_\theta^2}{r} + v_z\frac{\partial v_r}{\partial z}\right) = -\frac{\partial P}{\partial r} + \mu\left(\frac{\partial^2 v_r}{\partial r^2} + \frac{1}{r}\frac{\partial v_r}{\partial r} - \frac{v_r}{r^2} + \frac{1}{r^2}\frac{\partial^2 v_r}{\partial \theta^2} - \frac{2}{r^2}\frac{\partial v_\theta}{\partial \theta} + \frac{\partial^2 v_r}{\partial z^2}\right) + F_r$$

(M_θ)
$$\rho\left(\frac{\partial v_\theta}{\partial t} + v_r\frac{\partial v_\theta}{\partial r} + \frac{v_\theta}{r}\frac{\partial v_\theta}{\partial \theta} + \frac{v_r v_\theta}{r} + v_z\frac{\partial v_\theta}{\partial z}\right) = -\frac{1}{r}\frac{\partial P}{\partial \theta} + \mu\left(\frac{\partial^2 v_\theta}{\partial r^2} + \frac{1}{r}\frac{\partial v_\theta}{\partial r} - \frac{v_\theta}{r^2} + \frac{1}{r^2}\frac{\partial^2 v_\theta}{\partial \theta^2} + \frac{2}{r^2}\frac{\partial v_r}{\partial \theta} + \frac{\partial^2 v_\theta}{\partial z^2}\right) + F_\theta$$

(M_z)
$$\rho\left(\frac{\partial v_z}{\partial t} + v_r\frac{\partial v_z}{\partial r} + \frac{v_\theta}{r}\frac{\partial v_z}{\partial \theta} + v_z\frac{\partial v_z}{\partial z}\right) = -\frac{\partial P}{\partial z} + \mu\left(\frac{\partial^2 v_z}{\partial r^2} + \frac{1}{r}\frac{\partial v_z}{\partial r} + \frac{1}{r^2}\frac{\partial^2 v_z}{\partial \theta^2} + \frac{\partial^2 v_z}{\partial z^2}\right) + F_z$$

Energy Equation

(E)
$$\rho c_P\left(\frac{\partial T}{\partial t} + v_r\frac{\partial T}{\partial r} + \frac{v_\theta}{r}\frac{\partial T}{\partial \theta} + v_z\frac{\partial T}{\partial z}\right) = k\left[\frac{1}{r}\frac{\partial}{\partial r}\left(r\frac{\partial T}{\partial r}\right) + \frac{1}{r^2}\frac{\partial^2 T}{\partial \theta^2} + \frac{\partial^2 T}{\partial z^2}\right] + \dot{q} + \mu\Phi$$

where

$$\Phi = 2\left[\left(\frac{\partial v_r}{\partial r}\right)^2 + \left(\frac{1}{r}\frac{\partial v_\theta}{\partial \theta} + \frac{v_r}{r}\right)^2 + \left(\frac{\partial v_z}{\partial z}\right)^2\right] + \left(\frac{\partial v_\theta}{\partial r} - \frac{v_\theta}{r} + \frac{1}{r}\frac{\partial v_r}{\partial \theta}\right)^2 + \left(\frac{1}{r}\frac{\partial v_z}{\partial \theta} + \frac{\partial v_\theta}{\partial z}\right)^2 + \left(\frac{\partial v_r}{\partial z} + \frac{\partial v_z}{\partial r}\right)^2$$

that enters through the left side of the control volume is $\rho u^2 \, \Delta y$, where $\rho u \, \Delta y$ is the stream flowrate and u is the velocity of that stream in the x direction. On the other hand, entering through the bottom side of the control volume we see a stream of flowrate $\rho v \, \Delta x$, whereas the x velocity of the fluid carried by that stream is u. Therefore, the x-momentum flowrate that enters through the bottom side is $(\rho v \, \Delta x)u$.

The bottom right drawing of Fig. 5.3 shows the forces that act on the control volume, namely, the forces due to the normal stress σ_{xx}, the tangential stress τ_{xy}, and the per-unit-volume body force in the x direction, X. Note that the forces due to the normal and tangential stresses act on, and are proportional to, the respective surfaces of the control volume. The body force associated with X is proportional to the volume of the control volume. An example of body force is the buoyancy effect that serves as driving mechanism in all the natural convection configurations of Chapter 7.

It is a simple matter to add together all the "arrows" indicated on the two frames of Fig. 5.3, in other words, to substitute the indicated terms in their appropriate places in the momentum theorem, eq. (5.7), and to take into account eq. (5.6). Dividing the resulting equation by $\Delta x \, \Delta y$ and invoking the limit $(\Delta x \, \Delta y) \to 0$, we obtain [3]

$$\rho\left(\frac{\partial u}{\partial t} + u\frac{\partial u}{\partial x} + v\frac{\partial u}{\partial y} \right) = -\frac{\partial \sigma_{xx}}{\partial x} + \frac{\partial \tau_{xy}}{\partial y} + X \tag{5.8}$$

This equation holds at every point (x,y) in the flow field. The three terms on the left side represent the effects of x-momentum accumulation and flow through the point (x,y). The three terms on the right side represent, in order, the effects of the normal, shear, and body forces.

The next step in the development of the momentum theorem for the x direction is the empirical observation that a fluid packet can be deformed without resistance if, regardless of size, the deformation occurs infinitely slowly. The fluid packet poses an increasing resistance as it is being deformed faster. It is said that a fluid packet offers no resistance to a finite-size (infinitely slow) change of shape, and that, instead, it resists the *time rate* of the given change of shape. This observation is the basis for a set of constitutive relations that we borrow from the field of fluid mechanics [4]:

$$\sigma_{xx} - P = -2\mu \, \frac{\partial u}{\partial x} + \frac{2}{3}\mu\left(\frac{\partial u}{\partial x} + \frac{\partial v}{\partial y} \right) \tag{5.9}$$

$$\tau_{xy} = \mu\left(\frac{\partial u}{\partial y} + \frac{\partial v}{\partial x} \right) \tag{5.10}$$

These equations relate the stresses to the local velocity gradients. On the left side of eq. (5.9), the group $\sigma_{xx} - P$ is the "excess" normal stress, that is, the normal stress above the level of the local pressure P.

The coefficient μ is the *viscosity* of the fluid. The value of this physical property can be measured using eq. (5.10): its units are $N \cdot s/m^2$, or $kg/m \cdot s$. Equation (5.10) states that the shear stress is proportional to the rate of angular deformation of the fluid packet: fluids that obey this proportionality are called *Newtonian*. All the analyses reported in the convection chapters of this course

Table 5.3 The Governing Equations for Nearly Constant-Property (ρ, μ, k) Flow and Convection in Spherical Coordinates

Mass Conservation

$$(m) \quad \frac{1}{r}\frac{\partial}{\partial r}(r^2 v_r) + \frac{1}{\sin\phi}\frac{\partial}{\partial \phi}(v_\phi \sin\phi) + \frac{1}{\sin\phi}\frac{\partial v_\theta}{\partial \theta} = 0$$

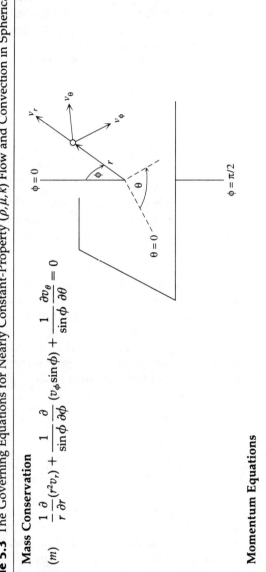

Momentum Equations

$$(M_r) \quad \rho\left(\frac{Dv_r}{Dt} - \frac{v_\phi^2 + v_\theta^2}{r}\right) = -\frac{\partial P}{\partial r} + \mu\left(\nabla^2 v_r - \frac{2v_r}{r^2} - \frac{2}{r^2}\frac{\partial v_\phi}{\partial \phi} - \frac{2v_\phi \cot\phi}{r^2} - \frac{2}{r^2 \sin\phi}\frac{\partial v_\theta}{\partial \theta}\right) + F_r$$

(M_ϕ) $\quad \rho\left(\dfrac{Dv_\phi}{Dt} + \dfrac{v_r v_\phi - v_\theta^2\cot\phi}{r}\right) = -\dfrac{1}{r}\dfrac{\partial P}{\partial\phi} + \mu\left(\nabla^2 v_\phi + \dfrac{2}{r^2}\dfrac{\partial v_r}{\partial\phi} - \dfrac{v_\phi}{r^2\sin^2\phi} - \dfrac{2\cos\phi}{r^2\sin^2\phi}\dfrac{\partial v_\theta}{\partial\theta}\right) + F_\phi$

(M_θ) $\quad \rho\left(\dfrac{Dv_\theta}{Dt} + \dfrac{v_\theta v_r + v_\phi v_\theta\cot\phi}{r}\right) = -\dfrac{1}{r\sin\phi}\dfrac{\partial P}{\partial\theta} + \mu\left(\nabla^2 v_\theta - \dfrac{v_\theta}{r^2\sin^2\phi} + \dfrac{2}{r^2\sin\phi}\dfrac{\partial v_r}{\partial\theta} + \dfrac{2\cos\phi}{r^2\sin^2\phi}\dfrac{\partial v_\phi}{\partial\theta}\right) + F_\theta$

where

$$\frac{D}{Dt} = \frac{\partial}{\partial t} + v_r\frac{\partial}{\partial r} + \frac{v_\phi}{r}\frac{\partial}{\partial\phi} + \frac{v_\theta}{r\sin\phi}\frac{\partial}{\partial\theta}$$

$$\nabla^2 = \frac{1}{r^2}\frac{\partial}{\partial r}\left(r^2\frac{\partial}{\partial r}\right) + \frac{1}{r^2\sin\phi}\frac{\partial}{\partial\phi}\left(\sin\phi\frac{\partial}{\partial\phi}\right) + \frac{1}{r^2\sin^2\phi}\frac{\partial^2}{\partial\theta^2}$$

Energy Equation

(E) $\quad \rho c_p\left(\dfrac{\partial T}{\partial t} + v_r\dfrac{\partial T}{\partial r} + \dfrac{v_\phi}{r}\dfrac{\partial T}{\partial\phi} + \dfrac{v_\theta}{r\sin\phi}\dfrac{\partial T}{\partial\theta}\right) = k\left[\dfrac{1}{r^2}\dfrac{\partial}{\partial r}\left(r^2\dfrac{\partial T}{\partial r}\right) + \dfrac{1}{r^2\sin\phi}\dfrac{\partial}{\partial\phi}\left(\sin\phi\,\dfrac{\partial T}{\partial\phi}\right) + \dfrac{1}{r^2\sin^2\phi}\dfrac{\partial^2 T}{\partial\theta^2}\right] + \dot{q} + \mu\Phi$

where

$$\Phi = 2\left[\left(\frac{\partial v_r}{\partial r}\right)^2 + \left(\frac{1}{r}\frac{\partial v_\phi}{\partial\phi} + \frac{v_r}{r}\right)^2 + \left(\frac{1}{r\sin\phi}\frac{\partial v_\theta}{\partial\theta} + \frac{v_r}{r} + \frac{v_\phi\cot\phi}{r}\right)^2\right] + \left[r\frac{\partial}{\partial r}\left(\frac{v_\phi}{r}\right) + \frac{1}{r}\frac{\partial v_r}{\partial\phi}\right]^2 +$$

$$\left[\frac{\sin\phi}{r}\frac{\partial}{\partial\phi}\left(\frac{v_\theta}{r\sin\phi}\right) + \frac{1}{r\sin\phi}\frac{\partial v_\phi}{\partial\theta}\right]^2 + \left[\frac{1}{r\sin\phi}\frac{\partial v_r}{\partial\theta} + r\frac{\partial}{\partial r}\left(\frac{v_\theta}{r}\right)\right]^2$$

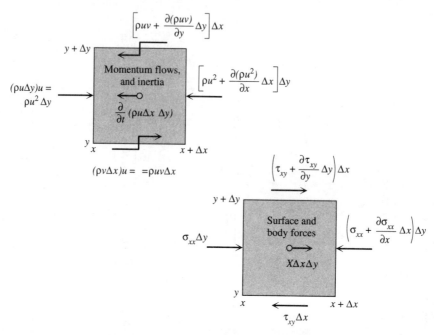

Figure 5.3 The development of the momentum equation for the x direction: the effects of momentum flows and inertia (top, left), and the surface and body forces (bottom, right).

contain eq. (5.10) as a basic assumption; consequently, all the fluids to which these analyses refer are Newtonian fluids.

The viscosity coefficient μ varies in general with the local temperature of the fluid. Several examples of the $\mu(T)$ dependence are exhibited in Appendixes C and D. When the temperature range spanned by the flow (e.g., the difference $T_w - T_\infty$) is sufficiently small, the viscosity can be treated as a constant in the analysis. In this limiting case the substitution of eqs. (5.9) and (5.10) into eq. (5.8) and the use of the constant-density mass conservation eq. (5.6) lead to

$$\rho\left(\frac{\partial u}{\partial t} + u\frac{\partial u}{\partial x} + v\frac{\partial u}{\partial y}\right) = -\frac{\partial P}{\partial x} + \mu\left(\frac{\partial^2 u}{\partial x^2} + \frac{\partial^2 u}{\partial y^2}\right) + X \tag{5.11}$$

This equation is the *momentum equation* for the flow of a constant-property fluid in the x direction.* The properties that have been modeled as constant are the density and the viscosity. The three-dimensional counterpart of eq. (5.11) is listed in Table 5.1. The same table shows also the corresponding forms of the constant-property momentum equations in the y and z directions of the Cartesian system. For example, the y-momentum equation can be deduced from the just-derived x-momentum equation by replacing u with v in all the derivatives, $\partial P/\partial x$ with $\partial P/\partial y$, and the per-unit-volume body force X with the volumetric body force in the y direction, Y.

The three momentum equations for flows in cylindrical coordinates are

---

*The momentum equations are known also as the *Navier–Stokes equations*, after the French engineer Claude-Louis-Marie-Henri Navier (1785–1836) and the British mathematician George Gabriel Stokes (1819–1903).

presented in Table 5.2, where v_r, v_θ, and v_z are the velocity components and F_r, F_θ, and F_z are the components of the body force per unit volume. The corresponding set of equations for spherical coordinates is presented in Table 5.3. This time, the components of velocity and body force per unit volume are v_r, v_ϕ, and v_θ and F_r, F_ϕ, and F_θ, respectively. The D/Dt and ∇^2 notations are also explained in Table 5.3.

5.2.3   The Energy Equation

In forced-convection configurations, the mass conservation and momentum equations are sufficient for determining the flow field, that is, the velocity distribution through the fluid. In such cases the fluid temperature distribution can be determined as an afterthought, by using the just-determined velocity distribution and a new equation (conservation principle): the first law of thermodynamics. In natural convection problems, on the other hand, the velocity and temperature distributions must be determined simultaneously* using the mass conservation principle, the momentum equations, and the first law of thermodynamics.

The derivation of the differential equation that accounts for the first law of thermodynamics at every point in the fluid is a lengthy analysis that can be found in a more advanced treatment of convection [e.g., Ref. [3], pp. 8–16]. In this course we review only the structure of this derivation (Fig. 5.4) and the most useful form of the resulting equation.

With reference to any control volume with inlet and outlet ports, heat and work transfer, and time-dependent energy inventory, the first law of thermodynamics requires (see, for example, Ref. [1], p. 24) that

$$\frac{\partial}{\partial t}(me) = \sum_{\text{in}} \dot{m}e - \sum_{\text{out}} \dot{m}e + \sum_i q_i - \sum_j w_j \tag{5.12}$$

In this equation the left side accounts for the accumulation of energy inside the control volume, in which e is the local specific *internal* energy.† On the right side the first two sums account for the inflows and outflows of energy that are associated with the mass flows across the control surface (i.e., through the "ports"). The third sum represents the net rate of heat transfer into the control volume, in such a way that i indicates the boundary point (or portion) that is crossed by the individual heat current q_i. Finally, the fourth sum on the right side of eq. (5.12) accounts for the net rate of work transfer delivered by the control volume to its environment. Note that while writing eq. (5.12) we are neglecting the contributions of kinetic energy and gravitational potential energy to the local specific energy of the fluid. This assumption is appropriate in the convection heat transfer problems addressed in this course.

It remains to apply eq. (5.12) to the infinitesimal control volume $\Delta x\,\Delta y$ of Fig. 5.2. This amounts to identifying the appropriate expressions that must be

---

*The velocity and temperature fields are coupled because the temperature distribution influences the buoyancy effect that drives the flow and because the flow affects the temperature distribution.
†The common thermodynamics notation for this property is u. In convection the symbol u is reserved for the velocity component in the x direction.

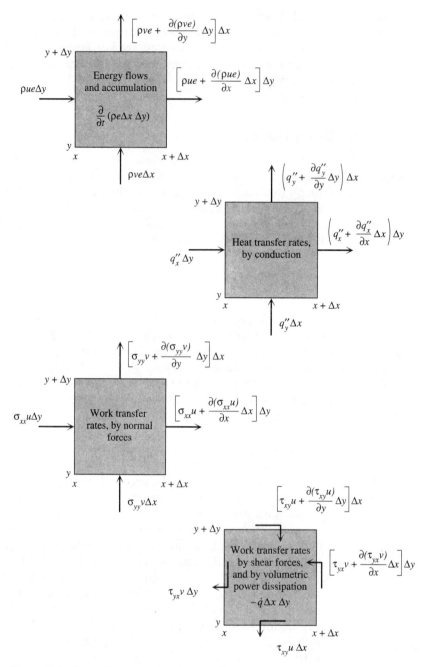

Figure 5.4 The various contributions to the statement that energy is conserved in the control volume $\Delta x\,\Delta y$ of Fig. 5.2.

substituted in place of the terms of eq. (5.12). The four sums listed on the right side of eq. (5.12) represent, in order, the effects of energy inflow (via fluid flow), energy outflow, heat transfer rate into the control volume, and work transfer rate out of the control volume.

The appropriate substitutions for eq. (5.12) are identified in four stages in Fig. 5.4. The first (top) drawing of the $\Delta x\,\Delta y$ system shows the effects due to energy accumulation, energy inflow, and energy outflow. This first drawing takes care of the left side of eq. (5.12) and the first two summations listed on the right side.

The second drawing in Fig. 5.4 identifies all the heat transfer rate interactions between the surrounding fluid and the system. For example, q_x'' is the conduction heat flux oriented in the x direction at the location (x,y) in the fluid. The corresponding heat flux in the next plane in the x direction can be approximated using a Taylor expansion:

$$q_{x+\Delta x}'' = q_x'' + \frac{\partial q_x''}{\partial x} \Delta x \tag{5.13}$$

The third drawing of the $\Delta x \Delta y$ system outlines the work transfer rate interactions associated with the *normal* stresses (σ_{xx} and σ_{yy}) and the pistonlike movement of fluid across the faces of the system. For example, entering through the left side of the control volume is the work transfer rate associated with the normal stress σ_{xx} and the volume swept per unit time, $u \Delta y$. This particular contribution represents work done by the surroundings on the $\Delta x \Delta y$ system: this is why the $\sigma_{xx} u \Delta y$ arrow points toward the system.

The bottom drawing of Fig. 5.4 presents the work transfer rate contributions made by the shear stresses that "work" along the four faces, and the contribution made by an internal (volumetric) mechanism of power dissipation (e.g., Joule heating). The latter is listed with the minus sign, $-\dot{q} \Delta x \Delta y$, because it represents power deposited by an external agent into the control volume. The factor $\dot{q}$ is the volumetric rate of heat generation encountered first in eq. (1.19).

Now, if we substitute the quantities of Fig. 5.4 into the first law, eq. (5.12), divide the resulting equation by $\Delta x \Delta y$, and invoke the constant-density model, we obtain [3]

$$\rho\left(\frac{\partial e}{\partial t} + u \frac{\partial e}{\partial x} + v \frac{\partial e}{\partial y}\right) = -\frac{\partial q_x''}{\partial x} - \frac{\partial q_y''}{\partial y} + \dot{q} + \mu\Phi \tag{5.14}$$

where Φ is shorthand for the *viscous dissipation function* for a constant-ρ fluid:

$$\Phi = \left(\frac{\partial u}{\partial y} + \frac{\partial v}{\partial x}\right)^2 + 2\left[\left(\frac{\partial u}{\partial x}\right)^2 + \left(\frac{\partial v}{\partial y}\right)^2\right] \tag{5.15}$$

The left side of eq. (5.14) can be restated in terms of the local temperature T and its gradients by executing two steps. First, the specific internal energy e can be eliminated based on the definition of specific enthalpy (i):*

$$e = i - \frac{1}{\rho} P \tag{5.16}$$

The specific enthalpy derivatives that result from this operation can be eliminated using the general differential form for a single-phase fluid (see, for example, Ref. [1], p. 187):

$$di = c_p dT + (1 - \beta T)\frac{1}{\rho} dP \tag{5.17}$$

---

*Recall that the modern thermodynamics notation for specific enthalpy is h. In the field of heat transfer, the h symbol is reserved for the convective heat transfer coefficient.

where c_P is not necessarily constant, and where β is the coefficient of volumetric thermal expansion, also called volume expansivity:

$$\beta = -\frac{1}{\rho}\left(\frac{\partial \rho}{\partial T}\right)_P \tag{5.18}$$

In the end, if ρ continues to be treated as *nearly constant*, eq. (5.6), these two steps transform the left side of eq. (5.14) into

$$\rho\left(\frac{\partial e}{\partial t} + u\frac{\partial e}{\partial x} + v\frac{\partial e}{\partial y}\right) = \rho c_P\left(\frac{\partial T}{\partial t} + u\frac{\partial T}{\partial x} + v\frac{\partial T}{\partial y}\right) -$$
$$\beta T\left(\frac{\partial P}{\partial t} + u\frac{\partial P}{\partial x} + v\frac{\partial P}{\partial y}\right) \tag{5.19}$$

The contribution made by the pressure terms is negligible relative to that of the temperature terms; therefore, we abandon the P terms at this stage. This decision leaves nothing but temperature-derivative terms on the left side of eq. (5.14). On the right side of the same equation, the heat fluxes q_x'' and q_y'' can be replaced with the respective Fourier laws, eq. (1.36x), and (1.36y), in which k is the thermal conductivity of the isotropic fluid.

In summary, the first law, eq. (5.14), assumes the form of a differential equation in terms of the fluid temperature T:

$$\rho c_P\left(\frac{\partial T}{\partial t} + u\frac{\partial T}{\partial x} + v\frac{\partial T}{\partial y}\right) = \frac{\partial}{\partial x}\left(k\frac{\partial T}{\partial x}\right) + \frac{\partial}{\partial y}\left(k\frac{\partial T}{\partial y}\right) + \dot{q} + \mu\Phi \tag{5.20}$$

This result constitutes the *energy equation* for the two-dimensional flow of a fluid with nearly constant properties (ρ, μ), variable conductivity (k), and prescribed internal heat generation ($\dot{q}$). The three-dimensional counterpart of eq. (5.20) in Cartesian coordinates is listed as eq. (E) in Table 5.1. It is instructive to compare this equation with the "conduction" equation for the same coordinate system, eq. (1.37): When the fluid comes to a complete rest ($u = v = 0$, $\Phi = 0$), the energy equation reduces to the equation that governs the phenomenon of pure conduction (thermal diffusion).

Tables 5.2 and 5.3 also contain the corresponding energy equations for constant-property (ρ, μ, k) convection in cylindrical and spherical coordinates. In almost all the convection problems considered in this course, the $\dot{q}$ term is absent (equal to zero), and the viscous dissipation effect ($\mu\Phi$) is negligible.

It is worth reviewing the constant-ρ approximation that led to eq. (5.6) and recognizing that it differs conceptually from the "incompressible substance model" of thermodynamics. The latter is considerably more restrictive than the "nearly constant" density model, eq. (5.6). For example, a compressible substance such as air can flow in such a way that eq. (5.6) is a very good approximation of eq. (5.5).

For the restrictive class of fluids that are "incompressible" from the thermodynamic point of view, eq. (1.12), the specific heat at constant pressure c_P can be replaced by the lone specific heat of the fluid, c, on the left side of eq. (5.20). Water, liquid mercury, and engine oil are examples of fluids for which this substitution is justified. There are even convection problems in which the moving media are actually solid (e.g., a roller and its substrate, in the zone of elastic contact—see Project 5.1). In such cases the $c_P = c$ substitution is permissible also; after all, this is why in the limit of pure conduction in a solid the energy eq. (1.37) has c as a factor in the energy-inertia term.

The specific heat that, conceptually speaking, does not belong on the left side of eq. (5.20) is the specific heat at constant volume, c_v. This observation is important because Fourier [5, 6], and later Poisson [7], who were the first to derive the energy equation for a convective flow, wrote c on the left side of eq. (5.20). They made this choice because their analyses were aimed specifically at *incompressible* fluids (liquids), for which c happens to have nearly the same value as c_v. Because of this choice, they did not have to account for the PdV-type work done by the fluid packet as it expands or contracts in the flow field. In the modern era, however, the use of c_v instead of c_P is an error (a misreading of the pioneering work) that continues to propagate through the convection literature (e.g., Ref. [8]).

The prethermodynamics (caloric conservation) origins of the science of convection heat transfer are also responsible for the "thermal energy equation" label that some prefer to attach to eq. (5.20). This terminology is sometimes used to stress (incorrectly) the conservation of "thermal" energy in eq. (5.20) as something distinct from "mechanical and thermal" energy. In classical thermodynamics, however, this distinction disappeared as soon as the first law of thermodynamics was enunciated, that is, as soon as the thermodynamic property "energy" was defined, which happened in the years 1850–1851 (see Ref. [1], pp. 31–33).

Equation (5.20) represents the first law of thermodynamics. This law proclaims the conservation of the sum of energy change (the property) and energy interactions (heat transfer *and* work transfer), subject of course to all the simplifying assumptions that have been made en route to eq. (5.20). The suggestion that mechanical effects (e.g., work transfer) are absent from eq. (5.20) is erroneous. The presence of c_P on the left side of the equation is a clear sign that each fluid packet expands, or contracts (i.e., it does PdV-type work) as it rides on the flow. The terms $\dot{q}$ and $\mu\Phi$ are work transfer rate terms also.

5.3 LAMINAR BOUNDARY LAYER OVER A PLANE WALL

5.3.1 The Velocity Boundary Layer

The simplest forced convection configuration to consider is the flow and heat transfer near a flat wall, Fig. 5.5. Sufficiently far away from the $y = 0$ plane the fluid is isothermal (T_∞) and isobaric (P_∞) and flows to the right with a uniform velocity (U_∞). We assume further that the leading edge of the flat plate is sharp enough so that the collision between the U_∞ flow and the leading edge does not disturb the more-or-less longitudinal (parallel) motion of every fluid packet that flows in the close vicinity of the $y = 0$ plane.

The flow field is therefore steady, two-dimensional, and *laminar*, that is, a sandwich of smooth fluid blades* of infinitesimal thickness, slipping past one another with different longitudinal speeds. In eq. (5.85) we shall discover that the flow pattern is indeed laminar at a given location x if the free-stream velocity U_∞ is sufficiently small, or, when U_∞ is given, if the downstream length x does not exceed a critical value.

According to eqs. (m), (M_x), and (M_y) of Table 5.1, the steady two-dimen-

---

*Indeed, the word "laminar" originates from the Latin noun *lamina*, which means a thin piece of metal (blade), or a thin sheet of wood.

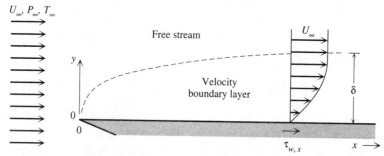

Figure 5.5 Velocity boundary layer in laminar flow near a plane wall.

sional flow in Fig. 5.5 is described by the velocity components $u(x,y)$ and $v(x,y)$ that, along with the pressure $P(x,y)$, must satisfy the system:

$$(m) \qquad \frac{\partial u}{\partial x} + \frac{\partial v}{\partial y} = 0 \tag{5.21}$$

$$(M_x) \qquad u\frac{\partial u}{\partial x} + v\frac{\partial u}{\partial y} = -\frac{1}{\rho}\frac{\partial P}{\partial x} + v\left(\frac{\partial^2 u}{\partial x^2} + \frac{\partial^2 u}{\partial y^2}\right) \tag{5.22}$$

$$(M_y) \qquad u\frac{\partial v}{\partial x} + v\frac{\partial v}{\partial y} = -\frac{1}{\rho}\frac{\partial P}{\partial y} + v\left(\frac{\partial^2 v}{\partial x^2} + \frac{\partial^2 v}{\partial y^2}\right) \tag{5.23}$$

The coefficient v (m^2/s) is the *kinematic viscosity* of the fluid:

$$v = \frac{\mu}{\rho} \tag{5.24}$$

This coefficient is a constant, because both ρ and μ have already been modeled as constant.

The momentum equations that apply to the present flow are considerably simpler than eqs. (5.22) and (5.23). Let δ be the characteristic size (length scale) of the transversal distance over which the horizontal velocity $u(x,y)$ changes from the free-stream value U_∞ to $u = 0$ right at the solid wall. The analytical order-of-magnitude definition of δ is therefore

$$\frac{\partial u}{\partial y} \sim \frac{U_\infty}{\delta} \tag{5.25}$$

The velocity profile "thickness" δ is a function of the downstream position x. Let us *assume* that δ is always negligible relative to x,

$$\delta << x \tag{5.26}$$

in other words, that the flow region of length x and thickness δ (the region between the wall and the undisturbed free stream) is *slender*. Following Prandtl [9],* in convection heat transfer a slender region of this kind is called a *boundary layer*.

---

*Ludwig Prandtl (1875–1953) was professor of applied mechanics at Göttingen University, Germany. Arguably the founder of modern fluid mechanics, he is best known for his boundary layer theory, wing theory, development of the wind tunnel, and research on supersonic flow and turbulence.

The slenderness inequality (5.26) allows us to simplify the x-momentum eq. (5.22) in two ways. First, the term $\partial^2 u/\partial x^2$ can be neglected in favor of the term $\partial^2 u/\partial y^2$, which is much larger. The respective scales of these terms are (review the method of scale analysis, Section 3.2.2)

$$\frac{\partial^2 u}{\partial x^2} \sim \frac{\left(\dfrac{\partial u}{\partial x}\right)_{x=x} - \left(\dfrac{\partial u}{\partial x}\right)_{x=0}}{x-0} \sim \frac{\dfrac{-U_\infty}{x} - 0}{x-0} \sim \frac{-U_\infty}{x^2} \tag{5.27}$$

$$\frac{\partial^2 u}{\partial y^2} \sim \frac{\left(\dfrac{\partial u}{\partial y}\right)_{y=\delta} - \left(\dfrac{\partial u}{\partial y}\right)_{y=0}}{\delta-0} \sim \frac{0 - \dfrac{U_\infty}{\delta}}{\delta-0} \sim \frac{-U_\infty}{\delta^2} \tag{5.28}$$

Dividing the absolute values of these scales, we find that

$$\frac{|\partial^2 u/\partial x^2|}{|\partial^2 u/\partial y^2|} \sim \left(\frac{\delta}{x}\right)^2 \ll 1 \tag{5.29}$$

In conclusion, the slenderness inequality (5.26) implies that the $\partial^2 u/\partial x^2$ term can be neglected inside the boundary layer.

The second simplification of the x-momentum equation stems from the fact that the pressure P does not vary appreciably across the thin boundary layer:

$$P(x,y) \cong P(x) \tag{5.30}$$

The validity of eq. (5.30) can be demonstrated based on the y-momentum equation, eq. (5.23) and the slenderness condition (5.26) (see Ref. [3], pp. 33–34). In other words, it can be shown that inside the boundary layer region the y-momentum equation reduces to eq. (5.30).

In the present problem the pressure inside the boundary layer, $P(x)$, must be the same as the pressure at the outer edge of the boundary layer, P_∞, which is a constant. This means that the pressure gradient term that appears in eq. (5.22) is zero:

$$\frac{\partial P}{\partial x} \cong \frac{dP}{dx} = \frac{dP_\infty}{dx} = 0 \tag{5.31}$$

Equations (5.29) and (5.31) indicate the two simplifications that result from the slenderness of the boundary layer region. Combining eqs. (5.22), (5.29), and (5.31) we obtain the lone momentum equation of the flat-plate boundary layer flow:

$$u \underbrace{\frac{\partial u}{\partial x} + v \frac{\partial u}{\partial y}}_{\text{Inertia}} = \underbrace{\nu \frac{\partial^2 u}{\partial y^2}}_{\text{Friction}} \tag{5.32}$$

Equation (5.32) is the only momentum equation because its derivation is based on both (M_x) and (M_y). In qualitative terms the momentum eq. (5.32) expresses a balance between the inertia (deceleration) of a fluid packet and the friction effect (restraining force) transmitted by the wall to the fluid packet via viscous diffusion in the y direction.

The flow distribution (u,v) can be determined by solving the mass and momentum eqs. (5.21) and (5.32). Term by term, the order-of-magnitude counterparts of these two equations are

$$(m) \qquad \frac{U_\infty}{x} \sim \frac{v}{\delta} \tag{5.33}$$

$$(M) \qquad U_\infty \frac{U_\infty}{x}, \; v \frac{U_\infty}{\delta} \sim v \frac{U_\infty}{\delta^2} \tag{5.34}$$

Note first that the two inertia scales on the left side of eq. (5.34) are of the same order of magnitude [cf. eq. (5.33)]; therefore, the momentum balance between inertia and friction is simply

$$\underbrace{U_\infty \frac{U_\infty}{x}}_{\text{Inertia}} \sim \underbrace{v \frac{U_\infty}{\delta^2}}_{\text{Friction}} \tag{5.35}$$

Equations (5.33) and (5.35) deliver the two unknown scales of the boundary layer flow:

$$\delta \sim \left(\frac{vx}{U_\infty} \right)^{1/2} \tag{5.36}$$

$$v \sim U_\infty \left(\frac{U_\infty x}{v} \right)^{-1/2} \tag{5.37}$$

This approximate solution contains also an order-of-magnitude answer to the chief engineering question regarding this flow, namely, the wall shear stress at the location x:

$$\tau_{w,x} = \mu \left(\frac{\partial u}{\partial y} \right)_{y=0} \sim \mu \frac{U_\infty}{\delta} \sim \rho U_\infty^2 \left(\frac{U_\infty x}{v} \right)^{-1/2} \tag{5.38}$$

The dimensionless alternative to reporting the local wall shear stress is the *local skin friction coefficient* $C_{f,x}$, which is defined as the ratio

$$C_{f,x} = \frac{\tau_{w,x}}{\frac{1}{2} \rho U_\infty^2} \tag{5.39}$$

Combining eqs. (5.38) and (5.39) and neglecting* the numerical factor $\frac{1}{2}$, we conclude that the local skin friction coefficient is approximately

$$C_{f,x} \sim \mathrm{Re}_x^{-1/2} \tag{5.40}$$

where Re_x is the *Reynolds number*† based on the downstream distance x:

$$\mathrm{Re}_x = \frac{U_\infty x}{v} \tag{5.41}$$

All these results hold true when the boundary layer is slender, as was assumed in the inequality (5.26): combining eq. (5.26) with the just-derived δ scale (5.36), we find that the boundary layer is indeed slender when $\mathrm{Re}_x^{1/2}$ is much greater than 1:

---

*Factors of order 1 can be neglected in an order-of-magnitude analysis.

†Osborne Reynolds (1842–1912) was professor of engineering at Owens College, in Manchester, England. Among his many contributions are the discovery of the critical velocity for the laminar–turbulent transition in pipe flow (the critical Reynolds number), and the thin-film theory of lubrication.

$$\mathrm{Re}_x^{1/2} \gg 1 \qquad\qquad (5.42)$$

Another implication of this inequality is that the boundary layer theory results hold for positions x sufficiently far downstream from the leading edge of the wall, so that $\mathrm{Re}_x^{1/2} \gg 1$. The same theory breaks down at $x = 0$ and at small enough x values (called the "tip region") where $\mathrm{Re}_x^{1/2} \lesssim 1$.

The scale analysis solution for the wall friction effect, eq. (5.40), can be improved by pursuing a more exact method of solution. One alternative is the integral method outlined in Sections 2.6.2 and 3.2.1. The step-by-step integral analysis of the δ-thin flow region forms the subject of Problem 5.11: the more refined $C_{f,x}$ formulas that are recommended by that analysis are listed in the problem statement.

Students who perform the integral analysis may find it inconsistent that the initial (or "starting," or "tip") condition $\delta = 0$ at $x = 0$ is necessary in the development of the solution for the boundary layer thickness function $\delta(x)$. The stipulation of a condition (any condition) at $x = 0$ appears to contradict the observation made under eq. (5.42), namely, that the boundary layer theory does not hold at $x = 0$ and near $x = 0$. In fact, there is no inconsistency. The boundary layer theory holds sufficiently far downstream, over that long stretch where $\mathrm{Re}_x^{1/2} \gg 1$ and the boundary layer is slender. The theory states (predicts) that over that downstream region the thickness *varies in a certain way*, $\delta(x)$, which turns out to be supported very well by experiments. That "certain" variation of $\delta(x)$ is such that when extrapolated to $x = 0$ it yields $\delta = 0$. In other words, the invocation of the tip condition $\delta(0) = 0$ is an important component of the theory, as the theory attempts to predict what happens *sufficiently far* from $x = 0$.

To repeat, the fact that the theoretical thickness varies as $x^{1/2}$ and the wall shear stress varies as $x^{-1/2}$ does not mean that the real thickness is zero and the real shear stress is infinite at $x = 0$. They are not. Examine the photographed tip region of the laminar boundary layer in vertical natural convection, Fig. 7.3*a*, and you will note that right at the tip δ is finite. An entirely new theory, not a boundary layer theory, is needed for predicting the behavior of the flow, friction, and heat transfer over the near-tip region.

The *exact solution* to eqs. (5.21) and (5.32) proceeds from the idea that, regardless of x, the u-vs.-y profiles are similar. They all start with the no-slip condition at the wall ($u = 0$) and approach $u = U_\infty$ as y takes values of order δ. The $u(x,y)$ distribution is summarized by the universal or "master" profile

$$u(x,y) = U_\infty \text{ function } \left[\frac{y}{\delta(x)}\right] \qquad\qquad (5.43)$$

where $\delta(x)$ is the scale determined in eq. (5.36). The transversal coordinate of this universal profile is therefore the *similarity variable* η:

$$\eta = \frac{y}{x} \mathrm{Re}_x^{1/2} \qquad\qquad (5.44)$$

The *similarity solution* for the laminar boundary layer flow was carried out by Blasius [10, 11], who replaced the unknowns $u(x,y)$ and $v(x,y)$ with a single unknown — the *streamfunction* $\psi(x,y)$ — the definition of which is

$$u = \frac{\partial \psi}{\partial y} \qquad v = -\frac{\partial \psi}{\partial x} \qquad\qquad (5.45)$$

It is easy to see that the continuity eq. (5.21) is satisfied identically by ψ and that the momentum eq. (5.32) becomes

$$\frac{\partial\psi}{\partial y}\frac{\partial^2\psi}{\partial x\,\partial y} - \frac{\partial\psi}{\partial x}\frac{\partial^2\psi}{\partial y^2} = \nu\frac{\partial^3\psi}{\partial y^3} \tag{5.46}$$

The appropriate boundary conditions to be satisfied by ψ are

$$\frac{\partial\psi}{\partial y} = 0 \quad \text{at} \quad y = 0 \qquad \text{(no-slip, } u = 0\text{)} \tag{5.47a}$$

$$\psi = 0 \quad \text{at} \quad y = 0 \qquad \text{(impermeable wall, } v = 0\text{)} \tag{5.47b}$$

$$\frac{\partial\psi}{\partial y} \to U_\infty \quad \text{as} \quad y \to \infty \qquad \text{(free stream, } u \to U_\infty\text{)} \tag{5.47c}$$

The problem (5.46)–(5.47) can be restated in terms of the similarity variable η, however, this transformation is beyond the level of this course, and what follows is only an outline of the method. Combining the similarity u profile (5.43) with the first of eqs. (5.45), we see that we must choose a streamfunction of the form

$$\psi(x,y) = (U_\infty\nu x)^{1/2}\,f(\eta) \tag{5.48}$$

The unknown function $f(\eta)$ is the similarity streamfunction profile; and, in view of eqs. (5.43) and (5.45), the derivative $df/d\eta$ is the similarity longitudinal velocity profile,

$$u = U_\infty\frac{df}{d\eta} \tag{5.49}$$

The $f(\eta)$ formulation of the flow problem (5.46)–(5.47) is

$$2f''' + ff'' = 0 \tag{5.50}$$

$$f' = f = 0 \quad \text{at } \eta = 0 \tag{5.51a, b}$$

$$f' \to 1 \quad \text{as } \eta \to \infty \tag{5.51c}$$

The numerical solution to this problem (see Ref. [3], p. 61, Problem 3) is presented here in terms of the velocity profile $df/d\eta$ (Fig. 5.6). A key result of the "Blasius solution" is the slope of the similarity velocity profile right at the wall, which has the following value:

$$\left(\frac{d^2f}{d\eta^2}\right)_{\eta=0} = 0.332 \tag{5.52}$$

This slope is needed for calculating the exact value of the wall shear stress, or the local skin friction coefficient:

$$C_{f,x} = \frac{\mu(\partial u/\partial y)_{y=0}}{\frac{1}{2}\rho U_\infty^2} = 2\left(\frac{d^2f}{d\eta^2}\right)_{\eta=0}\mathrm{Re}_x^{-1/2} \tag{5.53}$$

$$= 0.664\mathrm{Re}_x^{-1/2}$$

This exact result is not far off from the integral analysis estimate (Problem 5.11) and the back-of-the-envelope calculation that gave us eq. (5.40).

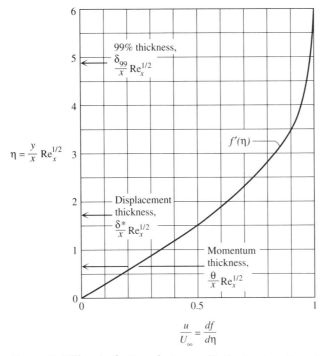

Figure 5.6 The similarity velocity profile for laminar boundary layer flow over a plane wall.

The *total* shear force* experienced by the plane wall of longitudinal length x is

$$\int_0^x \tau_{w,x} \, dx = x\bar{\tau}_{w,x} \tag{5.54}$$

where $\bar{\tau}_{w,x}$ is the x-averaged value of the local shear stress $\tau_{w,x}$. The average shear stress can be evaluated using eq. (5.53): that result can be arranged as a dimensionless *average skin friction coefficient*

$$\bar{C}_{f,x} = \frac{\bar{\tau}_{w,x}}{\frac{1}{2} \rho U_\infty^2} = 1.328 \mathrm{Re}_x^{-1/2} \tag{5.55}$$

Figure 5.6 shows that u increases smoothly in the transversal direction η (or y) and *approaches U_∞ asymptotically.* For this reason, the actual "thickness" of the velocity boundary layer must be defined based on a precise definition. There are three such definitions. First, there is the δ_{99} thickness, which is defined as the distance y where u reaches 99 percent of its maximum (free-stream) value:

$$u(x, \delta_{99}) = 0.99 U_\infty \tag{5.56}$$

---

*This force is per unit length in the direction normal to the plane of Fig. 5.5.

The numerical solution of Fig. 5.6 shows that $u = 0.99U_\infty$ at $\eta = 4.92$; therefore,

$$\delta_{99} = 4.92 \, x \, Re_x^{-1/2} \tag{5.57}$$

The other two definitions that are in use refer to the *displacement thickness* δ^* and the *momentum thickness* θ:

$$\delta^* = \int_0^\infty \left(1 - \frac{u}{U_\infty}\right)dy = 1.72 \, x \, Re_x^{-1/2} \tag{5.58}$$

$$\theta = \int_0^\infty \frac{u}{U_\infty}\left(1 - \frac{u}{U_\infty}\right)dy = 0.664 \, x \, Re_x^{-1/2} \tag{5.59}$$

Listed on the right side of these equations are the actual sizes revealed by the Blasius solution of Fig. 5.6. The physical meaning and the explanation of the names of δ^* and θ can be found in a more advanced treatment (e.g., Ref. [3], pp. 48–50). Worth noting is that all three thicknesses (δ_{99}, δ^*, and θ) confirm the validity of the order-of-magnitude estimate provided by scale analysis, eq. (5.36).

The laminar boundary layer region and the longitudinal velocity profile have been made visible using strings of tiny hydrogen bubbles in the experiments of Figs. 5.7 and 5.8. This flow visualization method works in water and consists of suspending a fine wire as cathode in the flow. Hydrogen bubbles are generated by electrolysis when voltage pulses are applied to the wire. When the wire is straight and the pulses occur at uniform intervals (as in Figs. 5.7 and 5.8), the hydrogen bubbles form a pattern of "time lines."

In Fig. 5.7 the wire is perpendicular to the free stream and is positioned at less than 1 mm upstream of the leading edge of the flat plate. By looking in the downstream direction, we see that the time lines are pushed away from the wall by the fluid that finds itself inside the boundary layer region (dark). The longitudinal flow inside the boundary layer region is made visible in Fig. 5.8, in

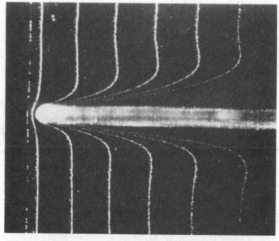

Figure 5.7 Visualization of the laminar boundary layers on the two sides of a flat plate by using the hydrogen bubble method: 0.01% salt water, $U_\infty = 0.6$ cm/s, plate thickness = 0.5 mm. (Nakayama et al. [12], with permission from Pergamon Press.)

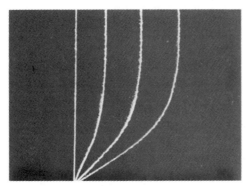

Figure 5.8 Visualization of the longitudinal velocity profile in the laminar boundary layer over a flat plate: 0.01% salt water, $U_\infty = 0.6$ cm/s, distance from the leading edge $x = 20$ cm. (Nakayama et al. [12], with permission from Pergamon Press.)

which the wire is perpendicular to the wall, and at $x = 20$ cm downstream from the leading edge.

Example 5.1

The Laminar Velocity Boundary Layer

Examine the flow conditions and geometry of the velocity profile photographed in Fig. 5.8 and estimate the following:

(a) The boundary layer thicknesses δ_{99}, δ^*, and θ.

(b) The time between two consecutive wire-voltage pulses that generate the lines of hydrogen bubbles.

(c) The local and wall-averaged shear stresses associated with the downstream position where the photograph of Fig. 5.8 was taken.

Solution.   (a) In the following order-of-magnitude calculations, it is reasonable to use the properties of distilled water at 25°C. We calculate in order

$$\text{Re}_x = \frac{U_\infty x}{\nu} = \frac{0.6 \frac{\text{cm}}{\text{s}} \, 20 \text{ cm}}{0.00894 \frac{\text{cm}^2}{\text{s}}} \cong 1340$$

$$\delta_{99} = 4.92 \, x \, \text{Re}_x^{-1/2} = 2.7 \text{ cm}$$

$$\delta^* = 1.72 \, x \, \text{Re}_x^{-1/2} = 0.94 \text{ cm}$$

$$\theta = 0.664 \, x \, \text{Re}_x^{-1/2} = 0.36 \text{ cm}$$

Figure 5.6 showed that the knee of the velocity profile is located at $y \sim \delta_{99}$. This means that the bends in the hydrogen bubble profiles of Fig. 5.8 occur at a distance of approximately 2.7 cm away from the wall.

(b) To calculate the time between two consecutive pulses in the generation of lines of hydrogen bubbles in Fig. 5.8, we note that

$$3\Delta x \sim \delta_{99}$$

where Δx is the distance between two consecutive lines in the free stream:

$$\Delta x = U_\infty \Delta t$$

The bubble-generating wire is therefore charged at intervals of order

$$\Delta t = \frac{\Delta x}{U_\infty} \sim \frac{\delta_{99}}{3U_\infty} = \frac{2.7 \text{ cm}}{3 \times 0.6 \text{ cm/s}} = 1.5 \text{ s}$$

(c) The local and wall-averaged shear stresses can be evaluated based on eqs. (5.53) and (5.55):

$$C_{f,x} = 0.664 \text{Re}_x^{-1/2} = 0.0181$$

$$\tau_{w,x} = \frac{1}{2} \rho U_\infty^2 C_{f,x}$$

$$= \frac{1}{2} \left[1 \frac{g}{cm^3} \left(0.6 \frac{cm}{s} \right)^2 0.0181 \right]$$

$$= 3.3 \times 10^{-3} \frac{g}{cm \cdot s^2} = 3.3 \times 10^{-4} \text{ N/m}^2$$

$$\overline{C}_{f,x} = 1.328 \text{ Re}_x^{-1/2} = 0.0362 = 2C_{f,x}$$

$$\overline{\tau}_{w,x} = 2\tau_{w,x} = 6.6 \times 10^{-4} \text{ N/m}^2$$

5.3.2   The Thermal Boundary Layer (Isothermal Wall)

The temperature distribution $T(x,y)$ near the isothermal wall (T_w) swept by the parallel flow U_∞ is outlined in Fig. 5.9. Let δ_T be the transversal length scale that represents the distance over which the fluid temperature makes the transition from the wall value T_w to the free-stream value T_∞. The thermal boundary layer thickness δ_T is defined therefore as the distance from the wall to the knee in the T-vs.-y curve:

$$\left| \frac{\partial T}{\partial y} \right| \sim \frac{\Delta T}{\delta_T} \tag{5.60}$$

where ΔT is the imposed temperature difference $\Delta T = T_w - T_\infty$.

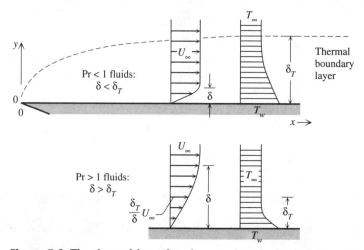

Figure 5.9 The thermal boundary layer in low-Pr fluids (top) and its relative thinness in high-Pr fluids (bottom).

Assume next that the x-long and δ_T-thin region is slender,

$$\delta_T \ll x \tag{5.61}$$

namely, that it is a *thermal boundary layer region*. Based on a reasoning analogous to the one used in the development of eq. (5.29), we conclude that in the steady two-dimensional energy eq. (E) of Table 5.1 the $\partial^2 T / \partial x^2$ term is negligible:

$$\underbrace{u\,\frac{\partial T}{\partial x} + v\,\frac{\partial T}{\partial y}}_{\text{Convection}} = \underbrace{\alpha\,\frac{\partial^2 T}{\partial y^2}}_{\substack{\text{Transversal}\\ \text{conduction}}} \tag{5.62}$$

This is the boundary layer-simplified form of the energy equation for the fluid that resides in the (x,δ_T) layer. The coefficient α (m²/s) is the thermal diffusivity of the fluid,

$$\alpha = \frac{k}{\rho c_P} \tag{5.63}$$

The unknown characteristic of the thermal boundary layer is its thickness δ_T. Note that as soon as δ_T is known, the wall heat flux can be evaluated [cf. eq. (1.56)]:

$$q''_{w,x} \sim k\frac{\Delta T}{\delta_T} \tag{5.64}$$

This would be an order-of-magnitude estimate of the relationship between wall heat flux and imposed temperature difference. The δ_T scale needed for this estimate can be determined analytically in the following two limits:

Thick Thermal Boundary Layer.   In this limit, the δ_T layer is "thick" relative to the velocity boundary layer thickness measured at the same x, Fig. 5.9 (top):

$$\delta_T \gg \delta \tag{5.65}$$

Inside the δ_T layer, the respective scales of the three terms of the energy eq. (5.62) are

$$u\,\frac{\Delta T}{x}, \; v\,\frac{\Delta T}{\delta_T} \sim \alpha\,\frac{\Delta T}{\delta_T^2} \tag{5.66}$$

According to eq. (5.33), the v scale outside the thin velocity boundary layer (and inside the δ_T layer) is $v \sim U_\infty \delta / x$. This means that the second term on the left side of eq. (5.66) is

$$v\,\frac{\Delta T}{\delta_T} \sim U_\infty\,\frac{\Delta T}{x}\,\frac{\delta}{\delta_T} \tag{5.67}$$

in which $\delta / \delta_T \ll 1$. The second term is therefore δ / δ_T times smaller than the first, and the left side of eq. (5.66) is dominated by $u\Delta T / x$. In conclusion, the convection $\sim$ conduction balance expressed by the energy equation is simply

$$u\,\frac{\Delta T}{x} \sim \alpha\,\frac{\Delta T}{\delta_T^2} \tag{5.68}$$

Now we must face the question of what is the scale of the longitudinal velocity u *inside the* δ_T-*thick layer*. In the case of a thermal boundary layer that is thicker than the velocity boundary layer, Fig. 5.9 (top) shows that $u \sim U_\infty$. Substituting this u estimate into eq. (5.68) we obtain immediately

$$\delta_T \sim x\mathrm{Pe}_x^{-1/2} \tag{5.69}$$

where Pe_x is the *Peclet number** based on the downstream position x:

$$\mathrm{Pe}_x = \frac{U_\infty x}{\alpha} \tag{5.70}$$

The thermal boundary layer thickness increases as $x^{1/2}$ in the downstream direction, as illustrated in Fig. 5.9. The Prandtl number (Pr) referred to on the drawing of Fig. 5.9 will be defined shortly, in eq. (5.74).

The local wall heat flux $q''_{w,x}$ can now be deduced from eq. (5.64). The result can be reported as a dimensionless *local Nusselt number*,† Nu_x, which is defined by

$$\mathrm{Nu}_x = \frac{q''_{w,x}}{\Delta T}\frac{x}{k} = \frac{h_x x}{k} \tag{5.71}$$

where h_x is the *local* heat transfer coefficient, $h_x = q''_{w,x}/\Delta T$. Combining eqs. (5.64) and (5.69), we draw the conclusion that the local Nusselt number is of order

$$\mathrm{Nu}_x \sim \mathrm{Pe}_x^{1/2} \quad (\mathrm{Pr} \ll 1) \tag{5.72}$$

This conclusion holds as long as the assumption $\delta_T > \delta$ is correct. Substituting eqs. (5.69) and (5.36) on the left and right sides of the inequality (5.65), we learn that the $\delta_T > \delta$ assumption is the same as writing

$$\alpha > \nu \tag{5.73}$$

The thermal boundary layer is therefore thicker than the velocity boundary layer when the fluid properties are such that the *Prandtl number*

$$\mathrm{Pr} = \frac{\nu}{\alpha} \tag{5.74}$$

is considerably smaller than 1. This restriction has been added to the right of eq. (5.72). Examples of low-Prandtl-number fluids are the liquid metals (mercury, lead, sodium, Appendix C).

Thin Thermal Boundary Layer. What changes when the thermal layer is thinner than the superimposed velocity layer, that is, when

$$\delta_T \ll \delta \tag{5.75}$$

---

*Jean-Claude-Eugène Péclet (1793–1857) was a French physicist who wrote an influential 1829 treatise on heat transfer and its applications, which was translated into English in 1843.

†Ernst Kraft Wilhelm Nusselt (1882–1957) was professor of theoretical mechanics at the Technical University of Munich. He pioneered the dimensional analysis of convective processes, the dimensionless correlation of experimental data, and the analysis of laminar film condensation.

is the scale of the longitudinal velocity u in the δ_T-thin region. According to the configuration shown in the lower part of Fig. 5.9, that u scale is

$$u \sim \frac{\delta_T}{\delta} U_\infty \qquad (5.76)$$

Combining this idea with the convection $\sim$ conduction balance that rules the δ_T-thin layer, eq. (5.68), and using also the previously determined velocity thickness δ, eq. (5.36), we obtain

$$\delta_T \sim x \mathrm{Pr}^{-1/3} \mathrm{Re}_x^{-1/2} \qquad (5.77)$$

The corresponding local heat flux and local Nusselt number follow from this conclusion and the definitions (5.64) and (5.71):

$$\mathrm{Nu}_x \sim \mathrm{Pr}^{1/3} \mathrm{Re}_x^{1/2} \qquad (\mathrm{Pr} \gg 1) \qquad (5.78)$$

These heat transfer results (δ_T, Nu_x) are valid when δ_T is indeed smaller than δ: combined, eqs. (5.36), (5.75), and (5.77) show that the thermal layer is thinner than the velocity layer when the fluid has a Prandtl number greater than 1. Most nonmetallic liquids (e.g., water, oils) have relatively large Pr values.

The most common gases (e.g., air, steam, CO_2) have Prandtl numbers of order 1. As a class, they fall right between the limits (a) and (b), which were discussed until now. It is instructive to verify that these two limits, eqs. (5.72) and (5.78), predict the same heat transfer result ($\mathrm{Nu}_x \sim \mathrm{Re}_x^{1/2}$) when the Prandtl number is constrained to a number of order 1.

A common characteristic of all the boundary layer thicknesses uncovered so far (δ, δ_T in $\mathrm{Pr} < 1$ fluids, and δ_T in $\mathrm{Pr} > 1$ fluids) is their monotonic increase as $x^{1/2}$ in the downstream direction. One property of the $x^{1/2}$ function is that it has infinite slope at $x = 0$. This feature is being stressed in Figs. 5.5 and 5.9 because it is almost never respected in the boundary layer sketches that appear in heat transfer books and journals. As noted in eq. (5.42), however, the boundary layer approximation does not hold in the immediate vicinity of the tip of the boundary layer region.

The heat transfer results (5.72) and (5.78) have been refined on the basis of more exact methods of solution. The integral method and samples of the Nu_x estimates produced by this method are illustrated in Problem 5.12. The exact solution based on the similarity formulation was carried out by Pohlhausen [13]. Of interest to the problem-solving component of this course are the "low-Pr" and "high-Pr" limits of this solution (see also Ref. [3], pp. 50–52):

$$\mathrm{Nu}_x \cong 0.564 \mathrm{Pe}_x^{1/2} \qquad (\mathrm{Pr} \lesssim 0.5) \qquad (5.79\mathrm{a})$$

$$\mathrm{Nu}_x \cong 0.332 \mathrm{Pr}^{1/3} \mathrm{Re}_x^{1/2} \qquad (\mathrm{Pr} \gtrsim 0.5) \qquad (5.79\mathrm{b})$$

The *total* heat transfer rate between the x-long wall and the adjacent flow, per unit length, in the direction normal to the plane of Fig. 5.9 is

$$\int_0^x q''_{w,x} dx = x \, \overline{q}''_{w,x} \qquad (5.80)$$

Equation (5.80) is the definition of the x-averaged wall heat flux $\overline{q}''_{w,x}$: this can be calculated by substituting eqs. (5.79a, b) into the Nu_x definition (5.71) and using the resulting $q''_{w,x}$ formulas in the integrand on the left side of eq. (5.80). The average heat flux $\overline{q}''_{w,x}$ obtained in this manner can be summarized in dimensionless form,

$$\overline{\mathrm{Nu}}_x = \frac{\overline{q}''_{w,x}}{\Delta T}\frac{x}{k} = \frac{\overline{h}_x x}{k} \qquad (5.81)$$

where $\overline{\mathrm{Nu}}_x$ is the *average or overall Nusselt number*, and $\overline{h}_x$ is the average heat transfer coefficient. The $\overline{\mathrm{Nu}}_x$ formulas that correspond* to the local Nusselt number asymptotes (5.79a,b) are

$$\overline{\mathrm{Nu}}_x = 1.128\mathrm{Pe}_x^{1/2} \qquad (\mathrm{Pr} \lesssim 0.5) \qquad (5.82a)$$

$$\overline{\mathrm{Nu}}_x = 0.664\mathrm{Pr}^{1/3}\mathrm{Re}_x^{1/2} \qquad (\mathrm{Pr} \gtrsim 0.5) \qquad (5.82b)$$

in which, it is worth noting, the numerical coefficients are twice as large as the coefficients of eqs. (5.79a,b). An average Nusselt number expression that covers the entire Prandtl number range was recommended by Churchill and Ozoe [14]:

$$\overline{\mathrm{Nu}}_x = \frac{0.928\mathrm{Pr}^{1/3}\mathrm{Re}_x^{1/2}}{[1 + (0.0207/\mathrm{Pr})^{2/3}]^{1/4}} \qquad (5.83)$$

It is valid when the Peclet number $\mathrm{Pe}_x = U_\infty x/\alpha$ is greater than approximately 100.

5.3.3 Nonisothermal Wall Conditions

The similarity solutions (5.79a,b) for heat transfer in laminar boundary layer flow refer to the heat flux from an *isothermal* wall (T_w) to an isothermal free stream (T_∞). There are many other wall-heating conditions that occur in practical situations, and, for some of these, heat transfer solutions have already been developed. Four results of this kind have been collected from Refs. [3,4,15,16] and arranged in Table 5.4. They were derived based on the *integral method*, and apply to gases and nonmetallic liquids, that is, to fluids with Pr values greater than approximately 0.5.

The first solution in Table 5.4 shows the heat flux through the isothermal section of a wall with unheated starting length. The effect of the unheated length x_0 is to decrease the heat flux to values below those of the fully isothermal wall, eq. (5.79b). The isothermal wall problem is, in fact, the special case $x_0 = 0$ of the first solution of Table 5.4.

---

*The average heat transfer coefficient that corresponds to a certain local heat transfer coefficient expression can be estimated rapidly by invoking the following theorem: If the local quantity (h_x) has a power law dependence on x,

$$h_x = Cx^n \qquad (a)$$

where C is a constant, then the quantity averaged from $x = 0$ to x (in this case $\overline{h}_x$) is simply

$$\overline{h}_x = \frac{h_x}{1 + n} \qquad (b)$$

To apply this theorem, it is necessary to identify the correct exponent in the power law (a). In eqs. (5.79a,b), for example, the exponent of h_x is $n = -\frac{1}{2}$; the denominator of eq. (b) becomes $1 - \frac{1}{2} = \frac{1}{2}$ and explains why the leading numerical coefficients in eqs. (5.82a,b) are twice as large as in eqs. (5.79a,b).

Table 5.4 Heat Transfer Results for Laminar Boundary Layer Flows Near Walls with Various Heating Conditions ($Pr \gtrsim 0.5$)

Unheated Starting Length

$$q''_{w,x} = 0.332 \frac{k(T_w - T_\infty)}{x} Pr^{1/3} Re_x^{1/2} \left[1 - \left(\frac{x_0}{x}\right)^{3/4} \right]^{-1/3}$$

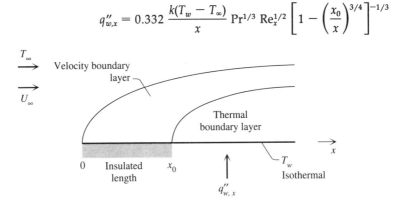

Nonuniform Wall Temperature

$$q''_{w,x} = 0.332 \frac{k}{x} Pr^{1/3} Re_x^{1/2} \int_0^x \frac{(dT_w/d\xi)d\xi}{[1 - (\xi/x)^{3/4}]^{1/3}}$$

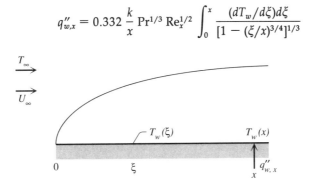

Uniform Heat Flux

$$T_w(x) - T_\infty = \frac{q''_w x}{0.453 k Pr^{1/3} Re_x^{1/2}}$$

(Continued)

Table 5.4, (continued)

Nonuniform Heat Flux

$$T_w(x) - T_\infty = \frac{0.623}{k\mathrm{Pr}^{1/3}\mathrm{Re}_x^{1/2}} \int_{\xi=0}^{x} \left[1 - \left(\frac{\xi}{x}\right)^{3/4}\right]^{-2/3} q_w''(\xi)\,d\xi$$

The second entry in Table 5.4 shows the wall heat flux distribution in the general case where the wall temperature is a certain function of longitudinal position, $T_w(\xi)$. The heat flux registered at the downstream position x is an integral result of the interaction between the wall and the fluid all along the wall section that precedes x, namely, $0 < \xi < x$. The wall with unheated starting length is clearly a simple case of the second solution listed in Table 5.4.

The last two entries in Table 5.4 refer to walls with specified heat flux. The uniform heat flux solution is a special case of the nonuniform wall heating configuration shown at the bottom of the table. In problems with specified wall heat flux, the unknown quantity is the wall–fluid temperature difference, $T_w(x) - T_\infty$, that is, the temperature variation along the wall. For example, the temperature of the wall with constant heat flux increases as $x^{1/2}$ in the downstream direction. Note also that the $T_w(x) - T_\infty$ formula listed for the wall with uniform heat flux agrees within a numerical factor of order 1 with the scale analysis result (5.78). In other words, the order-of-magnitude conclusions of Section 5.3.2 apply to a laminar boundary layer whose temperature T_w is not necessarily a constant.

5.3.4   Film Temperature

The wall friction and heat transfer results summarized in Section 5.3 are based conceptually on the constant property model adopted in Section 5.2. In real situations, fluid properties such as k, v, μ, and α are not constant, as they depend primarily on the local temperature in the flow field. It turns out that the constant property formulas describe sufficiently accurately the actual convective flows encountered in engineering applications, provided the maximum temperature variation experienced by the fluid $(T_w - T_\infty)$ is small relative to the *absolute* temperature level of the fluid $(T_w$, or T_∞, expressed in degrees Kelvin). In such cases the properties needed for calculating the various dimensionless groups (Re_x, Pe_x, Pr, $C_{f,x}$, Nu_x) can be evaluated at the average temperature of the fluid in the thermal boundary layer,

$$T = \tfrac{1}{2}\,(T_w + T_\infty) \tag{5.84}$$

This average is commonly recognized as the *film temperature* of the fluid, and is generally recommended for use in formulas of the constant property type. Worth keeping in mind is that there are special correlations in which the effect of temperature-dependent properties is taken into account by means of explicit correction factors. An example of this kind will be encountered in eq. (6.92).

Example 5.2

The Thermal Laminar Boundary Layer

The thin plate swept by laminar flow in Fig. 5.7 is a plate fin attached to a distant wall that is parallel to the plane of the figure. The length of the plate in the flow direction is 20 times greater than the plate thickness. The plate is isothermal, $T_w = 35\,°C$, and the water stream is somewhat colder, $T_\infty = 25\,°C$. Calculate the local heat flux at the trailing edge of the plate, the plate-averaged heat flux and heat transfer coefficient, and the total heat transfer rate between the plate and the water stream.

Solution.   The swept length of the plate fin is

$$x = 20 \times 0.5 \text{ mm} = 1 \text{ cm}$$

By evaluating the properties of water at the film temperature,

$$T = \frac{1}{2}\,(T_w + T_\infty) = \frac{1}{2}\,(35\,°C + 25\,°C) = 30\,°C$$

we first calculate the Reynolds number, to be sure that the entire boundary layer is laminar [see eq. (5.85)]:

$$\text{Re}_x = \frac{U_\infty x}{\nu} = 0.6\,\frac{\text{cm}}{\text{s}}\,1\text{ cm}\,\frac{\text{s}}{0.008\text{ cm}^2} = 75 \qquad \text{(laminar)}$$

The local heat flux at $x = 1$ cm can be estimated based on eq. (5.79b):

$$\text{Nu}_x = 0.332\,\text{Pr}^{1/3}\text{Re}_x^{1/2}$$

$$= 0.332(5.49)^{1/3}(75)^{1/2} = 5.07$$

$$\text{Nu}_x = \frac{q''_{w,x}}{T_w - T_\infty}\,\frac{x}{k} = 5.07$$

$$q''_{w,x} = 5.07(T_w - T_\infty)\,\frac{k}{x}$$

$$= 5.07\,(35-25)\text{ K }0.61\,\frac{\text{W}}{\text{m}\cdot\text{K}}\,\frac{1}{0.01\,\text{m}} = 3094 \text{ W/m}^2$$

The value of the heat flux averaged over the entire length x is furnished by eq. (5.82b):

$$\overline{\text{Nu}}_x = 0.664\,\text{Pr}^{1/3}\text{Re}_x^{1/2} = 10.14$$

$$\overline{q}''_{w,x} = 10.14\,(T_w - T_\infty)\,\frac{k}{x} = 6188 \text{ W/m}^2$$

The same values could have been calculated more rapidly by comparing eqs. (5.82b) and (5.79b) and noting that in the laminar boundary layer the x-averaged heat flux is twice the local heat flux evaluated at x:

$$\overline{Nu}_x = 2Nu_x = 10.14$$

$$\overline{q}''_{w,x} = 2q''_{w,x} = 6188 \text{ W/m}^2$$

$$\overline{h}_x = \frac{\overline{q}''_{w,x}}{T_w - T_\infty} = \frac{6188 \text{ W/m}^2}{10\,\text{K}} \cong 620 \text{ W/m}^2 \cdot \text{K}$$

The total heat transfer rate out of the plate fin follows after the observation that both sides of the plate are bathed by the water stream:

$$q' = 2x\,\overline{q}''_{w,x} = 2 \times 0.01 \text{ m } 6188 \text{ W/m}^2 = 124 \text{ W/m}$$

This heat transfer rate is expressed per unit length in the direction perpendicular to the plane of Fig. 5.7, that is, away from the wall that supports the fin.

5.4 TURBULENT BOUNDARY LAYER OVER A PLANE WALL

5.4.1 Transition from Laminar to Turbulent Flow

The laminar boundary layer flow discussed until now prevails in the leading section of the flow, that is, at small enough longitudinal distances from the leading edge (x), so that the x-based Reynolds number does not exceed a certain (critical) value:

$$\text{Re}_x \lesssim 5 \times 10^5 \qquad \text{(laminar)}$$
$$\text{Re}_x \gtrsim 5 \times 10^5 \qquad \text{(turbulent)} \tag{5.85}$$

The critical Reynolds number 5×10^5 is an approximate and much abbreviated way of recording the empirical observations of the laminar–turbulent transition along a plane wall. As illustrated in Fig. 5.10, the *transition section* occupies a finite range of x values (or a finite Re_x range), marked by a "beginning" and an "end" of transition. Furthermore, the position of the transition section along the wall depends strongly on the degree of smoothness of the free stream, that is, on the presence of flow disturbances (eddies) in the fluid outside the laminar boundary layer. The critical Reynolds number on the right side of eq. (5.85) can be as low as 2×10^4 when the free stream is strongly disturbed, and as high as 10^6 and even higher when the outer flow is extremely smooth.

The transition criterion can be expressed also in terms of a *local Reynolds number*, which is based on the local longitudinal velocity scale (U_∞) and the local thickness of the flow region. Taking the momentum thickness θ of eq. (5.59) as a measure of the transversal length scale of the velocity boundary layer, the

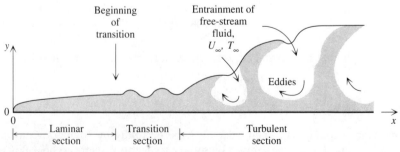

Figure 5.10 The succession of the laminar, transition, and turbulent sections in the boundary layer over a plane wall.

critical local Reynolds number $(U_\infty\theta/\nu)$ that corresponds to the observed transition range $2 \times 10^4 < \mathrm{Re}_x < 10^6$ is

$$\frac{U_\infty\theta}{\nu} \sim 94 - 660 \tag{5.86}$$

In conclusion, at transition the local Reynolds number has a value of order 10^2. It is shown in Appendix F that this local Reynolds number criterion governs also the other laminar–turbulent transitions of the flows treated in the remainder of this book.

Students are often confused that the boundary layer critical Reynolds number 5×10^5 is so much greater than the critical Reynolds number of only about 2000 for duct flow [see eq. (6.59) in the next chapter]. The main reason for this apparently huge discrepancy is that the two Reynolds numbers are defined differently. In boundary layer flow the critical Re is based on the *longitudinal* distance to the transition region, whereas in duct flow the critical Re is based on an appropriate *transversal* length scale of the duct [see "hydraulic diameter," eq. (6.28)]. When the boundary layer Reynolds number is rewritten in terms of the transversal length scale, it becomes the *local* Reynolds number and its critical value agrees much better with the critical value known for duct flow. The order-of-magnitude correlation of seemingly divergent critical numbers for transition in different flow configurations is one of the contributions of the local Reynolds number criterion (Appendix F).

One of the simplest and least expensive ways to visualize the transition to turbulence in the laminar boundary layer is by dropping a front-loaded piece of toilet paper through the air. Figure 5.11 shows the instantaneous shape of the

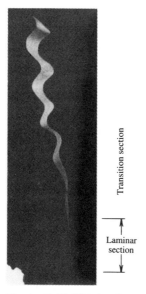

Transition section

Laminar
section

Figure 5.11 The laminar section and the beginning of transition in the air boundary layers over a 1.25-m piece of toilet paper falling to the ground: $U_\infty = 2.93$ m/s, ribbon width $= 11.4$ cm, ribbon mass $= 3.21$ g (Bejan [17]).

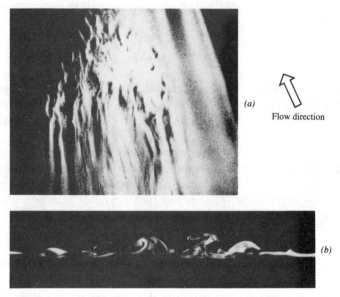

Figure 5.12 A well-developed turbulent spot viewed from above (*a*) and in cross section (*b*). (Perry et al. [18], with permission from Cambridge University Press.)

paper ribbon. The leading section is straight, as it is lined by laminar boundary layers on both sides. The transition region follows. The tissue paper is flexible enough to mimic the meandering, or buckling, shape of the flow in its first stages of transition.

The transition section in Fig. 5.11 appears two-dimensional only because of the narrowness of the paper ribbon. On a wide and rigid flat surface the transition is marked by local, three-dimensional deformations of the laminar flow. One such *turbulent spot* is illustrated in Fig. 5.12, in which the fluid is air and the visualization agent is smoke. The photographs were taken in the downstream section where Re_x was in the range $3.2 \times 10^4 - 5.7 \times 10^4$.

5.4.2 Time-Averaged Equations

In the boundary layer section situated immediately downstream from the transition section, Fig. 5.10, the elbows of the sinusoidal deformation become exaggerated, as they are sucked into the surrounding free stream. The deformed boundary layer is then rolled into eddies that tumble as they are swept down the wall. This chaotic tumbling flow constitutes the *turbulent section* of the boundary layer.

One way to visualize the difference between the laminar section and the turbulent section is to recall that the laminar section is ruled by a perfect balance between fluid inertia and by the transversal viscous diffusion of the friction effect of the wall, eq. (5.35). It can be said that in the laminar section the fluid is fully penetrated in the y direction by the effect of viscous diffusion. This effect requires a certain amount of time, and, as shown in Appendix F, in the transition section the boundary layer has become marginally too thick to allow viscous diffusion to traverse it completely.

In the transition section and, especially, in the subsequent turbulent section, the thickness of the flow regions penetrated by viscous diffusion is of the same order as the thickness of the laminar boundary layer just upstream of the transition section. Consequently, the turbulent section becomes a conglomerate of viscous layers rolled around pockets of inviscid fluid inhaled from the free stream. In Fig. 5.10 these inviscid regions are illustrated by means of white areas; they travel longitudinally with a velocity of order U_∞; therefore, when they make contact with the solid wall they generate local (and temporary) laminar layers similar to the leading laminar section of the boundary layer flow.

This description of the turbulent section betrays the influence of a modern trend in turbulence research, which recognizes the instantaneous large-scale structure of the turbulent flow field (see, for example, Chapter 7 of Ref. [3]). The classical approach to the analytical study of turbulent flows proceeds from an entirely different point of view. It begins with a time-averaging process—a smoothing out of the kinks—of the admittedly complicated turbulent flow field. It recognizes that a flow variable such as the longitudinal velocity component is a function of spatial position and time, $u(x,y,z,t)$. At a fixed location in the turbulent flow field, the u value fluctuates about a mean value $\bar{u}$ defined by the time-averaging operation

$$\bar{u} = \frac{1}{p} \int_0^p u\,dt \qquad (5.87)$$

which is illustrated in Fig. 5.13. The time-averaged value $\bar{u}$ is independent of time when the period p of the time-averaging operation exceeds the period of the slowest fluctuation exhibited by the actual variable u. The end result of this time-averaging procedure is the decomposition of the actual flow variable into a mean value plus a time-dependent correction (fluctuation) labeled u':

$$u(x,y,z,t) = \bar{u}(x,y,z) + u'(x,y,z,t) \qquad (5.88)$$

The same decomposition rule applies to the remaining variables of the flow field:

$$\begin{aligned} v &= \bar{v} + v' \qquad P = \bar{P} + P' \\ w &= \bar{w} + w' \qquad T = \bar{T} + T' \end{aligned} \qquad (5.89)$$

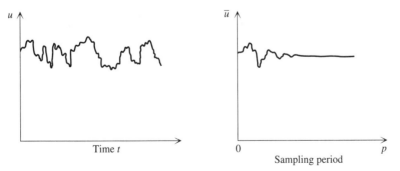

Figure 5.13 The behavior of an instantaneous quantity (u) in turbulent flow, and the calculation of the time-independent mean value $\bar{u}$ by using a long enough sampling period p.

The next step is the time averaging of the mass conservation, momentum, and energy equations assembled in Table 5.1. This step consists of substituting eqs. (5.88)–(5.89) into the governing equations, and then applying the time-averaging operation (5.87) to every term of the resulting equations. This analysis relies on a special set of algebraic rules that follow from eq. (5.87):

$$\overline{u'} = 0 \qquad\qquad \overline{\left(\dfrac{\partial u}{\partial x}\right)} = \dfrac{\partial \overline{u}}{\partial x}$$

$$\overline{u + v} = \overline{u} + \overline{v}$$

$$\overline{\overline{u}u'} = 0 \qquad\qquad \dfrac{\partial \overline{u}}{\partial t} = 0 \qquad\qquad (5.90)$$

$$\overline{uv} = \overline{u}\,\overline{v} + \overline{u'v'} \qquad \overline{\left(\dfrac{\partial u}{\partial t}\right)} = 0$$

$$\overline{u^2} = (\overline{u})^2 + \overline{(u')^2}$$

It is not difficult to demonstrate that in the end the governing equations reduce to the following [3]:

$$(m) \qquad \dfrac{\partial \overline{u}}{\partial x} + \dfrac{\partial \overline{v}}{\partial y} + \dfrac{\partial \overline{w}}{\partial z} = 0 \qquad\qquad (5.91)$$

$$(M_x) \qquad \overline{u}\,\dfrac{\partial \overline{u}}{\partial x} + \overline{v}\,\dfrac{\partial \overline{u}}{\partial y} + \overline{w}\,\dfrac{\partial \overline{u}}{\partial z} = -\dfrac{1}{\rho}\dfrac{\partial \overline{P}}{\partial x} + \nu\nabla^2\overline{u}$$

$$-\dfrac{\partial}{\partial x}(\overline{u'^2}) - \dfrac{\partial}{\partial y}(\overline{u'v'}) - \dfrac{\partial}{\partial z}(\overline{u'w'}) \qquad (5.92)$$

$$(M_y) \qquad \overline{u}\,\dfrac{\partial \overline{v}}{\partial x} + \overline{v}\,\dfrac{\partial \overline{v}}{\partial y} + \overline{w}\,\dfrac{\partial \overline{v}}{\partial z} = -\dfrac{1}{\rho}\dfrac{\partial \overline{P}}{\partial y} + \nu\nabla^2\overline{v}$$

$$-\dfrac{\partial}{\partial x}(\overline{u'v'}) - \dfrac{\partial}{\partial y}(\overline{v'^2}) - \dfrac{\partial}{\partial z}(\overline{v'w'}) \qquad (5.93)$$

$$(M_z) \qquad \overline{u}\,\dfrac{\partial \overline{w}}{\partial x} + \overline{v}\,\dfrac{\partial \overline{w}}{\partial y} + \overline{w}\,\dfrac{\partial \overline{w}}{\partial z} = -\dfrac{1}{\rho}\dfrac{\partial \overline{P}}{\partial z} + \nu\nabla^2\overline{w}$$

$$-\dfrac{\partial}{\partial x}(\overline{u'w'}) - \dfrac{\partial}{\partial y}(\overline{v'w'}) - \dfrac{\partial}{\partial z}(\overline{w'^2}) \qquad (5.94)$$

$$(E) \qquad \overline{u}\,\dfrac{\partial \overline{T}}{\partial x} + \overline{v}\,\dfrac{\partial \overline{T}}{\partial y} + \overline{w}\,\dfrac{\partial \overline{T}}{\partial z} = \alpha\nabla^2\overline{T}$$

$$-\dfrac{\partial}{\partial x}(\overline{u'T'}) - \dfrac{\partial}{\partial y}(\overline{v'T'}) - \dfrac{\partial}{\partial z}(\overline{w'T'}) \quad (5.95)$$

In these opening stages, we have treated the turbulent flow field as three-dimensional (u,v,w), because by their very nature turbulent flows are *instantaneously three-dimensional*, regardless of the simple shapes of the walls that may bound them. In the turbulent section of the boundary layer of Fig. 5.10, for example, the eddies can rotate not only in the plane of the figure but also into and out of that plane.

The turbulent boundary layer near a flat wall is *two-dimensional only as a time-averaged flow field*. This is why beginning with eqs. (5.91)–(5.95) we set

$\bar{w} = 0$ and $\partial( )/\partial z = 0$. The first consequence of this decision is that the z-momentum eq. (5.94) requires that $\partial \bar{P}/\partial z = 0$, in other words, that $\bar{P} = \bar{P}(x,y)$. Boundary layer theory and the y-momentum eq. (5.93) indicate further that $\bar{P}$ is mainly a function of x, which means that $\bar{P} = \bar{P}(x) = P_\infty$. And, since P_∞ is a constant in the free stream near a flat wall, the term $\partial \bar{P}/\partial x$ of eq. (5.92) is zero.

The second group of simplifications recommended by boundary layer theory (Section 5.3.1) is the neglect of the longitudinal derivatives $\partial( )/\partial x$ and $\partial^2( )/\partial x^2$ on the right side of eqs. (5.92)–(5.95). The final form of the boundary layer-simplified time-averaged equations for the turbulent section of the boundary layer is therefore

$$(m) \quad \frac{\partial \bar{u}}{\partial x} + \frac{\partial \bar{v}}{\partial y} = 0 \tag{5.96}$$

$$(M) \quad \bar{u}\frac{\partial \bar{u}}{\partial x} + \bar{v}\frac{\partial \bar{u}}{\partial y} = \frac{\partial}{\partial y}\left(v\frac{\partial \bar{u}}{\partial y} - \overline{u'v'}\right) \tag{5.97}$$

$$(E) \quad \bar{u}\frac{\partial \bar{T}}{\partial x} + \bar{v}\frac{\partial \bar{T}}{\partial y} = \frac{\partial}{\partial y}\left(\alpha\frac{\partial \bar{T}}{\partial y} - \overline{v'T'}\right) \tag{5.98}$$

The momentum and energy eqs. (5.97) and (5.98) must be compared with their laminar flow counterparts, eqs. (5.32) and (5.62), to appreciate the new features introduced by the time averaging of the turbulent flow. The products $\overline{u'v'}$ and $\overline{v'T'}$ survive the time-averaging operation, and increase by two the number of unknowns that could be determined by solving the system (5.96)–(5.98). The five unknowns in this three-equation system are $\bar{u}, \bar{v}, \bar{T}, \overline{u'v'}$, and $\overline{v'T'}$. The search for the two additional equations that are required to determine the five unknowns uniquely is activity recognized as *turbulence modeling*. To this we return briefly in the next section.

5.4.3 Eddy Diffusivities

It is customary to replace the time-averaged products $\overline{u'v'}$ and $\overline{v'T'}$ with the following expressions:

$$-\overline{u'v'} = \varepsilon_M \frac{\partial \bar{u}}{\partial y} \tag{5.99}$$

$$-\overline{v'T'} = \varepsilon_H \frac{\partial \bar{T}}{\partial y} \tag{5.100}$$

where the coefficient ε_M (m^2/s) is recognized as the *momentum eddy diffusivity* and ε_H (m^2/s) as the *thermal eddy diffusivity*, or the eddy diffusivity for heat. The momentum and energy equations become

$$(M) \quad \bar{u}\frac{\partial \bar{u}}{\partial x} + \bar{v}\frac{\partial \bar{u}}{\partial y} = \frac{\partial}{\partial y}\left[(v + \varepsilon_M)\frac{\partial \bar{u}}{\partial y}\right] \tag{5.101}$$

$$(E) \quad \bar{u}\frac{\partial \bar{T}}{\partial x} + \bar{v}\frac{\partial \bar{T}}{\partial y} = \frac{\partial}{\partial y}\left[(\alpha + \varepsilon_H)\frac{\partial \bar{T}}{\partial y}\right] \tag{5.102}$$

The ε_M and ε_H notation is recommended by the view that on the right side of eqs. (5.97)–(5.98) the time-averaged groups $(-\overline{u'v'})$ and $(-\overline{v'T'})$ are "eddy" contributions that augment the effects of molecular diffusion, which are represented respectively by $v(\partial \bar{u}/\partial y)$ and $\alpha(\partial \bar{T}/\partial y)$. Consider first the quantity placed inside the square brackets on the right side of eq. (5.101),

$$v\frac{\partial \bar{u}}{\partial y} + \varepsilon_M \frac{\partial \bar{u}}{\partial y} = \frac{1}{\rho}\left(\underbrace{\mu \frac{\partial \bar{u}}{\partial y}}_{\tau_{\text{mol}}} + \underbrace{\rho\varepsilon_M \frac{\partial \bar{u}}{\partial y}}_{\tau_{\text{eddy}}}\right) \tag{5.103}$$

where the terms indicated by τ_{mol} and τ_{eddy}, respectively, represent the usual (molecular) shear stress and the shear stress contribution made by the time-averaged effect of the eddies that act at the point to which eq. (5.101) applies. The sum of the molecular and eddy shear stresses is the *apparent shear stress*

$$\tau_{\text{app}} = \tau_{\text{mol}} + \tau_{\text{eddy}} \tag{5.104}$$

Similarly, the right side of the energy eq. (5.102) can be decomposed to see the molecular heat flux $-k(\partial \bar{T}/\partial y)$ and the eddy heat flux $-\rho c_P \varepsilon_H(\partial \bar{T}/\partial y)$:

$$\alpha \frac{\partial \bar{T}}{\partial y} + \varepsilon_H \frac{\partial \bar{T}}{\partial y} = -\frac{1}{\rho c_P}\left[\underbrace{\left(-k\frac{\partial \bar{T}}{\partial y}\right)}_{q''_{\text{mol}}} + \underbrace{\left(-\rho c_P \varepsilon_H \frac{\partial \bar{T}}{\partial y}\right)}_{q''_{\text{eddy}}}\right] \tag{5.105}$$

Both q''_{mol} and q''_{eddy} are defined as positive when pointing in the positive y direction, that is, away from the wall. Their sum represents the *apparent heat flux*

$$q''_{\text{app}} = q''_{\text{mol}} + q''_{\text{eddy}} \tag{5.106}$$

In conclusion, the momentum and energy eqs. (5.101) and (5.102) can be written in terms of the gradients of apparent shear stress and apparent heat flux:

$$(M) \qquad \bar{u}\frac{\partial \bar{u}}{\partial x} + \bar{v}\frac{\partial \bar{u}}{\partial y} = \frac{\partial}{\partial y}\left(\frac{\tau_{\text{app}}}{\rho}\right) \tag{5.107}$$

$$(E) \qquad \bar{u}\frac{\partial \bar{T}}{\partial x} + \bar{v}\frac{\partial \bar{T}}{\partial y} = \frac{\partial}{\partial y}\left(\frac{q''_{\text{app}}}{-\rho c_P}\right) \tag{5.108}$$

The net contribution of the analysis described between eqs. (5.99) and (5.108) is the replacement of the unknown quantities $\overline{u'v'}$ and $\overline{v'T'}$ with two alternative unknowns, ε_M and ε_H. These require the use of two additional equations (two assumptions), the simplest of which are described next.

To determine ε_M, imagine a packet of fluid that at some point in time is situated at a distance y away from the wall, in the turbulent boundary layer. The average longitudinal velocity at that location is $\bar{u}(x,y)$. Imagine, next, that this fluid packet rides on an eddy and migrates toward the wall to the new location $y - l$, where the mean velocity is $\bar{u}(x, y - l)$. The distance l is called the *mixing length* over which the fluid packet maintains its identity: this length is of the same order as the eddy diameter. The u' fluctuation induced by this migration at the new level $y - l$ is of the same order as

$$|u'| \sim \bar{u}(x,y) - \bar{u}(x,y-l) \sim l\frac{\partial \bar{u}}{\partial y} \tag{5.109}$$

Because of the wheel-like motion of the eddy, it can be argued [3] that the v'

fluctuation is of the same order as u', that is, the same as the peripheral speed of the eddy,

$$|v'| \sim l \frac{\partial \bar{u}}{\partial y} \qquad (5.110)$$

Substituting these u' and v' estimates into the ε_M definition (5.99), we obtain

$$\varepsilon_M \sim l^2 \left| \frac{\partial \bar{u}}{\partial y} \right| \qquad (5.111)$$

Measurements of the $\bar{u}$-vs.-y profile suggest that the mixing length l is proportional to the distance to the wall (like the diameter of the eddy that fits across the distance y):

$$l = \kappa y \qquad (5.112)$$

where $\kappa \cong 0.4$ is known as *von Karman's constant*. In summary, the mixing length model for the momentum eddy diffusivity reads [19]:

$$\varepsilon_M = \kappa^2 y^2 \left| \frac{\partial \bar{u}}{\partial y} \right| \qquad (5.113)$$

This is the simplest of many eddy diffusivity models that have been proposed. A compilation of these models is available in Ref. [3].

Regarding the thermal eddy diffusivity ε_H, the simplest model is the assumption that ε_H is approximately the same as ε_M. By analogy with the Prandtl number notation $\mathrm{Pr} = \nu/\alpha$, the *eddy diffusivity ratio*

$$\frac{\varepsilon_M}{\varepsilon_H} = \mathrm{Pr}_t \qquad (5.114)$$

is called the *turbulent Prandtl number*. Therefore, the ε_H model consists of writing that Pr_t is a constant approximately equal to 1. Measurements of the temperature distribution in turbulent boundary layers of $\mathrm{Pr} \gtrsim 1$ fluids recommend the value $\mathrm{Pr}_t \cong 0.9$.

5.4.4 Wall Friction

The writing of the momentum equation as eq. (5.101) or (5.107) allows us to view the turbulent boundary layer as a sandwich of two distinct regions. In the outer region the inertia of the flow [the left side of eq. (5.107)] is *finite* and negative, and the apparent shear stress decreases to zero as y reaches into the free stream. This feature is illustrated in Fig. 5.14. The outer layer is also called the wake region of the turbulent boundary layer.

Sufficiently close to the wall the inertia effect becomes negligible, and both sides of the momentum eq. (5.107) approach zero. Integrating in y, we reach the conclusion that in this inner layer τ_{app} is practically independent of y; in other words,

$$\tau_{\mathrm{app}} = \text{constant} = \tau_{w,x} \qquad (5.115)$$

where the wall shear stress $\tau_{w,x}$ is the value reached by τ_{app} right at the wall. (Note that ε_M and τ_{eddy} are zero at $y = 0$.) The inner layer is recognized also as the constant-τ_{app} region of the boundary layer.

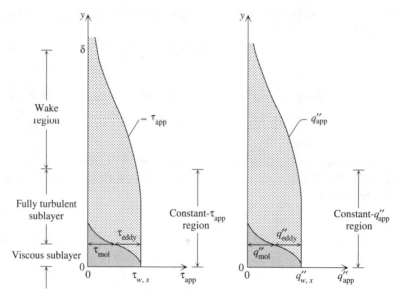

Figure 5.14 The structure of the turbulent boundary layer: composition of the apparent shear stress τ_{app} and the apparent heat flux q''_{app}.

Working now with eq. (5.115), in which τ_{app} is equal to the quantity contained between brackets on the right side of eq. (5.103), it is possible to derive an analytical expression for the velocity profile $\bar{u}$ in the constant-τ_{app} region. Figure 5.14 and eq. (5.113) show that immediately close to the wall, ε_M approaches zero and can be neglected in favor of ν. The region where $\nu \gg \varepsilon_M$ is called the *viscous sublayer* of the constant-τ_{app} region.

Immediately outside the viscous sublayer (and still inside the constant-τ_{app} region) resides the *fully turbulent sublayer*, in which ε_M is much greater than ν. In summary, the velocity distribution that is obtained for the constant-τ_{app} region by using eq. (5.115) and the mixing length model (5.113) is

$$u^+ = \begin{cases} y^+ & \text{(viscous sublayer, } \nu \gg \varepsilon_M) \\ \dfrac{1}{\kappa}\ln y^+ + B & \text{(fully turbulent sublayer, } \varepsilon_M \gg \nu) \end{cases} \tag{5.116}$$

in which u^+ and y^+ are the dimensionless "wall coordinates" defined by

$$u^+ = \frac{\bar{u}}{(\tau_{w,x}/\rho)^{1/2}} \qquad y^+ = \frac{y}{\nu}\left(\frac{\tau_{w,x}}{\rho}\right)^{1/2} \tag{5.117}$$

Measurements of the $u^+(y^+)$ distribution in the fully turbulent sublayer recommend the value $B \cong 5.5$ for the constant of integration appearing in the second of eqs. (5.116). By equating the u^+ values indicated by the two expressions stacked on the right side of eq. (5.116), we learn that the interface between the viscous sublayer and the fully turbulent sublayer is located at $y^+ \cong 11.6$. It has been shown [3] that this viscous sublayer thickness (namely, a y^+ thickness of order 10) is another manifestation of the local Reynolds number criterion for transition to turbulence, eq. (5.86) and Appendix F.

The physical meaning of $y^+ \sim 10$ as an approximate boundary between the viscous sublayer and the fully turbulent sublayer is illustrated in Fig. 5.15. The

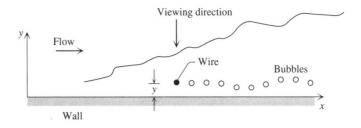

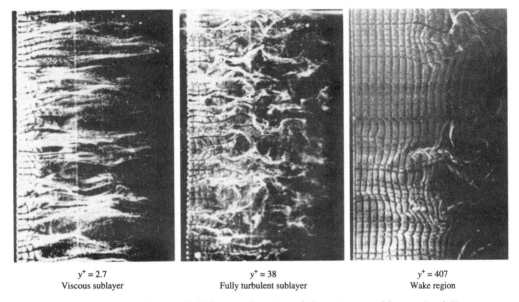

$y^+ = 2.7$	$y^+ = 38$	$y^+ = 407$
Viscous sublayer	Fully turbulent sublayer	Wake region

Figure 5.15 Hydrogen bubble visualization of the viscous sublayer, the fully turbulent sublayer, and the wake region. (Kline et al. [20], with permission from Cambridge University Press.)

photographs refer to the same water boundary layer, which flows from left to right and is being viewed from above. Hydrogen bubbles are released intermittently by a thin (0.02 mm diameter) platinum wire stretched perpendicular to the flow and parallel to the solid wall. The wire is visible along the left margin of each photograph. The distance between the wall and the wire is y.

In the first photograph ($y^+ = 2.7$) the bubbles are released into the viscous sublayer, generating streaks that are relatively smooth and parallel (i.e., unmixed). The second photograph ($y^+ = 38$) shows what happens when the bubbles are released from farther above the wall, in the fully turbulent sublayer. In this region the mixing is intense and three-dimensional. The third photograph refers to a large y^+ that places the bubble-generating wire outside the constant-τ_{app} region: In the wake region (Fig. 5.14, left) the flow is uniform except for periodic, large-scale eddies that roll between (and mix) the freestream and the constant-τ_{app} layer. These large eddies punch big holes into the regular grid formed by the hydrogen bubbles.

A simpler empirical $u^+(y^+)$ expression that approximates most of the curve represented by eq. (5.116) is the so-called Prandtl's $\frac{1}{7}$th power law [19]

$$u^+ = 8.7 \, (y^+)^{1/7} \qquad (5.118)$$

An equally compact formula for the wall shear stress $\tau_{w,x}$ is obtained by stretching the validity of eq. (5.118) all the way to the edge of the outer (wake) region, where $\bar{u} = U_\infty$ and $y = \delta$:

$$\frac{U_\infty}{(\tau_{w,x}/\rho)^{1/2}} = 8.7\left[\frac{\delta}{\nu}\left(\frac{\tau_{w,x}}{\rho}\right)^{1/2}\right]^{1/7} \tag{5.119}$$

This equation relates $\tau_{w,x}$ to the outer thickness of the turbulent boundary layer, δ. A second relation between $\tau_{w,x}$ and δ is the integral form of the full momentum eq. (5.101):

$$\frac{d}{dx}\int_0^\delta \bar{u}(U_\infty - \bar{u})dy = \frac{\tau_{w,x}}{\rho} \tag{5.120}$$

Using the $\frac{1}{7}$th power law (5.118) for $\bar{u}$ in the integrand, the system (5.119)–(5.120) can be solved for the wall shear stress and the boundary layer thickness:

$$\frac{\tau_{w,x}}{\rho U_\infty^2} = \frac{1}{2}C_{f,x} = 0.0296\left(\frac{U_\infty x}{\nu}\right)^{-1/5} \tag{5.121}$$

$$\bar{\tau}_{w,L} = 0.037\rho U_\infty^2 \mathrm{Re}_L^{-1/5} \tag{5.121'}$$

$$\frac{\delta}{x} = 0.37\left(\frac{U_\infty x}{\nu}\right)^{-1/5} \tag{5.122}$$

The derivation of the *average* shear stress $\bar{\tau}_{w,L}$, which is averaged over a wall of length L, is presented in eqs. (2) and (3) in Example 5.4 at the end of this section. Multiplied by the total area of the swept wall, the average shear stress permits the calculation of the total tangential drag force exerted by the flow on the plane wall.

The local skin friction coefficient formula (5.121) may be used at Reynolds numbers as high as $\mathrm{Re}_x \cong 10^8$. This formula shows that $\tau_{w,x}$ decreases as $x^{-1/5}$ in the downstream direction, that is, at a much slower rate than in the laminar section, eq. (5.53). This trend is illustrated in Fig. 5.16.

According to eq. (5.122), the boundary layer thickness increases as $x^{4/5}$,

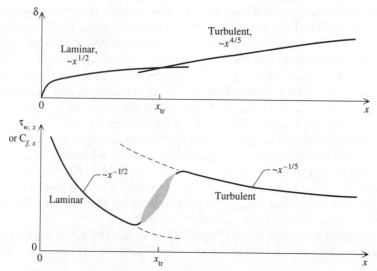

Figure 5.16 The behavior of the boundary layer thickness and wall shear stress in the laminar and turbulent sections of the boundary layer over a plane wall.

namely, almost linearly. This increase is considerably steeper than in the leading laminar section, where δ increases as $x^{1/2}$. This trend is presented also in Fig. 5.16. Unlike their purely theoretical counterparts in the laminar boundary layer, the analytical forms of the turbulent layer, specifically the exponents in $\tau_{w,x} \sim x^{-1/5}$ and $\delta \sim x^{4/5}$, are simply the result of having adopted the empirical curve fit (5.118) at the start of the analysis. In other words, different $u^+(y^+)$ expressions [e.g., eq. (5.116)] lead to different formulas for $\tau_{w,x}$, $\overline{\tau}_{w,L}$, and δ.

5.4.5 Heat Transfer

In $\text{Pr} \gtrsim 1$ fluids, the structure of the time-averaged temperature distribution in the turbulent boundary layer mimics that of the longitudinal velocity distribution. Figure 5.14 shows that sufficiently close to the wall the apparent heat flux q''_{app} is independent of the distance to the wall, y. This region is approximately of the same thickness as the constant-τ_{app} region sketched on the left side of the figure. In it, both sides of the energy eq. (5.102) or (5.108) are zero; therefore, integrating in y we obtain

$$q''_{\text{app}} = \text{constant} = q''_{w,x} \tag{5.123}$$

The local wall heat flux $q''_{w,x}$ is the special value of q''_{app} at $y = 0$, because right at the wall ε_H and q''_{eddy} are zero [review eq. (5.106) and the definitions under eq. (5.105)]. The constant-q''_{app} statement can be rewritten as

$$-(k + \rho c_P \varepsilon_H)\frac{\partial \overline{T}}{\partial y} = q''_{w,x} \tag{5.124}$$

It is possible to obtain an expression for the local heat transfer coefficient $h_x = q''_{w,x}/(T_w - T_\infty)$ by combining eq. (5.124) with the analogous equation recommended by the constant-τ_{app} condition (5.115):

$$(\mu + \rho \varepsilon_M)\frac{\partial \overline{u}}{\partial y} = \tau_{w,x} \tag{5.125}$$

Dividing these last two equations side by side,

$$\frac{\rho(\nu + \varepsilon_M)}{\rho c_P(\alpha + \varepsilon_H)}\frac{d\overline{u}}{d\overline{T}} = -\frac{\tau_{w,x}}{q''_{w,x}} \tag{5.126}$$

and assuming that $\nu = \alpha$ and $\varepsilon_M = \varepsilon_H$ (i.e., that $\text{Pr} = 1$ and $\text{Pr}_t = 1$), we obtain

$$\frac{1}{c_P}\frac{d\overline{u}}{d\overline{T}} = -\frac{\tau_{w,x}}{q''_{w,x}} \qquad (\text{Pr} = \text{Pr}_t = 1) \tag{5.127}$$

This can be integrated from the wall ($\overline{u} = 0$, $\overline{T} = T_w$ at $y = 0$) to a large enough y where $\overline{u} \cong U_\infty$ and $\overline{T} \cong T_\infty$, and the result is

$$\frac{U_\infty}{c_P(T_\infty - T_w)} = -\frac{\tau_{w,x}}{q''_{w,x}} \tag{5.128}$$

Rearranged, eq. (5.128) states that the local heat transfer coefficient $h_x = q''_{w,x}/(T_w - T_\infty)$ is proportional to the local wall shear stress $\tau_{w,x}$. This result can be nondimensionalized by defining the *local Stanton* number

---

*Thomas Edward Stanton (1865–1931) was professor of engineering at Bristol University College. He devoted his research to the relationship between heat transfer and fluid flow, the aerodynamic loads on solid structures, and the air cooling of internal combustion engines.

$$\mathrm{St}_x = \frac{h_x}{\rho c_p U_\infty} = \frac{q''_{w,x}}{\rho c_p U_\infty (T_w - T_\infty)} \tag{5.129}$$

or

$$\mathrm{St}_x = \frac{\mathrm{Nu}_x}{\mathrm{Pe}_x} = \frac{\mathrm{Nu}_x}{\mathrm{Re}_x \mathrm{Pr}} \tag{5.129'}$$

and by recognizing the $\frac{1}{2}C_{f,x}$ definition listed as the first of eqs. (5.121). The dimensionless form of eq. (5.128),

$$\mathrm{St}_x = \tfrac{1}{2}C_{f,x} \qquad (\mathrm{Pr} = \mathrm{Pr}_t = 1) \tag{5.130}$$

is known best as *the Reynolds analogy* between wall friction and heat transfer, in recognition of Osborne Reynolds' essay [21], in which he argued that a relationship between wall shear stress and heat flux must exist.

The preceding analysis shows that the Reynolds analogy holds strictly for $\mathrm{Pr} = 1$ and $\mathrm{Pr}_t = 1$. For fluids with molecular Prandtl numbers (Pr) different than 1, Colburn [22] proposed an empirical correlation that displays $\mathrm{Pr}^{2/3}$ as a factor:

$$\mathrm{St}_x \mathrm{Pr}^{2/3} = \frac{1}{2}C_{f,x} \qquad (\mathrm{Pr} \gtrsim 0.5) \tag{5.131}$$

This can be restated in terms of Nu_x and Re_x, in accordance with eqs. (5.121) and (5.129'):

$$\begin{aligned}
\mathrm{Nu}_x &= \tfrac{1}{2}C_{f,x}\mathrm{Re}_x \mathrm{Pr}^{1/3} \\
&= 0.0296\,\mathrm{Re}_x^{4/5}\mathrm{Pr}^{1/3} \qquad (\mathrm{Pr} \gtrsim 0.5)
\end{aligned} \tag{5.131'}$$

Equation (5.131) is known as *the Colburn* analogy* between wall friction and heat transfer, relative to which the Reynolds analogy (5.130) represents the special case of a $\mathrm{Pr} = 1$ fluid.

Although eq. (5.131') was developed for an isothermal wall, it works satisfactorily when the wall heat flux is uniform. The Nu_x value for a wall with uniform heat flux is only 4 percent greater than the value furnished by eq. (5.131'). Note further that when the wall heat flux is uniform, the local Nusselt number is defined by $\mathrm{Nu}_x = q''_w x / k[T_w(x) - T_\infty]$. Regardless of the thermal boundary condition that may exist at the wall, the reference temperature for evaluating the properties in eqs. (5.131) and (5.131') is the average film temperature $(\overline{T}_w + T_\infty)/2$, where $\overline{T}_w$ is the x-averaged wall temperature.

A theoretical derivation of Colburn's empirical formula (5.131) was offered in 1984 (Ref. [3], pp. 252–256), based on the view that the time-averaged wall shear and heat flux are dominated by an array of discrete regions of direct contact between the free-stream fluid (U_∞, T_∞) and the wall. According to this view, the shear flow under each contact spot is *laminar*, with characteristics similar to the leading laminar section of the boundary layer. An additional contribution of this theory is the proof that the Colburn analogy (5.131) applies only to $\mathrm{Pr} \gtrsim 0.5$ fluids. According to the same theory, for low-Pr fluids the factor

---

*Allan Philip Colburn (1904–1955) was professor of chemical engineering at the University of Delaware. His research focused on the condensation of water vapor and the analogy between heat, mass, and momentum transfer.

$Pr^{2/3}$ is replaced on the left side of eq. (5.131) by the group $cPr^{1/2}$, where c is a constant of order 10.

In summary, the local heat transfer coefficient h_x (or St_x) may be calculated using eq. (5.131), for which $\frac{1}{2}C_{f,x}$ is given by eq. (5.121). The proportionality between St_x and $\frac{1}{2}C_{f,x}$ means that the x dependence of the local heat transfer coefficient is similar to that of the wall shear stress. Figure 5.17 shows that in the turbulent section h_x decreases as $x^{-1/5}$ in the downstream direction. This behavior departs significantly from the $h_x \sim x^{-1/2}$ decrease that prevails in the laminar section of the boundary layer.

Consider finally the *average* heat transfer coefficient $\overline{h}_L$ for the wall of total length L shown in Fig. 5.17. This quantity is needed to calculate the total heat transfer rate through the wall, per unit length, in the direction normal to the (x,y) plane, namely, $q' = \overline{h}_L(T_w - T_\infty)L$. The value of $\overline{h}_L$ depends on how the transition length x_{tr} compares with the total length L:

$$\overline{h}_L = \frac{1}{L}\left(\int_0^{x_{tr}} \underset{\substack{\text{eq. (5.79b)}}}{h_{x,\,\text{laminar}}} \; dx + \int_{x_{tr}}^{L} \underset{\substack{\text{eqs. (5.131)}\\\text{and (5.121)}}}{h_{x,\,\text{turbulent}}} \; dx \right) \tag{5.132}$$

Using the h_x formulas indicated under each integrand, we find that the Nusselt number based on $\overline{h}_L$ and L depends on the longitudinal transition Reynolds number $(Re_{x,tr} = U_\infty x_{tr}/\nu)$:

$$\overline{Nu}_L = \frac{\overline{h}_L L}{k} = 0.664 Pr^{1/3}\, Re_{x,tr}^{1/2} + 0.037 Pr^{1/3}\,(Re_L^{4/5} - Re_{x,tr}^{4/5}) \tag{5.133}$$

Taking the right side of eq. (5.85) as representative of the transition Reynolds number, $Re_{x,tr} \cong 5 \times 10^5$, the above $\overline{Nu}_L$ formula becomes

$$\overline{Nu}_L = 0.037 Pr^{1/3}(Re_L^{4/5} - 23,550) \qquad (Pr \gtrsim 0.5) \tag{5.134}$$

In view of the ingredients used in its derivation, this formula is valid for $5 \times 10^5 < Re_L < 10^8$ and $Pr \gtrsim 0.5$. At Reynolds numbers Re_L smaller than 5×10^5, the entire length L is covered by laminar boundary layer flow, and the $\overline{Nu}_L$ formula is given by eq. (5.82b). A method of correlating a certain set of laminar *and* turbulent $\overline{Nu}_L$ data with a single formula over the entire Pr and Re_L ranges has been described by Churchill [23].

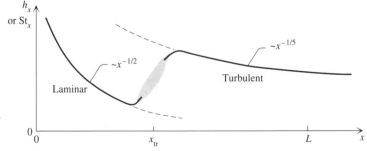

Figure 5.17 The behavior of the local heat transfer coefficient in the laminar and turbulent sections of the boundary layer. (Compare this figure with Fig. 5.16.)

Example 5.3

Heat Transfer, Turbulent Boundary Layer, Plane Surface

The flat (tabular) iceberg shown in the figure drifts over the ocean, as it is driven by the wind that blows over the top. The iceberg may be modeled as a block of frozen fresh water at $0°C$. The temperature of the surrounding seawater is $10°C$, and the relative velocity between it and the iceberg is 10 cm/s. The length of the iceberg in the direction of drift is $L = 100$ m.

The relative motion between the seawater and the flat bottom of the iceberg produces a boundary layer of length L. The $10°C$ temperature difference across this boundary layer drives a certain heat flux into the bottom surface of the iceberg. This heating effect causes the steady erosion (thinning) of the flat piece of ice. If $H(t)$ is the instantaneous height of the ice slab, calculate the ice melting rate dH/dt averaged over the swept length of the iceberg.

Solution. The following calculations are by definition approximate, because they are based on the rather crude model that the bottom surface of the iceberg is plane and that the drift is steady. This is why it is sufficient to approximate the properties of the seawater as those of fresh water at the film temperature of $5°C$:

$$\nu \cong 0.015 \text{ cm}^2/\text{s} \qquad k \cong 0.57 \text{ W/m·K} \qquad \text{Pr} \cong 11.13$$

First, we calculate the Reynolds number, to determine the prevailing flow regime:

$$\text{Re}_L = \frac{U_\infty L}{\nu} = 10 \frac{\text{cm}}{\text{s}} \, 100 \text{ m} \, \frac{\text{s}}{0.015 \text{ cm}^2}$$

$$= 6.7 \times 10^5 \qquad \text{(turbulent flow)}$$

Next, starting with eq. (5.134) we calculate, in order, the average Nusselt number, the L-averaged heat transfer coefficient, and the L-averaged heat flux that impinges on the bottom of the iceberg:

$$\overline{\text{Nu}}_L = 0.037\text{Pr}^{1/3}(\text{Re}_L^{4/5} - 23{,}550)$$

$$= 0.037(11.13)^{1/3}(2.89 \times 10^5 - 23{,}550) = 2.19 \times 10^4$$

$$\overline{h}_L = \frac{k}{L}\overline{\text{Nu}}_L = 0.57 \frac{\text{W}}{\text{m·K}} \frac{1}{100 \text{ m}} 2.19 \times 10^4 =$$

$$\cong 125 \text{ W/m}^2\text{·K}$$

$$\overline{q}''_{w,L} = \overline{h}_L \Delta T = 125 \frac{\text{W}}{\text{m}^2\text{·K}} 10°C$$

$$\cong 1250 \text{ W/m}^2$$

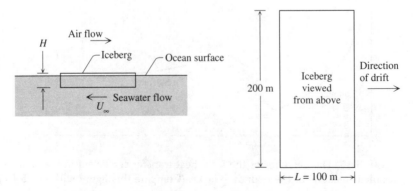

Figure E5.3

The average heat flux $\bar{q}''_{w,L}$ is absorbed by the melting process (review Section 4.7),

$$\bar{q}''_{w,L} = \rho h_{sf} \frac{dH}{dt}$$

where h_{sf} is the latent heat of melting for ice,

$$h_{sf} = 333.4 \text{ kJ/kg}$$

The average melting rate is therefore

$$\frac{dH}{dt} = \frac{\bar{q}''_{w,L}}{\rho h_{sf}} = 1250 \frac{W}{m^2} \frac{(0.1 \text{ m})^3}{1 \text{ kg}} \frac{kg}{333.4 \text{ kJ}}$$

$$= 3.75 \times 10^{-6} \text{ m/s} = 1.35 \text{ cm/h}$$

Example 5.4

Wall Friction, Turbulent Boundary Layer, Plane Surface

Seen from above, the flat iceberg described in the preceding example and Fig. E5.3 drifts stably in the direction of its shorter side, $L = 100$ m. The side oriented across the direction of drift is 200 m long. Calculate the time-averaged total drag force exerted by the water on the bottom surface of the iceberg.

Solution.   The total drag force is the area integral of all the shear stresses of type $\tau_{w,x}$,

$$F_D = \bar{\tau}_{w,L} A \tag{1}$$

where A is the swept area, $(100 \text{ m})(200 \text{ m}) = 2 \times 10^4 \text{ m}^2$, and $\bar{\tau}_{w,L}$ is the L-averaged shear stress

$$\bar{\tau}_{w,L} = \frac{1}{L} \int_0^L \tau_{w,x} dx \tag{2}$$

Assuming that the Reynolds number is high enough so that L is covered almost entirely by turbulent flow, we can use eq. (5.121) in the above integral; the result is

$$\bar{\tau}_{w,L} = 0.037 \rho U_\infty^2 \text{Re}_L^{-1/5} \tag{3}$$

In the present example this formula yields

$$\bar{\tau}_{w,L} = 0.037 \frac{1 \text{ kg}}{(0.1 \text{ m})^3} \left(10 \frac{cm}{s}\right)^2 (6.7 \times 10^6)^{-1/5}$$

$$\cong 0.016 \text{ kg/m} \cdot \text{s}^2 = 0.016 \text{ N/m}^2$$

Therefore, the total drag force on the bottom surface is

$$F_D = 0.016 \frac{N}{m^2}(2 \times 10^4 \text{ m}^2) = 320 \text{ N} \cong 72 \text{ lbf}$$

For steady drift this bottom drag force must be balanced by a driving drag force associated with the air flow over the top of the iceberg.

5.5   OTHER EXTERNAL FLOWS

5.5.1   Single Cylinder in Cross-Flow

In the preceding two sections we uncovered the most basic aspects of laminar and turbulent convection by focusing on the simplest wall geometry — the flat plate. These aspects are present also in more complicated configurations and account for the largest portion of the external convection literature and handbooks. The present section is a problem solving-oriented review of some of the simplest external convection results that have been obtained for nonplane walls.

Consider the heat transfer between a long cylinder oriented across a fluid stream of uniform velocity U_∞ and temperature T_∞, Fig. 5.18. The temperature of the cylindrical surface is uniform, T_w. There are many heat transfer correlations for this configuration but, generally speaking, they are not in very good agreement with the experimental data. One correlation that is based on data from many independent sources was developed by Churchill and Bernstein [24],

$$\overline{Nu}_D = 0.3 + \frac{0.62 Re_D^{1/2} Pr^{1/3}}{[1 + (0.4/Pr)^{2/3}]^{1/4}} \left[1 + \left(\frac{Re_D}{282,000} \right)^{5/8} \right]^{4/5} \tag{5.135}$$

where $\overline{Nu}_D = \overline{h}D/k$. This formula holds for all values of Re_D and Pr, provided the Peclet number $Pe_D = Re_D Pr$ is greater than 0.2. In the intermediate Reynolds number range $7 \times 10^4 < Re_D < 4 \times 10^5$, eq. (5.135) predicts $\overline{Nu}_D$ values that can be 20 percent smaller than those furnished by direct measurement. The physical properties needed for calculating Nu_D, Pr, and Re_D are evaluated at the film temperature $(T_\infty + T_w)/2$. Equation (5.135) applies also to a cylinder with uniform heat flux, in which case the average heat transfer coefficient $\overline{h}$ is based on the perimeter-averaged temperature difference between the cylindrical surface and the free stream.

In flows slow enough so that $Pe_D < 0.2$, more accurate than eq. (5.135) is a formula due to Nakai and Okazaki [25]:

$$\overline{Nu}_D = \frac{1}{0.8237 - 0.5 \ln (Pe_D)} \tag{5.136}$$

This agrees well with experimental measurements conducted in air; however, it has not been tested for a wide range of the Prandtl numbers.

The single cylinder in cross-flow has been studied extensively also from a fluid mechanics standpoint. The first portion of the surrounding flow that

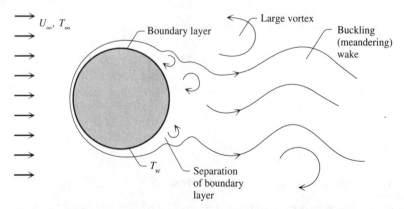

Figure 5.18 Single cylinder (or sphere) in cross-flow, and some of the features of the surrounding flow.

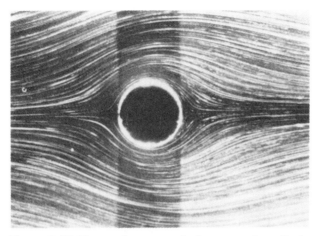

Figure 5.19 The nearly symmetric flow around a cylinder when the Reynolds number is small: $Re_D = 1.1$, glycerine–water solution, $U_\infty = 0.2$ cm/s, $D = 1$ cm. (Nakayama et al. [12], with permission from Pergamon Press.)

becomes turbulent as the Reynolds number $Re_D = U_\infty D/\nu$ increases is the wake. When the Reynolds number is of order 1 or smaller, the flow is nearly symmetric about the transversal diameter of the cylinder. A flow of this type is illustrated in Fig. 5.19, in which it is not easy to tell whether the fluid flows to the right or to the left (correct answer: to the right). This long-exposure photograph shows the path followed by particles of aluminum powder sprinkled on the surface of the liquid that flows through an open channel. The cylinder rises as a vertical "tree trunk" in this stream.

Eddies of size comparable with the cylinder itself begin to "shed" periodically when $Re_D \sim 40$: this is another manifestation of the local Reynolds number criterion of transition to turbulence (Appendix F). The meandering wake is visualized by means of lines of hydrogen bubbles in Fig. 5.20.

The wake becomes increasingly turbulent as Re_D increases; however, the shedding of vortices of diameter comparable with D remains a feature of the wake over most of the $10^2 < Re_D < 10^7$ range. The frequency $f_v(s^{-1})$ with which the vortices shed is roughly proportional to the flow velocity [26]:

$$f_v \sim 0.21 \frac{U_\infty}{D} \tag{5.137}$$

The vortex-shedding frequency is important in the design of cylinders in cross-flow (e.g., tubes in a heat exchanger) because it may induce and sustain vibrations in the mechanical structure. The various flow transitions, regimes, and slight departures from eq. (5.137) have been reviewed by Lienhard [26].

The total, time-averaged drag force F_D exerted by the flow on the cylinder can be calculated with the help of Fig. 5.21. Plotted on the ordinate is the *drag coefficient* C_D, which is a dimensionless way of expressing the drag force [27]:

$$C_D = \frac{F_D/A}{\frac{1}{2}\rho U_\infty^2} \tag{5.138}$$

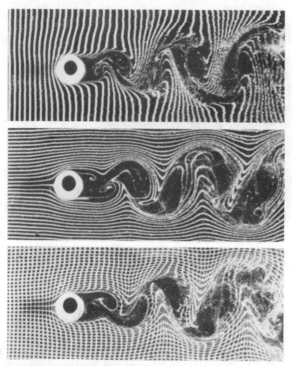

Figure 5.20 The meandering wake behind a cylinder in cross-flow: $Re_D = 170$, water, $U_\infty = 2.6$ cm/s, $D = 0.8$ cm; time lines, top; streak lines, middle; time lines and streak lines, bottom. (Nakayama et al. [12], with permission from Pergamon Press.)

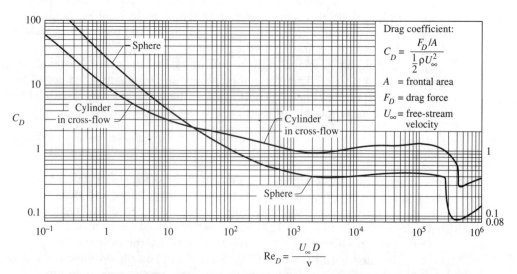

Figure 5.21 Drag coefficients of a smooth sphere and a single smooth cylinder in cross-flow. (Drawn with permission, after the data compiled in H. Schlichting, *Boundary Layer Theory*, translated by J. Kestin, 4th edition, McGraw-Hill, New York, 1960, p. 16.)

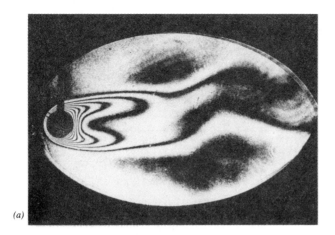

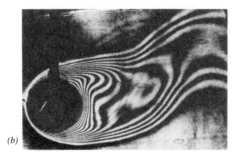

Figure 5.22 Isotherms around a cylinder suspended in a cross-flow of air: (*a*) $Re_D = 100$, $T_w = 55°C$, $T_\infty = 18.5°C$; (*b*) $Re_D = 150$, $T_w = 90°C$, $T_\infty = 18.5°C$. (Sallam and Mitunaga [29], courtesy of Dr. Akiharu Mitunaga, Osaka University.)

The area A is the *frontal* area of the cylinder (as seen by the approaching stream), namely, $A = LD$, if L is the length of the cylinder. The drag coefficients of several other body shapes have been summarized by Simiu and Scanlan [28].

The temperature distribution around a cylinder in cross-flow is illustrated by the interferometer photographs shown in Fig. 5.22. The Reynolds number range is comparable with that of Fig. 5.20. The isothermal bands show the thermal boundary layer that coats the cylinder, and the undulating wake. The wake temperature decreases in the downstream direction. The periodic shedding of vortices from the upper and lower downstream portions of the cylinder is especially visible in Fig. 5.22*b*.

Example 5.5

Single Cylinder in Cross-Flow

A pin fin of average temperature 100°C extends into a 1-m/s uniform air stream of temperature 200°C. The pin fin diameter is 4 mm. Calculate the average heat transfer coefficient between the pin fin and the surrounding air. Estimate also the drag force per unit of fin length.

Solution. The air properties are first evaluated at the film temperature of $(100°C + 200°C)/2 = 150°C$:

$$v = 0.288 \text{ cm}^2/\text{s} \qquad k = 0.036 \text{ W/m·K}$$

$$\rho = 0.846 \text{ kg/m}^3 \qquad \text{Pr} = 0.69$$

The average heat transfer coefficient can now be calculated by using eq. (5.135):

$$\text{Re}_D = \frac{U_\infty D}{v} = 1 \frac{\text{m}}{\text{s}} 0.004 \text{ m} \frac{\text{s}}{0.288 \times 10^{-4} \text{ m}^2}$$

$$= 138.9$$

$$\overline{\text{Nu}}_D = 0.3 + 5.66 \ (1 + 0.0086)^{4/5} \cong 6$$

$$\bar{h} = \overline{\text{Nu}}_D \frac{k}{D} = 6 \times 0.036 \frac{\text{W}}{\text{m·K}} \frac{1}{0.004 \text{ m}}$$

$$= 54 \text{ W/m}^2 \cdot \text{K}$$

For the drag force calculation we turn to Fig. 5.21, which shows that $C_D \cong 1.7$ when $\text{Re}_D \cong 140$. If L is the length of the fin, the frontal area is $A = DL$, and the resulting drag force per unit fin length is

$$\frac{F_D}{L} = C_D D \frac{1}{2} \rho U_\infty^2$$

$$= 1.7 \times 0.004 \text{ m} \frac{1}{2} 0.846 \frac{\text{kg}}{\text{m}^3} \left(1 \frac{\text{m}}{\text{s}} \right)^2$$

$$\cong 0.003 \text{ N/m}$$

5.5.2 Sphere

For the average heat transfer coefficient between an isothermal spherical surface (T_w) and an isothermal free stream (U_∞, T_∞), Fig. 5.18, Whitaker [30] proposed the correlation

$$\overline{\text{Nu}}_D = 2 + (0.4\text{Re}_D^{1/2} + 0.06\text{Re}_D^{2/3})\text{Pr}^{0.4} \left(\frac{\mu_\infty}{\mu_w} \right)^{1/4} \tag{5.139}$$

This relation has been tested for $0.71 < \text{Pr} < 380$, $3.5 < \text{Re}_D < 7.6 \times 10^4$, and $1 < (\mu_\infty/\mu_w) < 3.2$. All the physical properties in eq. (5.139) are evaluated at the free-stream temperature T_∞, except μ_w, which is the viscosity evaluated at the surface temperature, $\mu_w = \mu(T_w)$. It is worth noting that the no-flow limit of this formula, $\overline{\text{Nu}}_D = 2$, agrees with the pure conduction estimate for steady radial conduction between the spherical surface and the motionless, infinite conducting medium that surrounds it (Section 2.3). Equation (5.139) applies also to spherical surfaces with uniform heat flux, with the understanding that in such cases $\overline{\text{Nu}}_D$ is based on the surface-averaged temperature difference between the sphere and the surrounding stream, $\overline{\text{Nu}}_D = \bar{h}D/k = q_w'' D/k(\overline{T}_w - T_\infty)$.

Figure 5.21 shows the dimensionless drag coefficient for a sphere suspended in a uniform stream. The total, time-averaged drag force F_D experienced by the holder of the sphere can be calculated with eq. (5.138), in which the frontal area this time is $A = (\pi/4)D^2$.

The regimes that are exhibited by the flow around the sphere are similar to those encountered in the case of a cylinder in cross-flow. An air flow example is

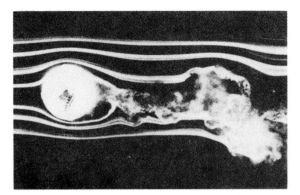

Figure 5.23 The flow around a baseball traveling at 53 miles/h and spinning at 630 rpm. (Photograph by Prof. F. N. M. Brown, University of Notre Dame, courtesy of Prof. T. J. Mueller, University of Notre Dame.)

visualized with smoke in Fig. 5.23, which shows the presence of vortex shedding (i.e., a buckling, meandering wake) from a baseball in a 53 miles/h wind tunnel. The sinuous wake is visible even though the baseball (a "curve ball") rotates at 630 rpm about an axis perpendicular to the figure.

5.5.3   Other Body Shapes

The single cylinder and the sphere discussed in the preceding two subsections are the simplest geometries of bodies immersed in a uniform flow of different temperature. The heat transfer literature contains a wealth of analogous results for bodies of other shapes. Some of these formulas have been reviewed critically by Yovanovich [31], who also proposed a universal correlation for *spheroids*, that is, bodies of nearly spherical shape.

As illustrated in the lower part of Fig. 7.16, a spheroid can be obtained by rotating an ellipse about one of the semiaxes. The spheroid geometry is characterized by two dimensions (C, B), where C is the semiaxis aligned with the free stream U_∞. As length scale for the definition of the Reynolds and Nusselt numbers, Yovanovich chose the square root of the spheroid surface A,

$$\mathcal{L} = A^{1/2} \tag{5.140}$$

therefore

$$\mathrm{Re}_\mathcal{L} = \frac{U_\infty \mathcal{L}}{\nu} \quad \text{and} \quad \overline{\mathrm{Nu}}_\mathcal{L} = \frac{\bar{h}\mathcal{L}}{k} \tag{5.141}$$

The universal correlation developed by Yovanovich [31] is

$$\overline{\mathrm{Nu}}_\mathcal{L} = \overline{\mathrm{Nu}}_\mathcal{L}^0 + \left[0.15 \left(\frac{p}{\mathcal{L}} \right)^{1/2} \mathrm{Re}_\mathcal{L}^{1/2} + 0.35 \mathrm{Re}_\mathcal{L}^{0.566} \right] \mathrm{Pr}^{1/3} \tag{5.142}$$

where p is the maximum (equatorial) perimeter of the spheroid, perpendicular to the flow direction U_∞. The constant $\overline{\mathrm{Nu}}_\mathcal{L}^0$ is the overall Nusselt number in the no-flow (pure conduction) limit: representative values of this constant are listed in Table 7.1, which is to be read in conjunction with Fig. 7.16. Equation (5.142) is recommended for $0 < \text{Re}_L < 2 \times 10^5$, $\text{Pr} > 0.7$, and $0 < C/B < 5$. The physical properties involved in the definitions of $\overline{\text{Nu}}_L$ and Re_L are evaluated at the film temperature.

5.5.4 Arrays of Cylinders in Cross-Flow

A considerably more complicated geometry is that of a large number of regularly spaced (parallel) cylinders in cross-flow. As indicated in Fig. 5.24, this geometry is characterized by the cylinder diameter D, the longitudinal spacing of two consecutive rows (longitudinal pitch, X_l), and the transversal spacing of two consecutive cylinders (transversal pitch, X_t). It is assumed that the array is wide enough, that is, enough cylinders exist in each row, so that the top and bottom boundaries (the shroud) do not affect the overall flow and heat transfer characteristics of the array.

In the field of heat exchanger design, arrays such as those of Fig. 5.24 form when *tube banks* or *tube bundles* are placed perpendicular to a stream of fluid. Assume, for example, that this stream is warm. A second stream of cold fluid flows through the many tubes of the bundle, and, because of the heat transfer interaction between each tube and the external cross-flow, the second stream picks up the enthalpy lost by the warm stream. This two-stream arrangement, and the hardware that allows the streams to experience heat transfer without mixing with one another, is an example of a *heat exchanger* (see Chapter 9).

The two most common types of arrays are the cases where the cylinders are *aligned* one behind the other in the direction of flow (Fig. 5.24, left), and where the cylinders are *staggered* (Fig. 5.24, right). Aligned cylinders form rectangles with the centers of their cross sections, whereas staggered cylinders form isosceles triangles. The flow patterns have been visualized in Fig. 5.25 by sprinkling aluminum powder on the surface of a water channel flow in which the cylinders act as islands.

An enormous amount of work has been published on the heat transfer performance of banks of cylinders in cross-flow: the most up-to-date review of this work was presented by Zukauskas [33]. The overall Nusselt number formulas presented below are based on Zukauskas' recommendations, where

$$\overline{\text{Nu}}_D = \frac{\overline{h}D}{k} \tag{5.143}$$

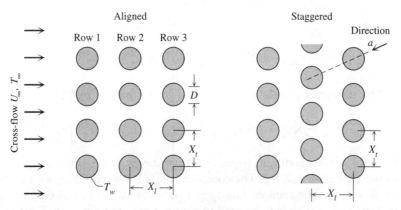

Figure 5.24 Banks of cylinders in cross-flow: aligned array (left) vs. staggered array (right).

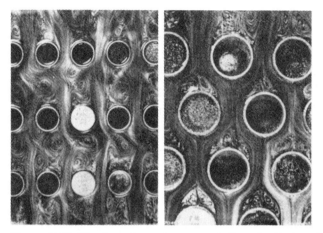

Figure 5.25 The flow through arrays of aligned cylinders (left) and staggered cylinders (right): water, $U_\infty = 3$ cm/s, $D = 4.2$ cm; the flow direction is upward in the photographs (Wallis [32]).

and $\bar{h}$ is the heat transfer coefficient averaged over all the cylindrical surfaces in the array. The total area of these surfaces is $nm\pi DL$, where n is the number of rows, m is the number of cylinders in each row (across the flow direction), and L is the length of the array in the direction perpendicular to the plane of Fig. 5.24.

For *aligned* arrays of cylinders, the array-averaged Nusselt number is anticipated within ± 15 percent by

$$\overline{Nu}_D = 0.9 C_n Re_D^{0.4} Pr^{0.36} \left(\frac{Pr}{Pr_w} \right)^{1/4}, \qquad Re_D = 1 - 10^2$$

$$= 0.52 C_n Re_D^{0.5} Pr^{0.36} \left(\frac{Pr}{Pr_w} \right)^{1/4}, \qquad = 10^2 - 10^3 \qquad (5.144)$$

$$= 0.27 C_n Re_D^{0.63} Pr^{0.36} \left(\frac{Pr}{Pr_w} \right)^{1/4}, \qquad = 10^3 - (2 \times 10^5)$$

$$= 0.033 C_n Re_D^{0.8} Pr^{0.4} \left(\frac{Pr}{Pr_w} \right)^{1/4}, \qquad = (2 \times 10^5) - (2 \times 10^6)$$

where C_n is a function of the total number of rows in the array (Fig. 5.26). The Reynolds number Re_D is based on the average velocity through the narrowest cross section formed by the array, that is, the *maximum* average velocity U_{max},

$$Re_D = \frac{U_{max}D}{\nu} \qquad (5.145)$$

In the case of aligned cylinders, the narrowest flow cross section forms in the plane that contains the centers of all the cylinders of one row. The conservation of mass through such a plane requires (see Fig. 5.24, left) that

$$U_\infty X_t = U_{max}(X_t - D) \qquad (5.146)$$

All the physical properties except Pr_w in eqs. (5.144) are evaluated at the *mean* temperature of the fluid that flows through the spaces formed between the cylinders. The mean (or "bulk") temperature is a concept defined in Section

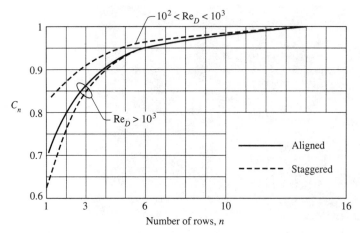

Figure 5.26 The effect of the number of rows on the array-averaged Nusselt number for banks of cylinders in cross-flow. (Drawn after Zukauskas [33].)

6.2.2, in which the cross-flow of Fig. 5.24 can be viewed as a stream that flows through the "duct" constituted by all the spaces between cylinders. The denominator Pr_w is the Prandtl number evaluated at the temperature of the cylindrical surface, $Pr_w = Pr(T_w)$.

Figure 5.26 shows that the number of rows has an effect on the array-averaged Nusselt number only when n is less than approximately 16. In the $n < 16$ range, C_n and $\overline{Nu}_D$ increase as more rows are added to the array. This effect is analogous to the observation that the individual $\bar{h}$ value of a cylinder positioned in the front row is lower than that of the cylinder situated behind it. The front row cylinder is coated by a relatively smooth boundary layer formed by the undisturbed incoming stream U_∞, whereas a downstream cylinder benefits from the heat transfer "augmentation" effect provided by the eddies of the turbulent wake created by the preceding cylinder.

The effect of the turbulent wake is visualized in Fig. 5.27, where the array is staggered and made up of parallel plates. The growing interaction between the shed vortices and the downstream plates as Re increases is blamed for the generation of noise in plate fin heat exchangers.

For *staggered* arrays, the following relations approximate $\overline{Nu}_D$ within ± 15 percent [33]:

$$\overline{Nu}_D = 1.04 C_n Re_D^{0.4} Pr^{0.36} \left(\frac{Pr}{Pr_w}\right)^{1/4}, \qquad Re_D = 1-500$$

$$= 0.71 C_n Re_D^{0.5} Pr^{0.36} \left(\frac{Pr}{Pr_w}\right)^{1/4}, \qquad = 500-10^3 \qquad (5.147)$$

$$= 0.35 C_n Re_D^{0.6} Pr^{0.36} \left(\frac{Pr}{Pr_w}\right)^{1/4} \left(\frac{X_t}{X_l}\right)^{0.2}, \qquad = 10^3-(2 \times 10^5)$$

$$= 0.031 C_n Re_D^{0.8} Pr^{0.36} \left(\frac{Pr}{Pr_w}\right)^{1/4} \left(\frac{X_t}{X_l}\right)^{0.2}, \qquad = (2 \times 10^5)-(2 \times 10^6)$$

The observations made in connection with eqs. (5.144) apply here as well. A new effect is the role played by the aspect ratio of the isosceles triangle, X_t/X_l,

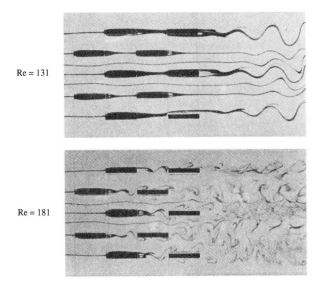

Re = 131

Re = 181

Figure 5.27 Staggered array of plate fins, and the effect of the turbulent wake on the downstream plates as Re increases: dye injection in water, $Re = U_{max}\delta/\nu$, $\delta = 2$ mm. (Mochizuki and Yagi [34], courtesy of Prof. S. Mochizuki, Tokyo University of Agriculture and Technology.)

which is felt at relatively large Reynolds numbers. The Reynolds number continues to be based on U_{max}, eq. (5.145); however, which flow cross section is the narrowest depends on the slenderness of the isosceles triangle. For example, in the staggered array shown on the right side of Fig. 5.24, the narrowest flow area occurs in the vertical plane drawn through the centers of one row of cylinders. In the other extreme, where X_l is considerably smaller than X_t, the strangling of the flow may occur through the area aligned with the sloped direction labeled a.

Means for calculating the drag experienced by a bundle of cylinders in cross-flow are presented in Section 9.5.4. The aggregate effect of the drag forces of all the cylinders is the pressure drop experienced by the stream that bathes the bundle.

5.5.5   Turbulent Jets

There are many flows and convection phenomena in which the solid boundaries are totally absent or irrelevant. Prime examples are the momentum-driven jets discussed below and the buoyancy-driven plumes treated in Ref. [3]. These flows are particularly important in environmental applications, where, instead of the usual heat transfer problem (Section 1.2), the chief question is to what extent the stream mixes with and spreads through the surrounding fluid. Since the flow regime is usually turbulent and the solid boundaries are absent, these flows are recognized under the general title of *free turbulent flows*.

In this section we consider the simplest and most basic of the nonbuoyant free turbulent flows, namely, the round turbulent jet (Fig. 5.28). A more extensive coverage of this class of convection problems can be found in Chapter 8 of Ref. [3]. The jet of Fig. 5.28 issues from a round nozzle of inner diameter D_0 with a cross section-averaged velocity U_0 and a uniform nozzle temperature T_0. The

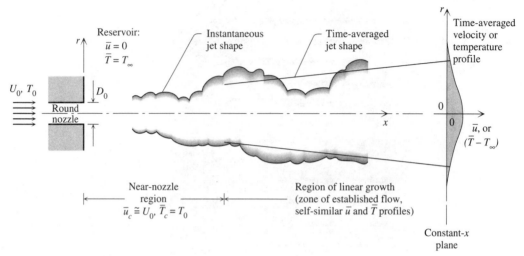

Figure 5.28 Turbulent, nonbuoyant round jet mixing with a stagnant reservoir of different temperature.

jet flows into a stationary infinite reservoir containing the same type of fluid as the jet itself. The temperature of the reservoir fluid, T_∞, is, in general, different than the bulk temperature in the plane of the nozzle, T_0.

Experiments with turbulent jets indicate that the time-averaged velocity field $\bar{u}(r,x)$ has two distinct regions. In the near-nozzle region of axial length of approximately six nozzle diameters, the centerline velocity $\bar{u}_c = \bar{u}(0,x)$ is practically independent of x. In this first region the eddy mixing that operates across the imaginary cylindrical surface between jet and reservoir has not had time to reach the jet centerline. The time-averaged centerline temperature $\bar{T}_c = \bar{T}(0,x)$ is also constant and equal to T_0 in this region.

The jet velocity and temperature distributions described below refer exclusively to the second region of the jet, which develops at x values greater than approximately $6D_0$. In this downstream region it is observed that the radial thickness of the $\bar{u}(r,x)$ and $\bar{T}(r,x)$ profiles increases almost linearly with x. This key feature is illustrated in Fig. 5.28 along with the characteristic radial variation (bell-shaped profile) of both $\bar{u}$ and $\bar{T}$. The two regions of the jet are visible also in Fig. 5.29, which is worth comparing with Fig. 5.28.

An analytical solution for the time-averaged flow and temperature field can be developed by exploiting the observation that the downstream region of the jet is shaped like a cone. This observation leads, among other things, to the conclusion that the eddy diffusivity ε_M is a constant inside this region of the jet [3]. More practical than the analytical solution is a set of Gaussian expressions that correlate the experimental measurements furnished by many investigators [35]:

$$\bar{u} = \bar{u}_c \exp\left[-\left(\frac{r}{b}\right)^2\right], \qquad b \cong 0.107x \tag{5.148}$$

$$\bar{T} - \bar{T}_\infty = (\bar{T}_c - T_\infty)\exp\left[-\left(\frac{r}{b_T}\right)^2\right], \qquad b_T \cong 0.127x \tag{5.149}$$

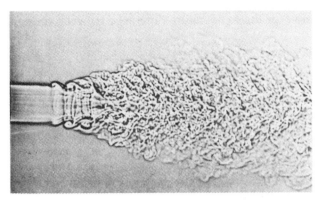

Figure 5.29 Turbulent round jet of carbon dioxide discharging into air: $D_0 \cong 0.64$ cm, $U_0 \cong 39$ m/s. (Photograph by Profs. F. Landis and A. H. Shapiro, courtesy of Prof. A. H. Shapiro, Massachusetts Institute of Technology.)

The linear growth of the jet mixing region is visible in the empirical expressions listed above for the transversal length scales $b(x)$ and $b_T(x)$. The numerical coefficients 0.107 and 0.127 are accurate within ± 3 percent, and are based on experiments conducted mainly in air and water, that is, in Pr ~ 1 fluids.

The solution, eqs. (5.148) and (5.149), is complete only when the x-dependent *centerline* quantities $\bar{u}_c$ and $\bar{T}_c - T_\infty$ are known. These follow from two important theorems, the proofs of which are proposed as problems at the end of this chapter. The first theorem states that the total longitudinal momentum in any constant-x cut through the jet is conserved (i.e., is independent of x):

$$2\pi \int_0^\infty \rho \bar{u}^2 r\, dr = \text{constant} = \rho U_0^2 \frac{\pi}{4} D_0^2 \tag{5.150}$$

The value of the constant was evaluated by carrying out the left-side integral in the plane of the nozzle, where $\bar{u} = U_0$. Combined, eqs. (5.148) and (5.150) deliver the needed formula for the centerline velocity,

$$\bar{u}_c \cong 6.61 \frac{U_0 D_0}{x} \tag{5.151}$$

Noteworthy is the fact that the centerline velocity decreases as x^{-1} in the downstream direction and that it is equal to the mean nozzle velocity U_0 at $x \cong 6.61 D_0$ downstream from the nozzle. This location marks the beginning of the linear growth region of the turbulent jet, Fig. 5.28.

The second theorem states that the energy flow through any constant-x plane of the jet is conserved:

$$2\pi \int_0^\infty \rho c_p \bar{u}(\bar{T} - T_\infty) r\, dr = \text{constant} = \rho c_p U_0 (T_0 - T_\infty)\frac{\pi}{4} D_0^2 \tag{5.152}$$

Again, the constant listed on the right side was determined by performing the integral in the plane of the nozzle, where $\bar{u} = U_0$ and $\bar{T} = T_0$. Using both eqs. (5.148) and (5.149) in the integrand on the left side of eq. (5.152), we obtain the wanted expression for the temperature difference between the centerline and the reservoir,

$$\overline{T}_c - T_\infty \cong 5.65 \frac{(T_0 - T_\infty)D_0}{x} \tag{5.153}$$

This result shows that the excess centerline temperature decreases as x^{-1} in the downstream direction. Together, eqs. (5.149) and (5.153) describe the extent to which the hotness or coldness of the jet has spread into the otherwise isothermal reservoir.

When a jet is oriented perpendicular to a solid surface, the spot that is struck by the jet is characterized by a relatively high heat transfer coefficient. Arrays of circular or two-dimensional (curtain) *impinging jets* can be used to generate an enhanced heat transfer coefficient over a wider area. Some of the recent literature on impinging gas jets was reviewed by Martin [36]. The application of impinging water jets to the cooling of high-speed and high-temperature metallurgical processes was described recently by Zumbrunnen et al. [37].

REFERENCES

1. A. Bejan, *Advanced Engineering Thermodynamics*, Wiley, New York, 1988.

2. W. C. Reynolds and H. C. Perkins, *Engineering Thermodynamics*, 2nd edition, McGraw-Hill, New York, 1977, p. 348.

3. A. Bejan, *Convection Heat Transfer*, Wiley, New York, 1984.

4. W. M. Rohsenow and H. Y. Choi, *Heat, Mass and Momentum Transfer*, Prentice-Hall, Englewood Cliffs, NJ, 1961, p. 48.

5. J. B. J. Fourier, Mémoire d'analyse sur le mouvement de la chaleur dans les fluides, *Memoires de l'Academie Royale des Sciences de l'Institut de France*, Didot, Paris, 1833, pp. 507–530 (presented on Sept. 4, 1820).

6. J. B. J. Fourier, *Oevres de Fourier*, G. Darboux, editor, Vol. 2, Gauthier-Villars, Paris, 1890, pp. 595–614.

7. S. D. Poisson, *Théorie Mathématique de la Chaleur*, Paris, 1835, Chapter 4, p. 86.

8. A. Russell, The convection of heat from a body cooled by a stream of fluid, *Phil. Mag.*, sixth series, Vol. 20, July–December 1910, pp. 591–610.

9. L. Prandtl, Über Flüssigkeitsbewegung bei sehr kleiner Reibung, Proc. 3rd Int. Math. Congr., Heidelberg, 1904; also NACA TM 452, 1928.

10. H. Blasius, Grenzschichten in Flüssigkeiten mit kleiner Reibung, *Z. Math. Phys.*, Vol. 56, 1908, pp. 1–37; also NACA TM 1256.

11. H. Schlichting, *Boundary Layer Theory*, translated by J. Kestin, 4th ed., McGraw-Hill, New York, 1960, pp. 116–123.

12. Y. Nakayama, W. A. Woods, D. G. Clark, and the Japan Society of Mechanical Engineers, Eds., *Visualized Flow*, Pergamon Press, Oxford, 1988.

13. E. Pohlhausen, Der Wärmeaustausch zwischen festen Körpern und Flüssigkeiten mit kleiner Reibung und kleiner Wärmeleitung, *Z. Angew. Math. Mech.*, Vol. 1, 1921, pp. 115–121.

14. S. W. Churchill and H. Ozoe, Correlations for laminar forced convection with uniform heating in flow over a plate and in developing and fully developed flow in a tube, *J. Heat Transfer*, Vol. 95, 1973, pp. 78–84.

15. W. M. Kays and M. E. Crawford, *Convective Heat and Mass Transfer*, McGraw-Hill, New York, 1980, p. 151.

16. E. R. G. Eckert, *Introduction to the Transfer of Heat and Mass*, McGraw-Hill, New York, 1959.

17. A. Bejan, The meandering fall of paper ribbons, *Phys. Fluids*, Vol. 25, 1982, pp. 741–742.

18. A. E. Perry, T. T. Lim, and E. W. Teh, A visual study of turbulent spots, *J. Fluid Mechanics*, Vol. 104, 1981, pp. 387–405.

19. L. Prandtl, *Essentials of Fluid Dynamics*, Blackie & Son, London, 1969, p. 117.

20. S. J. Kline, W. C. Reynolds, F. A. Schraub, and P. W. Runstadler, The structure of turbulent boundary layers, *J. Fluid Mechanics*, Vol. 30, 1967, pp. 741–773.

21. O. Reynolds, On the extent and action of the heating surface for steam boilers, *Proc. Manchester Lit. Phil. Soc.*, Vol. 14, 1874, pp. 7–12.

22. A. P. Colburn, A method for correlating forced convection heat transfer data and a comparison with fluid friction, *Trans. Am. Inst. Chem. Eng.*, Vol. 29, 1933, pp. 174–210; reprinted in *Int. J. Heat Mass Transfer*, Vol. 7, 1964, pp. 1359–1384.

23. S. W. Churchill, A comprehensive correlating equation for forced convection from flat plates, *A.I.Ch.E. J.*, Vol. 22, 1976, pp. 264–268.

24. S. W. Churchill and M. Bernstein, A correlating equation for forced convection from gases and liquids to a circular cylinder in crossflow, *J. Heat Transfer*, Vol. 99, 1977, pp. 300–306.

25. S. Nakai and T. Okazaki, Heat transfer from horizontal circular wire at small Reynolds and Grashof numbers—I Pure convection, *Int. J. Heat Mass Transfer*, Vol. 18, 1975, pp. 387–396.

26. J. H. Lienhard, *A Heat Transfer Textbook*, 2nd edition, Prentice-Hall, Englewood Cliffs, NJ, 1987, p. 346.

27. H. Schlichting, *Boundary Layer Theory*, translated by J. Kestin, 4th edition, McGraw-Hill, New York, 1960, p. 16.

28. E. Simiu and R. H. Scanlan, *Wind Effects on Structures*, 2nd edition, Wiley, New York, 1986, pp. 143–152.

29. T. M. Sallam and A. Mitunaga, Visualization and computational studies on the formation and stability of Karman vortex street, *Flow Visualization III*, W. J. Yang, Ed., Hemisphere, Washington, DC, 1983, pp. 353–358.

30. S. Whitaker, Forced convection heat transfer correlations for flow in pipes, past flat plates, single cylinders, single spheres, and flow in packed beds and tube bundles, *AIChE J*, Vol. 18, 1972, pp. 361–371.

31. M. M. Yovanovich, General expression for forced convection heat and mass transfer from isopotential spheroids, Paper No. AIAA 88-0743, AIAA 26th Aerospace Sciences Meeting, Reno, Nevada, January 11–14, 1988.

32. R. P. Wallis, Photographic study of fluid flow between banks of tubes, *Engineering*, Vol. 148, 1939, pp. 423–426.

33. A. A. Zukauskas, Convective heat transfer in cross flow, Chapter 6 in S. Kakac, R. K. Shah, and W. Aung, Eds., *Handbook of Single-Phase Convective Heat Transfer*, Wiley, New York, 1987.

34. S. Mochizuki and Y. Yagi, Characteristics of vortex shedding in plate arrays, *Flow Visualization II*, W. Merzkirch, Ed., Hemisphere, Washington, DC, 1980, pp. 99–103.

35. H. B. Fischer, E. J. List, R. C. Y. Koh, J. Imberger, and N. H. Brooks, *Mixing in Inland and Coastal Waters*, Academic Press, New York, 1979, Chapter 9.

36. H. Martin, Heat and mass transfer between impinging gas jets and solid surfaces, *Advances in Heat Transfer*, Vol. 13, 1977.

37. D. A. Zumbrunnen, R. Viskanta, and F. P. Incropera, The effect of surface motion on forced convection film boiling heat transfer, *J. Heat Transfer*, Vol. 111, 1989, pp. 760–766.

38. A. Bejan, Theory of rolling contact heat transfer, *J. Heat Transfer*, Vol. 111, 1989, pp. 257–263.

39. E. M. Sparrow, Reexamination and correction of the critical radius for radial heat conduction, *AIChE J*, Vol. 16, p. 149, 1970.

40. L. D. Simmons, Critical thickness of insulation accounting for variable convection coefficient and radiation loss, *J. Heat Transfer*, Vol. 98, 1976, pp. 150–152.

PROBLEMS

Basic Principles of Convection

5.1 Retrace the steps between eqs. (5.8) and (5.11). Show that the derivation of eq. (5.11) requires the use of the constant-μ and nearly constant-ρ models.

5.2 Perform the two-step analysis outlined in connection with eqs. (5.16) and (5.17), and show that the left side of the energy equation, eq. (5.14), can be expressed in terms of temperature and pressure gradients, as in eq. (5.19).

Laminar Boundary Layers

5.3 The free-stream water temperature in the laminar boundary layer of Fig. 5.8 in the text is $T_\infty = 25°C$. The temperature of the wall is $T_w = 24°C$. By using the data listed in the figure caption and the properties of 25°C water, calculate the local wall heat flux at $x = 20$ cm and the heat flux averaged over the wall section of length x.

5.4 Assume that the entire swept length of the thin plate photographed in Fig. 5.7 is $x = 10$ cm (only the leading 0.5 cm or so is shown). The water free-stream temperature is 30°C, and the uniform temperature of the plate is 20°C. By using the properties of distilled water at the appropriate film temperature, calculate

(a) The average heat transfer coefficient, the average heat flux, and the total heat transfer rate received by the plate; and

(b) The total drag force experienced by the plate.

5.5 Assume that the thin plate photographed in Fig. 5.7 is an electrical resistance that is being cooled on both sides by the laminar boundary layer flow of water. The width of the resistance (the length swept by the flow) is $x = 10$ cm. Assume further that the water is pure and that its temperature is $T_\infty = 20°C$. The heat flux is distributed uniformly over the swept length, $q''_w = 1000$ W/m². Calculate

(a) The plate temperature at its trailing edge (x), and

(b) The x-averaged temperature of the plate.

5.6 Consider the laminar boundary layer formed by the flow of 10°C water over a 10°C flat wall of length L. Show that the total shear force experienced by the wall and the mechanical power P spent on dragging the wall through the fluid is proportional to $v^{1/2}$.

(a) The dissipated drag power described above refers to the case in which the wall is as cold as the free-stream water. Show that if the wall is heated isothermally so that its temperature rises to 90°C, the dissipated power decreases by 35 percent. In other words, show that $P_h/P_c = 0.65$, where h and c refer to the hot-wall and cold-wall conditions.

(b) Compare the power savings due to heating the wall $(P_c - P_h)$ with the electrical power needed to heat the wall to 90°C. How fast must the water flow be so that the savings in fluid-friction power dissipation become greater than the electrical power invested in heating the wall? How short must the swept length L be so that the boundary layer remains laminar while the power savings $P_c - P_h$ exceed the heat input to the wall?

5.7   The wind blows at 0.5 m/s parallel to the short side of a flat roof with the rectangular area 10 m × 20 m. The roof temperature is 40°C, and the temperature of the air free stream is 20°C. Calculate the total force experienced by the roof. Estimate also the total heat transfer rate by laminar forced convection, from the roof to the atmosphere.

5.8   Make a qualitative sketch of how the local heat flux varies along an isothermal wall bathed by a laminar boundary layer of total length L. Use $q''_{w,x}$ on the ordinate and x on the abscissa. On the same sketch, draw a horizontal line at the level that would correspond to the heat flux averaged over the entire length of the plate, $\bar{q}''_{w,L}$. Determine analytically

   (a) The position x where the local heat flux matches the value of the L-averaged heat flux, and

   (b) The relationship between the midpoint local flux and the L-averaged value, that is, the ratio $q''_{w,L/2}/\bar{q}''_{w,L}$.

5.9   Consider the sharp-edged entrance to a round duct of diameter D. The laminar boundary layer that forms over the duct length L is much thinner than the duct diameter. The temperature difference between the duct wall (isothermal) and the inflowing stream is ΔT. The longitudinal inlet velocity of the stream is U_∞. Derive expressions for the total force F experienced by the duct section of length L, and the total heat transfer rate from the duct wall to the stream, q. In the end, show that q and F are proportional:

$$\frac{q}{F} = \mathrm{Pr}^{-2/3}\frac{c_P\Delta T}{U_\infty} \qquad (\mathrm{Pr}\gtrsim 0.5)$$

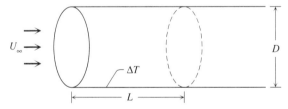

Figure P5.9

5.10   A stream of 20°C water enters a duct the wall temperature of which is uniform and equal to 50°C. The water inlet velocity is 5 cm/s. The duct cross section is a 20 cm × 20 cm square. Assume that the thickness of the boundary layer that lines the inner surface of the duct is much smaller than 20 cm, and calculate

   (a) The local heat transfer coefficient at $x = 1$ m downstream from the mouth,

   (b) The total heat transfer rate between the duct section of length $x = 1$ m and the water stream, and

   (c) The velocity boundary layer thickness (δ_{99}) at $x = 1$ m. Verify in this way the validity of the assumption that δ_{99} is much smaller than the duct width.

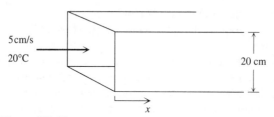

Figure P5.10

5.11 Perform the integral analysis of the laminar velocity boundary layer illustrated in Fig. 5.5 and follow these steps:

(a) Show that the momentum equation, eq. (5.32), is equivalent to

$$\frac{\partial(u^2)}{\partial x} + \frac{\partial(uv)}{\partial y} = v\frac{\partial^2 u}{\partial y^2}$$

(b) Integrate this new momentum equation across the boundary layer and show that

$$\frac{d}{dx}\int_0^\delta u(U_\infty - u)dy = v\left(\frac{\partial u}{\partial y}\right)_{y=0}$$

where $\delta(x)$ is the unknown thickness of the boundary layer.

(c) Assume a reasonable shape for the u-vs.-y profile,

$$\frac{u}{U_\infty} = m(n) \qquad \text{where } n = \frac{y}{\delta(x)}$$

and $m(n)$ is one of the functions listed in the left column of the table below. Show that the integral analysis produces the following $\delta(x)$ and $C_{f,x}$ estimates [3]:

Assumed Profile		$\dfrac{\delta}{x}\text{Re}_x^{1/2}$	$C_{f,x}\text{Re}_x^{1/2}$
Linear:	$m = n$	3.46	0.577
Parabolic:	$m = \dfrac{n}{2}(3 - n^2)$	4.64	0.646
Sine arc:	$m = \sin\left(\dfrac{\pi n}{2}\right)$	4.8	0.654

5.12 Carry out the integral analysis of the thermal boundary layer near the isothermal wall of Fig. 5.9 (bottom), that is, when the fluid has a large Prandtl number. Execute the following steps:

(a) Show that the boundary layer energy equation, eq. (5.62), is equivalent to

$$\frac{\partial(uT)}{\partial x} + \frac{\partial(vT)}{\partial y} = \alpha\frac{\partial^2 T}{\partial y^2}$$

(b) Integrate the energy equation across the thermal boundary layer, and arrive at the "integral" energy equation

$$\frac{d}{dx}\int_0^{\delta_T} u(T_\infty - T)dy = \alpha\left(\frac{\partial T}{\partial y}\right)_{y=0}$$

(c) Assume the parabolic temperature profile

$$\frac{T_w - T}{T_w - T_\infty} = \frac{p}{2}(3 - p^2) \qquad \text{where } p = \frac{y}{\delta_T}$$

and note that the corresponding velocity boundary layer thickness $\delta(x)$ is listed in the middle column of the table in the preceding problem statement. Note also that since $\text{Pr} \gg 1$, you can use the approximation $\delta \gg \delta_T$. Show finally that the local Nusselt number produced by this integral analysis is

$$\text{Nu}_x = 0.331 \text{Pr}^{1/3} \text{Re}_x^{1/2}$$

5.13 Consider the boundary layers δ_T and δ near an isothermal wall, Fig. 5.9 (top). In the low-Pr limit δ is negligible relative to δ_T, therefore the flow in the δ_T-thick region is uniform and parallel:

$$u = U_\infty \qquad \text{and} \qquad v = 0$$

(a) Show that in this limit the energy equation for the thermal boundary layer has the same form as eq. (4.22) of Section 4.3.1. Recognize further the analogy between the boundary conditions in your problem and those of the unsteady conduction problem of Section 4.3.1.

(b) Note the different notations employed in the two problems, and deduce from eq. (4.42) the closed-form expression for the temperature distribution $T(x,y)$ in the low-Pr thermal boundary layer.

(c) Determine analytically the local heat flux and the local Nusselt number; in other words, prove the validity of eq. (5.79a).

5.14 It is important to note that the aveage Nusselt number $\overline{\text{Nu}}_x$ is *not* the result of averaging the local Nusselt number Nu_x over the downstream length of the boundary layer, that is,

$$\overline{\text{Nu}}_x \neq \frac{1}{x} \int_0^x \text{Nu}_x dx'$$

The proper definition of $\overline{\text{Nu}}_x$ is eq. (5.81). It is the average wall heat flux $\overline{q}''_{w,x}$ (or the average heat transfer coefficient $\overline{h}_x$) that is the result of averaging the local quantity $q''_{w,x}$ (or h_x) over the distance x [review eq. (5.80)]. Keep this observation in mind as you derive the $\overline{\text{Nu}}_x$ expressions (5.82a, b), by starting with the local Nusselt number formulas listed as eqs. (5.79a, b).

5.15 The plane wall shown in the figure is swept by the laminar boundary layer flow of an isothermal fluid (T_∞, U_∞) with Prandtl number greater than 0.5. Deposited on the surface of this wall is a narrow strip of metallic film that runs parallel to the leading edge of the wall, that is, in the direction normal to the plane of the

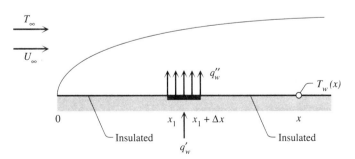

Figure P5.15

figure. As part of an electrical circuit, the strip generates Joule heating at the rate q'_w(W/m), or as the heat flux $q''_w = q'_w/\Delta x$ = constant. The width of the strip is much smaller than the distance to the leading edge, $\Delta x \ll x_1$.

Consult Table 5.4 and determine analytically the wall temperature distribution over the unheated downstream portion $x > x_1$. In other words, determine the "thermal wake" effect of the strip conductor. Assume that the entire Joule heating effect q'_w can only escape through the fluid side of the metallic strip. In other words, assume that the wall side of the strip is adiabatic.

5.16 Use the formula for the nonuniform heat flux case (the fourth in Table 5.4), set $q''_w(\xi)$ = constant, and derive in this way the temperature distribution along a wall with uniform heat flux (the third in Table 5.4).

5.17 A plane wall of length L is cooled by the laminar boundary layer flow of a fluid with $Pr \gtrsim 0.5$. The wall is heated electrically so that it releases the uniform heat flux q'' over the front half of its swept length, $0 < x < L/2$. The trailing half $L/2 < x < L$ is without heat transfer. Consult Table 5.4 and determine analytically the wall–fluid temperature difference at the trailing edge, $T_w(L) - T_\infty$.

A simpler (approximate) approach would be to assume that the total heat transfer rate described above ($q''L/2$) is distributed uniformly over the entire length L. Determine the trailing edge temperature difference, compare it with the previous estimate, and comment on the accuracy of this approximate approach.

5.18 An isothermal flat strip is swept by a parallel stream of water with a temperature of 20°C and free-stream velocity of 0.5 m/s. The width of the strip, $L = 1$ cm, is parallel to the flow. The temperature difference between the strip and the free stream is $\Delta T = 1$°C. Calculate the L-averaged shear stress $\bar\tau_{w,L}$ and the L-averaged heat flux $\bar{q}''_{w,L}$ between the strip and the water flow.

5.19 An unspecified fluid with $Pr = 17.8$ flows parallel to a flat isothermal wall and develops a laminar boundary layer along the wall. It is known that at a certain location x along the wall the local skin friction coefficient is equal to 0.008. Calculate the value of the local Nusselt number at the same location.

5.20 It is proposed to estimate the uniform velocity U_∞ of a stream of air of temperature 20°C, by measuring the temperature of a thin metallic blade that is heated and inserted parallel to U_∞ in the air stream. The width of the blade (i.e., the dimension aligned with U_∞) is $L = 2$ cm. The blade is considerably longer in the direction normal to the attached figure; therefore, the boundary layer flow that develops is two-dimensional. The blade is heated volumetrically by an electric current, so that 0.03-W electrical power is dissipated in each square centimeter of metallic blade. It is assumed that the blade is so thin that the effect of heat conduction through the blade (in the x direction) is negligible. A temperature sensor mounted on the trailing edge of the blade reads $T_w = 30$°C. Calculate the free stream velocity U_∞ that corresponds to this reading.

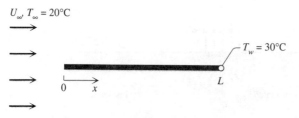

Figure P5.20

Turbulent Boundary Layers

5.21 Demonstrate that the time-averaged version of the mass conservation equation, eq. (m) of Table 5.1, is eq. (5.91). In the derivation of eq. (5.91), use the velocity decomposition relationships (5.88) and (5.89), and the appropriate algebraic rules listed as eqs. (5.90).

5.22 Derive the time-averaged form of the energy equation in Cartesian coordinates, eq. (5.95), by starting with eq. (E) of Table 5.1, in which $\dot{q} = 0$ and $\mu\Phi = 0$. In this analysis, use the decomposition formulas (5.88) and (5.89) and the time-averaging rules (5.90).

5.23 Derive the two-curve expression for the longitudinal velocity profile $u^+(y^+)$ in turbulent boundary layer flow near a flat wall, eq. (5.116). Begin with the constant-τ_{app} condition, eq. (5.115), and in the fully turbulent sublayer ($\varepsilon_M >> \nu$) use the mixing-length model (5.113).

5.24 Derive the formulas for skin friction coefficient and boundary layer thickness for turbulent flow over a flat plate, eqs. (5.121) and (5.122). Base this derivation on the $\frac{1}{7}$th power law profile (5.118). As starting condition for evaluating the left-side integral of eq. (5.120), use $\delta = 0$ at $x = 0$.

5.25 (a) Show that the following general relationships exist between the Stanton, Nusselt, Reynolds, Peclet, and Prandtl numbers in boundary layer flow:

$$St_x = \frac{Nu_x}{Re_x Pr} = \frac{Nu_x}{Pe_x}$$

(b) Show that the Colburn analogy (5.131) applies also to the *laminar* section of the boundary layer near an isothermal wall if the fluid has a Prandtl number in the range $Pr \gtrsim 0.5$.

5.26 Evaluate the average heat transfer coefficient for a flat wall of length L, with laminar and turbulent flow on it, eq. (5.132). Show that in $Pr \gtrsim 0.5$ fluids the average Nusselt number $\overline{Nu_L}$ is given by eq. (5.133).

5.27 The laminar section of the paper ribbon falling in Fig. 5.11 has the approximate length $x \cong 0.3$ m. The air temperature is $20°C$.

(a) Calculate the Re_x value for the end of the laminar section and compare it with the transition criterion recommended in the text.

(b) Imagine that the ribbon is straight (rigid) and covered with laminar boundary layers over its entire length. Assume also that the terminal velocity of free fall has been reached. Calculate this velocity and compare it with the measured value listed in the caption of Fig. 5.11. Why is the calculated value larger?

5.28 Examine again the information listed in the caption of Fig. 5.11 and imagine that the paper ribbon remains straight and vertical as it falls to the ground. The air temperature is $20°C$.

(a) Calculate the terminal free-fall velocity of the ribbon, $U_{\infty,turb}$, by assuming that the boundary layers are turbulent over the entire length L.

(b) Had we assumed instead that the flow is laminar over L, the answer to part (a) would have been $U_{\infty, lam} = 11.7$ m/s. Compare your $U_{\infty,turb}$ estimate with the measured value (Fig. 5.11 caption) and $U_{\infty,lam}$, and comment on which method of calculation is more accurate and why.

5.29 Water flows with the velocity $U_\infty = 0.2$ m/s parallel to a plane wall. The following calculations refer to the position $x = 6$ m measured downstream from the leading edge. The water properties can be evaluated at $20°C$.

(a) A probe is to be inserted in the viscous sublayer to the position represented by $y^+ = 2.7$. Calculate the actual spacing y (mm) between the probe and the wall.

(b) Calculate the boundary layer thickness δ, and compare this value with the estimate based on the assumption that the length x is covered by turbulent boundary layer flow.

(c) Calculate the heat transfer coefficient averaged over the length x.

5.30   It is proposed here to reduce the drag experienced by the hull of a ship by heating the hull to a high enough temperature so that the viscosity of the water in the boundary layer decreases. Evaluate the merit of this proposal in the following steps:

(a) Model the hull as a flat wall of length L that is swept by turbulent boundary layer flow. Show that the power spent on dragging the wall through the water is proportional to $v^{1/5}$.

(b) Assume that the hull temperature is raised to 90°C, while the water free-stream temperature is 10°C. Calculate the relative decrease in the drag power, a decrease that is caused by the heating of the wall.

(c) Compare the decrease (savings) in drag power with the electrical power that would be needed to maintain a wall temperature of 90°C. Show that when the ship speed is of order 10 m/s, the savings in drag power are much smaller than the power used for heating the wall.

5.31   The flat iceberg described in Examples 5.3 and 5.4 in the text is driven by the air drag felt by its upper surface. The iceberg drifts steadily when the air drag is balanced by the bottom-surface seawater drag calculated in Example 5.4. The relative speed between seawater and iceberg is 10 cm/s. Calculate the corresponding wind velocity when the atmospheric air temperature is 40°C.

5.32   Air flows with velocity 3.24 m/s over the top surface of the flat iceberg discussed in Example 5.3. The air temperature outside the boundary layer is 40°C, while the ice surface temperature is 0°C. The length of the iceberg in the direction of air flow is $L = 100$ m. The ice latent heat of melting is $h_{sf} = 333.4$ kJ/kg.

Calculate the L-averaged heat flux deposited by the air flow into the upper surface of the iceberg (model this surface as flat). Calculate, in millimeters per hour, the rate of melting caused by this heat flux, that is, the erosion (thinning) of the ice slab.

5.33   The large faces of the tall building shown in the figure are swept by a parallel wind with $U_\infty = 15$ m/s and $T_\infty = 0°C$. Calculate the heat transfer coefficient for the outer surface of a window situated on a large face, at $x = 60$ m downstream from the leading surface of the building. Evaluate all the air properties at 0°C. Do you expect the calculated heat transfer coefficient to be smaller or larger than the true value?

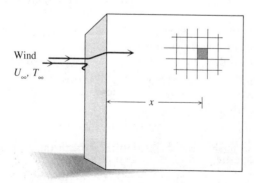

Figure P5.33

Other External Flows

5.34 The cold wind forces the air to flow with a velocity of 4 cm/s through the hair strands in the fur of a bear. The hair strand is approximately cylindrical, with a diameter of 21×10^{-6} m. The mean temperature of the air that sweeps the hair strands is 10°C. Calculate the average coefficient for heat transfer between the individual hair strand and the surrounding air.

5.35 You are fishing while wading in a 75-cm-deep river that flows at 1 m/s. The immersed portion of each bare leg can be approximated as a vertical cylinder with a diameter of 15 cm. The river velocity is approximately uniform, and the water temperature is 10°C.

(a) Calculate the horizontal drag force experienced by one leg.

(b) Compare the drag force with the weight (in air) of the immersed portion of one leg.

(c) Determine the average heat transfer coefficient between the wetted skin and the river water.

(d) After a long enough wait, the temperature of the wetted skin drops to 11°C. Calculate the instantaneous heat transfer rate that escapes into the river through one leg.

5.36 The baseballs used in the major leagues have an average diameter of 7.4 cm and an average weight of 145 g. The distance between the pitcher's mound and home plate is 18.5 m. The pitcher throws an 80 miles/h fastball. The rotation of the ball and its motion in the vertical direction are negligible.

(a) Calculate the drag force experienced by the fastball.

(b) Show that the calculated drag force is comparable with the weight of the ball.

(c) Estimate the final horizontal velocity of the baseball as it reaches the catcher's mitt.

5.37 By holding and rubbing the ball in his hand, the pitcher warms the leather cover of the baseball to 30°C. The outside air temperature is 20°C and the ball diameter is 7 cm. The pitcher throws the ball at 50 miles/h to the catcher, who is stationed 18.5 m away.

(a) Assume that the ball surface temperature remains constant and calculate the heat transferred from the ball to the surrounding air during the throw.

(b) Calculate the temperature drop experienced by the leather cover to account for the heat transfer calculated in part (a). Assume that the thickness of the layer of leather that experiences the air cooling effect is comparable with the conduction penetration depth $\delta \sim (\alpha t)^{1/2}$, where α is the thermal diffusivity of leather. Validate in this way the correctness of the constant surface temperature assumption made in part (a).

5.38 An electrical light bulb for the outdoors is approximated well by a sphere with a diameter of 6 cm. It is being swept by a 2 m/s wind, while its surface temperature is 60°C. The air temperature is 10°C. Calculate the rate of forced-convection heat transfer from the outer surface of the light bulb to the atmosphere.

5.39 Rely on the results of Section 2.3 to prove that in the pure conduction limit the overall Nusselt number for a sphere immersed in a motionless conducting medium is $\overline{Nu}_D = 2$. In other words, demonstrate that the $Re_D = 0$ limit of eq. (5.139) is correct. Show also that in the same limit $\overline{Nu}_{\mathcal{L}} = 3.545$, for which the $\mathcal{L}$ scale is defined in eq. (5.140).

5.40 During the cooling and hardening phase of its manufacturing process, a glass bead with a diameter of 0.5 mm is dropped from a height of 10 m. The bead falls through still air of temperature 20°C. The properties of the bead material are the same as those listed for window glass in Appendix B.

(a) Calculate the "terminal" velocity of the free-falling bead, that is, the velocity when its weight is balanced by the air drag force. Calculate also the approximate time that is needed by the bead to achieve this velocity, and show that the bead travels most of the 10-m height at terminal velocity.

(b) Calculate the average heat transfer coefficient between the bead surface and the surrounding air, when the bead travels at terminal velocity and when its surface temperature is 500°C. Treat the bead as a lumped capacitance and estimate its temperature at the end of the 10-m fall. Assume that its initial temperature was 500°C.

5.41   Hot air with an average velocity $U_\infty = 2$ m/s flows across a bank of 4 cm-diameter cylinders in an array with $X_l = X_t = 7$ cm. Assume that the air bulk temperature is 300°C, and that the cylinder wall temperature is 30°C. The array has 20 rows in the direction of flow. Calculate the average heat transfer coefficient when the cylinders are

(a) Aligned, and

(b) Staggered.

Comment on the relative heat transfer augmentation effect associated with staggering the cylinders.

5.42   Combine the empirical turbulent jet velocity profile (5.148) with the momentum conservation theorem (5.150) to determine the centerline jet velocity (5.151). In this way, prove that the centerline velocity decreases as x^{-1}, from a distance $x \cong 6.61 D_0$ downstream from the nozzle, where $\bar{u}_c = U_0$.

5.43   Demonstrate that the time-averaged excess temperature on the centerline of a turbulent jet decreases as x^{-1}, in accordance with eq. (5.153). In this derivation, rely on the Gaussian profiles (5.148) and (5.149), the centerline velocity (5.151), and the energy conservation theorem (5.152).

5.44   Guided by Table 5.2., it would not be difficult to show that the boundary layer-simplified mass and momentum equations for a time-averaged turbulent jet (Fig. 5.28) are

$$(m) \quad \frac{\partial \bar{u}}{\partial x} + \frac{1}{r}\frac{\partial}{\partial r}(r\bar{v}) = 0$$

$$(M) \quad \bar{u}\frac{\partial \bar{u}}{\partial x} + \bar{v}\frac{\partial \bar{u}}{\partial r} = \frac{1}{r}\frac{\partial}{\partial r}\left[r(v + \varepsilon_M)\frac{\partial \bar{u}}{\partial r}\right]$$

Rely on these equations to prove the correctness of eq. (5.150), which states that the cross section integrated longitudinal momentum of the jet is conserved (independent of x). Use part (a) of Problem 5.11 as a hint. Does the longitudinal momentum theorem (5.150) apply also to a laminar jet?

5.45   The boundary layer-simplified energy equation for a time-averaged turbulent round jet (Fig. 5.28) is

$$(E) \quad \bar{u}\frac{\partial \bar{T}}{\partial x} + \bar{v}\frac{\partial \bar{T}}{\partial r} = \frac{1}{r}\frac{\partial}{\partial r}\left[r(\alpha + \varepsilon_H)\frac{\partial \bar{T}}{\partial r}\right]$$

Rely on this equation and the mass conservation eq. (m) listed in the preceding problem to prove the validity of the theorem (5.152). Step (a) of Problem 5.12 can be used as a guide. The theorem (5.152) states that the flow of energy through any constant-x cross section is conserved. Does this theorem apply also to a laminar round jet?

PROJECTS

5.1   The Heat Transfer Rate Between Two Bodies in Rolling Contact

When two elastic cylinders are pressed against one another, they make contact over a strip of width L. This width is assumed known. In general, it depends on the elastic properties and radii of the two cylinders, and on the force with which one cylinder is pressed against the other. The radii of the cylinders are much larger than the contact width L.

Two cylindrical bodies have different temperatures, T_1 and T_2, and roll past one another without slip. The peripheral velocity U, with which both bodies pass through the frame of reference attached to the contact region, is known. The objective of this project is to show how the "fluid" boundary layer method of this chapter can be used to calculate the heat transfer rate between two *solids* [38].

(a) Assume that the interface temperature T_0 (unknown) is uniform, that is, independent of x. Write down the expression for the local heat flux q_x'' by noting that the "flow" of each solid through its respective thermal boundary layer region (with constant U) is similar to that of a fluid with extremely small Prandtl number.

(b) Show that the interface temperature depends on the physical properties of the two solids in the following manner:

$$T_0 = \frac{r}{1+r}T_1 + \frac{1}{1+r}T_2 \quad \text{with} \quad r = \frac{(\rho c k)_1^{1/2}}{(\rho c k)_2^{1/2}}$$

(c) Derive the following expression for the L-averaged heat flux between the two bodies:

$$\bar{q}'' = \frac{1.128}{1+r}k_1(T_1 - T_2)\left(\frac{U}{\alpha_1 L}\right)^{1/2}$$

(d) How fast must the cylinders roll for these analytical results to be valid?

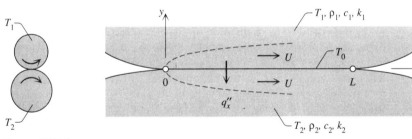

Figure PR5.1

5.2   Optimum Plate Fin Geometry for Maximum Heat Transfer Rate Subject to Fixed Fin Volume

The thin-plate fin shown in the figure is bathed by a uniform flow (U_∞, T_∞) that is oriented parallel to the dimension b. The resulting boundary layers (length b) are laminar. The heat transfer through the fin itself can be described as unidirectional, in the sense that the fin temperature is a function of x only. The fin dimensions are such that $L \gg b \gg t$. The plate thickness t is fixed.

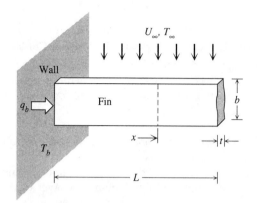

Figure PR5.2

You are asked to design (i.e., select the dimensions of) this plate fin so that the heat transfer rate through its base (q_b) is maximum when the volume of fin material is fixed [3]:

$$V = bLt = \text{constant}$$

Note that there is only one degree of freedom in this design problem, namely, the choice of b or L. Express your result in the following dimensionless form:

$$\frac{b_{\text{opt}}}{t} = \text{function} \left(\frac{V}{t^3}, \frac{k_f}{k}, \text{Pr}, \frac{U_\infty t}{v} \right)$$

where k_f, v, and Pr are respectively the fluid thermal conductivity, kinematic viscosity and Prandtl number. The thermal conductivity of the fin material is k, while b_{opt} is the optimum width of the plate. Determine the fin length that corresponds to this optimum design, L_{opt}, and the resulting base heat transfer maximum, $q_{b,\text{max}}$.

5.3 Combined Forced Convection and Conduction: Critical Insulation Radius

A key ingredient in the derivation of the critical insulation radius in Section 2.4 was the assumption that the heat transfer coefficient is constant, that is, independent of diameter. When h is due to forced convection, this assumption breaks down. Correlations of type (5.135) for the cylinder, and (5.139) for the sphere show that the overall Nusselt number generally increases with the Reynolds number. In fact, in a sufficiently narrow range of Re_D values, these correlations are approximated by power laws of the type $\text{Nu}_D = C\,\text{Re}_D^n$, or

$$\frac{hD}{k_{\text{fluid}}} = C \left(\frac{U_\infty D}{v} \right)^n$$

where C is a constant, and n is a number between 0 and 1 (usually in the vicinity of 0.5). This power-law approximation shows that h varies as D^{n-1}.

(a) Show that the critical insulation radius of a cylinder in cross-flow is given by

$$\frac{r_{o,c}}{r_i} = \left(\frac{nk}{r_i h_*} \right)^{1/n}$$

where h_* is a reference heat transfer coefficient evaluated with eq. (5.135) for $D = 2r_i$, and k is the insulation conductivity.

(b) To understand how the above critical radius differs from the classical estimate provided by eq. (2.44), consider insulating a bare pipe characterized by $r_i = 4$

cm and $T_i = 100°C$, which is exposed to a 2 m/s cross-flow of 20°C air. The insulation material would be polystyrene. Calculate the $r_{o,c}$ value based on part (a), and compare it with the classical $r_{o,c}$ based on $h = h_*$.

(c) By repeating the method of part (a), show that the critical insulation radius of a sphere is given by

$$\frac{r_{o,c}}{r_i} = \left(\frac{(n+1)k}{r_i h_*}\right)^{1/n}$$

More general treatments of the critical insulation radius when h is not a constant have been presented by Sparrow [39] and Simmons [40]. In this book, this topic is pursued based on simpler formulations in this project and, later, in Projects 7.4 and 10.5.

6
INTERNAL FORCED CONVECTION

6.1 LAMINAR FLOW THROUGH A DUCT

6.1.1 The Flow Entrance Region

In this chapter we turn our attention to forced convection configurations in which the flow is "internal," that is, surrounded by the wall of a duct at a different temperature. As a first example, consider the laminar flow through the round tube of diameter D shown in Fig. 6.1. The mouth of the tube is located in the plane $x = 0$, and x is the longitudinal position measured in the downstream direction.

Throughout this chapter we shall write U for the *mean velocity*, that is, the longitudinal velocity averaged over the duct cross section A:

$$U = \frac{1}{A} \int_A u \, dA \qquad (6.1)$$

In the round tube of Fig. 6.1, for example, A is equal to $\pi r_o{}^2$, where r_o is the outer radius of the flow region (the inner radius of the tube wall). Equation (6.1) is a general definition of U. Note that regardless of the shape of the duct cross section, U represents also the uniform (sluglike) distribution of longitudinal velocity through the entrance to the duct.

We assume that the mean velocity U is constant or nearly constant along the duct. According to the principle of mass conservation, this assumption is valid when the fluid density ρ is constant or nearly constant along the duct. It is particularly good in the case of liquids (e.g., water, oils) and fairly good for gases that experience only moderate temperature changes while flowing through the duct. In this case "moderate" means that the change in temperature from inlet to outlet is at least ten times smaller than the absolute (Kelvin) temperature of the gas. On the other hand, if the density variation is significant, the gas stream accelerates or decelerates along the duct, depending on whether it is heated or

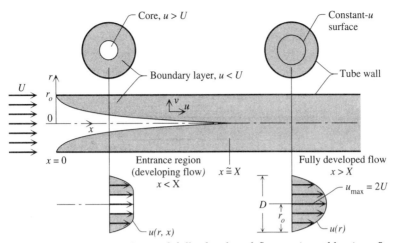

Figure 6.1 Entrance region and fully developed flow region of laminar flow through a tube.

cooled. The complications that arise when ρ and U are not sufficiently constant along the duct are treated in Section 9.5.3.

The uniform longitudinal velocity U assumed across the $x = 0$ plane in Fig. 6.1 is also an approximation. For example, if the flow originates from a large reservoir positioned to the left of the $x = 0$ plane, the entering fluid flows longitudinally only if it is close to the tube centerline. Closer to the rim, the velocity of the entering fluid has a negative radial component. These and other entrance effects form the subject of Section 9.5.2.

Immediately downstream from the mouth, the flow develops a boundary layer region that coats the inner surface of the tube wall. The thickness of this velocity boundary layer, δ, initially grows as $x^{1/2}$. For this reason we must identify the longitudinal position X where the velocity boundary layer thickness δ has grown to the same size as the tube radius r_o. In other words, X is the longitudinal position where the inflowing stream has had enough time to feel the effect of wall friction (transversal viscous diffusion) all the way to its centerline.

The *flow entrance length* X divides the flow into two important regions: the *flow entrance* region situated between the mouth and $x \sim X$, and the so-called *fully developed flow* region* located downstream from $x \sim X$. The size of the entrance length X can be estimated quite rapidly using the results of Section 5.3.1. As a measure of the thickness of the boundary layer that exists in the entrance region, we can use an approximate version of eq. (5.57):

$$\delta(x) \cong 5x \left(\frac{Ux}{\nu} \right)^{-1/2} \tag{6.2}$$

The order-of-magnitude definition of the entrance length X is the statement

$$\delta \cong r_o \quad \text{when} \quad x \cong X \tag{6.3}$$

---

*The flow entrance region and the fully developed flow region are known also as the *hydrodynamic* entrance region and the *hydrodynamically* fully developed region.

Combining this with eq. (6.2), we conclude that the geometric aspect ratio of the flow entrance region, X/D, is proportional to the Reynolds number based on tube diameter ($Re_D = UD/\nu$):

$$\frac{X}{D} \cong 10^{-2} Re_D \tag{6.4}$$

The more exact and more frequently used estimate of the flow entrance length is

$$\frac{X}{D} \cong 0.05 Re_D \tag{6.4'}$$

In the case of a duct whose cross section is not circular, in eq. (6.4') the diameter D is replaced by the hydraulic diameter D_h, which will be defined later in eq. (6.28).

The "developing" flow that takes place in the entrance region has two subregions of its own — the annular boundary layer and the core that surrounds the centerline. The fluid that once entered the tube with the velocity U is slowed down by the wall in the boundary layer ($u < U$). The conservation of the flowrate $\rho \pi r_o^2 U$ in each cross section requires longitudinal velocities larger than U in the core region. Indeed, in the next section we learn that the centerline fluid accelerates from $u = U$ at the mouth to $u = 2U$ on the centerline of the fully developed flow region.

These features are evident in the photograph shown in Fig. 6.2. The flow in the entrance region was visualized using the hydrogen bubble method. The bubble-generating wire was positioned right at the mouth, across the inflowing stream. Lines of hydrogen bubbles were generated at equal time intervals. These lines show the growth of the boundary layer regions and the downstream acceleration of the core region.

6.1.2 The Fully Developed Flow Region

The flow distribution in the fully developed region of the tube flow can be determined based on a relatively simple analysis. For this we must begin with the equations that govern the conservation of mass and momentum in the cylindrical system of coordinates (r, θ, x) attached to the tube centerline. Assuming a flow with nearly constant density (ρ) and viscosity (μ), and replacing z

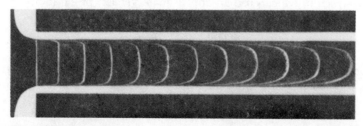

Figure 6.2 Laminar flow in the entrance region of a tube with circular cross section: water, $U = 6$ cm/s, $D = 2.7$ cm. (Nakayama et al. [1], with permission from Pergamon Press.)

with x, v_z with u, and v_r with v in the equations of Table 5.2, we find that the three equations that govern the flow in the steady state are

$$\frac{1}{r}\frac{\partial}{\partial r}(rv) + \frac{\partial u}{\partial x} = 0 \tag{6.5}$$

$$\rho\left(v\frac{\partial v}{\partial r} + u\frac{\partial v}{\partial x}\right) = -\frac{\partial P}{\partial r} + \mu\left(\frac{\partial^2 v}{\partial r^2} + \frac{1}{r}\frac{\partial v}{\partial r} - \frac{v}{r^2} + \frac{\partial^2 v}{\partial x^2}\right) \tag{6.6}$$

$$\rho\left(v\frac{\partial u}{\partial r} + u\frac{\partial u}{\partial x}\right) = -\frac{\partial P}{\partial x} + \mu\left(\frac{\partial^2 u}{\partial r^2} + \frac{1}{r}\frac{\partial u}{\partial r} + \frac{\partial^2 u}{\partial x^2}\right) \tag{6.7}$$

In writing these equations we have left out the derivatives with respect to θ, because the tube wall is such that the flow field does not change as θ changes. It is said that the flow exhibits "θ symmetry." Equations (6.5), (6.6), and (6.7) assume a dramatically simpler form if we recognize that in the tube flow region of length x and thickness D, the mass conservation eq. (6.5) requires

$$\frac{v}{D} \sim \frac{U}{x} \tag{6.8}$$

The transversal velocity scale v is therefore

$$v \sim \frac{UD}{x} \tag{6.9}$$

This scale shows that v becomes indeed negligible relative to U as x increases, that is, as the flow proceeds deeper into the fully developed region:

$$v = 0 \quad \text{(fully developed flow)} \tag{6.10}$$

From eq. (6.10) follow two additional conclusions. First, from eq. (6.5) we learn that in the fully developed region $\partial u/\partial x$ must also be zero, which means that

$$u = u(r) \tag{6.11}$$

Second, by substituting $v = 0$ into the r-momentum eq. (6.6), we find that $\partial P/\partial r = 0$; therefore,

$$P = P(x) \tag{6.12}$$

In view of the last three conclusions, eqs. (6.10), (6.11), and (6.12), the longitudinal momentum eq. (6.7) assumes the much simpler form (note the disappearance of the partial derivative signs)

$$\frac{dP}{dx} = \mu\left(\frac{d^2 u}{dr^2} + \frac{1}{r}\frac{du}{dr}\right) \tag{6.13}$$

where the left side is, at best, a function of x and the right side is a function of r. In conclusion, both sides of eq. (6.13) must be equal to the same constant; therefore, the *entire* system (6.5)–(6.7) reduces to

$$\frac{dP}{dx} = \text{constant} = \mu\left(\frac{d^2 u}{dr^2} + \frac{1}{r}\frac{du}{dr}\right) \tag{6.14}$$

Two equations (two equal signs) appear in eq. (6.14). The first indicates that the pressure varies linearly along the fully developed section of the tube flow. If the entrance length X is much smaller than the overall length of the tube, L, and if ΔP is the pressure difference (pressure "drop") maintained between the mouth and the downstream end of the tube, then the constant alluded to in eq. (6.14) is clearly

$$\frac{dP}{dx} = -\frac{\Delta P}{L} \tag{6.15}$$

The second equation* listed in eq. (6.14) can be solved for the longitudinal velocity distribution $u(r)$ in the fully developed region. The two boundary conditions that are demanded by this analysis are

$$u = 0 \quad \text{at} \quad r = r_o \quad \text{(no slip)} \tag{6.16a}$$

$$\frac{du}{dr} = 0 \quad \text{at} \quad r = 0 \quad \text{(symmetry)} \tag{6.16b}$$

The solution can be derived as an exercise (Problem 6.1):

$$u = \frac{r_o^2}{4\mu}\left(-\frac{dP}{dx}\right)\left[1 - \left(\frac{r}{r_o}\right)^2\right] \tag{6.17}$$

Substituted into the U definition (6.1), this solution yields also

$$U = \frac{r_o^2}{8\mu}\left(-\frac{dP}{dx}\right) \tag{6.18}$$

which shows that the average velocity (or the flowrate $\rho\pi r_o^2 U$) is *proportional* to the longitudinal pressure gradient. The $u(r)$ distribution (6.17) can then be rewritten as

$$u = 2U\left[1 - \left(\frac{r}{r_o}\right)^2\right] \tag{6.19}$$

to show that the velocity along the centerline of the tube is exactly twice as large as the tube-averaged velocity U. The fully developed flow distribution (6.19) is known as the *Hagen–Poiseuille flow*, after the first two investigators who reported it [2, 3].

In a completely analogous way, the laminar flow into the space of thickness D formed between two parallel walls (Fig. 6.3) can be divided into an entrance region followed by a fully developed region. With reference to the (x,y) system of coordinates and boundary conditions defined in Fig. 6.3, the governing equations for the fully developed region reduce to

$$\frac{dP}{dx} = \text{constant} = \mu\frac{d^2u}{dy^2} \tag{6.20}$$

The resulting velocity profile $u(y)$ is parabolic (Problem 6.2),

$$u = \frac{3}{2}U\left[1 - \left(\frac{y}{D/2}\right)^2\right] \tag{6.21}$$

$$U = \frac{D^2}{12\mu}\left(-\frac{dP}{dx}\right) \tag{6.22}$$

---

*The equation associated with the second equal sign.

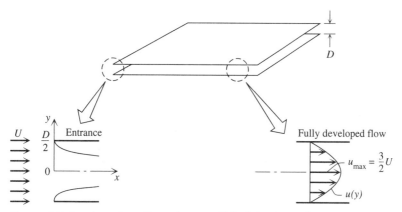

Figure 6.3 Laminar flow through a parallel-plate channel.

and the maximum (midplane) velocity is only 50 percent larger than the cross section-averaged value U. The flowrate is proportional to the longitudinal pressure gradient $-dP/dx$, which drives the flow.

The Hagen–Poiseuille flow in a parallel-plate channel can be seen by looking at the right (downstream) extremity of the flow shown in Fig. 6.4. That photograph illustrates also the development of the flow in the entrance region. Vertical wires generate lines of hydrogen bubbles at six different locations along the channel. The time interval between two consecutive lines of bubbles generated by the same wire is fixed. It is clear that at the downstream end of the channel the shape of each line of bubbles resembles the parabolic profile indicated by eq. (6.21).

6.1.3 Friction Factor and Pressure Drop

The two fully developed flow examples exhibited in eqs. (6.19) and (6.21) show that the longitudinal velocity u is not a function of the longitudinal position x. For this reason, the wall shear effect τ_w is also independent of x. In the case of Hagen–Poiseuille flow through a round tube, Fig. 6.1 (right side), the τ_w constant is easily evaluated from eq. (6.19):

$$\tau_w = \mu \left(-\frac{du}{dr} \right)_{r=r_o} = 4\mu \frac{U}{r_o} \tag{6.23}$$

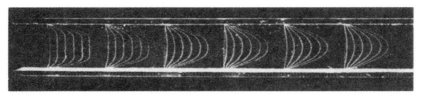

Figure 6.4 Laminar flow in the entrance region of a parallel-plate channel: water, $U = 3.2$ cm/s, $D = 2$ cm. (Nakayama et al. [1], with permission from Pergamon Press.)

It is customary to express the τ_w value in dimensionless form, by constructing the *Fanning friction factor**

$$f = \frac{\tau_w}{\frac{1}{2}\rho U^2} \qquad (6.24)$$

Worth noting are the similarities between this new group and the local skin friction coefficient defined in eq. (5.39). This time, the role of velocity unit is played by the tube-averaged velocity U, and, unlike in eq. (5.39), the resulting dimensionless group (f) is x-independent. Combining eqs. (6.23) and (6.24), we find that the friction factor for fully developed tube flow is inversely proportional to the Reynolds number based on mean velocity and tube diameter ($\mathrm{Re}_D = UD/\nu$):

$$f = \frac{16}{\mathrm{Re}_D} \qquad \text{(fully developed tube flow)} \qquad (6.25)$$

Either in f or τ_w form, the wall friction information can be used to deduce the overall pressure difference ΔP that must be maintained across the duct of length L to drive the flow of average velocity U. For this final step of the calculation, consider the flow through the straight duct shown in the upper-left corner of Fig. 6.5. The unspecified geometry of the duct cross section is described by the flow cross-sectional area A and by the wetted perimeter of the cross section, p.

---

*The Fanning friction factor f should not be confused with the Darcy–Weisbach [4] friction factor f_D, which is four times larger than f:

$$f_D = 4f$$

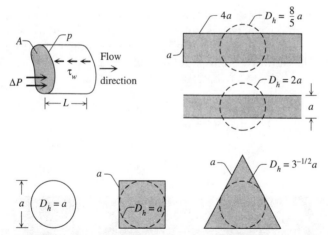

Figure 6.5 Force balance over the flow control volume (the upper-left corner) and five duct cross-sectional shapes and their hydraulic diameters. The cross sections are drawn to scale in such a way that they all have the same hydraulic diameter (after Ref. [5]).

When the length L is much larger than the entrance length estimated in Section 6.1.1, the inner surface of the duct is covered by a wall shear stress τ_w that does not vary with the longitudinal position. In the round-tube and parallel-plate examples of Figs. 6.1 and 6.3, τ_w is also uniformly distributed along the wetted perimeter of the cross section. In a duct with a less regularly shaped cross section (e.g., triangular), τ_w varies around the cross section periphery in such a way that its smallest values occur in the corners. For this reason, in the force balance suggested by the first drawing of Fig. 6.5 the term τ_w represents the wall shear stress *averaged* around the periphery of length p. The product $\tau_w pL$ is, therefore, the total friction force experienced by the tube wall. The force balance on the control volume AL requires that

$$\Delta P\, A = \tau_w p L \tag{6.26}$$

In terms of the friction factor f, the pressure drop given by eq. (6.26) reads

$$\Delta P = f \frac{L}{A/p} \frac{1}{2} \rho U^2 \tag{6.27}$$

Equations (6.26) and (6.27) hold for either laminar or turbulent flow, provided the length L contains fully developed flow. Note further that the denominator A/p has units of length; for example, the A/p value of a circular cross section of diameter D happens to be equal to $D/4$. It makes sense, then, to define $4A/p$ as the *hydraulic diameter* D_h of a cross section that is not necessarily circular,

$$D_h = \frac{4A}{p} \tag{6.28}$$

and to regard D_h as a characteristic transversal length scale of the flow through the unspecified cross section. In this way, the pressure drop formula (6.27) becomes

$$\Delta P = f \frac{4L}{D_h} \frac{1}{2} \rho U^2 \tag{6.27'}$$

Figure 6.5 shows several of the more common duct cross sections and their respective hydraulic diameters. These cross sections have been drawn to scale in such a way that they all have the same hydraulic diameter [5]. For example, in the case of a round cross section of diameter D, the hydraulic diameter is equal to the actual diameter, $D_h = 4(\pi D^2/4)(\pi D) = D$. On the other hand, in a parallel-plate channel of spacing S and width W (i.e., flow cross section $S \times W$), the hydraulic diameter is two times greater than the spacing, $D_h = 4(SW)/(2W) = 2S$.

The first column of Table 6.1 shows the friction factor formulas that apply to fully developed laminar flow through the special cross sections illustrated in Fig. 6.5. The general form of these friction factor formulas is

$$f = \frac{C}{\mathrm{Re}_{D_h}} \tag{6.29}$$

where Re_{D_h} is the Reynolds number based on hydraulic diameter, $U D_h / \nu$. The constant C appearing in the numerator depends on the shape of the duct cross

Table 6.1 Friction Factors (f), Cross Section Shape Numbers (B), and Nusselt Numbers (Nu_D) for Hydrodynamically and Thermally Fully Developed Duct Flows [5]

Cross Section Shape	$f\mathrm{Re}_{D_h}$	B	$\mathrm{Nu}_{D_h} = hD_h/k$ Uniform q_w''	Uniform T_w
60° triangle	13.3	0.605	3	2.35
square	14.2	0.785	3.63	2.89
circle	16	1	4.364	3.66
a by $4a$ rectangle	18.3	1.26	5.35	4.65
rectangle	24	1.57	8.235	7.54
rectangle (One side insulated)	24	1.57	5.385	4.86

section. It was pointed out in Ref. [5] that C increases monotonically with a new dimensionless parameter, the *cross section shape number*

$$B = \frac{\pi D_h^2/4}{A} \tag{6.30}$$

whose purpose it is to measure the deviation of the actual shape from the circular shape. Note that $B = 1$ corresponds to a round tube. The C values of Table 6.1 are approximated within 4 percent by the expression

$$C \cong 16 \exp(0.294B^2 + 0.068B - 0.318) \tag{6.31}$$

Example 6.1

Laminar Flow, Round Tube

Consider the laminar flow of 20°C water through a tube with the diameter $D = 2.7$ cm. The mean velocity of the stream is $U = 6$ cm/s. Calculate the pressure drop per unit length $\Delta P/L$ in the fully developed region. Determine also the extent of the flow entrance length, and compare this result with the length of the flow region photographed in Fig. 6.2.

Solution. We begin with the calculation of the Reynolds number, to verify that the flow is indeed laminar:

$$\text{Re}_D = \frac{UD}{v}$$

$$= 6\,\frac{\text{cm}}{\text{s}}\,2.7\,\text{cm}\,\frac{\text{s}}{0.01\,\text{cm}^2} = 1620 \qquad \text{(laminar)}$$

$$f = \frac{16}{\text{Re}_D} = \frac{16}{1620} = 0.0099 \tag{6.25}$$

$$\frac{\Delta P}{L} = f\frac{4}{D}\frac{1}{2}\rho U^2 \tag{6.27'}$$

$$= 0.0099\,\frac{4}{2.7\,\text{cm}}\,\frac{1}{2}\,1\frac{\text{g}}{\text{cm}^3}\left(6\,\frac{\text{cm}}{\text{s}}\right)^2$$

$$= 0.264\frac{\text{g}}{\text{cm}^2\,\text{s}^2} = 2.64\,\frac{\text{N/m}^2}{\text{m}}$$

$$X \cong 0.05D\text{Re}_D$$

$$= (0.05 \times 2.7\,\text{cm})\,(1620) = 2.2\,\text{m}$$

The length/diameter ratio of the flow region photographed in Fig. 6.2 is roughly 7.3/1. Recalling that $D = 2.7$ cm, we conclude that the length of the photographed region (namely, 7.3×2.7 cm $\cong 20$ cm) is about one tenth of the actual entrance region of the flow.

6.2   HEAT TRANSFER IN LAMINAR FLOW

6.2.1   The Thermal Entrance Region

The fluid mechanics covered in the preceding section is a prerequisite for the analysis of the heat transfer between the duct wall and the fluid stream. We continue to assume that the duct is long enough so that the flow is fully developed over most of the length L. This is why in Fig. 6.6 the flow entrance

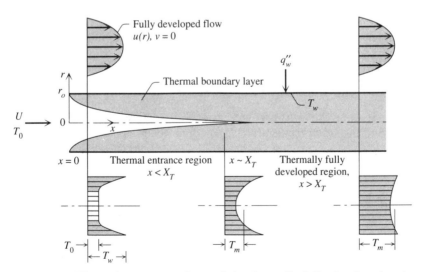

Figure 6.6 Thermal entrance region and the thermally fully developed region of fully developed flow through a tube.

region is not shown and the fully developed velocity profile u is drawn to emphasize that it does not change along the duct.

The entering stream has a uniform initial temperature T_0 at $x = 0$. Heat transfer occurs between the duct wall and the fluid, and this effect is represented by the local wall heat flux q_w'' and the local wall temperature T_w. In this opening subsection we say nothing more about q_w'' and T_w, except that, in general, they can both be functions of longitudinal position. Two special types of wall heating conditions will be analyzed soon, namely, the wall with uniform heat flux (Section 6.2.3) and the wall with uniform temperature (Section 6.2.4).

The wall heat transfer effect penetrates the flow gradually. It is felt first in a thin *thermal* boundary layer that coats the wall. The thermal boundary layer grows in the downstream direction and eventually becomes as thick as the duct itself. The approximate longitudinal position $x \sim X_T$ where the thermal boundary layer loses its identity divides the duct length into two distinct regions: (1) a *thermal entrance region* $(0 < x < X_T)$ in which the shape of the transversal temperature profile "develops" (i.e., where the T profile changes from one x to the next), and (2) a *thermally fully developed region* $(x > X_T)$ where the shape of the temperature profile is preserved. In tube flow, for example, the unchanging shape of the fully developed temperature profile is similar to that of the meniscus at the end of a liquid column in a capillary. This shape is preserved (i.e., it remains a "meniscus") even though its amplitude* may vary in the longitudinal direction.

A scale analysis similar to the one used in the derivation of the flow entrance length X, eq. (6.4), can be used to estimate the thermal entrance length X_T [5]. The result is

$$\frac{X_T}{D_h} \cong 0.05 \mathrm{Re}_{D_h} \mathrm{Pr} \tag{6.32}$$

It has been demonstrated [5] that this estimate holds for both high-Pr and low-Pr fluids, and is insensitive to whether the flow entrance length X is longer or shorter than the thermal entrance length X_T. In fluids with Prandtl numbers that do not differ greatly from 1 (e.g., air), the lengths X and X_T are of the same order of magnitude: if X is negligible relative to L, then X_T is negligible also. We return to the heat transfer characteristics of the thermal entrance region in Figs. 6.8 and 6.10.

6.2.2   The Thermally Fully Developed Region

For reasons that will become clear while reading eq. (6.38), it is convenient to define the *mean temperature* T_m of the stream as

$$T_m = \frac{1}{UA} \int_A uT \, dA \tag{6.33}$$

This quantity is known also as the "bulk" temperature of the stream, and it is an integral part of the *bulk flow model* that is used routinely in the thermodynamic

---

*In the case of Fig. 6.6, the amplitude is the difference between the temperature at the wall and the minimum or maximum temperature on the duct centerline. The figure shows that in the thermally fully developed region $(x > X_T)$ the amplitude changes while the shape—the meniscus—is preserved.

analysis of open (flow) systems [6]. The temperature T_m is a cross-sectional *weighted* average of the local fluid temperature T: The role of weighting factor is played by the longitudinal velocity u, which is zero at the wall and large in the central portion of the duct cross section.

Similarly, it is convenient to define the *mean temperature difference* between the wall and the stream as

$$\Delta T = T_w - T_m \tag{6.34}$$

in which both T_w and T_m can vary in the longitudinal direction. In problems of internal convection (duct flow), the mean temperature difference is used as temperature difference scale in the definition of the wall–fluid heat transfer coefficient h, which generally can be a function of x:

$$h = \frac{q_w''}{T_w - T_m} = \frac{q_w''}{\Delta T} \tag{6.35}$$

Consider now the problem of determining the heat transfer coefficient for a *hydrodynamically* and *thermally* fully developed flow through a straight duct of unspecified cross-sectional shape (A, p). The fact that the flow is hydrodynamically developed means that the u profile is x-independent and that the order of magnitude of u is the mean value U. The thermal development of the temperature field means that the transversal length scale of the temperature profile is the same as the duct size D_h. For example, in a tube of radius r_o full thermal development means that

$$\left(\frac{\partial T}{\partial r}\right)_{r = r_o} \sim \frac{\Delta T}{r_o} \tag{6.36}$$

An important relationship between the local wall heat flux q_w'' and the local change in mean fluid temperature, dT_m/dx, follows from the first law of thermodynamics, as applied to the steady flow through the duct control volume of length dx and cross-sectional area A:

$$\int_A \rho u (i_{x+dx} - i_x) dA = q_w'' p \, dx \tag{6.37}$$

The integral expresses the difference between the enthalpy outflow (at $x + dx$) and the enthalpy inflow (at x). The right side of eq. (6.37) represents the rate of heat transfer into the control volume, through the lateral area $p \, dx$. In accordance with the assumption of negligible local pressure changes, which was invoked in the derivation of the energy equation [review eq. (5.17)], we can replace the specific enthalpy change, $i_{x+dx} - i_x$, with $c_p dT$:

$$\rho c_p d \left(\int_A u T \, dA \right) = q_w'' p \, dx \tag{6.38}$$

On the left side of eq. (6.38) we now recognize the integral used in the definition of the mean temperature; in this way, we arrive at the conclusion that the local mean temperature gradient is proportional to the local wall heat flux:

$$\frac{dT_m}{dx} = \frac{p}{A} \frac{q_w''}{\rho c_p U} \tag{6.39}$$

In the special case of a round tube, this conclusion reads

$$\frac{dT_m}{dx} = \frac{2}{r_o} \frac{q_w''}{\rho c_p U} \qquad \text{(round tube)} \qquad (6.39')$$

It is instructive to show that the first-law statement (6.39) and the condition of full thermal development, eq. (6.36), are sufficient for estimating the order of magnitude of the heat transfer coefficient h. According to Table 5.2 and the cylindrical coordinates and velocity components defined in Fig. 6.6, the steady-state energy equation for fully developed flow and temperature in a round tube is

$$\rho c_p u \frac{\partial T}{\partial x} = k \frac{1}{r} \frac{\partial}{\partial r} \left(r \frac{\partial T}{\partial r} \right) \qquad (6.40)$$

This equation expresses the balance between the heat conducted from the wall radially into the fluid (the right side) and the enthalpy carried away longitudinally by the stream (the left side). The orders of magnitude of the two sides of this balance can be evaluated using, respectively, eqs. (6.39') and (6.36):

$$\rho c_p U \frac{1}{r_o} \frac{q_w''}{\rho c_p U} \sim k \frac{\Delta T}{r_o^2} \qquad (6.41)$$

From this we draw the conclusion that the heat transfer coefficient, $h = q_w''/\Delta T$, must be x-independent:

$$h \sim \frac{k}{r_o} \qquad \text{(constant)} \qquad (6.42)$$

The internal flow heat transfer coefficient is commonly nondimensionalized as a Nusselt number based on hydraulic diameter:

$$\text{Nu}_{D_h} = \frac{h D_h}{k} \qquad (6.43)$$

In a round tube, D_h is the same as the tube diameter D; therefore, the Nu_D estimate that corresponds to eq. (6.42) is

$$\text{Nu}_D \sim 1 \qquad \text{(constant)} \qquad (6.44)$$

In conclusion, the Nusselt number for hydrodynamically and thermally fully developed flow is a constant whose order of magnitude is 1. This conclusion means also that the local heat transfer coefficient is independent of longitudinal position, even though q_w'' and ΔT may vary with x. It has been shown in Ref. [5] that the constant-h conclusion (6.42) implies also that the fluid temperature distribution $T(r,x)$ must obey a function of the type

$$T(r,x) = T_w(x) - [T_w(x) - T_m(x)] \, \phi \left(\frac{r}{r_o} \right) \qquad (6.45)$$

where ϕ is a function of r only. Equation (6.45) is usually postulated as the definition of what is meant by thermally fully developed flow, that is, the condition that must be met by the temperature profile to be "fully" developed. In this subsection we started from a considerably less abstract definition, eq. (6.36), and arrived not only at eq. (6.45) but also at the Nusselt number estimate (6.44).

6.2.3   Uniform Wall Heat Flux

The Nusselt number for fully developed flow through a tube with uniform wall heat flux can be determined analytically by solving the energy eq. (6.40). This equation acquires a simpler form, which stems from the observation that when q''_w is constant, eqs. (6.35) and (6.42) require that

$$\frac{dT_w}{dx} = \frac{dT_m}{dx} \tag{6.46}$$

Differentiating with respect to x the fully developed temperature profile (6.45), and using eq. (6.46), we learn further that

$$\frac{\partial T}{\partial x} = \frac{dT_w}{dx} = \frac{dT_m}{dx} \tag{6.47}$$

In conclusion, instead of $\partial T/\partial x$ on the left side of the energy eq. (6.40), we are entitled to use dT_m/dx, which according to eq. (6.39') is a *constant*. This conclusion is illustrated graphically by the three-dimensional surface $T(r,x)$ presented in Fig. 6.7. Finally, on the right side of the energy eq. (6.40), we replace T with the expression listed in eq. (6.45), so that in the end the energy equation reads

$$2\frac{u}{U}\frac{q''_w}{kr_o\Delta T} = -\frac{1}{r}\frac{d}{dr}\left(r\,\frac{d\phi}{dr}\right) \tag{6.48}$$

In the denominator on the left side, ΔT is shorthand for the temperature difference $T_w(x) - T_m(x)$.

The dimensionless Hagen–Poiseuille profile u/U is given by eq. (6.19). The problem of finding $T(r,x)$ reduces to solving eq. (6.48) for $\phi(r)$, subject to the boundary conditions $\phi(r_o) = 0$ and $\phi'(0) = 0$. The result of this operation is

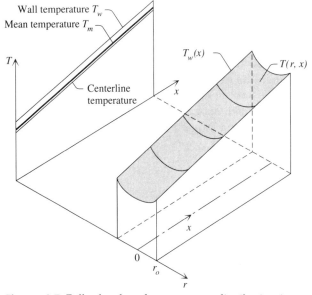

Figure 6.7 Fully developed temperature distribution in a tube with uniform heat flux.

$$\phi(r) = \frac{q_w'' r_o}{k \Delta T} \left[\frac{3}{4} - \left(\frac{r}{r_o} \right)^2 + \frac{1}{4} \left(\frac{r}{r_o} \right)^4 \right] \tag{6.49}$$

Combined with eq. (6.45), this $\phi(r)$ solution represents the temperature distribution $T(r,x)$ subject to the still undetermined factor* $q_w'' r_o / k \Delta T$ that appears in eq. (6.49). This factor is determined from the condition that $T(r,x)$ must obey the T_m definition (6.33). In terms of ϕ, the mean temperature definition (6.33) reads

$$\frac{1}{UA} \int_A u \phi \, dA = 1 \tag{6.50}$$

where $A = \pi r_o^2$. The substitution of eqs. (6.19) and (6.49) into this integral condition leads to the conclusion that the Nusselt number is indeed a constant of order 1 [cf. eq. (6.44)]:

$$\mathrm{Nu}_D = \frac{48}{11} = 4.364 \qquad (\text{constant } q_w'') \tag{6.51}$$

The third column of Table 6.1 lists the corresponding Nu values for five additional duct cross-sectional shapes. Except for the bottom entry (parallel-plate channel with one side insulated), the Nusselt numbers increase monotonically as the shape number B increases [5]. In the triangular, square, and rectangular cases the wall temperature varies along the perimeter of the cross section when the heat flux is uniform. In such cases the heat transfer coefficient is defined by $h = q_w'' / [\overline{T}_w(x) - T_m(x)]$, where $\overline{T}_w(x)$ is the wall temperature averaged over the entire perimeter. Additional examples of Nu values for fully developed flow through ducts with uniform heat flux can be found in Shah and London's monograph [4] and, later here, in Table 6.2.

Figure 6.8 shows the complete behavior of the Nusselt number along a tube with uniform heat flux. The figure was drawn based on data reported in Refs. [4] and [7]. The Nu_D constant determined in eq. (6.51) is reached when the flow finds itself in the thermally fully developed region. (Note that x is measured starting from the mouth of the tube.) In the thermal entrance region, the local Nusselt number based on tube diameter ($\mathrm{Nu}_D = q_w'' D / k \Delta T$) is a function of x, indicating that, unlike in the fully developed region, the temperature difference $\Delta T = T_w - T_m$ is not constant. It can be verified that toward the extreme left of Fig. 6.8 the Nu_D (or ΔT) behavior is consistent with the behavior seen in a thermal boundary layer near a plane wall with uniform heat flux, Table 5.4.

A closed-form expression that covers both the entrance and fully developed regions of Fig. 6.8 was developed by Churchill and Ozoe [8]:

$$\frac{\mathrm{Nu}_D}{4.364 \, [1 + (\mathrm{Gz}/29.6)^2]^{1/6}}$$

$$= \left[1 + \left(\frac{\mathrm{Gz}/19.04}{[1 + (\mathrm{Pr}/0.0207)^{2/3}]^{1/2}[1 + (\mathrm{Gz}/29.6)^2]^{1/3}} \right)^{3/2} \right]^{1/3} \tag{6.52}$$

---

*This factor is the unknown of the problem, as it is equal to $\mathrm{Nu}_D/2$, where $\mathrm{Nu}_D = q_w'' D / k \Delta T$.

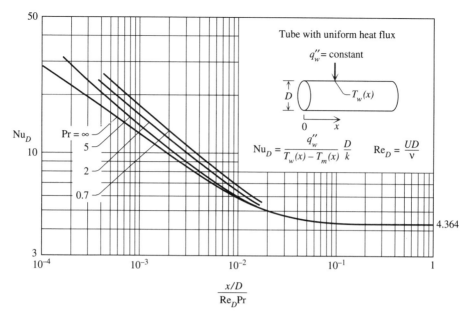

Figure 6.8 The Nusselt number for laminar flow through a tube with uniform wall heat flux.

The new dimensionless group Gz is the *Graetz number*[*]

$$Gz = \frac{\pi D^2 U}{4\alpha x} = \frac{\pi}{4}\left(\frac{x/D}{Re_D Pr}\right)^{-1} \tag{6.53}$$

which is inversely proportional to the abscissa parameter used in Fig. 6.8. The formula (6.52) agrees within 5 percent with experimental and numerical data for $Pr = 0.7$ and $Pr = 10$, and has the correct asymptotic behavior for large and small Gz and Pr.

6.2.4 Isothermal Wall

When the flow is hydrodynamically and thermally fully developed and when the tube wall is isothermal, the mean temperature of the stream varies exponentially toward the plateau value T_w as x increases. (See the projection on the vertical plane to the left in Fig. 6.9.) Such a relationship is required by eq. (6.39') because, after writing $q_w'' = h\,[T_w - T_m(x)]$, that equation reads

$$\frac{dT_m}{dx} = \frac{2h}{r_o \rho c_P U}[T_w - T_m(x)] \tag{6.54}$$

---

[*]Leo Graetz (1856–1941) was professor of physics at the University of Munich. His writings covered a very wide domain, from heat transfer (conduction, convection, and radiation) to mechanics (elasticity, friction) and electromagnetism.

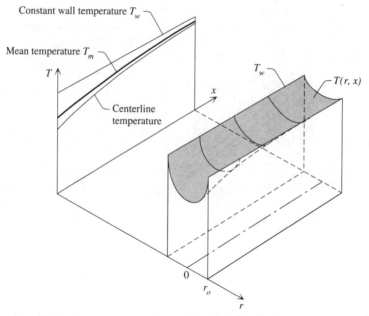

Figure 6.9 Fully developed temperature distribution in a tube with constant wall temperature.

This equation can be integrated from a reference position $x = x_1$ where $T_m = T_{m,1}$, by noting that h must be a constant [cf. eq. (6.42)]:

$$\frac{T_w - T_m(x)}{T_w - T_{m,1}} = \exp\left[-\frac{2h(x - x_1)}{r_o \rho c_p U}\right] \tag{6.55}$$

The heat transfer coefficient h, or the Nusselt number $Nu_D = hD/k$, can be determined by following the analytical steps outlined in the preceding subsection. This time, however, the resulting equation for $\phi(r)$ is more complicated and has to be integrated iteratively or numerically [5]. The result is

$$Nu_D = 3.66 \qquad (\text{constant } T_w) \tag{6.56}$$

The constancy of Nu_D implies that when the wall is isothermal, *both* ΔT and the wall heat flux q_w'' vary exponentially in the downstream direction:

$$q_w''(x) = 3.66\frac{k}{D}(T_w - T_{m,1})\exp\left[-3.66\,\frac{\alpha(x - x_1)}{r_o^2 U}\right] \tag{6.57}$$

Table 6.1 lists the Nu_D constants for other cross-sectional shapes of ducts with isothermal wall when the laminar flow is both hydrodynamically and thermally fully developed. The table shows again that duct cross sections with large shape numbers B have correspondingly large fully developed flow Nusselt numbers.

Figure 6.10 shows the Nusselt number in the thermal entrance and fully developed regions of a tube with isothermal wall. The curves were drawn based on data furnished by Shah and London [4] and Hornbeck [7]. The D-based Nusselt number approaches the 3.66 value as the stream proceeds into the thermally fully developed region. In the thermal entrance region Nu_D varies as $x^{-1/2}$, indicating that the thermal resistance between wall and stream is domi-

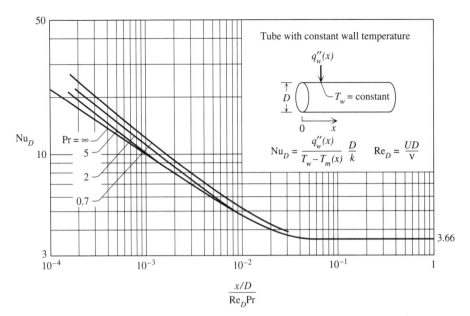

Figure 6.10 The Nusselt number for laminar flow through a tube with constant wall temperature.

nated by a thermal boundary layer of the type described by eq. (5.79). It is worth keeping in mind the difference between the D-based Nusselt number, eq. (6.43), and the Nu_x definition (5.71).

Table 6.2 shows the fully developed values of the friction factor and the Nusselt number in a duct with regular polygonal cross section. Some of the cases covered by this table are covered also by Table 6.1. The less than 1 percent discrepancies that exist between the data covered by both tables (e.g., the square cross section) is representative of the discrepancies found between results reported by different investigators in the literature. Note again that in the noncircular cases with uniform heat flux, the wall temperature changes along the perimeter. In these cases the heat transfer coefficient is based on the perimeter-averaged wall temperature $\overline{T}_w(x)$, namely, $h = q_w''/[\overline{T}_w(x) - T_m(x)]$.

Table 6.2 Friction Factors and Nusselt Numbers for Heat Transfer to Laminar Flow Through Ducts with Regular Polygonal Cross Sections[a]

| | $f\mathrm{Re}_{D_h}$ | Nu_{D_h} | | | |
| | | Uniform Heat Flux | | Isothermal Wall | |
Cross Section	Fully Developed Flow	Fully Developed Flow	Slug Flow	Fully Developed Flow	Slug Flow
Square	14.167	3.614	7.083	2.980	4.926
Hexagon	15.065	4.021	7.533	3.353	5.380
Octagon	15.381	4.207	7.690	3.467	5.526
Circle	16	4.364	7.962	3.66	5.769

[a]The data are from Asako et al. [9].

Table 6.2 also shows that the thermally developed Nu_D value is considerably smaller in fully developed flow than in "slug flow." The latter refers to the flow of a fluid with extremely small Prandtl number ($\mathrm{Pr} \to 0$), in which the viscosity is so much smaller than the thermal diffusivity that the longitudinal velocity profile remains uniform over the cross section, $u = U$, like the velocity distribution of a solid slug. The persistence of slug flow in the $\mathrm{Pr} \to 0$ limit, even at a large enough x where the temperature profile is fully developed, can be expected from the relationship between the flow and thermal developing lengths.

$$\frac{X_T}{X} \cong \mathrm{Pr} \tag{6.58}$$

This relationship follows from eqs. (6.4′) and (6.32). It says that when Pr is very small, the flow entrance distance X is much longer than the thermal entrance length X_T, and that when $X_T < x < X$, the flow distribution in the thermally fully developed region is approximately sluglike.

Example 6.2

Heat Transfer, Laminar Flow, Round Tube

A stream of room-temperature water is being heated as it flows through a pipe with uniform wall heat flux, $q_w'' = 0.1 \ \mathrm{W/cm^2}$. The flow is hydrodynamically and thermally fully developed. The mass flowrate of the stream is $\dot{m} = 10 \ \mathrm{g/s}$ and the pipe radius is $r_o = 1 \ \mathrm{cm}$. The properties of room-temperature water are approximately $\mu = 0.01 \ \mathrm{g/cm \cdot s}$ and $k = 0.006 \ \mathrm{W/cm \cdot K}$. Calculate (a) the Reynolds number based on diameter ($D = 2r_o$) and mean velocity (U), (b) the heat transfer coefficient (h), and (c) the difference between the local wall temperature and the local mean (bulk) temperature.

Solution. (a) The D-based Reynolds number is

$$\mathrm{Re}_D = \frac{\rho U D}{\mu}$$

for which the product ρU can be determined from the given mass flowrate:

$$\rho U = \frac{\dot{m}}{\pi D^2 / 4} = \frac{10 \, \mathrm{g}}{\mathrm{s}} \frac{4}{\pi} \frac{1}{(2 \, \mathrm{cm})^2}$$

$$= 3.18 \ \mathrm{g/cm^2 \cdot s}$$

The Reynolds number is therefore

$$\mathrm{Re}_D = 3.18 \, \frac{\mathrm{g}}{\mathrm{cm^2 \cdot s}} \, 2 \ \mathrm{cm} \, \frac{\mathrm{cm \cdot s}}{0.01 \ \mathrm{g}}$$

$$= 637 \qquad \text{(laminar regime)}$$

(b) The heat transfer coefficient follows from the definition of the Nusselt number for tube flow,

$$\mathrm{Nu}_D = \frac{hD}{k} = 4.364$$

Therefore,

$$h = \mathrm{Nu}_D \, \frac{k}{D} = 4.364 \times 0.006 \, \frac{\mathrm{W}}{\mathrm{cm \cdot K}} \frac{1}{2 \, \mathrm{cm}} = 0.0131 \ \mathrm{W/cm^2 \cdot K}$$

(c) From the definition of h in internal flow, $q_w'' = h(T_w - T_m)$, we draw the conclusion that

$$T_w - T_m = \frac{q_w''}{h} = \frac{0.1 \text{ W}}{\text{cm}^2} \frac{\text{cm}^2 \cdot \text{K}}{0.0131 \text{ W}}$$

$$= 7.64 \text{ K} = 7.64°C$$

6.3 TURBULENT FLOW

6.3.1 Transition, Entrance Region, and Fully Developed Flow

Consider first the turbulent flow through a straight pipe of inner diameter D. Experimental observations indicate that the fully developed laminar flow, eq. (6.19), ceases to exist if the D-based Reynolds number exceeds the *approximate* level of (Appendix F)

$$\text{Re}_D \cong 2000 \qquad (6.59)$$

The turbulent regime takes over when Re_D exceeds approximately 2300. It must be noted, however, that the laminar regime can survive at considerably higher Reynolds numbers (for example, in the range $\text{Re}_D \sim 10^4 - 10^5$) if the pipe wall is exceptionally smooth and the supply of fluid to the pipe is steady and free of disturbances. As soon as a disturbance occurs, however, the laminar flow disappears and is replaced by the meandering procession of eddies of several sizes known as turbulent duct flow. The usual condition of the internal surface of ducts is sufficiently rough and their operation sufficiently "noisy" so that the transition is summarized adequately by the criterion (6.59). The same criterion applies to ducts with other cross-sectional shapes, provided Re_D is replaced by the Reynolds number based on hydraulic diameter, Re_{D_h}.

The onset of turbulence in a pipe has been visualized by the dye injection method in Fig. 6.11, in which Re_D is in the vicinity of 2000. At higher Reynolds numbers the flow pattern has the features illustrated in Fig. 6.12. In this second set of photographs the water flows in an open channel (parallel-walls channel with free top surface). Most of the fluid meanders at high speed through the

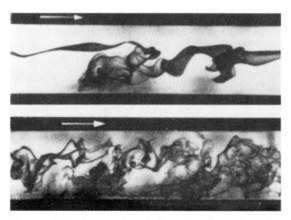

Figure 6.11 Onset of turbulence in flow through a straight pipe. (Photograph by B. Dubs published in Eck [10], with permission from Springer-Verlag, Heidelberg.)

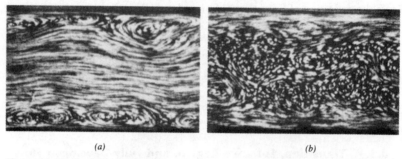

(a)   (b)

Figure 6.12 Turbulent water flow in an open channel, photographed by a camera that travels with velocity V in the downstream direction: (a) $V \cong$ velocity of the layers (eddies) near the walls; (b) $V \cong$ centerline velocity (Prandtl and Tietjens [11], with permission from McGraw-Hill Book Company).

core of the channel while bumping into and rebounding from the walls. These collisions lead to the formation of slower eddy-filled layers of fluid along both walls: these layers serve as "lubricant" for the relative motion between the fast core fluid and the stationary walls.

In terms of the time-averaged flow variables discussed in Section 5.4.2, the turbulent pipe flow is described by the mean longitudinal velocity component $\bar{u}$, the radial component $\bar{v}$, the pressure $\bar{P}$, and the temperature $\bar{T}$ (Fig. 6.13). The time averaging of the governing equations for a flow in cylindrical coordinates with θ symmetry yields the following system [5]:

$$\frac{\partial \bar{u}}{\partial x} + \frac{1}{r}\frac{\partial}{\partial r}(r\bar{v}) = 0 \tag{6.60}$$

$$\bar{u}\frac{\partial \bar{u}}{\partial x} + \bar{v}\frac{\partial \bar{u}}{\partial r} = -\frac{1}{\rho}\frac{d\bar{P}}{dx} + \frac{1}{r}\frac{\partial}{\partial r}\left[(v + \varepsilon_M)r\frac{\partial \bar{u}}{\partial r}\right] \tag{6.61}$$

$$\bar{u}\frac{\partial \bar{T}}{\partial x} + \bar{v}\frac{\partial \bar{T}}{\partial r} = \quad + \frac{1}{r}\frac{\partial}{\partial r}\left[(\alpha + \varepsilon_H)r\frac{\partial \bar{T}}{\partial r}\right] \tag{6.62}$$

The eddy diffusivities ε_M and ε_H emerge after an argument analogous to the one presented in Section 5.4.3. They contribute to modeling the apparent shear stress and the apparent radial heat flux as sums of molecular and eddy contributions:

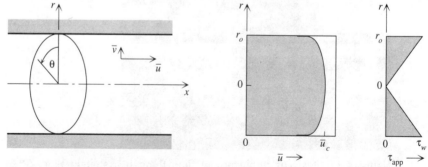

Figure 6.13 Longitudinal velocity profile and apparent shear stress distribution in fully developed turbulent flow through a pipe.

$$\tau_{\text{app}} = -\mu \frac{\partial \bar{u}}{\partial r} - \rho \varepsilon_M \frac{\partial \bar{u}}{\partial r} \tag{6.63}$$

$$q''_{\text{app}} = \underbrace{k \frac{\partial \bar{T}}{\partial r}}_{\text{Molecular}} + \underbrace{\rho c_p \varepsilon_H \frac{\partial \bar{T}}{\partial r}}_{\text{Eddy}} \tag{6.64}$$

As in Figs. 6.6, 6.8, and 6.10, we continue to define the radial heat flux positive when oriented toward the centerline.

It is also observed that the turbulent flow becomes hydrodynamically *and* thermally fully developed after a relatively short distance from the entrance to the tube,

$$\frac{X}{D} \cong 10 \cong \frac{X_T}{D} \tag{6.65}$$

This full-development criterion is particularly applicable to fluids with Prandtl numbers of order 1. It is easy to verify that the turbulent entrance length (6.65) is much shorter than the would-be laminar estimate (6.4′) when $Re_D > 2000$. According to Section 6.1.2, in fully developed flow, $\bar{v} = 0$. Therefore, $\bar{u} = \bar{u}(r)$, $\bar{P} = \bar{P}(x)$, and the momentum and energy eqs. (6.61) and (6.62) reduce to

$$0 = -\frac{1}{\rho}\frac{d\bar{P}}{dx} - \frac{1}{\rho r}\frac{d}{dr}(r\tau_{\text{app}}) \tag{6.66}$$

$$\bar{u}\frac{\partial \bar{T}}{\partial x} = \frac{1}{\rho c_p r}\frac{\partial}{\partial r}(r q''_{\text{app}}) \tag{6.67}$$

In the remainder of this subsection we focus only on the implications of the momentum eq. (6.66). The longitudinal pressure gradient is a constant proportional to the wall shear stress τ_w:

$$-\frac{d\bar{P}}{dx} = 2\frac{\tau_w}{r_o} \tag{6.68}$$

This is a rewriting of the force balance (6.26), in which $\Delta P/L = -d\bar{P}/dx$, $A = \pi r_o^2$, and $p = 2\pi r_o$. Eliminating the pressure gradient between eqs. (6.66) and (6.68) and integrating the resulting equation away from the centerline (where symmetry requires $\tau_{\text{app}} = 0$), we obtain

$$\frac{\tau_{\text{app}}}{\tau_w} = \frac{r}{r_o} \tag{6.69}$$

In conclusion, the apparent shear stress increases linearly toward the wall value τ_w, and, in a sufficiently thin flow region close to the wall,

$$\tau_{\text{app}} \cong \tau_w \tag{6.70}$$

In this way we rediscover the constant-τ_{app} wall region that was discussed in some detail in Section 5.4.4. All the features of the constant-τ_{app} region (Fig. 5.14) apply here as well. The longitudinal velocity distribution obeys the universal law $u^+ = u^+(y^+)$ presented in eq. (5.116), where

$$u^+ = \frac{\bar{u}}{(\tau_w/\rho)^{1/2}} \qquad y^+ = \frac{y}{\nu}\left(\frac{\tau_w}{\rho}\right)^{1/2} \tag{6.71}$$

and $y = r_o - r$ is the short distance measured *away from the wall*. The velocity profile is remarkably flat over most of the central portion of the cross section: this feature was intentionally exaggerated in the $\bar{u}(r)$ diagram of Fig. 6.13. Of use in the friction factor derivation presented below is the $\frac{1}{7}$th power law (5.118), which approximates well the $u^+(y^+)$ function,

$$u^+ \cong 8.7 \, (y^+)^{1/7} \tag{6.72}$$

6.3.2   Friction Factor and Pressure Drop

In problems of internal convection we use the friction factor f defined by eq. (6.24) as a dimensionless measure of the wall shear stress. The cross section-averaged velocity U used in this definition is [cf. eq. (6.1)]

$$U = \frac{1}{\pi r_o^2} \int_0^{2\pi} d\theta \int_0^{r_o} \bar{u} \, r \, dr \tag{6.73}$$

A compact friction factor formula can be derived by applying eq. (6.72) at the pipe centerline ($r = 0$), where $\bar{u}$ reaches its centerline value $\bar{u}_c$:

$$\frac{\bar{u}_c}{(\tau_w/\rho)^{1/2}} \cong 8.7 \left[\frac{r_o}{\nu} \left(\frac{\tau_w}{\rho} \right)^{1/2} \right]^{1/7} \tag{6.74}$$

The f definition (6.24) provides an additional relationship between f and τ_w/ρ:

$$\left(\frac{\tau_w}{\rho} \right)^{1/2} = U \left(\frac{f}{2} \right)^{1/2} \tag{6.75}$$

The relation between the cross section-averaged velocity U and the centerline velocity $\bar{u}_c$ follows from the definition (6.73), for which $\bar{u}$ is provided by eq. (6.72). It is not difficult to show that the result of combining eqs. (6.72) through (6.75) is the friction factor formula

$$f \cong 0.079 \text{Re}_D^{-1/4} \qquad 2 \times 10^3 < \text{Re}_D < 2 \times 10^4 \tag{6.76}$$

Compared with the curve drawn for "smooth pipes" in Fig. 6.14, the friction factor formula (6.76) is fairly accurate in the specified range, $2 \times 10^3 < \text{Re}_D < 2 \times 10^4$, where $\text{Re}_D = UD/\nu$. An empirical relation that holds at higher Reynolds numbers is

$$f \cong 0.046 \text{Re}_D^{-1/5} \qquad 2 \times 10^4 < \text{Re}_D < 10^6 \tag{6.77}$$

An implicit expression for $f^{1/2}$ that covers the entire range of eqs. (6.76) and (6.77) is the *Karman–Nikuradse relation* [13]:

$$\frac{1}{f^{1/2}} = 1.737 \ln(f^{1/2} \text{Re}_D) - 0.396 \tag{6.78}$$

Figure 6.14 also shows the position occupied by the laminar flow friction factor formula $f = 16/\text{Re}_D$, which was derived in eq. (6.25). The use of $4f$ is worth noting on the ordinate of this figure, which is widely known as *Moody's chart* [12]. Across the $\text{Re}_D \cong 2000$ transition, the friction factor executes a jump of a factor of order 10 from the laminar level to the corresponding turbulent level. A similar observation can be made regarding the behavior of the local skin friction coefficient $C_{f,x}$ along a plane wall with boundary layer flow, Fig. 5.16.

For fully developed turbulent flow through ducts with *cross sections other*

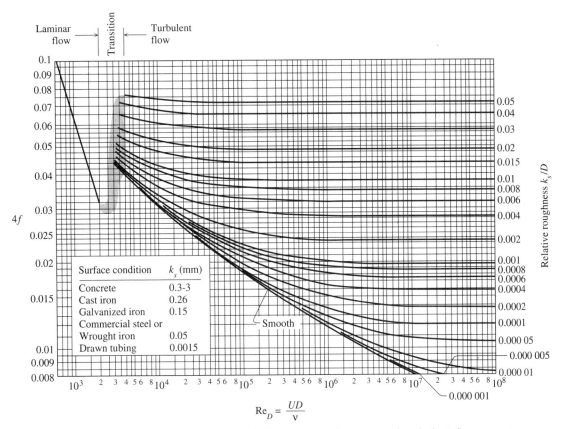

Figure 6.14 Friction factor for fully developed laminar and turbulent flow in a pipe. (Drawn after Moody [12].)

than round, eqs. (6.76) and (6.77) and Moody's chart continue to apply, provided Re_D is replaced by the Reynolds number based on hydraulic diameter, Re_{D_h}. In a duct with noncircular cross section, the time-averaged wall shear stress τ_w does not have a constant value around the periphery of the cross section. Consequently, in the friction factor definition (6.24), τ_w represents the wall shear stress averaged over the perimeter of the duct cross section.

Ducts with rough internal surfaces have higher friction factors and pressure drops than their counterparts with perfectly polished walls. The effect of wall roughness is presented in Fig. 6.14 by means of the dimensionless ratio k_s/D, where the length scale k_s (mm) is Nikuradse's "sand roughness" size [14]. Experimentally, a k_s value can be attached to a given type of surface, as illustrated in the table attached to Fig. 6.14. In general, the friction factor for fully developed turbulent flow through a duct with rough internal surface approaches a constant in the "fully rough" limit; that is, as Re_D becomes sufficiently large,

$$f = \left[1.74 \ln \left(\frac{D}{k_s} \right) + 2.28 \right]^{-2} \tag{6.79}$$

In this subsection we focused on the calculation of the friction factor f in the case of fully developed flow in a straight pipe, or a duct with different cross-sectional shape. The overall *pressure drop* registered between the ends of a duct of length L can then be calculated by applying eqs. (6.27) and (6.27').

6.3.3   Heat Transfer Coefficient

In fully developed turbulent flow through a straight duct, the heat transfer coefficient can be calculated using a Stanton number formula analogous to the Colburn relationship (5.131). This analogy exists because sufficiently close to the duct wall the apparent heat flux is nearly constant, that is, similar to the distribution plotted in Fig. 5.14 for a turbulent boundary layer.

The existence of the constant-q''_{app} layer near the wall of the duct can be demonstrated by using the pipe flow energy eq. (6.62). Noting that in fully developed flow $\bar{v} = 0$, and using eq. (6.64), the energy equation becomes

$$\rho c_p \bar{u} \frac{\partial \bar{T}}{\partial x} = \frac{1}{r} \frac{\partial}{\partial r} (r q''_{app}) \tag{6.80}$$

This can be integrated over a central cross-sectional disc of radius r:

$$2\pi \int_0^r \rho c_p \bar{u} \frac{\partial \bar{T}}{\partial x} r \, dr = 2\pi r q''_{app} \tag{6.81}$$

where q''_{app} is the apparent heat flux at the radial distance r. Note that in accordance with eq. (6.64), the heat flux is positive when pointing toward the centerline (i.e., away from the pipe wall). The special form of eq. (6.81) at the wall ($r = r_o$) is

$$\int_0^{r_o} \rho c_p \bar{u} \frac{\partial \bar{T}}{\partial x} r \, dr = r_o q''_w \tag{6.82}$$

Dividing eqs. (6.81) and (6.82) side by side, we obtain a relationship that resembles eq. (6.69),

$$\frac{q''_{app}}{q''_w} = M \frac{r}{r_o} \tag{6.83}$$

except for the factor M, which is shorthand for

$$M = \frac{\dfrac{1}{r^2} \displaystyle\int_0^r \bar{u} \frac{\partial \bar{T}}{\partial x} r \, dr}{\dfrac{1}{r_o^2} \displaystyle\int_0^{r_o} \bar{u} \frac{\partial \bar{T}}{\partial x} r \, dr} \tag{6.84}$$

In Section 6.2.3 we learned that when the wall heat flux q''_w is independent of x, the longitudinal temperature gradient ($\partial \bar{T}/\partial x$ in this case) is independent of r; therefore, M assumes the simpler form

$$M = \frac{\dfrac{1}{r^2} \displaystyle\int_0^r \bar{u} \, r \, dr}{\dfrac{1}{r_o^2} \displaystyle\int_0^{r_o} \bar{u} \, r \, dr} \tag{6.85}$$

Furthermore, since the turbulent flow $\bar{u}$ profile is relatively flat (Fig. 6.13), the effect of $\bar{u}(r)$ is weak on the right side of eq. (6.85). In conclusion, the M factor is a weak function of radius and does not deviate much from the value 1. This means that the apparent heat flux increases *almost* linearly in the radial direction [cf. eq. (6.83)]:

$$\frac{q''_{app}}{q''_w} \cong \frac{r}{r_o} \tag{6.86}$$

This distribution can be visualized by replacing τ_{app} with q''_{app} and τ_w with q''_w on the extreme-right drawing shown in Fig. 6.13. Equation (6.86) shows finally that sufficiently close to the wall $(r \lesssim r_o)$, the apparent heat flux is nearly constant:

$$q''_{app} \cong q''_w \tag{6.87}$$

This conclusion is the starting point—the basis—of analyses [5] that led to Stanton number formulas similar to eq. (5.131), except that in the case of duct flow the longitudinal velocity scale is the cross section-averaged velocity U, not U_∞:

$$St = \frac{h}{\rho c_p U} \tag{6.88}$$

Another distinction to be made between eqs. (6.88) and (5.129) is that in fully developed flow both h and St are independent of x. Furthermore, in duct flow we speak of friction factors f, eq. (6.24), instead of local skin friction coefficients; therefore, the equivalent Colburn formula for the Stanton number is [15]

$$St = \frac{\frac{1}{2}f}{Pr^{2/3}} \tag{6.89}$$

This formula holds for $Pr \gtrsim 0.5$ and is to be used in conjunction with the Moody chart (Fig. 6.14), which supplies the value of the friction factor. It applies to ducts of various cross-sectional shapes, with wall surfaces having various degrees of roughness. For example, in the special case of a pipe with smooth internal surface, we can combine eq. (6.89) with eq. (6.77) to derive the Nusselt number formula

$$Nu_D = \frac{hD}{k} = 0.023 Re_D^{4/5} Pr^{1/3} \tag{6.90}$$

which, in accordance with eq. (6.77), holds in the range $2 \times 10^4 < Re_D < 10^6$.

There are many formulas that, in one way or another, improve on the accuracy with which Colburn's eq. (6.90) predicts actual measurements. The most recent review of these formulas has been published by Bhatti and Shah [16]. The most popular formula is a correlation due to Dittus and Boelter [17],

$$Nu_D = 0.023 Re_D^{4/5} Pr^n \tag{6.91}$$

which was developed for $0.7 \le Pr \le 120$, $2500 \le Re_D \le 1.24 \times 10^5$, and $L/D > 60$. The Prandtl number exponent is $n = 0.4$ when the fluid is being heated $(T_w > T_m)$, and $n = 0.3$ when the fluid is being cooled $(T_w < T_m)$. All the physical properties needed for the calculation of Nu_D, Re_D, and Pr are to be evaluated at the bulk temperature T_m. The maximum deviation between experimental data and values predicted using eq. (6.91) is of the order of 40 percent.

For applications in which the temperature influence on properties is significant, Sieder and Tate's [18] modification of eq. (6.90) is recommended:

$$\mathrm{Nu}_D = 0.027 \mathrm{Re}_D^{4/5} \mathrm{Pr}^{1/3} \left(\frac{\mu}{\mu_w} \right)^{0.14} \tag{6.92}$$

The correlation is valid for $0.7 < \mathrm{Pr} < 16{,}700$ and $\mathrm{Re}_D > 10^4$. The effect of temperature-dependent properties is taken into account by evaluating all the properties (except μ_w) at the mean temperature of the stream, T_m. The viscosity μ_w is evaluated at the wall temperature, $\mu_w = \mu(T_w)$.

The most accurate of the correlations of type (6.90) through (6.92) is a formula due to Gnielinski [16,19],

$$\mathrm{Nu}_D = \frac{(f/2)(\mathrm{Re}_D - 10^3)\mathrm{Pr}}{1 + 12.7(f/2)^{1/2}(\mathrm{Pr}^{2/3} - 1)} \tag{6.93}$$

in which the friction factor is supplied by Fig. 6.14. It is accurate within ± 10 percent in the range $0.5 < \mathrm{Pr} < 10^6$ and $2300 < \mathrm{Re}_D < 5 \times 10^6$. Like eqs. (6.90)–(6.92), the Gnielinski correlation (6.93) can be used both in constant-q_w'' and constant-T_w applications. Two simpler alternatives to eq. (6.93) are [19]

$$\mathrm{Nu}_D = 0.0214(\mathrm{Re}_D^{0.8} - 100)\mathrm{Pr}^{0.4}$$
$$0.5 \leq \mathrm{Pr} \leq 1.5,\ 10^4 \leq \mathrm{Re}_D \leq 5 \times 10^6 \quad (6.93')$$

$$\mathrm{Nu}_D = 0.012(\mathrm{Re}_D^{0.87} - 280)\mathrm{Pr}^{0.4}$$
$$1.5 \leq \mathrm{Pr} \leq 500,\ 3 \times 10^3 \leq \mathrm{Re}_D \leq 10^6 \quad (6.93'')$$

The preceding results refer to gases and liquids, that is, to the range $\mathrm{Pr} \gtrsim 0.5$. For liquid metals, the most accurate formulas are those of Notter and Sleicher [20],

$$\mathrm{Nu}_D = 6.3 + 0.0167\,\mathrm{Re}_D^{0.85}\mathrm{Pr}^{0.93} \quad (q_w'' = \text{constant}) \tag{6.94}$$
$$\mathrm{Nu}_D = 4.8 + 0.0156\,\mathrm{Re}_D^{0.85}\mathrm{Pr}^{0.93} \quad (T_w = \text{constant}) \tag{6.95}$$

These are valid for $0.004 < \mathrm{Pr} < 0.1$ and $10^4 < \mathrm{Re}_D < 10^6$, and are based on both computational and experimental data. All the properties used in eqs. (6.94) and (6.95) are evaluated at the mean temperature T_m.

One peculiarity of the mean temperature of the stream is that it varies with

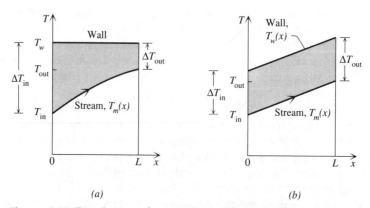

(a)   *(b)*

Figure 6.15 Distribution of temperature along a duct: (a) isothermal wall; (b) wall with uniform heat flux.

the position along the duct, $T_m(x)$. This variation is linear in the case of constant-q_w'', and exponential when the duct wall is isothermal (review Figs. 6.8 and 6.10). To simplify the recommended evaluation of the physical properties at the T_m temperature, it is convenient to choose as representative mean temperature the average value:

$$T_m = \tfrac{1}{2}(T_{in} + T_{out}) \tag{6.96}$$

In this definition, T_{in} and T_{out} are the bulk temperatures of the stream at the duct inlet and outlet, respectively (Fig. 6.15).

Example 6.3

Pressure Drop, Turbulent Flow, Annular Space

The inner pipe of a downhole coaxial heat exchanger used for geothermal energy extraction has a diameter of 16 cm. The pipe material is commercial steel. At a certain location along this pipe, the mean temperature of the water stream that flows through it is 80°C. The water flowrate is 100 tons/h. Calculate the frictional pressure drop per unit pipe length $(\Delta P/L)$ experienced by the water stream at that location.

Solution. The frictional pressure drop per unit length can be calculated with eq. (6.27), which can be rewritten as

$$\frac{\Delta P}{L} = \frac{f}{D_h} 2\rho U^2 \tag{6.27'}$$

The right side of this expression asks us to evaluate, in order,

$$U = \frac{\dot{m}}{\rho A} = 100 \frac{10^3 \, \text{kg}}{3600 \, \text{s}} \frac{\text{cm}^3}{0.9718 \, \text{g}} \frac{1}{(\pi/4)(16)^2 \, \text{cm}^2}$$

$$= 1.42 \text{ m/s}$$

$$\text{Re} = \frac{UD}{\nu} = 142 \frac{\text{cm}}{\text{s}} \, 16 \, \text{cm} \frac{\text{s}}{0.00366 \, \text{cm}^2}$$

$$= 6.2 \times 10^5 \quad \text{(turbulent flow)}$$

The properties ρ and ν have been evaluated at the mean temperature of 80°C.

Next, Fig. 6.14 delivers the value of the friction factor, provided we also know the roughness parameter k_s/D. In the case of a commercial steel surface, this is

$$\frac{k_s}{D} = \frac{0.05 \, \text{mm}}{160 \, \text{mm}} = 3.1 \times 10^{-4}$$

Therefore, Fig. 6.14 recommends $4f \cong 0.016$, or

$$f \cong 0.004$$

The right side of eq. (6.27') can now be evaluated numerically:

$$\frac{\Delta P}{L} = \frac{0.004}{16 \, \text{cm}} 2 \times 0.9718 \frac{\text{g}}{\text{cm}^3} (142)^2 \frac{\text{cm}^2}{\text{s}^2}$$

$$= 9.8 \frac{\text{g}}{\text{cm}^2 \cdot \text{s}^2} = 98 \frac{\text{N/m}^2}{\text{m}} = 9.7 \times 10^{-4} \, \text{atm/m}$$

This $\Delta P/L$ value may not look impressive; however, for a pipe that reaches to a depth of 1 km into the earth's crust, it indicates a frictional pressure drop of approximately 1 atm.

6.4 THE TOTAL HEAT TRANSFER RATE

A primary objective of Sections 6.2 and 6.3 has been the proper evaluation of the heat transfer coefficient between the duct wall and the stream. We learned, for example, that in fully developed laminar and turbulent duct flows, h is independent of longitudinal position. The heat transfer coefficient is essential in the calculation of the *total* heat transfer rate q(W) that is picked up by the stream as it travels the entire length L of the duct. The heat transfer rate q can be expected to be proportional to the h constant, to the total duct surface swept by the stream ($A_w = pL$), and to an "effective" temperature difference labeled for the time being ΔT_{lm}:

$$q = hA_w \Delta T_{lm} \tag{6.97}$$

Isothermal Wall. The magnitude of the effective temperature difference ΔT_{lm} depends on how the actual wall–stream temperature difference varies along the duct, $T_w(x) - T_m(x)$. Consider first the case where the wall temperature T_w is constant, as shown earlier in Fig. 6.9, and now in Fig. 6.15a. In fully developed laminar or turbulent flow, the temperature difference

$$\Delta T = T_w - T_m \tag{6.98}$$

decreases exponentially in the downstream direction, between a certain value at the tube inlet and a smaller value at the tube outlet,

$$\Delta T_{in} = T_w - T_{in} \qquad \Delta T_{out} = T_w - T_{out} \tag{6.99}$$

The effective temperature difference ΔT_{lm} falls somewhere between the extremes ΔT_{in} and ΔT_{out}. Its precise value can be determined by deriving the q formula (6.97) in a rigorous manner based on thermodynamic analysis. From the point of view of the stream as an elongated control volume, the total heat transfer rate through the duct wall is

$$q = \dot{m}c_P(T_{out} - T_{in}) \tag{6.100}$$

Figure 6.15a shows that the bulk temperature excursion $T_{out} - T_{in}$ is the same as the difference $\Delta T_{in} - \Delta T_{out}$; therefore, an alternative to eq. (6.100) is

$$q = \dot{m}c_P(\Delta T_{in} - \Delta T_{out}) \tag{6.101}$$

It remains to determine the relationship between the heat capacity flowrate $\dot{m}c_P$ and the group hA_w that appears on the right side of eq. (6.97). For this we use eq. (6.39), in which $q_w'' = h(T_w - T_m)$; therefore,

$$\frac{dT_m}{T_w - T_m} = \frac{hp}{A\rho c_P U}dx \tag{6.102}$$

The special form acquired by this differential equation in the case of a duct of circular cross section is listed as eq. (6.54). Assuming constant A, p, and c_P, we integrate eq. (6.102) from the inlet ($T_m = T_{in}$ at $x = 0$) all the way to the outlet ($T_m = T_{out}$ at $x = L$), and obtain

$$\ln\frac{T_w - T_{in}}{T_w - T_{out}} = \frac{hpL}{\rho A U c_P} \tag{6.103}$$

In this equation we recognize the inlet and outlet temperature differences (6.99), the mass flow rate $\rho A U = \dot{m}$, and the total duct area $pL = A_w$. Therefore, an alternative form of eq. (6.103) is

$$\ln \frac{\Delta T_{in}}{\Delta T_{out}} = \frac{h A_w}{\dot{m} c_p} \tag{6.103$'$}$$

The proper definition of the ΔT_{lm} factor adopted in eq. (6.97) becomes clear as we eliminate $\dot{m} c_p$ between eqs. (6.103$'$) and (6.101):

$$q = h A_w \frac{\Delta T_{in} - \Delta T_{out}}{\ln\left(\dfrac{\Delta T_{in}}{\Delta T_{out}}\right)} \tag{6.104}$$

in other words,

$$\Delta T_{lm} = \frac{\Delta T_{in} - \Delta T_{out}}{\ln\left(\dfrac{\Delta T_{in}}{\Delta T_{out}}\right)} \tag{6.105}$$

Because of the ΔT-averaging operation prescribed by eq. (6.105), the ΔT_{lm} factor is recognized as the *log-mean temperature difference* between the wall and the stream. When the wall and inlet temperatures are specified, eq. (6.104) expresses the relationship between the total heat transfer rate q, the total duct surface conductance $h A_w$, and the outlet temperature of the stream. Alternatively, eqs. (6.103$'$)–(6.105) can be combined to express the total heat transfer rate in terms of the inlet temperatures, mass flowrate, and duct surface thermal conductance $h A_w$,

$$q = \dot{m} c_p \Delta T_{in} \left[1 - \exp\left(-\frac{h A_w}{\dot{m} c_p}\right)\right] \tag{6.105$'$}$$

It is important to note that it was not necessary to assume that h is independent of x while integrating eq. (6.102). In cases where the heat transfer coefficient varies longitudinally, $h(x)$, the h factor on the right side of eqs. (6.103) and (6.103$'$) represents the L-averaged heat transfer coefficient, namely, $\int_L h(x)\,dx / L$.

Uniform Wall Heating.   It will be demonstrated in Section 9.3 that the applicability of eq. (6.104) is considerably more general than what is suggested by Fig. 6.15a. For example, when the heat transfer rate q is distributed uniformly along the duct, the temperature difference ΔT does not vary with the longitudinal position (review Section 6.2.3). This case is illustrated in Fig. 6.15b, where it was again assumed that A, p, h, and c_p are independent of x. Geometrically, it is evident that the "effective" value labeled ΔT_{lm} can only be the same as the constant ΔT recorded all along the duct,

$$\Delta T_{lm} = \Delta T_{in} = \Delta T_{out} \tag{6.106}$$

It is not difficult to show that eq. (6.106) is indeed a special case of eq. (6.105), namely, the limit $\Delta T_{in}/\Delta T_{out} \rightarrow 1$.

REFERENCES

1. Y. Nakayama, W. A. Woods, D. G. Clark, and the Japan Society of Mechanical Engineers, Eds., *Visualized Flow*, Pergamon Press, Oxford, 1988.

2. G. Hagen, Über die Bewegung des Wassers in engen zylindrischen Röhren, *Pogg. Ann.*, Vol. 46, 1839, pp. 423–442.

3. J. Poiseuille, Récherches expérimentales sur le mouvement des liquides dans les tubes de très petits diamètres, *C. R. Acad. Sci.*, Vol. 11, 1840, pp. 961–967, 1041–1048; Vol. 12, 1841, pp. 112–115.

4. R. K. Shah and A. L. London, *Laminar Flow Forced Convection in Ducts*, Academic Press, New York, 1978, p. 40.

5. A. Bejan, *Convection Heat Transfer*, Wiley, New York, 1984, p. 77.

6. A. Bejan, *Advanced Engineering Thermodynamics*, Wiley, New York, 1988, p. 71.

7. R. W. Hornbeck, An all-numerical method for heat transfer in the inlet of a tube, *Am. Soc. Mech. Eng.* Paper No. 65-WA/HT-36, 1965.

8. S. W. Churchill and H. Ozoe, Correlations for forced convection with uniform heating in flow over a plate and in developing and fully developed flow in a tube, *J. Heat Transfer*, Vol. 95, 1973, pp. 78–84.

9. Y. Asako, H. Nakamura, and M. Faghri, Developing laminar flow and heat transfer in the entrance region of regular polygonal ducts, *Int. J. Heat Mass Transfer*, Vol. 31, 1988, pp. 2590–2593.

10. B. Eck, *Technische Strömungslehre*, Springer-Verlag, Berlin, 1966, p. 103.

11. L. Prandtl and O. G. Tietjens, *Applied Hydro- and Aeromechanics*, translated by J. P. Den Hartog, McGraw-Hill, New York, 1934.

12. L. F. Moody, Friction factors for pipe flows, *Trans. ASME*, Vol. 66, 1944, pp. 671–684.

13. W. M. Kays and H. C. Perkins, Forced convection, internal flow in ducts, in W. M. Rohsenow and J. P. Hartnett, Eds., *Handbook of Heat Transfer*, Section 7, McGraw-Hill, New York, 1973.

14. J. Nikuradse, Strömungsgesetze in rauhen Rohren, *VDI-Forschungsh.*, Vol. 361, 1933, pp. 1–22.

15. A. P. Colburn, A method for correlating forced convection heat transfer data and a comparison with fluid friction, *Trans. Am. Inst. Chem. Eng.*, Vol. 29, 1933, pp. 174–210; reprinted in *Int. J. Heat Mass Transfer*, Vol. 7, 1964, pp. 1359–1384.

16. M. S. Bhatti and R. K. Shah, Turbulent and transition flow convective heat transfer in ducts, Chapter 4 in S. Kakac, R. K. Shah and W. Aung, *Handbook of Single-Phase Convective Heat Transfer*, Wiley, New York, 1987.

17. P. W. Dittus and L. M. K. Boelter, Heat transfer in automobile radiators of the tubular type, *Univ. California Pub. Eng.*, Vol. 2, No. 13, pp. 443–461, Oct. 17, 1930; reprinted in *Int. Comm. Heat Mass Transfer*, Vol. 12, 1985, pp. 3–22.

18. E. N. Sieder and G. E. Tate, Heat transfer and pressure drop of liquids in tubes, *Ind. Eng. Chem.*, Vol. 28, 1936, pp. 1429–1436.

19. V. Gnielinski, New equations for heat and mass transfer in turbulent pipe and channel flow, *Int. Chem. Eng.*, Vol. 16, 1976, pp. 359–368.

20. R. H. Notter and C. A. Sleicher, A solution to the turbulent Graetz problem III. Fully developed and entry region heat transfer rates, *Chem. Eng. Sci.*, Vol. 27, 1972, pp. 2073–2093.

21. A. Bejan, *Entropy Generation through Heat and Fluid Flow*, Wiley, New York, 1982, p. 60.

22. A. Bejan, The fundamentals of sliding contact melting and friction, *J. Heat Transfer*, Vol. 111, 1989, pp. 13–20.

23. J. L. Lage and J. S. Lim, Office conversation, October 1990.

24. A. Bejan and E. Sciubba, The optimal spacing of parallel plates cooled by forced convection, *Int. J. Heat and Mass Transfer*, Vol. 35, 1992, to appear.

25. D. B. Tuckerman and R. F. W. Pease, High-performance heat sinking for VLSI, *IEEE Electron Device Letters*, Vol. EDL-2, 1981, pp. 126–129.

26. R. W. Knight, J. S. Goodling, and D. J. Hall, Optimal thermal design of forced convection heat sinks—analytical, *Journal of Electronic Packaging*, Vol. 113, 1991, pp. 313–321.

27. A. Bar-Cohen and A. D. Kraus, eds., *Advances in Thermal Modeling of Electronic Components and Systems*, Vol. 2, ASME Press, New York, 1990.

PROBLEMS

Laminar Flow

6.1 Derive the parabolic velocity distribution (6.17)–(6.18) for fully developed laminar flow through a tube of radius r_o. In other words, solve eq. (6.14) subject to the boundary conditions (6.16a,b).

6.2 Consider the fully developed laminar flow through the parallel-plate channel shown in Fig. 6.3. Determine the velocity profile (6.21)–(6.22) by solving eq. (6.20) subject to boundary conditions identified on the figure.

6.3 The friction factor for fully developed laminar flow through a straight duct with regular hexagonal cross section is (Table 6.2)

$$f = \frac{15.065}{\text{Re}_{D_h}}$$

Compare the constant 15.065 with the approximate value that can be anticipated using eq. (6.31).

6.4 Consider the laminar flow of 5°C water through the parallel-plate channel photographed in Fig. 6.4. The plate-to-plate spacing is $D = 2$ cm and the mean velocity is $U = 3.2$ cm/s. Calculate the pressure drop per unit length $\Delta P/L$ in the fully developed region. Estimate also the flow entrance length X. Compare X with the length of the flow region photographed in Fig. 6.4.

6.5 Water flows with the mean velocity 6 cm/s through a 2.7-cm-diameter pipe. The pipe wall is isothermal, $T_w = 10°C$, and the water inlet temperature is 30°C. The total length of the pipe is $L = 10$ m. Calculate the outlet mean temperature of the stream, by assuming that the flow is fully developed and that the water properties can be evaluated at 20°C. Verify that the flow is in the laminar regime.

6.6 Water is heated as it flows through a stack of parallel metallic blades. The blade-to-blade spacing is $D = 1$ cm and the mean velocity through each channel is $U = 3.2$ cm/s. Each blade is heated electrically so that both sides of the blade release together 1600 W/m² into the water. Assuming that the water properties can be evaluated at 50°C, and that the flow is thermally fully developed,

(a) Verify that the flow is laminar.

(b) Calculate the mean temperature difference between the blade and the water stream.

(c) Calculate the rate of temperature increase along the channel.

(d) Develop a feeling for how long the channel must be so that the assumption that the flow is thermally fully developed is valid.

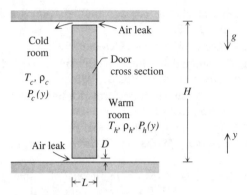

Figure P6.7

6.7 The air flow through the gaps formed at the top and bottom of a closed door is driven by the local air pressure difference between the two sides of the door. The door separates two isothermal rooms at different temperatures, T_c and T_h. In each room the pressure distribution is purely hydrostatic, $P_c(y)$ and $P_h(y)$, and the height-averaged pressure is the same on both sides of the door.

(a) Assume that the air flow through each gap is laminar and fully developed. In terms of the geometric parameters indicated in the figure, show that the air flowrate through one gap is

$$\dot{m} = (\rho_c - \rho_h)\frac{gD^3 WH}{24\nu L}$$

where W is the door width in the direction perpendicular to the plane of the figure. Show further that the net convection heat transfer rate from the warm room to the cold room, through the two gaps, is

$$q = \dot{m}c_P(T_h - T_c)$$

(b) Given are $T_c = 10°C$, $T_h = 30°C$, $D = 0.5$ mm, $L = 5$ cm, $H = 2.2$ m, and $W = 1.5$ m. Calculate $\dot{m}$ and q, and comment on how these quantities react to an increase in the gap thickness D.

6.8 A stream of water is heated with uniform heat flux in a pipe with a diameter of 2 cm and length of 2.86 m. The stream enters the pipe with the uniform velocity 10 cm/s and temperature $T_{in} = 10°C$. The mean temperature at the pipe outlet is $T_{out} = 20°C$. The water properties can be estimated at the average bulk temperature $(T_{in} + T_{out})/2$.

(a) Calculate the thermal entrance length and compare it with the length of the pipe. Is the stream thermally fully developed at the pipe outlet?

(b) Estimate the local Nusselt number Nu_D at $x = 2.86$ m in four ways: (1) graphically, based on Fig. 6.8; (2) by assuming that the temperature profile is fully developed; (3) by assuming that the thermal boundary layer extends over the entire length of the pipe wall; and (4) based on the correlation (6.52)–(6.53). Comment on the relative merit (accuracy) of the four methods.

(c) Calculate the wall heat flux and the local wall temperature at the pipe outlet. Develop also an estimate for the wall temperature averaged over the entire length of the pipe.

6.9 Determine the Nusselt number for fully developed laminar flow through a tube with uniform heat flux. Begin with the energy eq. (6.48), integrate that equation twice, and invoke finally the mean temperature definition (6.50).

6.10   The design of the cooling jacket for the apparatus shown in the figure calls for the use of a copper tube of inner diameter $D = 0.5$ cm and total length $L = 4$ m. Through this tube, cooling water flows with the mean velocity $U = 10$ cm/s. The tube is soldered to the side of the apparatus, the temperature of which is 30°C. Neglecting the thermal resistance due to pure conduction across the copper wall of the tube, we can model the tube wall as isothermal with the temperature $T_w = 30$°C. The water stream enters the tube with the mean temperature $T_1 = 20$°C. In the calculations proposed below, the physical properties of water can be evaluated at 25°C. Assume also that the tube is sufficiently straight, that is, neglect the effect of bends.

(a) Verify that the flow is hydrodynamically fully developed over most of the tube length. Calculate the pressure drop across the tube.

(b) Verify also that the flow is thermally fully developed along most of the tube. Calculate the mean outlet temperature of the water stream, and the total heat transfer rate extracted by the stream from the tube wall (i.e., from the apparatus to which the tube is attached).

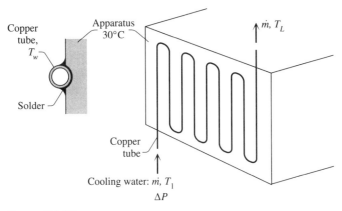

Figure P6.10

6.11   A highly viscous fluid is forced to flow through a straight pipe of inner radius r_o. The effect of friction (viscous shearing) tends to warm up the fluid as it advances through the pipe. This effect is offset by the cooling provided all along the pipe wall, which is isothermal ($T_w = $ constant). The flow is hydrodynamically and thermally fully developed. The energy equation for a fluid with constant properties reduces in this case to

$$0 = k\frac{1}{r}\frac{d}{dr}\left(r\frac{dT}{dr}\right) + \mu\Phi$$

where Φ is the viscous dissipation function

$$\Phi = \left(\frac{du}{dr}\right)^2$$

and $u(r)$ is the Hagen–Poiseuille velocity profile.

(a) Determine the temperature distribution through the fluid, $T(r)$.

(b) If q is the total heat transfer rate (i.e., the cooling effect) through the wall of a pipe of length L, show that

$$q = \frac{\dot{m}\,\Delta P}{\rho}$$

where ΔP is the pressure difference that drives the mass flowrate $\dot{m}$ through the same pipe length L.

Turbulent Flow

6.12   The criterion for transition to turbulence in duct flow, eq. (6.59), is essentially the same as the criterion for transition in boundary layer flow, eq. (5.85). Demonstrate this equivalence by noting that when the flow is at the laminar/turbulent threshold in the fully developed region, it is also undergoing transition at the end of the entrance region (i.e., at $x \sim X$: see the figure). Rely on this observation as you start with eq. (6.59) and derive a critical Re_x criterion that reproduces approximately eq. (5.85). The equivalence of these and other transition criteria forms the subject of Appendix F.

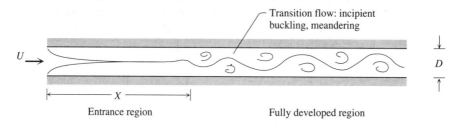

Figure P6.12

6.13   Consider the flow of a certain fluid through a tube of fixed diameter D and length L. The mass flowrate $\dot{m}$ is also fixed. The only change that may occur is the switch from laminar to turbulent flow, because the Reynolds number Re_D happens to be in the vicinity of 2000. In either regime, the flow is fully developed. Calculate the change in the required pumping power as the laminar flow is replaced by turbulent flow [21].

6.14   A stream of air (Pr = 0.72) is heated in fully developed flow through a pipe of diameter D (fixed) and uniform heat flux q_w'' (fixed). Since the Reynolds number Re_D happens to be equal to 2500, there is some uncertainty with regard to the flow regime that prevails in the air stream. Calculate the change experienced by the local temperature difference $T_w - T_m$ as the flow regime switches from laminar to turbulent.

6.15   Water flows at the rate of 0.5 kg/s through a 10-m-long pipe with an inside diameter of 2 cm. It is being heated with uniform wall heat flux at the rate of 5×10^4 W/m². Evaluate the water properties at 20°C, assume that the flow and temperature fields are fully developed, and calculate:

(a) The pressure drop over the entire pipe length,

(b) The heat transfer coefficient based on the Colburn analogy (6.89),

(c) The heat transfer coefficient based on the Dittus–Boelter correlation (6.91),

(d) The difference between the wall temperature and the local mean water temperature, and

(e) The temperature increase experienced by the mean water temperature in the longitudinal direction, from the inlet to the outlet.

6.16   Atmospheric pressure steam condenses on the outside of a metallic tube and maintains the tube wall temperature uniform at 100°C. The interior of the tube is cooled by a stream of 1 atm air with a mean velocity of 5 m/s and an inlet temperature of 30°C. The tube inside diameter is 4 cm. Assume that the flow and temperature distributions across the tube are fully developed, and calculate

(a) The heat transfer coefficient,

(b) The length of the tube, if the outlet mean temperature of the air stream is 90°C, and

(c) The flow and thermal entrance lengths. Is the assumption of fully developed flow and heat transfer justified?

6.17 Water is being heated in a straight pipe with an inside diameter of 2.5 cm. The heat flux is uniform, $q_w'' = 10^4$ W/m^2, and the flow and temperature fields are fully developed. The local difference between the wall temperature and the mean temperature of the stream is 4°C. Calculate the mass flowrate of the water stream, and verify that the flow is turbulent. Evaluate the properties of water at 20°C.

6.18 Derive eq. (6.69), which shows that the apparent shear stress in fully developed turbulent pipe flow increases linearly in the radial direction.

6.19 Tap water of temperature 20°C flows through a straight pipe 1 cm in diameter, with a mean velocity of 1 m/s. Verify that the flow regime is turbulent, and calculate the friction factor f for fully developed flow. If the dimensionless thickness of the viscous sublayer of the constant-τ_{app} region of the flow is equal to $y^+ \cong 11.6$, what is the actual thickness y (mm) of the viscous sublayer?

6.20 Show that in fully developed turbulent flow through a pipe, in which the velocity distribution is given by eq. (6.72), the relationship between the centerline velocity $\bar{u}_c$ and the mean velocity U is

$$\bar{u}_c = 1.224U$$

Rely on this result and eqs. (6.74)–(6.75) to derive your version of the friction factor formula (6.76).

6.21 Derive Colburn's formula (6.90) for fully developed flow in a straight tube by combining eqs. (6.89) and (6.77). This derivation is based on the assumption that eq. (6.90) is accurate above $Re_D \cong 2 \times 10^4$. Derive an alternative Nu_D formula that is more accurate at lower Reynolds numbers, in the range $2 \times 10^3 < Re_D < 2 \times 10^4$.

Total Heat Transfer Rate

6.22 A water chiller circulates a water stream of 0.1 kg/s through a pipe immersed in a bath of crushed ice and water. For this reason, the pipe wall temperature may be modeled as constant, $T_w = 0$°C. The inlet temperature of the water stream is $T_1 = 40$°C. The pipe is long enough so that the flow is hydrodynamically and thermally fully developed over most of the pipe length L. The mean temperature at the pipe outlet is $T_L = 6$°C.

(a) Calculate the total heat transfer rate between the stream and the 0°C bath.

(b) Let D be the inner diameter of the pipe discussed until now. Consider the option of using a different pipe with the diameter $D_1 = D/2$. The mass flowrate, and the inlet and outlet temperatures of the water stream are fixed by design. What is the required length of the new pipe (L_1)?

6.23 Consider the definition of the log-mean temperature difference, eq. (6.105), in an application in which the end differences ΔT_{in} and ΔT_{out} are equal (e.g., Fig. 6.15b). Show that in the limit $\Delta T_{in}/\Delta T_{out} \to 1$, eq. (6.105) approaches the ΔT_{lm} values listed in eq. (6.106).

6.24 One method of extracting the energy contained in a geothermal reservoir consists of using the "downhole coaxial heat exchanger" shown in the figure. The underground temperature increases almost linearly with depth (see next problem). The stream $\dot{m} = 100$ tons/h of cold water is pumped downward through the annular space of outer diameter $D_o = 22$ cm and inner diameter $D_i = 16$ cm. In this portion of its circuit, the stream is heated by contact with the increasingly warmer rock material, across the wall of diameter D_o.

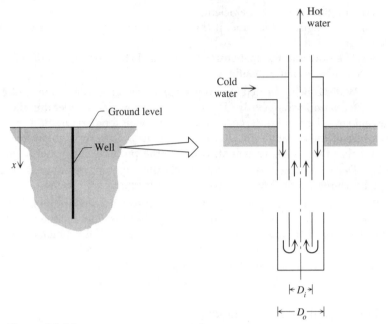

Figure P6.24

After reaching the lower extremity of the well, the heated stream returns to the surface, by flowing through the inner pipe. A very effective layer of insulation is built into the wall of diameter D_i, which separates the downflowing cold stream from the upflowing hot stream.

(a) Consider only the downflow through the annular space, and assume that the depth (x) to which your calculations apply is such that the mean temperature of the stream is 80°C. The wetted surfaces of the annular space are made of commercial steel. Calculate the frictional pressure drop per unit length experienced by the stream at that depth.

(b) Calculate the temperature difference ΔT between the outer wall of the annulus and the mean temperature of the stream. Again, the depth x is such that the mean temperature of the stream is 80°C. Known also is the mean temperature gradient $dT_m/dx = 200°C/km$, that is, the rate of temperature increase with depth.

6.25 Review the method of extracting energy from hot dry rock, which is illustrated in Problem 4.24. The fracture can be modeled as a parallel-plate channel of thickness $D = 1$ cm, length parallel to the flow $L = 100$ m, and width $W = 100$ m. The wetted rock surfaces are isothermal at $T_w = 200°C$. The water inlet temperature is $T_{in} = 30°C$, and the flowrate is $\dot{m} = 10$ kg/s. The water properties are approximately the same as those of saturated liquid at 200°C. Calculate

(a) The flow and thermal entrance lengths,

(b) The pressure drop across the length L, and

(c) The total heat transfer rate between the rock surfaces and the water stream, that is, the rate of energy extraction from the hot rock.

6.26 Air at 300°C and 2 m/s approaches a bundle of 4 cm-diameter tubes arranged in a staggered array ($X_t = X_l = 7$ cm), with 21 rows and 6 or 5 tubes per row (i.e., across the flow). Each tube is 3 m long, and its wall temperature is maintained at 30°C by water flowing in the tube. The bundle-averaged heat transfer coefficient between tubes and air stream is 62 W/m²·K. The parallel side walls of the air duct

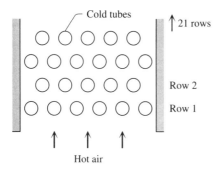

Figure P6.26

are insulated. Calculate the outlet temperature of the air stream, and the total heat transfer rate absorbed by the tube bundle.

6.27 The metallic blade shown in the figure is an electric conductor that must be cooled by forced convection in a channel with insulated walls, with spacing D and length L. The blade and the channel are sufficiently long in the direction perpendicular to the figure. The pressure difference across the arrangement is fixed, ΔP, and the flow on either side of the blade is laminar and fully developed. The inlet temperature of the coolant is T_0.

The designer's objective is to lower the blade temperature as much as possible. The total heat transfer rate from the blade to the fluid, through both sides of the blade, is fixed by electrical design. What is the best position that the blade should occupy in the channel — right in the middle, or closer to one of the side walls?

For simplicity, assume that the blade is isothermal (T_w). Assume also that on either side of the blade the group $hA_w/\dot{m}c_P$ is sufficiently greater than 1 so that the outlet temperature of the stream is approximately equal to T_w. The blade thickness is negligible relative to D.

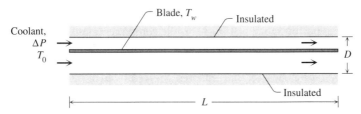

Figure P6.27

6.28 The figure shows a simplified model of an electronic circuit board cooled by a laminar fully developed flow in a parallel-wall channel of fixed length L. The walls of the channel are insulated. The board substrate has a sufficiently high thermal conductivity so that the board temperature T_w may be assumed uniform in the longitudinal direction. The pressure difference that drives the flow is fixed, ΔP, and the fluid inlet temperature is T_0.

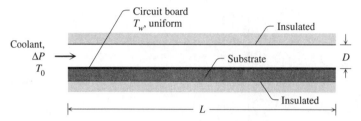

Figure P6.28

The channel spacing D must be selected in such a way that the thermal conductance $q/(T_w - T_0)$ is maximum. In this ratio, q is the total heat transfer rate removed by the stream from the board. Show that the optimal spacing is given by

$$\frac{D_{opt}}{L} = 2.70\left(\frac{\mu\alpha}{\Delta P \cdot L^2}\right)^{1/4}$$

and that the corresponding maximum thermal conductance or average heat transfer coefficient is

$$\left(\frac{\overline{q}''}{T_w - T_0}\right)_{max}\frac{L}{k} = 0.693\left(\frac{\Delta P \cdot L^2}{\mu\alpha}\right)^{1/4}$$

The dimensionless group that emerged on the right side deserves to be called *the pressure drop number*

$$\Pi = \frac{\Delta P \cdot L^2}{\mu\alpha}$$

It plays the leading role in forced convection through ducts with a specified pressure difference across the duct, as in the next problem and in Project 6.3.

6.29   The electronic circuit board shown in the figure is thin and long enough to be modeled as a surface with uniform heat flux q''. The heat generated by the circuitry is removed by the fully developed laminar flow channeled by the board and a parallel wall above it. That wall and the underside of the board are insulated. The length L is specified, and the inlet temperature of the coolant is T_0.

The circuit board reaches its highest temperature (T_h) at the trailing edge, that is, in the plane of the outlet. That temperature ceiling is fixed by electrical design, otherwise the performance of the electronic components incorporated in the board will deteriorate. The designer would like to build as much circuitry and as many components into the board as possible. This objective is equivalent to seeking a board and channel design that ensures the removal of the largest rate of heat generated by the board ($q''L$). The lone degree of freedom is the selection of the spacing D.

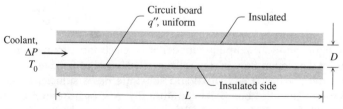

Figure P6.29

(a) Maximize the heat transfer rate removed by the stream, and show that the optimal design is characterized by

$$\frac{D_{opt}}{L} = 3.14\left(\frac{\mu\alpha}{\Delta P \cdot L^2}\right)^{1/4}$$

$$\left(\frac{q''}{T_h - T_0}\right)_{max}\frac{L}{k} = 0.644\left(\frac{\Delta P \cdot L^2}{\mu\alpha}\right)^{1/4}$$

(b) Compare this optimal design with the results of the preceding problem (with $T_w = T_h$), in which the board was made isothermal by bonding it to a high conductivity substrate. Why is the maximum heat transfer rate higher when the board is isothermal? Is the increase in heat transfer significant enough to justify the use of a high-conductivity substrate?

PROJECTS

6.1   Coaxial Heat Exchanger for Geothermal Energy Extraction

The single-stream, coaxial heat exchanger described in Problem 6.24 brings up a fundamental design question regarding the diameter of the inner pipe, D_i. The thickness of the wall of diameter D_i is assumed negligible. If D_i is much smaller than D_o, the stream is "strangled" as it flows upward through the inner pipe. Conversely, when D_i is nearly the same as D_o, the flow is impeded by the narrowness of the annular space. In both extremes the overall pressure drop that must be overcome by the pump is excessive. Clearly, when D_o is fixed, there exists an optimal inner diameter D_i (or an optimal ratio D_i/D_o) such that the total pressure drop experienced by the stream is minimum.

(a) Determine this optimal D_i/D_o ratio in the large Reynolds number limit of the turbulent regime, Fig. 6.14, where the friction factors for the annular space (f_a) and for the upflow through the inner pipe (f_i) are both constant. For simplicity, assume that $f_a = f_i$.

(b) Consider next the regime in which the flow is laminar both through the annular space and through the inner pipe. Assume that the friction factor for the annular space is approximately equal to the friction factor for flow between two parallel plates positioned $(D_o - D_i)/2$ apart. Calculate the optimal D_i/D_o ratio for minimum total pressure drop, and show that this result is almost the same as the result obtained in part (a).

6.2   Melting and Lubrication by Sliding Contact

A block of a certain crystalline substance at the melting point T_m is pushed with the force F_n (N/m) against another solid that moves to the right with the velocity U. The sliding solid is relatively warmer ($T_m + \Delta T$) and causes steady melting at the lower surface of the T_m block. The liquid that is generated by this melting process fills the gap δ and "lubricates" the relative motion between the two solids [22].

The objective of this project is to illustrate a "parallel-plate channel" with heat transfer, where the analysis is considerably more interesting than what we saw in Sections 6.1 and 6.2. In this channel the liquid enters "from above" by crossing the $y = \delta$ interface. Furthermore, the channel spacing δ and the melting rate V are not fixed by design: they are both consequences of the imposed vertical force F_n and temperature difference ΔT.

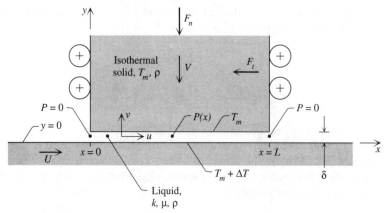

Figure PR6.2

(a) Assume that the relative motion gap is so narrow that the effects of convection and longitudinal conduction can be neglected in the energy equation for the liquid flow. That equation reduces to

$$\frac{\partial^2 T}{\partial y^2} = 0$$

Determine the temperature distribution in the liquid and show that the melting rate and the gap thickness are related by

$$V\delta = \frac{k\,\Delta T}{\rho h_{sf}}$$

In this way, you are proving that the gap thickness δ is not a function of longitudinal position x.

(b) In the same small-δ limit, the effects of liquid inertia and longitudinal viscous diffusion are negligible in the x and y momentum equations. In a way similar to the discussion that led to eq. (6.13) in the text, the momentum equation reduces to

$$\frac{dP}{dx} = \mu\frac{\partial^2 u}{\partial y^2}$$

where $P = P(x)$ is the liquid pressure and $u = u(x,y)$ is the liquid longitudinal velocity. Determine $u(x,y)$ as a function of dP/dx, U, and δ.

(c) To discover the actual distribution of pressure $P(x)$, determine sequentially the liquid flowrate

$$Q = \int_0^\delta u\,dy$$

and the result of integrating from $y = 0$ to $y = \delta$ the mass conservation equation

$$\frac{\partial u}{\partial x} + \frac{\partial v}{\partial y} = 0$$

Show that the pressure distribution is parabolic and that the maximum pressure occurs in the middle of the swept length ($x = L/2$).

(d) Relate your pressure distribution result to the vertical force maintained from above. Show that the melting rate V can be calculated with the formula

$$V = \left(\frac{F_n}{\mu}\right)^{1/4}\left(\frac{k\,\Delta T}{\rho h_{sf}L}\right)^{3/4}$$

(e) The melting block is maintained vertical and guided downward by the mechanism (rollers) shown in the figure. Derive an expression for the horizontal force felt by this mechanism, F_t. Show that the coefficient of friction $\mu_f = F_t/F_n$ decreases as the vertical force F_n increases:

$$\mu_f = U\left(\frac{\mu}{F_n}\right)^{3/4}\left(\frac{ph_{sf}L}{k\Delta T}\right)^{1/4}$$

6.3 Optimal Spacing Between Electronic Circuit Boards

A large number of parallel electronic circuit boards must be cooled by forced convection in a space whose height H is fixed. The pressure difference ΔP that drives the flow is also fixed. The fluid has the inlet temperature T_∞, and its Prandtl number is larger than 0.5. Each board can be modeled as an isothermal flat plate with negligible thickness. The board temperature T_0 is set by electronic operational constraints.

In the design of the electronic package there are incentives to mount as much circuitry as possible in the $H \times L$ space, provided the board temperature does not exceed T_0. Since each circuit generates heat, this design objective translates into the maximization of the total rate of heat transfer in the $H \times L$ space, q', from all the boards to the forced flow. The heat transfer rate q' is expressed per unit length normal to the plane $H \times L$. You are asked to determine the optimal board-to-board spacing D that accomplishes this objective, namely, the optimal number of boards ($n \cong H/D$) that must be installed [23,24].

Procedure. A cost-effective approach that allows you to see how D affects the design consists of analyzing the two limits associated with changing D:

(a) Assume that as $D \to 0$, each channel becomes slender enough so that the flow is laminar and fully developed, and the fluid outlet temperature is nearly equal to T_0. Show that in this limit the total heat transfer rate removed by the fluid is proportional to D^2.

(b) Consider the opposite limit in which D is large enough so that each board is lined by distinct boundary layers. Note that the imposed ΔP is related to the L-averaged shear stress experienced by each surface. Assume that the flow is laminar, and calculate the average longitudinal velocity and the heat transfer through each L-long surface. Show that in this limit the total heat transfer rate removed by the fluid is proportional to $D^{-2/3}$.

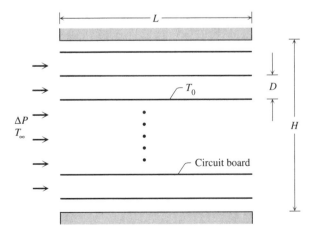

Figure PR6.3

(c) Obtain an approximate estimate for the optimal spacing by intersecting the asymptotic results (a) and (b):

$$\frac{D_{opt}}{L} \sim 2.73 \left(\frac{\mu\alpha}{\Delta P \cdot L^2} \right)^{1/4}$$

Illustrate numerically how the property group $(\mu\alpha)^{1/4}$ changes the dimension D_{opt}. For example, compare the optimal spacing of an air-cooled package with that of a package cooled by liquid Freon 12 when T_0 is of order 100°C.

(d) Determine the order of magnitude of the total heat transfer rate q' and the "density" of heat-generating circuitry, $q'/HL(T_0 - T_\infty)$, that corresponds to D_{opt}. Comment on how the geometry (H, L) and the choice of coolant influence these important design quantities.

6.4 Optimal Fin Thickness for Water-Cooled Integrated Circuit Chip

The figure shows the cross section through a bundle of microscopic water channels that are intended to serve as a heat sink in the substrate of an integrated circuit [25]. Their job is to remove the heat generated by the circuit. The channels are etched into the high conductivity substrate (silicon) and are capped with a cover plate that is a relatively poor thermal conductor. Water is pumped through the channels, and the flow is laminar and fully developed.

Several ways of optimizing the geometry of this compact heat sink are described in Refs. [25–27]. In this project we consider only one question, namely, what is the optimal fin thickness (t) that maximizes the heat transfer rate from the substrate (T_w) to the water flow (local bulk temperature T_f)? We conduct the optimization in the cross section, that is, at each location along the stream.

Assume that the heat transfer from T_w to T_f occurs mainly through the fins, in other words, that the heat transfer through the unfinned portions of the substrate is negligible. For the purpose of estimating the heat transfer coefficient only, assume that $L \gg D$, and that the wetted surfaces are almost isothermal along L. Assume further that the thermal resistance posed by the substrate of thickness S is negligible. Derive a formula for the total heat transfer rate removed by all the channel streams, q'. This quantity is expressed per unit length in the direction of flow. Arrange your result in the following dimensionless form,

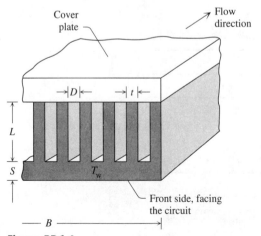

Figure PR6.4

$$Q = \text{function}\left(b, \frac{t}{D}\right)$$

where Q is the dimensionless heat transfer rate group

$$Q = \frac{q'}{k_w(T_w - T_f)B/L}$$

and b is shorthand for

$$b = \frac{L}{D}\left(\text{Nu}\,\frac{k_f}{k_w}\right)^{1/2}$$

In this notation k_f, k_w, and Nu are the fluid thermal conductivity, the fin thermal conductivity, and the Nusselt number (constant) for fully developed flow and heat transfer in the individual channel.

To find the optimal fin thickness for maximum q' when all the other parameters are fixed, you can maximize Q numerically with respect to t/D for every b. In this way you can show that the optimal fin thickness is approximately equal to the channel spacing when $b \gtrsim 2$. On the other hand, when $b \lesssim 1$ the optimal fin thickness is proportional to (i.e., a certain fraction of) the fin length L.

6.5 The Positioning of a Heat Generating Board Inside a Parallel-Plate Channel

The electronic circuit board shown in the figure must be cooled by forced convection in a parallel plate channel of spacing D, length L, and width W perpendicular to the plane of the figure. The channel walls are insulated. The coolant (air) is forced to flow through the channel by the pressure difference ΔP, which is maintained by a fan. All the surfaces may be modelled as smooth. The channel is sufficiently slender so that the flow is laminar and fully developed in both subchannels, this is, on both sides of the board.

The board substrate has the thermal conductivity k_w and thickness t, which is negligible relative to D. The two surfaces of the board are loaded equally and uniformly with electronics: the constant heat generation rate per unit board surface is q''. It is important to note, however, that the heat fluxes removed by the two streams generally are not equal, because of the conduction heat transfer across the board. The temperatures of the two board surfaces (T_1, T_2) increase in the downstream direction, and reach their highest levels at the trailing edge, $x = L$.

How would you mount the board inside the channel to lower the highest board temperature as much as possible? Would you place it in the middle of the channel, or closer to one of the side walls? This is a fundamental design question that requires careful analysis ("gut feelings" are not necessarily correct, here or anywhere else). You will discover that the correct answer depends on the degree to which the board substrate is a good thermal conductor in the transversal direction.

To simplify the analysis, assume that the local bulk temperature of each stream is nearly the same as the temperature of the neighboring spot on the board surface bathed by that stream. In other words, assume that the local temperature difference between the stream and the board surface is considerably smaller than the temperature rise from $x = 0$ to $x = L$ along the board surface. Derive expressions for the surface temperature distributions $T_1(x)$ and $T_2(x)$, and try to minimize the larger of the two trailing edge temperatures, $T_1(L)$ or $T_2(L)$. It is recommended to carry out this analysis in terms of the following dimensionless parameters:

$$\xi = \frac{x}{L} \qquad y = \frac{D_1}{D} \qquad 1 - y = \frac{D_2}{D}$$

$$\theta_1 = (T_1 - T_0)\frac{\rho c_p \Delta P D^3}{12 \mu L^2 q''} \qquad \theta_2 = (T_2 - T_0)\frac{\rho c_p \Delta P D^3}{12 \mu L^2 q''}$$

$$B = 12 \frac{k_w}{k} \frac{\mu \, \alpha \, L^2}{\Delta P \cdot D^3 \, t}$$

The dimensionless number B expresses the relative size of the thermal conductance of the board substrate. The optimal board location (y) will be a function of B.

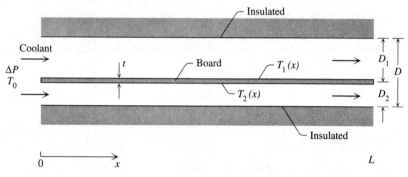

Figure PR 6.5

7
NATURAL CONVECTION

7.1 WHAT DRIVES THE NATURAL CONVECTION FLOW?

In natural convection, or free convection, the fluid flows "naturally" (by itself), as it is driven by the effect of buoyancy. This effect is distributed throughout the fluid, and is associated with the general tendency of fluids to expand (or, in special cases,* to contract), when heated at constant pressure. The layer that feels the warm vertical wall in Fig. 7.1 becomes lighter than the rest of the fluid. Its lightness forces it to flow upward, to sweep the wall and to collect heat transfer from the wall in a manner that reminds us of the boundary layer of Fig. 5.5. This time, however, the flow is a vertical jet parallel to the wall, whereas the fluid situated far from the wall is stagnant.

For those who remember the lessons of classical thermodynamics, specifically, the impossibility of a perpetual motion machine of the first kind, it is instructive to show what thermodynamic principle is responsible for the fluid motion in natural convection. Consider the evolution of a small fluid packet Δm (a closed thermodynamic system) as it proceeds clockwise along the circuit drawn in Fig. 7.1. While it is in the close vicinity of the heated vertical wall, the fluid packet is *heated* by thermal diffusion from the wall. At the same time, Δm *expands* because it rises to altitudes where the hydrostatic pressure imposed by the distant fluid reservoir is lower.

Mass conservation requires that the upflow of the wall jet be complemented by a sufficient downflow of the cold reservoir fluid. For this reason, the fluid packet Δm returns eventually to the heated wall, by flowing downward (and very slowly) through the reservoir. In the latter, the fluid packet is *cooled* and *compressed* while it descends to levels of increasingly larger pressure.

In summary, the Δm system executes a cycle the processes of which are arranged in the sequence heating — expansion — cooling — compression. This is the classical sequence of a work-producing cycle. The finite work put out by Δm is invested in accelerating and increasing the kinetic energy inventory of Δm to

---

*One example of this kind is water at nearly atmospheric pressures, in the 0–4°C temperature range.

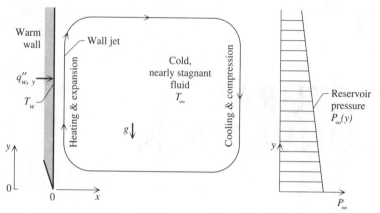

Figure 7.1 Wall jet driven by buoyancy along a heated wall, and pressure distribution in the reservoir of stagnant fluid.

the point where any additional work is dissipated fully by the effect of friction between Δm and the other fluid packets with which Δm comes in contact.

Looking at the entire flow field as an ensemble of fluid packets of type Δm, we see the "wheel" of a heat engine driven by the temperature difference $T_w - T_\infty$. The Carnot work potential of this heat engine is dissipated entirely in the "fluid brake" that is distributed throughout the fluid, that is, in the interfaces between all the fluid packets that experience relative movement. The existence of a work-producing potential in natural convection can be demonstrated by inserting a propeller across the path followed by Δm, especially in the wall jet section where Δm travels the fastest.

7.2   BOUNDARY LAYER FLOW ALONG A VERTICAL WALL

7.2.1   The Boundary Layer-Simplified Equations

From a heat transfer standpoint, the challenge is again to determine the local heat transfer coefficient

$$h_y = \frac{q''_{w,y}}{T_w - T_\infty} = \frac{-k(\partial T/\partial x)_{x=0}}{T_w - T_\infty} \tag{7.1}$$

The two-dimensional frame $x-y$ is oriented in the usual way that Cartesian systems are drawn, with y pointing upward and x pointing horizontally. The gravitational acceleration g points in the negative y direction.

The numerator appearing on the right side of eq. (7.1) invites us to consider the problem of determining the temperature distribution in the fluid that makes contact with the wall. The four equations that govern the two-dimensional flow and temperature field can be found in Table 5.1, in which the body force components are $X = 0$ and $Y = -\rho g$:

$$(m) \qquad \frac{\partial u}{\partial x} + \frac{\partial v}{\partial y} = 0 \tag{7.2}$$

$$(M_x) \qquad \rho\left(u\,\frac{\partial u}{\partial x} + v\,\frac{\partial u}{\partial y}\right) = -\frac{\partial P}{\partial x} + \mu\left(\frac{\partial^2 u}{\partial x^2} + \frac{\partial^2 u}{\partial y^2}\right) \tag{7.3}$$

$$(M_y) \qquad \rho\left(u\,\frac{\partial v}{\partial x} + v\,\frac{\partial v}{\partial y}\right) = -\frac{\partial P}{\partial y} + \mu\left(\frac{\partial^2 v}{\partial x^2} + \frac{\partial^2 v}{\partial y^2}\right) - \rho g \tag{7.4}$$

$$(E) \qquad u\,\frac{\partial T}{\partial x} + v\,\frac{\partial T}{\partial y} = \alpha\left(\frac{\partial^2 T}{\partial x^2} + \frac{\partial^2 T}{\partial y^2}\right) \tag{7.5}$$

Starting with Fig. 7.2, we consider only the flow region immediately adjacent to the vertical wall. The only assumption we make in connection with the wall temperature T_w is that it is *different* than the temperature of the fluid reservoir, T_∞. More specialized models of the wall heating mechanism will be adopted in Sections 7.2.3 (isothermal wall) and 7.2.5 (uniform heat flux).

The laminar boundary layer region is brought into view by Fig. 7.3. The photographs show the vertical boundary layers that form on the two sides of a metallic sheet that is heated uniformly by an electric current. In the laboratory, the surface with uniform heat flux is easier to construct than a surface with uniform temperature. The photographs (interferograms) were generated with an optical apparatus (interferometer) that first splits the light beam into two, passing the first beam through the boundary layer region (perpendicular to Fig. 7.3) and the second beam through an isothermal sample of the same fluid. Farther along, the two beams are recombined and projected on a screen, or directly on film. Dark and light fringes form on the screen, in relation to the density gradient encountered by the beam that passes through the boundary layer region.

The thickness of the near-wall region, δ_T, is defined as the transversal

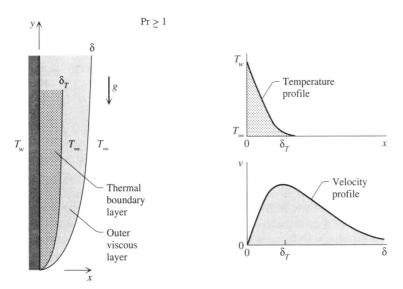

Figure 7.2 High Prandtl number fluids: the structure of the thermal and velocity boundary layers over a heated vertical wall.

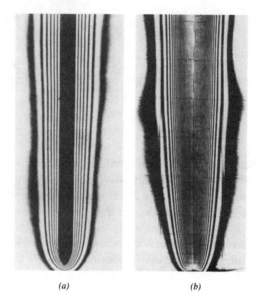

(a)   (b)

Figure 7.3 Interferometer photographs showing the laminar natural convection boundary layers over vertical surfaces with uniform heat flux: (a) Plate suspended in air (Eckert and Soehngen [1]), and (b) metallic foil suspended in gaseous nitrogen at 17.1 atm. The boundary layer thickness has been magnified six times (Gebhart [2], courtesy of Prof. B. Gebhart, University of Pennsylvania).

(horizontal) distance over which the temperature changes from its wall value, T_w, to a value comparable to that of the fluid reservoir, T_∞. The order-of-magnitude definition of δ_T is, therefore,

$$\left(-\frac{\partial T}{\partial x}\right)_{x=0} \sim \frac{\Delta T}{\delta_T} \tag{7.6}$$

where $\Delta T = T_w - T_\infty$. Next, we assume that this wall layer of thickness δ_T and height y is *slender*, namely, that it is a *thermal boundary layer* in the sense of Section 5.3.2,

$$\delta_T << y \tag{7.7}$$

The boundary layer feature (7.7) has the power to simplify the governing equations in two ways. First, all the longitudinal curvature terms $\partial^2/\partial y^2$ can be neglected in favor of the transversal curvature terms of type $\partial^2/\partial x^2$. This simplification occurs on the right side of eqs. (7.3) through (7.5), that is, in a total of three places.

Second, in a way that parallels the conclusion (5.30) of Chapter 5, it can be shown that the transversal momentum eq. (7.3) implies that the pressure P does not vary appreciably across the δ_T-thin region; in other words,

$$P(x,y) \cong P(y) = P_\infty(y) \tag{7.8}$$

According to the right side of Fig. 7.1, the distribution of pressure in the reservoir is hydrostatic; therefore,

$$\frac{dP_\infty}{dy} = -\rho_\infty g \tag{7.9}$$

In view of eqs. (7.8) and (7.9), it is easy to see that the longitudinal pressure gradient $\partial P/\partial y$ in eq. (7.4) can be replaced by the quantity $-\rho_\infty g$, in which ρ_∞ is the density of reservoir fluid.

To summarize the two simplifications made possible by the slenderness of the thermal boundary layer, eq. (7.7), we record the so-called *boundary layer-simplified* forms of the momentum and energy equations:

$$(M) \quad \rho\left(u\,\frac{\partial v}{\partial x} + v\,\frac{\partial v}{\partial y} \right) = \mu\,\frac{\partial^2 v}{\partial x^2} + (\rho_\infty - \rho)\,g \tag{7.10}$$

$$(E) \quad u\,\frac{\partial T}{\partial x} + v\,\frac{\partial T}{\partial y} = \alpha\,\frac{\partial^2 T}{\partial x^2} \tag{7.11}$$

There is only one momentum equation now—eq. (7.10)—because the other one, (M_x), eq. (7.3), was invoked to write eq. (7.8), that is, to eliminate P as one of the unknowns of the problem.*

The temperature T will appear explicitly in the momentum equation if we make use of the thermodynamic *equation of state* $\rho = \rho(T,P)$. Expanding the density function as a Taylor series for small departures from the reference density,

$$\rho_\infty = \rho\,(T_\infty,\, P_0) \tag{7.12}$$

we obtain

$$\rho \cong \rho_\infty + \left(\frac{\partial \rho}{\partial T}\right)_P (T - T_\infty) + \left(\frac{\partial \rho}{\partial P}\right)_T (P - P_0) + \cdots \tag{7.13}$$

where P_0 is a reference pressure level (e.g., the pressure at the bottom of the reservoir, $P_{\infty,0}$). It turns out that in most of the natural convection configurations that are encountered in thermal engineering, the pressure correction term on the right side of eq. (7.13) is negligible relative to the temperature correction term; therefore, eq. (7.13) reduces to

$$\rho \cong \rho_\infty - \rho_\infty \beta\,(T - T_\infty) + \cdots$$
$$\cong \rho_\infty\,[1 - \beta\,(T - T_\infty) + \cdots] \tag{7.14}$$

where β is the coefficient of volumetric thermal expansion defined in eq. (5.18). In an ideal gas, for example, β is equal to $1/T$, where T is the *absolute* temperature expressed in degrees Kelvin or Rankine. Also worth noting is the fact that eq. (7.14) is an adequate linear approximation of the $\rho = \rho(T,P)$ equation of

---

*Note that in the original system of four equations, eqs. (7.2) through (7.5), the four unknowns were u, v, P, and T.

state *only when the departures from the reference density* $\rho_\infty(T_\infty, P_0)$ *are sufficiently small*, that is, when the temperature correction is small enough so that

$$1 >> \beta (T - T_\infty) \tag{7.15}$$

The writing of ν as shorthand for μ/ρ_∞ does not mean that in an *actual application* the fluid properties are to be evaluated at the reservoir temperature T_∞. It turns out that when the fluid properties vary with the temperature across the boundary layer region, reasonable agreement between predictions based on this theory and actual measurements is achieved when the fluid properties are evaluated at the film temperature $(T_w + T_\infty)/2$.

Substituting the ρ expression (7.14) on both sides of the momentum eq. (7.10), we obtain

$$\rho_\infty [1 - \beta(T - T_\infty)] \left(u \frac{\partial v}{\partial x} + v \frac{\partial v}{\partial y} \right) = \mu \frac{\partial^2 v}{\partial x^2} + \rho_\infty \beta (T - T_\infty)g \tag{7.16}$$

In the square brackets on the left side, the $\beta(T - T_\infty)$ term can be neglected (left out) in favor of 1, in accordance with the observation (7.15). Dividing both sides of the equation by ρ_∞, and writing $\nu = \mu/\rho_\infty$ for the kinematic viscosity of the fluid, we arrive at a momentum equation that displays T in the buoyancy term:

$$(M) \qquad \underbrace{u \frac{\partial v}{\partial x} + v \frac{\partial v}{\partial y}}_{\text{Inertia}} = \underbrace{\nu \frac{\partial^2 v}{\partial x^2}}_{\text{Friction}} + \underbrace{g\beta (T - T_\infty)}_{\text{Buoyancy}} \tag{7.17}$$

The linearization of the $\rho = \rho(T,P)$ equation of state, eq. (7.14), or the rewriting of the momentum eq. (7.10) as eq. (7.17), is recognized as the *Oberbeck–Boussinesq approximation* [3,4] or, more succinctly, as the Boussinesq* approximation. The presence of the temperature in the buoyancy term of the momentum eq. (7.17) "couples" the flow to the temperature field, and vice versa. In summary, the boundary layer-simplified equations are three: the mass conservation eq. (7.2), which remained unchanged, the momentum eq. (7.17), and the energy eq. (7.11).

7.2.2 Scale Analysis of the Laminar Regime

The simplest approach to calculating the local heat transfer coefficient h_y is the order-of-magnitude analysis. According to the second h_y expression in eq. (7.1), and in view of eq. (7.6), the chief heat transfer unknown is

$$h_y \sim \frac{k}{\delta_T} \tag{7.18}$$

The problem reduces to figuring out the size of the *thermal* boundary layer thickness δ_T. Let u and v represent the orders of magnitude (scales) of the horizontal and vertical velocity components *inside* the flow region of thickness

---

*(Valentin-)Joseph Boussinesq (1842–1929) was professor of physics and experimental mechanics at the Faculty of Sciences in Paris. He made many contributions to the theory of elasticity and magnetism, and was the first to attack the subject of turbulent flow in a fundamental theoretical manner. In 1877 he pioneered the use of the time-averaged Navier–Stokes equations in the analytical study of turbulent flows.

δ_T and height y. The mass, momentum, and energy eqs. (7.2), (7.17), and (7.11) require, in order,

$$(m) \qquad \frac{u}{\delta_T} \sim \frac{v}{y} \qquad (7.19)$$

$$(M) \qquad u\frac{v}{\delta_T}, v\frac{v}{y} \quad \sim \quad v\frac{v}{\delta_T^2}, g\beta\Delta T \qquad (7.20)$$

$$(E) \qquad u\frac{\Delta T}{\delta_T}, v\frac{\Delta T}{y} \quad \sim \quad \alpha\frac{\Delta T}{\delta_T^2} \qquad (7.21)$$

The mass conservation proportionality (7.19) implies that the first two scales (the convection scales) on the left side of the energy eq. (7.21) are of the same order of magnitude, $v\Delta T/y$. Therefore, the energy equation (7.21) reduces to

$$(E) \qquad v\frac{\Delta T}{y} \quad \sim \quad \alpha\frac{\Delta T}{\delta_T^2} \qquad (7.22)$$

Convection   Transversal
conduction

that is, a balance between the heat conducted horizontally, from the wall into the thermal boundary layer, and the enthalpy carried upward by the vertical stream.

Similarly, the mass conservation proportionality $u/\delta_T \sim v/y$ guarantees that the first two terms (the inertia scales) on the left side of the momentum eq. (7.20) are of the same order of magnitude, v^2/y. The momentum equation describes the competition between only three scales,

$$(M) \qquad \frac{v^2}{y} \qquad v\frac{v}{\delta_T^2} \qquad g\beta\Delta T \qquad (7.23)$$

Inertia   Friction   Buoyancy

among which the driving force (the buoyancy scale) is never negligible (without it, there would be no flow). The remaining effects—inertia and friction—oppose the driving effect of buoyancy. The question is which flow-restraining effect plays the dominant role, inertia or friction? To answer this question, consider the two limiting possibilities described below.

(a) Buoyancy Balanced by Friction. In this limit, the momentum eq. (7.23) reduces to

$$(M) \qquad v\frac{v}{\delta_T^2} \sim g\beta\Delta T \qquad (7.24)$$

Equations (7.24), (7.22), and (7.19) form a system of approximate algebraic equations containing three unknowns, u, v, and δ_T. The solving of this system is straightforward, and the results are

$$u \sim \frac{\alpha}{y}\,\mathrm{Ra}_y^{1/4} \qquad (7.25a)$$

$$v \sim \frac{\alpha}{y}\,\mathrm{Ra}_y^{1/2} \qquad (7.25b)$$

$$\delta_T \sim y \mathrm{Ra}_y^{-1/4} \tag{7.25c}$$

in which the dimensionless group labeled Ra_y is the *Rayleigh number** based on the local altitude y:

$$\mathrm{Ra}_y = \frac{g\beta\,(T_w - T_\infty)y^3}{\alpha\nu} \tag{7.26}$$

Equation (7.25c) shows that the Rayleigh number Ra_y is a number of the same order of magnitude as the slenderness ratio y/δ_T raised to the fourth power. This is why Ra_y has characteristically large numerical values in actual calculations. Equation (7.25c) also delivers the answer to the engineering question that motivated this analysis, namely, eq. (7.18):

$$h_y \sim \frac{k}{y}\,\mathrm{Ra}_y^{1/4} \tag{7.27}$$

This conclusion is traditionally summarized in dimensionless form, by defining the *local Nusselt number*:

$$\mathrm{Nu}_y = \frac{h_y y}{k} \tag{7.28}$$

so that eq. (7.27) is analogous to writing

$$\mathrm{Nu}_y \sim \mathrm{Ra}_y^{1/4} \tag{7.29}$$

When are the results (7.25) and (7.29) valid? They are valid when the assumed buoyancy $\sim$ friction balance holds, namely, whenever the inertia scale v^2/y is negligible relative to either the buoyancy scale or the friction scale, for example,

$$\frac{v^2}{y} < \nu\,\frac{v}{\delta_T^2} \tag{7.30}$$

After using the v and δ_T scales listed in eqs. (7.25b) and (7.25c), this inequality becomes $\alpha < \nu$, or $1 < \mathrm{Pr}$. In conclusion, the results (7.25) and (7.29) are valid in fluids with Prandtl numbers of order 1 or greater. Put another way, the thermal boundary layer of a $\mathrm{Pr} \gtrsim 1$ fluid is ruled by a momentum balance between the effects of buoyancy and friction.

When the kinematic viscosity is greater than the thermal diffusivity, the thermal boundary layer flow of thickness δ_T and vertical velocity v entrains (drags along) an additional layer that contains nearly isothermal fluid. It can be shown using the method of Section 5.3.1 that the thickness of this "outer layer" is of order [5]

$$\delta \sim y \mathrm{Ra}_y^{-1/4} \mathrm{Pr}^{1/2} \tag{7.31}$$

---

*Lord Rayleigh (John William Strutt, 1842–1919) was professor of natural philosophy in the Royal Institution of Great Britain. Next to his seminal work in sound theory, optics, electrodynamics and elasticity, in heat transfer he is known for his work on hydrodynamic stability and cellular convection in a fluid layer heated from below (Bénard convection, Section 7.4.3).

The outer layer is indeed thicker than the inner (thermal) boundary layer because, according to eqs. (7.25c) and (7.31),

$$\frac{\delta}{\delta_T} \sim \text{Pr}^{1/2} > 1 \tag{7.32}$$

The structure and relative position of the thermal and vertical velocity profiles in $\text{Pr} \gtrsim 1$ fluids are illustrated in Fig. 7.2. It is important to note that *the velocity profile has two length scales*, the distance from the wall to the velocity peak (δ_T) and the distance to the nearly stagnant reservoir (δ). Combining eqs. (7.25c) and (7.31) with slenderness criteria of type (7.7), we learn that the δ_T and δ layers are indeed "boundary layers" if, respectively,

$$\text{Ra}_y^{1/4} > 1 \qquad \text{Ra}_y^{1/4}\text{Pr}^{-1/2} > 1 \tag{7.33a, b}$$

In $\text{Pr} \gtrsim 1$ fluids, the second of these criteria is more stringent than the first, as $\text{Ra}_y^{1/4}$ must exceed the order of magnitude of $\text{Pr}^{1/2}$. This is why natural convection boundary layer flows are often referred to as "high Rayleigh number" flows.

(b) Buoyancy Balanced by Inertia. When the contribution made by the friction term is negligible in the momentum balance (7.23), what is left reads

$$(M) \qquad \frac{v^2}{y} \sim g\beta\Delta T \tag{7.34}$$

This equation and eqs. (7.19) and (7.22) are sufficient for determining the unknown scales of the δ_T-thin region:

$$u \sim \frac{\alpha}{y}(\text{Ra}_y\text{Pr})^{1/4} \tag{7.35a}$$

$$v \sim \frac{\alpha}{y}(\text{Ra}_y\text{Pr})^{1/2} \tag{7.35b}$$

$$\delta_T \sim y(\text{Ra}_y\text{Pr})^{-1/4} \tag{7.35c}$$

These results bring to light the new dimensionless group (Ra_yPr), whose noteworthy feature is that it does not contain the kinematic viscosity v in the denominator:

$$\text{Ra}_y\text{Pr} = \frac{g\beta(T_w - T_\infty)y^3}{\alpha^2} \tag{7.36}$$

This group is recognized also as the *Boussinesq number*, and is labeled Bo_y. In the present treatment, the Ra_yPr label will be used to stress that the shift from the results (7.25a–c) to the new set of results (7.35a–c) occurs when the Prandtl number changes.

The local Nusselt number Nu_y can now be evaluated by combining eqs. (7.18), (7.28), and (7.35c):

$$\text{Nu}_y \sim (\text{Ra}_y\text{Pr})^{1/4} \tag{7.37}$$

The results (7.35a–c) and (7.37) apply when the friction effect is negligible in the momentum balance, that is, when the opposite of the inequality (7.30)

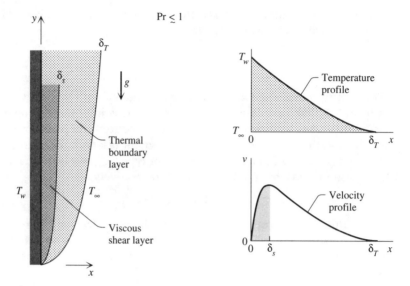

Figure 7.4 Low Prandtl number fluids: structure of the thermal and velocity boundary layers over a heated vertical wall.

prevails. Substituting the v and δ_T scales (7.35b) and (7.35c) into the inequality leads to $\alpha > v$, or to the conclusion that the present results apply to Pr $\lesssim 1$ fluids.

The structure of the temperature and vertical velocity profiles across the boundary layer flow of a low-Pr fluid is illustrated in Fig. 7.4. The wall jet is as thick as the temperature profile (δ_T). Immediately adjacent to the wall, there is a thin shear layer in which the effect of friction is important. It can be demonstrated that the thickness of this shear layer (i.e., the distance to the velocity peak) is of the order [5]

$$\delta_s \sim y \left(\frac{\mathrm{Ra}_y}{\mathrm{Pr}} \right)^{-1/4} \tag{7.38}$$

which shows that δ_s is smaller than δ_T:

$$\frac{\delta_s}{\delta_T} \sim \mathrm{Pr}^{1/2} < 1 \tag{7.39}$$

The dimensionless group ($\mathrm{Ra}_y/\mathrm{Pr}$) is known as the *Grashof number:*[*]

$$\mathrm{Gr}_y = \frac{\mathrm{Ra}_y}{\mathrm{Pr}} = \frac{g\beta(T_w - T_\infty)y^3}{v^2} \tag{7.40}$$

Contrary to the impression left by the presence of Pr when the Grashof number is written as $\mathrm{Ra}_y/\mathrm{Pr}$, the Grashof number — the right side of eq. (7.40) — does not contain the thermal diffusivity α in its denominator.

---

[*]Franz Grashof (1826–1893) was professor of mechanical engineering at the University of Karlsruhe.

The results developed in this limit of low Prandtl numbers are valid provided that the δ_s and δ_T layers are slender. These requirements lead respectively to the following inequalities:

$$\left(\frac{Ra_y}{Pr}\right)^{1/4} > 1 \qquad (Ra_y Pr)^{1/4} > 1 \qquad\qquad (7.41a, b)$$

The second of these inequalities is more restrictive because Pr is a number of order 1 or smaller. The slenderness of the boundary layer structure is assured if $Ra_y > Pr^{-1}$, that is, when the Rayleigh number is sufficiently large.

The limits (a) and (b) analyzed in this section correspond to the two Prandtl number limits $Pr \gtrsim 1$ and $Pr \lesssim 1$, respectively. At the intersection of these two limiting sets of results lies the class of fluids with Prandtl numbers of order 1 (e.g., air). Note that the two limiting Nu_y formulas, eqs. (7.29) and (7.37), give the same result when Pr is of order 1 and fixed.

7.2.3   Isothermal Wall (Laminar Flow)

More exact methods of solution have the power to predict the dimensionless coefficients of order 1 that are now missing from the right side of the Nu_y formulas (7.29) and (7.37). For the flow near an isothermal wall, the similarity solution can be determined by first defining the similarity variable [5]

$$\eta = \frac{x}{y Ra_y^{-1/4}} \qquad\qquad (7.42)$$

in which the reference transversal length scale is the thermal boundary layer thickness of a high-Pr fluid, eq. (7.25c). The corresponding similarity vertical velocity profile recommended by eq. (7.25b) is

$$G\,(\eta, Pr) = \frac{v}{(\alpha/y)\, Ra_y^{1/2}} \qquad\qquad (7.43)$$

Introducing the streamfunction $\psi(x,y)$ defined by

$$u = \frac{\partial \psi}{\partial y} \qquad \text{and} \qquad v = -\frac{\partial \psi}{\partial x} \qquad\qquad (7.44)$$

it can be shown that the similarity velocity profile (7.43) requires a similarity streamfunction of the form

$$F(\eta, Pr) = \frac{\psi}{\alpha Ra_y^{1/4}} \qquad\qquad (7.45)$$

where $G = -dF/d\eta$. Finally, the similarity temperature profile is the function

$$\theta\,(\eta, Pr) = \frac{T - T_\infty}{T_w - T_\infty} \qquad\qquad (7.46)$$

in which the denominator $T_w - T_\infty = \Delta T$ is a constant.

The similarity form of the energy and momentum eqs. (7.11) and (7.17) is obtained in two steps: first, by eliminating u and v in favor of the respective ψ derivatives (7.44); and, second, by eliminating x, y, ψ, and T in favor of η, F, and θ using eqs. (7.42), (7.45), and (7.46). The two equations that emerge from this analysis are [5]

$$(E) \qquad \frac{3}{4}F\theta' = \theta'' \tag{7.47}$$

$$(M) \qquad \frac{1}{\text{Pr}}\left(\frac{1}{2}F'^2 - \frac{3}{4}FF''\right) = -F''' + \theta \tag{7.48}$$

The boundary conditions that must be satisfied by the functions F and θ are

$F = 0$	at	$\eta = 0$	(impermeable wall, $u = 0$)	(7.49a)
$F' = 0$	at	$\eta = 0$	(no slip, $v = 0$)	(7.49b)
$\theta = 1$	at	$\eta = 0$	(isothermal wall, $T = T_w$)	(7.49c)
$F' \to 0$	as	$\eta \to \infty$	($v = 0$ in the fluid reservoir)	(7.49d)
$\theta \to 0$	as	$\eta \to \infty$	($T = T_\infty$ in the fluid reservoir)	(7.49e)

Figure 7.5 shows the resulting similarity profiles for temperature and vertical velocity near the isothermal wall. The curves were drawn in this manner by

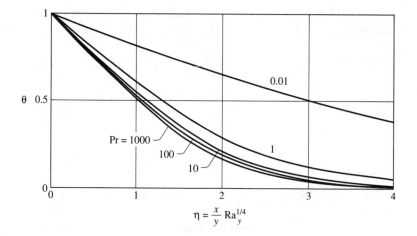

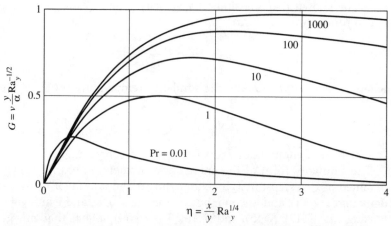

Figure 7.5 The similarity temperature and velocity profiles for laminar natural convection along a vertical isothermal wall (Bejan [5]).

Bejan [5] using the numerical data reported by Ostrach [6], who had solved the equivalent of the similarity problem (7.47)–(7.49e) by using a different set of dimensionless variables. Bejan [5] chose the set of dimensionless variables that reflect the correct scales of the thermal boundary layer region in $\text{Pr} \gtrsim 1$ fluids. This choice was intentional, so that the many curves would collapse onto the same curve in the fluid region that corresponds to the thermal boundary layer.

Indeed, the upper graph shows that the temperature profiles fall on the same curve when the Prandtl number is large. This behavior is due to the use of the δ_T scale in the definition of η, eq. (7.42). For the same reason, the near-wall portions of the vertical velocity profiles approach the same curve as Pr becomes large. (Review the location of the v maximum relative to $x \sim \delta_T$ in Fig. 7.2.)

The use of the correct v scale for $\text{Pr} \gtrsim 1$ fluids in the G definition (7.43) is the reason why the maximum G values are consistently of order 1 in the lower graph of Fig. 7.5. The outer "tails" of the G profiles extend toward progressively larger η values as Pr increases, in accordance with the outer thickness scaling (7.32). Both the temperature and the vertical velocity profiles deviate from the behavior described thus far if Pr becomes smaller than 1. This departure should be expected in view of the use of the δ_T and v scales of the large-Pr fluids in the similarity formulation of the problem, eqs. (7.42) and (7.43).

This last observation is stressed by the $\text{Pr} = 0.01$ curves drawn on the two graphs of Fig. 7.5, which appear to contradict the sketch made on the right side of Fig. 7.4. In the latter, the outer thicknesses of the thermal and velocity boundary layers clearly have the same length scale. In Fig. 7.5, even though the outer thickness of the velocity profile (G) has the same scale as the outer thickness of the temperature profile (θ), this is obscured somewhat by the fact that the dimensionless G *amplitude* is only a fraction of the θ amplitude when $\text{Pr} = 0.01$.

The local Nusselt number revealed by the similarity solution of Fig. 7.5 is [cf. eqs. (7.28) and (7.1)]

$$\text{Nu}_y = \frac{h_y y}{k} = \frac{y}{k \Delta T}\left(-k \frac{\partial T}{\partial x}\right)_{x=0}$$

$$= \left(-\frac{d\theta}{d\eta}\right)_{\eta=0} \text{Ra}_y^{1/4} \tag{7.50}$$

The proportionality between Nu_y and $\text{Ra}_y^{1/4}$ implies that when the vertical wall is isothermal, the local heat flux $q''_{w,y}$ varies as $y^{-1/4}$. The similarity temperature gradient at the wall, $(d\theta/d\eta)_{\eta=0}$, is a function of Prandtl number because Pr appears as a parameter in the momentum eq. (7.48). That function is approximated with 0.5 percent by the Pr-dependent expression listed on the right side of the equal sign (after Ref. [7]):

$$\text{Nu}_y = 0.503 \left(\frac{\text{Pr}}{\text{Pr} + 0.986\text{Pr}^{1/2} + 0.492}\right)^{1/4} \text{Ra}_y^{1/4} \tag{7.51}$$

This expression covers the entire range of Prandtl numbers. Its two asymptotes,

$$\text{Nu}_y = 0.503\,\text{Ra}_y^{1/4} \quad (\text{Pr} \gg 1) \tag{7.52a}$$

$$\text{Nu}_y = 0.600(\text{Ra}_y\text{Pr})^{1/4} \quad (\text{Pr} \ll 1) \tag{7.52b}$$

confirm the validity of the order-of-magnitude conclusions (7.29) and (7.37), respectively.

The *average* or *overall Nusselt number* $\overline{\mathrm{Nu}}_y$ for a wall of height y is defined as

$$\overline{\mathrm{Nu}}_y = \frac{\overline{h}_y y}{k} = \frac{\overline{q}''_{w,y}}{T_w - T_\infty}\frac{y}{k} \tag{7.53}$$

In this expression $\overline{q}''_{w,y}$ is the wall heat flux averaged from $y = 0$ to y, and

$$q'_{w,y} = \overline{q}''_{w,y}y \tag{7.54}$$

is the total heat transfer rate through the wall, per unit length in the direction perpendicular to the plane of Fig. 7.2, for example. Or, if W is the width of the wall in the direction perpendicular to Fig. 7.2, the total heat transfer rate through the wall $q_{w,y}$ (watts) is $q_{w,y} = q'_{w,y}W = \overline{q}''_{w,y}Wy$. The wall-averaged Nusselt number $\overline{\mathrm{Nu}}_y$ is therefore a dimensionless version of the total heat transfer rate $q'_{w,y}$, namely, $\overline{\mathrm{Nu}}_y = q'_{w,y}/k\Delta T$. The $\overline{\mathrm{Nu}}_y$ formula that corresponds to eq. (7.51) is

$$\overline{\mathrm{Nu}}_y = 0.671 \left(\frac{\mathrm{Pr}}{\mathrm{Pr} + 0.986\mathrm{Pr}^{1/2} + 0.492}\right)^{1/4}\mathrm{Ra}_y^{1/4} \tag{7.55}$$

and, in the special case of air or any other gas with $\mathrm{Pr} = 0.72$,

$$\overline{\mathrm{Nu}}_y = 0.517\mathrm{Ra}_y^{1/4} \qquad (\mathrm{Pr} = 0.72) \tag{7.55'}$$

An alternative correlation for the entire Pr range in the laminar regime will be presented in eq. (7.62), that is, after the discussion of the transition from laminar flow to turbulent flow, eq. (7.56).

Looking back at the material covered in Sections 7.2.1 through 7.2.3, we focused in detail on natural convection along a vertical isothermal wall because it is the most basic (and oldest) of all the natural convection boundary layer configurations. The vertical wall with uniform heat flux will be treated in Section 7.2.5. The history of the progress made in the study of this flow was recounted recently by Martin [8]. The first similarity formulation and partial solution were reported by Pohlhausen, in association with Schmidt and Beckmann, in 1930 [9]. The scales and double-layer structures presented in Section 7.2.2 were first discovered by Kuiken [10, 11]. The scale analysis of the same problem was presented independently by Bejan [5].

Example 7.1

Laminar Natural Convection Boundary Layer

The door of a kitchen oven is a vertical rectangular area 0.5 m tall and 0.65 m wide. The external surface of the oven door is at 40°C, while the room air is at 20°C. Calculate the natural convection heat transfer rate from the door to the ambient air.

Solution.   The film temperature of the air in the boundary layer is $(T_w + T_\infty)/2 = 30°C$, and the air properties that will be needed are

$$\mathrm{Pr} = 0.72 \qquad k = 0.026\,\frac{\mathrm{W}}{\mathrm{m}\cdot\mathrm{K}} \qquad \frac{g\beta}{\alpha\nu} = \frac{90.7}{\mathrm{cm}^3\cdot\mathrm{K}}$$

The calculation begins with the Rayleigh number based on the door height $H = 50$ cm:

$$\mathrm{Ra}_H = \frac{g\beta}{\alpha\nu} H^3 \, (T_w - T_\infty)$$

$$= \frac{90.7}{\mathrm{cm^3 \cdot K}} (50 \text{ cm})^3 (40 - 20) \text{ K} = 2.27 \times 10^8 \qquad \text{(laminar)}$$

$$\overline{\mathrm{Nu}}_H = 0.517 \mathrm{Ra}_H^{1/4} \qquad\qquad\qquad\qquad\qquad\qquad (7.55')$$

$$= 0.517(2.27 \times 10^8)^{1/4} = 63.44$$

$$\overline{\mathrm{Nu}}_H = \frac{\overline{q''_w}H}{\Delta T k}$$

$$\overline{q''_w} = (63.44 \times 20\,\mathrm{K})\left(0.026 \, \frac{\mathrm{W}}{\mathrm{m \cdot K}}\right) \frac{1}{0.5\,\mathrm{m}} = 66 \text{ W/m}^2$$

The door area is $A = 0.5 \text{ m} \times 0.65 \text{ m} = 0.325 \text{ m}^2$, and this means that the natural convection heat transfer rate to the room air is

$$q = \overline{q''_w}A$$

$$= 66 \, \frac{\mathrm{W}}{\mathrm{m}^2} \, (0.325 \text{ m}^2) = 21.4 \text{ W}$$

7.2.4   Transition and the Effect of Turbulence on Heat Transfer

The boundary layer flow discussed until now remains laminar if y is small enough so that the Rayleigh number Ra_y does not exceed a certain critical value (Fig. 7.6). Until very recently it was thought that the transition to turbulent flow

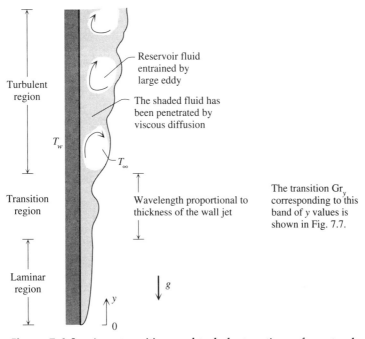

Figure 7.6 Laminar, transition, and turbulent sections of a natural convection boundary layer along a vertical wall.

occurs at the y position where $Ra_y \sim 10^9$, regardless of the value of the Prandtl number (e.g., Ref. [12]). This established view was questioned by Bejan and Lage [13], who showed that it is the *Grashof number of order 10^9* (i.e., not the Rayleigh number of order 10^9) that marks the transition in all fluids:

$$Gr_y \sim 10^9 \qquad (10^{-3} \leq Pr \leq 10^3) \qquad (7.56)$$

This universal transition criterion can be expressed also in terms of the Rayleigh number, by recalling that $Ra_y = Gr_y Pr$:

$$Ra_y \sim 10^9 Pr \qquad (10^{-3} \leq Pr \leq 10^3) \qquad (7.57)$$

It is supported very well by the experimental observations [14–22] reviewed in Fig. 7.7, and, in great detail, in Ref. [13]. The $Gr_y \sim 10^9$ transition criterion coincides with the traditional criterion $Ra_y \sim 10^9$ only in the case of fluids with Prandtl numbers of order 1 (e.g., air). In the liquid–metal range of Prandtl numbers ($Pr \sim 10^{-3}$–10^{-2}), for example, the Grashof number criterion (7.56) means that the actual transition Rayleigh number is of order 10^6–10^7, which is well below the often mentioned threshold of 10^9.

This is a good opportunity to extend the Pr range of the transition observations assembled in Fig. 7.7 and Ref. [13], by adding the observations reported by Lloyd et al. [23]. In the current context, Lloyd et al.'s experiments indicated that the laminar boundary layer regime expires somewhere above $Ra_y \sim 3 \times 10^{11}$

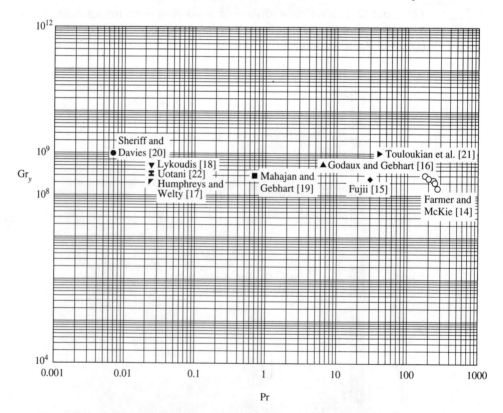

Figure 7.7 The Grashof number of order 10^9, as a universal criterion for transition to turbulence along a vertical wall with natural convection boundary layer flow (Bejan and Lage [13]).

when $Pr = 2000$. This agrees very well with the transition criterion (7.56), as $Gr_y = Ra_y/Pr \sim (3 \times 10^{11})/2000 = 1.5 \times 10^8$.

In conclusion, the laminar flow formulas (7.53)–(7.55) apply when $Gr_y \lesssim 10^9$ or $Ra_y \lesssim 10^9 Pr$. When the Rayleigh number exceeds the order of magnitude of $10^9 Pr$, the Nusselt number can be calculated with eq. (7.61), which is discussed later in this section.

Instructive at this stage is the observation that, deep down, the transition criterion (7.56) is another manifestation of the *local Reynolds number criterion* for transition to turbulence, which appears to be universally applicable (Appendix F). For example, if the present flow (wall jet) has a Prandtl number of order 1, its local Reynolds number has the δ_T of eq. (7.25c) as transversal length scale and the v of eq. (7.25b) as longitudinal velocity scale:

$$Re \sim \frac{\delta_T v}{\nu} \sim \frac{y Ra_y^{-1/4}}{\nu} \frac{\alpha}{y} Ra_y^{1/2}$$

$$\sim Ra_y^{1/4}/Pr \tag{7.58}$$

In this last expression, Ra_y can be replaced by Gr_y because $Pr \sim 1$:

$$Re \sim Gr_y^{1/4} \qquad (Pr \sim 1) \tag{7.59}$$

From eq. (7.56) we learn again that at transition the local Reynolds number has a value of order 10^2:

$$Re \sim (10^9)^{1/4} = 178 \qquad (Pr \sim 1) \tag{7.60}$$

The transition to turbulence is illustrated by the two interferograms shown in Fig. 7.8. The vertical surface is heated with uniform flux, and the laminar boundary layer is disturbed by a horizontal ribbon that moves periodically in and out. Combined with the vertical motion of the boundary layer, this mechanical disturbance generates a wave that propagates upward with the fluid. The photograph on the right shows that when the disturbance frequency is so high that the wavelength is much smaller than the boundary layer thickness, the disturbance is attenuated and the boundary layer remains laminar. In the photograph on the left we see that a disturbance the wavelength of which is comparable with the boundary layer thickness is amplified downstream and triggers the transition to turbulence. In Appendix F we can see that the proportionality between the wavelength of the buckled shape (meander) and the stream thickness is a common characteristic of all slender laminar flows during transition.

The heat transfer rate from a vertical wall in the presence of turbulence in the boundary layer has been measured experimentally and correlated as a function $\overline{Nu}_y(Ra_y, Pr)$. It was found that in the turbulent regime $\overline{Nu}_y$ is approximately proportional to $Ra_y^{1/3}$: this dependence differs from the $\overline{Nu}_y \sim Ra_y^{1/4}$ proportionality that characterizes laminar flow in high-Pr fluids.

An empirical correlation that reports the wall-averaged Nusselt number $\overline{Nu}_y$ for the *entire* Rayleigh number range—laminar, transition, and turbulent— was constructed by Churchill and Chu [25]:

$$\overline{Nu}_y = \left\{ 0.825 + \frac{0.387 Ra_y^{1/6}}{[1 + (0.492/Pr)^{9/16}]^{8/27}} \right\}^2 \tag{7.61}$$

This correlation holds true for $10^{-1} < Ra_y < 10^{12}$ and for all Prandtl numbers.

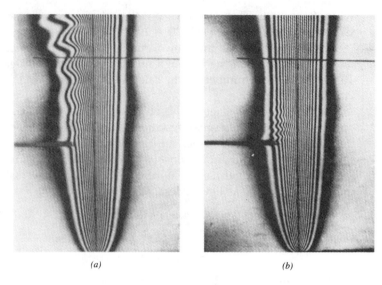

(a)   (b)

Figure 7.8 Interferometer photographs showing the transition to turbulence in the laminar natural convection boundary layer along a surface heated with uniform flux: (a) Undamped disturbance, and (b) damped disturbance. (Polymeropoulos and Gebhart [24], with permission from Cambridge University Press.)

The physical properties used in the definition of $\overline{\mathrm{Nu}}_y$, Ra_y, and Pr are evaluated at the film temperature $(T_w + T_\infty)/2$. In the case of air $(\mathrm{Pr} = 0.72)$, eq. (7.61) reduces to

$$\overline{\mathrm{Nu}}_y = (0.825 + 0.325\,\mathrm{Ra}_y^{1/6})^2 \qquad (\mathrm{Pr} = 0.72) \tag{7.61'}$$

In the *laminar* range, $\mathrm{Gr}_y < 10^9$, a correlation that represents the experimental data more accurately than eq. (7.61) is [25]

$$\overline{\mathrm{Nu}}_y = 0.68 + \frac{0.67\,\mathrm{Ra}_y^{1/4}}{[1 + (0.492/\mathrm{Pr})^{9/16}]^{4/9}} \tag{7.62}$$

$$\overline{\mathrm{Nu}}_y = 0.68 + 0.515\,\mathrm{Ra}_y^{1/4} \qquad (\mathrm{Pr} = 0.72) \tag{7.62'}$$

These correlations are an alternative to eqs. (7.55) and (7.55') especially in the low Rayleigh number limit where the boundary layer (slender flow) approximation loses its appeal.

7.2.5   Uniform Wall Heat Flux

When the vertical wall is heated uniformly, $q''_w = $ constant, the wall temperature T_w increases monotonically in the y direction. In the laminar regime, the temperature difference $T_w - T_\infty$ increases as $y^{1/5}$; this relationship follows from eqs. (7.29) and (7.37), which have general applicability in laminar natural convection boundary layer flow. In the case of a high-Pr fluid, for example, eq. (7.29) states that

$$\frac{q_w''}{[T_w(y) - T_\infty]}\frac{y}{k} \sim \left[\frac{g\beta(T_w - T_\infty)y^3}{\alpha v}\right]^{1/4} \tag{7.63}$$

This result can be rearranged to show that $T_w - T_\infty$ is indeed proportional to $y^{1/5}$.

Alternatively, eq. (7.63) can be rewritten so that the right side no longer contains the temperature difference $T_w - T_\infty$:

$$\frac{q''}{[T_w(y) - T_\infty]}\frac{y}{k} \sim \left(\frac{g\beta q_w'' y^4}{\alpha v k}\right)^{1/5} \tag{7.64}$$

The group formed inside the brackets on the right side is the Rayleigh number based on constant heat flux,

$$\mathrm{Ra}_y^* = \frac{g\beta q_w'' y^4}{\alpha v k} \tag{7.65}$$

while the left side of eq. (7.64) continues to show the local Nusselt number defined in eq. (7.28). In conclusion, in laminar high-Pr flow Nu_y is of the same order as $\mathrm{Ra}_y^{*1/5}$. This idea is confirmed by the similarity solution for the uniform heat flux case [26], which is fitted well by the expression

$$\mathrm{Nu}_y \cong 0.616 \left(\frac{\mathrm{Pr}}{\mathrm{Pr} + 0.8}\right)^{1/5} \mathrm{Ra}_y^{*1/5} \tag{7.66}$$

In fluids of the air–water Prandtl number range, the transition to turbulence occurs in the vicinity of $\mathrm{Ra}_y^* \sim 10^{13}$. For calculating the local and wall-averaged Nusselt numbers, Vliet and Liu [27] recommend the following formulas:

$$\left.\begin{array}{l} \mathrm{Nu}_y = 0.6\mathrm{Ra}_y^{*1/5} \\[2mm] \overline{\mathrm{Nu}}_y = 0.75\mathrm{Ra}_y^{*1/5} \end{array}\right\} \quad \text{laminar, } 10^5 < \mathrm{Ra}_y^* < 10^{13} \tag{7.67}$$

$$\left.\begin{array}{l} \mathrm{Nu}_y = 0.568\ \mathrm{Ra}_y^{*0.22} \\[2mm] \overline{\mathrm{Nu}}_y = 0.645\mathrm{Ra}_y^{*0.22} \end{array}\right\} \quad \text{turbulent, } 10^{13} < \mathrm{Ra}_y^* < 10^{16} \tag{7.68}$$

The average Nusselt number $\overline{\mathrm{Nu}}_y$ is based on the wall-averaged temperature difference $\overline{T_w - T_\infty}$. In particular, for calculations of heat transfer to *air*, Vliet and Ross [28] recommend

$$\mathrm{Nu}_y = 0.55\mathrm{Ra}_y^{*1/5} \quad \text{(laminar)} \tag{7.69a}$$

$$\mathrm{Nu}_y = 0.17\mathrm{Ra}_y^{*1/4} \quad \text{(turbulent)} \tag{7.69b}$$

A correlation that is valid for all Rayleigh and Prandtl numbers has been proposed by Churchill and Chu [25]:

$$\overline{\mathrm{Nu}}_y = \left\{0.825 + \frac{0.387\mathrm{Ra}_y^{1/6}}{[1 + (0.437/\mathrm{Pr})^{9/16}]^{8/27}}\right\}^2 \tag{7.70}$$

In this expression, Ra_y is based on the *y-averaged* temperature difference, namely $\overline{T_w} - T_\infty$. This correlation is almost identical to the one recommended for isothermal walls, eq. (7.61). For air at normal (room) conditions, the correlation (7.70) reduces to the simpler formula

$$\overline{\mathrm{Nu}}_y = (0.825 + 0.328\mathrm{Ra}_y^{1/6})^2 \quad (\mathrm{Pr} = 0.72) \tag{7.70'}$$

The high Rayleigh number asymptote of this last formula is

$$\overline{Nu}_y \cong 0.107 Ra_y^{1/3} \qquad (Pr = 0.72 \text{ and } Ra_y > 10^{10}) \tag{7.70''}$$

Equations (7.70)–(7.70″) can be restated in terms of the flux Rayleigh number Ra_y^* by noting the substitution $Ra_y = Ra_y^*/\overline{Nu}_y$. For example, the high Rayleigh number asymptote for air (7.70″) becomes

$$\overline{Nu}_y \cong 0.187 \ Ra_y^{*1/4} \qquad (Pr = 0.72 \text{ and } Ra_y^* > 10^{12}) \tag{7.70'''}$$

7.3 OTHER EXTERNAL FLOW CONFIGURATIONS

7.3.1 Thermally Stratified Fluid Reservoir

There are many instances in which the fluid reservoir that bathes the heated or cooled object is not isothermal, as was assumed throughout Section 7.2. Environmental applications provide good examples of quiescent fluid reservoirs that exhibit a temperature variation—a stratification—in the vertical direction. The core fluid in an enclosure heated from the side (Section 7.4.2) and the bulk of the air in a sealed room are just two examples.

The insert in the upper-right corner of Fig. 7.9 shows the assumed *linear* distribution of temperature in a fluid reservoir adjacent to an *isothermal* vertical wall of height H. The dimensionless stratification parameter,

$$b = \frac{\Delta T_{max} - \Delta T_{min}}{\Delta T_{max}} \tag{7.71}$$

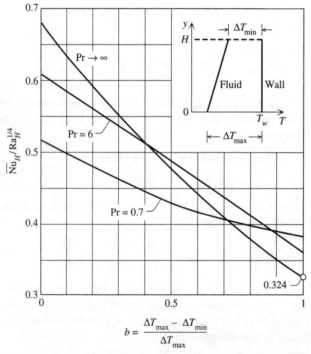

Figure 7.9 Average Nusselt number for laminar flow along an isothermal wall in contact with a linearly stratified fluid reservoir.

varies between the isothermal reservoir limit ($b = 0$) and the limit of maximum stratification ($b = 1$), where the uppermost layer of the reservoir is as warm as the heated wall. The temperature differences ΔT_{min} and ΔT_{max} are the wall–reservoir differences measured at $y = H$ and $y = 0$, respectively.

Figure 7.9 shows the average Nusselt number in the laminar regime. Both $\overline{Nu}_H$ and Ra_H are based on height and the *maximum* temperature difference,

$$\overline{Nu}_H = \frac{\overline{q}''_{w,H}}{\Delta T_{max}} \frac{H}{k} \qquad Ra_H = \frac{g\beta \Delta T_{max} H^3}{\alpha \nu} \qquad (7.72a, b)$$

In the laminar range, the Nusselt number obeys the relationship

$$\overline{Nu}_H = f(b,Pr)\, Ra_H^{1/4} \qquad (7.73)$$

for which the function $f(b,Pr)$ is displayed in the figure. It is clear that f decreases significantly as the stratification parameter b increases from 0 to 1. The Prandtl number has a less pronounced effect especially when it is considerably greater than 1. The curves drawn for $Pr = 0.7$ and $Pr = 6$ are based on Chen and Eichhorn's [29] results, which were obtained using the local nonsimilarity method [30]. The $Pr \rightarrow \infty$ curve was developed by Bejan [5] based on the integral method.

7.3.2 Inclined Walls

Figure 7.10 shows four possible configurations in which a plane wall is inclined relative to the vertical direction. The angle between the plane and the vertical direction, ϕ, is restricted to the range $-60° < \phi < 60°$ (horizontal walls are discussed in the next subsection). In the cases labeled (a) and (d)—heated wall tilted upward, and cooled wall tilted downward—the effect of the angle ϕ is to thicken the tail end of the boundary layer, and to give the wall jet a tendency to separate from the wall. The opposite effect is illustrated in cases (b) and (c), where the wall jet is squeezed against the wall until it flows over the trailing edge.

In the boundary layer analysis of the flows of Fig. 7.10 it is found that the momentum equation is analogous to eq. (7.17), except that $g \cos \phi$ replaces g in the buoyancy term. The group $g \cos \phi$ is the gravitational acceleration component that is oriented parallel to the wall. For this reason, the heat transfer rate in the *laminar* regime along an isothermal wall can be calculated with eq. (7.62), provided the Rayleigh number Ra_y is based on $g \cos \phi$:

$$Ra_y = \frac{(g\cos\phi)\beta(T_w - T_\infty)y^3}{\alpha\nu} \qquad (7.74)$$

Similarly, for laminar flow over a plate with uniform heat flux the Nusselt number can be calculated with eqs. (7.67) and (7.69a), in which

$$Ra_y^* = \frac{(g\cos\phi)\beta q''_w y^4}{\alpha\nu k} \qquad (7.75)$$

In the *turbulent* regime it was found that the heat transfer measurements are correlated better using g instead of $g \cos \phi$ in the Rayleigh number grouping [31]. Therefore, eq. (7.61) is recommended for isothermal plates, in which Ra_y follows the usual definition, eq. (7.26). For inclined walls with *uniform heat* flux, the Nusselt number is given by eqs. (7.68) and (7.69b), and the flux Rayleigh

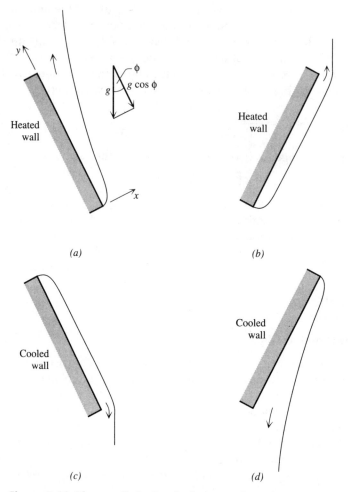

Figure 7.10 Plane walls inclined relative to the vertical direction.

number Ra_y^* is defined by eq. (7.65). Experimental data for both laminar and turbulent natural convection on inclined plates can be found also in Lloyd et al. [23].

Vliet [31] also showed that the tilt angle ϕ has a noticeable effect on the location of the laminar–turbulent transition when the *uniform flux* wall is oriented as in cases (a) and (d) of Fig. 7.10. The flux Rayleigh numbers tabulated below mark the beginning and the end of the transition region in water experiments (Pr $\cong 6.5$):

ϕ	Ra_y^*
0°	5×10^{12}–10^{14}
30°	3×10^{10}–10^{12}
60°	6×10^{7}–6×10^{9}

The Ra_y^* spread between the beginning and end of transition covers almost two orders of magnitude. This observation reinforces the approximate character of the threshold value $Ra_y^* \sim 10^{13}$ used in eqs. (7.67) and (7.68).

The effect of wall inclination on transition along an *isothermal* wall was documented by Lloyd and Sparrow [32] for water (Pr ~ 6). Once again, the transition Rayleigh number $Ra_y = g\beta y^3 \Delta T / \alpha\nu$ decreases as the tilt angle ϕ increases:

ϕ	Ra_y
0°	8.7×10^8
20°	2.5×10^8
45°	1.7×10^7
60°	7.7×10^5

7.3.3   Horizontal Walls

The flow changes its character as the tilt angle ϕ increases beyond the moderate values considered in Fig. 7.10. Two flow types are encountered in the extreme where the plane wall becomes horizontal (Fig. 7.11). When the wall is heated and faces upward, or when it is cooled and faces downward, the flow leaves the boundary layer as a vertical plume rooted in the central region of the wall. When the temperature difference is sufficiently large, the heated fluid rises from all over the surface. As illustrated in Fig. 7.12, this flow is intermittent and consists of balls of heated fluid (called *thermals*) that rise and meander through the colder fluid [33].

In cases where the surface is hot and faces downward, or when it is cold and faces upward, the boundary layer covers the entire surface and the flow spills over the edges. This situation is illustrated in the lower half of Fig. 7.11. Of course, a two-sided plate, hot or cold, will have flow of one type on the top side and flow of the other type on the bottom side.

Average Nusselt number measurements for several horizontal-plate configu-

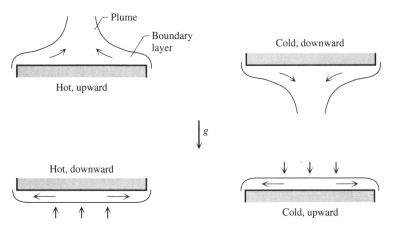

Figure 7.11 Horizontal surfaces with central plume (top row) and without plume flow (bottom row).

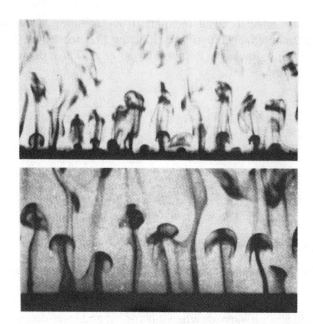

Figure 7.12 The intermittent formation and rise of balls of heated fluid (thermals) from a horizontal surface facing upward in a colder pool of water. (Sparrow et al. [33], with permission from Cambridge University Press.)

rations have been correlated by defining the *characteristic length* of the plane surface [34]:

$$L = \frac{A}{p} \qquad (7.76)$$

In this definition, A is the area of the plane surface and p is the perimeter of A. In the case of a disc of diameter D, for example, $L = D/4$. The $\overline{\mathrm{Nu}}_L$ formulas listed below are valid for Prandtl numbers greater than 0.5. The average Nusselt number $\overline{\mathrm{Nu}}_L$ and the Rayleigh number Ra_L are both based on L.

For hot surfaces facing upward, or cold surfaces facing downward (Fig. 7.9, top), the Nusselt number varies as follows [35]:

$$\overline{\mathrm{Nu}}_L = 0.54 \mathrm{Ra}_L^{1/4} \qquad (10^4 < \mathrm{Ra}_L < 10^7) \qquad (7.77a)$$

$$\overline{\mathrm{Nu}}_L = 0.15 \mathrm{Ra}_L^{1/3} \qquad (10^7 < \mathrm{Ra}_L < 10^9) \qquad (7.77b)$$

The corresponding correlation for hot surfaces facing downward or cold surfaces facing upward (Fig. 7.11, bottom) is (cf. Ref. [12], p. 548)

$$\overline{\mathrm{Nu}}_L = 0.27 \mathrm{Ra}_L^{1/4} \qquad (10^5 < \mathrm{Ra}_L < 10^{10}) \qquad (7.78)$$

Equations (7.77)–(7.78) are written for isothermal surfaces. The same correlations can be used for uniform flux surfaces, noting that in those cases $\overline{\mathrm{Nu}}_L$ and Ra_L would be based on the *L-averaged* temperature difference between the surface and the surrounding fluid. The flux Rayleigh number Ra_L^* can be made to appear on the right side of these correlations by noting one more time the substitution $\mathrm{Ra}_L = \mathrm{Ra}_L^* / \overline{\mathrm{Nu}}_L$.

Worth keeping in mind is that the heat transfer and fluid mechanics literature contains many theoretical (similarity, integral) solutions for laminar natural convection over horizontal walls. Many of these have been reviewed by Gebhart et al. [36]. Their common feature is the prediction of a proportionality between $\overline{\mathrm{Nu}}_L$ and $\mathrm{Ra}_L^{1/5}$, which, as demonstrated by eqs. (7.77)–(7.78), agrees only approximately with direct measurements.

Example 7.2

Hot Plate Facing Upward in a Colder Fluid

The temperature of the horizontal plate in the experiment of Fig. 7.12 is $T_w = 43.1°C$, while the temperature of the water pool is $T_\infty = 23.6°C$. The plate is a square with side $a = 8.9$ cm. Calculate the rate at which the plate must be heated electrically to generate the thermals illustrated in the figure.

Solution.   The film temperature is $(43.1°C + 23.6°C)/2 = 33.4°C$ and, after interpolating linearly in Appendix C, we find that the relevant water properties are

$$k = 0.62\frac{W}{m \cdot K} \qquad \frac{g\beta}{\alpha v} = \frac{28{,}982}{cm^3 \cdot K}$$

The characteristic length of the upward facing square of side a is

$$L = \frac{A}{p} = \frac{a^2}{4a} = \frac{8.9 \text{ cm}}{4} = 2.23 \text{ cm}$$

with the corresponding Rayleigh number

$$\mathrm{Ra}_L = \frac{g\beta}{\alpha v} L^3 (T_w - T_\infty)$$

$$= \frac{28{,}982}{cm^3 \cdot K} (2.23 \text{ cm})^3 (43.1 - 23.6) \text{ K} = 6.23 \times 10^6$$

The appropriate heat transfer correlation is eq. (7.77a):

$$\overline{\mathrm{Nu}}_L = 0.54\mathrm{Ra}_L^{1/4}$$

$$= 0.54 (6.23 \times 10^6)^{1/4} = 26.97$$

$$\overline{h} = \overline{\mathrm{Nu}}_L \frac{k}{L}$$

$$= 26.97 \times 0.62 \frac{W}{m \cdot K} \frac{1}{0.0223\,m} = 750 \text{ W/m}^2 \cdot K$$

$$q = \overline{h}a^2 (T_w - T_\infty)$$

$$= 750\frac{W}{m^2 \cdot K} (0.089 \text{ m})^2 (43.1 - 23.6) \text{ K} = 116 \text{ W}$$

7.3.4   Horizontal Cylinder

The natural convection flow around an isothermal cylinder positioned horizontally in a fluid reservoir, Fig. 7.13, is similar to the flow along a vertical surface. The only difference is the fact that now the wall is curved, and instead of the wall height y (or H) the vertical dimension is the cylinder diameter D. These similarities explain why the heat transfer correlation for horizontal cylinders [25],

$$\overline{\mathrm{Nu}}_D = \left\{0.6 + \frac{0.387\mathrm{Ra}_D^{1/6}}{[1 + (0.559/\mathrm{Pr})^{9/16}]^{8/27}}\right\}^2 \qquad (7.79)$$

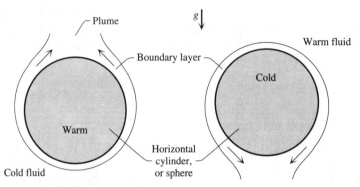

Figure 7.13 Horizontal cylinder or sphere immersed in a fluid at a different temperature.

has the same form as the vertical wall correlation (7.61). Equation (7.79) is valid for $10^{-5} < \text{Ra}_D < 10^{12}$ and the entire Prandtl number range. The average Nusselt number and the Rayleigh number are both based on diameter:

$$\overline{\text{Nu}_D} = \frac{\bar{q}_{w,D}''}{\Delta T}\frac{D}{k} \qquad \text{Ra}_D = \frac{g\beta\Delta T D^3}{\alpha v} \qquad \text{(7.80a, b)}$$

The air boundary layer around a horizontal cylinder is illustrated in Fig. 7.14. The boundary layer is laminar because the Grashof number $\text{Gr}_D = 2.45 \times 10^5$ is considerably smaller than the order of magnitude 10^9 associated with the transition to turbulence. The experimental conditions in which the photograph of Fig. 7.14 was made are the subject of Example 7.3.

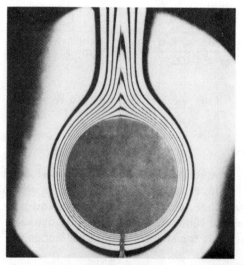

Figure 7.14 Interferometer photograph showing the laminar boundary layer around a horizontal cylinder in air. The cylinder diameter is 6 cm, the length perpendicular to the photograph is 60 cm, and the temperature difference between the cylinder and the distant air is 9°C. (Courtesy of Professor U. Grigull, Technical University of Munich.)

7.3.5   Sphere

The flow around a sphere suspended in a pool at a different temperature has the general features outlined in Fig. 7.13. The vertical dimension of the spherical body is its diameter D, on which both $\overline{\mathrm{Nu}}_D$ and Ra_D are based [cf. eqs. (7.80a, b)]. Heat transfer data are correlated well by the formula [37]

$$\overline{\mathrm{Nu}}_D = 2 + \frac{0.589\mathrm{Ra}_D^{1/4}}{[1 + (0.469/\mathrm{Pr})^{9/16}]^{4/9}} \tag{7.81}$$

in which the appropriate ranges are $\mathrm{Pr} \gtrsim 0.7$ and $\mathrm{Ra}_D < 10^{11}$.

7.3.6   Vertical Cylinder

There are three surfaces for heat transfer in the vertical cylinder geometry, the top and bottom discs, which can be handled according to Section 7.3.3, and the lateral surface. The boundary layer flow that develops over the lateral surface is illustrated in Fig. 7.15. When the boundary layer thickness δ_T is much smaller than the cylinder diameter D, the curvature of the lateral surface does not play a role, and the Nusselt number can be calculated with the vertical wall formulas (7.61)–(7.62). Note that if H is the height of the cylinder and if $\mathrm{Pr} \gtrsim 1$, the $\delta_T < D$ criterion requires

$$\frac{D}{H} > \mathrm{Ra}_H^{-1/4} \tag{7.82}$$

The above inequality and the simplified heat transfer calculation recommended by it may be regarded as the "thick cylinder" limit. The opposite limit is illustrated on the right side of Fig. 7.15. An integral heat transfer solution that accounts for the effect of wall curvature in the laminar regime was developed by LeFevre and Ede [38],

$$\overline{\mathrm{Nu}}_H = \frac{4}{3}\left[\frac{7\mathrm{Ra}_H\mathrm{Pr}}{5\,(20 + 21\mathrm{Pr})}\right]^{1/4} + \frac{4\,(272 + 315\mathrm{Pr})H}{35\,(64 + 63\mathrm{Pr})D} \tag{7.83}$$

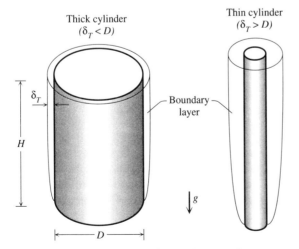

Figure 7.15 Vertical cylinders with natural convection boundary layer flow over the lateral surface.

where $\overline{\mathrm{Nu}}_H = \overline{h}H/k$, $\mathrm{Ra}_H = g\beta\Delta TH^3/\alpha v$, $\overline{h}$ is the wall-averaged heat transfer coefficient, and ΔT is the temperature difference between the surface and the fluid reservoir.

7.3.7   Other Immersed Bodies

In Section 7.3 we have reviewed several of the simplest and most common body shapes that are encountered in calculations of external natural convection. It turns out that the natural convection heat transfer from bodies of other, less regular shapes follows $\overline{\mathrm{Nu}}$–Ra relationships that are similar to what we have seen thus far. This similarity and the general need for fewer and simpler formulas continue to stimulate research on a universal heat transfer correlation for immersed bodies of various shapes.

The simplest formula of this kind was proposed by Lienhard [39],

$$\overline{\mathrm{Nu}}_l \cong 0.52\mathrm{Ra}_l^{1/4} \tag{7.84}$$

where $\overline{\mathrm{Nu}}_l = \overline{h}l/k$, $\mathrm{Ra}_l = g\beta\Delta Tl^3/\alpha v$ and $\overline{h}$ is the heat transfer coefficient averaged over the entire surface of the body. Lienhard's length l, on which both Nu_l and Ra_l are based, is the distance traveled by the boundary layer fluid while in contact with the body. In the case of a horizontal cylinder, for example, $l = \pi D/2$. Equation (7.84) should be accurate within 10 percent, provided that $\mathrm{Pr} \gtrsim 0.7$ and the Rayleigh number is sufficiently large so that the boundary layer is thin.

Sparrow and Ansari [40] tested eq. (7.84) in an experiment in which a vertical cylinder of equal height and diameter ($H = D$, hence $l = 2D$) was suspended in air. They found that eq. (7.84) generally underpredicts the measured $\overline{\mathrm{Nu}}_l$ (Ra_l) data. The error was of order 30 percent at $\mathrm{Ra}_l \sim 10^4$ and less than 8 percent when Ra_l exceeded 10^6. Their experimental data for the vertical cylinder with $H = D$ in air was correlated very closely by $\overline{\mathrm{Nu}}_D = 0.775\mathrm{Ra}_D^{0.208}$ when $\mathrm{Ra}_D > 1.4 \times 10^4$, in which D is the length scale used for defining the Nusselt and Rayleigh numbers.

Yovanovich [41] developed a correlation that covers the entire laminar range, $0 < \mathrm{Ra}_{\mathcal{L}} < 10^8$, that is, including the limit of pure conduction $\mathrm{Ra}_{\mathcal{L}} \to 0$. As length scale, he used the square root of the entire surface of the immersed body,

$$\mathcal{L} = A^{1/2} \tag{7.85}$$

and defined $\overline{\mathrm{Nu}}_{\mathcal{L}} = \overline{h}\mathcal{L}/k$ and $\mathrm{Ra}_{\mathcal{L}} = g\beta\Delta T\mathcal{L}^3/\alpha v$. The correlation contains two constants,

$$\overline{\mathrm{Nu}}_{\mathcal{L}} = \overline{\mathrm{Nu}}_{\mathcal{L}}^0 + \frac{0.67G_{\mathcal{L}}\mathrm{Ra}_{\mathcal{L}}^{1/4}}{[1 + (0.492/\mathrm{Pr})^{9/16}]^{4/9}} \tag{7.86}$$

namely, the conduction limit Nusselt number $\overline{\mathrm{Nu}}_{\mathcal{L}}^0$, and the geometric parameter $G_{\mathcal{L}}$. The latter is a weak function of body shape, aspect ratio, and orientation in the gravitational field. Table 7.1 lists the values of these two constants for the bodies and orientations shown in Fig. 7.16. These values do not vary appreciably; therefore, a general expression based on the average values of $\overline{\mathrm{Nu}}_{\mathcal{L}}^0$ and $G_{\mathcal{L}}$, and valid for $\mathrm{Pr} \gtrsim 0.7$, is [41]

$$\overline{\mathrm{Nu}}_{\mathcal{L}} \cong 3.47 + 0.51\,\mathrm{Ra}_{\mathcal{L}}^{1/4} \tag{7.87}$$

Table 7.1 Constants for Yovanovich's Correlation [41] for Laminar Natural Convection Heat Transfer from Immersed Bodies, Fig. 7.16

Body Shape	$\overline{Nu}^0_{\mathcal{L}}$	$G_{\mathcal{L}}$
Sphere	3.545	1.023
Bisphere	3.475	0.928
Cube 1	3.388	0.951
Cube 2	3.388	0.990
Cube 3	3.388	1.014
Vertical cylinder[a]	3.444	0.967
Horizontal cylinder[a]	3.444	1.019
Cylinder[a] at 45°	3.444	1.004
Prolate spheroid ($C/B = 1.93$)	3.566	1.012
Oblate spheroid ($C/B = 0.5$)	3.529	0.973
Oblate spheroid ($C/B = 0.1$)	3.342	0.768

[a]Short cylinder, $H = D$.

A more extensive correlation that covers the conduction, laminar, and turbulent regimes was developed by Hassani and Hollands [42]. This correlation employs two length scales, one of which is $\mathcal{L}$, eq. (7.85). It covers the Rayleigh number range $0-10^{14}$. Experimental and numerical results for heat transfer from horizontal disks and rings were published recently by Sahraoui et al. [43].

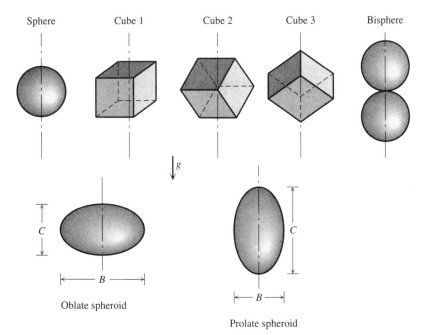

Figure 7.16 Various shapes and orientations of isothermal bodies immersed in a fluid at a different temperature (Table 7.1).

Example 7.3

Horizontal Cylinder, Laminar Natural Convection

The horizontal cylinder photographed in Fig. 7.14 has the diameter $D = 6$ cm and length $L = 60$ cm. Its surface is isothermal at $T_w = 34°C$, while the distant air that surrounds it has the temperature $T_\infty = 25°C$. Verify numerically that the boundary layer is laminar and that it has a thickness compatible (of the same order) with the one revealed by the photograph. Calculate the heat transfer rate released by the cylindrical surface.

Solution.   The film temperature of the air boundary layer is $(34°C + 25°C)/2 \cong 30°C$; therefore, the physical properties that will be needed are

$$Pr = 0.72 \qquad k = 0.026 \, \frac{W}{m \cdot K} \qquad \frac{g\beta}{\alpha v} = \frac{90.7}{cm^3 \cdot K}$$

The Rayleigh and Grashof numbers based on diameter,

$$Ra_D = \frac{g\beta}{\alpha v} D^3 \, (T_w - T_\infty)$$

$$= \frac{90.7}{cm^3 \cdot K} \, (6 \text{ cm})^3 \, (34 - 25) \text{ K} = 1.76 \times 10^5$$

$$Gr_D = \frac{Ra_D}{Pr} = \frac{1.76 \times 10^5}{0.72} = 2.45 \times 10^5 < 10^9$$

confirm that the boundary layer is laminar. The thermal boundary layer thickness is approximately

$$\delta_T \sim D Ra_D^{-1/4} = 6 \text{ cm } (1.76 \times 10^5)^{-1/4}$$

$$= 0.3 \text{ cm}$$

which shows that the slenderness ratio δ_T/D is of order $1/20$. This ratio is compatible with what we see in Fig. 7.14, where the δ_T-thin layer is the innermost region in which the dark lines are nearly equidistant.

To calculate the total heat transfer rate released by the cylinder, we turn our attention to eq. (7.79), in which we substitute $Pr = 0.72$:

$$\overline{Nu}_D = (0.6 + 0.322 Ra_D^{1/6})^2 = 9.06$$

$$\overline{q}''_{w,D} = \overline{Nu}_D \, \frac{\Delta T}{D} \, k$$

$$= 9.06 \, \frac{9 \text{ K}}{0.06 \text{ m}} 0.026 \frac{W}{m \cdot K} = 35.3 \, \frac{W}{m^2}$$

$$A = \pi D L$$

$$= \pi \, (0.06 \text{ m}) \, (0.6 \text{ m}) = 0.113 \text{ m}^2$$

$$q = \overline{q}''_{w,D} A$$

$$= 35.3 \, \frac{W}{m^2} \, (0.113 \text{ m}^2) = 4 \text{ W}$$

7.4   INTERNAL FLOW CONFIGURATIONS

7.4.1   Vertical Channels

We now consider problems in which the walls surround the buoyant fluid. The simplest configuration of this kind is the vertical gap formed between two heated plates, Fig. 7.17. In many instances this is an adequate model for the air space between two vertically mounted circuit boards in an electronic package or for the space between two vertical plate fins. The plates are modeled as isothermal (T_w), and so is the fluid situated *outside* the channel (T_∞). For the sake of better illustration, it is assumed that the plates are heated ($T_w > T_\infty$) and that the fluid rises through the chimney formed between them.

The channel has two length scales: the height H and the plate-to-plate spacing L. When the thermal boundary layers that coat each plate (δ_T) are considerably thinner than the plate-to-plate spacing, the heat transfer rate from the plates to the channel fluid can be calculated with the single-wall formulas (7.61)–(7.62) or, if they are modeled as uniform flux, with eqs. (7.67)–(7.70). In $\text{Pr} \gtrsim 1$ fluids, the "wide channel" limit represented by $\delta_T < L$ is represented alternatively by the inequalities

$$\frac{L}{H} > \text{Ra}_H^{-1/4} \quad \text{or} \quad \frac{L}{H} > \text{Ra}_L^{-1} \quad\quad (7.88\text{a, b})$$

Of special interest is the "narrow channel" limit in which the above inequalities break down. Figure 7.17 shows that when the channel is narrow enough, the velocity profiles merge into a single profile that resembles that of Hagen–

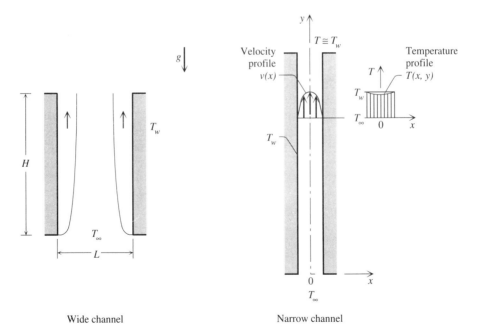

Wide channel   Narrow channel

Figure 7.17 Vertical channel with isothermal walls; the top and bottom ends communicate with a reservoir of isothermal fluid.

Poiseuille flow (Fig. 6.3, right side). Even though the temperature of the fluid that enters through the bottom opening is equal to T_∞, when the channel is narrow and tall enough the fluid temperature $T(x,y)$ is nearly the same as the plate temperature T_w. We record this observation by means of an inequality between temperature differences:

$$T_w - T(x,y) < T_w - T_\infty \tag{7.89}$$

In the language of Section 6.2.2, the above inequality means that the duct flow that proceeds vertically through the channel is thermally fully developed over most of the channel height H. Assuming further that the Prandtl number is of order 1 (e.g., air) or greater, the inequality (7.89) means also that the channel flow is hydrodynamically fully developed. Therefore, with reference to the momentum eq. (7.17), which applies at every point in the channel, the hydrodynamic development of the flow means the flow is purely *vertical* ($u = 0$) and, consequently, $\partial v/\partial y = 0$. [Review eqs. (6.10)–(6.14), and keep in mind that in the present problem the longitudinal direction is y.] In conclusion, eq. (7.17) reduces to

$$0 = v \frac{d^2 v}{dx^2} + g\beta(T - T_\infty) \tag{7.90}$$

and, in view of eq. (7.89),

$$\frac{d^2 v}{dx^2} \cong -\frac{g\beta}{v}(T_w - T_\infty) = \text{constant} \tag{7.91}$$

Important to note is that eq. (7.90) follows from eq. (7.17), which was developed for an external flow with $dP_\infty/dy = -\rho_\infty g$, eq. (7.9). In the present "internal" geometry, $\partial P/\partial y$ is a *constant* equal to dP/dy because the flow is assumed to be fully developed. [Review eqs. (6.12) and (6.14).] It is because both ends of the channel are open that the dP/dy constant inside the channel is equal to (i.e., controlled by) the hydrostatic pressure gradient of the surroundings, $dP/dy = dP_\infty/dy = -\rho_\infty g$.

Equation (7.91) is equivalent to the momentum eq. (6.20) of forced convection through a parallel-plate channel. The role of the forced-convection constant $(1/\mu)dP/dx$ is now played by the natural-convection constant $-g\beta\Delta T/v$, where $\Delta T = T_w - T_\infty$. Solving eq. (7.91) subject to the no-slip conditions that are evident in Fig. 7.17, we obtain the parabolic velocity distribution

$$v = \frac{g\beta\Delta T L^2}{8v}\left[1 - \left(\frac{x}{L/2}\right)^2\right] \tag{7.92}$$

and the vertical mass flowrate per unit length in the direction normal to Fig. 7.17,

$$\dot{m}' = \int_{-L/2}^{L/2} \rho v\, dx = \frac{\rho g\beta\Delta T L^3}{12v} \tag{7.93}$$

An important observation is that in fully developed laminar flow the vertical velocity and mass flowrate *are independent of the channel height H*. Although this may seem strange, it is a consequence of the fact that the net longitudinal (i.e., vertical) pressure gradient that drives the chimney flow is independent of H. By reason of the analogy between eqs. (7.91) and (6.20), that net pressure gradient is $-\rho g\beta(T_w - T_\infty)$.

The total heat transfer rate extracted by the $\dot{m}'$ stream from the two vertical surfaces that touch it is (again, per unit length normal to the plane of Fig. 7.17)

$$q' = \dot{m}'c_P(T_w - T_\infty) = \frac{\rho g \beta c_P (\Delta T)^2 L^3}{12\nu} \tag{7.94}$$

This result can be nondimensionalized by defining the channel-averaged heat flux (note that the channel has two sides)

$$\bar{q}'' = \frac{q'}{2H} \tag{7.95}$$

and the average Nusselt number $\overline{Nu}_H$. Hence,

$$\overline{Nu}_H = \frac{\bar{q}''}{\Delta T}\frac{H}{k} = \frac{1}{24}\,Ra_L \tag{7.96}$$

The right side of eq. (7.96) follows from using eq. (7.94). This $\overline{Nu}_H$ result is valid when the inequality (7.89) prevails over most of the channel: the two sides of that inequality are respectively represented by the scales

$$\frac{\bar{q}''L}{k} < \Delta T \tag{7.97}$$

Dividing by ΔT and using the last of eqs. (7.96), we arrive at

$$Ra_L < \frac{H}{L} \tag{7.98}$$

which is the opposite of the wide-channel criterion (7.88b). In conclusion, the $\overline{Nu}_H$ result (7.96) holds true in the narrow-channel limit represented by the inequality (7.98).

Results similar to eq. (7.96) have been obtained in the narrow-channel limit of vertical ducts with other cross-sectional shapes (Table 7.2). In each case the ratio $\overline{Nu}_H/Ra_{D_h}$ is a constant, D_h being the hydraulic diameter of the particular cross section, and $\overline{Nu}_H$ is the channel-averaged Nusselt number defined by the

Table 7.2 Average Nusselt Numbers for Chimney Flow[a] in the Narrow-Channel Limit $H/D_h > Ra_{D_h}$

Cross Section Shape	$\overline{Nu}_H/Ra_{D_h}$
Parallel plates	$\dfrac{1}{192}$
Circular	$\dfrac{1}{128}$
Square	$\dfrac{1}{113.6}$
Equilateral triangle	$\dfrac{1}{106.4}$

[a]From Bejan [5], p. 157

first of eqs. (7.96). The height of the channel in all cases is H. Many other results for laminar and turbulent flow and for mixed (forced and natural) convection in vertical channels have been reviewed by Aung [44].

7.4.2 Enclosures Heated from the Side

An important class of internal natural convection problems are the flows induced in enclosed spaces that are subjected to temperature differences in the horizontal direction. A classical example of this type is the circulation of air in the slot of a double-pane window. Figure 7.18 shows that the circulation consists of a vertical jet that rises along the heated wall and of a cold jet that descends along the cooled wall. The circulation is completed by two horizontal streams that proceed in counterflow along the top and bottom walls. It is assumed that the flow is practically two-dimensional, in other words, that the enclosure is sufficiently wide in the direction normal to the plane of Fig. 7.18.

A vast amount of research has been devoted during the 1970s to predicting the total heat transfer rate between the differentially heated walls of the enclosure (T_h, T_c). The general trends revealed by this research have been sorted out relatively recently. Therefore, it pays to begin this discussion with Fig. 7.19, which shows that in $\mathrm{Pr} \gtrsim 1$ fluids the flow pattern depends on the Rayleigh number $\mathrm{Ra}_H = g\beta(T_h - T_c)H^3/\alpha\nu$ and the geometric aspect ratio H/L.

The cavity is "wide" when the vertical thermal boundary layer thickness δ_T is smaller than the horizontal dimension L. The condition $\delta_T < L$ leads to the inequality (7.88a), which is represented by the rising diagonal on Fig. 7.19. To the right of this diagonal the vertical boundary layers are distinct and, to a large degree, the overall heat transfer rate can be predicted with the formulas of Section 7.2.4. (Note that in the present problem the temperature difference between one wall and the "fluid reservoir" would be $\Delta T/2$, where $\Delta T = T_h - T_c$.) Consequently, in the laminar regime we expect a proportionality between the average Nusselt number and $\mathrm{Ra}_H^{1/4}$, and no effect due to L. The Berkovsky–

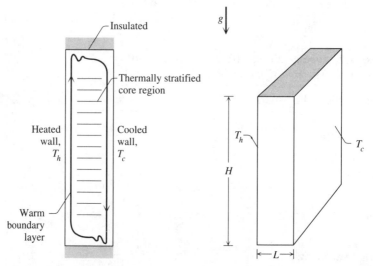

Figure 7.18 Enclosure filled with fluid and heated and cooled along the two lateral walls.

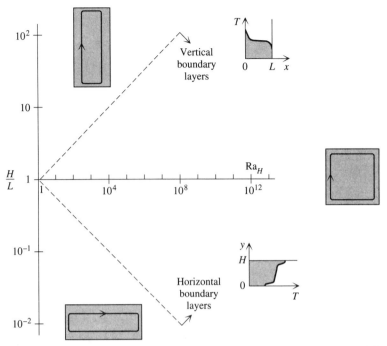

Figure 7.19 Flow regimes for natural convection in enclosures heated from the side and filled with $Pr \gtrsim 1$ fluid (Bejan [5], p. 166).

Polevikov correlations* recommended by Catton [45] deviate only slightly from this rule:

$$\overline{Nu}_H = 0.22 \left(\frac{Pr}{0.2 + Pr} Ra_H \right)^{0.28} \left(\frac{L}{H} \right)^{0.09}$$

$$2 < \frac{H}{L} < 10, \; Pr < 10^5, \; Ra_H < 10^{13} \quad (7.99)$$

$$\overline{Nu}_H = 0.18 \left(\frac{Pr}{0.2 + Pr} Ra_H \right)^{0.29} \left(\frac{L}{H} \right)^{-0.13}$$

$$1 < \frac{H}{L} < 2, \; 10^{-3} < Pr < 10^5, \; 10^3 < \frac{Pr}{0.2 + Pr} Ra_H \left(\frac{L}{H} \right)^3 \quad (7.100)$$

where $\overline{Nu}_H = q''H/k\,\Delta T$. These correlations are valid in the indicated (H/L, Pr, Ra_H) domain, and in the "wide" cavity limit, eq. (7.88a).

In the opposite extreme, $L/H < Ra_H^{-1/4}$, the cavity is too narrow to allow distinct vertical thermal boundary layers. Instead, the boundary layers are squeezed together, giving birth to a linear temperature variation in the horizontal direction. This feature occurs at large values of H/L, that is, to the left of the rising diagonal of Fig. 7.19. The linear temperature profile means that in this

---

*In Ref. [45] the Berkovsky–Polevikov correlations show a deceptively strong L/H effect, because the Nusselt and Rayleigh numbers in that paper were based on L as length scale, namely, $\overline{Nu}_L$ and Ra_L.

limit the heat transfer rate from T_h to T_c is ruled by conduction and, therefore, that $\overline{\mathrm{Nu}}_H$ approaches the constant H/L:

$$\overline{\mathrm{Nu}}_H = \overline{q}'' \, \frac{H}{k \Delta T} \rightarrow \left(k \, \frac{\Delta T}{L} \right) \frac{H}{k \Delta T} = \frac{H}{L} \tag{7.101}$$

The preceding discussion referred to square or tall enclosures, where $H/L \geq 1$. The flows in shallow enclosures (Fig. 7.19, $H/L < 1$) can be divided into flows with distinct horizontal jets and flows without distinct jets along the top and bottom walls. The overall Nusselt number is reported in Fig. 7.20, which is based on an integral analysis of the natural circulation in the entire cavity [46]. Note that at sufficiently low Rayleigh numbers Ra_H, the Nusselt number $\overline{\mathrm{Nu}}_H$ approaches again the pure conduction constant H/L. This limit corresponds to the wedge-shaped domain situated to the left of the descending diagonal in Fig. 7.19.

A completely theoretical solution for the flow and temperature field is possible [47] when the side walls are heated and, respectively, cooled with uniform heat flux q''. In the boundary layer regime, the temperature increases *linearly* in the vertical direction along the heated wall, the cooled wall, and in the core region:

$$\frac{\partial T}{\partial y} = 0.0425 \, \frac{\alpha v}{g \beta H^4} \left(\frac{H}{L} \right)^{4/9} \mathrm{Ra}_H^{*8/9} \quad \text{(constant)} \tag{7.102}$$

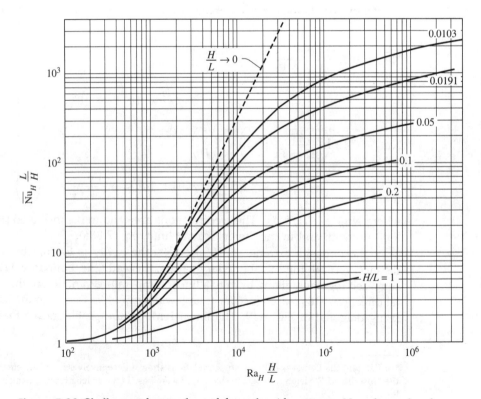

Figure 7.20 Shallow enclosures heated from the side: average Nusselt number for $\mathrm{Pr} \gtrsim 1$ fluids (Bejan and Tien [46]).

Since the temperature increases at the same rate in the vertical direction along both walls, the wall-to-wall temperature difference is a *constant* at every level, $T_h(y) - T_c(y) = \Delta T = $ constant. The theoretical solution for the average Nusselt number $\overline{\mathrm{Nu}}_H = q''H/(\Delta Tk)$ in the boundary layer regime in $\mathrm{Pr} \gtrsim 1$ fluids is [47]

$$\overline{\mathrm{Nu}}_H = 0.34 \mathrm{Ra}_H^{*2/9} \left(\frac{H}{L}\right)^{1/9} \tag{7.103a}$$

where $\mathrm{Ra}_H^* = g\beta H^4 q''/(\alpha \nu k)$. If, instead of the flux Rayleigh number Ra_H^*, we use the ΔT-based Rayleigh number $\mathrm{Ra}_H = \mathrm{Ra}_H^*/\overline{\mathrm{Nu}}_H = g\beta \Delta T H^3/(\alpha \nu)$, then eq. (7.103a) becomes

$$\overline{\mathrm{Nu}}_H = 0.25 \mathrm{Ra}_H^{2/7} \left(\frac{H}{L}\right)^{1/7} \tag{7.103b}$$

It is important that, since $\mathrm{Ra}_H^{2/7} = \mathrm{Ra}_H^{0.286}$, this theoretical alternative reproduces almost all the features of the empirical correlations (7.99)–(7.100) recommended for enclosures with isothermal side walls. This reinforces the seemingly universal observation that the heat transfer correlations developed for a system with isothermal walls apply reasonably well to the uniform flux configuration, provided the Ra_H number is based then on the wall-averaged temperature difference, $\mathrm{Ra}_H = \mathrm{Ra}_H^*/\overline{\mathrm{Nu}}_H$.

There are many applications in which the enclosed fluid departs from the single-enclosure model used throughout this subsection (e.g., rooms in buildings, solar collectors, lakes). A more advanced model for the description of such systems is the association of two enclosures communicating laterally through a doorway, window, or corridor, or over an incomplete dividing wall. The *partially divided enclosure* model has only recently come under scrutiny [48–50].

7.4.3 Enclosures Heated from Below

The fundamental difference between enclosures heated from the side (Fig. 7.18) and enclosures heated from below (Fig. 7.21) is that in enclosures heated from the side a buoyancy-driven flow is present as soon as a very small temperature difference $(T_h - T_c)$ is imposed between the two side walls. By contrast, in enclosures heated from below, the imposed temperature difference must exceed a finite *critical* value before the first signs of fluid motion and convective heat transfer are detected.

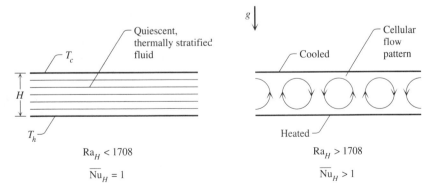

Figure 7.21 Horizontal fluid layer held between two parallel walls and heated from below.

When the enclosure is sufficiently long and wide in the horizontal direction, the condition for the onset of convection is expressed by the critical Rayleigh number [51]

$$\text{Ra}_H \gtrsim 1708 \tag{7.104}$$

where $\text{Ra}_H = g\beta(T_h - T_c)H^3/(\alpha\nu)$. As suggested in Fig. 7.21, immediately above $\text{Ra}_H \cong 1708$ the flow consists of counterrotating two-dimensional rolls the cross sections of which are almost square. This flow pattern is commonly recognized as *Bénard cells*, or *Bénard convection*, in honor of H. Bénard who reported the first investigation of this phenomenon in 1900. The cellular flow becomes considerably more complicated as Ra_H exceeds by one or more orders of magnitude the critical (convection onset) value. The two-dimensional rolls break up into three-dimensional cells, which appear hexagonal in shape when viewed from above (Fig. 7.22). At even higher Rayleigh numbers, the cells multiply (become narrower) and, eventually, the flow becomes oscillatory and turbulent. The hierarchy of flow regimes and various transitions in Bénard convection has been reviewed by Busse [53].

The heat transfer effect of the cellular flow is to augment the net heat transfer rate in the vertical direction, that is, to increase it above the pure conduction rate that would prevail in the absence of fluid motion. The dimensionless number that measures this augmentation effect is the average Nusselt number based on the vertical dimension H, namely, $\overline{\text{Nu}}_H = \bar{q}''H/(k\,\Delta T)$. Experimental heat transfer measurements in the range $3 \times 10^5 < \text{Ra}_H < 7 \times 10^9$ support the correlation [54]

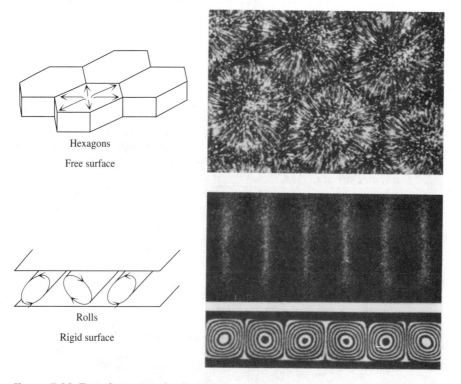

Hexagons

Free surface

Rolls

Rigid surface

Figure 7.22 Two-dimensional rolls and three-dimensional hexagonal cells in a fluid layer heated from below. (Oertel [52], with permission from G. Braun, Karlsruhe.)

$$\overline{Nu}_H = 0.069 Ra_H^{1/3} Pr^{0.074} \tag{7.105}$$

The physical properties needed for calculating $\overline{Nu}_H$, Ra_H, and Pr are evaluated at the average fluid temperature $(T_h + T_c)/2$. Equation (7.105) holds when the horizontal layer is sufficiently wide so that effect of the short vertical sides is minimal.

The experimental data correlated by eq. (7.105) line up in such a way that the exponent of Ra_H in that correlation increases slightly as Ra_H increases. The exponent is closer to 0.29 at the low Ra_H end of the correlation (roughly, when $Ra_H \lesssim 10^8$). Its value becomes practically equal to one-third at higher Rayleigh numbers, in other words $\overline{Nu}_H \sim Ra_H^{1/3}$ when Ra_H exceeds approximately 10^8. The proportionality $\overline{Nu}_H \sim Ra_H^{1/3}$ persists as Ra_H increases above the range covered by eq. (7.105).

It is important to note that the proportionality $\overline{Nu}_H \sim Ra_H^{1/3}$ expressed by eq. (7.105) means that the actual heat transfer rate (q'') from T_h to T_c is *independent* of the layer thickness H. It has been shown that the $\overline{Nu}_H \sim Ra_H^{1/3}$ proportionality can be predicted theoretically by performing the scale analysis of a single roll (cell) of the Bénard convection pattern [5]. An even simpler alternative derivation is presented in Example 7.5.

At subcritical Rayleigh numbers, $Ra_H \lesssim 1708$, the fluid is quiescent and the temperature decreases linearly from T_h to T_c. The heat transfer rate across the fluid layer is by pure conduction; therefore, $\overline{Nu}_H = 1$.

The convection onset criterion (7.104)—the value 1708 on the right side— refers strictly to an infinite horizontal layer with rigid (no-slip) and isothermal top and bottom boundaries. Similar criteria, with critical Rayleigh numbers of the order of 10^3, hold for other horizontal layer configurations (i.e., combinations of top and bottom boundary conditions).

Example 7.4

The Origin of Thermals Rising from a Heated Surface

We are in a position to estimate the time interval between the formation of two successive balls of heated fluid (thermals) that rise from the same spot on the hot surface photographed in Fig. 7.12. As indicated earlier in Example 7.2, the bottom surface and water pool tempera-

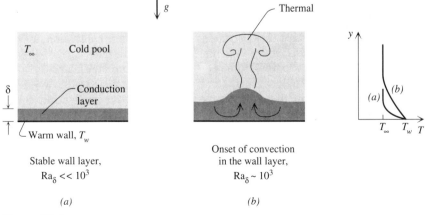

Stable wall layer,
$Ra_\delta \ll 10^3$

(a)

Onset of convection
in the wall layer,
$Ra_\delta \sim 10^3$

(b)

Figure E7.4

tures are 43.1°C and 23.6°C. The following order-of-magnitude calculation is based on the view that the surface heats by conduction a thin layer of water. The thickness of this layer increases in time until its thickness-based Rayleigh number reaches the critical level for the onset of convection. Thermals are the aftermath of the onset of convection in the conduction layer that grew over the surface.

Solution. The water properties that will be used are evaluated at the film temperature of 33.4°C:

$$\alpha = 0.00148 \ \frac{cm^2}{s} \qquad \frac{g\beta}{\alpha\nu} = \frac{28,982}{cm^3 \cdot K}$$

Assume that in the beginning the water is motionless. The water that comes in contact with the hot surface develops a "conduction" layer the thickness of which increases as (see Chapter 4)

$$\delta \sim (\alpha t)^{1/2} \tag{1}$$

This δ-tall layer is heated from below and, in accordance with a criterion similar to eq. (7.104), it becomes unstable when its Rayleigh number based on height exceeds the order of magnitude 10^3:

$$Ra_\delta \sim 10^3 \tag{2}$$

$$\frac{g\beta}{\alpha\nu} \delta^3 (T_w - T_\infty) \sim 10^3 \tag{3}$$

$$\delta^3 \sim 10^3 \ \frac{cm^3 \cdot K}{28,982} \ \frac{1}{(43.1 - 23.6)K} = 0.0018 \ cm^3 \tag{4}$$

or, approximately, $\delta \sim 1$ mm. In view of eq. (1), this result means that

$$\alpha t \sim 0.0146 \ cm^2 \tag{5}$$

$$t \sim \frac{0.0146 \ cm^2}{0.00148 \ cm^2/s} \cong 10 \ s \tag{6}$$

In conclusion, thermals will rise from the same region of the heated surface at time intervals of the order of 10 seconds.

Example 7.5

Turbulent Bénard Convection: Why $\overline{Nu}_H$ Increases as $Ra_H^{1/3}$, and Why the Heat Transfer Rate Is Independent of H

When the Rayleigh number Ra_H is orders of magnitude greater than the critical value, convection in the bottom-heated fluid layer is turbulent. The core of the fluid layer is practically at the average temperature $(T_h + T_c)/2$, while temperature drops of size $(T_h - T_c)/2$ occur across thin fluid layers that line the two horizontal walls. Rely on the scales of the thermals analyzed in the preceding example to derive analytically an approximate substitute for the heat transfer correlation (7.105).

Solution. In the high Rayleigh number regime, the turbulence over most of the horizontal layer of thickness H is caused by the meandering thermals that rise from the heated bottom and those that fall from the cooled top. In this way, the turbulent core of the layer is sandwiched between two δ-thin conduction layers of the type described in the preceding example. There, we learned that the δ thickness is such that, in an order of magnitude sense,

$$Ra_\delta \sim 10^3 \tag{1}$$

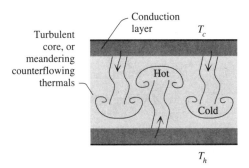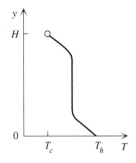

Figure E7.5

This is equivalent to writing

$$\text{Ra}_H \sim 10^3 \left(\frac{H}{\delta}\right)^3 \tag{2}$$

where Ra_H is the usual Rayleigh number, $\text{Ra}_H = g\beta\Delta T H^3/(\alpha\nu)$, and $\Delta T = T_h - T_c$. The heat transfer from T_h to T_c is impeded by layers of thickness δ (not H); therefore,

$$\overline{q''} \sim k\,\frac{\Delta T}{\delta} \tag{3}$$

and

$$\overline{\text{Nu}}_H = \frac{\overline{q''}}{k\,\Delta T/H} \sim \frac{H}{\delta} \tag{4}$$

By eliminating H/δ between eqs. (2) and (4), we obtain

$$\overline{\text{Nu}}_H \sim 10^{-1}\text{Ra}_H^{1/3} \tag{5}$$

It is indeed rewarding to see that this simple theoretical result reproduces almost all the features of the empirical correlation recommended for turbulent (high Ra_H) Bénard convection calculations, eq. (7.105). This back-of-the-envelope derivation is also a good reminder that the actual heat transfer rate does not depend on H [see eq. (3)].

7.4.4   Inclined Enclosures

Let us consider now the tilted enclosure configuration of Fig. 7.23, in which L is the distance measured in the direction of the imposed temperature difference $T_h - T_c$. The angle made with the horizontal direction, τ, is defined such that in the range $0° < \tau < 90°$ the heated surface is positioned *below* the cooled surface. The enclosure heated from the side, Fig. 7.18, is clearly the $\tau = 90°$ case of the inclined enclosure of Fig. 7.23. Note also that the $\tau = 0°$ case represents the rectangular enclosure heated from below (Fig. 7.21), except that now the distance between the differentially heated walls is labeled L.

The inclination angle τ has a dramatic effect on the flow and heat transfer characteristics of the enclosure. As τ decreases from $180°$ to $0°$, the heat transfer mechanism switches from pure conduction at $\tau = 180°$ to single-cell convection at $\tau = 90°$ and, finally, to Bénard convection at $\tau = 0°$. The conduction-referenced Nusselt number $\overline{\text{Nu}}_L = \overline{q''}L/(k\,\Delta T)$ rises from the pure conduction level $\overline{\text{Nu}}_L(180°) = 1$ to a maximum $\overline{\text{Nu}}_L$ value near $\tau = 90°$. As τ decreases below $90°$, the Nusselt number decreases and passes through a local

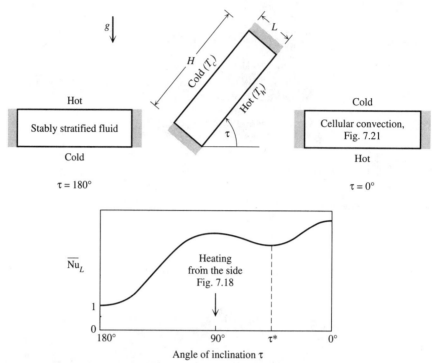

Figure 7.23 The effect of tilt angle on the heat transfer rate and flow pattern in an enclosure with a temperature difference imposed between two opposing walls.

minimum at a special tilt angle $\tau = \tau^*$, which is a function of the geometric aspect ratio of the enclosure [55]:

H/L	1	3	6	12	>12
τ^*	25°	53°	60°	67°	70°

As τ continues to decrease below $\tau = \tau^*$, the heat transfer rate rises toward another maximum associated with the Bénard convection regime, $\overline{\mathrm{Nu}}_L(0°)$.

The $\overline{\mathrm{Nu}}_L(\tau)$ curve is illustrated qualitatively in the lower part of Fig. 7.23. According to Refs. [56–58] and the review article Ref. [45], the various portions of the curve are represented well by the following formulas:

$$\overline{\mathrm{Nu}}_L(\tau) = 1 + [\overline{\mathrm{Nu}}_L(90°) - 1]\sin\tau \qquad 180° > \tau > 90° \qquad (7.106a)$$

$$\overline{\mathrm{Nu}}_L(\tau) = \overline{\mathrm{Nu}}_L(90°)(\sin\tau)^{1/4} \qquad 90° > \tau > \tau^* \qquad (7.106b)$$

$$\overline{\mathrm{Nu}}_L(\tau) = \left[\frac{\overline{\mathrm{Nu}}_L(90°)}{\overline{\mathrm{Nu}}_L(0°)}(\sin\tau^*)^{1/4}\right]^{\tau/\tau^*} \qquad \tau^* > \tau > 0° \text{ and } H/L < 10 \qquad (7.106c)$$

$$\overline{\mathrm{Nu}}_L(\tau) = 1 + 1.44\left(1 - \frac{1708}{\mathrm{Ra}_L\cos\tau}\right)^*\left[1 - \frac{(\sin 1.8\tau)^{1.6} \times 1708}{\mathrm{Ra}_L\cos\tau}\right] +$$

$$\left[\left(\frac{\mathrm{Ra}_L\cos\tau}{5830}\right)^{1/3} - 1\right]^* \qquad \tau^* > \tau > 0° \text{ and } H/L > 10 \qquad (7.106d)$$

The quantities contained between the parentheses with asterisk, ( )*, must be set equal to zero if they become negative. The Rayleigh number is based on the distance between the differentially heated walls, $\mathrm{Ra}_L = g\beta(T_h - T_c)L^3/(\alpha\nu)$.

7.4.5 Annular Space Between Horizontal Cylinders

The flow generated between concentric horizontal cylinders at different temperatures (T_i, T_o) has features similar to the circulation in an enclosure heated from the side. As illustrated in Figs. 7.24 and 7.25, in the laminar regime two counterrotating kidney-shaped cells are positioned symmetrically about the vertical plane drawn through the cylinder centerline. Because the two cells are essentially being heated and cooled from the side, the overall heat transfer correlation should have the features of eqs. (7.99) and (7.100). The role of vertical dimension in this case can be played by either of the two diameters D_i and D_o: when D_i is sensibly smaller than D_o, the overall heat transfer rate is determined ("throttled") by the smallness of D_i rather than D_o.

These expectations are confirmed by the following new interpretation* of a correlation developed by Raithby and Hollands [60],

$$q' \cong \frac{2.425k(T_i - T_o)}{[1 + (D_i/D_o)^{3/5}]^{5/4}} \left(\frac{\mathrm{Pr}\,\mathrm{Ra}_{D_i}}{0.861 + \mathrm{Pr}} \right)^{1/4} \qquad (7.107)$$

This q' (W/m) expression refers to the total heat transfer rate between the two cylinders, per unit length in the direction normal to the plane of Fig. 7.24. The Rayleigh number is based on the cylinder-to-cylinder temperature difference and on the *inner* diameter, $\mathrm{Ra}_{D_i} = g\beta(T_i - T_o)D_i^3/(\alpha\nu)$.

Equation (7.107) agrees within ±10 percent with experimental data at moderate (laminar regime) Rayleigh numbers as high as 10^7. The low Rayleigh number limit of the validity of eq. (7.107) occurs where the thickest free convection boundary layer would become larger than the transversal dimension of the annular cavity:

$$D_o\mathrm{Ra}_{D_o}^{-1/4} > D_o - D_i \qquad (7.108)$$

---

*The writing and use of the original correlation [60] is a bit more complicated, because it was patterned after the pure conduction formula (2.32), and its Rayleigh number was based on the thickness of the annular space $(D_o - D_i)/2$.

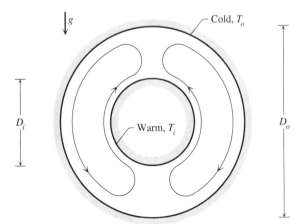

Figure 7.24 Natural convection pattern in the annular space between horizontal concentric cylinders at different temperatures.

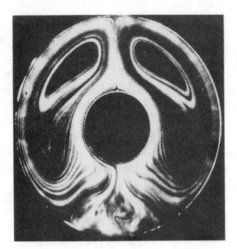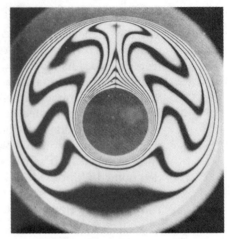

Figure 7.25 Air natural convection between two concentric horizontal cylinders. The photograph on the left shows the flow pattern (streamlines) visualized with cigarette smoke. The interferometer photograph on the right illustrates the temperature field, or isotherms. (Grigull and Hauf [59]; courtesy of Professor U. Grigull, Technical University of Munich.)

The left side of this inequality shows the thermal boundary layer thickness along (i.e. inside) the outer cylinder, which — it is easy to see* — is greater than the corresponding thickness associated with the inner cylinder, $D_i \mathrm{Ra}_{D_i}^{-1/4}$. When the Rayleigh number is small enough so that the inequality (7.108) is satisfied, the heat transfer mechanism approaches the pure conduction regime, for which the recommended formula is eq. (2.32). Another way of making certain that the Rayleigh number is large enough for eq. (7.107) to apply is to calculate q' twice, using eqs. (7.107) and (2.32), and to retain the larger of the two q' values. This alternative is based on the idea that the true heat transfer rate q' cannot be smaller than the pure conduction estimate based on eq. (2.32).

Equation (7.107) has been tested extensively in the range $\mathrm{Pr} > 0.7$. It should also hold in the low Prandtl number range, because the function $[\mathrm{Pr}/(0.861 + \mathrm{Pr})]^{1/4}$ is known to account properly for the Pr effect at both ends of the Pr spectrum. Compare, for example, eq. (7.107) with eq. (7.55). All the physical properties that appear on the right side of eq. (7.107) are evaluated at the average temperature $(T_i + T_o)/2$. A lengthier correlation that covers a wider Rayleigh number range, including the turbulent regime, was developed by Kuehn and Goldstein [61].

7.4.6   Annular Space Between Concentric Spheres

The natural circulation in the annulus between two concentric spheres has the approximate shape of a doughnut. In fact, at small enough Rayleigh numbers in

---

*Recall that in laminar boundary layer natural convection along (or around) a surface (flat or curved) of a certain height, the boundary layer thickness increases as the height raised to the power $\frac{1}{4}$. The boundary layer thickness scale is equal to the height times the Rayleigh number (based on height) raised to the power $-\frac{1}{4}$. This observation stood behind the laminar flow correlations assembled in Section 7.3.

the laminar regime, a vertical cut through the center of the two spheres reveals two kidney-shaped flow patterns similar to what we saw in Fig. 7.24.

A relatively compact expression for the total heat transfer rate q (W) between the two spherical surfaces can be derived from a correlation developed by Raithby and Hollands [60]:

$$q \cong \frac{2.325 k D_i \, (T_i - T_o)}{[1 + (D_i/D_o)^{7/5}]^{5/4}} \left(\frac{\Pr \mathrm{Ra}_{D_i}}{0.861 + \Pr} \right)^{1/4} \tag{7.109}$$

where, again, $\mathrm{Ra}_{D_i} = g\beta(T_i - T_o)D_i^3/(\alpha\nu)$. The argument for employing the Rayleigh number based on D_i is the same as in the opening paragraph to the preceding section. Equation (7.109) holds over the entire Prandtl number range, and the physical properties are evaluated at the average temperature $(T_i + T_o)/2$. The similarities between eq. (7.109) and eq. (7.107) are worth contemplating.

Equation (7.109) is accurate within ± 10 percent in the laminar range, at Rayleigh numbers as high as 10^7. When the Rayleigh number is so low that the inequality (7.108) applies, eq. (7.109) is no longer valid: the formula to use in this limit is the pure conduction expression (2.39). As a safety check, it is a good idea to always calculate q based on both formulas, eqs. (7.109) and (2.39), and to retain the larger of the two values. Note that the ratio $q_{7.109}/q_{2.39}$ indicates the degree of heat transfer enhancement (augmentation) that is due to the presence of convection inside the cavity. In this sense, the q ratio is similar to the conduction-referenced Nusselt number $\overline{\mathrm{Nu}}_H$ encountered in fluid layers heated from below, eq. (7.105).

All the topics and configurations assembled in this chapter refer exclusively to *pure* natural convection, that is, to flows driven solely by the effect of buoyancy. Even more numerous in nature are the situations where the flow is the result of a combination of natural and forced convection. Think, for example, of the room air flow near a warm window at the time when the air-conditioning system blows a charge of new (cleaner and cooler) air through the room. These more complicated flows are examples of *mixed convection*: their fundamentals are covered in the more advanced courses on convection (e.g., Ref. [5]). The current progress on mixed convection is described by Aung [44] for internal flow, by Chen and Armaly [62] for external flow, and by Zubair and Kadaba [63] for external unsteady flow.

REFERENCES

1. E. R. G. Eckert and E. Soehngen, Studies in heat transfer, U.S. Air Force Technical Report 5747, Dec. 1948.
2. B. Gebhart, *Heat Transfer*, 2nd edition, McGraw-Hill, New York, 1971, p. 321.
3. A. Oberbeck, Über die Wärmeleitung der Flüssigkeiten bei Berücksichtigung der Strömungen infolge von Temperaturdifferenzen, *Ann. Phys. Chem.*, Vol. 7, 1879, pp. 271–292.
4. J. Boussinesq, *Theorie Analytique de la Chaleur*, Vol. 2, Gauthier-Villars, Paris, 1903.
5. A. Bejan, *Convection Heat Transfer*, Wiley, New York, 1984, p. 116.
6. S. Ostrach, An analysis of laminar free-convection flow and heat transfer about a flat plate parallel to the direction of the generating body force, NACA TN 2635, 1952.

7. E. J. LeFevre, Laminar free convection from a vertical plane surface, *Proc. Ninth Int. Congr. Appl. Mech.*, Brussels, Vol. 4, 1956, pp. 168–174.

8. B. W. Martin, An appreciation of advances in natural convection along an isothermal vertical surface, *Int. J. Heat Mass Transfer*, Vol. 27, 1984, pp. 1583–1586.

9. E. Schmidt and W. Beckmann, Das Temperatur- und Geschwindigkeitsfeld vor einer Wärme abgebenden senkrechten Platte bei natürlicher Konvektion, *Tech. Mech. Thermo-Dynam.*, Vol. 1, 1930, pp. 391–406.

10. H. K. Kuiken, An asymptotic solution for large Prandtl number free convection, *J. Engng. Math.*, Vol. 2, 1968, pp. 355–371.

11. H. K. Kuiken, Free convection at low Prandtl numbers, *J. Fluid Mech.*, Vol. 37, 1969, pp. 785–798.

12. F. P. Incropera and D. P. DeWitt, *Fundamentals of Heat and Mass Transfer*, 3rd edition, Wiley, New York, 1990, pp. 539–540.

13. A. Bejan and J. L. Lage, The Prandtl number effect on the transition in natural convection along a vertical surface, *J. Heat Transfer*, Vol. 112, 1990, pp. 787–790.

14. W. P. Farmer and W. T. McKie, Natural convection from a vertical isothermal surface in oil, ASME Paper No. 64-WA/HT-12, 1964.

15. T. Fujii, Experimental studies of free convection heat transfer, *Bulletin of JSME*, Vol. 2, No. 8, 1959, pp. 555–558.

16. R. Godaux and B. Gebhart, An experimental study of the transition of natural convection flow adjacent to a vertical surface, *Int. J. Heat Mass Transfer*, Vol. 17, 1974, pp. 93–107.

17. W. W. Humphreys and J. R. Welty, Natural convection with mercury in a uniformly heated vertical channel during unstable laminar and transitional flow, *AIChE J*, Vol. 21, 1975, pp. 268–274.

18. P. S. Lykoudis, Private communication, 1989.

19. R. L. Mahajan and B. Gebhart, An experimental determination of transition limits in a vertical natural convection flow adjacent to a surface, *J. Fluid Mechanics*, Vol. 91, 1979, pp. 131–154.

20. N. Sheriff and N. W. Davies, Sodium natural convection from a vertical plate, *Heat Transfer 1978*, Hemisphere, Washington, DC, Vol. 5, 1978, pp. 131–136.

21. Y. S. Touloukian, G. A. Hawkins, and M. Jakob, Heat transfer by free convection from heated vertical surfaces to liquids, *Trans. ASME*, Vol. 70, 1948, pp. 13–23.

22. M. Uotani, Natural convection heat transfer in thermally stratified liquid metal, *J. Nuclear Science and Technology*, Vol. 24, No. 6, 1987, pp. 442–451.

23. J. R. Lloyd, E. M. Sparrow, and E. R. G. Eckert, Laminar, transition and turbulent natural convection adjacent to inclined and vertical surfaces, *Int. J. Heat Mass Transfer*, Vol. 15, 1972, pp. 457–473.

24. C. E. Polymeropoulos and B. Gebhart, Incipient instability in free convection laminar boundary layers, *J. Fluid Mechanics*, Vol. 30, 1967, pp.225–239.

25. S. W. Churchill and H. H. S. Chu, Correlating equations for laminar and turbulent free convection from a vertical plate, *Int. J. Heat Mass Transfer*, Vol. 18, 1975, pp. 1323–1329.

26. E. M. Sparrow and J. L. Gregg, Laminar free convection from a vertical plate with uniform surface heat flux, *Trans. ASME*, Vol. 78, 1956, pp. 435–440.

27. G. C. Vliet and C. K. Liu, An experimental study of turbulent natural convection boundary layers, *J. Heat Transfer*, Vol. 91, 1969, pp. 517–531.

28. G. C. Vliet and D. C. Ross, Turbulent natural convection on upward and downward facing inclined heat flux surfaces, *J. Heat Transfer*, Vol. 97, 1975, pp. 549–555.

29. C. C. Chen and R. Eichhorn, Natural convection from a vertical surface to a thermally stratified medium, *J. Heat Transfer*, Vol. 98, 1976, pp. 446–451.

30. W. J. Minkowycz and E. M. Sparrow, Local nonsimilar solutions for natural convection on a vertical cylinder, *J. Heat Transfer*, Vol. 96, 1974, pp. 178–183.

31. G. C. Vliet, Natural convection local heat transfer on constant-heat-flux inclined surfaces, *J. Heat Transfer*, Vol. 91, 1969, pp. 511–516.

32. J. R. Lloyd and E. M. Sparrow, On the instability of natural convection flow on inclined plates, *J. Fluid Mechanics*, Vol. 42, 1970, pp. 465–470.

33. E. M. Sparrow, R. B. Husar, and R. J. Goldstein, Observations and other characteristics of thermals, *J. Fluid Mechanics*, Vol. 41, 1970, pp. 793–800.

34. R. J. Goldstein, E. M. Sparrow, and D. C. Jones, Natural convection mass transfer adjacent to horizontal plates, *Int. J. Heat Mass Transfer*, Vol. 16, 1973, pp. 1025–1035.

35. J. R. Lloyd and W. R. Moran, Natural convection adjacent to horizontal surfaces of various planforms, ASME Paper 74-WA/HT-66, 1974.

36. B. Gebhart, Y. Jaluria, R. L. Mahajan, and B. Sammakia, *Buoyancy-Induced Flows and Transport*, Hemisphere, New York, 1988.

37. S. W. Churchill, Free convection around immersed bodies, E. U. Schlünder, Ed., *Heat Exchanger Design Handbook*, Section 2.5.7, Hemisphere, New York, 1983.

38. E. J. LeFevre and A. J. Ede, Laminar free convection from the outer surface of a vertical circular cylinder, *Proc. Ninth Int. Congr. Appl. Mech.*, Brussels, Vol. 4, 1956, pp. 175–183.

39. J. H. Lienhard, On the commonality of equations for natural convection from immersed bodies, *Int. J. Heat Mass Transfer*, Vol. 16, 1973, pp. 2121–2123.

40. E. M. Sparrow and M. A. Ansari, A refutation of King's rule for multi-dimensional external natural convection, *Int. J. Heat Mass Transfer*, Vol. 26, 1983, pp. 1357–1364.

41. M. M. Yovanovich, On the effect of shape, aspect ratio and orientation upon natural convection from isothermal bodies of complex shape, ASME HTD-Vol. 82, 1987, pp. 121–129.

42. A. V. Hassani and K. G. T. Hollands, On natural convection heat transfer from three-dimensional bodies of arbitrary shape, *J. Heat Transfer*, Vol. 111, 1989, pp. 363–371.

43. M. Sahraoui, M. Kaviany, and H. Marshall, Natural convection from horizontal disks and rings, *J. Heat Transfer*, Vol. 112, 1990, pp. 110–116.

44. W. Aung, Mixed convection in internal flow, Chapter 15 in S. Kakac, R. K. Shah, and W. Aung, Eds., *Handbook of Single-Phase Convective Heat Transfer*, Wiley, New York, 1987.

45. I. Catton, Natural convection in enclosures, *6th Int. Heat Transfer Conf., Toronto, 1978*, Vol. 6, 1979, pp. 13–43.

46. A. Bejan and C. L. Tien, Laminar natural convection heat transfer in a horizontal cavity with different end temperatures, *J. Heat Transfer*, Vol. 100, 1978, pp. 641–647.

47. S. Kimura and A. Bejan, The boundary layer natural convection regime in a rectangular cavity with uniform heat flux from the side, *J. Heat Transfer*, Vol. 106, 1984, pp. 98–103.

48. M. W. Nansteel and R. Greif, An investigation of natural convection in enclo-

sures with two and three dimensional partitions, *Int. J. Heat Mass Transfer*, Vol. 27, 1984, pp. 561–571.

49. R. Jetli, S. Acharya, and E. Zimmerman, Influence of baffle location on natural convection in a partially divided enclosure, *Numerical Heat Transfer*, Vol. 10, 1986, pp. 521–536.

50. J. Neymark, C. R. Boardman III, A. T. Kirkpatrick, and R. Anderson, High Rayleigh number natural convection in partially divided air and water filled enclosures, *Int. J. Heat Mass Transfer*, Vol. 32, 1989, pp. 1671–1679.

51. A. Pellew and R. V. Southwell, On maintained convection motion in a fluid heated from below, *Proc. Royal Soc.*, Vol. A176, 1940, pp. 312–343.

52. H. Oertel, Jr., Thermal instabilities, *in* J. Zierep and H. Oertel, Jr., Eds., *Convective Transport and Instability Phenomena*, G. Braun, Karlsruhe, 1982, p. 10.

53. F. H. Busse, Non-linear properties of thermal convection, *Rep. Prog. Phys.*, Vol. 41, 1978, pp. 1929–1967.

54. S. Globe and D. Dropkin, Natural convection heat transfer in liquids confined by two horizontal plates and heated from below, *J. Heat Transfer*, Vol. 81, 1959, pp. 24–28.

55. J. N. Arnold, I. Catton, and D. K. Edwards, Experimental investigation of natural convection in inclined rectangular regions of differing aspect ratios, *J. Heat Transfer*, Vol. 98, 1976, pp. 67–71.

56. J. N. Arnold, P. N. Bonaparte, I. Catton, and D. K. Edwards, Proc. 1974 Heat Transfer Fluid Mech. Inst., Stanford University Press, Stanford, CA, 1974.

57. P. S. Ayyaswamy and I. Catton, The boundary layer regime for natural convection in a differentially heated tilted rectangular cavity, *J. Heat Transfer*, Vol. 95, 1973, pp. 543–545.

58. K. G. T. Hollands, T. E. Unny, G. D. Raithby, and L. J. Konicek, Free convection heat transfer across inclined air layers, *J. Heat Transfer*, Vol. 98, 1976, pp. 189–193.

59. U. Grigull and W. Hauf, Natural convection in horizontal cylindrical annuli, Proc. 3rd Int. Heat Transfer Conf., Vol. 2, 1966, pp. 182–195.

60. G. D. Raithby and K. G. T. Hollands, A general method of obtaining approximate solutions to laminar and turbulent free convection problems, *Advances in Heat Transfer*, Vol. 11, 1975, pp. 265–315.

61. T. H. Kuehn and R. J. Goldstein, Correlating equations for natural convection heat transfer between horizontal circular cylinders, *Int. J. Heat Mass Transfer*, Vol. 19, 1976, pp. 1127–1134.

62. T. S. Chen and B. F. Armaly, Mixed convection in external flow, Chapter 14 in S. Kakac, R. K. Shah, and W. Aung, Eds., *Handbook of Single-Phase Convective Heat Transfer*, Wiley, New York, 1987.

63. S. M. Zubair and P. V. Kadaba, Similarity transformations for boundary layer equations in unsteady mixed convection, *Int. Comm. Heat Mass Transfer*, Vol. 17, 1990, pp. 215–226.

64. A. Bejan and Z. Zhang, Natural convection melting of a slab of ice, in R. A. Granger and K. R. Read, Eds., *Experiments in Thermodynamics and Heat Transfer* (to be published).

65. W. S. Janna and G. S. Jakubowski, Heat transfer analysis for ice objects melting in air, ASME HTD-Vol. 143, 1990, pp. 1–8.

66. A. Bar-Cohen and W. M. Rohsenow, Thermally optimum spacing of vertical, natural convection cooled, parallel plates, *J. Heat Transfer*, Vol. 106, 1984, pp. 116–123.

67. F. P. Incropera, Convection heat transfer in electronic equipment cooling, *J. Heat Transfer*, Vol. 110, 1988, pp. 1097–1111.
68. G. P. Peterson and A. Ortega, Thermal control of electronic equipment and devices, *Advances in Heat Transfer*, Vol. 20, 1990, pp. 181–314.
69. J. S. Lim, A. Bejan and J. H. Kim, The optimal thickness of a wall with convection on one side, *Int. J. Heat Mass Transfer*, Vol. 35, 1992 (to be published).

PROBLEMS

Laminar Vertical Boundary Layer Flow

7.1 **(a)** Show that in the case of a fluid that can be modeled as an ideal gas, the linearized equation of state (7.13) becomes

$$\frac{\rho}{\rho_\infty} \cong 1 - \frac{T - T_\infty}{T_\infty} + \frac{P - P_0}{P_0} + \cdots$$

(b) Consider an application in which the fluid is air ($T_\infty = 20°C$, $P_0 = 1$ atm), the vertical heated wall is 4 m tall, and the wall temperature is approximately 40°C. The pressure excursion $P - P_0$ is of the same order as the hydrostatic pressure difference between floor and ceiling (Fig. 7.1, right side). Numerically, show that the term $(P - P_0)/P_0$ is relatively negligible in the ρ/ρ_∞ formula derived in part (a).

7.2 Let $\delta(y)$ be the outer thickness of the vertical velocity profile for laminar natural convection along a vertical wall. Show that the integral form of the boundary layer-simplified momentum eq. (7.17) is

$$\frac{d}{dy}\int_0^\delta v^2\,dx = -\nu\left(\frac{\partial v}{\partial x}\right)_{x=0} + \int_0^\delta g\beta\,(T - T_\infty)dx$$

7.3 Let δ_T be the thermal boundary layer thickness of the buoyant jet along a vertical wall. Show that the integral form of the boundary layer-simplified energy eq. (7.11) is

$$\frac{d}{dy}\int_0^{\delta_T} v(T - T_\infty)dx = -\alpha\left(\frac{\partial T}{\partial x}\right)_{x=0}$$

In this derivation, keep in mind that the vertical velocity v at the outer edge of the thermal boundary layer ($x = \delta_T$) is not necessarily zero.

7.4 Consider the boundary layer flow near a vertical isothermal wall (T_w) surrounded by an isothermal fluid reservoir (T_∞). Assume the following temperature and vertical velocity profiles:

$$T - T_\infty = \Delta T\left(1 - \frac{x}{\delta}\right)^2$$

$$v = V\frac{x}{\delta}\left(1 - \frac{x}{\delta}\right)^2$$

where $\Delta T = T_w - T_\infty$. Determine the functions $\delta(y)$ and $V(y)$ by solving the integral momentum and energy equations listed in the preceding two problem statements. Show that the local Nusselt number that corresponds to this solution is

$$Nu_y = \frac{q''_{w,y}}{\Delta T}\frac{y}{k} = 0.508\left(1 + \frac{20}{21Pr}\right)^{-1/4}Ra_y^{1/4}$$

7.5   The door of a household refrigerator is 1 m tall and 0.6 m wide. The temperature of its external surface is 20°C while the room air temperature is 30°C. Verify that the natural convection air boundary layer that descends along the door is laminar, and calculate the heat transfer rate absorbed by the door.

7.6   An immersion heater for water consists of a thin vertical plate shaped as a rectangle of height 8 cm and length 15 cm. The plate is heated electrically and maintained at 55°C, while the average temperature in the surrounding water tank is 15°C. The plate is bathed by water on both sides. Show that the natural convection boundary layer that rises along the plate is laminar. Later, calculate the total heat transfer rate released by the heater into the water pool.

7.7   For the buoyant jet along a vertical wall with uniform heat flux q_w'', assume the following temperature and velocity profiles:

$$T - T_\infty = (T_w - T_\infty)\left(1 - \frac{x}{\delta}\right)^2$$

$$v = V\frac{x}{\delta}\left(1 - \frac{x}{\delta}\right)^2$$

Solve the integral boundary layer-simplified equations for momentum and energy listed in Problems 7.2 and 7.3, and show that the local Nusselt number is

$$\text{Nu}_y = \frac{q_w''}{T_w(y) - T_\infty}\frac{y}{k} = 0.616\left(\frac{0.8}{\text{Pr}} + 1\right)^{-1/5}\text{Ra}_y^{*1/5}$$

7.8   One way to visualize the $y^{1/4}$-dependence of the thickness of the laminar natural convection boundary layer is to execute the experiment shown in the figure. The vertical isothermal wall, $T_w = 20$°C, is placed in contact with an isothermal pool of paraffin, $T_\infty = 35$°C. Since the solidification point of this paraffin is $T_m = 27.5$°C, the wall becomes covered with a thin layer of solidified paraffin.

   Show that under steady-state conditions the thickness of the solidified layer, L, is proportional to the laminar boundary layer thickness, that is, it increases in the downward direction as $y^{1/4}$. Calculate L numerically and plot to scale the $L(y)$ shape of the solidified layer. The relevant properties of liquid paraffin are $k_f = 0.15$ W/m·K, $\beta = 8.5 \times 10^{-4}$ K^{-1}, $\alpha = 9 \times 10^{-4}$ cm^2/s, and Pr = 55.9. The thermal conductivity of solid paraffin is $k_s = 0.36$ W/m·K. The overall height of the isothermal wall is $H = 10$ cm.

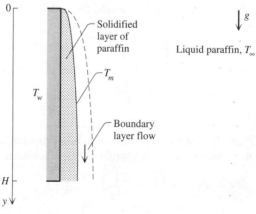

Figure P7.8

Turbulent Vertical Boundary Layer Flow

7.9 One 0.5 m $\times$ 0.5 m vertical wall of a parallelipipedic water container is heated uniformly by an array of electrical strip heaters mounted on its back. The total heat transfer rate furnished by these heaters is 1000 W. The water temperature is 25°C. Calculate the height-averaged temperature of the heated surface. Later, verify that most of the boundary layer that rises along this surface is turbulent.

7.10 A small box filled with water must be designed (sized) for the purpose of simulating in the laboratory the natural convection of air in a 3-m-tall room. The average temperature of the room air is 25°C while the temperature of the cold vertical wall is 15°C.

The designer must select a certain vertical dimension (height H) for the water enclosure so that the Rayleigh number $g\beta\,\Delta TH^3/(\alpha\nu)$ of the water flow matches the Rayleigh number of the air flow in the actual room. In the water experiment the average water temperature and the cold-wall temperature are 25°C and 15°C, respectively. What is the required height H of the water apparatus?

7.11 **(a)** Consider the natural convection boundary layer flow of water along a vertical wall with uniform heat flux q_w''. The Rayleigh number based on heat flux, Ra_y^*, is known. Show that

$$Ra_y = \frac{Ra_y^*}{\overline{Nu}_y}$$

where $\overline{Nu}_y$ is the average Nusselt number (based on average temperature difference, $\overline{T}_w - T_\infty$) and Ra_y is the Rayleigh number based on $\overline{T}_w - T_\infty$.

(b) It is observed that the transition to turbulence takes place in the vicinity of $Ra_y^* \sim 10^{13}$. Calculate the corresponding value of the Rayleigh number based on temperature difference, Ra_y, and compare the Ra_y value with the one anticipated based on eq. (7.57).

Other External Flow Configurations

7.12 The downward trend exhibited by the $\overline{Nu}_H/Ra_H$ curves of Fig. 7.9 can be anticipated by means of a very simple analysis. Let ΔT_{avg} be the average temperature difference between the isothermal wall and the stratified reservoir. Assume a large Prandtl number ($Pr \to \infty$), and estimate the Nusselt number by using eq. (7.55), in which both $\overline{Nu}_y$ and Ra_y are based on ΔT_{avg} (note also that $y = H$). Multiply and divide this relationship by ΔT_{max} to reshape it in terms of the $\overline{Nu}_H$ and Ra_H groups defined in eqs. (7.72a, b). Show that this relationship becomes

$$\overline{Nu}_H = 0.671\left(1 - \frac{b}{2}\right)^{5/4} Ra_H^{1/4}$$

and that it approximates within 13 percent the values suggested by the $Pr \to \infty$ curve of Fig. 7.9.

7.13 The air in a room is thermally stratified so that its temperature increases by 5°C for each 1-m rise in altitude. Facing the room air is a 10°C vertical window, which is 1 m tall and 0.8 m wide. The room air temperature at the same level as the window midheight is 20°C. Assume that the natural convection boundary layer that falls along the window is laminar, and calculate the heat transfer rate through the window.

7.14 A horizontal disc with a diameter of 8.7 cm and temperature of 43.1°C is immersed facing upward in a pool of 23.6°C water. Calculate the total heat transfer rate released by the surface into the water.

7.15 A long, rectangular metallic blade has width $H = 4$ cm and temperature $T_w = 40°C$. It is surrounded on both sides by atmospheric air at $T_\infty = 20°C$. The long side of the blade is always horizontal. Calculate the total heat transfer rate per unit of blade length, when the short side of its rectangular shape (H) is
(a) Vertical
(b) Inclined at 45° relative to the vertical
(c) Horizontal
Comment on the effect that blade orientation has on the total heat transfer rate.

7.16 The external surface of a spherical container with the diameter $D = 3$ m has a temperature of 10°C. The container is surrounded by 30°C air, which is motionless except in the immediate vicinity of the container.
(a) Calculate the total heat transfer rate absorbed by the spherical container.
(b) It is proposed to replace the spherical container with one shaped as a horizontal cylinder with the diameter $d = 1.5$ m. This new container would have the same volume as the old one. Calculate the total heat transfer rate absorbed by the cylindrical container.
(c) Which container design would you choose if your objective is to prevent the warming of the liquid stored inside the container?

7.17 A cubic block of metal is immersed in a pool of 20°C water and is oriented as "Cube 1" in Fig. 7.16. The cube has a 2-cm side and an instantaneous temperature of 80°C. Calculate the average heat transfer coefficient between the cube and the water, by using
(a) Lienhard's method, eq. (7.84),
(b) Yovanovich's method, eq. (7.86), and
(c) Yovanovich's simplified formula (7.87).
Comment on the agreement between these three estimates.

7.18 A block of ice has the shape of a parallelepiped with the dimensions indicated in the figure. The block is oriented in such a way that two of its long surfaces are horizontal. It is surrounded by 20°C air from all sides.
    Calculate the total heat transfer rate between the ambient air and the ice block, q (W), by estimating the heat transfer through each face. Calculate also the ice melting rate that corresponds to the total heat transfer rate. The ice latent heat of melting is $h_{sf} = 333.4$ kJ/kg.

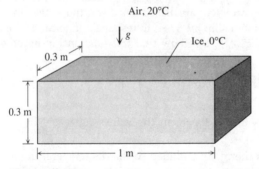

Figure P7.18

7.19 Consider again the natural convection heat transfer from air to the ice block described in the preceding problem. Obtain a "quick" estimate of the total heat transfer rate q (W) by using Lienhard's formula (7.84). The length l on which $\overline{Nu}_l$ and Ra_l are based can be approximated as the half-perimeter of the smaller (0.3 m $\times$ 0.3 m) cross section of the ice block (see Fig. P7.18).

7.20 An electrical wire of diameter $D = 1$ mm is suspended horizontally in air of temperature 20°C. The Joule heating of the wire is responsible for the heat generation rate $q' = 0.01$ W/cm, per unit length in the axial direction. The wire can be modeled as a cylinder with isothermal surface. Sufficiently far from the wire, the ambient air is motionless.

Calculate the temperature difference that is established between the wire and the ambient air. (*Note*: This calculation requires a trial-and-error procedure; expect a relatively small Rayleigh number.)

7.21 The large-diameter cylindrical reservoir shown in the figure is perfectly insulated and filled with water. Two horizontal tubes with an outer diameter of 4 cm are positioned at the same level in the vicinity of the reservoir centerline. The temperatures of the tube walls are maintained at $T_1 = 30$°C and $T_2 = 20$°C by internal water streams of appropriate (controlled) temperature.

Calculate the heat transfer rate from the hot tube to the cold tube via the water reservoir, by assuming that the tube-to-tube spacing is wide enough so that the boundary layers that coat the tubes do not touch (this case is illustrated in the figure). Assume further that the film-temperature properties of the two boundary layers are equal to the properties evaluated at the average temperature of the water reservoir, T_∞. Begin with the calculation of T_∞, and recognize the centrosymmetry of the flow pattern, that is, the symmetry about the centerline of the large reservoir.

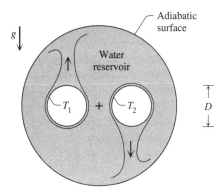

Figure P7.21

7.22 A hot dog is immersed horizontally in a stagnant pool of 95°C water. The temperature of the hot dog is uniform at 20°C. The shape of the hot dog is approximated well by a cylinder with a diameter of 1.7 cm and length of 10 cm. Assume that the natural convection boundary layer flow has had time to develop around the hot dog surface.

(a) Calculate the average heat transfer coefficient and the rate of heat transfer absorbed by the hot dog.

(b) As the time increases and the hot dog warms up, does the hot dog temperature remain uniform over the hot dog cross section? In other words, would it be a good idea to model the hot dog as a lumped capacitance?

7.23 Cake batter is poured to fill completely a horizontal shallow pan that is shaped as a square. The side of the square is 46 cm (measured internally, i.e., touched by the batter). The room air is motionless and at 28°C, while the initial temperature of the batter is 22°C. Calculate the rate of heat transfer through the top surface, into the batter. Assuming that the pan is not put in the oven too soon, will the calculated heat transfer rate increase or decrease as time passes?

Internal Flow Configurations

7.24 Consider the narrow-channel limit of a vertical duct with isothermal wall T_w, height H, cross-sectional area A, wetted perimeter p, and hydraulic diameter $D_h = 4A/p$. The top and bottom ends of the duct communicate with an isothermal fluid reservoir (T_∞), Fig. 7.17. By assuming fully developed laminar flow and reviewing Sections 6.1.2 and 6.1.3, show that the average vertical velocity through the duct is

$$V = \frac{g\beta(T_w - T_\infty)D_h^2}{2\nu f \mathrm{Re}_{D_h}}$$

where $\mathrm{Re}_{D_h} = D_h V/\nu$. Show further that the average Nusselt number $\overline{\mathrm{Nu}}_H = \overline{q''}H/(k\,\Delta T)$ is proportional to the Rayleigh number based on hydraulic diameter,

$$\overline{\mathrm{Nu}}_H = \frac{1}{8(f\mathrm{Re}_{D_h})}\,\mathrm{Ra}_{D_h}$$

where the group $f\mathrm{Re}_{D_h}$ is a constant listed in Table 6.1. Use this last formula for the purpose of verifying the validity of the numerical values assembled in Table 7.2.

7.25 The *single-pane window problem* consists of estimating the heat transfer rate through the vertical glass layer shown in the figure. The window separates two air reservoirs of temperatures T_h and T_c. Assuming constant properties, laminar boundary layers on both sides of the glass, and a *uniform* glass temperature T_w, show that the average heat flux $\overline{q''}$ from T_h to T_c obeys the relationship

$$\frac{\overline{q''}}{T_h - T_c}\frac{H}{k} = 0.217\left[\frac{g\beta(T_h - T_c)H^3}{\alpha\nu}\right]^{1/4}$$

Use eq. (7.55) as starting point in this analysis, and neglect the thermal resistance due to pure conduction across the glass layer itself.

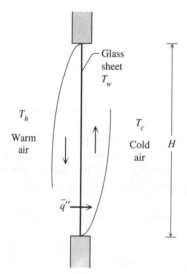

Figure P7.25

7.26 Consider again the single-pane window described in the preceding problem, and model the heat flux through the glass layer as uniform, q''. Starting with the height-averaged version of eq. (7.66) for air, show that the relationship between q'' and the overall temperature difference $T_h - T_c$ is

$$\frac{q''}{T_h - T_c} \frac{H}{k} = 0.252 \left[\frac{g\beta(T_h - T_c)H^3}{\alpha v} \right]^{1/4}$$

Compare this result with the formula recommended by the uniform-T_w model in the preceding problem, and you will get a feel for the "certainty" with which you can calculate the total heat transfer rate through the window.

7.27 The *double-pane window problem* consists of determining the H-averaged heat flux through the system, when the overall temperature difference $T_h - T_c$ is specified (see figure). If the glass-to-glass spacing is wide enough to house *distinct* laminar boundary layers, it is possible to approximate the double-pane system as a sandwich of two of the single-pane windows treated in the preceding problem. Use the formula listed above for the single-pane window with uniform heat flux, and show that the average heat flux through the double-pane window system is given approximately by

$$\frac{q''}{T_h - T_c} \frac{H}{k} \cong 0.106 \left[\frac{g\beta(T_h - T_c)H^3}{\alpha v} \right]^{1/4}$$

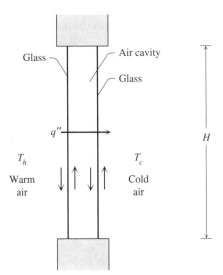

Figure P7.27

7.28 In the drawing that accompanies Problem 4.3 a small bottle containing 60°C-temperature milk is immersed in a pool of 10°C tap water. The height of the bottle wall wetted on the inside by milk and on the outside by tap water is $H = 6$ cm. Calculate the H-averaged *overall* heat transfer coefficient between the warm milk and the cold tap water. In this calculation recognize that the two sides of the vertical wall are lined by natural convection boundary layers, and that the "overall" heat transfer coefficient is related to the individual heat transfer coefficients calculated for the two sides of the vertical surface.

7.29 An interesting application of the air flow in enclosures heated from the side is shown in the figure. The space between the vertical surfaces of the double wall is divided into many cells by means of inclined partitions. In the house heating arrangement illustrated here, the function of this wall is to allow the transfer of heat from left to right (when the outer surface is heated by the sun) and to prevent the transfer of heat in the opposite direction (during night time).

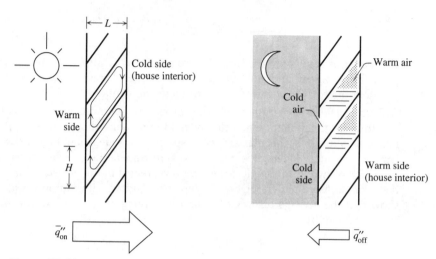

Figure P7.29

In the "on" mode, the heated portion of each cell is positioned below the cooled portion. The enclosed air rises at one end and sinks at the other and completes one roll. In the "off" mode, the heated air remains trapped above the cold air, and the circulation is inhibited.

Estimate the average heat flux through the double wall in the two modes. Model the flow during the "on" mode as the single-roll circulation inside a square cavity heated from the side, $H = L = 10$ cm. Assume that the heat transfer during the "off" mode is by pure conduction. In both cases the temperature difference between the vertical surfaces of the double wall is 20°C. Evaluate all the air properties at 20°C.

PROJECTS

7.1 The Natural Convection Melting of a Slab of Ice

Objective. The effect of heat transfer by boundary layer natural convection over a vertical wall can be visualized and measured by experimenting with thin slabs of ice suspended vertically in still air [64]. The uneven distribution of heat flux is demonstrated by the uneven thinning of the ice slab. The instantaneous flowrate of meltwater collected under the dripping ice is a measure of the overall heat transfer rate from the ambient to the isothermal surfaces of the slab. An additional objective of this project is to show that laboratory apparatuses can be built quite inexpensively, often by using kitchen utensils. This experiment teaches a group of students to critically evaluate each others' data, and to pool all their findings into a comprehensive report that may have engineering significance.

Apparatus. The heart of the apparatus is a vertical slab of ice that is suspended by means of a string in still air. The manufacture of the ice slab and its suspension and the maintenance of a nearly motionless and isothermal ambient are the critical aspects of the apparatus construction.

An inexpensive way of producing ice slabs of one or more sizes is to use a flat-bottom baking pan (or cookie sheet) placed horizontally in the freezer of a household refrigerator. Desirable ice slab qualities are

(a) A temperature close to the melting point, 0°C, in other words, a minimum degree of solid subcooling; and

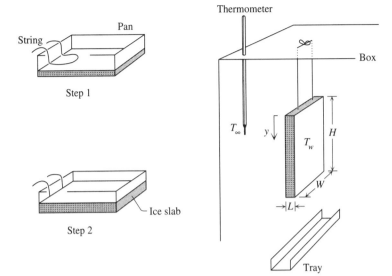

Figure PR7.1a

(b) A minimum amount of trapped air bubbles and other defects (cracks, bulges) in the free surface.

The first feature is enhanced by using a freezer the temperature setting of which reads relatively warm. The second feature is more difficult to attain, as it requires the sequential freezing of thin layers of water no deeper than approximately 0.5 cm.

Figure PR7.1a shows the main steps that may be taken to produce one ice slab. In the first step, the slab is frozen (built up) to its half-thickness, $L/2$. At this point the suspension string (total length ~1 m) is looped and placed flat over the free surface of the frozen slab. The two ends of the string hang over the side of the tray, by intersecting that edge of the ice slab which in the actual experiment will face upward.

In the second step the remainder of the slab thickness L is built up through the freezing of additional water layers. For a slab of final thickness $L = 1$ cm, it is sufficient to freeze a single 0.5-cm-deep water layer in the first step, followed by another 0.5-cm-deep layer in the second step.

The baking pan dictates the large dimensions of the ice slab, namely, the height H and the width W. The very slow process of natural convection melting — the actual experiment — begins by suspending the ice slab in still air. An inexpensive type of enclosure that prevents the forced convection of air is a cardboard box. The suspension of the ice slab inside this box is achieved by passing the two ends of the string through two holes in the ceiling of the box and tying them into a knot on the outside.

Positioned under the ice slab is a sheet metal tray (trough) that catches the falling water droplets. This tray is inclined relative to the horizontal, so that the collected liquid passes through an opening in the side of the cardboard box, and falls into a graduated beaker or measuring glass column.

Procedure. The quantity that is measured during this experiment is the meltwater flowrate $\dot{m}$. This measurement is obtained by monitoring the rise of the water level in the graduated beaker. It is a relatively simple measurement, as the melting rate is expectedly slow. It is a good idea to wait at least 30 minutes before measuring $\dot{m}$, to allow the transient conduction inside the slab to run its course and bring the entire slab volume to the melting point.

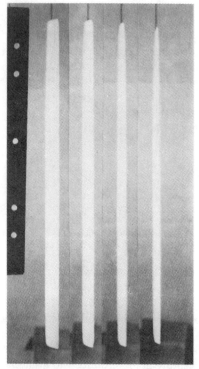

Figure PR7.1*b*

The other quantity of interest is the temperature of the air maintained in the box, T_∞ (°C). This air is not sealed off completely, because of the slab suspension device and the opening used for the collection of meltwater. The function of this "leaky" box is simply to prevent the action of forced air currents in the vicinity of the vertical ice surface. Such currents are driven on and off by the ventilation and air-conditioning system of the laboratory.

The slow leaks of air into and out of the box have the beneficial effect of regulating the bulk temperature of the nearly motionless air contained inside the box. This can be measured with one or more temperature sensors (e.g., thermometers), and the Rayleigh number can be calculated,

$$\text{Ra}_H = \frac{g\beta(T_\infty - T_w)H^3}{\alpha v \cdot} \tag{1}$$

by noting that $T_w = 0°C$, and $\beta/\alpha v$ is a group of air properties evaluated at the film temperature $(T_\infty + T_w)/2$.

Explanation.   The instructional value of this very simple experiment lies in the visual observations afforded by the melting slab (Fig. PR7.1*b*) and in the effort of predicting (anticipating theoretically) the melting rate $\dot{m}$. The four photographs juxtaposed in Fig. PR7.1*b* were taken at 25, 70, 115, and 160 minutes from the start of the melting process. The ruler suspended on the left side of the figure indicates inches.

Two long windows cut into the side of the box, and covered with plastic wrapping material (of the kind used in the kitchen) allow the experimentalist to look in the direction parallel to the large surfaces of area $H \times W$. As time passes, you will observe that the slab melts unevenly: near its upper edge, the slab becomes considerably narrower than over the remainder of its height.

This observation finds an explanation in the argument that the change of phase at the slab surface is driven mainly by the local heat flux from the T_∞ air to the T_w surface across a boundary layer of cold air that descends along the wet ice surface:

$$q''(y) = \rho_w h_{sf} \frac{d}{dt} \left(\frac{L(y)}{2} \right) \tag{2}$$

In this equation, $\rho_w \cong 1$ g/cm^3 is the density of water at $0°$C and $h_{sf} = 333.4$ J/g is the latent heat of melting of ice. The uneven shape of the instantaneous slab half-thickness $L(y)/2$ is the time integral effect of the local free-convection heat flux $q''(y)$. Recall that in the laminar regime, q'' decreases as $y^{-1/4}$ in the downstream direction.

The meltwater flowrate $\dot{m}$ can be predicted by noting its geometric definition,

$$\dot{m} = \rho_w HW \left(\frac{\overline{dL}}{dt} \right) \tag{3}$$

where $\overline{dL/dt}$ is the rate of slab thinning averaged over the height H. Combined, eqs. (2) and (3) yield

$$\dot{m} = 2W \frac{k}{h_{sf}} (T_\infty - T_w) \, \overline{\mathrm{Nu}_H} \tag{4}$$

where $\overline{\mathrm{Nu}_H}$ is the overall Nusselt number based on the height averaged heat flux $\overline{q''}$:

$$\overline{\mathrm{Nu}_H} = \frac{\bar{h}H}{k} = \frac{\overline{q''}H}{(T_\infty - T_w)k} \tag{5}$$

In eqs. (4) and (5), the symbol k is the thermal conductivity of air evaluated at the film temperature. The overall Nusselt number can be calculated using the appropriate formula $\overline{\mathrm{Nu}_H}(\mathrm{Ra}_H, \mathrm{Pr})$ for laminar boundary layer natural convection over a vertical plane wall.

It is likely that the experimentally measured melting rate $\dot{m}$ will be greater than the value anticipated using eq. (4). This discrepancy should be commented on. It may be caused, for example, by the effect of melting along the four narrow surfaces of the slab, and by the radiation heat transfer between the slab and the room-temperature wall of the cardboard box. The radiation effect can be estimated in an order-of-magnitude sense by using the two-surface gray enclosure theory (Section 10.4.4).

Of particular interest in ice storage applications and meteorology is the relationship between the melting rate and the bulk temperature of the surrounding air. The present experiment reveals this information as a relationship between $\dot{m}$ and Ra_H. It is considerably easier to vary the Rayleigh number by changing the vertical dimension of the ice slab, as opposed to changing the imposed temperature difference $T_\infty - T_w$.

The $\dot{m} - \mathrm{Ra}_H$ curve can be pinpointed by several students (or groups of students) who monitor the melting of ice slabs with several different slab heights H. Worth noting is that a single baking pan can produce ice slabs for two different Ra_H cases, depending on which side of the large rectangular surface is oriented vertically during the experiment. By using ice slabs with heights H in the range $0.25 - 1.25$ m, it is possible to vary Ra_H by two orders of magnitude in the range $10^7 - 10^9$.

It is recommended that at the end of the experimental session each student report his or her experimental point $(\dot{m}, \mathrm{Ra}_H)$ on one line in the table below. After the session and before handing in the individual experimental report, the student should analyze the completed table. One question to pursue is whether the $\dot{m}$ (Ra_H) curve is anticipated adequately by the theory of eqs. (2)–(5). Furthermore, by plotting the $\dot{m}$ and Ra_H data in dimensionless logarithmic coordinates M and Ra_H, where

$$M = \frac{\dot{m}h_{sf}}{Wk(T_\infty - T_w)} \qquad (6)$$

it is possible to tell if M is proportional to $Ra_H^{1/4}$, in accordance with laminar boundary layer theory.

This final exercise gives each student a global perspective on the success of the experimental session, for example, on who agrees with whom, and why. It also teaches that disagreements between two or more experimentalists are very much part of the game, and, if these disagreements cannot be reasoned away, it is time to devise a better theory!

The melting of ice objects of other shapes can be studied in a similar manner. Heat transfer correlations for the melting of objects of diverse shapes in air have been developed by Janna and Jakubowski [65].

Student's Name	H (cm)	W (cm)	T_∞ (°C)	$\dot{m}$ (g/h)	M	Ra_H

7.2   Optimal Spacing Between Vertical Fins

The heat exchanger tube mounted horizontally in the back of a refrigerator is to be fitted with a number of vertical rectangular plate fins of height H, width W, and fin-to-fin spacing D. The fin surface may be modeled as isothermal at temperature T_0; the ambient fluid is air at temperature T_∞.

The purpose of this design work is to maximize the total heat transfer rate q between the tube of length L and the ambient air (Bejan [5], p. 157). The only design variable is the number of fins, that is, fin-to-fin spacing D. The parameters H, W, T_0, T_∞, and L are assumed given, and the thickness of the fin plate is considered negligible. Also negligible is the direct heat transfer between the unfinned portions of the tube and the ambient.

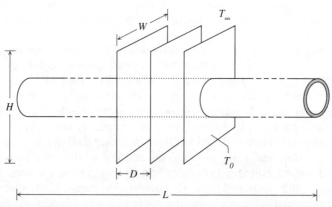

Figure PR7.2

Procedure. The simplest way to approach this optimization problem is by considering the two possible extremes:

(a) A sufficiently small D, so that the air (chimney) flow through the fin-to-fin spacing is fully developed, eq. (7.94); and

(b) A sufficiently large D, so that the thickness of the air boundary layer on one side of the fin is smaller than D; in this second limit the heat transfer rate from one vertical surface can be calculated with eq. (7.55').

Determine analytically the total heat transfer rate q in these two extremes. Rely on the intersection of these two asymptotes to determine approximately the optimal spacing D for which q is maximum:

$$\frac{D_{\text{opt}}}{H} \sim 2.3 \text{Ra}_H^{-1/4}$$

I first composed and solved the problem proposed in this project as one of several original fundamental research problems in my 1984 graduate course on convection (Bejan [5], p. 157). In the same year the same problem was treated using a different approach by Bar-Cohen and Rohsenow [66], with application to the natural convection cooling of electronic equipment. Reviews of the literature on the fundamentals of electronic equipment cooling have been published by Incropera [67] and Peterson and Ortega [68].

Note also that this optimal spacing design question is the natural-convection counterpart of the forced-convection question posed as Project 6.3.

7.3 The Tapered Glass Window: An Energy Saving Feature in Buildings?

Evaluate the proposal to install windows with glass panes of nonuniform thickness as a means of minimizing the heat leak from a room to the cold outside air. This proposal was stimulated by the thought that since the room-side heat transfer coefficient is higher near the top of the window (at the start of the descending boundary layer), it is there that a thicker glass layer can have the greatest effect on reducing the local heat flux [69]. This thought points toward the tapered glass design illustrated on the left side of the figure. The thickness of the glass layer has been exaggerated to make the notation clearer.

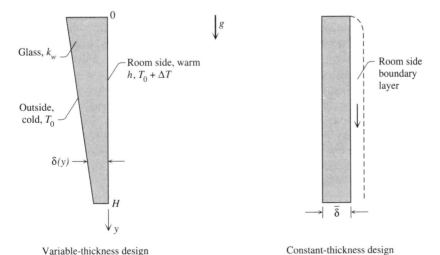

Variable-thickness design   Constant-thickness design

Figure PR7.3

One way to evaluate the merit of the proposed design is to compare the total heat leak through the window, q' (W/m), with the corresponding heat leak through a constant-thickness glass window. The amount of glass used in the variable-thickness design is the same as in the constant-thickness (or reference) design. For the tapered glass window assume that the glass thickness decreases linearly in the downward direction:

$$\delta = \bar{\delta} + b \left(\frac{1}{2} - \xi \right)$$

In this expression, $\xi = y/H$, and the taper parameter $b = -d\delta/d\xi$ is a design variable that must be determined optimally. The H-averaged thickness $\bar{\delta}$ is fixed because the window height H and the glass volume are fixed.

For the heat transfer coefficient on the room-air side assume a y dependence consistent with that found in laminar natural convection boundary layers,

$$h = h_{\min} \xi^{-1/4}$$

where $h_{\min}$ is the smallest h value that occurs at the bottom of the window, where the room-side boundary layer is the thickest. The heat transfer coefficient on the outside of the glass layer is sufficiently large so that the temperature of that surface is equal to the atmospheric temperature.

To have access to a bird's-eye view of the merit of the proposed design relative to the reference design, determine numerically the ratio q'/q'_{ref} as a function of two dimensionless groups, the taper parameter

$$S = \frac{H}{\bar{\delta}} \left(-\frac{d\delta}{dy} \right)$$

and the bottom-end Biot number

$$\text{Bi} = \frac{h_{\min} \bar{\delta}}{k_w}$$

Determine the best taper parameter S for the smallest q'/q'_{ref} ratio for a fixed Bi. By means of a numerical example, determine the Bi range in which a common window is likely to operate. Comment on the practicality of the heat leak reduction promised by the tapered glass design.

7.4 Combined Natural Convection and Conduction: Critical Insulation Radius

In the derivation of the critical insulation radius for a cylinder (Section 2.4), it was assumed that the average heat transfer coefficient h is a constant. We learned in this chapter that in natural convection h depends not only on the vertical size of the body but also on the wall-fluid temperature difference.

Reconsider the heat transfer through a layer of insulation wrapped around a bare horizontal cylinder of radius r_i, length L, and temperature T_i. The outer radius of the layer of insulation is r_o, the insulation conductivity is k, and the temperature of the surrounding fluid reservoir is T_∞. Assume that the natural convection boundary layer is laminar so that the overall Nusselt number based on $D = 2r_o$ is proportional to $\text{Ra}_D^{1/4}$. In other words, write

$$\frac{hD}{k_{\text{fluid}}} = C \left[\frac{g\beta (T_o - T_\infty) D^3}{\alpha \nu} \right]^{1/4}$$

where C is a constant and T_o is the temperature at the outer radius of the insulation layer.

(a) Develop a method for calculating numerically the total heat transfer rate through the insulation q, as a function of the outer radius r_o. Since this will require using a trial-and-error procedure, it is advisable to employ the following dimensionless variables

$$Q = \frac{q}{2\pi kL \, (T_i - T_\infty)} \qquad \text{Bi} = \frac{h_* \, r_i}{k}$$

$$\rho = \frac{r_o}{r_i} \qquad \theta = \frac{T_o - T_\infty}{T_i - T_\infty}$$

In the Bi definition, h_* is the average heat transfer coefficient in the case where the insulation is absent ($D = 2r_i$, $T_o = T_i$).

(b) For one or more Bi values smaller than 0.5, determine numerically the curve Q vs. ρ. Report the critical insulation radius ($\rho_c = r_{o,c}/r_i$) for which Q reaches its maximum.

(c) Compare the calculated ρ_c values with those predicted by the classical constant-h model of eq. (2.44). How small must the Biot number be for eq. (2.44) to be sufficiently accurate?

7.5 The Suppression of Bénard Convection by Means of a Horizontal Partition

The thermal insulation capability of a horizontal layer of fluid is impaired if natural convection currents are present. As shown in Fig. 7.21, the heat transfer coefficient is lower when convection is absent, and the transfer of heat from the bottom wall to the top wall is by pure conduction.

Consider the design of a thermal insulation that consists of a horizontal layer of fluid of thickness H and bottom-to-top temperature difference $T_h - T_c = \Delta T$. These two parameters, H and ΔT, happen to be large enough so that convection currents would form in the fluid. To suppress the formation of these currents, it is proposed to install a horizontal partition at some level between the bottom wall and the top wall. What is the optimal level at which the partition should be installed?

To simplify your analysis, assume that the partition can be modeled as an isothermal wall with a temperature between the bottom wall temperature and the top wall temperature. Assume further that convection currents are absent above and below the partition. Find the optimal partition level by maximizing the overall temperature difference ΔT for which this state of pure conduction can be preserved.

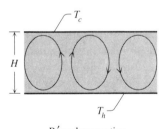

Bénard convection

Conduction on both sides
of the partition

Figure PR 7.5

8

CONVECTION WITH
CHANGE OF PHASE

8.1 CONDENSATION HEAT TRANSFER

8.1.1 Laminar Film on a Vertical Surface

The common thread of the convection problems treated in Chapters 5, 6, and 7 is that the fluid—the convective medium—remains in its original single-phase state, in spite of the heating or cooling that it experiences. In the present chapter we take a look at convection phenomena in which the fluid undergoes a change of phase. Early cases of phase change under the influence of heat transfer were the melting and solidification processes discussed in Section 4.7, as examples of time-dependent unidirectional conduction. The phase change phenomena detailed in the present chapter are all caused by convection.

Perhaps the simplest convection phase change process is the condensation of a vapor on a cold vertical surface, Fig. 8.1. The film of condensate that forms on the vertical surface can have three distinct regions. The *laminar* section is near the top, where the film is the thinnest. The film thickness increases in the downward direction, as more and more of the surrounding vapor condenses on the exposed surface of the film. There comes a point where the film becomes thick enough to show the first signs of transition to a nonlaminar flow regime. In this *wavy* flow region the visible surface of the film shows a sequence of regular ripples. Finally, if the wall extends sufficiently far downward, the film enters and remains in the *turbulent* region, where the ripples appear irregular in both space and time.

The "transitions" from one flow regime to the next will be described in more precise terms in the discussion of Fig. 8.5. At this stage it is instructive to see the similarity between the three-regime sequence of the vertical film (Fig. 8.1, left) and the flow regimes of a single-phase natural convection boundary layer (rotate Fig. 7.6 by 180°). Indeed, the flow and development of the laminar film described in this section are conceptually analogous to what we learned in Sections 7.2.2 and 7.2.3. The laminar film of condensate is, in fact, a laminar boundary layer flow, where the vertical surface is so cold that it causes the change of phase (condensation) of the descending boundary layer flow.

On the right side of Fig. 8.1 we see that even in the laminar film region,

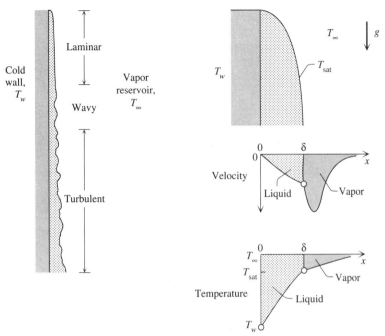

Figure 8.1 Flow regimes of the film of condensate on a cooled vertical surface.

which is the simplest of the three regions, the flow of the liquid film generally interacts with the descending boundary layer of cooled vapor. The temperature of the liquid–vapor interface is the saturation temperature that corresponds to the local pressure along the wall, T_{sat}. The saturation temperature is sandwiched between the temperature of the isothermal vapor reservoir, T_∞, and the wall temperature T_w. Through the shear stress at the liquid–vapor interface, the descending jet of vapor aids the downward flow of the liquid film. The vapor in the descending jet is colder than the vapor reservoir and warmer than the liquid in the film attached to the wall.

This two-phase flow is considerably more complicated in the wavy and turbulent sections of the wall. To make matters worse, in applications where the film is sufficiently long to exhibit all three regimes, the overall heat transfer rate from the vapor reservoir to the wall is dominated by the contributions made by the wavy and turbulent sections. The same can be said about the total rate of condensation, which, as we shall learn in eq. (8.21), is proportional to the overall heat transfer rate from the vapor to the vertical wall.

For these reasons, in this opening section we focus on the simplest region of the film — the laminar region. We do this to define the proper terminology and to demonstrate analytically the relationship between the condensation of vapor at the interface, the vertical flow of the growing film of liquid, and the heat transfer rate absorbed by the cold vertical wall. Later, we shall rely on this set of fundamentals as we review the recommended results for the calculation of condensation and heat transfer rates in more complex configurations.

Consider the two-dimensional laminar film sketched in Fig. 8.2, in which the distance y measures the downward length of the film. This flow is considerably simpler than the one seen in Fig. 8.1, because this time the entire reservoir of

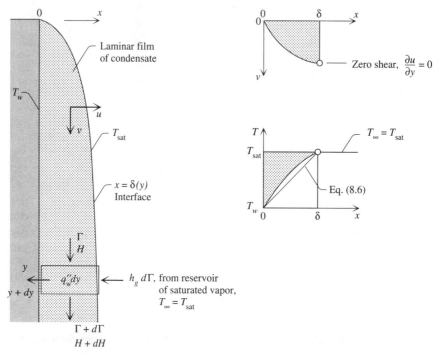

Figure 8.2 Laminar film of condensate, supplied by a reservoir of stationary saturated vapor.

vapor is isothermal at the saturation pressure, T_{sat}. The merit of this simplification is that it allows us to focus exclusively on the flow of the liquid film and to neglect (as much as is permissible) the movement of the nearest layers of vapor.

The analysis of the flow of liquid begins with the steady-state version of the momentum equations (M_x) and (M_y) of Table 5.1, which in the case of a *slender* film (i.e., in boundary layer-type flow) reduce to the single equation

$$\rho_l \left(u \frac{\partial v}{\partial x} + v \frac{\partial v}{\partial y} \right) = -\frac{dP}{dy} + \mu_l \frac{\partial^2 v}{\partial x^2} + \rho_l g \tag{8.1}$$

The last term on the right side represents the body force experienced by each small packet of liquid. Because of the slenderness of the film, the vertical pressure gradient in the liquid is the same as the hydrostatic pressure gradient in the outside vapor* [1], $dP/dy = \rho_v g$. Equation (8.1) can now be rewritten to show that the net sinking force felt by the liquid is, in general, resisted by a combination of the effects of friction and inertia:

$$\rho_l \underbrace{\left(u \frac{\partial v}{\partial x} + v \frac{\partial v}{\partial y} \right)}_{\text{Inertia}} = \underbrace{\mu_l \frac{\partial^2 v}{\partial x^2}}_{\text{Friction}} + \underbrace{g(\rho_l - \rho_v)}_{\text{Sinking effect}} \tag{8.2}$$

---

*Review the "second" simplification of the boundary layer momentum equations in Chapter 5, specifically Eqs. (5.30) and (5.31).

We continue the analysis by assuming that the inertia effect is small when compared with the effect of friction, and we set the left side of eq. (8.2) equal to zero. The resulting equation can be integrated twice in x and subjected to the conditions of no slip at the wall ($v = 0$ at $x = 0$) and zero shear at the liquid–vapor interface ($\partial v / \partial x = 0$ at $x = \delta$). The solution for the vertical liquid velocity profile is

$$v(x,y) = \frac{g}{\mu_l}(\rho_l - \rho_v)\delta^2 \left[\frac{x}{\delta} - \frac{1}{2}\left(\frac{x}{\delta}\right)^2\right] \qquad (8.3)$$

in which the film thickness is an unknown function of longitudinal position, $\delta(y)$.

Related to the parabolic velocity profile (8.3) is the local mass flowrate through a cross section of the film,

$$\Gamma(y) = \int_0^\delta \rho_l u \, dx = \frac{g\rho_l}{3\mu_l}(\rho_l - \rho_v)\delta^3 \qquad (8.4)$$

The mass flowrate is traditionally labeled Γ (kg/s·m), and is expressed per unit length in the direction normal to the plane of Fig. 8.2. A less unusual symbol for this quantity would be $\dot{m}'$. The downward velocity and the flowrate are obviously proportional to the sinking effect $g(\rho_l - \rho_v)$; they are also inversely proportional to the liquid viscosity.

The film thickness $\delta(y)$ can be determined by invoking the first law of thermodynamics in connection with the control volume $\delta \times dy$ shown on the lower-left side of Fig. 8.2. Entering this control volume from the right is the saturated vapor stream $d\Gamma$, whose enthalpy flowrate is $h_g d\Gamma$. The vertical enthalpy inflow (W/m) associated with the mass flowrate Γ is

$$H = \int_0^\delta \rho_l v [h_f - c_{P,l}(T_{\text{sat}} - T)] dx \qquad (8.5)$$

where the quantity in the square brackets is the local specific enthalpy (kJ/kg) of the liquid at the point (x,y). Since the liquid is slightly subcooled ($T < T_{\text{sat}}$), its specific enthalpy is smaller than the specific enthalpy of saturated liquid, h_f. As was done originally by Nusselt [2], we assume that the local temperature T is distributed approximately linearly across the film,

$$\frac{T_{\text{sat}} - T}{T_{\text{sat}} - T_w} \cong 1 - \frac{x}{\delta} \qquad (8.6)$$

and, after using eqs. (8.3) and (8.6) in the integral (8.5), we obtain

$$H = \left[h_f - \frac{3}{8}c_{P,l}(T_{\text{sat}} - T_w)\right]\Gamma \qquad (8.7)$$

Finally, in accordance with the linear temperature profile assumption (8.6), we note that the heat flux absorbed by the solid boundary of the control volume (i.e., the wall) is

$$q_w'' \cong k_l \frac{T_{\text{sat}} - T_w}{\delta} \qquad (8.8)$$

The steady-state form of the first law of thermodynamics for the $\delta \times dy$ system is

$$0 = H - (H + dH) + h_g d\Gamma - q''_w dx \tag{8.9}$$

or, after using eqs. (8.7) and (8.8),

$$\frac{k_l}{\delta}(T_{\text{sat}} - T_w)dx = \underbrace{\left[h_{fg} + \frac{3}{8}c_{P,l}(T_{\text{sat}} - T_w) \right]}_{h'_{fg}} d\Gamma \tag{8.10}$$

Starting with this equation, we use h'_{fg} as shorthand for the *augmented* latent heat of condensation, which includes the proper latent heat, h_{fg}, and a contribution accounting for the cooling of the fresh condensate to temperatures below T_{sat}. Combined with the Γ expression (8.4), this equation becomes

$$\frac{k_l \nu_l (T_{\text{sat}} - T_w)}{h'_{fg} g(\rho_l - \rho_v)} dy = \delta^3 d\delta \tag{8.11}$$

and, after integrating from $y = 0$ where $\delta = 0$,

$$\delta(y) = \left[x\frac{4k_l \nu_l (T_{\text{sat}} - T_w)}{h'_{fg} g(\rho_l - \rho_v)} \right]^{1/4} \tag{8.12}$$

In conclusion, the thickness of the laminar film increases as the longitudinal length raised to the power $\frac{1}{4}$, that is, in the same way as the thickness (thermal, or velocity) of the vertical laminar boundary layer in a single-phase fluid, eqs. (7.25c) and (7.31). Knowing $\delta(y)$, we can calculate, in order, the local heat transfer coefficient

$$h_y = \frac{q''_w}{T_{\text{sat}} - T_w} = \frac{k_l}{\delta} = \left[\frac{k_l^3 h'_{fg} g(\rho_l - \rho_v)}{4y\nu_l(T_{\text{sat}} - T_w)} \right]^{1/4} \tag{8.13}$$

the average* heat transfer coefficient for a film of height L,

$$\bar{h}_L = \frac{h_{y=L}}{1 + (-1/4)} = \frac{4}{3}h_{y=L} \tag{8.14}$$

and the overall Nusselt number based on the L-averaged heat transfer coefficient,

$$\overline{\text{Nu}}_L = \frac{\bar{h}_L L}{k_l} = 0.943 \left[\frac{L^3 h'_{fg} g(\rho_l - \rho_v)}{k_l \nu_l (T_{\text{sat}} - T_w)} \right]^{1/4} \tag{8.15}$$

Regarding the physical meaning of the dimensionless group formed on the right side of eq. (8.15), we note that it is nearly equal to the geometric slenderness ratio of the liquid film [cf. eq. (8.12)]:

$$\frac{L}{\delta(L)} = 0.707 \left[\frac{L^3 h'_{fg} g(\rho_l - \rho_v)}{k_l \nu_l (T_{\text{sat}} - T_w)} \right]^{1/4} \tag{8.16}$$

In numerical applications, the liquid properties that appear in these formulas are best evaluated at the average film temperature $(T_w + T_{\text{sat}})/2$. The latent heat of condensation h_{fg} is found in thermodynamic tables of saturated-state

---

*Review the footnote on page 244.

properties and takes the value that corresponds to the phase change temperature T_{sat}. Rohsenow [3] refined the preceding analysis by avoiding the linear profile assumption (8.6), and by performing an integral analysis of the temperature distribution across the film. He found a temperature profile whose curvature increases with the degree of liquid subcooling, $c_{P,l}(T_{sat} - T_w)$. In place of the modified latent heat h'_{fg} defined under eq. (8.10), Rohsenow recommended using

$$h'_{fg} = h_{fg} + 0.68c_{P,l}(T_{sat} - T_w) \tag{8.17}$$

This expression is recommended also for calculations involving the wavy and turbulent flow regimes. It can be rewritten as

$$h'_{fg} = h_{fg}(1 + 0.68\text{Ja}) \tag{8.18}$$

in which the *Jakob* number* Ja is a relative measure of the degree of subcooling experienced by the liquid film:

$$\text{Ja} = \frac{c_{P,l}(T_{sat} - T_w)}{h_{fg}} \tag{8.19}$$

This group is analogous to the Stefan number defined in eq. (4.119), which in melting and solidification problems accounts for the degree of liquid superheating or solid subcooling.

To summarize, the total heat transfer rate absorbed by the wall, per unit length in the direction normal to the plane of Fig. 8.2, is

$$q' = \overline{h}_l L(T_{sat} - T_w)$$
$$= k_l(T_{sat} - T_w)\overline{\text{Nu}_L} \tag{8.20}$$

The total (bottom-end) flowrate $\Gamma(L)$ can be calculated by substituting $y = L$ in what results from combining eqs. (8.4) and (8.12). In this way it is easy to show that the total condensation rate $\Gamma(L)$ is proportional to the total cooling rate provided by the vertical wall:

$$\Gamma(L) = \frac{q'}{h'_{fg}} = \frac{k_l}{h'_{fg}}(T_{sat} - T_w)\overline{\text{Nu}_L} \tag{8.21}$$

The global eqs. (8.20) and (8.21) hold true for the entire film, not only for the laminar section. Rewritten as $q' = \Gamma(L)h_{fg}(1 + 0.68\text{Ja})$, eq. (8.21) shows also that the cooling rate q' increases with both the latent heat h_{fg} and the degree of liquid subcooling, Ja. This trend should have been expected, because the cooling provided by the wall causes both the condensation of vapor at the $x = \delta$ interface, and the cooling of the newly formed liquid to temperatures below T_{sat}.

The physical properties needed for evaluating the Nusselt number of eq. (8.15) are listed in the next section in Table 8.2 and in Appendix C. Worth noting is that in many instances ρ_l is much greater than ρ_v, so that $\rho_l - \rho_v$ can safely be replaced with ρ_l.

---

*Max Jakob (1879–1955) was a leading figure in German heat transfer engineering. He emigrated to the United States in 1936, and became a professor at the Illinois Institute of Technology. Through his research on boiling and condensation and his famous treatise on *Heat Transfer* (Wiley, 1949), he played a major role in the shaping of heat transfer as a core discipline in modern mechanical engineering.

The laminar film results discussed until now were derived by Nusselt [2] based on the assumption that the effect of inertia is negligible in the momentum balance (8.2). The complete momentum equation was used by Sparrow and Gregg [4] in a similarity formulation of the same problem. Their solution for $\overline{Nu}_L$ falls below Nusselt's eq. (8.15) when the Prandtl number is smaller than 0.03 and the Jakob number is greater than 0.01.

In a subsequent analysis, Chen [5] abandoned the assumption of zero shear at the liquid–vapor interface (Fig. 8.2, right) while retaining the effect of inertia in the momentum equation. The vapor was assumed saturated and stagnant *sufficiently far* from the interface. Next to the interface, the vapor is dragged downward by the falling film of condensate and forms a velocity boundary layer that bridges the gap between the downward velocity of the interface and the zero velocity of the outer vapor (see the small detail above Fig. 8.3).

Chen's chart [5] for calculating the overall Nusselt number $\overline{Nu}_L$ is reproduced in Fig. 8.3. Especially at low Prandtl numbers, the $\overline{Nu}_L$ values read off Fig. 8.3 are smaller than those furnished by Sparrow and Gregg's solution [4], and agree better with experimental data. The lower $\overline{Nu}_L$ values are due to the additional restraining effect that the vapor drag has on the downward acceleration of the liquid film.

The scale analysis of the laminar film condensation problem (Ref. [1], pp. 146–151) showed that the fall of the liquid film is restrained by friction when $Pr_l > Ja$, and by inertia when $Pr_l < Ja$. The group that marks the transition from one type of flow to the other is the ratio* Pr_l/Ja. Indeed, if we use the group Pr_l/Ja on the abscissa of Fig. 8.4, the low-Pr_l information of Fig. 8.3 is correlated well by the single curve shown in Fig. 8.4.

---

*Recall that in single-phase natural convection (Figs. 7.2 and 7.3) the parameter that describes this transition is the Prandtl number alone.

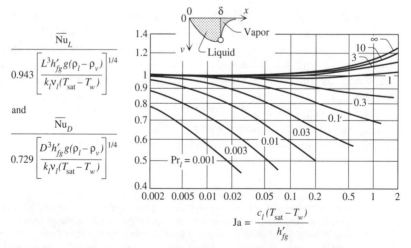

Figure 8.3 Effect of Prandtl number on heat transfer from a laminar film of condensate on a vertical wall [5] or on a single horizontal cylinder.

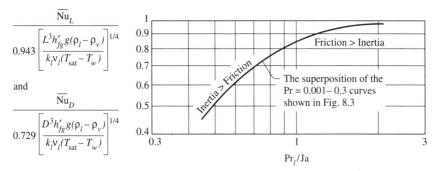

$$\frac{\overline{\mathrm{Nu}}_L}{0.943\left[\dfrac{L^3 h'_{fg} g(\rho_l - \rho_v)}{k_l \nu_l (T_{sat} - T_w)}\right]^{1/4}}$$

and

$$\frac{\overline{\mathrm{Nu}}_D}{0.729\left[\dfrac{D^3 h'_{fg} g(\rho_l - \rho_v)}{k_l \nu_l (T_{sat} - T_w)}\right]^{1/4}}$$

Figure 8.4 Transition from inertia-restrained to friction-restrained film condensation on a vertical wall or on a single horizontal cylinder [1].

Example 8.1

Laminar Film Condensation, Vertical Surface

A plane vertical wall of temperature 60°C faces a space filled with stagnant saturated steam at atmospheric pressure. The height of the wall is $L = 2$ m. Assume that the film is laminar, and calculate the rate at which steam condenses on the vertical surface.

Solution.   The relevant properties of water at atmospheric pressure and film temperature $(100°C + 60°C)/2 = 80°C$ are (see Appendix C)

$$c_{P,l} = 4.196 \frac{\mathrm{kJ}}{\mathrm{kg \cdot K}} \qquad \rho_l = 972 \frac{\mathrm{kg}}{\mathrm{m}^3}$$

$$k_l = 0.67 \frac{\mathrm{W}}{\mathrm{m \cdot K}} \qquad \nu_l = 3.66 \times 10^{-7} \frac{\mathrm{m}^2}{\mathrm{s}}$$

The latent heat of condensation at atmospheric pressure (or at 100°C) is $h_{fg} = 2257$ kJ/kg. The Jakob number is small,

$$\mathrm{Ja} = \frac{c_{P,l}(T_{sat} - T_w)}{h_{fg}} = 0.074$$

and this means that h'_{fg} is nearly the same as h_{fg}:

$$h'_{fg} = h_{fg}(1 + 0.68\mathrm{Ja}) = 1.05 h_{fg}$$
$$= 2370.6 \,\mathrm{kJ/kg}$$

The total (bottom-end) condensation rate $\Gamma(L)$ is given by eq. (8.21). The overall Nusselt number $\overline{\mathrm{Nu}}_L$ can be calculated using eq. (8.15) by assuming that the film is laminar over its entire height:

$$\frac{L^3 h'_{fg} g(\rho_l - \rho_v)}{k_l \nu_l (T_{sat} - T_w)} = \frac{(8\,\mathrm{m}^3)\left(2370.6\dfrac{\mathrm{kJ}}{\mathrm{kg}}\right)\left(9.81\dfrac{\mathrm{m}}{\mathrm{s}^2}\right)\left(972\dfrac{\mathrm{kg}}{\mathrm{m}^3}\right)}{\left(0.67\dfrac{\mathrm{W}}{\mathrm{m \cdot K}}\right)\left(3.66 \times 10^{-7}\dfrac{\mathrm{m}^2}{\mathrm{s}}\right)(100 - 60)\mathrm{K}} = 1.844 \times 10^{16}$$

$$\overline{\mathrm{Nu}}_L = 0.943(1.844 \times 10^{16})^{1/4} = 10{,}988 \qquad\qquad (8.15)$$

$$\Gamma(L) = \frac{k_l}{h'_{fg}}(T_{sat} - T_w)\overline{Nu}_L$$

$$= \frac{0.67 W}{m \cdot K} \frac{kg}{2370.6 \, kJ}(40 \, K)(10{,}988) \qquad (8.21)$$

$$= 0.124 \, kg/m \cdot s$$

In Example 8.2 we learn that the laminar film assumption is inappropriate and that most of the film is wavy and turbulent. For the same reason, the actual condensation rate will be higher than the value calculated above.

8.1.2 Turbulent Film on a Vertical Surface

The liquid film becomes wavy and, farther downstream, turbulent when the order of magnitude of its local Reynolds number is greater than 10^2 (see Appendix F). The local Reynolds number of the liquid film can be constructed in the form of the group $\rho_l \bar{u} \delta / \mu_l$, in which δ is the local thickness, and $\bar{u}$ is the scale representative of the local downward velocity. And since the product $\rho_l \bar{u} \delta$ is of the same order of magnitude as the local liquid mass flowrate Γ, the local Reynolds number can be expressed also as the ratio Γ/μ_l. This is why in the field of condensation heat transfer the local Reynolds number of the liquid film has historically been defined as

$$Re_y = \frac{4}{\mu_l}\Gamma(y) \qquad (8.22)$$

The flowrate $\Gamma(y)$ and the Reynolds number Re_y increase in the downstream direction. Experimental observations of the condensate indicate that the laminar section of the film expires in the general vicinity of $Re_y \sim 30$. The film can be described as wavy in the segment corresponding approximately to $30 \lesssim Re_y \lesssim 1800$. Farther downstream, the film appears turbulent. The succession of these flow regimes is illustrated on the abscissa of Fig. 8.5.

Experiments have revealed also that the heat transfer rate in the wavy and turbulent sections is considerably larger than the estimate based on the laminar film analysis, eq. (8.15) and Fig. 8.3. The sizeable record of experimental data and correlations on condensation heat transfer in the wavy and turbulent regimes was reviewed recently by Chen, Gerner, and Tien [6] and Chun and Kim [7]. Chen et al. [6] developed and recommended the following correlation for the *average* heat transfer coefficient for an L-tall film that may have wavy and turbulent regions

$$\frac{\bar{h}_L}{k_l}\left(\frac{\nu_l^2}{g}\right)^{1/3} = [Re_L^{-0.44} + (5.82 \times 10^{-6})Re_L^{0.8}Pr_l^{1/3}]^{1/2} \qquad (8.23)$$

Figure 8.5 stresses that this correlation applies only above $Re_L \sim 30$. Equation (8.23) agrees within ± 10 percent with measurements in experiments where the vapor was stagnant (or slow enough) so that the effect of shear at the interface was negligible. Below $Re_L \sim 30$, the recommended average heat transfer formula is eq. (8.15), which, when $\rho_l \gg \rho_v$, can be projected on Fig. 8.5 as the line

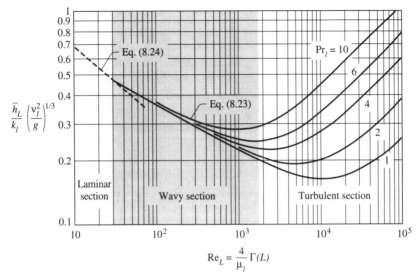

Figure 8.5 The L-averaged heat transfer coefficient for laminar, wavy, and turbulent film condensation on a vertical surface.

$$\frac{\bar{h}_L}{k_l}\left(\frac{v_l^2}{g}\right)^{1/3} = 1.468\mathrm{Re}_L^{-1/3} \tag{8.24}$$

The usual engineering unknown in the vertical film condensation problem is the total condensation rate $\Gamma(L)$ or, alternatively, Re_L. Unfortunately, this unknown influences both sides of eq. (8.23), or both the ordinate and abscissa parameters of Fig. 8.5. Instead of the trial-and-error procedure that goes with using eq. (8.23) or Fig. 8.5, it is more convenient to rewrite the ordinate parameter of Fig. 8.5 as

$$\frac{\bar{h}_L}{k_l}\left(\frac{v_l^2}{g}\right)^{1/3} = \frac{\mathrm{Re}_L}{B} \tag{8.25}$$

where B is a new dimensionless group that is proportional to the physical quantities that, when increasing, tend to augment the condensation rate, namely, L and $T_{\mathrm{sat}} - T_w$:

$$B = L(T_{\mathrm{sat}} - T_w)\frac{4k_l}{\mu_l h'_{fg}}\left(\frac{g}{v_l^2}\right)^{1/3} \tag{8.26}$$

The B group can be viewed as the *driving parameter for film condensation.* Equation (8.25) is a consequence of the global statements (8.20)–(8.21). It allows us to rewrite eqs. (8.23) and (8.24) as

$$B = \mathrm{Re}_L[\mathrm{Re}_L^{-0.44} + (5.82 \times 10^{-6})\mathrm{Re}_L^{0.8}\mathrm{Pr}_l^{1.3}]^{-1/2} \tag{8.27}$$

$$B = 0.681\mathrm{Re}_L^{4/3} \tag{8.28}$$

Figure 8.6 displays this information by using the unknown Re_L on the abscissa and the driving parameter B on the ordinate. We now see directly how the condensation rate and the bottom-end Reynolds number increase monotonically as the driving parameter increases. The condensation rate increases faster when the film length is dominated by the turbulent regime.

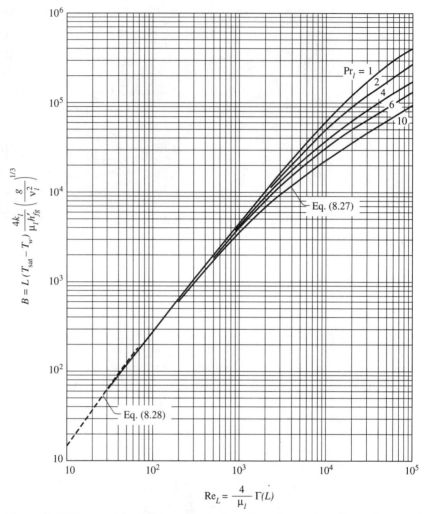

Figure 8.6 Film condensation on a vertical surface: the total condensation rate (or Re_L) as a function of the condensation driving parameter B.

Example 8.2

Turbulent Film Condensation, Vertical Surface

A plane vertical wall of temperature 60°C faces a space filled with stagnant saturated steam at atmospheric pressure. The wall height is $L = 2$ m. Calculate the rate at which steam condenses on the vertical surface.

Solution. The properties of water at atmospheric pressure and film temperature (100°C + 60°C)/2 = 80°C are (see Example 8.1 and Appendix C) as follows:

$$c_{Pl} = 4.196 \frac{kJ}{kg \cdot K} \qquad \mu_l = 3.55 \times 10^{-4} \frac{kg}{m \cdot s}$$

$$k_l = 0.67 \frac{W}{m \cdot K} \qquad \nu_l = 3.66 \times 10^{-7} \frac{m^2}{s}$$

The vertical axis label reads: $B = L(T_{sat} - T_w) \frac{4k_l}{\mu_l h'_{fg}} \left(\frac{g}{\nu_l^2}\right)^{1/3}$

Horizontal axis: $Re_L = \frac{4}{\mu_l} \dot{\Gamma}(L)$

Curves labeled $Pr_l = 1, 2, 4, 6, 10$. Eq. (8.27) and Eq. (8.28) marked.

$$h'_{fg} = 2370.6 \frac{\text{kJ}}{\text{kg}} \qquad \text{Pr}_l = 2.23$$

For calculating the condensation rate $\Gamma(L)$ we have two options: Fig. 8.6 and, depending on the Re_L range, eqs. (8.27)–(8.28). In either case we must begin by calculating the driving parameter

$$B = L(T_{\text{sat}} - T_w) \frac{4k_l}{\mu_l h'_{fg}} \left(\frac{g}{v_l^2} \right)^{1/3}$$

$$= 2 \text{ m } (100-60)\text{K} \left(\frac{4 \times 0.67 \dfrac{\text{W}}{\text{m} \cdot \text{K}}}{3.55 \times 10^{-4} \dfrac{\text{kg}}{\text{m} \cdot \text{s}} 2370.6 \dfrac{\text{kJ}}{\text{kg}}} \right) \left(\frac{9.81 \dfrac{\text{m}}{\text{s}^2}}{(3.66)^2 \, 10^{-14} \dfrac{\text{m}^4}{\text{s}^2}} \right)^{1/3}$$

$$= 10{,}659$$

On the graphic route, we enter $B = 10{,}659$ and $\text{Pr}_l = 2.23$ in Fig. 8.6, and read $\text{Re}_L \cong 2200$, which shows that the film is dominated by its turbulent section. The corresponding condensation rate is

$$\Gamma(L) = \frac{\mu_l}{4} \text{Re}_L = \frac{1}{4} \left(3.55 \times 10^{-4} \frac{\text{kg}}{\text{m} \cdot \text{s}} \right) 2200$$

$$= 0.195 \, \text{kg/m} \cdot \text{s}$$

Numerically, we can enter the same B and Pr_l values in eq. (8.27) to obtain more precisely

$$\text{Re}_L = 2177 \qquad \text{and} \qquad \Gamma(L) = 0.193 \text{ kg/m} \cdot \text{s}$$

This condensation rate is 56 percent greater than the estimate obtained in Example 8.1, by assuming that the film is laminar. This comparison shows the extent to which the turbulence of the actual film augments the condensation rate that would have occurred had the flow remained laminar.

8.1.3 Film Condensation in Other Configurations

The vertical wall results described until now hold true not only for flat surfaces (Fig. 8.7, left) but also for curved vertical surfaces on which the condensate film is sufficiently thin. For the vertical cylindrical surface (internal or external) shown on the right side of Fig. 8.7, the film is "thin" when the order of magnitude of its thickness is smaller than the cylinder diameter.

In the case of a plane wall inclined at an angle θ with respect to the vertical direction (Fig. 8.8, left), the gravitational acceleration component that acts along the surface is $g \cos \theta$. Condensation heat transfer results for the inclined wall can be obtained by replacing g with $g \cos \theta$ in the results reported until now for a vertical plane wall.

More complicated surface shapes are those that are curved in such a way that the tangential component of gravity varies along the flowing film of condensate. One example is the *spherical surface* shown on the right side of Fig. 8.8. If the film is laminar all around, then the diameter-averaged heat transfer coefficient is given by [8, 9]

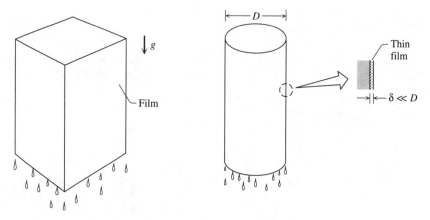

Figure 8.7 Vertical surfaces whose films of condensate can be regarded as plane.

$$\overline{\mathrm{Nu}}_D = \frac{\overline{h}_D D}{k_l} = 0.815\left[\frac{D^3 h'_{fg} g(\rho_l - \rho_v)}{k_l \nu_l (T_{\mathrm{sat}} - T_w)}\right]^{1/4} \tag{8.29}$$

The laminar film condensation on the surface of a *single horizontal cylinder* of diameter D (Fig. 8.9, upper left) was first analyzed by Nusselt [2], who relied on the same simplifying assumptions as in the vertical wall analysis detailed in Section 8.1.1. The formula for the average heat transfer coefficient under a condensate film that is laminar all around the cylinder is [8].

$$\overline{\mathrm{Nu}}_D = \frac{\overline{h}_D D}{k_l} = 0.729\left[\frac{D^3 h'_{fg} g(\rho_l - \rho_v)}{k_l \nu_l (T_{\mathrm{sat}} - T_w)}\right]^{1/4} \tag{8.30}$$

Worth noting is that this expression is similar to the laminar film relationships for the sphere, eq. (8.29), and the vertical wall, eq. (8.15). For the horizontal cylinder and the sphere, the diameter D plays the role of vertical dimension in the same sense that L measures the height of the vertical plane wall in eq. (8.15). The physical properties needed in eqs. (8.29) and (8.30) and other formulas presented in this section are to be evaluated at the film temperature $(T_w + T_{\mathrm{sat}})/2$.

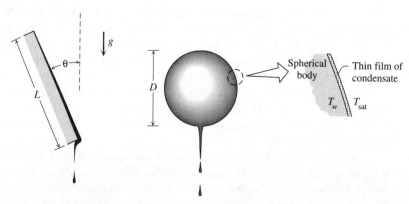

Figure 8.8 Film condensation on an inclined plane surface (left side) and on a spherical surface (right side).

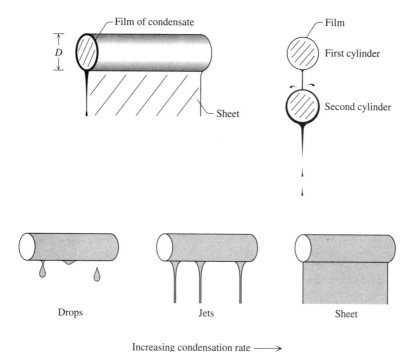

Figure 8.9 Film condensation on a single horizontal cylinder and on a vertical column ($n = 2$) of horizontal cylinders of the same size. Lower drawing shows the effect of the condensation rate on the type of flow that impinges on the next cylinder.

The Prandtl number effect on laminar film condensation on a single horizontal cylinder was documented by Sparrow and Gregg [10] and Chen [11]. The latter took into account also the effect of interfacial shear, and found that the Pr_l effect is described fairly well by the curves plotted in Fig. 8.3. Note the alternative $\overline{Nu}_D$ meaning of the ordinates of Figs. 8.3 and 8.4.

Analogous to the analysis that yields the single-cylinder formula (8.30), the laminar film analysis of a *vertical column* of n *horizontal cylinders* (Fig. 8.9, upper right) leads to

$$\overline{Nu}_{D,n} = \frac{\overline{h}_{D,n}D}{k_l} = 0.729 \left[\frac{D^3 h'_{fg} g (\rho_l - \rho_v)}{n k_l \nu_l (T_{sat} - T_w)} \right]^{1/4} \qquad (8.31)$$

The heat transfer coefficient $\overline{h}_{D,n}$ has been averaged over *all* the cylindrical surfaces, so that the total heat transfer rate per unit of cylinder length is $q' = \overline{h}_{D,n} n \pi D (T_{sat} - T_w)$. By comparing the right sides of eqs. (8.30) and (8.31), we note that the average heat transfer coefficient of the n-tall column is generally smaller than that of the single cylinder:

$$\overline{h}_{D,n} = \frac{\overline{h}_D}{n^{1/4}} \qquad (8.32)$$

The $\overline{h}_{D,n}$ values that are found experimentally are usually greater than the values calculated based on eq. (8.31). This augmentation effect can be attributed to the splashing caused by the sheet or droplets of condensate as they impinge on the next cylinder. The lower part of Fig. 8.9 shows that the condensation rate influences the type of flow that falls on the next cylinder, namely, drops, jets, or sheet, as the condensation rate increases. An additional factor can be the condensation that takes place on the sheet (or droplets) between two consecutive cylinders [11], because the falling sheet is at an average temperature below T_{sat} [i.e., it is subcooled; review the discussion under eq. (8.5)]. Finally, if each tube is slightly tilted or bowed (due to its weight or to a defect in the assembly), the condensate runs longitudinally along the tube and drips only from its lowest region. In such cases most of the length of the next cylinder is not affected by the condensate generated by the preceding cylinder.

When the cooled surface is perfectly horizontal and faces upward, Fig. 8.10 top, the condensate flows away from the central region and spills over the edges [12]. In the case of a long *horizontal strip* of width L, the average heat transfer

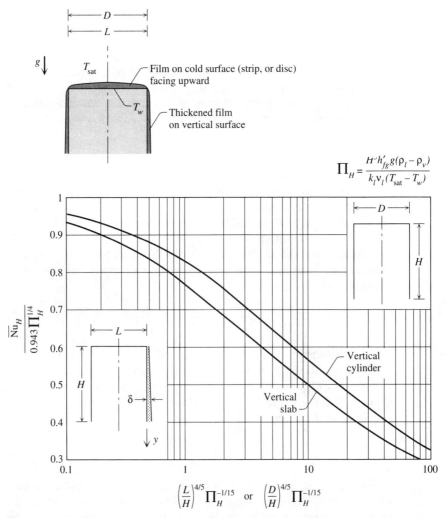

$$\Pi_H = \frac{H^{\scriptscriptstyle\sim} h'_{fg}\, g(\rho_l - \rho_v)}{k_l \nu_l (T_{sat} - T_w)}$$

Figure 8.10 Film of condensate on a horizontal strip of width L, or a horizontal disc of diameter D (after Bejan [12]).

coefficient is given by a formula similar to eq. (8.15), except that the exponent of the dimensionless group on the right side is $\frac{1}{5}$:

$$\overline{Nu}_L = \frac{\overline{h}_L L}{k_l} = 1.079 \left[\frac{L^3 h'_{fg} g (\rho_l - \rho_v)}{k_l \nu_l (T_{sat} - T_w)}\right]^{1/5} \tag{8.33}$$

The average heat transfer coefficient for an upward facing *disc* with free edges is similar [12]:

$$\overline{Nu}_D = \frac{\overline{h}_D D}{k_l} = 1.368 \left[\frac{D^3 h'_{fg} g (\rho_l - \rho_v)}{k_l \nu_l (T_{sat} - T_w)}\right]^{1/5} \tag{8.34}$$

Chapman [13] pointed out that a fully analytical solution for the film on the horizontal two-dimensional strip [14] replaces the 1.079 factor with 1.083 on the right side of eq. (8.33).

The corresponding formula for any other surface whose shape is somewhere between the "very long" shape of the strip and the "round" shape of the disc can be deduced from eqs. (8.33) and (8.34) by using the concept of *characteristic length* of eq. (7.76) (see Problem 8.11). Equations (8.33) and (8.34) are based on an analysis and set of assumptions of the same type as the ones outlined in Section 8.1.1.

These results may be used to estimate the contribution made by the horizontal "roof" surface to the total condensation rate on a three-dimensional body. The horizontal surface contributes to the total condensation rate in two ways: directly, through the flowrate estimated based on eqs. (8.33) and (8.34); and, indirectly, by thickening the film that coats the vertical lateral surface [12]. In other words, when the condensate collected on the top surface spills over the edge, the vertical surface heat transfer coefficient is smaller than the value that would be calculated based on eq. (8.15).

This effect is documented in the lower part of Fig. 8.10, in which, unlike in eq. (8.15), the height of the vertical surfaces is labeled H. The symbol Π_H is defined on the figure and is used as shorthand for the dimensionless group that emerged on the right side of eq. (8.15). Figure 8.10 shows that the roof condensate inhibits* the condensation on the vertical surfaces when the abscissa parameter exceeds the order of magnitude of 1.

All the film condensation processes described until now are examples of *natural convection*, because in all cases the flow is being driven by gravity. Considerably more complicated are the condensation processes where the vapor is forced to flow over the cooled surface. In these processes the vapor and the condensate film interact across their mutual interface. The forced flow of vapor tends to drag the liquid in its direction, and the overall condensation heat transfer process is one of natural convection mixed with *forced convection*.

One example of this kind is the film condensation on the outside of a *horizontal cylinder in cross-flow*, Fig. 8.11, left. The surface heat transfer coefficient depends on the free-stream velocity of the vapor, U_∞, as well as on gravity [15]:

$$\frac{\overline{h}_D D}{k_l} = 0.64 Re_D^{1/2} \left[1 + \left(1 + 1.69 \frac{g h'_{fg} \mu_l D}{U_\infty^2 k_l (T_{sat} - T_w)}\right)^{1/2}\right]^{1/2} \tag{8.35}$$

---

*Note that the ordinate values in Fig. 8.10 are less than 1.

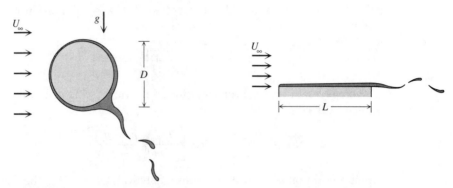

Figure 8.11 Film condensation on a horizontal cylinder in cross-flow (left), and on a flat plate parallel to the flow (right).

The Reynolds number in this expression is based on the kinematic viscosity of the liquid, $Re_D = U_\infty D / v_l$. Equation (8.35) holds for Reynolds numbers up to 10^6. It is easy to see that in the limit of negligible gravitational effect, the right side of eq. (8.35) approaches $0.64 Re_D^{1/2}$. In the opposite extreme, when the vapor stream slows to a halt, eq. (8.35) becomes identical to eq. (8.30) which, after all, represents the pure natural convection limit of the process of Fig. 8.11.

The results for laminar film condensation on a *flat plate in a parallel stream* of saturated vapor (Fig. 8.11, right) are represented well by the single formula [16]

$$\frac{\bar{h}_L L}{k_l} = 0.872 Re_L^{1/2} \left[\frac{1.508}{(1 + Ja/Pr_l)^{3/2}} + \frac{Pr_l}{Ja} \left(\frac{\rho_v \mu_v}{\rho_l \mu_l} \right)^{1/2} \right]^{1/3} \tag{8.36}$$

In this expression the Reynolds number is again based on the liquid viscosity, $Re_L = U_\infty L / v_l$, and the Jakob number is the same as the one defined in eq. (8.19). Equation (8.36) has the proper asymptotic behavior and has been tested in the ranges $(\rho_l \mu_l / \rho_v \mu_v)^{1/2} \sim 10-500$ and $Ja/Pr_l \sim 0.01-1$.

Inside a *vertical cylinder* with cocurrent vapor flow, Fig. 8.12, the downward

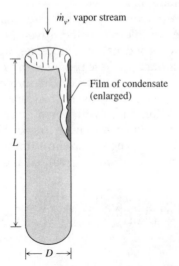

$\dot{m}_v$, vapor stream

Film of condensate (enlarged)

Figure 8.12 Condensation in a vertical tube with cocurrent flow of vapor.

progress of the liquid is aided by the vapor that flows through the core of the cross section. The liquid film is therefore thinner than in the absence of downward vapor flow, and the L-averaged heat transfer coefficient $\bar{h}_L$ and the condensation rate are greater. Chen et al. [6] reviewed the experimental information available on this configuration and proposed a correlation that here can be rearranged in the following way:

$$\frac{\bar{h}_L}{k_l}\left(\frac{v_l^2}{g}\right)^{1/3} = \left[\mathrm{Re}_L^{-0.44} + (5.82 \times 10^{-6})\mathrm{Re}_L^{0.8}\mathrm{Pr}_l^{1.3}\right.$$

$$\left. + (3.27 \times 10^{-4})\frac{\mathrm{Pr}_l^{1.3}}{D^2}\left(\frac{v_l^2}{g}\right)^{2/3}\left(\frac{\mu_v}{\mu_l}\right)^{0.156}\left(\frac{\rho_l}{\rho_v}\right)^{0.78}\frac{\mathrm{Re}_L^{0.4}\mathrm{Re}_t^{1.4}}{(1.25 + 0.39\ \mathrm{Re}_L/\mathrm{Re}_t)^2}\right]^{1/2}$$

$$(8.37)$$

This form is similar to eq. (8.23), because the only difference between the present configuration (Fig. 8.12) and that of Fig. 8.7 is the presence of the core flow of vapor. Indeed, the third group on the right side of eq. (8.37) accounts for the increase in $\bar{h}_L$ that is due to the interfacial shear between the vapor and the liquid film. The Reynolds number Re_L is defined according to eq. (8.22). The "terminal" Reynolds number Re_t is based not on the actual flowrate $\Gamma(L)$ but on $\dot{m}_v/\pi D$, in which $\dot{m}_v$ is the total flowrate of the vapor that enters through the top of the tube. In other words, the terminal Reynolds number Re_t is the maximum value approached by Re_L as a greater fraction of the original vapor stream is converted into liquid at the bottom of the tube (note that $\mathrm{Re}_L < \mathrm{Re}_t$).

The tube orientation is no longer a factor when *the vapor stream is fast enough* so that the last term overwhelms the others on the right side of eq. (8.37). In this limit the gravitational effect is negligible, and the average heat transfer coefficient can be calculated with the simpler formula

$$\frac{\bar{h}_L D}{k_l} = 0.0181\mathrm{Pr}_l^{0.65}\left(\frac{\mu_v}{\mu_l}\right)^{0.078}\left(\frac{\rho_l}{\rho_v}\right)^{0.39}\frac{\mathrm{Re}_L^{0.2}\mathrm{Re}_t^{0.7}}{1.25 + 0.39\ \mathrm{Re}_L/\mathrm{Re}_t} \qquad (8.38)$$

The rate of condensation inside a *horizontal tube* with fast vapor flow (Fig. 8.13, left) can be calculated based on eq. (8.38). In this limit the liquid film coats

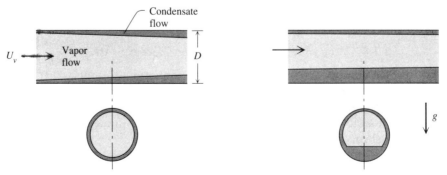

Figure 8.13 Condensation as annular film in a tube with *fast* vapor flow (left), and accumulation of condensate at the bottom of a horizontal tube with *slow* vapor flow (right).

the perimeter of the cross section uniformly. When the vapor flow is slow, the liquid flow favors the lower region of the tube cross section (Fig. 8.13, right). Chato [17] found that when the vapor flow Reynolds number is small,

$$\frac{\rho_v U_v D}{\mu_v} < 3.5 \times 10^4 \tag{8.39}$$

the condensation process is dominated by natural convection, and that $\bar{h}_D$ is given by an expression similar to eq. (8.30):

$$\frac{\bar{h}_D D}{k_l} = 0.555 \left[\frac{D^3 h'_{fg} g(\rho_l - \rho_v)}{k_l \nu_l (T_{sat} - T_w)} \right]^{1/4} \tag{8.40}$$

It is worth noting that in eq. (8.40) the effect of the buildup of condensate in the longitudinal direction has not been taken into account.

Procedures for calculating the heat transfer and condensation rates in several other configurations are described in Refs. [8, 18, 19]. For example, in cases where the vapor is *superheated*, $T_\infty > T_{sat}$, Rohsenow [19] recommends replacing h'_{fg} with a slightly larger quantity, h''_{fg}, which accounts also for the cooling experienced by the vapor en route to its saturation temperature at the interface:

$$h''_{fg} = h'_{fg} + c_{P,v}(T_\infty - T_{sat}) \tag{8.41}$$

A common feature of all the configurations discussed until now is that the vapor is pure, that is, it contains nothing but the substance that eventually condenses into the liquid film. When the "vapor" is a mixture containing not only the condensing species but also one or more *noncondensable* gases, the heat transfer coefficient is significantly lower than when the noncondensable gases are absent. The condensation rate is lower because the condensing species must first diffuse through the concentration boundary layer* that coats the gas side of the interface. The condensing species must first overcome the mass transfer resistance posed by the concentration boundary layer. This process and its effect on the condensation rate are described in Refs. [20, 21].

The film condensation process can be complicated further by transient effects, such as a sudden change in wall temperature, vapor flow, or wall orientation. Note that all the results reviewed until now refer to the steady state, in which the heat transfer and condensation rates are time-independent. A good overview of the results developed for situations in which the condensation process is time-dependent was presented by Flik and Tien [22], who also developed a general analysis for this class of problems.

8.1.4 Dropwise and Direct-Contact Condensation

The condensate distributes itself as a continuous thin film on the cooled surface only when the liquid "wets" the solid. This happens when the surface tension between the liquid and the solid material is sufficiently small, for example, when the solid surface is clean (grease free), as in the condensation of steam on a clean metallic surface.

When the surface tension is large, the condensate coalesces into a multitude of droplets of many sizes. In time, each droplet grows as more vapor condenses

---

*This concept is discussed in detail in Section 11.4.1.

on its exposed surface. The formation of each droplet is initiated at a point of surface imperfection (pit, scratch) called *nucleation site*. There comes a time when the tangential pull of gravity, or shear force exerted by the vapor stream, dislodges the droplet and carries it downstream. The moving droplet devours the smaller droplets found in its path, creating in this way a clean trail ready for the initiation of a new generation of droplets of the smallest size.

Since the condensation rate is the highest in the absence of condensate on the surface (film or droplets), the periodic cleaning performed by the large drops renews finite-size regions of the surface for the restart of the time-dependent condensation process, Fig. 8.14. This *surface renewal process* is the main reason why dropwise condensation is a highly effective mechanism. The heat transfer coefficient for dropwise condensation is approximately ten times greater than the corresponding coefficient estimated based on the assumption that the condensate forms a continuous film.

In the design of condensers, whose function it is to cool a vapor stream and to convert it into liquid, there is a great incentive to promote the breakup of the condensate film into drops. This can be accomplished by (1) coating the solid surface with an organic substance (e.g., oil, wax, kerosene, oleic acid), (2) injecting nonwetting chemicals into the vapor so that they will be deposited on the condenser surface, and (3) coating the surface with a polymer of low surface energy (e.g., teflon, silicone) or with a noble metal (e.g., gold, silver), Fig. 8.15 [23]. In methods (1) and (2), unfortunately, the "promoter" material wears off, as it is gradually removed by the scraping action of the droplet

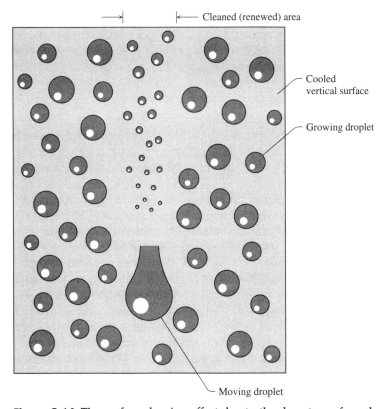

Figure 8.14 The surface cleaning effect due to the departure of one large drop.

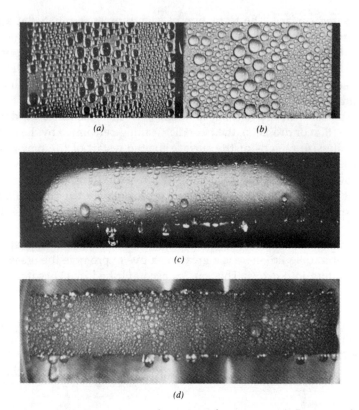

Figure 8.15 Dropwise condensation of steam on: (a) Copper-nickel coated with Nedox, (b) titanium coated with Nedox, (c) copper tube coated with fluoroacrylic, (d) copper tube coated with Nostick. (Courtesy of Prof. P. J. Marto, Naval Postgraduate School; see also Holden et al. [23].)

movement. In method (3), fluorocarbon coatings such as teflon have good surface characteristics but a relatively low thermal conductivity. If the coating is thicker than about 20 μm, its conduction resistance tends to offset the heat transfer augmentation effect owing to dropwise condensation on the vapor side of the coating.

As a fundamental mechanism of convection with change of phase, the phenomenon of dropwise condensation is complicated by its intermittent time-dependent character, the dominant effect of surface tension (drop size and shape), and the uncertainty associated with the location of nucleation sites and the time when the largest droplet will start its movement downstream. For all these reasons, a unifying theory of dropwise condensation has not been developed. Reviews of the experimental information on the performance of surfaces with promoters of dropwise condensation can be found in Refs. [24–26].

The film and drop condensation mechanisms are two examples of what is generally referred to as *surface condensation*. In both cases the condensate adheres to a solid surface that is being cooled by an external entity (from the back side). The mechanism of *direct-contact condensation* is conceptually different because the solid surface is absent here and the cooling effect is provided by the large pool of subcooled liquid through which bubbles of condensing vapor rise. Surface tension plays an important role in determining the size, shape, and life of each vapor bubble. The progress on direct-contact condensation has been reviewed most recently in Refs. [27, 28].

8.2   BOILING HEAT TRANSFER

8.2.1   Pool Boiling Regimes

In this section we turn our attention to the mechanism of boiling heat transfer, which occurs when the temperature of a solid surface is sufficiently higher than the saturation temperature of the liquid with which it comes in contact. The solid–liquid heat transfer is accompanied by the transformation of some of the heated liquid into vapor, and by the formation of distinct vapor bubbles, jets, and films. The vapor and the surrounding packets of heated liquid are carried away by the effect of buoyancy (natural convection, or pool boiling) or by a combination of buoyancy and the forced flow of liquid that may be sweeping the solid heater (mixed convection, or flow boiling).

Boiling is therefore the short name for convective heat transfer with change of phase (liquid → vapor) when a *liquid is being heated* by a sufficiently hot surface. In this way, boiling emerges as the reverse of the condensation phenomenon discussed in the first half of this chapter, where the change of phase (vapor → liquid) was caused by the *cooling of a vapor* by contact with a sufficiently cold surface.

We begin with the case of *pool boiling*, Fig. 8.16, in which the heater surface (T_w) is immersed in a pool of initially stagnant liquid (T_l). The basic heat transfer engineering question consists again of pinpointing the relationship between surface heat flux q_w'' and temperature difference $T_w - T_{sat}$, where T_{sat} is the saturation temperature of the liquid. When, as shown on the left side of Fig. 8.16, the liquid mass is at a temperature below saturation (i.e., subcooled, $T_l < T_{sat}$), boiling is confined to a layer in the immediate vicinity of the heater surface. The vapor bubbles collapse (recondense) as they rise through the subcooled liquid. When the liquid pool is at the saturation temperature (Fig. 8.16, right), the vapor generated at the heater surface reaches the free surface of the pool. In what follows it is assumed that the liquid in the pool is all saturated ($T_l = T_{sat}$).

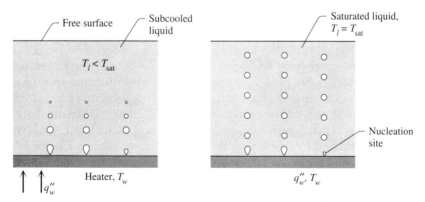

Figure 8.16 Nucleate pool boiling of a subcooled liquid (left) and a saturated liquid (right).

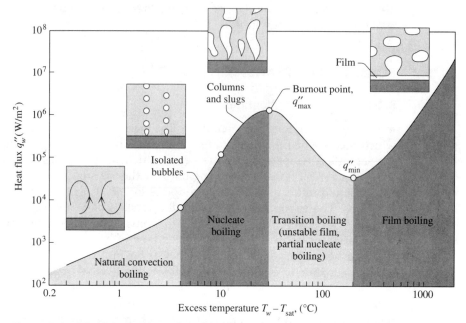

Figure 8.17 The four regimes of pool boiling in water at atmospheric pressure.

Figure 8.17 shows the main features of the "boiling curve," or the relationship between q_w'' and the "excess" temperature $(T_w - T_{sat})$. This particular curve corresponds to the pool boiling of water at atmospheric pressure; however, its "roller coaster" shape is a characteristic of the curves describing the pool boiling of other liquids. The nonmonotonic relationship between heat flux and excess temperature is due to the various forms (bubbles, film) that the newly generated vapor takes in the vicinity of the heater surface. The peculiar shape of the boiling curve is the basis for distinguishing between several pool boiling *regimes*.

The transition from one regime to the next can be seen by reading Fig. 8.17 from left to right. This corresponds to a boiling experiment in which the heater surface temperature is increased monotonically and the resulting heat flux is measured. The experimental setups that can be used to trace the boiling curve are taken up at the end of this section.

The first interesting aspect of the curve is that at very low excess temperatures (in water, at $T_w - T_{sat} \lesssim 4\,°C$), the heat transfer occurs without the appearance of bubbles on the heater surface. In this regime the near-surface liquid becomes superheated and rises in the form of *natural convection* currents to the free surface of the pool. If the heater surface is horizontal and large, and if the liquid pool is shallow, the convection currents are similar to the cellular (Bénard) flow shown on the right side of Fig. 7.21. The relationship between q_w'' and $T_w - T_{sat}$ depends on the shape and orientation of the immersed heater, and can be determined by employing the formulas assembled in Chapter 7.

Proceeding toward larger excess temperatures, the next regime is that of *nucleate boiling*. This is characterized by the generation of vapor at a number of favored spots on the surface, which are called *nucleation sites*. A probable nucleation site is a tiny crack in the surface in which the trapped liquid is surrounded by a relatively large heater area per unit of liquid volume. At the low end of the nucleate boiling curve, the boiling process consists of isolated

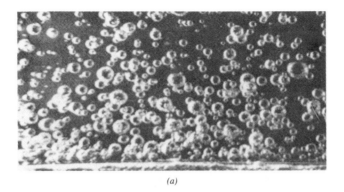

(a)

(b)

Figure 8.18 Nucleate boiling of water at atmospheric pressure: (a) Isolated bubbles, $q''_w = 1.2 \times 10^5$ W/m²; (b) columns and slugs, $q''_w = 3.7 \times 10^5$ W/m². (Photographs by Zmola [29], reproduced in Moissis and Berenson [30].)

bubbles (Fig. 8.18a). At higher temperatures, the bubble frequency increases, the nucleation sites multiply, and the isolated bubbles interact and are replaced by slugs and columns of vapor (Fig. 8.18b).

The formation of more and more vapor in the vicinity of the surface has the effect of gradually insulating the surface against the T_{sat}-cold liquid. This effect is responsible for the gradual decrease of the slope of the nucleate boiling part of the curve and for its expiration at the point of maximum (peak) heat flux q''_{max}. The latter is also called *critical heat flux*, and in water it is of the order of 10^6 W/m². The excess temperature at this point is approximately 30°C.

The next regime is the most peculiar, because the heat flux actually decreases as the excess temperature $T_w - T_{sat}$ continues to increase. This trend is a reflection of the fact that increasingly greater portions of the heater surface become coated with a continuous film of vapor. The vapor is unstable and is intermittently replaced by nucleate boiling. This regime is called *transition boiling*: it expires at the point* of minimum heat flux q''_{min}, where the excess temperature has become just large enough to sustain a stable vapor film on the heater surface. In water at 1 atmosphere, the minimum heat flux is in the $10^4 - 10^5$ W/m² range and occurs at an excess temperature in the 100–200°C range.

---

*The corresponding temperature of the surface is called the *Leidenfrost temperature*, or *Leidenfrost point*, in honor of Johann Gottlob Leidenfrost, a German medical doctor who studied the evaporation of liquid droplets on hot surfaces. A portion of his 1756 treatise *De Aquae Communis Nonnullis Qualitatibus Tractatus* was translated from Latin [31].

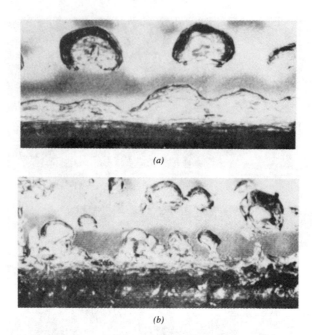

(a)

(b)

Figure 8.19 Film boiling of isopropanol on a horizontal tube: (a) $T_w - T_{sat} = 138°C$, $D = 0.64$ cm; (b) $T_w - T_{sat} = 138°C$, $D = 1.7$ cm. (Photographs by Breen and Westwater [32].)

At even larger excess temperatures, the vapor film covers the entire surface and the heat flux q_w'' resumes its monotonic increase with $T_w - T_{sat}$. The vapor film is illustrated in Fig. 8.19. Radiation heat transfer across the film plays a progressively greater role as the excess temperature increases. This high-temperature mode is called the *film boiling* regime. It persists until T_w reaches the melting point of the surface material, that is, until the meltdown (burnout) of the heater surface. If, as in the case of a platinum surface, the melting point is very high (namely, 2042 K), the film boiling portion of the boiling curve can extend to heat fluxes above the critical q_{max}'' of the nucleate boiling regime.

The tracing of the boiling curve from left to right in Fig. 8.17 was based on the assumption that the excess temperature $T_w - T_{sat}$ can be controlled and increased monotonically. An experiment in which the heater temperature control is possible is shown on the left side of Fig. 8.20. The heater is a horizontal tube immersed in a pool of liquid. The heater surface temperature is controlled by a preheated stream that flows through the tube. In this temperature-controlled experiment, the boiling curve can be traced in either direction by gradually increasing or decreasing the excess temperature.

An alternative setup is the power-controlled experiment shown on the right side of Fig. 8.20. The heater is a horizontal cylinder (wire) stretched in a pool of liquid. The heat flux q_w'' is controlled by the experimentalist who measures the power dissipated in the electrical resistance posed by the wire. In this experiment the shape of the emerging boiling curve depends on whether the power (q_w'') is increased or decreased.

When the power increases monotonically, the experimentalist observes the

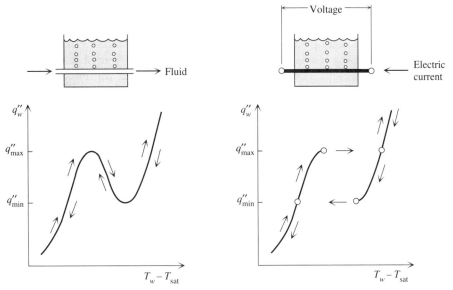

Figure 8.20 The pool boiling curve in a temperature-controlled experiment (left), and in a power-controlled experiment (right).

transition from the natural convection regime to the several forms of the nucleate boiling regime. As the imposed heat flux increases slightly above the critical value q''_{max}, the wire temperature increases abruptly (and dramatically) to the value associated with the film boiling portion of the boiling curve. In most cases this new temperature would be above the melting point of the surface material, and the wire burns up. This is why the peak of the nucleate boiling portion of the curve is often called the *burnout point*. The catastrophic event that can occur at heat fluxes comparable with and greater than q''_{max} is why in power-controlled applications of boiling heat transfer (e.g., nuclear reactors, electrical resistance heaters) it is advisable to operate at heat fluxes safely smaller than q''_{max}.

The power-controlled pool boiling experiment can be run in reverse by decreasing the heat flux. In that case the excess temperature decreases along the film boiling portion of the curve all the way down to the Leidenfrost temperature. As the heat flux is lowered slightly below the minimum heat flux of film boiling, q''_{min}, the vapor film collapses, isolated bubbles form, and the wire temperature drops to the low level associated with the nucleate boiling regime.

To summarize, during heat flux-controlled boiling the transition boiling regime is inaccessible, and certain portions of the boiling curve can be reached while varying q''_w in only one direction. For example, the nucleate boiling regime in the vicinity of the point of maximum heat flux can be established only by increasing the heat flux, starting from a sufficiently low level. Although much less important in practice, the film boiling regime in the vicinity of q''_{min} can be achieved only by decreasing the heat flux, starting from a sufficiently high level. It is said that q''_w-controlled boiling is an example of *hysteresis,** a phenomenon

---

*From the Greek words *hysteresis* (a deficiency) and *hysterein* (to lag behind, to fall short).

that depends not only on the imposed condition (q_w'') but also on its previous history, in this instance the previous value of q_w''.

The boiling curve was first determined by Prof. S. Nukiyama [33] of Tohoku University (Sendai, Japan), who employed the q_w''-controlled method illustrated on the right side of Fig. 8.20. The existence of the missing transition boiling portion of the curve was demonstrated based on T_w-controlled experiments by Drew and Mueller [34].

8.2.2   Nucleate Boiling and Peak Heat Flux

The most important regime of the entire curve displayed in Fig. 8.17 is the regime of nucleate boiling, because here the boiling heat transfer coefficient

$$h = \frac{q_w''}{T_w - T_{\text{sat}}} \tag{8.42}$$

reaches characteristically large values. These cover the range $10^3 - 10^5$ W/m²·K, as we saw in the beginning of this course at the top of Fig. 1.12.

An enormous volume of research has been devoted to the measurement and correlation of the nucleate boiling heat transfer coefficient. One of the earliest and most successful correlations, which has withstood the test of time, was proposed in 1952 by Rohsenow [35]:

$$T_w - T_{\text{sat}} = \frac{h_{fg}}{c_{P,l}} \text{Pr}_l^s C_{sf} \left[\frac{q_w''}{\mu_l h_{fg}} \left(\frac{\sigma}{g(\rho_l - \rho_v)} \right)^{1/2} \right]^{1/3} \tag{8.43}$$

This correlation applies to clean surfaces, and, as an engineering approximation, is insensitive to the shape and orientation of the surface. It depends on two empirical constants, C_{sf} and s, which are listed in Table 8.1. The dimensionless factor C_{sf} accounts for the particular combination of liquid and surface material, whereas the Prandtl number exponent s differentiates only between water and other liquids. The subscripts l and v denote saturated liquid and saturated vapor, and indicate the temperature (T_{sat}) at which the properties are evaluated.

The symbol σ (N/m) denotes the surface tension of the liquid in contact with its own vapor. Representative values of this physical property have been collected in Table 8.2 along with other data needed for boiling heat transfer calculations. The surface tension plays an important role in the growth of each vapor bubble, and this role is being recognized in the theories aimed at predicting the nucleate boiling curve [41]. For example, in Problem 8.13 it is shown that the radius of a spherical vapor bubble in mechanical equilibrium is given by $2\sigma/(P_v - P_l)$, where P_v and P_l are the pressures inside and outside the bubble.

Written as eq. (8.43), Rohsenow's nucleate boiling correlation can be used to calculate the excess temperature $T_w - T_{\text{sat}}$ when the heat flux q_w'' is known. The calculated excess temperature agrees within ±25 percent with experimental data. In the reverse case in which the excess temperature is specified, eq. (8.43) can be rewritten as

$$q_w'' = \mu_l h_{fg} \left(\frac{g(\rho_l - \rho_v)}{\sigma} \right)^{1/2} \left[\frac{c_{P,l}(T_w - T_{\text{sat}})}{\text{Pr}_l^s C_{sf} h_{fg}} \right]^3 \tag{8.44}$$

to calculate the unknown heat flux. In this instance the calculated q_w'' agrees within a factor of 2 with actual heat flux measurements.

In conclusion, eqs. (8.43) and (8.44) provide only an approximate, engineer-

Table 8.1 Empirical Constants for Rohsenow's Nucleate Pool Boiling Correlations (8.43) and (8.44) [35, 36]

Liquid – Surface Combination	C_{sf}	s
Water – copper		
Polished	0.013	1.0
Scored	0.068	1.0
Emery polished, paraffin treated	0.015	1.0
Water – stainless steel		
Ground and polished	0.008	1.0
Chemically etched	0.013	1.0
Mechanically polished	0.013	1.0
Teflon pitted	0.0058	1.0
Water – brass	0.006	1.0
Water – nickel	0.006	1.0
Water – platinum	0.013	1.0
CCl_4 – copper	0.013	1.7
Benzene – chromium	0.010	1.7
n-Pentane – chromium	0.015	1.7
n-Pentane – copper		
Emery polished	0.0154	1.7
Emery rubbed	0.0074	1.7
Lapped	0.0049	1.7
n-Pentane – nickel		
Emery polished	0.013	1.7
Ethyl alcohol – chromium	0.0027	1.7
Isopropyl alcohol – copper	0.0025	1.7
35% K_2CO_3 – copper	0.0054	1.7
50% K_2CO_3 – copper	0.0027	1.7
n-Butyl alcohol – copper	0.0030	1.7

ing-type estimate of the true position of the nucleate boiling curve. One reason for this is the S shape taken by the nucleate boiling curve on the logarithmic grid of Fig. 8.17: this shape departs from the straight line that would correspond to eq. (8.44). Another reason is the potential effect of surface roughness, which tends to increase the number of active nucleation sites. In artificially roughened surfaces, for example, the heat flux can be one order of magnitude greater than the q_w'' value furnished by eq. (8.44).

For calculations involving the critical or peak heat flux on a large horizontal surface, the recommended relation is [42]

$$q_{max}'' = 0.149 h_{fg} \rho_v^{1/2} [\sigma g(\rho_l - \rho_v)]^{1/4} \tag{8.45}$$

The analytical form of this expression has a theoretical foundation, having been first proposed based on dimensional analysis by Kutateladze [43] and on the hydrodynamic stability of vapor columns by Zuber [44]. In both theories the place of the recommended numerical factor of 0.149 in eq. (8.45) was occupied originally by 0.131.

The peak heat flux formula (8.45) is clearly independent of the surface material. It applies to a sufficiently large surface whose linear length is consider-

Table 8.2 Surface Tension and Other Physical Properties Needed for Calculating Boiling and Condensation Heat Transfer Rates[a]

Fluid	T_{sat} (K)	T_{sat} (°C)	P[b] (10^5 N/m²)	ρ_l (kg/m³)	ρ_v (kg/m³)	h_{fg} (kJ/kg)	σ (N/m)
Ammonia	223	−50	0.409	702	0.38	1417	0.038
	300	27	10.66	600	8.39	1158	0.020
Ethanol	351	78	1.013	757	1.44	846	0.018
Helium	4.2	−269	1.013	125	16.9	20.42	10^{-4}
Hydrogen	20.3	−253	1.013	70.8		442	0.002
Lithium	600	327	4.2×10^{-9}	503		22340	0.375
	800	527	9.6×10^{-6}	483	10^{-6}	21988	0.348
Mercury	630	357	1.013	12740	3.90	301	0.417
Nitrogen	77.3	−196	1.013	809	4.61	198.4	0.0089
Oxygen	90.2	−183	1.013	1134		213.1	0.013
Potassium	400	127	1.84×10^{-7}	814	2.2×10^{-7}	2196	0.110
	800	527	0.0612	720	0.037	2042	0.083
Refrigerant 12	243	−30	1.004	1488	6.27	165.3	0.016
Refrigerant 22	200	−73	0.166	1497	0.87	252.8	0.024
	250	−23	2.174	1360	9.64	221.9	0.016
	300	27	10.96	1187	46.55	180.1	0.007
Sodium	500	227	7.64×10^{-7}	898	4.3×10^{-7}	4438	0.175
	1000	727	0.1955	776	0.059	4022	0.130
Water	323	50	0.1235	988	0.08	2383	0.068
	373	100	1.0133	958	0.60	2257	0.059
	423	150	4.758	917	2.55	2114	0.048
	473	200	15.54	865	7.85	1941	0.037
	523	250	39.73	799	19.95	1716	0.026
	573	300	85.81	712	46.15	1405	0.014

[a] Adapted from Refs. [37–40].

[b] Note that the standard atmospheric pressure is nearly the same as the pressure of 10^5 N/m² (i.e., 1 bar), specifically, 1 atm = 1.0133×10^5 N/m².

ably greater than the characteristic size of the vapor bubble. Equation (8.45) can be used also for a sufficiently large horizontal cylinder by replacing the 0.149 factor with 0.116 [45]. When the size of the heater is comparable with, or smaller than, the bubble size, the peak heat flux depends also on the size and geometry of the heater. The peak heat flux can be calculated with a formula similar to eq. (8.45), which contains an additional geometric correction factor [42]. Overall, the peak heat flux is *relatively insensitive to the shape and orientation of the heater surface*; therefore, eq. (8.45) provides an adequate order-of-magnitude estimate of q''_{max} when more specific correlations are not available.

The maximum heat flux of eq. (8.45) depends strongly on the pressure that prevails in the liquid pool. The pressure effect is brought into this relation through both h_{fg} and σ. Near the critical pressure, for example, the peak heat flux approaches zero because both h_{fg} and $\rho_l - \rho_v$ approach zero. The relationship between q''_{max} and pressure is not monotonic: in the case of water, q''_{max} increases as the pressure rises to about 70 atm, and then decreases to zero as the pressure approaches the critical point pressure of 218.2 atm.

Regarding the effect of the gravitational acceleration, the nucleate boiling correlations presented in this and the next section are valid in the range $10-1$ m/s². Despite the proportionality between q''_w and $g^{1/2}$ suggested by eq. (8.44), it has been found that the gravitational acceleration has a considerably weaker effect on the nucleate boiling heat flux.

Example 8.3

Nucleate Boiling, Critical Heat Flux

A cylindrical heating element with a diameter of 1 cm and length of 30 cm is immersed horizontally in a pool of saturated water at atmospheric pressure. The cylindrical surface is plated with nickel. Calculate the heat flux q''_w and total heat transfer rate from the cylinder to the water pool, q_w, when the surface temperature is $T_w = 108°C$. Calculate also the critical heat flux q''_{max} and compare this value with q''_w and the approximate layout of the boiling curve shown in Fig. 8.17.

Solution.   Using Fig. 8.17 as a guide, we see that the excess temperature of $108°C - 100°C = 8°C$ is associated with the nucleate boiling regime. The wall-averaged heat flux can be evaluated using eq. (8.44), for which the pertinent properties of water at $100°C$ are (Table 8.2 and Appendix C)

$$\rho_l = 958 \text{ kg/m}^3 \qquad\qquad \sigma = 0.059 \text{ N/m}$$
$$\rho_v = 0.6 \text{ kg/m}^3 \qquad\qquad h_{fg} = 2257 \text{ kJ/kg}$$
$$c_{P,l} = 4.216 \text{ kJ/kg·K} \qquad\qquad \text{Pr}_l = 1.78$$
$$\mu_l = 2.83 \times 10^{-4} \text{ kg/s·m}$$

From Table 8.1 we collect the values of the empirical constants $C_{sf} = 0.006$ and $s = 1$. The right side of eq. (8.44) can be calculated in modules,

$$\frac{g(\rho_l - \rho_v)}{\sigma} \cong \frac{g\rho_l}{\sigma} = \frac{\left(9.81\dfrac{m}{s^2}\right)\left(958\dfrac{kg}{m^3}\right)}{0.059\dfrac{1}{m}\dfrac{kg·m}{s^2}}$$

$$= 1.593 \times 10^5 \text{ 1/m}^2$$

$$\frac{c_{P,l}(T_w - T_{sat})}{\text{Pr}_l^s C_{sf} h_{fg}} = \frac{4.216\dfrac{kJ}{kg·K}(108-100)K}{1.78 \times 0.006 \times \left(2257\dfrac{kJ}{kg}\right)}$$

$$= 1.40$$

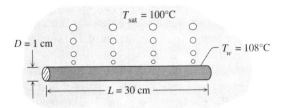

Figure E8.3

and the nucleate boiling heat flux becomes

$$q_w'' = \left(2.83 \times 10^{-4}\frac{kg}{s \cdot m}\right)\left(2257\frac{kJ}{kg}\right)\left(1.593 \times 10^5\frac{1}{m^2}\right)^{1/2}(1.40)^3$$

$$\cong 7 \times 10^5 \; W/m^2$$

(8.44)

The total heat transfer rate q_w is proportional to the exposed cylindrical area; therefore,

$$q_w = \pi D L q_w'' = \pi(0.01 \text{ m})(0.3 \text{ m})\left(7 \times 10^5\frac{W}{m^2}\right)$$

$$\cong 6600 \; W$$

The critical (peak) heat flux can be estimated using eq. (8.45), in which $\rho_l - \rho_v \cong \rho_l$:

$$q_{max}'' \cong 0.149 h_{fg}\rho_v^{1/2}(\sigma g \rho_l)^{1/4}$$

$$= \left(0.149 \times 2257\frac{kJ}{kg}\right)(0.6)^{1/2}\frac{kg^{1/2}}{m^{3/2}}\left[\left(0.059\frac{N}{m}\right)\left(9.81\frac{m}{s^2}\right)\left(958\frac{kg}{m^3}\right)\right]^{1/4}$$

(8.45)

$$= 1.26 \times 10^6 \; W/m^2$$

$$q_{max} = \pi D L q_{max}'' = \pi(0.01 \text{ m})(0.3 \text{ m})\left(1.26 \times 10^6\frac{W}{m^2}\right)$$

$$\cong 11900 \; W$$

The critical heat flux is 80 percent greater than the nucleate boiling heat flux calculated in the first part of this example. The behavior of the same cylindrical heating element in the film boiling regime will be analyzed in Example 8.4.

8.2.3   Film Boiling and Minimum Heat Flux

The outstanding feature of the film boiling regime is the continuous layer of vapor (typically, 0.2–0.5 mm thick) that separates the heat surface from the rest of the liquid pool. The minimum heat flux q_{min}'' is registered at the lowest heater temperature where the film is still continuous and stable, Fig. 8.17. The recommended correlation for the minimum heat flux on a sufficiently large horizontal plane surface is

$$q_{min}'' = 0.09 h_{fg}\rho_v \left[\frac{\sigma g(\rho_l - \rho_v)}{(\rho_l + \rho_v)^2}\right]^{1/4}$$

(8.46)

An interesting feature of this correlation is that q_{min}'' does not depend on the excess temperature $T_w - T_{sat}$. The analytical form of eq. (8.46) was discovered by Zuber [44], who analyzed the stability of the horizontal vapor–liquid interface of the film. In the field of fluid mechanics, the unstable wavy shape that can be assumed by the horizontal interface between a heavy fluid (above) and lighter fluid (below) is called *Taylor instability* [46]. The 0.09 numerical coefficient in eq. (8.46) was determined by Berenson [47] based on experimental data. For minimum-flux film boiling on a horizontal cylinder, the right side of eq. (8.46) is multiplied by an additional factor to account for the cylinder radius [48].

The minimum heat flux calculated with eq. (8.46) agrees within 50 percent with laboratory measurements at low and moderate pressures. The accuracy of this correlation deteriorates as the pressure increases. The surface roughness has only a negligible effect on the minimum heat flux (of the order of 10 percent), because the surface asperities are cushioned by the film against the liquid.

For the rising portion of the film boiling curve (Fig. 8.17), the correlations that have been developed have the same analytical form as the formulas encountered in our study of film condensation. For example, the formula for the average heat transfer coefficient on a *horizontal cylinder* [49],

$$\frac{\bar{h}_D D}{k_v} = 0.62 \left[\frac{D^3 h'_{fg} g \, (\rho_l - \rho_v)}{k_v \nu_v \, (T_w - T_{sat})} \right]^{1/4} \tag{8.47}$$

is similar to the film condensation formula (8.30). One important difference is that in eq. (8.47) the transport properties are those of vapor (k_v, ν_v), because this time the film is occupied by vapor. The similarity between film boiling and film condensation is also geometric, as can be seen by comparing Fig. 8.21 with the upper left side of Fig. 8.9. The corresponding formula for film boiling on a *sphere* is [42]

$$\frac{\bar{h}_D D}{k_v} = 0.67 \left[\frac{D^3 h'_{fg} g \, (\rho_l - \rho_v)}{k_v \nu_v \, (T_w - T_{sat})} \right]^{1/4} \tag{8.48}$$

In eqs. (8.47) and (8.48) the augmented latent heat of vaporization h'_{fg} accounts also for the superheating of the fresh vapor to temperatures above the saturation temperature [49]:

$$h'_{fg} = h_{fg} + 0.4 c_{P,v} \, (T_w - T_{sat}) \tag{8.49}$$

The vapor properties k_v, ν_v, ρ_v, and $c_{P,v}$ are best evaluated at the average film temperature $(T_w + T_{sat})/2$. Equations (8.47) and (8.48) further state that the heat transfer coefficient $\bar{h}_D$ is proportional to $(T_w - T_{sat})^{-1/4}$, which means that during film boiling the heat flux q''_w is proportional to $(T_w - T_{sat})^{3/4}$.

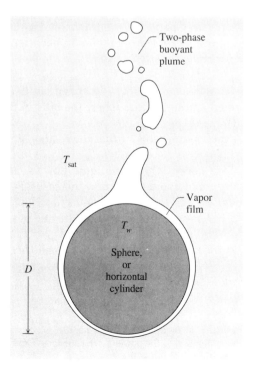

Figure 8.21 The film boiling regime on a sphere or horizontal cylinder.

As the heater temperature increases, the effect of direct thermal radiation across the film contributes more and more to the overall heat transfer rate from the heater to the liquid pool. Bromley [49] showed that the thermal radiation effect can be incorporated into an "effective" average heat transfer coefficient $\bar{h}$,

$$\bar{h} \cong \bar{h}_D + \frac{3}{4}\, \bar{h}_{\text{rad}} \qquad (\text{when } \bar{h}_D > \bar{h}_{\text{rad}}) \tag{8.50}$$

for which $\bar{h}_D$ is furnished by eqs. (8.47) and (8.48) and $\bar{h}_{\text{rad}}$ is the radiation heat transfer coefficient

$$\bar{h}_{\text{rad}} = \frac{\sigma \epsilon_w\, (T_w^4 - T_{\text{sat}}^4)}{T_w - T_{\text{sat}}} \tag{8.51}$$

In water, the effect of thermal radiation begins to be felt as $T_w - T_{\text{sat}}$ increases above the 550–660°C range [50]. When $\bar{h}_{\text{rad}}$ is comparable with or greater than $\bar{h}_D$, Bromley's [49] recommended rule for the effective heat transfer coefficient $\bar{h}$ is

$$\bar{h} = \bar{h}_D \left(\frac{\bar{h}_D}{\bar{h}} \right)^{1/3} + \bar{h}_{\text{rad}} \qquad (\text{when } \bar{h}_D \lesssim \bar{h}_{\text{rad}}) \tag{8.52}$$

Sparrow [51] showed that a rigorous analysis of combined convection and radiation in the vapor film leads to results that match within a few percentage points the values calculated based on Bromley's rule (8.52). It can be shown that the simpler eq. (8.50) follows from eq. (8.52) as $\bar{h}_{\text{rad}}/\bar{h}_D \to 0$.

It is important to note also that the σ factor in eq. (8.51) is the Stefan–Boltzmann constant, $\sigma = 5.669 \times 10^{-8}$ W/m²·K⁴, which should not be confused with the symbol used for surface tension. In the same equation, ϵ_w is the emissivity of the heater surface; and, numerically, the temperatures (T_w, T_{sat}) *must be expressed in degrees Kelvin*. In Chapter 10 we learn that eq. (8.51) can be derived from the more general formula for the net radiation heat transfer across a narrow gap, eq. (10.84), by assuming that the emissivity of the liquid surface is equal to 1.

All the pool-boiling heat transfer correlations described until now apply when the pool contains saturated liquid. In cases where the bulk of the liquid is subcooled (e.g., Fig. 8.16, left), the degree of *liquid subcooling* ($T_{\text{sat}} - T_l$) constitutes an additional parameter that complicates further the relationship between the actual heat flux q_w'' and the excess temperature $T_w - T_{\text{sat}}$. On the natural convection portion of the boiling curve, where the liquid flow is single phase, the heat flux increases if the degree of liquid subcooling increases. For example, when the heater is small enough so that the natural convection flow is laminar, the heat flux increases* as $(T_w - T_l)^{5/4}$, or, in terms of the degree of subcooling, as $[(T_w - T_{\text{sat}}) + (T_{\text{sat}} - T_l)]^{5/4}$. The subcooling parameter $T_{\text{sat}} - T_l$ has a relatively negligible effect on q_w'' in the nucleate boiling regime, while both q_{max}'' and q_{min}'' increase linearly with $T_{\text{sat}} - T_l$. The effect of liquid subcooling is the most pronounced in the film boiling regime.

---

*In Chapter 7 we learned that in the laminar regime the Nusselt number is proportional to the overall temperature difference (Rayleigh number) to the power $\frac{1}{4}$, which means that the heat flux is proportional to the overall temperature difference to the power $\frac{5}{4}$.

Example 8.4

Film Boiling, Contribution Due to Radiation

The surface temperature of the cylindrical heating element described in Example 8.3 is raised to $T_w = 300°C$. The water pool is saturated at 1 atm, and the dimensions of the horizontal cylinder are $D = 1$ cm and $L = 30$ cm. Calculate the average heat transfer coefficient, the heat flux, and the total heat transfer rate from the heater to the water pool. The emissivity of the heater surface is $\epsilon_w = 0.8$. Compare these heat transfer results with those obtained for nucleate boiling (Example 8.3), and with the boiling curve outlined in Fig. 8.17.

Solution. According to the approximate position of the boiling curve shown in Fig. 8.17, the present excess temperature $(300°C - 100°C = 200°C)$ corresponds to the low-flux end of the film boiling portion of the curve. To calculate the heat flux,

$$q_w'' = \bar{h}\,(T_w - T_{sat})$$

we must use eqs. (8.48)–(8.51), for which the physical properties of steam are evaluated at the average temperature of the film, $(300°C + 100°C)/2 = 200°C = 473$ K (Appendix D):

$$\rho_v = 0.46 \text{ kg/m}^3 \qquad c_{P,v} = 1.982 \text{ kJ/kg·K}$$
$$v_v = 3.55 \times 10^{-5} \text{ m}^2/\text{s} \qquad k_v = 0.0334 \text{ W/m·K}$$

From Table 8.2 we further collect

$$\rho_l = 958 \text{ kg/m}^3 \qquad h_{fg} = 2257 \text{ kJ/kg}$$

and, using eq. (8.49), we can calculate the augmented latent heat of vaporization:

$$h_{fg}' = h_{fg} + 0.4\, c_{P,v}\,(T_w - T_{sat})$$

$$= 2257\, \frac{\text{kJ}}{\text{kg}} + \left(0.4 \times 1.982\, \frac{\text{kJ}}{\text{kg·K}}\right)(300 - 100)\text{ K}$$

$$= 2416 \text{ kJ/kg}$$

The heat transfer coefficient for pure convection, $\bar{h}_D$, follows in two steps from eq. (8.48), in which $\rho_l - \rho_v \cong \rho_l$:

$$\frac{\bar{h}_D D}{k_v} \cong 0.62 \left[\frac{D^3 h_{fg}' g \rho_l}{k_v v_v (T_w - T_{sat})} \right]^{1/4}$$

$$= 0.62 \left[\frac{10^{-6}\text{ m}^3\; 2416\, \dfrac{\text{kJ}}{\text{kg}}\; 9.81\, \dfrac{\text{m}}{\text{s}^2}\; 958\, \dfrac{\text{kg}}{\text{m}^3}}{0.0334\, \dfrac{\text{W}}{\text{m·K}}\; 3.55 \times 10^{-5}\, \dfrac{\text{m}^2}{\text{s}}\;(300 - 100)\text{ K}} \right]^{1/4}$$

$$= 0.62\,(9.57 \times 10^7)^{1/4} = 61.3$$

$$\bar{h}_D = 61.3\, \frac{k_v}{D} = 61.3\, \frac{0.0334\text{ W}}{\text{m·K}}\, \frac{1}{0.01\text{ m}}$$

$$= 204.8 \text{ W/m}^2\text{·K}$$

The correction due to radiation can be evaluated based on eq. (8.51), in which $\epsilon_w = 0.8$, and where T_w and T_{sat} are *absolute* temperatures:

$$\bar{h}_{rad} = \frac{\sigma \epsilon_w (T_w^4 - T_{sat}^4)}{T_w - T_{sat}} = \frac{5.669 \times 10^{-8}\text{ W}}{\text{m}^2\text{·K}^4}\,0.8\,\frac{(573.15^4 - 373.15^4)\text{ K}^4}{(573.15 - 373.15)\text{ K}}$$

$$= 20.1 \text{ W/m}^2\text{·K}$$

Since $\bar{h}_{rad} < \bar{h}_D$, we can use eq. (8.50) to calculate, in order,

$$\bar{h} = \bar{h}_D + \frac{3}{4} \bar{h}_{rad} = 220 \text{ W/m}^2 \cdot \text{K}$$

$$q''_w = \bar{h} \, (T_w - T_{sat}) = 220 \, \frac{\text{W}}{\text{m}^2 \cdot \text{K}} \, 200 \text{ K}$$

$$\cong 4.4 \times 10^4 \text{ W/m}^2$$

Note that this heat flux level agrees in an order-of-magnitude sense with the lower end of the film boiling curve in Fig. 8.17. The total heat transfer rate from the cylindrical element to the pool of water is, finally,

$$q_w = \pi D L q''_w = \pi (0.01 \text{ m})(0.3 \text{ m}) \left(4.4 \times 10^4 \, \frac{\text{W}}{\text{m}^2} \right)$$

$$\cong 414 \text{ W}$$

8.2.4 Flow Boiling

The preceding material referred to pool boiling, that is, to boiling in a bulk-stationary volume of liquid (Fig. 8.16). The boiling heat transfer process is considerably more complicated (and more difficult to correlate) in situations where the liquid is forced to flow past the heater. In nucleate flow boiling, for example, Fig. 8.22, the actual heat transfer rate is due to a combination of two closely interrelated effects: (1) the bubble formation and motion near the surface, and (2) the direct sweeping of the heater surface by the liquid itself. The heat transfer mechanism is a combination of two basic ones: the nucleate pool boiling of Section 8.2.2 and a forced convection phenomenon of the kind treated in Chapters 5 and 6.

There is no general, definitive method of correlating flow boiling data. The progress in this direction has been reviewed by Rohsenow [52] and, more recently, by Whalley [53]. In one of the earliest and simplest methods, Rohsenow showed that the experimental data on nucleate boiling with convection are represented adequately by the additive formula

$$q'' = q''_w + q''_c \tag{8.53}$$

In this expression, q''_w is the nucleate pool boiling heat flux calculated based on eq. (8.44) and the assumption that the bulk of the liquid is stationary. The second term, q''_c, is the single-phase convection heat flux to the liquid, $q''_c = h_c (T_w - T_l)$, for which the convection heat transfer coefficient h_c can be estimated

Figure 8.22 Turbulent boundary layer in water that flows from left to right and boils over a flat surface: $U_\infty = 0.52$ m/s, $q''_w = 4.8 \times 10^5$ W/m². (J. H. Lienhard, *A Heat Transfer Textbook*, 2nd edition, 1987, p. 444. Reprinted by permission of Prentice Hall, Englewood Cliffs, NJ. Photograph courtesy of Prof. J. H. Lienhard, University of Houston.)

by employing the results listed in Chapters 5 and 6 or, in the case of a significant natural convection effect, Chapter 7. In particular, for nucleate boiling in duct flow, Rohsenow recommends calculating h_c by replacing the coefficient 0.023 with 0.019 in the Dittus–Boelter correlation (6.91). The superposition formula (8.53) works best when the flowing liquid is subcooled and the generation of vapor near the heater surface is not excessive.

Correlations for the peak heat flux q''_{max} on a cylinder in cross-flow have been developed by Lienhard and Eichhorn [54] and Kheyrandish and Lienhard [55]. The combined process of film boiling in the presence of forced convection was documented by Bromley et al. [56]. The current progress on film boiling and transition boiling was reviewed by Sakurai [57] and, respectively, Dhir [58].

REFERENCES

1. A. Bejan, *Convection Heat Transfer*, Wiley, New York, 1984, pp. 33–34.
2. W. Nusselt, Die Oberflächenkondensation des Wasserdampfes, *Zeitschrift des Vereines deutscher Ingenieure*, Vol. 60, 1916, pp. 541–569.
3. W. M. Rohsenow, Heat transfer and temperature distribution in laminar-film condensation, *Trans. ASME*, Vol. 78, 1956, pp. 1645–1648.
4. E. M. Sparrow and J. L. Gregg, A boundary-layer treatment of laminar-film condensation, *J. Heat Transfer*, Vol. 81, 1959, pp. 13–18.
5. M. M. Chen, An analytical study of laminar film condensation: Part 1—Flat plates, *J. Heat. Transfer*, Vol. 83, 1961, pp. 48–54.
6. S. L. Chen, F. M. Gerner, and C. L. Tien, General film condensation correlations, *Experimental Heat Transfer*, Vol. 1, 1987, pp. 93–107.
7. M. H. Chun and K. T. Kim, Assessment of the new and existing correlations for laminar and turbulent film condensations on a vertical surface, *Int. Comm. Heat Mass Transfer*, Vol. 17, 1990, pp. 431–441.
8. V. K. Dhir and J. H. Lienhard, Laminar film condensation on plane and axisymmetric bodies in non-uniform gravity, *J. Heat Transfer*, Vol. 93, 1971, pp. 97–100.
9. J. H. Lienhard, *A Heat Transfer Textbook*, 2nd edition, Prentice Hall, Englewood Cliffs, NJ, 1987, p. 400.
10. E. M. Sparrow and J. L. Gregg, Laminar condensation heat transfer on a horizontal cylinder, *J. Heat Transfer*, Vol. 81, 1959, pp. 291–296.
11. M. M. Chen, An analytical study of laminar film condensation: Part 2—Single and multiple horizontal tubes, *J. Heat Transfer*, Vol. 83, 1961, pp. 55–60.
12. A. Bejan, Film condensation on an upward facing plate with free edges, *Int. J. Heat Mass Transfer*, Vol. 34, 1991, pp. 578–582.
13. A. J. Chapman, Private communication, April 11, 1991.
14. J. V. Clifton and A. J. Chapman, Condensation of a pure vapor on a finite-size horizontal plate, ASME Paper No. 67-WA/HT-18, 1967.
15. I. G. Shekriladze and V. I. Gomelauri, Theoretical study of laminar film condensation of flowing vapour, *Int. J. Heat Mass Transfer*, Vol. 9, 1966, pp. 581–591.
16. J. W. Rose, A new interpolation formula for forced-convection condensation on a horizontal surface. *J. Heat Transfer*, Vol. 111, 1989, pp. 818–819.
17. J. C. Chato, Laminar condensation inside horizontal and inclined tubes, *J ASHRAE*, Vol. 4, February 1962, pp. 52–60.

18. J. G. Collier, *Convective Boiling and Condensation*, 2nd edition, McGraw-Hill, New York, 1981.

19. W. M. Rohsenow, Film condensation, Section 12A in *Handbook of Heat Transfer*, W. M. Rohsenow and J. P. Hartnett, Eds., McGraw-Hill, New York, 1973.

20. E. M. Sparrow and S. H. Lin, Condensation in the presence of a noncondensible gas, *J. Heat Transfer*, Vol. 86, 1963, pp. 430–436.

21. W. J. Minkowycz and E. M. Sparrow, Condensation heat transfer in the presence of non-condensibles, interfacial resistance, superheating, variable properties, and diffusion, *Int. J. Heat Mass Transfer*, Vol. 9, 1966, pp. 1125–1144.

22. M. I. Flik and C. L. Tien, An approximate analysis for general film condensation transients, *J. Heat Transfer*, Vol. 111, 1989, pp. 511–517.

23. K. M. Holden, A. S. Wanniarachchi, P. J. Marto, D. H. Boone, and J. W. Rose, The use of organic coatings to promote dropwise condensation of steam, *J. Heat Transfer*, Vol. 109, 1987, pp. 768–774.

24. P. Griffith, Dropwise condensation, Section 12B in *Handbook of Heat Transfer*, W. M. Rohsenow and J. P. Hartnett, Eds., McGraw-Hill, New York, 1973.

25. H. Merte, Jr., Condensation heat transfer, *Advances in Heat Transfer*, Vol. 9, 1973, pp. 181–272.

26. I. Tanasawa, Advances in condensation heat transfer, *Advances in Heat Transfer*, Vol. 21, 1991, pp. 55–139.

27. S. Sideman and D. Moalem-Maron, Direct contact condensation, *Advances in Heat Transfer*, Vol. 15, 1982, pp. 227–281.

28. F. Kreith and R. F. Boehm, Eds., *Direct Contact Heat Transfer*, Hemisphere, Washington, DC, 1988.

29. P. C. Zmola, Ph.D. Thesis, Purdue University, June 1950.

30. R. Moissis and P. J. Berenson, On the hydrodynamic transitions in nucleate boiling, *J. Heat Transfer*, Vol. 85, 1963, pp. 221–229.

31. C. Wares. On the fixation of water in diverse fire, *Int. J. Heat Mass Transfer*, Vol. 9, 1966, pp. 1153–1166.

32. B. P. Breen and J. W. Westwater, Effect of diameter of horizontal tubes on film boiling heat transfer, *Chem. Engng. Progress*, Vol. 58, No. 7, pp. 67–72.

33. S. Nukiyama, The maximum and minimum values of the heat Q transmitted from metal to boiling water under atmospheric pressure, *J. Jap. Soc. Mech. Eng.*, Vol. 37, 1934, pp. 367–374; English translation in *Int. J. Heat Mass Transfer*, Vol. 9, 1966, pp. 1419–1433.

34. T. B. Drew and C. Mueller, Boiling, *Trans. AIChE*, Vol. 33, 1937, pp. 449–473.

35. W. M. Rohsenow, A method for correlating heat transfer data for surface boiling of liquids, *Trans. ASME*, Vol. 74, 1952, pp. 969–976.

36. R. I. Vachon, G. H. Nix, and G. E. Tanger, Evaluation of constants for the Rohsenow pool-boiling correlation, *J. Heat Transfer*, Vol. 90, 1968, pp. 239–247.

37. P. E. Liley, Thermophysical properties, Chapter 22 in S. Kakac, R. K. Shah, and W. Aung, Eds., *Handbook of Single-Phase Convective Heat Transfer*, Wiley, New York, 1987.

38. *ASHRAE Handbook of Fundamentals*, ASHRAE, New York, 1981.

39. R. H. Kropschot, B. W. Birmingham, and D. B. Mann, Eds., *Technology of Liquid Helium*, NBS Monograph 111, Washington, DC, October 1968.

40. D. K. Edwards, V. E. Denny, and A. F. Mills, *Transfer Processes*, 2nd edition, Hemisphere, Washington, DC, 1979.

41. W. M. Rohsenow, What we don't know and do know about nucleate pool boiling heat transfer, ASME HTD-Vol. 104, Vol. 2, 1988, pp. 169–172.

42. J. H. Lienhard and V. K. Dhir, Extended hydrodynamic theory of the peak and minimum pool boiling heat fluxes, NASA CR-2270, July 1973.

43. S. S. Kutateladze, On the transition to film boiling under natural convection, *Kotloturbostroenie*, No. 3, 1948, p. 10.

44. N. Zuber, On the stability of boiling heat transfer, *Trans. ASME*, Vol. 80, 1958, pp. 711–720.

45. K. H. Sun and J. H. Lienhard, The peak pool boiling heat flux on horizontal cylinders, *Int. J. Heat Mass Transfer*, Vol. 13, 1970, pp. 1425–1439.

46. C. S. Yih, *Fluid Mechanics*, McGraw-Hill, New York, 1969, pp. 439–440.

47. P. J. Berenson, Film boiling heat transfer for a horizontal surface, *J. Heat Transfer*, Vol. 83, 1961, pp. 351–358.

48. J. H. Lienhard and P. T. Y. Wong, The dominant unstable wavelength and minimum heat flux during film boiling on a horizontal cylinder, *J. Heat Transfer*, Vol. 86, 1964, pp. 220–226.

49. A. L. Bromley, Heat transfer in stable film boiling, *Chem. Eng. Progress*, Vol. 46, 1950, pp. 221–227.

50. M. R. Duignan, G. A. Greene, and T. F. Irvine, Jr., Film boiling heat transfer to large superheats from a horizontal flat surface, *J. Heat Transfer*, Vol. 113, 1991, pp. 266–268.

51. E. M. Sparrow, The effect of radiation on film-boiling heat transfer, *Int. J. Heat Mass Transfer*, Vol. 7, 1964, pp. 229–238.

52. W. M. Rohsenow, Boiling, Section 13 in *Handbook of Heat Transfer*, W. M. Rohsenow and J. P. Hartnett, Eds., McGraw-Hill, New York, 1973.

53. P. B. Whalley, *Boiling, Condensation and Gas-Liquid Flow*, Clarendon Press, Oxford, 1987, Chapters 16, 17, and 20.

54. J. H. Lienhard and R. Eichhorn, Peak boiling heat flux on cylinders in a cross flow, *Int. J. Heat Mass Transfer*, Vol. 19, 1976, pp. 1135–1142.

55. K. Kheyrandish and J. H. Lienhard, Mechanisms of burnout in saturated and subcooled flow boiling over a horizontal cylinder, ASME-AIChE National Heat Transfer Conference, Denver, CO, August 4–7, 1985.

56. A. L. Bromley, N. R. LeRoy, and J. A. Robbers, Heat transfer in forced convection film boiling, *Ind. Eng. Chem.*, Vol. 45, 1953, pp. 2639–2646.

57. A. Sakurai, Film boiling heat transfer, Proc. 9th Int. Heat Transfer Conf., Jerusalem, 1990, Vol. 1, pp. 157–186.

58. V. K. Dhir, Nucleate and transition boiling heat transfer under pool and external flow conditions, Proc. 9th Int. Heat Transfer Conf., Jerusalem, 1990, Vol. 1, pp. 129–155.

59. D. Poulikakos, Interaction between film condensation on one side of a vertical wall and natural convection on the other side, *J. Heat Transfer*, Vol. 108, 1986, pp. 560–566.

PROBLEMS

Condensation on Vertical Surfaces

8.1 Rely on the formulas developed for a vertical laminar film, and demonstrate that the total cooling rate provided by the wall is proportional to the total rate of condensation, $q' = h'_{fg}\Gamma(L)$.

8.2 Consider the control volume drawn around the entire film of height L shown in the figure, and make no assumption concerning the flow regimes that may be present inside the control volume. The vertical wall is isothermal, T_w. To the right of the film of condensate the vapor is stagnant and saturated. Show that in this general configuration the total wall cooling rate is proportional to the condensate mass flowrate, $q' = h'_{fg}\Gamma(L)$.

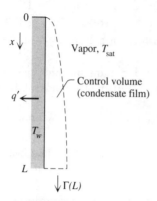

Figure P8.2

8.3 Show that regardless of the flow regime, the length L of a vertical film of condensate is related to the average heat transfer coefficient $\bar{h}_L$ and the total condensation rate $\Gamma(L)$ by the general formula

$$L = \frac{h'_{fg}\Gamma(L)}{(T_{sat} - T_w)\bar{h}_L}$$

Use this formula to derive eq. (8.24), which holds only for a *laminar* vertical film, when $\rho_l \gg \rho_v$.

8.4 Demonstrate that the local Reynolds number Re_x along a laminar vertical film of condensate is equal to

$$Re_x = \frac{8}{3}\frac{u_{max}(x)\,\delta(x)}{\mu_l}$$

where $u_{max}(x)$ is the downward velocity of the liquid–vapor interface. [Review the zero-shear boundary condition discussed above eq. (8.3).]

8.5 Saturated vapor condenses on a cold vertical slab of height L. Both sides of the slab are covered by laminar films of condensate. A single horizontal cylinder of diameter D, and at the same temperature as the slab, is immersed in the same saturated vapor. For what special diameter D will the total condensation rate on the cylinder equal the total condensation rate produced by the slab?

8.6 **(a)** Saturated steam at 1 atm condenses on a vertical wall of temperature 80°C and height 1 m. Assume that the condensate forms a laminar film, and calculate the average heat transfer coefficient, the condensation rate, and the film Reynolds number at the bottom of the wall.

(b) Is the film laminar over its entire height? If not, recalculate the quantities of part (a) by relying on the chart given in Fig. 8.6. Compare the new condensation rate estimate with the value calculated in part (a), and you will see the condensation augmentation effect of the film waviness.

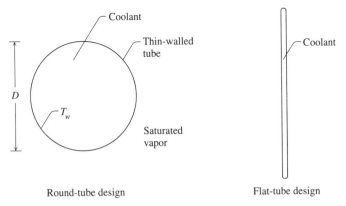

Figure P8.7

Condensation in Other Configurations

8.7 The horizontal thin-walled tube shown in the figure is cooled by an internal fluid of temperature T_w. The tube is immersed in a stagnant atmosphere of saturated vapor, which condenses in laminar film fashion on the outer cylindrical surface.

It is proposed to increase the total condensation rate by flattening the tube cross section into the shape shown on the right side of the figure. Calculate the percent increase in condensation flowrate associated with this design change.

8.8 A plane rectangular surface of width $L = 1$ m, length $Z \gg L$, and temperature $T_w = 80°C$ is suspended in saturated steam of temperature $100°C$. When this surface is oriented in such a way that L is aligned with the vertical (i.e., as in Problem 8.6), the steam condenses on it at the rate 0.063 kg/s·m. The purpose of this exercise is to show how the condensation rate decreases when the surface becomes tilted relative to the vertical direction.

(a) Calculate the condensation rate when the L width makes a 45° degree angle with the vertical and the Z length is aligned with the horizontal. Determine also the film Reynolds number and the flow regime.

(b) Assume that the surface is perfectly horizontal facing upward (as the top surface in Fig. 8.10) and that the film of condensate is laminar. Calculate the condensation rate and the Reynolds number of the liquid film spilling over one edge, and verify the validity of the laminar film assumption.

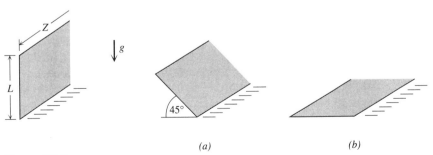

(a)   *(b)*

Figure P8.8

8.9 Atmospheric-pressure saturated steam condenses on the outside of a horizontal tube of wall temperature $T_w = 60°C$ and outer diameter $D = 2$ cm. Assume that the condensate forms a laminar film, and calculate the mass flowrate of condensate dripping from the bottom of the tube. Calculate also the film Reynolds number, and in this way prove the validity of the laminar flow assumption.

8.10 **(a)** The bank of horizontal tubes shown on the left side of the figure is surrounded by 100°C saturated steam, which condenses on the outside of each tube. The tube surface is maintained at 60°C by a cold fluid that flows through each tube in the direction perpendicular to the plane of the figure. Assuming that the condensate film is laminar, calculate the total mass flowrate of condensate per unit length of tube bank.

(b) In a competing design, the same bundle of tubes appears rotated by 90°, as shown on the right side of the figure. Calculate the total condensate mass flowrate in this new design, and comment on the effect of the 90° rotation.

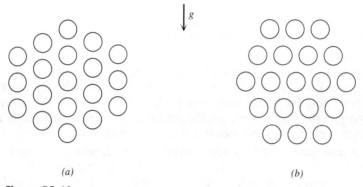

(a)   (b)

Figure P8.10

8.11 The average heat transfer coefficients for film condensation on an upward facing strip and disc (Fig. 8.10) are listed in eqs. (8.33) and (8.34). Rewrite each of these formulas by using as length scale the characteristic length of the surface, eq. (7.76),

$$L_c = \frac{A}{p}$$

where A and p are the area and perimeter of the surface, respectively. In this way, show that the average heat transfer coefficient of any other surface whose shape is somewhere between the "very long" limit (the strip) and the "round" limit (the disc) is given by the approximate formula

$$\overline{Nu}_{L_c} = \frac{\overline{h}L_c}{k_l} \cong 0.8 \left[\frac{L_c^3 h'_{fg} g\, (\rho_l - \rho_v)}{k_l \nu_l\, (T_{sat} - T_w)} \right]^{1/5}$$

8.12 **(a)** Saturated steam at 1 atm condenses as a laminar film on a metallic horizontal tube at 80°C. The tube outside diameter is 4 cm. Calculate the condensation rate and verify that the laminar film assumption is adequate.

(b) It is proposed to coat the tube with a 0.5-mm layer of teflon to achieve dropwise condensation on the exposed surface. Assume that the average heat transfer coefficient for dropwise condensation is ten times greater than the value calculated in part (a). Calculate the new condensation rate, and explain why it is not greater than when the film was laminar.

(c) How thin must the teflon coating be if it is to increase the condensation rate?

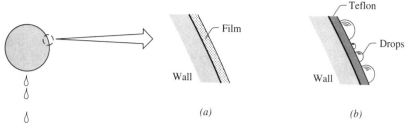

Figure P8.12

Nucleate Boiling

8.13 Consider the spherical vapor bubble of radius r shown in the figure. The pressure and temperature inside the bubble (P_v, T_v) are slightly above the pressure and temperature in the liquid (P_l, T_l). The liquid is saturated, $T_l = T_{sat}$.

   (a) Invoke the mechanical equilibrium of one hemispherical control volume, and show that the bubble radius varies inversely with the pressure difference:

$$r = \frac{2\sigma}{P_v - P_l}$$

   (b) Rely on the Clausius–Clapeyron relationship $dP/dT = h_{fg}/(Tv_{fg})$ to show that the bubble radius also varies inversely with the temperature difference:

$$r = \frac{2\sigma T_{sat}}{h_{fg}\rho_v(T_v - T_{sat})}$$

   (c) Calculate the radius of a steam bubble with $T_v - T_{sat} = 2$ K in water at $T_{sat} = 100°C$.

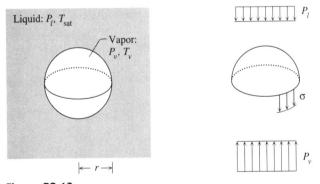

Figure P8.13

8.14 Consider the water on nickel nucleate boiling calculations outlined in Example 8.3. Obtain an estimate for the excess temperature $T_w - T_{sat}$ at critical heat flux conditions by equating the nucleate boiling heat flux q''_w with the calculated peak heat flux q''_{max}. Compare your estimate with the actual excess temperature at peak heat flux (Fig. 8.17), and explain why your ($T_w - T_{sat}$) value is smaller.

8.15 Estimate the C_{sf} constant that corresponds to the nucleate boiling portion of the curve shown in Fig. 8.17. Use Rohsenow's correlation (8.43) and a point (q''_w,

$T_w - T_{sat}$) in the vicinity of the transition from the isolated bubbles regime to the regime of columns and slugs.

8.16 The vacuum insulation around a spherical liquid helium vessel breaks down (develops an air leak) and allows the heat flux $q_w'' = 10^3$ W/m² to land on the external surface of the vessel. An amount of saturated liquid helium at atmospheric pressure boils at the bottom of the vessel. Calculate the excess temperature $T_w - T_{sat}$ by assuming nucleate boiling with $C_{sf} = 0.02$ and $s = 1.7$. Compare the heat leak q_w'' with the peak heat flux for nucleate boiling in the pool of liquid helium.

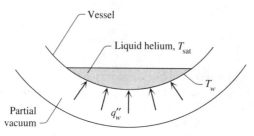

Figure P8.16

8.17 Water boils in the pressurized cylindrical vessel shown in the figure. The steam relief valve is set in such a way that the pressure inside the vessel is 4.76×10^5 N/m². The bottom surface is made out of copper (polished), and its temperature is maintained at $T_w = 160°$C. Assume nucleate boiling, and calculate the total heat transfer rate from the bottom surface to the boiling water. Later, verify the correctness of the nucleate boiling assumption.

The calculation of the time when there is no liquid left in the vessel is the subject of the next problem.

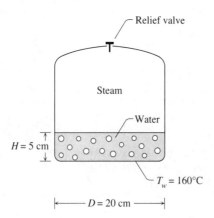

Figure P8.17

8.18 The cylindrical vessel described in the preceding problem has an inner diameter of 20 cm. The depth of the original amount of liquid is 5 cm, and the pressure is maintained at 4.76×10^5 N/m². The nucleate boiling heat transfer rate to the liquid (calculated in the preceding problem) is $q_w = 12.44$ kW.

(a) Estimate the time needed to evaporate all the liquid. Base your estimate on the simple relation $q_w = \dot{m} h_{fg}$, which is routinely recommended.

(b) The $q_w = \dot{m} h_{fg}$ relation is valid only approximately and is incorrect from a

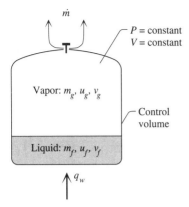

Figure P8.18

thermodynamic standpoint. With reference to the control volume defined by the pressurized vessel, show that the correct proportionality between q_w and $\dot{m}$ is, in fact,

$$q_w = \dot{m}\left(h_g - \frac{u_f v_g - u_g v_f}{v_g - v_f}\right)$$

where u, v, $()_g$, and $()_f$ are the usual thermodynamics symbols for specific internal energy, specific volume, saturated vapor, and saturated liquid, respectively. Note that the pressure (or temperature) and the total volume V remain constant. Numerically, show that the quantity arrived at in the round brackets deviates from h_{fg}, assumed in part (a), as the saturated liquid–vapor mixture approaches the critical point.

8.19 The bottom of a shallow pan is a 20-cm-diameter disc made out of mechanically polished stainless steel. It is used to boil water at atmospheric pressure while the temperature of the bottom surface is 108°C.

(a) Calculate the heat flux supplied by the bottom surface to the water pool. Consult Fig. 8.17 to anticipate the boiling regime, and later compare the calculated q_w'' value with the one plotted on the figure.

(b) Estimate also the heat transfer coefficient and the total heat transfer rate to the water pool. Compare the calculated h value with the range of values indicated in Fig. 1.12.

Film Boiling

8.20 Let δ be the average thickness of the vapor film in the film boiling regime depicted in Fig. 8.21. When the heat transfer across this film is dominated by convection, the heat flux is roughly the same as the conduction heat flux across a vapor layer of thickness δ:

$$q_w'' \sim k_v \frac{T_w - T_{sat}}{\delta}$$

Use this idea and the numerical data of Example 8.4 to evaluate the film thickness δ.

8.21 In a power-controlled pool boiling experiment a horizontal cylindrical heater is immersed in saturated water at atmospheric pressure. The peak heat flux is 10^6 W/m². The power is increased slightly above this level, and the nucleate boiling regime is replaced abruptly by film boiling (Fig. 8.20, right). Estimate the excess temperature in this new regime by assuming that radiation is the dominant (i.e., the only) mode of heat transfer across the film. Assume also $\epsilon_w = 1$. Compare your estimate with the value read off Fig. 8.17. Will the actual excess temperature be larger or smaller than this pure-radiation estimate?

8.22 Water boils in the film boiling regime on a horizontal surface with an excess temperature of 200°C. Calculate the minimum heat flux when

(a) $T_{sat} = 100°C$, and

(b) $T_{sat} = 300°C$,

and comment on the behavior of q''_{min} as the saturated water pool approaches the critical point.

PROJECTS

8.1 Film Condensation Caused by Natural Convection Cooling

A vertical wall of height $L = 50$ cm separates a volume of 100°C saturated steam from the atmosphere of temperature $T_\infty = 20°C$. The steam side of the wall is coated by a thin film of condensate while the air side is swept by a free-convection upflow of heated air. Neglect the thermal resistance posed by the wall thickness, and for simplicity assume that the wall temperature T_w is uniform. Calculate

(a) The heat transfer rate from the steam to the atmosphere, q' (W/m);

(b) The mass flowrate of the condensate generated on the steam side of the wall, namely, $\Gamma(L)$; and

(c) The average wall temperature T_w.

Before starting these calculations, examine Fig. 1.12 and note that, in general, the thermal resistance on the air side is orders of magnitude greater than on the condensing side. This observation will allow you to set approximately $T_w \cong 100°C$ in all the calculations that refer to natural convection on the air side. The *conjugate* condensation and natural convection heat transfer mechanism analyzed in this project is discussed in more general and rigorous terms by Poulikakos [59].

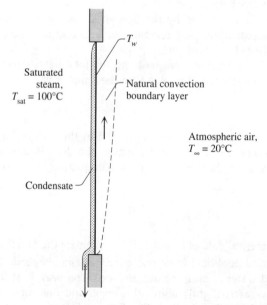

Figure PR8.1

8.2 Film Boiling Cooling of a Lumped Capacitance

Film boiling occurs on the surface of a sphere of temperature $T_w = 354°C$ and diameter $D = 2$ cm, which is plunged in a bath of saturated water at atmospheric pressure. The sphere is made out of polished copper.

(a) Calculate the heat transfer coefficient due to convection, $\bar{h}_D$, the correction due to radiation, $\bar{h}_{rad}$, and the total heat transfer rate from the copper ball to the water pool. For the emissivity of the copper surface assume $\epsilon_w = 0.05$.

(b) Model the copper sphere as a lumped capacitance, and calculate its new temperature 10 s after the start of film boiling. Consult Fig. 4.5, and demonstrate that the lumped capacitance model applies in the present case.

9

HEAT
EXCHANGERS

9.1 CLASSIFICATION OF HEAT EXCHANGERS

In the convection phenomena discussed thus far (Chapters 5 to 8), our primary goal has been to determine the relationship between the heat transfer rate and the driving temperature difference. In most instances this operation was reduced to calculating the proper heat transfer coefficient. In the present chapter we have the opportunity to use this heat transfer coefficient information in the greater effort of designing an actual apparatus called a *heat exchanger*.

A heat exchanger is a device, or piece of hardware, the function of which is to promote the transfer of heat between two or more entities at different temperatures. In most cases, and especially in the examples considered in this chapter, the heat exchanging entities are two streams of fluid. To prevent stream-to-stream mixing, the two fluids are separated by solid walls that, together, constitute the *heat transfer surface*, or *heat exchanger surface*. In some heat exchangers a solid heat transfer surface is not necessary, because of the natural immiscibility of the two streams or the separation (stratification) of the two fluids in the gravitational field. In such cases the heat transfer between the two fluids occurs through their mutual interface, and the apparatus is called a *direct contact heat exchanger*.

It must be recognized from the outset that the heat exchanger is a multifaceted engineering system whose design involves not only the calculation of the heat transfer rate across the heat exchanger surface but also the pumping power needed to circulate the two streams through the various flow passages, the geometric layout of the flow pattern (the wrapping and weaving of one stream around and through the other), the construction of the actual hardware, and the ability to disassemble the apparatus for periodic cleaning. Over the past 100 years, the analysis, design, and manufacturing of heat exchangers has grown into a separate discipline in thermal engineering, the current status of which is best described in the heat exchanger monographs of Refs. [1–4]. In the present chapter, we can only examine the heat transfer roots of this discipline.

The other aspect that makes the study of heat exchangers challenging is their diversity. There are many types of heat exchangers, and there is even more than one way to categorize these types. According to Shah's classification [5], one

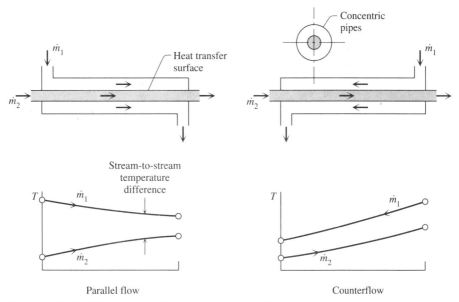

Figure 9.1 Double-pipe parallel flow and counterflow heat exchangers.

way to distinguish between the various types is by looking at the flow arrangement. Figures 9.1 and 9.2 show the three main flow configurations known as *parallel flow* (or cocurrent flow), *counterflow* (or countercurrent flow), and *cross-flow*. In parallel flow, the two inlet ports are positioned at the same end of the heat exchanger, where the stream-to-stream temperature difference is the greatest. In the counterflow scheme, the stream-to-stream temperature difference is more evenly distributed along the heat exchanger.

The two cross-flow arrangements shown in Fig. 9.2 show that the design can also vary with respect to the degree of lateral mixing that is experienced by each stream inside the channel. The lateral mixing effect can be inhibited by installing longitudinal corrugations that divide the "unmixed" stream into many ministreams that flow in parallel. In the absence of longitudinal partitions, the stream can mix transversally in each channel cross section: Whether the stream can be regarded as truly "mixed" depends on the design of the channel (width, length), the presence of turbulence, and the degree to which the mixing effect can propagate downstream [6]. The aluminum core for an air-to-air cross-flow heat exchanger with both streams unmixed is shown in Fig. 9.3.

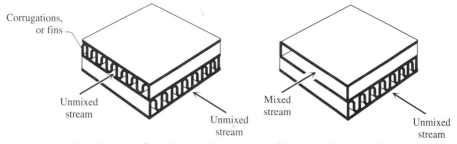

Figure 9.2 Plate fin cross-flow heat exchangers, and the use of longitudinal corrugations to prevent the transversal mixing of the stream.

Figure 9.3 The aluminum core of an air-to-air cross-flow heat exchanger. (Courtesy of Mr. D. P. Shatto, AKG of America, Inc., Mebane, North Carolina.)

The flow arrangements of Figs. 9.1 and 9.2 are all *single-pass* schemes, because in each case the stream passes only once through the heat exchanger control volume. Examples of *multipass* arrangements are given in the second and third cases illustrated in Fig. 9.4.

Another way to differentiate between various heat exchanger designs is to consider their construction. Perhaps the simplest design is the *double pipe* arrangement used in the two examples of Fig. 9.1. In the concentric pipes arrangement, the streams are separated by the wall of the inner pipe, the outer or inner surface of which plays the role of "heat transfer surface."

More complicated is the *shell-and-tube* heat exchanger illustrated in Fig. 9.5, in which the inner stream flows through not one but several tubes. The outer stream is confined by a large-diameter vessel (the "shell"). Transversal baffles

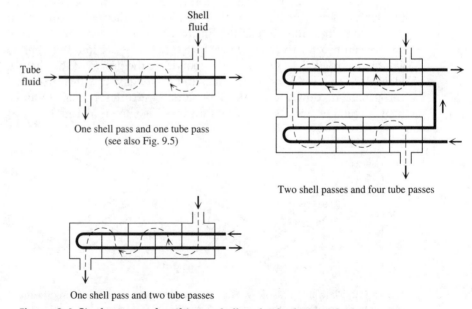

Figure 9.4 Single-pass and multipass shell-and-tube heat exchangers.

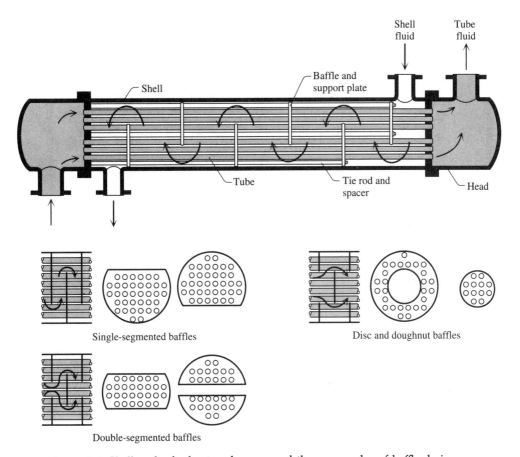

Figure 9.5 Shell-and-tube heat exchanger, and three examples of baffle design.

force the outer stream to flow across the tubes, augmenting in this way the overall heat transfer coefficient between the two fluids. Three of the more common baffle designs are also illustrated in Fig. 9.5. Worth noting is that the shell-and-tube heat exchanger of Fig. 9.5 is of the same type as the single-pass heat exchanger illustrated as the first case in Fig. 9.4, where, for simplicity, only one of the tubes was drawn. Note also that the direction of gravity was not taken into account when the flow directions were indicated with arrows in Figs. 9.1, 9.4, 9.5 and, later, in Figs. 9.16 to 9.20 and 9.22 to 9.28.

Another construction type is the *plate fin* heat exchanger, in which each channel is defined by two parallel plates separated by fins or spacers. Examples of plate fin heat exchangers are given in Figs. 9.6 and 9.7. The fins are connected to the parallel plates by tight mechanical fit, gluing, soldering, brazing, welding, or extrusion. Alternating passages are connected in parallel to end chambers (called headers) and form one side (i.e., one stream) of the heat exchanger. Fins are employed on both sides in gas-to-gas applications. In gas-to-liquid applications, fins are needed only on the gas side because there the heat transfer coefficient is lower. If present on the liquid side, the fins play a structural (stiffening) role. In bar-and-plate heat exchangers, solid parallel bars are used to seal the edges of each passage.

Figure 9.8 shows two examples of *tube fin* heat exchangers, also called finned tube exchangers. These are directed to gas-to-liquid applications such as car

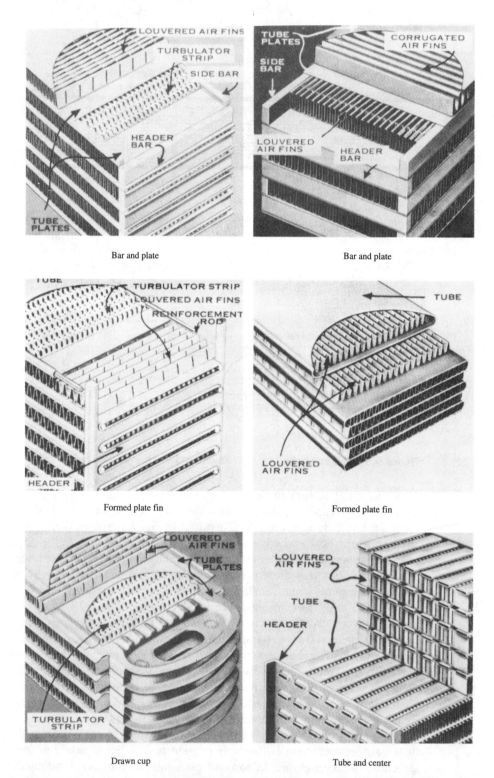

Figure 9.6 Plate fin heat exchangers. (Shah [5], courtesy of Dr. R. K. Shah, Harrison Division, General Motors Corporation, Lockport, New York.)

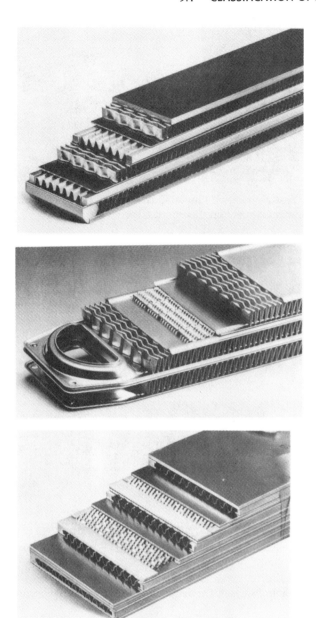

Figure 9.7 Plate fin heat exchangers. (Courtesy of Mr. D. P. Shatto, AKG of America, Inc., Mebane, North Carolina.)

radiators. The liquid flows through the tubes, which can accommodate relatively high pressures. Tube fin heat exchangers are less compact than plate fin heat exchangers; in other words, they have less heat transfer area per unit volume. Six examples of continuous fins that are used in tube fin heat exchangers are presented in Fig. 9.9.

The fins can be continuous, as in the case of the unmixed streams of Fig. 9.2, or interrupted, as on the left side of Fig. 9.10. Many fin shapes and orientations have been developed. In the cross-flow heat exchanger shown on the right side

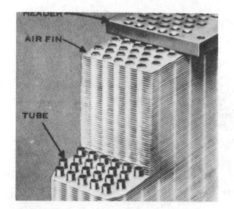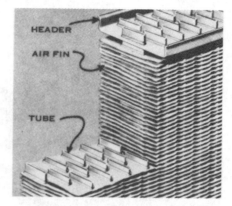

Figure 9.8 Tube fin heat exchangers. (Shah [5], courtesy of Dr. R. K. Shah, Harrison Division, General Motors Corporation, Lockport, New York.)

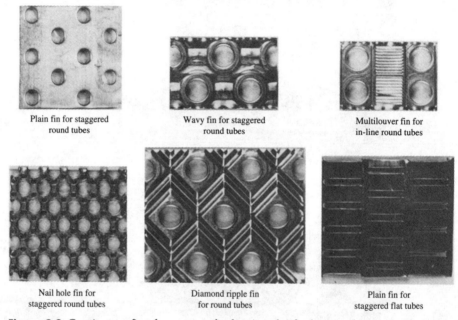

Plain fin for staggered round tubes

Wavy fin for staggered round tubes

Multilouver fin for in-line round tubes

Nail hole fin for staggered round tubes

Diamond ripple fin for round tubes

Plain fin for staggered flat tubes

Figure 9.9 Continuous fins for arrays of tubes in tube fin heat exchangers. (Shah [5], courtesy of Dr. R. K. Shah, Harrison Division, General Motors Corporation, Lockport, New York.)

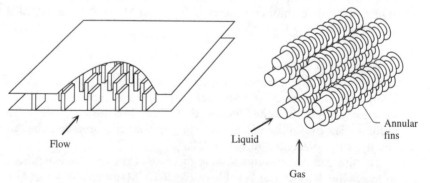

Flow

Liquid

Gas

Annular fins

Figure 9.10 Parallel-plate channel with interrupted fins (left) and bank of finned tubes in cross-flow (right).

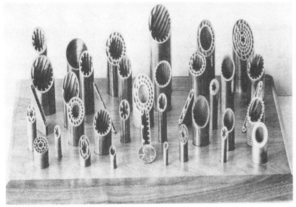

Figure 9.11 Pin fins (top) and tubes with internal fins (bottom). (Shah [5], courtesy of Dr. R. K. Shah, Harrison Division, General Motors Corporation, Lockport, New York.)

of Fig. 9.10, the outer surface of each tube is fitted with equidistant annular fins of constant thickness. In this arrangement the finned side of the tube wall usually faces a gaseous stream (mixed, or partially mixed), while the inner surface of each tube (unfinned) faces a liquid stream. Once again, the fins are needed more on the gas side of the tube wall, because the bare-wall heat transfer coefficient on the gas side is smaller than on the liquid side. Additional examples of finned surfaces are given in Fig. 9.11.

Finally, heat exchangers can also be categorized with respect to their degree of compactness. The plate fin and tube fin designs of Figs. 9.2 through 9.9 have more heat transfer area per unit volume than the corresponding designs without fins. Figure 9.12 shows that a high heat transfer area density corresponds to a small hydraulic diameter for the passages wetted by the streams. *Compact heat exchangers* have heat transfer area densities in excess of 700 m^2/m^3 [5] and are essential in applications where the size and weight of the heat exchanger is an important design constraint (e.g., automobiles, naval and airborne power plants, and air conditioning systems).

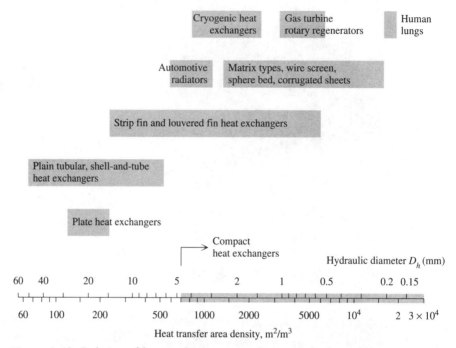

Figure 9.12 Ordering of heat exchangers according to their degree of compactness. (Drawn based on data presented by Shah [5].)

9.2 OVERALL HEAT TRANSFER COEFFICIENT

The heat exchanger surface is generally more complicated than the three-resistance sandwich analyzed earlier in Section 2.1.3 and Fig. 2.3. In addition to the conduction resistance of the wall and the convective film resistances on the two sides of the wall, the heat exchanger wall may be complicated by fins and the gradual accumulation of a layer of oxides and other deposits on all the surfaces exposed to fluids. This general configuration is illustrated in Fig. 9.13, where the

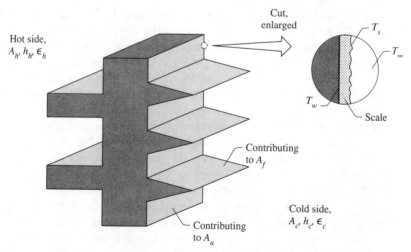

Figure 9.13 Heat exchanger surface with fins and scale on both sides.

deposited layer is called *scale*. The time-dependent process by which this layer is formed is called *scaling*, or *fouling*.

To understand the makeup of the overall thermal resistance formula for the heat exchanger surface, consider only one side of the wall. The total area of this side, A, is the sum of the area contributed by the exposed surfaces of the fins, A_f, plus the area of the unfinned portions of the wall, A_u:

$$A = A_f + A_u \tag{9.1}$$

Assuming now that the heat transfer coefficient has the same value h on both A_f and A_u, the total heat transfer rate through the surface A is

$$q = \eta h A_f (T_w - T_\infty) + h A_u (T_w - T_\infty) \tag{9.2}$$

The temperature T_w refers to the unfinned portions of the wall, or the base of each fin, while T_∞ is the bulk temperature of the fluid. The first term on the right side of eq. (9.2) represents the heat transfer contribution made by the fins, the factor η being the fin efficiency defined in eq. (2.115). The fin efficiency can be calculated based on formulas such as eq. (2.116) and the charts of Figs. 2.14 and 2.16.

The second term accounts for the direct heat transfer through the bare portions of the wall. The total heat transfer rate expression (9.2) can be rearranged so that q is proportional to the total heat transfer area A,

$$q = \left(\eta \frac{A_f}{A} + \frac{A_u}{A} \right) h A (T_w - T_\infty)$$
$$= \epsilon h A (T_w - T_\infty) \tag{9.3}$$

leading in this way to the definition of the *overall surface efficiency* factor

$$\epsilon = \eta \frac{A_f}{A} + \frac{A_u}{A} = 1 - \frac{A_f}{A}(1 - \eta) \tag{9.4}$$

This factor should not be confused with the overall projected surface effectiveness ϵ_0, which was defined in eq. (2.79).

When the surface A is covered with a layer of solid debris, the total heat transfer rate q must overcome not only the convective thermal resistance determined above, $(\epsilon h A)^{-1}$, but also the thermal resistance of the deposited layer:

$$r_s = \frac{T_w - T_s}{q''} \tag{9.5}$$

The r_s resistance refers to the scale that covers only one unit of the total area A, while T_s is the temperature of the fluid side of the scale, Fig. 9.13. It is not difficult to show that when the scale is present, the total heat transfer rate is given by an expression similar to eq. (9.3),

$$q = \epsilon h_e A (T_w - T_\infty) \tag{9.6}$$

where the *effective heat transfer coefficient* h_e accounts for two effects — the conduction thermal resistance across the scale (r_s) and the convective film resistance $(1/h)$ on the fluid side of the scale,

$$\frac{1}{h_e} = r_s + \frac{1}{h} \tag{9.7}$$

The r_s resistance is called *fouling factor* and has the units $m^2 \cdot K/W$. Table 9.1 shows a set of the most up-to-date recommendations concerning the r_s values to be used in overall heat transfer coefficient calculations [7]. The order of magnitude of the convective heat transfer coefficient [h in eq. (9.7)] depends on the flow arrangement and can be read off Fig. 1.12.

A worthwhile observation is that in eq. (9.7) it has been assumed that the size of the area crossed by the heat flux does not change as the flux travels the thickness of the scale. This assumption holds true as long as the scale is thin in comparison with the body that it covers (e.g., heat exchanger tube). Otherwise, the area variation must be taken into account, for example, by modeling the scale as a shell around the tube [cf. eq. (2.33)].

In summary, according to eq. (9.6) the total thermal resistance for one side of the heat exchanger surface is $(\epsilon h_e A)^{-1}$. Let the subscripts h and c represent the two sides (hot and cold) of the heat exchanger surface shown in Fig. 9.13. If $R_{t,w}$ is the conduction resistance of the wall itself, the total thermal resistance posed by the heat exchanger surface is

Table 9.1 Representative Values of the Fouling Factor r_s ($m^2 \cdot K/W$)[a]

Temperature of Heating Medium Water Temperature	Up to 115°C 50°C or less		115–205°C Above 50°C	
Water Velocity	1 m/s and less	Over 1 m/s	1 m/s and less	Over 1 m/s
Water types				
Distilled water	0.0001	0.0001	0.0001	0.0001
Sea water	0.0001	0.0001	0.0002	0.0002
Brackish water	0.0004	0.0002	0.0005	0.0004
City or well water	0.0002	0.0002	0.0004	0.0004
River water, average	0.0005	0.0004	0.0007	0.0005
Hard water	0.0005	0.0005	0.0009	0.0009
Treated boiler feed water	0.0002	0.0001	0.0002	0.0002
Liquids				
Liquid gasoline, oil, and liquefied petroleum gases		0.0002–0.0004		
Vegetable oils		0.0005		
Caustic solutions		0.0004		
Refrigerants, ammonia		0.0002		
Methanol, ethanol, and ethylene glycol solutions		0.0004		
Gases				
Natural gas		0.0002–0.0004		
Acid gas		0.0004–0.0005		
Solvent vapors		0.0002		
Steam (non-oil bearing)		0.0001		
Steam (oil bearing)		0.0003–0.0004		
Compressed air		0.0002		
Ammonia		0.0002		

[a]Constructed based on the data compiled in Ref. [7].

$$\frac{1}{U_c A_c} = \frac{1}{\epsilon_h h_{e,h} A_h} + R_{t,w} + \frac{1}{\epsilon_c h_{e,c} A_c} \tag{9.8}$$

On the left side, the overall heat transfer coefficient U_c is said to be based on the cold-side area A_c. Alternatively, the left side of eq. (9.8) can be labeled $1/(U_h A_h)$, in which U_h is the overall heat transfer coefficient based on the hot-side surface A_h:

$$\frac{1}{U_c A_c} \equiv \frac{1}{U_h A_h} \tag{9.9}$$

Before conducting a numerical calculation of the overall thermal resistance, it is a good idea to verify that all the resistance terms are significant (large enough) on the right side of eq. (9.8). The representative order of magnitude of the overall heat transfer coefficient in various two-fluid heat exchangers is listed in Table 9.2.

The wall resistance $R_{t,w}$ can be calculated using the methods of Chapter 2, for example, based on eq. (2.7) for a plane wall of thickness L, area A, and conductivity k, and eq. (2.33) for a tube wall of radii r_o and r_i, length l, and conductivity k. This method of calculating $R_{t,w}$ is based on the assumption that the conduction through the wall is unidirectional. (This conflicts somewhat with the wall sketched on the left side of Fig. 9.13, because there the size and density of the fins were exaggerated intentionally.) The overall heat transfer coefficient formula for a plane wall with convective heat transfer on both sides, eq. (2.17), is the special form taken by the more general eq. (9.8) in the case where the fins are absent and both surfaces are clean.

Table 9.2 Representative Orders of Magnitude of the Overall Heat Transfer Coefficient (After Refs. [2, 8])

Hot Fluid	Cold Fluid	$U(W/m^2 \cdot K)$
Water	Water	1000–2500
Ammonia	Water	1000–2500
Gases	Water	10–250
Light organics[a]	Water	370–730
Heavy organics[b]	Water	25–370
Steam	Water	1000–3500
Steam	Ammonia	1000–3500
Steam	Gases	25–250
Steam	Light organics	500–1000
Steam	Heavy organics	30–300
Light organics	Light organics	200–400
Heavy organics	Heavy organics	50–200
Light organics	Heavy organics	50–200
Heavy organics	Light organics	150–300

[a]Organic liquids with viscosities below 0.0005 kg/s·m.
[b]Organic liquids with viscosities greater than 0.001 kg/s·m.

Example 9.1

Plane Wall, Pin Fins On One Side

The heat exchanger surface shown in the figure separates a stream of hot liquid ($T_h = 100°C$, $h_h = 200$ W/m²·K) from a stream of cold gas ($T_c = 30°C$, $h_c = 10$ W/m²·K). This surface is made of a slab of thickness $t = 0.8$ cm, which has the frontal area of a square with side $H = 0.5$ m. To offset the effect of the small heat transfer coefficient h_c, the area of the cold side was increased by adding a number of pin fins arranged in a square pattern. The dimensions of the wall, the individual fin, and the square array are indicated directly on the figure. The wall and the fins are made of a metal whose conductivity is $k = 40$ W/m·K. The effect of fouling is negligible on both sides of the wall; in other words,

$$h_{e,h} \cong h_h \qquad h_{e,c} \cong h_c$$

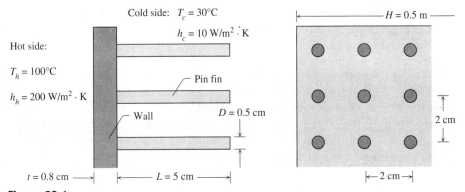

Figure E9.1

Calculate the overall thermal resistance of the heat exchanger surface ($1/U_h A_h$ or $1/U_c A_c$) and the total heat transfer rate q.

Solution. We begin with the geometry of the heat exchanger surface. The hot side is unfinned, therefore its total area is simply

$$A_h = (0.5 \text{ m})^2 = 0.25 \text{ m}^2$$

On the cold side we see one pin fin on each square with side 2 cm, in other words, one fin per 4 cm². The total number of fins is therefore

$$n = \frac{(0.5 \text{ m})^2}{4 \text{ cm}^2/\text{fin}} = \frac{0.25 \text{ m}^2}{0.0004 \text{ m}^2/\text{fin}} = 625 \text{ fins}$$

The total unfinned area on the right side is the wall area left between the roots of the fins:

$$A_u = H^2 - n\pi \left(\frac{D}{2}\right)^2$$

$$= (0.5 \text{ m})^2 - 625\pi \left(\frac{0.005 \text{ m}}{2}\right)^2 = 0.238 \text{ m}^2$$

To calculate the finned area A_f, we first model each fin (actual length L) as a slightly longer one with insulated tip (length L_c). This larger length is given by Harper and Brown's approximation, eq. (2.114):

$$L_c = L + \frac{D}{4} = 5.25 \text{ cm}$$

The finned area is the sum of the lateral (cylindrical) areas contributed by each fin:

$$A_f = n\pi D L_c$$

$$= 625\pi(0.005 \text{ m})(0.0525 \text{ m}) = 0.515 \text{ m}^2$$

Therefore, the total area of the cold side is

$$A_c = A_f + A_u = 0.753 \text{ m}^2$$

Next, to calculate the cold surface efficiency ϵ_c, we first estimate the efficiency of each pin fin [cf. eq. (2.116)],

$$\eta = \frac{\tanh(mL_c)}{mL_c}$$

in which the dimensionless group mL_c has the value

$$mL_c = \left(\frac{h_c P}{kA}\right)^{1/2} L_c = \left(\frac{h_c \pi D}{k(\pi/4)D^2}\right)^{1/2} L_c$$

$$= 2\left(\frac{h_c}{kD}\right)^{1/2} L_c$$

$$= 2\left(\frac{10 \text{ W}}{\text{m}^2 \cdot \text{K}} \frac{\text{m} \cdot \text{K}}{40 \text{ W}} \frac{1}{0.005 \text{ m}}\right)^{1/2} 0.0525 \text{ m} = 0.742$$

The fin efficiency is, then,

$$\eta = \frac{\tanh(0.742)}{0.742} = 0.85$$

and this translates into an overall cold-surface efficiency of

$$\epsilon_c = \eta \frac{A_f}{A_c} + \frac{A_u}{A_c}$$

$$= 0.85 \frac{0.515 \text{ m}^2}{0.753 \text{ m}^2} + \frac{0.238 \text{ m}^2}{0.753 \text{ m}^2} = 0.897$$

The overall thermal resistance posed by the heat exchanger surface is

$$\frac{1}{U_h A_h} = \frac{1}{h_h A_h} + \frac{t}{kA_h} + \frac{1}{\epsilon_c h_c A_c}$$

$$= \left(\frac{1}{200 \times 0.25} + \frac{0.008}{40 \times 0.25} + \frac{1}{0.897 \times 10 \times 0.753}\right) \frac{\text{K}}{\text{W}}$$

$$= (0.02 + 0.0008 + 0.148) \frac{\text{K}}{\text{W}} = 0.169 \text{ K/W}$$

This calculation shows that the cold-side resistance (0.148 K/W) accounts for 88 percent of the overall resistance (0.169 K/W), and that the resistance of the plane wall of thickness t is negligible. Finally, the total heat transfer rate through the entire arrangement is

$$q = U_h A_h (T_h - T_c) = \frac{(100 - 30) \text{ K}}{0.169 \text{ K/W}} = 415 \text{ W}$$

9.3   THE LOG-MEAN TEMPERATURE DIFFERENCE METHOD

9.3.1   Parallel Flow

In this and the next two sections we analyze the relationship between the total heat transfer rate q, the heat transfer area A, and the inlet and outlet temperatures of the two streams.

Consider first the *parallel flow* heat exchanger shown in Fig. 9.14. The abscissa shows the "length" of the heat exchanger, or the heat transfer area that is sandwiched between and swept by the two streams. Control volumes of width dA are drawn around each of the streams. The local stream-to-stream temperature difference is $T_h - T_c$, where T_h and T_c represent the bulk temperatures of the "hot" and "cold" streams, respectively. The heat transfer rate that crosses the area element dA is

$$dq = (T_h - T_c)U_A\, dA \tag{9.10}$$

where U_A is the overall heat transfer coefficient at this particular position along the heat exchanger surface. In addition, with reference to each of the control volumes, the first law of thermodynamics for steady flow requires that

$$dq = -C_h\, dT_h \tag{9.11}$$

$$dq = C_c\, dT_c \tag{9.12}$$

The symbol C (W/K) is shorthand for the product $\dot{m}c_P$ (i.e., the capacity rate) of each stream. Worth keeping in mind is that by writing the first-law statements (9.11) and (9.12) we are assuming either that the two fluids are ideal gases or that the pressure drop along each stream is thermodynamically negligible so that the specific enthalpy change experienced by each fluid can be approximated by $di \cong c_P\, dT$.

An equation that describes the variation of $T_h - T_c$ along the heat transfer surface A can be obtained by using eqs. (9.11) and (9.12),

$$d(T_h - T_c) = dT_h - dT_c = -\frac{dq}{C_h} - \frac{dq}{C_c}$$

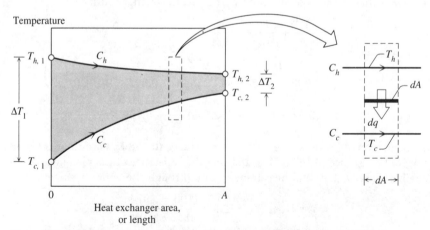

Figure 9.14 The temperature distribution in a parallel flow heat exchanger.

$$= -\left(\frac{1}{C_h} + \frac{1}{C_c}\right) dq \tag{9.13}$$

and by eliminating dq based on eq. (9.10):

$$d(T_h - T_c) = -\left(\frac{1}{C_h} + \frac{1}{C_c}\right)(T_h - T_c) U_A \, dA \tag{9.14}$$

The separation of variables is achieved by dividing both sides by $T_h - T_c$; after integrating from $A = 0$ to any A, we obtain

$$\ln \frac{\Delta T_2}{\Delta T_1} = -\left(\frac{1}{C_h} + \frac{1}{C_c}\right) UA \tag{9.15}$$

where ΔT_1 and ΔT_2 are the stream-to-stream temperature differences in the front section and the back section of the parallel-flow heat exchanger (see Fig. 9.14):

$$\Delta T_1 = T_{h,1} - T_{c,1} \qquad \Delta T_2 = T_{h,2} - T_{c,2} \tag{9.16}$$

The overall heat transfer coefficient U in eq. (9.15) is the result of *averaging* U_A over the entire area A. As a special case, $U = U_A$ when U_A is constant along the heat transfer area.

What we have achieved in eq. (9.15) is a relationship between the heat exchanger "size" (UA), the inlet and outlet temperatures (or ΔT_1 and ΔT_2), and the two capacity rates (C_h, C_c). The total stream-to-stream heat transfer rate q can be made to appear in this relationship by first integrating the first-law statements (9.11) and (9.12) along each stream:

$$q = -C_h (T_{h,2} - T_{h,1}) \tag{9.17}$$

$$q = C_c (T_{c,2} - T_{c,1}) \tag{9.18}$$

These integrations are based on the assumption that each C is a constant, in other words, that the c_p of each fluid does not vary appreciably as the stream travels the entire length of the heat exchanger. Together, eqs. (9.17) and (9.18) show that

$$\frac{1}{C_h} + \frac{1}{C_c} = \frac{1}{q}(-T_{h,2} + T_{h,1} + T_{c,2} - T_{c,1})$$

$$= \frac{1}{q}(\Delta T_1 - \Delta T_2) \tag{9.19}$$

Therefore, eq. (9.15) can also be written as

$$\ln \frac{\Delta T_2}{\Delta T_1} = -\frac{1}{q}(\Delta T_1 - \Delta T_2) \, UA \tag{9.20}$$

In conclusion, the proportionality between the total heat transfer rate q and the overall thermal conductance of the heat exchanger surface is

$$q = UA \, \Delta T_{lm} \tag{9.21}$$

where ΔT_{lm} is the *log-mean temperature difference* notation encountered in eq. (6.105):

$$\Delta T_{lm} = \frac{\Delta T_1 - \Delta T_2}{\ln \dfrac{\Delta T_1}{\Delta T_2}} = \frac{\Delta T_2 - \Delta T_1}{\ln \dfrac{\Delta T_2}{\Delta T_1}} \tag{9.22}$$

9.3.2   Counterflow

Equation (9.21) holds true not only for the parallel flow heat exchanger ana-lyzed above but also for a *counterflow heat exchanger,* for which ΔT_{lm} is given by eq. (9.22) with

$$\Delta T_1 = T_{h,in} - T_{c,out} \quad \text{and} \quad \Delta T_2 = T_{h,out} - T_{c,in} \tag{9.23}$$

All the flow arrangements to which the ΔT_{lm} heat transfer relation (9.21) applies are summarized in Fig. 9.15. It is easy to see now that the single-stream examples treated earlier in Fig. 6.15 are special cases of the more general result derived here in the form of eq. (9.21).

The upper right frame of Fig. 9.15 shows that in the counterflow arrange-ment the stream-to-stream temperature difference is more *uniform* along the heat exchanger, certainly more uniform than in the other three arrangements. The temperature difference is also *smaller* than in the parallel flow arrangement if the only difference between the two upper frames in Fig. 9.15 is the direction of the C_h stream (i.e., if the two heat exchangers have the same size, capacity rates, and fluid inlet temperatures). For this reason, the counterflow heat ex-changer is *thermodynamically more efficient* (it generates less entropy) than the corresponding parallel flow heat exchanger [10].

9.3.3   Other Flow Arrangements

In *cross-flow* and *multipass* arrangements, the ratio q/UA is a more complicated function of $T_{h,1}$, $T_{h,2}$, $T_{c,1}$, and $T_{c,2}$, that is, more complicated than the ΔT_{lm} group identified on the right side of eq. (9.21). For such arrangements, the heat transfer rate formula can be shaped in a way that mimics eq. (9.21),

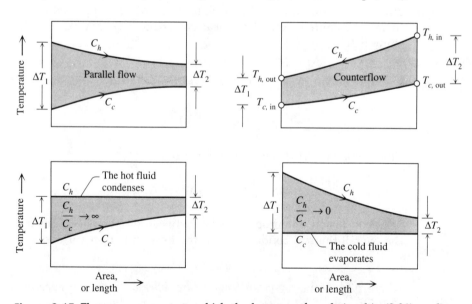

Figure 9.15 Flow arrangements to which the heat transfer relationship (9.21) applies.

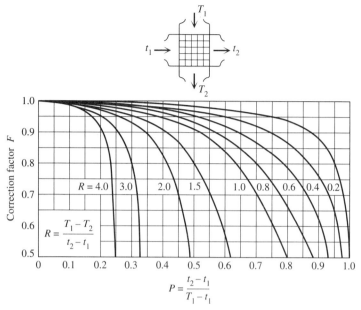

Figure 9.16 Correction factor F for cross-flow (single-pass) heat exchangers in which both streams remain unmixed [9].

$$q = UA\,\Delta T_{lm}F \tag{9.24}$$

in which the new *correction factor F* is a function of two dimensionless parameters P and R. The definitions of these parameters are indicated directly on Figs. 9.16 through 9.20, which show a sample of Bowman et al.'s [9] charts for the calculation of the correction factor. The ΔT_{lm} value that is to be substituted in eq.

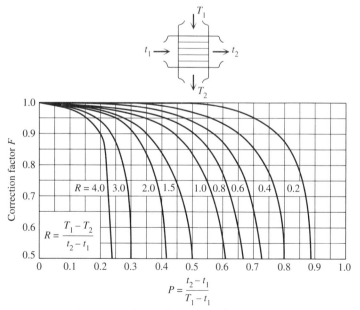

Figure 9.17 Correction factor F for cross-flow (single-pass) heat exchangers in which one stream is mixed and the other unmixed [9].

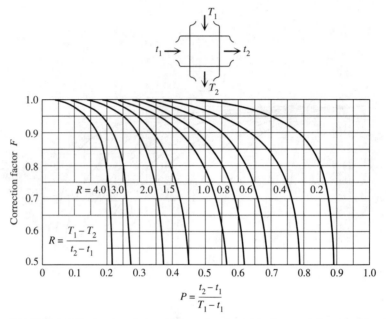

Figure 9.18 Correction factor F for cross-flow (single-pass) heat exchangers in which both streams are mixed [9].

(9.24) must be estimated by imagining that the flow arrangement is that of *counterflow*, that is, by using eqs. (9.23).

While reviewing the charts of Figs. 9.16 through 9.20, it is worth noting that F is generally less than 1, and that F approaches 1 as either P or R approaches the value 0. In either of these limits, eq. (9.24) approaches the limit represented

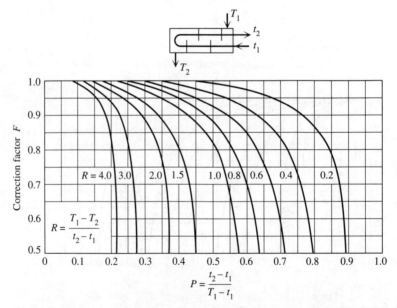

Figure 9.19 Correction factor F for shell-and-tube heat exchangers with one shell pass and any multiple of two tube passes (2, 4, 6, etc., tube passes) [9].

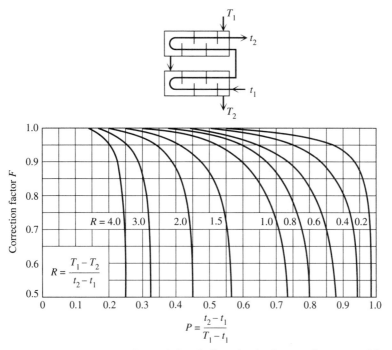

Figure 9.20 Correction factor F for shell-and-tube heat exchangers with two shell passes and any multiple of four tube passes (4, 8, 12, etc., tube passes) [9].

by eq. (9.21). Physically, the limit $P \rightarrow 0$ corresponds to a heat exchanger in which the stream represented by the temperatures t_1 and t_2 experiences a change of phase (i.e., it condenses or evaporates; therefore, $t_1 = t_2$ when the pressure drop is sufficiently small). In the other limit, $R \rightarrow 0$, the condensing or evaporating stream is the one represented by T_1 and T_2. Since the temperature of a condensing or evaporating stream does not change along the heat exchanger surface, that stream acts as if its assumed (single-phase) capacity rate C is infinite. This observation is made also graphically in the lower half of Fig. 9.15. Note also that $F = 1$ in all the cases illustrated in Fig. 9.15.

Another observation is that the temperature difference ratio R is, in fact, the ratio of the capacity rates of the two streams. To see this, assign the label C_t to the capacity rate of the $t_1 \rightarrow t_2$ stream and the label C_T to the capacity rate of the second stream, $T_1 \rightarrow T_2$. Under the assumption that the exterior of the heat exchanger enclosure is sufficiently well insulated with respect to the environment, the first law of thermodynamics for the entire heat exchanger reduces to

$$C_t (t_2 - t_1) = C_T (T_1 - T_2) \tag{9.25}$$

which means also that

$$\frac{C_t}{C_T} = \frac{T_1 - T_2}{t_2 - t_1} \equiv R \tag{9.26}$$

To summarize, the relationship between the total heat transfer rate q and the size of the heat exchanger area UA is given by eq. (9.21), or more generally by eq. (9.24). This relation *must always be used in conjunction with the first-law statements* of type (9.17) and (9.18), because the four inlet and outlet temperatures (which make up ΔT_{lm}) are interrelated and depend on the ratio of the two capacity rates, C_h/C_c. Indeed, one drawback of the ΔT_{lm}-type formula for the total heat transfer rate q is that, at first reading, it gives the impression that q does not depend on the capacity rates of the two streams.

Example 9.2

Calculating the Needed Heat Exchanger Area by the ΔT_{lm} Method

A hot water stream of flowrate $\dot{m}_h = 1$ kg/s is to be cooled from 90°C to 60°C in a heat exchanger, by contact with a larger stream of cold water, $\dot{m}_c = 2$ kg/s. The inlet temperature of the cold stream is 40°C.

Calculate the heat exchanger area A needed for accomplishing this task. The overall heat transfer coefficient (based on A) is known, $U = 1000$ W/m²·K. Consider first the counterflow arrangement and later the cross-flow scheme in which both fluids are mixed.

Solution. Regardless of the flow arrangement, the outlet temperature of the cold stream is dictated by the conservation of the total heat transfer rate between the two streams:

$$q = \dot{m}_h c_{P,h} \left(T_{h,in} - T_{h,out} \right) = \dot{m}_c c_{P,c} \left(T_{c,out} - T_{c,in} \right)$$

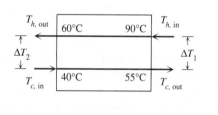

Counterflow   Cross-flow

Figure E9.2

Recognizing that for water $c_{P,h} \cong c_{P,c}$, we obtain

$$T_{c,out} = T_{c,in} + \frac{\dot{m}_h}{\dot{m}_c} \left(T_{h,in} - T_{h,out} \right)$$

$$= 40°C + \frac{1}{2} \left(90°C - 60°C \right) = 55°C$$

Consider first the *counterflow* arrangement, in which the end temperature differences are

$$\Delta T_1 = 35°C \quad \text{and} \quad \Delta T_2 = 20°C$$

The log-mean temperature difference will have a value somewhere between 35°C and 20°C;

$$\Delta T_{lm} = \frac{\Delta T_1 - \Delta T_2}{\ln\dfrac{\Delta T_1}{\Delta T_2}} = \frac{(35 - 20)°C}{\ln\dfrac{35}{20}} = 26.8°C$$

The needed heat exchanger area can be calculated by using eq. (9.21):

$$A_{\text{counterflow}} = \frac{q}{U\,\Delta T_{lm}} = \frac{\dot{m}_h c_{P,h}(T_{h,\text{in}} - T_{h,\text{out}})}{U\,\Delta T_{lm}}$$

$$= \frac{\left(1\,\dfrac{\text{kg}}{\text{s}}\right)\left(4.19\,\dfrac{10^3\text{J}}{\text{kg K}}\right)(90 - 60)^\circ\text{C}}{\left(1000\,\dfrac{\text{W}}{\text{m}^2\cdot\text{K}}\right)(26.8^\circ\text{C})} = 4.69\ \text{m}^2$$

As an alternative, consider the *cross-flow* arrangement in which both fluids are mixed. In the attached sketch, the inlet and outlet temperatures have been relabeled in accordance with the notation employed in Fig. 9.18. It is easy to calculate the dimensionless parameters employed in Fig. 9.18,

$$P = \frac{t_2 - t_1}{T_1 - t_1} = \frac{55 - 40}{90 - 40} = 0.3$$

$$R = \frac{T_1 - T_2}{t_2 - t_1} = \frac{90 - 60}{55 - 40} = 2 \quad \left(\text{note, again, that } R = \frac{C_c}{C_h}\right)$$

and to read on the ordinate of the graph

$$F \cong 0.885$$

The log-mean temperature difference is the value calculated for the *counterflow* arrangement [see the comment under eq. (9.24)]:

$$\Delta T_{lm} = 26.8^\circ\text{C}$$

Therefore, the new area requirement is [cf. eq. (9.24)]

$$A_{\text{cross-flow}} = \frac{q}{U\,\Delta T_{lm}F} = \frac{\dot{m}_h c_{P,h}(T_{h,\text{in}} - T_{h,\text{out}})}{U\,\Delta T_{lm}F}$$

$$= \frac{\left(1\,\dfrac{\text{kg}}{\text{s}}\right)\left(4.19\,\dfrac{10^3\text{J}}{\text{kg}\cdot\text{K}}\right)(90 - 60)^\circ\text{C}}{\left(1000\,\dfrac{\text{W}}{\text{m}^2\cdot\text{K}}\right)(26.8^\circ\text{C})(0.885)} = 5.30\ \text{m}^2$$

In conclusion, the area required by the cross-flow arrangement is 13 percent larger than the area required by the counterflow heat exchanger.

Summary.   The calculation of the area needed for the cross-flow arrangement was a bit more complicated than in the case of the counterflow, because of the graphic procedure needed for estimating the correction factor F. It is as if in counterflow we began the calculation by setting $F = 1$ and then continued with the same steps as those employed for the cross-flow scheme. In summary, the numerical calculation of the area requirement involved the following steps:

1. The parameters P and R were calculated based on the known inlet and outlet temperatures.
2. The correction factor F was read off the appropriate chart.
3. The log-mean temperature difference ΔT_{lm} was calculated based on eq. (9.22).
4. The area A was determined finally by using eq. (9.24) and by invoking the first law to replace q in that equation.

Example 9.3

Calculating the Outlet Temperatures and the Heat Transfer Rate by the ΔT_{lm} Method

The counterflow heat exchanger shown in the sketch has a heat transfer area $A = 10$ m^2 and a corresponding overall heat transfer coefficient $U = 500$ W/m$^2 \cdot$K. It is used to cool 1.5 kg/s of hot oil initially at 110°C, by contact with a 0.5 kg/s stream of cold water whose inlet temperature is 15°C. The respective c_P values of oil and water are 2.25 kJ/kg·K and 4.18 kJ/kg·K, respectively.

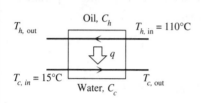

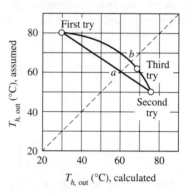

Figure E9.3

To summarize, known in this problem are A, U, C_h, C_c, $T_{h,in}$, and $T_{c,in}$. Calculate the two outlet temperatures and the total heat transfer rate q.

Solution. The capacity rates of the two streams are

$$C_h = \dot{m}_h c_{P,h} = 1.5 \frac{\text{kg}}{\text{s}} \, 2.25 \frac{\text{kJ}}{\text{kg} \cdot \text{K}} = 3.38 \text{ kW/K}$$

$$C_c = \dot{m}_c c_{P,c} = 0.5 \frac{\text{kg}}{\text{s}} \, 4.18 \frac{\text{kJ}}{\text{kg} \cdot \text{K}} = 2.09 \text{ kW/K}$$

To calculate the total heat transfer rate q, we need ΔT_{lm}. Here, we run into difficulty because, with only the two inlet temperatures being specified, we cannot calculate ΔT_{lm} directly. We are forced to *assume* (i.e., guess) the value of one of the outlet temperatures. Later, after calculating q, we will check the correctness of the initial guess, and, if necessary, we shall correct it. The numerical work makes more sense if we keep the following steps in mind:

1. Assume $T_{h,out} = 80$°C, and by recognizing the first law statement

$$C_h(T_{h,in} - T_{h,out}) = C_c(T_{c,out} - T_{c,in})$$

calculate the remaining outlet temperature:

$$T_{c,out} = T_{c,in} + \frac{C_h}{C_c}(T_{h,in} - T_{h,out})$$

$$= 15°C + \frac{3.38}{2.09}(110 - 80)°C$$

$$= 63.52°C$$

2. Calculate ΔT_{lm}:

$$\Delta T_1 = T_{h,out} - T_{c,in} = 80°C - 15°C = 65°C$$

$$\Delta T_2 = T_{h,in} - T_{c,out} = 110°C - 63.52°C = 46.48°C$$

$$\Delta T_{lm} = \frac{(65 - 46.48)°C}{\ln \dfrac{65}{46.48}} = 55.22°C$$

3. Determine q from the ΔT_{lm} expression (9.21):

$$q = UA \, \Delta T_{lm}$$

$$= \left(500 \, \frac{W}{m^2 \cdot K}\right)(10 \text{ m}^2)(55.22°C) = 2.76 \times 10^5 \text{ W}$$

4. Calculate $T_{h,out}$ and compare it with the value assumed in the first step,

$$q = C_h(T_{h,in} - T_{h,out})$$

$$T_{h,out} = T_{h,in} - \frac{q}{C_h}$$

$$= 110°C - \frac{2.76 \times 10^5 \text{ W}}{3.38 \times 10^3 \, \dfrac{W}{K}}$$

$$= 28.31°C$$

The $T_{h,out}$ value calculated above is considerably smaller than the assumed value (80°C). The next move, of course, is to go back to step 1 and repeat the procedure by starting with a better guess for $T_{h,out}$. The trouble is, will this "better" guess have to be a $T_{h,out}$ value greater or smaller than 80°C?

To see the correct choice, it pays to review the preceding calculation in the reverse sequence in which it was made. In this way, the conclusion that the calculated $T_{h,out}$ is too low can be attributed to the fact that q is too high. In turn, q is too high because ΔT_{lm} is too large, and that means that the assumed value $T_{h,out}$ is too large also. In conclusion, for the second try we begin with a lower guess for the $T_{h,out}$ value, and here are the results:

1. $T_{h,out} = 50°C$ (assumed)

   $T_{c,out} = 95.86°C$

2. $\Delta T_1 = 35°C$

   $\Delta T_2 = 14.14°C$

$$\Delta T_{lm} = \frac{(35 - 14.14)°C}{\ln \dfrac{35}{14.14}} = 23.02°C$$

3. $q = 1.15 \times 10^5 \text{ W}$

4. $T_{h,out} = 110°C - \dfrac{1.15 \times 10^5 \text{ W}}{3.38 \times 10^3 \text{ W/K}}$

   $= 76°C$ (calculated)

This time the calculated $T_{h,out}$ is greater than the assumed value; therefore, for the third try we'll assume a $T_{h,out}$ value greater than the preceding guess. The accuracy of this third guess can be improved considerably by drawing the sketch shown on the right side of the figure (or by performing the equivalent calculations on the computer). Each of the two tries executed until now is represented by one point. Connecting these points with a straight line and intersecting this line with the rising diagonal,

$$T_{h,out} \text{ (assumed)} = T_{h,out} \text{ (calculated)}$$

we obtain the point labeled a. In conclusion, $T_{h,out} \cong 61\,°C$ is a good candidate for the third guess, and the corresponding results are

1. $T_{h,out} = 61\,°C$ (assumed)

   $T_{c,out} = 94.24\,°C$

2. $\Delta T_1 = 46\,°C$

   $\Delta T_2 = 15.76\,°C$

   $\Delta T_{lm} = 28.23\,°C$

3. $q = 1.41 \times 10^5$ W

4. $T_{h,out} = 68.24\,°C$ (calculated)

The agreement between the calculated and assumed $T_{h,out}$ values has improved. For an even better estimate, we connect with an arc of parabola the three points that now appear on the graph. This arc intersects the rising diagonal at point b, or $T_{h,out} \cong 63.5\,°C$. The last (fourth) try yields, in order,

1. $T_{h,out} = 63.5\,°C$ (assumed)

   $T_{c,out} = 90.2\,°C$

2. $\Delta T_1 = 48.5\,°C$

   $\Delta T_2 = 19.8\,°C$

   $\Delta T_{lm} = 32.04\,°C$

3. $q = 1.6 \times 10^5$ W

4. $T_{h,out} = 62.6\,°C$ (calculated)

What we have discovered in this example is that when the area (A, U), capacity rates, and inlet temperatures are specified, the calculation of q by the ΔT_{lm} method requires a *trial-and-error procedure*. This procedure is even lengthier for a heat exchanger like those of Figs. 9.16 to 9.20, for which the correction factor F is generally less than 1. In such a case an integral part of step 1 is also the calculation of R, P, and F, and the correct ΔT_{lm} formula for use in step 2 is eq. (9.24). Fortunately, this type of heat exchanger problem can be solved in one shot (i.e., without trial and error) by applying the effectiveness–NTU method, which is described next.

9.4 THE EFFECTIVENESS – *NTU* METHOD

9.4.1 Effectiveness and Limitations Posed by the Second Law

An alternative to calculating the total heat transfer rate q—a method that brings out the individual effects of not only the total thermal conductance UA but also that of the capacity rates C_h and C_c—begins with defining two new dimensionless groups [11–13]. The first is the *number of heat exchanger (heat transfer) units, NTU*:

$$NTU = \frac{UA}{C_{min}} \tag{9.27}$$

in which C_{min} is the smaller of the two capacity rates,

$$C_{min} = \min(C_h, C_c) \qquad (9.28)$$

The second dimensionless group is the heat exchanger *effectiveness* ϵ, which is defined as the ratio between the actual heat transfer rate q and the thermodynamic "ceiling value" of the heat transfer rate that could take place between the two streams, q_{max}:

$$\epsilon = \frac{\text{actual heat transfer rate}}{\text{maximum heat transfer rate}} = \frac{q}{q_{max}} \qquad (9.29)$$

To see what is meant by q_{max}, consider the counterflow heat exchanger temperature distribution shown in Fig. 9.21. The actual heat transfer rate q can be written either as $C_h(T_{h,in} - T_{h,out})$, or as $C_c(T_{c,out} - T_{c,in})$. Imagine now that we increase the size of the heat exchanger UA, in an attempt to increase q while holding the two inlet temperatures ($T_{h,in}$, $T_{c,in}$) fixed. As we do this, the stream-to-stream temperature differences decrease everywhere along the heat exchanger, and this means that the two temperature distribution curves migrate toward one another (i.e., the curves become increasingly steeper). In other words, $T_{c,out}$ migrates toward the stationary $T_{h,in}$ and $T_{h,out}$ migrates toward the fixed temperature level $T_{c,in}$. Note that in Fig. 9.21 the *steeper* of the two temperature distribution curves is the one that corresponds to the *smaller* of the two capacity rates ($C_{min} = C_c$).

As these changes take place, the first outlet temperature to equal (i.e., "bump into") its facing inlet temperature is that of the stream with the lower capacity rate, namely, C_c, and $T_{c,out} = T_{h,in}$ in Fig. 9.21. Beyond this stage, the outlet temperature of the cold stream, $T_{c,out}$, remains at the $T_{h,in}$ temperature level no matter how large a UA we build into the heat exchanger. Otherwise, with $T_{c,out}$ going above $T_{h,in}$, we would be looking at a picture in which the two tempera-

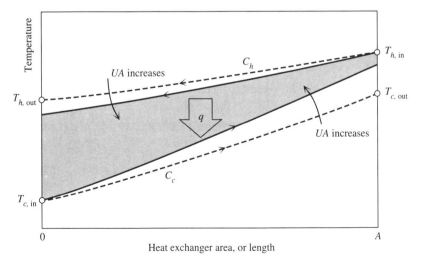

Figure 9.21 The effect of the heat transfer conductance UA on the temperature distributions inside a counterflow heat exchanger.

ture distributions would cross somewhere inside the heat exchanger. A heat exchanger with crossing temperature distribution curves is physically impossible (a violation of the second law of thermodynamics), because in the heat exchanger zone where the T_h curve happens to be below the T_c curve, the local heat transfer rate dq would be required to flow "uphill," from T_h to T_c (Ref. [10], p. 535).

In summary, since the actual heat transfer rate is $C_{min}(T_{c,out} - T_{c,in})$, and since $T_{c,in}$ is fixed while $T_{c,out}$ cannot exceed $T_{h,in}$, the ceiling value for q corresponds to the limit where the temperature excursion experienced by the C_{min} stream reaches its largest value, in this example, $T_{h,in} - T_{c,in}$,

$$q_{max} = C_{min}(T_{h,in} - T_{c,in}) \tag{9.30}$$

Although we derived this formula by analyzing the behavior of a counterflow heat exchanger, its validity as a definition for q_{max} is general. On the right side of eq. (9.30), the temperature difference $T_{h,in} - T_{c,in}$ is simply the temperature difference between the two inlet ports, that is, the *largest* temperature difference between two states that belong to the two streams that make up the heat exchanger.

9.4.2 Parallel Flow

The total heat transfer rate relation developed for a parallel flow heat exchanger in Section 9.3.1 can be restated in terms of effectiveness and NTU. Consider again the parallel flow temperature distributions shown in Fig. 9.14, and note that in this case the effectiveness ϵ is equal to

$$\epsilon = \frac{C_h(T_{h,1} - T_{h,2})}{C_{min}(T_{h,1} - T_{c,1})} = \frac{C_c(T_{c,2} - T_{c,1})}{C_{min}(T_{h,1} - T_{c,1})} \tag{9.31}$$

To be more specific, let us assume that the smaller of the two capacity rates is C_c, namely, that

$$C_c = C_{min} \quad \text{and} \quad C_h = C_{max} \tag{9.32}$$

The NTU definition (9.27) and eq. (9.18) allow us to rewrite eq. (9.20) as

$$\ln \frac{\Delta T_2}{\Delta T_1} = -NTU \left(1 - \frac{T_{h,2} - T_{h,1}}{T_{c,2} - T_{c,1}}\right) \tag{9.33}$$

Next, by eliminating q between eqs. (9.17) and (9.18), we conclude that

$$\frac{T_{h,2} - T_{h,1}}{T_{c,2} - T_{c,1}} = -\frac{C_c}{C_h} = -\frac{C_{min}}{C_{max}} \tag{9.34}$$

or that eq. (9.33) is the same as

$$\ln \frac{\Delta T_2}{\Delta T_1} = -NTU \left(1 + \frac{C_{min}}{C_{max}}\right) \tag{9.35}$$

The second part of this analysis focuses on the left side of eq. (9.35), which can be rewritten sequentially as

$$\frac{\Delta T_2}{\Delta T_1} = \frac{T_{h,2} - T_{c,2}}{T_{h,1} - T_{c,1}} = \frac{T_{h,2} - T_{c,2} + T_{h,1} - T_{c,1} - (T_{h,1} - T_{c,1})}{T_{h,1} - T_{c,1}}$$

$$= 1 + \frac{T_{h,2} - T_{c,2} - T_{h,1} + T_{c,1}}{T_{h,1} - T_{c,1}}$$

$$= 1 + \frac{-(T_{c,2} - T_{c,1}) + T_{h,2} - T_{h,1}}{\frac{1}{\epsilon}(T_{c,2} - T_{c,1})} \qquad (9.36)$$

$$= 1 + \epsilon\left(-1 + \frac{T_{h,2} - T_{h,1}}{T_{c,2} - T_{c,1}}\right) = 1 + \epsilon\left(-1 - \frac{C_{min}}{C_{max}}\right)$$

Substituting this last expression on the left side of eq. (9.35), we arrive at the expression for the effectiveness of a parallel flow heat exchanger:

$$\epsilon_{\text{parallel flow}} = \frac{1 - \exp[-NTU(1 + C_{min}/C_{max})]}{1 + C_{min}/C_{max}} \qquad (9.37)$$

This expression can be turned inside out to produce a formula for calculating the required *NTU* for an application in which the effectiveness and the capacity rate ratio are specified:

$$NTU_{\text{parallel flow}} = -\frac{\ln[1 - \epsilon(1 + C_{min}/C_{max})]}{1 + C_{min}/C_{max}} \qquad (9.38)$$

It is easy to verify that eqs. (9.37) and (9.38) are also obtained at the end of an analysis in which the reverse of the assumption (9.32) is made, that is, when $C_h = C_{min}$ and $C_c = C_{max}$. In conclusion, eqs. (9.37) and (9.38) represent the effectiveness–*NTU* relationship of a parallel flow heat exchanger, regardless of which stream (hot or cold) has the smaller capacity rate. The same relationship is shown graphically in Fig. 9.22. Two limiting cases of this relationship are

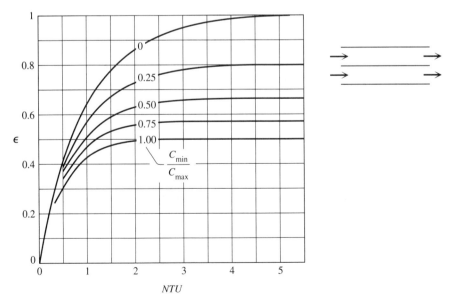

Figure 9.22 The effectiveness of a parallel flow heat exchanger. (Kays and London [4], with permission from McGraw-Hill Book Company.)

worth recording. When $C_{min} = C_{max}$, the capacity rates are said to be *balanced* and eq. (9.37) reduces to

$$\epsilon_{\text{parallel flow}} = \frac{1}{2}[1 - \exp(-2NTU)] \qquad (C_{min} = C_{max}) \qquad (9.39)$$

The opposite limit, $C_{min}/C_{max} = 0$, represents a heat exchanger in which one of the streams (C_{max}) undergoes a phase change at nearly constant pressure,

$$\epsilon = 1 - \exp(-NTU) \qquad \left(\frac{C_{min}}{C_{max}} = 0\right) \qquad (9.40)$$

9.4.3   Counterflow

Based on a completely analogous analysis, which this time would be focused on the counterflow arrangement of Fig. 9.21, we can write the effectiveness definition

$$\epsilon = \frac{C_h\,(T_{h,in} - T_{h,out})}{C_{min}\,(T_{h,in} - T_{c,in})} = \frac{C_c\,(T_{c,out} - T_{c,in})}{C_{min}\,(T_{h,in} - T_{c,in})} \qquad (9.41)$$

and derive the effectiveness–NTU relation,

$$\epsilon_{\text{counterflow}} = \frac{1 - \exp[-NTU(1 - C_{min}/C_{max})]}{1 - (C_{min}/C_{max})\exp[-NTU(1 - C_{min}/C_{max})]} \qquad (9.42)$$

or, conversely, at the NTU–effectiveness relation,

$$NTU_{\text{counterflow}} = \frac{\ln\left(\dfrac{1 - \epsilon\,C_{min}/C_{max}}{1 - \epsilon}\right)}{1 - C_{min}/C_{max}} \qquad (9.43)$$

This relationship is presented graphically in Fig. 9.23. The two extremes of this relationship are represented by the *balanced* counterflow heat exchanger,

$$\epsilon_{\text{counterflow}} = \frac{NTU}{1 + NTU} \qquad (C_{min} = C_{max}) \qquad (9.44)$$

and by the heat exchanger in which the C_{max} stream experiences a change of phase at nearly constant pressure:

$$\epsilon = 1 - \exp(-NTU) \qquad \left(\frac{C_{min}}{C_{max}} = 0\right) \qquad (9.45)$$

Note that the effectiveness formula in this second extreme, eq. (9.45), is exactly the same as that for parallel flow heat exchangers, eq. (9.40). This is caused by the fact that when the temperature of the condensing or evaporating stream (C_{max}) does not vary along the heat exchanger surface, the direction in which this stream flows loses its significance. This is why in the lower part of Fig. 9.15 the flow direction is not indicated on the temperature distribution curves of the streams that condense or evaporate.

9.4.4   Other Flow Arrangements

Figures 9.24 to 9.28 display the effectiveness–NTU curves for five other flow arrangements. This selection of cross-flow and multi-pass arrangements is the

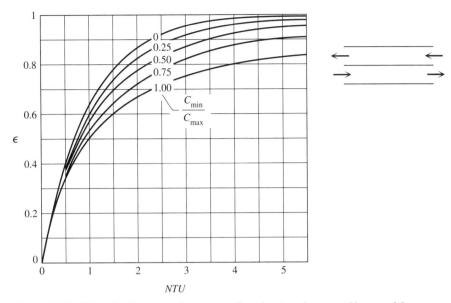

Figure 9.23 The effectiveness of a counterflow heat exchanger. (Kays and London [4], with permission from McGraw-Hill Book Company.)

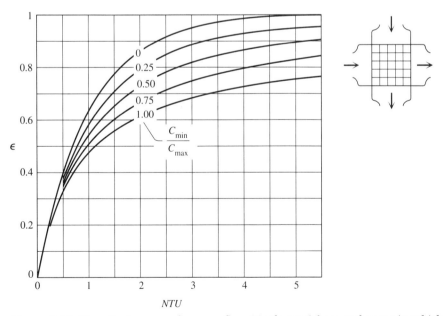

Figure 9.24 The effectiveness of a cross-flow (single-pass) heat exchanger in which both streams remain unmixed. (Kays and London [4], with permission from McGraw-Hill Book Company.)

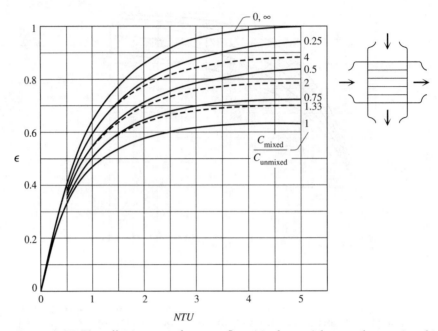

Figure 9.25 The effectiveness of a cross-flow (single-pass) heat exchanger in which one stream is mixed and the other is unmixed. (Kays and London [4], with permission from McGraw-Hill Book Company.)

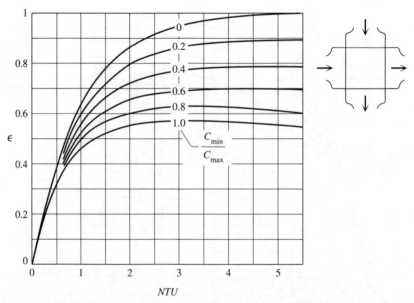

Figure 9.26 The effectiveness of a cross-flow (single-pass) heat exchanger in which both streams are mixed. (Kays and London [4], with permission from McGraw-Hill Book Company.)

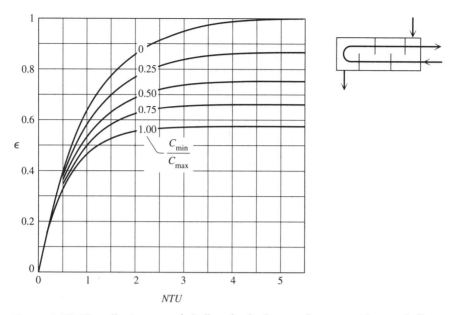

Figure 9.27 The effectiveness of shell-and-tube heat exchangers with one shell pass and any multiple of two tube passes (2, 4, 6, etc., tube passes). (Kays and London [4], with permission from McGraw-Hill Book Company.)

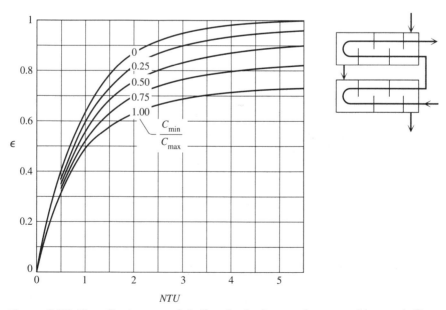

Figure 9.28 The effectiveness of shell-and-tube heat exchangers with two shell passes and any multiple of four tube passes (4, 8, 12, etc., tube passes). (Kays and London [4], with permission from McGraw-Hill Book Company.)

same as one covered by the ΔT_{lm} correction factor charts of Figs. 9.16 to 9.20. The analytical expressions that stand behind the curves plotted in Figs. 9.24 to 9.28 can be found in Ref. [4].

The effectiveness–NTU relationships and graphs presented in this entire section remind us that, in general, the effectiveness depends not only on NTU and the capacity rate ratio but also on the flow arrangement:

$$\epsilon = \text{function}\left(NTU, \frac{C_{min}}{C_{max}}, \text{flow arrangement}\right) \qquad (9.46)$$

Example 9.4

Calculating the Necessary Heat Exchanger Area by the ε–NTU Method

To see the details and merits of the ε–NTU method relative to those of the ΔT_{lm} method, consider again the problem that we solved in Example 9.2. It dealt with the design of a heat exchanger in which a hot water stream had to be cooled to a certain final temperature by contact with a stream of cold water. The overall heat transfer coefficient, the two flowrates, the two inlet temperatures, and the outlet temperature of the hot stream were specified:

$$U = 1000 \text{ W/m}^2\cdot\text{K} \qquad T_{h,in} = 90°\text{C}$$
$$\dot{m}_h = 1\text{kg/s} \qquad T_{h,out} = 60°\text{C}$$
$$\dot{m}_c = 2 \text{ kg/s} \qquad T_{c,in} = 40°\text{C}$$

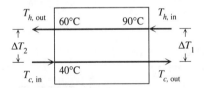

Figure E9.4

Consider only the counterflow arrangement. Calculate the heat exchanger area that is necessary to accomplish the task specified in the problem statement.

Solution.   The unknown A appears in the NTU definition (9.27); therefore, we will also need the value of NTU and, even before that, the value of the effectiveness ε. It is more instructive, however, if we construct the calculation of A as a sequence of three principal steps:

1. In the first step we calculate the effectiveness by using one of the two formulas listed in eq. (9.41), where

$$C_{min} = C_h \qquad \text{and} \qquad C_{max} = C_c$$

The first of these formulas is more convenient, because we know all the needed temperatures (in the second formula, we would first have to calculate $T_{c,out}$):

$$\epsilon = \frac{C_h}{C_{min}}\frac{T_{h,in} - T_{h,out}}{T_{h,in} - T_{c,in}} = \frac{90 - 60}{90 - 40} = 0.6$$

2. The next objective is to calculate NTU. For this we need the capacity rate ratio

$$\frac{C_{min}}{C_{max}} = \frac{C_h}{C_c} = \frac{(\dot{m}c_P)_h}{(\dot{m}c_P)_c} \cong \frac{\dot{m}_h}{\dot{m}_c} = \frac{1}{2}$$

in which we assumed that $c_P \cong 4.19$ kJ/kg·K applies to both water streams. The number of heat transfer units follows from eq. (9.43):

$$NTU = \frac{\ln\left[(1 - 0.6\frac{1}{2})/(1 - 0.6)\right]}{1 - \frac{1}{2}} = 1.12$$

3. In the last step we recognize the *NTU* definition (9.27), from which we pull out *A*:

$$A = NTU \frac{C_{min}}{U} = NTU \frac{(\dot{m}c_P)_h}{U}$$

$$= 1.12 \frac{(1 \text{ kg/s})(4.19 \times 10^3 \text{ J/kg·K})}{10^3 \text{ W/m}^2 \cdot \text{K}} = 4.69 \text{ m}^2$$

This area is, of course, the same as the answer obtained by the ΔT_{lm} method in Example 9.2. The ϵ–*NTU* method was a bit more direct, because, unlike the ΔT_{lm} method, it did not require the calculation of three auxiliary quantities—ΔT_{lm}, the cold-stream outlet temperature $T_{c,out}$, and the stream-to-stream heat transfer rate q.

Example 9.5

Calculating the Outlet Temperatures and Heat Transfer Rate by the ϵ–*NTU* Method

It is much more advantageous to use the ϵ–*NTU* method instead of the ΔT_{lm} method when the surface (A, U), capacity rates (C_c, C_h), and inlet temperatures $(T_{c,in}, T_{h,in})$ are specified. To see this, let us reconsider the problem described in Example 9.3, which was solved by a trial-and-error method.

Known in this problem are the flow arrangement—counterflow—and that a stream of hot oil is cooled by a stream of cold water:

Oil	Water
$\dot{m}_h = 1.5$ kg/s	$\dot{m}_c = 0.5$ kg/s
$c_{P,h} = 2.25$ kJ/kg·K	$c_{P,c} = 4.18$ kJ/kg·K
$T_{h,in} = 110°C$	$T_{c,in} = 15°C$

The characteristics of the heat exchanger surface are also known

$$A = 10 \text{ m}^2 \qquad U = 500 \text{ W/m}^2 \cdot \text{K}$$

Determine the two outlet temperatures and the total stream-to-stream heat transfer rate q.

Solution.   The chief unknown is q, because if we know q we can calculate each of the outlet temperatures by writing the first law individually for each stream. The total heat transfer rate q is the "actual" rate in the definition of effectiveness ϵ, eqs. (9.29) and (9.30). Therefore, working backward, we must calculate ϵ and, before that, the number of heat transfer units *NTU*. This reasoning suggests that the structure of the numerical solution consists of the following four steps:

1. In the first step we identify the larger and smaller capacity rates, the capacity rate ratio, and the number of heat transfer units:

$$C_h = (\dot{m}c_P)_{oil} = \left(1.5\,\frac{kg}{s}\right)\left(2.25\,\frac{kJ}{kg \cdot K}\right) = 3.38\text{ kW/K}$$

$$C_c = (\dot{m}c_P)_{water} = \left(0.5\,\frac{kg}{s}\right)\left(4.18\,\frac{kJ}{kg \cdot K}\right) = 2.09\text{ kW/K}$$

$$C_{max} = C_h \quad \text{and} \quad C_{min} = C_c$$

$$\frac{C_{min}}{C_{max}} = \frac{2.09}{3.38} = 0.618$$

$$NTU = \frac{UA}{C_{min}} = \left(500\,\frac{W}{m^2 \cdot K}\right)(10\text{ m}^2)\left(\frac{K}{2.09 \times 10^3\text{ W}}\right) = 2.392$$

2. We now have all the necessary ingredients for calculating ϵ using eq. (9.42):

$$\epsilon = \frac{1 - \exp[-2.392(1 - 0.618)]}{1 - 0.618\exp[-2.392(1 - 0.618)]} = 0.796$$

3. The total heat transfer rate follows from eqs. (9.29) and (9.30), which after eliminating q_{max} yields

$$q = \epsilon C_{min}(T_{h,in} - T_{c,in})$$

$$= 0.796 \times 2.09\frac{10^3\text{ W}}{K}(110 - 15)\text{ K} = 1.58 \times 10^5\text{ W}$$

4. In this final step we draw a control volume around each stream, write the first law, and calculate the respective outlet temperatures:

Water:
$$q = C_c(T_{c,out} - T_{c,in})$$

$$T_{c,out} = T_{c,in} + \frac{q}{C_c}$$

$$= 15°C + \frac{1.58 \times 10^5\text{ W}}{2.09 \times 10^3\text{ W/K}} = 90.6°C$$

Oil:
$$q = C_h(T_{h,in} - T_{h,out})$$

$$T_{h,out} = T_{h,in} + \frac{q}{C_h}$$

$$= 110°C - \frac{1.58 \times 10^5\text{ W}}{3.38 \times 10^3\text{ W/K}} = 63.2°C$$

In conclusion, by using the $\epsilon-NTU$ method it was possible to calculate the unknowns of the problem directly by avoiding the trial-and-error procedure that was required by the ΔT_{lm} method. The $\epsilon-NTU$ method is not only more direct but it is also more accurate. Here is a side-by-side comparison of the answers provided by the two methods, and keep in mind that the ΔT_{lm}-method results are based on no less than four iterations:

Result	$\epsilon-NTU$ Method	ΔT_{lm} Method (4 Iterations, Example 9.3)
q	1.58×10^5 W	1.6×10^5 W
$T_{c,out}$	90.6°C	90.2°C
$T_{h,out}$	63.2°C	62.6°C

9.5   PRESSURE DROP

9.5.1   Pumping Power

In an important way, the preceding discussion of the heat transfer characteristics of heat exchangers represents only half of the picture. This discussion has involved the two capacity rates (C_{max}, C_{min}), which influence the effectiveness directly through relations of type (9.46) and indirectly through the definition of NTU, eq. (9.27), and through the value of U. Recall that, in general, the area-averaged overall heat transfer coefficient depends on the flow velocities on the two sides of the heat transfer surface.

The second big item in the description of a heat exchanger deals with the mechanical power that is required for the purpose of forcing each stream to flow through its heat exchanger passage. This consideration leads ultimately to the calculation of the *total pressure drop* ΔP that is experienced by the stream across the passage. For example, if the stream $\dot{m}$ carries a liquid that may be modeled thermodynamically as incompressible, the power required by an adiabatic pump is [10]

$$\dot{W}_p = \frac{1}{\eta_p} \frac{\dot{m}}{\rho} \Delta P \tag{9.47}$$

In this expression, ΔP is the pressure rise through the pump, ρ is the liquid density, and η_p is the isentropic efficiency of the pump. The group $\dot{m}\Delta P/\rho$ represents the minimum (or isentropic) power requirement.

Likewise, when the stream carries a gas that may be modeled as ideal (with constants R and c_P), a monotonic relationship also exists between the power required by a compressor and the pressure rise experienced by $\dot{m}$ through the compressor [10]:

$$\dot{W}_c = \frac{1}{\eta_c} \dot{m} c_P T_{in} \left[\left(\frac{P_{out}}{P_{in}} \right)^{R/c_P} - 1 \right] \tag{9.48}$$

The subscripts in and out represent the compressor inlet and outlet conditions, and the temperature T_{in} is expressed in degrees Kelvin. Equation (9.48) refers to a single-stage adiabatic compressor with isentropic efficiency η_c.

The pressure drop through the heat exchanger passage is generally a complicated function of flow parameters and passage geometry. The pressure drop over a sufficiently straight portion of the passage ΔP_s may be evaluated with the method of Section 6.1.3, provided that the density does not vary appreciably along the passage:

$$\Delta P_s = f \frac{4L}{D_h} \frac{1}{2} \rho V^2 \tag{9.49}$$

We recognize in this expression the passage length L, the hydraulic diameter D_h, the mean velocity V, and the friction factor f. Unlike in Chapter 6, where we wrote U for the mean longitudinal velocity through the duct, here we write V to avoid any confusion with the overall heat transfer coefficient. When ρ is not constant, the ΔP_s expression includes a correction for the pressure difference required to accelerate the stream while inside the heat exchanger, eq. (9.61).

The total pressure drop ΔP consists of ΔP_s plus contributions made by other geometric features (sudden contractions, enlargements, bends, protuberances) that are encountered in the flow passage. The pressure drop contribution made by each feature of this kind is taken into account as the product between the dynamic pressure $\frac{1}{2}\rho V^2$ and a particular *pressure loss coefficient*, which can be found in the heat exchanger literature [1–4]. This procedure is illustrated below.

9.5.2 Abrupt Contraction and Enlargement

Consider first the contraction experienced by the stream in Fig. 9.29, as the passage cross section changes from A_a to A_b. The sharpness of this change in cross section may cause the separation of the flow upstream as well as downstream of the step change. The eddies that result are visible in the photograph of Fig. 9.29. They are a clear sign of the *irreversibility* of the flow in the direction $a \rightarrow b$.

In the absence of irreversibility, the total pressure $P + \frac{1}{2}\rho V^2$ would be conserved from a to b. In the present case this quantity decreases in proportion with the obstructive role of the abrupt contraction:

$$(P_a + \tfrac{1}{2}\rho_a V_a^2) - (P_b + \tfrac{1}{2}\rho_b V_b^2) = K_c \tfrac{1}{2}\rho_b V_b^2 \tag{9.50}$$

The *contraction loss coefficient* K_c is dimensionless. We can rewrite eq. (9.50) by noting that at constant ρ, mass conservation requires $V_a A_a = V_b A_b$:

$$P_a - P_b = (1 - \sigma_{a\text{-}b}^2) \tfrac{1}{2}\rho_b V_b^2 + K_c \tfrac{1}{2}\rho_b V_b^2 \tag{9.51}$$

In this equation, $\sigma_{a\text{-}b}$ represents the passage contraction ratio,

$$\sigma_{a\text{-}b} = \frac{A_b}{A_a} \tag{9.52}$$

The dash lines in Figs. 9.30 and 9.31 show the contraction loss coefficient for the flow entering a bundle of parallel tubes or parallel-plate channels. In each instance the sudden contraction occurs as the stream crosses the plane of the entrance to the channels. In other words, the original cross section of the stream is the same as the frontal area of the bundle, whereas the final stream cross section is the sum of all the channel cross sections. The abscissa parameter in these figures is the contraction ratio defined in eq. (9.52) which, in general, is defined by

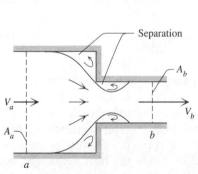
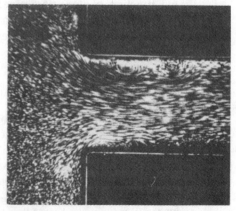

Figure 9.29 Abrupt contraction in a two-dimensional flow passage. (Photograph by Eck [14], with permission from Springer-Verlag.)

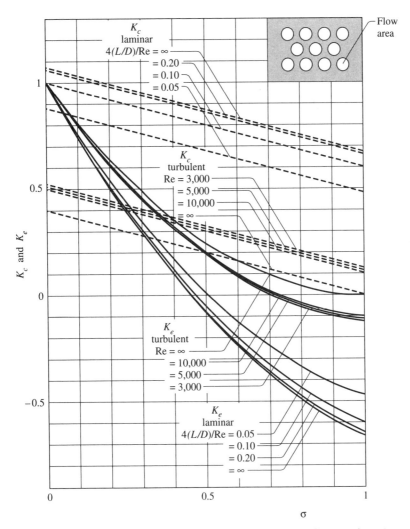

Figure 9.30 Abrupt contraction and enlargement loss coefficients for a heat exchanger core with multiple circular-tube passages (Kays and London [15]).

$$\sigma = \frac{\text{flow cross-sectional area}}{\text{frontal area}} \qquad (9.53)$$

The Reynolds number Re is based on the hydraulic diameter of one channel and on the mean flow velocity through the *narrower* channel (e.g., V_b in Fig. 9.29).

The irreversible flow through an abrupt enlargement of the passage is characterized similarly by introducing the *enlargement loss coefficient K_e*. With reference to the left side of Fig. 9.32, this coefficient is defined by writing the pressure *rise* experienced by the stream through the enlargement of the passage cross section:

$$P_d - P_c = (1 - \sigma_{c-d}^2) \tfrac{1}{2} \rho_c V_c^2 - K_e \tfrac{1}{2} \rho_c V_c^2 \qquad (9.54)$$

The passage enlargement ratio σ_{c-d} is defined as in eq. (9.53),

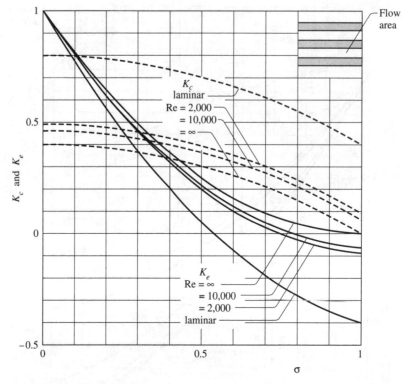

Figure 9.31 Abrupt contraction and enlargement loss coefficients for a heat exchanger core with multiple parallel-plate passages (Kays [16]).

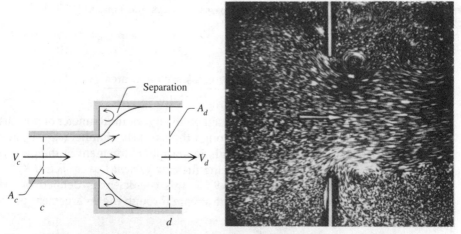

Figure 9.32 Abrupt enlargement in a two-dimensional flow passage (left) and the separated flow downstream from an orifice (right). (Photograph by Eck [14], with permission from Springer-Verlag.)

$$\sigma_{c\text{-}d} = \frac{A_c}{A_d} \tag{9.55}$$

while V_c represents the mean velocity in the narrower portion of the channel. Figure 9.32 shows that the flow separates downstream from the step increase in flow cross-sectional area: in actual flows, one separation region is almost always larger than the other, that is, unlike in the axially symmetric sketch of Fig. 9.32 (left). Figures 9.30 and 9.31 provide the K_e values for the sudden enlargement from a heat exchanger core (tubes or flat channels) to a flow with cross section equal to the frontal or backward facing area of the core.

The pressure losses calculated on the basis of K_c and K_e add up in passages where both effects—contraction and enlargement—are present. Examples in which contraction is followed by enlargement are presented in Figs. 9.33 and 9.34. In the former, the flow direction (left $\rightarrow$ right) is made evident by the asymmetry between contraction and enlargement with respect to flow separation (compare the left sides of Figs. 9.29 and 9.32).

Figure 9.34 shows the flow in and out of a heat exchanger core such as the multiple tubes and multiple parallel-plate channels sketched in the upper right corners of Figs. 9.30 and 9.31. An even earlier example was the tube side of the shell-and-tube heat exchanger presented in Fig. 9.5. The pressure drop during contraction ($a \rightarrow b$) and the pressure rise during enlargement ($c \rightarrow d$) are illustrated in the lower part of the figure. These can be calculated with eqs. (9.51) and (9.54). The pressure drop along the straight passage itself is $\Delta P_s = P_b - P_c$. The *total* pressure drop experienced by the stream is

$$\Delta P = P_a - P_d$$
$$= \underbrace{(P_a - P_b)}_{\text{eq. (9.51)}} + \underbrace{(P_b - P_c)}_{\substack{\Delta P_s \\ \text{eq. (9.61)}}} - \underbrace{(P_d - P_c)}_{\text{eq. (9.54)}} \tag{9.56}$$

In special cases where the density ρ is essentially constant from a to d (i.e., when $V_b = V_c = V$) and where the contraction ratio equals the enlargement ratio ($\sigma_{a\text{-}b} = \sigma_{c\text{-}d}$), eq (9.56) reduces to

$$\Delta P = K_c \tfrac{1}{2}\rho V^2 + \Delta P_s + K_e \tfrac{1}{2}\rho V^2 \tag{9.57}$$

The three components of the total ΔP represent, in order, the contraction loss, the straight-duct pressure drop, and the enlargement loss.

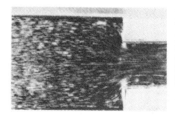 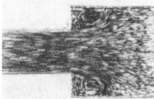

Figure 9.33 Water flow through a two-dimensional channel with sudden contraction followed by sudden enlargement. The width of the narrower passage is 0.4 m. The mean longitudinal velocity in that passage is 15 cm/s (left photo) and 20 cm/s (right photo). (Nakayama et al. [17], with permission from Pergamon Press.)

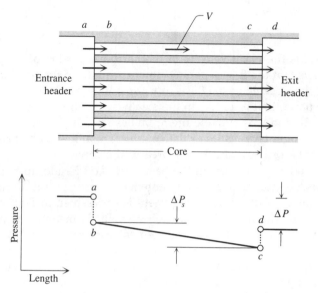

Figure 9.34 The cross section contraction and enlargement experienced by a stream flowing through the core of a heat exchanger, and the associated pressure distribution.

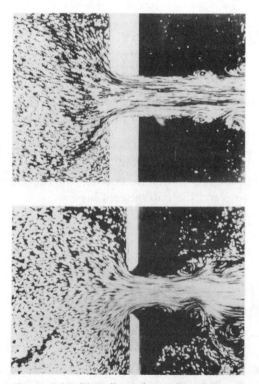

Figure 9.35 The effect of rounding the entrance to water flow through a two-dimensional 3-cm-wide slit. The mean velocity through the slit is 14 cm/s. (Nakayama et al. [17], with permission from Pergamon Press.)

One way to decrease the pressure losses associated with contractions, enlargements, and bends is to *smooth* these geometric features, that is, to make them less abrupt. Figure 9.35 relies on the flow through an orifice to show a comparison between a sharp contraction and a rounded entrance. When the contraction is gradual, the jet that forms is wider: such a jet is less likely to experience separation in the entrance to a duct. In other words, the postcontraction separation that is visible on the right side of Fig. 9.29 can be diminished through the proper rounding of the entrance to the channel.

Figure 9.36 shows that separation and large swirls are likely to occur in bends, because the naturally curved stream is pinched as it tries to make the corner. The flow is considerably smoother and the separation regions are smaller when the elbow is rounded. Pressure loss coefficients for this and other flow restrictions can be found in Refs. [1–4].

9.5.3   Acceleration and Deceleration

While developing the special-case formula (9.57), we assumed that the fluid density does not vary appreciably between inlet and outlet. This is a good approximation for liquids such as water and oil. It is not always adequate for gases. For example, when the gas is heated while flowing through the heat exchanger passage, its outlet density ρ_{out} is smaller than the density at the inlet, ρ_{in}. If the inlet and outlet have the same cross-sectional area A_c, the inlet

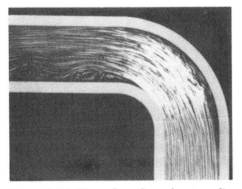

Figure 9.36 Water flow through a two-dimensional elbow 2 cm wide. The mean velocity is 10 cm/s. (Nakayama et al. [17], with permission from Pergamon Press.)

velocity is smaller than the outlet velocity, $V_{in} < V_{out}$. Conserved at every position along the flow passage is the mass flowrate $\dot{m} = \rho V A$, or the *mass velocity*

$$G = \frac{\dot{m}}{A} = \rho_{in} V_{in} = \rho_{out} V_{out} \tag{9.58}$$

When the flow is turbulent, the velocity distribution across the passage is relatively *uniform* (slug profile), so that the longitudinal momentum of the stream is approximately $\dot{m}V$. This increases from $\dot{m}V_{in}$ to $\dot{m}V_{out}$ because the gas accelerates from V_{in} to V_{out}. The change in longitudinal momentum $\dot{m}(V_{out} - V_{in})$ must be balanced by (charged to) the pressure difference applied between the two ends of the passage:

$$\dot{m}(V_{out} - V_{in}) = \Delta P_{acc} A_c \tag{9.59}$$

In view of eq. (9.58), we conclude that the pressure difference required to accelerate the stream from inlet to outlet is

$$\Delta P_{acc} = G^2 \left(\frac{1}{\rho_{out}} - \frac{1}{\rho_{in}} \right) \tag{9.60}$$

In conclusion, when the density is not constant, the pressure drop across the straight portion of the passage is equal to the wall friction effect (9.49) plus the acceleration effect (9.60):

$$\Delta P_s = f \frac{4L}{D_h} \tfrac{1}{2} \rho V^2 + G^2 \left(\frac{1}{\rho_{out}} - \frac{1}{\rho_{in}} \right) \tag{9.61}$$

In the frictional term on the right side, the density is averaged between inlet and outlet, $\rho = (\rho_{in} + \rho_{out})/2$, which means that $V = G/\rho$. The second term on the right side holds when the velocity does not vary appreciably across the passage. In the opposite case, for example, in laminar flow through a tube, the second term must be multiplied by $\tfrac{4}{3}$ to account for the parabolic velocity distribution.

When the gas is being cooled, ρ_{out} is greater than ρ_{in}, and ΔP_{acc} is negative. In this case the deceleration of the stream works toward decreasing the overall pressure difference that must be maintained across the passage, ΔP. It is important to keep in mind that when the gas density varies, the contraction–enlargement relationships to use are eqs. (9.51) and (9.54), not eq. (9.57).

9.5.4   Tube Bundles in Cross-Flow

The flow of a stream perpendicular to an array of cylinders formed the subject of Section 5.5.4. There we learned how to calculate the heat transfer rate between the stream and a given number of cylinders. Arrays of aligned cylinders and staggered cylinders were described in Figs. 5.24 and 5.25. In the present context the cross-flow and the array of cylinders (tubes with internal flow) constitute a cross-flow heat exchanger. The pressure drop experienced by the cross-flow is proportional to the number of tube rows counted in the flow direction, n_l:

$$\Delta P = n_l f \chi \tfrac{1}{2} \rho V_{max}^2 \tag{9.62}$$

The dimensionless factors f and χ are presented in Fig. 9.37 for *aligned arrays*.

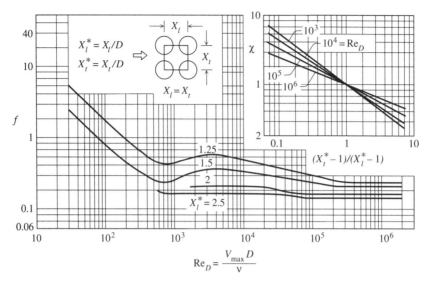

Figure 9.37 Arrays of aligned tubes: the coefficients f and χ for the pressure drop formula (9.62). (Zukauskas [18], with permission from John Wiley & Sons, Inc.)

Each array of this kind is described by the longitudinal pitch X_l and the transversal pitch X_t, or by their dimensionless counterparts

$$X_l^* = \frac{X_l}{D} \qquad X_t^* = \frac{X_t}{D} \tag{9.63}$$

where D is the tube outside diameter. The f curves in Fig. 9.37 correspond to a square array ($X_l = X_t$), while the χ correction factor accounts for arrangements in which $X_l \neq X_t$. Note that $\chi = 1$ for the square arrangement. The Reynolds number $\mathrm{Re}_D = V_{max}D/v$ is based on the maximum mean velocity V_{max}, which occurs in the smallest spacing between two adjacent tubes in a row perpendicular to the flow direction.

Figure 9.38 contains the corresponding f and χ information for *staggered arrays*. The f curves have been drawn for the equilateral triangle arrangement $[X_t = X_d$ or $X_l = (3^{1/2}/2)X_t]$. The upper right insert shows that the χ value plays the role of correction factor in instances where the tube centers do not form equilateral triangles. The Reynolds number is again based on the mean velocity through the narrowest tube-to-tube spacing, V_{max}.

Equation (9.62) and Figs. 9.37 and 9.38 are valid when $n_l > 9$. They were constructed based on extensive experiments in which the test fluids were air, water, and several oils [18].

9.5.5 Compact Heat Exchanger Surfaces

The pressure drop and heat transfer in passages with even more complicated geometries have also been measured and catalogued in dimensionless form. A representative sample is shown in Fig. 9.39, which refers to the passage external to a bundle of finned tubes. The fluid (usually a gas) flows perpendicular to the tubes. The heat transfer surface area is increased by the fins, which occupy

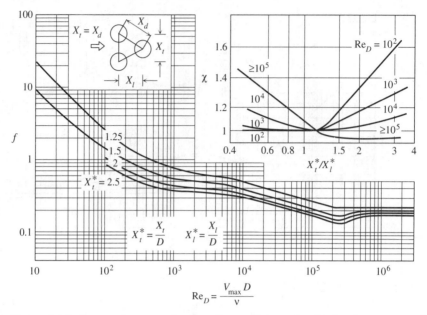

Figure 9.38 Arrays of staggered tubes: the coefficients f and χ for the pressure drop formula (9.62). (Zukauskas [18], with permission from John Wiley & Sons, Inc.)

much of the space that would have been left open between the tubes. Geometries of this kind—and there are many variations [3, 4]—have been designed to increase the heat transfer area per unit volume (heat transfer area density):

$$\alpha = \frac{A}{\mathcal{V}} \tag{9.64}$$

A heat exchanger passage with a large α value exhibits a high degree of compactness. With $\alpha = 446$ m^2/m^3, the example illustrated in Fig. 9.39 almost qualifies as a compact heat exchanger, which according to Fig. 9.12 requires $\alpha > 700$ m^2/m^3. The symbol for area density α should not be confused with the symbol used for thermal diffusivity.

In addition to α, the other key parameters that describe the geometry of the flow passage are

$$\sigma = \frac{A_c}{A_{f_r}} = \frac{\text{minimum free-flow area}}{\text{frontal area}} \tag{9.65}$$

$$\frac{A_f}{A} = \frac{\text{fin area}}{\text{total heat transfer area}} \tag{9.66}$$

and the hydraulic diameter

$$D_h = 4\,\frac{A_c L}{A} \tag{9.67}$$

In this last definition L is the length of the heat exchanger in the direction of flow. It is useful to take another look at the D_h definition for a duct with constant cross-sectional geometry, eq. (6.28), and to recognize it as a special case of eq. (9.67). The *minimum* free-flow area A_c refers to the smallest cross section

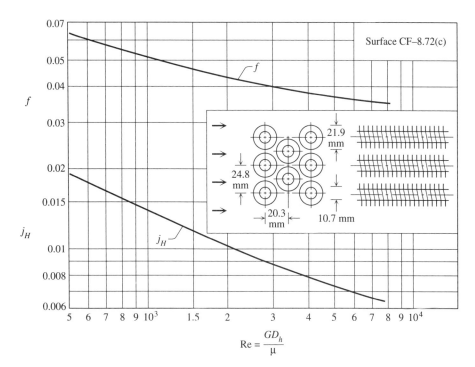

Tube outside diameter = 10.7 mm
Fin pitch = 343 per meter
Flow passage hydraulic diameter D_h = 4.43 mm
Fin thickness, average (fins are tapered slightly) = 0.48 mm, copper
Free-flow area/frontal area, σ = 0.494
Heat transfer area/total volume, α = 446 m²/m³
Fin area/total area, A_f/A = 0.876

Figure 9.39 Pressure drop and heat transfer data for the flow passage through a bundle of staggered finned tubes (London et al. [19]; also in Kays and London [4]).

encountered by the fluid. In Fig. 9.39 this area occurs between two adjacent tubes aligned on the vertical: A_c is coplanar with the two centerlines. The dimensionless parameter σ accounts for the contraction and enlargement experienced by the stream. The fin area/total area ratio A_f/A is an ingredient in the formula for the overall surface efficiency ϵ, eq. (9.4).

The mass flowrate of the stream that flows through this complicated passage is conserved in the direction of flow:

$$\dot{m} = \rho A_{fr} V \qquad (9.68)$$

By this definition, V is the mean velocity based on the frontal area. On the other hand, the mass velocity G is defined based on the *maximum* mean velocity, that is, the velocity through the minimum free-flow area:

$$G = \rho V_{\max} = \frac{\dot{m}}{A_c} \qquad (9.69)$$

where $A_c = \sigma A_{fr}$. The Reynolds number used on the abscissa of Fig. 9.39 is also based on the maximum velocity:

$$\text{Re} = \frac{V_{\max}D_h}{\nu} = \frac{GD_h}{\mu} \tag{9.70}$$

The total pressure drop across a heat exchanger core constructed as shown in the example of Fig. 9.39, namely, the difference between the pressure on the left (the inlet) and the pressure on the right (the outlet), is given by [4]

$$\Delta P = \frac{G^2}{2\rho_{\text{in}}} \left[f \frac{A}{A_c} \frac{\rho_{\text{in}}}{\rho} + (1 + \sigma^2) \left(\frac{\rho_{\text{in}}}{\rho_{\text{out}}} - 1 \right) \right] \tag{9.71}$$

In this equation ρ is the average density evaluated at the temperature averaged between the inlet and the outlet, $(T_{\text{in}} + T_{\text{out}})/2$. The average density can also be estimated by averaging the fluid specific volume $(1/\rho)$ between the inlet and outlet values,

$$\frac{1}{\rho} = \frac{1}{2} \left(\frac{1}{\rho_{\text{in}}} + \frac{1}{\rho_{\text{out}}} \right) \tag{9.72}$$

or

$$\rho = \frac{2 \, \rho_{\text{in}} \rho_{\text{out}}}{\rho_{\text{in}} + \rho_{\text{out}}} \tag{9.73}$$

The friction factor f that multiplies the first term in the square brackets of eq. (9.71) has been found experimentally and plotted in Fig. 9.39. The f factor accounts for fluid friction against solid walls *and* for the entrance and exit losses. The second term in the square brackets accounts for the acceleration or deceleration of the stream. This contribution is negligible when the density is essentially constant along the passage.

It is important to note that the term multiplied by f in the ΔP formula (9.71) is, as expected, proportional to the length of the flow passage, as $A/A_c = 4L/D_h$ according to eq. (9.67). It is not difficult to show that when the volume $\mathcal{V}$ is specified, the ratio A/A_c needed in eq. (9.71) can be estimated by writing

$$\frac{A}{A_c} = \frac{\alpha \mathcal{V}}{\sigma A_{fr}} \tag{9.74}$$

Charts such as Fig. 9.39 also contain the information necessary for calculating the average heat transfer coefficient h for the particular flow passage. This is expressed in nondimensional terms as the *Colburn* j_H *factor*,

$$j_H \doteq \text{St Pr}^{2/3} \tag{9.75}$$

in which the Stanton number is based on G (i.e., on maximum velocity):

$$\text{St} = \frac{h}{Gc_P} = \frac{h}{\rho c_P V_{\max}} \tag{9.76}$$

In Fig. 9.39 and many like it for other compact surfaces, factors j_H and f exhibit almost the same dependence on Reynolds number. Here, the ratio $j_H/(f/2)$ decreases only by one third (from 0.6 to 0.4) as Re increases from 500 to 800. The observation that even in a complicated passage the ratio $j_H/(f/2)$ is relatively constant and of order of magnitude 1 is important. It agrees qualitatively with the Colburn analogy for a duct with constant cross-sectional geometry, eq. (6.89).

The calculation of the heat transfer between the two fluids that interact in cross-flow in Fig. 9.39 also requires an estimate of the average heat transfer coefficient for the internal surface of the tubes. This estimate can be obtained by following the method outlined in Chapter 6. The complex procedure of selecting a certain heat exchanger (surface type, size) subject to various constraints (e.g., volume, pressure drop) is known as *heat exchanger design*. Several classical and modern books have been devoted to this important area of thermal engineering practice [1–4, 20–22].

The newer work on the design and optimization of heat exchangers is rooted in thermodynamics, and seeks to minimize the destruction of useful energy (exergy, or availability) in the heat exchanger. This work is described in Refs. [10] and [23]. The destruction of useful energy is due to thermodynamic irreversibility, or the generation of entropy through one-way processes such as heat transfer (always from hot to cold) and fluid flow (always against friction). For this reason the thermodynamic approach to heat exchanger design is also known as irreversibility minimization, or entropy generation minimization. This methodology has shown that certain dimensions and operating conditions can be selected optimally, often on the basis of simple formulas, so that the overall irreversibility of the heat exchanger is minimum.

Example 9.6

Compact Heat Exchanger, Pressure Drop, and Heat Transfer Coefficient

Air enters at 1 atm and 300°C the core of a finned tube heat exchanger of the type shown in Fig. 9.39. The air flows at the rate of 1500 kg/h perpendicular to the tubes and exits with a mean temperature of 100°C. The core is 0.5 m long, with a 0.25-m² frontal area. Calculate the total pressure drop between the air inlet and outlet, and the average heat transfer coefficient on the air side.

Solution. The air density at the inlet and outlet is listed in Appendix D:

$$\rho_{in} = 0.616 \text{ kg/m}^3 \qquad \rho_{out} = 0.946 \text{ kg/m}^3$$

The other relevant properties of air at the average temperature (300 °C + 100°C)/2 = 200°C are

$$\rho = 0.746 \text{ kg/m}^3 \qquad\qquad \text{Pr} = 0.68$$
$$\mu = 2.58 \times 10^{-5} \text{ kg/s·m} \qquad c_P = 1.025 \text{ kJ/kg·K}$$

The pressure drop is furnished by eq. (9.71), for which we calculate, in order,

$$\frac{A}{A_c} = \frac{\alpha \mathcal{V}}{\sigma A_{fr}} = \frac{\alpha}{\sigma} L = \frac{446 \text{m}^2/\text{m}^3}{0.494}(0.5\,\text{m}) = 451.4$$

$$A_c = \sigma A_{fr} = 0.494 \times 0.25 \text{ m}^2 = 0.124 \text{ m}^2$$

$$G = \frac{\dot{m}}{A_c} = \frac{1500 \text{ kg}}{3600 \text{ s}} \frac{1}{0.124 \text{ m}^2} = 3.36 \text{ kg/m}^2\text{·s}$$

$$D_h = 4\frac{L}{A/A_c} = 4\frac{0.5 \text{ m}}{451.4} = 4.43 \text{ mm} \qquad \text{(listed also in Fig. 9.39)}$$

$$\text{Re} = \frac{GD_h}{\mu} = 3.36 \frac{\text{kg}}{\text{m}^2\text{·s}} \frac{0.00443\text{m}}{2.58 \times 10^{-5} \text{ kg/s·m}} = 577$$

$$f \cong 0.061 \quad \text{(Fig. 9.39, at Re} = 577)$$

$$\Delta P = \left(3.36 \; \frac{\text{kg}}{\text{m}^2 \cdot \text{s}}\right)^2 \frac{1}{2 \times 0.616 \; \text{kg/m}^3} \left[(0.061 \times 451.4) \; \frac{0.616}{0.746}\right.$$

$$\left. + (1 + 0.494^2)\left(\frac{0.616}{0.946} - 1\right)\right]$$

$$= 204.4 \; \text{N/m}^2$$

The heat transfer coefficient emerges out of the j_H value read off Fig. 9.39:

$$j_H \cong 0.018$$

$$\text{St} = j_H \text{Pr}^{-2/3} \cong 0.018(0.68)^{-2/3} = 0.0233$$

$$h = \text{St} \; Gc_P = \left(0.0233 \times 3.36 \; \frac{\text{kg}}{\text{m}^2 \cdot \text{s}}\right)\left(1.025 \; \frac{10^3 \text{J}}{\text{kg} \cdot \text{K}}\right) \cong 80 \; \text{W/m}^2 \cdot \text{K}$$

REFERENCES

1. S. Kakac, A. E. Bergles, and F. Mayinger, Eds., *Heat Exchangers: Thermal-Hydraulic Fundamentals and Design*, Hemisphere, Washington, DC, 1981.

2. R. K. Shah and A. C. Mueller, Heat exchangers, Chapter 4 in W. M. Rohsenow, J. P. Hartnett, and E. N. Ganic, Eds., *Handbook of Heat Transfer Applications*, 2nd edition, McGraw-Hill, New York, 1985.

3. A. P. Fraas, *Heat Exchanger Design*, 2nd edition, Wiley, New York, 1989.

4. W. M. Kays and A. L. London, *Compact Heat Exchanges*, 3rd edition, McGraw-Hill, New York, 1984.

5. R. K. Shah, Classification of heat exchangers, pp. 9–46 in Ref. [1].

6. M. A. DiGiovanni and R. L. Webb, Uncertainty in effectiveness–*NTU* calculations for crossflow heat exchangers, *Heat Transfer Engineering*, Vol. 10, No. 3, 1989, pp. 61–70.

7. J. M. Chenoweth, Final report of the HTRI/TEMA joint committee to review the fouling section of the TEMA Standards, *Heat Transfer Engineering*, Vol. 11, No. 1, 1990, pp. 73–107.

8. D. Q. Kern, *Process Heat Transfer*, McGraw-Hill, New York, 1950.

9. R. A. Bowman, A. C. Mueller, and W. M. Nagle, Mean temperature difference in design, *Trans. ASME*, Vol. 62, 1940, pp. 283–294.

10. A. Bejan, *Advanced Engineering Thermodynamics*, Wiley, New York, 1988, p. 632.

11. W. Nusselt, A new heat transfer formula for cross-flow, *Technische Mechanik und Thermodynamik*, Vol. 12, 1930, pp. 417–422.

12. H. Ten Broeck, Multipass exchanger calculations, *Ind. Eng. Chem.*, Vol. 30, 1938, pp. 1041–1042.

13. W. M. Kays, A. L. London, and D. W. Johnson, *Gas Turbine Plant Heat Exchangers*, American Society of Mechanical Engineers, New York, 1951.

14. B. Eck, *Technische Strömungslehre*, 7th ed., Springer-Verlag, Berlin, 1966.

15. W. M. Kays and A. L. London, Convective heat-transfer and flow-friction behavior of small cylindrical tubes—circular and rectangular cross sections, *Trans. ASME*, Vol. 74, 1952, pp. 1179–1189.

16. W. M. Kays, Loss coefficients for abrupt changes in flow cross-section with low

Reynolds number flow in single and multiple-tube systems, *Trans. ASME*, Vol. 72, 1950, pp. 1067–1074.

17. Y. Nakayama, W. A. Woods, D. G. Clark, and the Japan Society of Mechanical Engineers, Eds., *Visualized Flow*, Pergamon Press, Oxford, 1988.

18. A. Zukauskas, Convective heat transfer in cross flow, Chapter 6 in S. Kakac, R. K. Shah, and W. Aung, Eds., *Handbook of Single-Phase Convective Heat Transfer*, Wiley, New York, 1987.

19. A. L. London, W. M. Kays, and D. W. Johnson, Heat-transfer and flow-friction characteristics of some compact heat-exchanger surfaces, *Trans. ASME*, Vol. 74, 1952, pp. 1167–1178.

20. E. U. Schlünder, Ed., *Heat Exchanger Design Handbook*, Vols. 1–5, Hemisphere, New York, 1983.

21. F. W. Schmidt and A. J. Willmott, *Thermal Energy Storage and Regeneration*, Hemisphere, New York, 1981.

22. R. F. Boehm, *Design Analysis of Thermal Systems*, Wiley, New York, 1987.

23. A. Bejan, *Entropy Generation through Heat and Fluid Flow*, Wiley, New York, 1982.

24. J. S. Lim, A. Bejan, and J. H. Kim, Thermodynamics of exergy extraction from fractured hot dry rock, *Int. J. Heat and Fluid Flow*, Vol. 13, 1992, pp. 71–77.

PROBLEMS

Overall Heat Transfer Coefficient

9.1 Assume that a scale layer of resistance r_s covers the entire area A of one side of the heat exchanger surface shown in Fig. 9.13. Defining $h_s = 1/r_s$ as an equivalent scale heat transfer coefficient (across the temperature difference $T_w - T_s$), repeat the analysis contained in eqs. (9.1) through (9.3) to arrive at the total heat transfer rate formula (9.6). In other words, prove that the effective heat transfer coefficient for one side of the heat exchanger surface accounts for the two effects shown on the right side of eq. (9.7).

9.2 The heat exchanger surface shown in the figure is exposed to a stream of city water on the hot side and a stream of compressed air on the cold side. The thermal conductivity of the wall and pin fin material is 50 W/m·K. The heat transfer coefficients (h_h, h_c) are distributed uniformly over their respective surfaces. The total area of the hot side of the wall is 1 m².

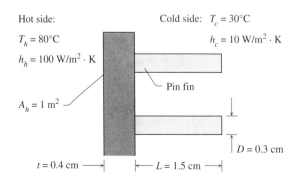
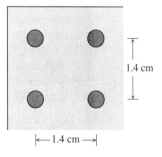

Figure P9.2

(a) Calculate the total heat transfer rate through the heat exchanger surface, q_a, by taking into account the effect of fouling on both sides of the surface. On which side is the effect of fouling insignificant?

(b) Repeat the calculation of the total heat transfer rate (q_b) by assuming that the pin fins are absent from the cold side of the surface. Compare q_b and q_a, and comment on the extent to which the finning of the cold side augments (enhances) the total heat transfer rate.

9.3 Consider the heat exchanger surface illustrated in the figure. The cold side is flat and of size $A_c = 0.4$ m². The hot side of the wall is covered with plate fins arranged in a staggered array. The geometric dimensions and thermal conditions are indicated directly on the drawing. The wall and the fins have the same thermal conductivity, $k = 20$ W/m·K. The effect of fouling is negligible on both sides of the heat exchanger surface.

Calculate the overall thermal resistance $(U_c A_c)^{-1}$ and the total heat transfer rate through the heat exchanger surface. Show also that the heat transfer through each plate fin can be modeled as one-dimensional conduction (review Section 2.7.4).

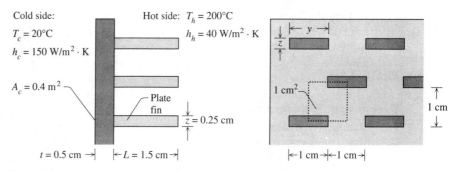

Figure P9.3

9.4 Hot air flows through the finned tubes dimensioned on Fig. 9.39 in the text, while cold water flows inside the tubes. The heat transfer coefficients on the hot and cold sides of the surface are $h_h = 80$ W/m²·K and $h_c = 1000$ W/m²·K, respectively. The tube wall and the fins are made of aluminum; their average temperature is 200°C. The scale resistances are negligible on both sides of the surface.

(a) Assume that the fin thickness is negligible, and prove that the area ratio is given by

$$\frac{A_c}{A_h} \cong \frac{D_i}{D_o}\left(1 - \frac{A_{f,h}}{A_h}\right)$$

In this relationship D_i and D_o are the inner and outer diameters of the bare tube and $A_{f,h}$ is the contribution the fins make to A_h. Note that $(A_f/A)_h$ is listed under the chart of Fig. 9.39.

(b) Calculate the overall surface efficiency for the hot side, ϵ_h.

(c) Determine the overall heat transfer coefficient based on the cold side, U_c, and later deduce the overall heat transfer coefficient based on the hot side, U_h.

The Log-Mean Temperature Difference Method

9.5 A 1 kg/s stream of hot water is cooled from 90°C to 60°C in a parallel flow heat exchanger in which the cooling agent is a 2 kg/s stream of water with the initial temperature of 40°C. The overall heat transfer coefficient between the two streams is constant and known, $U = 1000$ W/m²·K.

(a) Calculate by the ΔT_{lm} method the size of the required heat exchanger area A.

(b) Note that in the first part of Example 9.2, the same stream was cooled in a counterflow heat exchanger arrangement, requiring a total heat transfer area of 4.69 m^2. Compare this area with your answer to part (a) and comment on which of the two heat exchangers (parallel flow or counterflow) is preferable.

9.6 A cross-flow heat exchanger in which both streams are unmixed has the function to cool a 1 kg/s stream of hot water from 90°C to 60°C. The cold stream carries 2 kg/s of water with an initial (inlet) temperature of 40°C. The overall heat transfer coefficient is $U = 1000$ W/m²·K.

(a) Calculate by the ΔT_{lm} method the required heat transfer area A.

(b) Compare this result with the area needed by a cross-flow heat exchanger with both fluids mixed (see the second part of Example 9.2), and comment on the effect of stream mixing on the size of the required heat exchanger surface.

9.7 Show that the total heat transfer rate between the two streams of a counterflow heat exchanger is also given by eq. (9.21) in which ΔT_{lm} assumes the form listed in eqs. (9.22) and (9.23). In other words, derive eq. (9.21) while analyzing the counterflow heat exchanger shown in the upper right quadrant of Fig. 9.15. As a guide, use the analysis listed between eqs. (9.10) and (9.22) in the text.

9.8 Consider the *parallel flow* oil cooler in which an oil stream of 1.5 kg/s is cooled by a water stream of 0.5 kg/s. The oil inlet temperature is 110°C, and the water inlet temperature is 15°C. The respective specific heats c_P of oil and water are, respectively, 2.25 kJ/kg·K and 4.18 kJ/kg·K. The total heat transfer area is $A = 10$ m^2, while the overall heat transfer coefficient is $U = 500$ W/m²·K.

(a) Calculate the two outlet temperatures and the total stream-to-stream heat transfer rate by using the ΔT_{lm} method. As a starting guess for the trial-and-error procedure use $T_{h,out} = 70$°C for the outlet temperature of the oil stream.

(b) The *counterflow* alternative to this parallel flow design was analyzed in Example 9.3. Compare the heat transfer rates and oil outlet temperatures promised by the two designs, and comment on the relative attractiveness of the counterflow arrangement.

The Effectiveness – *NTU* Method

9.9 The effectiveness–*NTU* relation for a parallel flow heat exchanger, eq. (9.37), was derived in the text by assuming that $C_c = C_{min}$ and $C_h = C_{max}$. Rederive the effectiveness–*NTU* relationship by making the opposite assumption, $C_c = C_{max}$ and $C_h = C_{min}$. Prove in this way that eqs. (9.37) and (9.38) are valid regardless of whether the hot stream or the cold stream has the smaller capacity rate.

9.10 The job of a counterflow oil cooler is to lower the temperature of a 2 kg/s stream of oil from a 110°C inlet to an outlet temperature of 50°C. The oil specific heat is 2.25 kJ/kg·K. The coolant is a 1 kg/s stream of cold water with an inlet temperature of 20°C. The specific heat of water is 4.18 kJ/kg·K and the overall heat transfer coefficient has the value $U = 400$ W/m²·K.

(a) Calculate the necessary heat exchanger area A by the ϵ–*NTU* method.

(b) Later, calculate the total stream-to-stream heat transfer rate and the water outlet temperature.

9.11 The heat transfer surface of a counterflow heat exchanger is characterized by $U = 600$ W/m²·K and $A = 10$ m^2. The hot side is bathed by a 1 kg/s stream of hot water with an inlet temperature of 90°C. On the cold side, the surface is cooled by a 4 kg/s stream of water with an inlet temperature of 20°C. Calculate the stream-to-stream heat transfer rate and the two outlet temperatures by using the ϵ–*NTU* method.

9.12   The specified job of a heat exchanger is to lower the temperature of a 1 kg/s stream of hot water from 80°C to 40°C. The available coolant is a 1 kg/s stream of cold water, with an inlet temperature of 20°C. The overall stream-to-stream heat transfer coefficient is $U = 800 \ W/m^2 \cdot K$. Calculate the required heat transfer area A by assuming that the two streams are oriented in

(a) Parallel flow (provide a physical or thermodynamic explanation for why this flow arrangement will not work), and

(b) Counterflow.

9.13   The pores of the granular material shown on the left side of the figure are filled by a fluid with thermal diffusivity α. Such a combination of solid and fluid is known as a *saturated porous medium* (see the end of Appendix B). The fluid flows through the spaces between the grains with an imposed average velocity V. A temperature difference is maintained across the saturated porous medium, and as a result heat is transferred through it.

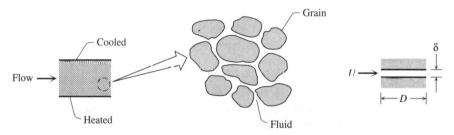

Figure P9.13

Consider now the sample enlarged on the right side of the figure. Generally speaking, there is always a difference between the local bulk temperature of the fluid and the temperature of the most immediate grain. When the grains and the interstitial spaces are sufficiently small, this local temperature difference can be neglected, and the saturated porous medium is described (approximately) by a single temperature at each point. A critical question, then, is how small the grains and fluid spaces should be for this local thermal equilibrium approximation to be valid.

(a) Let D and δ represent the sizes of the grain and the fluid channel. As fluid channel in your analysis use the parallel-plate model illustrated in the figure. Review the $\epsilon - NTU$ charts exhibited in the text, and note that the notion of good thermal contact between fluid and solid walls is associated with the range $NTU \gg 1$. Rely on this observation to arrive at the following criterion for local fluid–grain thermal equilibrium:

$$\frac{U\delta}{\alpha} << \frac{D}{\delta}$$

(b) In a certain laboratory apparatus, the grains are spheres of diameter 0.4 cm, the fluid is water at roughly 25°C, and the average water velocity through the pores is 0.1 cm/s. The spheres are packed in such a way that $\delta \sim 0.1$ cm is a good estimate for the average fluid channel thickness. Determine numerically whether the water and the spheres are in local thermal equilibrium, that is, whether the criterion derived in part (a) is satisfied.

9.14   The figure shows the low temperature (preheating) zone of a long reheating oven used in the manufacture of commercial steel plate. The steel plate enters the oven at 100°C, and is heated in counterflow by a stream of hot gas with an inlet temperature of 300°C. The gas channel is 30 m long, with a ceiling-to-plate spacing

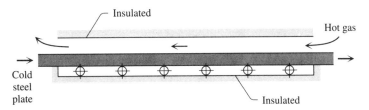

Figure P9.14

of 15 cm. The gas mean velocity is 2 m/s. The gas properties are very similar to those of air.

The steel plate is continuous, and has a thickness of 10 cm. Its motion is slow: one spot on the plate spends a total of 2 hours inside the oven. This is called the residence time. The oven is sufficiently wide in the direction perpendicular to the figure. The top and bottom surfaces of the oven are well insulated.

(a) Calculate the gas outlet temperature, and the final (outlet) temperature of the steel plate.

(b) Estimate the needed residence time if the final temperature of the steel plate is to be 200°C.

9.15 The heat exchanger analyses presented in the text are based on the assumption that all of the heat transfer occurs internally (between two streams), and that the heat exchanger enclosure is perfectly insulated with respect to its environment. When this assumption is not valid, the stream-to-stream heat transfer depends on the environmental temperature and on the properties of the external wall of the heat exchanger.

To understand this dependence, consider a counterflow heat exchanger of length L in which the cold stream is isothermal at T_c (e.g., a liquid boiling at nearly constant pressure). The hot stream is single-phase, and enters the heat exchanger with the high temperature T_h. The hot stream engulfs the cold stream, so that the wall of the heat exchanger separates the hot stream of local bulk temperature $T(x)$, from the environment of temperature T_0. The external wall has the constant thickness t, thermal conductivity k, and a geometry that does not depend on the longitudinal coordinate x. The figure shows a symmetric (round) geometry for simplicity. The outer wall is relatively "thin," that is, its thickness is smaller than the effective radius of the cross section. Assume that the temperatures of the external and internal surfaces of the outer wall are equal to T_0 and, respectively, $T(x)$. Note further that $T_h > T_c > T_0$. The overall stream-to-stream heat transfer coefficient is known, and is based on the internal surface of the external wall.

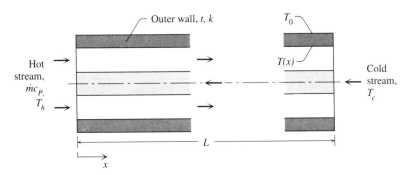

Figure P9.15

Derive expressions for the temperature distribution along the hot stream, the heat transfer rate received by the cold stream, and the heat transfer through the outer wall. Show that these results approach the more familiar forms (given in the text) for the limit where the wall insulation becomes perfect.

Pressure Drop

9.16 The core of a shell-and-tube heat exchanger contains 50 tubes, each with an inside diameter of 2 cm and length of 5 m. The header is shaped as a disc with a diameter of 30 cm. Air at 1 atm flows through the tubes with a total flowrate of 500 kg/h. The air inlet and outlet temperatures are 100°C and 300°C, respectively. Calculate the total pressure drop across the tube bundle by evaluating, in order, the pressure drops due to friction in the tube, acceleration, abrupt contraction, and abrupt enlargement. In the end, compare the frictional pressure drop estimate with the total pressure drop.

9.17 The two-dimensional water flow photographed on the left side of Fig. 9.33 has a mean velocity of 15 cm/s through the narrower channel. The width of the narrower channel is 0.4 m. The water temperature is 25°C.

(a) Determine the pressure drop through the abrupt contraction shown in the photograph.

(b) How long must the narrower channel be if its pressure drop due to wall friction is to equal the pressure drop calculated in part (a)?

9.18 A 5000 kg/h stream of 500°C air at atmospheric pressure is cooled in a shell-and-tube heat exchanger. The air flows through 300 parallel tubes. Each tube has an internal diameter of 1.5 cm and a length of 3 m. The header is built in such a way that the total cross-sectional area of the tubes is equal to 60 percent of the header (frontal) area. The outlet temperature of the air stream is 100°C. Calculate the pressure differences associated with the abrupt contraction, friction, and deceleration in the straight tubes, and, finally, with enlargement. Explain why the total pressure drop is smaller than the pressure drop due to friction in the straight tube.

9.19 The core of a cross-flow heat exchanger employs a bank of staggered *bare* tubes with a longitudinal pitch of 20.3 mm and transverse pitch of 24.8 mm. The outer diameter of each tube is 10.7 mm. Air flows perpendicular to the bare tubes. The frontal area seen by the air stream is a 0.5 m × 0.5 m square. The length of the heat exchanger core is 0.5 m. The air mass flowrate is 1500 kg/h, and the air properties may be evaluated at 200°C and 1 atm.

(a) Calculate the air pressure drop across the core of the heat exchanger.

(b) The equivalent calculation was performed in Example 9.6 for the case where the tubes were finned. Compare the ΔP calculated in part (a) with the ΔP calculated in Example 9.6, and comment on the effect of finning on the total pressure drop.

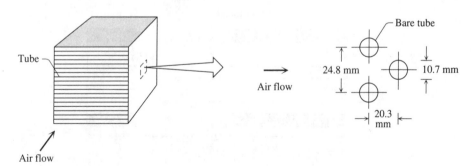

Figure P9.19

9.20 **(a)** Show that the following relationship exists between hydraulic diameter, heat transfer area density, and contraction ratio:

$$D_h = 4 \frac{\sigma}{\alpha}$$

(b) The scale drawn along the baseline of Fig. 9.12 was constructed by assuming a certain value for σ. Calculate that value.

(c) Derive eq. (9.74) and show that

$$\frac{A}{A_c} = \frac{4L}{D_h}$$

9.21 In a cross-flow heat exchanger, air flows across a bundle of tubes with dimensions $D = 5$ cm and $X_t = X_l = 9$ cm. The air velocity averaged over the frontal area of the bundle is 3 m/s. There are 21 rows of tubes counted in the direction of air flow. The average air temperature inside the heat exchanger is 100°C. Calculate the air pressure drop caused by the tubes, by assuming that the tubes are **(a)** aligned, and **(b)** staggered. Compare the two pressure drops, and comment on the effect of staggering the tubes in the array.

PROJECTS

9.1 Batch Heating Processes

Many industrial processes require the heating of a liquid as a *batch*, not as the steady *stream* that has been assumed in all the configurations discussed until now. The batch can be heated by contact with an immersed heat exchanger coil, or by mixing with a stream that has been heated outside the tank that holds the batch. In either case, the batch temperature increases with time. Batch heating processes are time dependent, which distinguishes them from all the heat exchange schemes considered in Chapter 9.

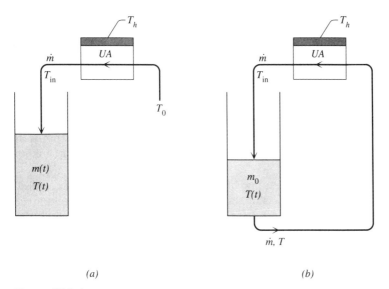

(a)   (b)

Figure PR9.1

Two of the simplest batch heating schemes are outlined in the figure. Assume first that the liquid (batch and stream) is water at constant pressure. The initial mass of the cold batch is m_0, and the initial temperature is T_0.

In scheme (a), the heated stream $\dot{m}$ (constant) is discharged into the batch beginning with the time $t = 0$. The stream and the batch mix instantly to the time-dependent temperature $T(t)$ and mass inventory $m(t)$. The stream is heated externally in a heat exchanger of size UA. The hot side of the heat exchanger is at constant temperature T_h.

In scheme (b), the stream $\dot{m}$ has its origin in the batch, as it is drawn from the bottom of the vessel at the same rate ($\dot{m}$) at which it is poured in. As a consequence, the batch mass inventory does not change, $m = m_0$, even though the temperature of the well-mixed liquid changes, $T(t)$. The stream is heated in the same external heat exchanger.

Consider now the following challenge. You must choose the scheme, (a) or (b), that would allow you to raise the batch temperature T the fastest. All the other parameters (m_0, $\dot{m}$, T_0, T_h, UA) are fixed and assumed known.

The easiest way to begin making progress on answering this question is by assuming that the external heat exchanger is sufficiently large so that $NTU \gg 1$ and $\epsilon \cong 1$ in eq. (9.40). This assumption leads to the all-important simplification that $T_{in} \cong T_h$.

9.2 Heat Transfer Surface Between a Liquid Stream and a Solid, with Time-Dependent Conduction in the Solid

The objective of this project is to determine the temperature history of the water and the adjacent rock layers situated at the bottom of a well for the extraction of energy from a region of hot dry rock. This method of energy extraction formed also the subject of Problem 4.24. It consists of creating a fracture (crack) in the rock, and circulating cold water through the crack. In time, the temperature of the rock adjacent to the crack drops as a result of time-dependent conduction.

As the simplest model [24] for the "heat exchanger surface" between the water-filled space and the rock, assume that the conduction through the rock is unidirectional. Assume also that the crack is more like a small cave, that is, wide and short enough so that the water stream is very well mixed to a unique instantaneous temperature $T_{out}(t)$. This unique temperature is equal also to the temperature of the outgoing water stream. The water inlet temperature T_{in} is fixed.

This model is outlined in the attached figure, in which only one half of the system (one side of the crack) is shown. The rock surface in contact with the water stream is A. The heat transfer coefficient h between the rock surface $T_0(t)$ and the water bulk temperature $T_{out}(t)$ is constant.

The initial temperature of the rock is uniform, T_∞. Beginning at $t = 0$, the stream of cold water is forced through the crack, and the wetted rock is cooled by conduction. In time, all the temperatures decrease, esspecially the hot-water delivery temperature $T_{out}(t)$, which is the most important engineering unknown in this problem.

Procedure. Write the first law of thermodynamics for the water control volume in the steady state, by neglecting the thermal inertia of the crack water inventory. Note also the continuity of heat flux across the wetted surface (A, T_0). Finally, write the conduction equation, the initial condition and the boundary conditions for $T(y,t)$, or time-dependent conduction in the semi-infinite rock.

Simplify the conduction problem by using the *integral method*. Assume the following exponential-decay profile for the rock temperature

$$\theta = (\theta_0 - 1) \exp\left(-\frac{\eta}{\delta}\right) + 1$$

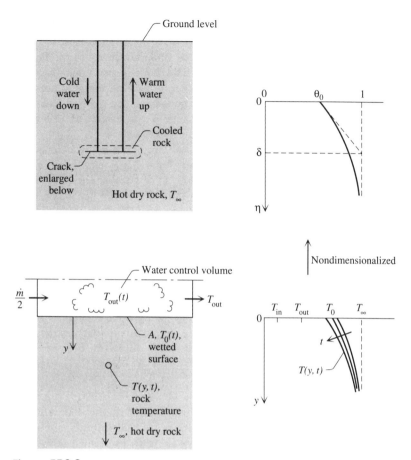

Figure PR9.2

for which the dimensionless variables are defined by

$$\theta = \frac{T - T_{\text{in}}}{T_\infty - T_{\text{in}}} \qquad \theta_{\text{out}} = \frac{T_{\text{out}} - T_{\text{in}}}{T_\infty - T_{\text{in}}}$$

$$\theta_0 = \frac{T_0 - T_{\text{in}}}{T_\infty - T_{\text{in}}} \qquad \eta = \frac{y}{k/h}$$

In the η definition, k is the thermal conductivity of the rock. The parameter δ is a dimensionless number proportional to the effective thickness of the layer of rock that has been cooled by contact with the cold water. We can expect δ to increase approximately as $(\alpha t)^{1/2}$, where α is the rock thermal diffusivity. Complete the integral analysis by integrating the conduction equation from the rock surface ($\eta = 0$) to the distant layers that are still uncooled ($\eta \to \infty$).

Solve the resulting system of equations analytically or numerically. Report your results as dimensionless curves for $\theta_{\text{out}}(\tau)$, $\theta_0(\tau)$ and $\delta(\tau)$, where τ is the dimensionless time

$$\tau = t\,\frac{h^2\alpha}{k^2}$$

There will be one θ_{out}, θ_0, or δ curve for each value of $NTU = hA/(\dot{m}c)$.

9.3 Counterflow in a Vertical Channel Connecting Two Fluid Reservoirs at Different Temperatures

Consider the vertical narrow parallel-plate channel shown in the figure. The channel and the two reservoirs contain a fluid with nearly constant properties. The reservoir temperatures (T_h, T_c) are known. When the temperature of the bottom reservoir is higher than the temperature of the top reservoir (and if the fluid expands upon heating at constant pressure, $\beta > 0$), warm fluid tends to rise and cold fluid tends to fall through the vertical channel.

The simplest type of channel flow that can occur is the balanced counterflow sketched in the figure. If the regime is laminar and fully developed, the counterflow can be viewed as a "straight" flow made up of parallel sheets of fluid that move relative to one another. As is shown on the right side of the figure, each branch of the counterflow can be modeled (i.e., approximated) as Hagen-Poiseuille flow through a $D/2$-wide channel. The stream-to-stream bulk temperature difference (ΔT) and the stream-to-stream heat flux are constant (independent of altitude), even though the temperatures of both streams decrease with altitude.

An important issue in cryogenic engineering is the calculation of the vertical heat transfer rate between the two reservoirs. This "heat leak" is equal to $q' = \dot{m}' c_P \Delta T$, where $\dot{m}'$ is the mass flowrate of one branch of the counterflow. Both q' and $\dot{m}'$ are expressed per unit length perpendicular to the plane $H \times D$. The vertical heat leak q' is a convective heat transfer rate. For example, looking across the opening in the bottom of the cold reservoir, q' represents the difference between the upflow of enthalpy due to the warm branch, $\dot{m}' c_P (T_c + \Delta T)$, and the enthalpy downflow associated with the cold branch, $\dot{m}' c_P T_c$.

To determine q' analytically, you may consider these steps:

(a) Note that, if motionless, the two branches of the counterflow (two fluid columns) would have different hydrostatic pressure distributions. For the pressures of the two streams to match everywhere along H, they must flow fast enough so that their individual frictional pressure drops eliminate the hydrostatic pressure mismatch. Show that the frictional pressure drop experienced by each stream must be equal to

$$\Delta P = \tfrac{1}{2} \rho \beta \Delta T g H$$

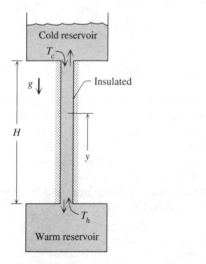

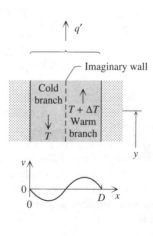

(b) Rely on the channel flow and heat transfer information developed in Chapter 6, and on the ϵ-NTU relation for a balanced counterflow heat exchanger, to arrive at the stream-to-stream bulk temperature difference

$$\frac{\Delta T}{T_h - T_c} \cong 1 - 1034\,\frac{H/D}{\text{Ra}}$$

and the vertical heat transfer rate,

$$\frac{q'}{k(T_h - T_c)} \cong \frac{1}{192}\,\text{Ra}\left(1 - 1034\,\frac{H/D}{\text{Ra}}\right)^2$$

where Ra is the Rayleigh number based on D and reservoir-to-reservoir temperature difference,

$$\text{Ra} = \frac{g\beta(T_h - T_c)D^3}{\alpha v}$$

(c) Show that the fluid starts moving in the vertical channel when the temperature difference $(T_h - T_c)$ and D are large enough so that

$$\text{Ra} > 1034\,\frac{H}{D}$$

Comment on the similarities between this condition and the criterion for the onset of Bénard convection (nearly square rolls, Fig. 7.21, right side).

(d) Apply the local Reynolds number criterion (F.2) to the $D/2$-wide stream, and show that the flow remains laminar as long as

$$\text{Ra} - 1034\,\frac{H}{D} \lesssim 2 \times 10^4\,\text{Pr}$$

9.4 The Best Way of Wrapping a Finite Amount of Insulation on a Duct

A hot stream originally at the temperature T_h flows through a pipe suspended in a cold environment of temperature T_0. The stream temperature $T(x)$ decreases in the longitudinal direction, because of the heat transfer that takes place from $T(x)$ to T_0 everywhere along the pipe. The stream behaves in the same way as in a counterflow (or parallel flow) heat exchanger in which the other stream is isothermal (T_0) because of change of phase (boiling) or a very large capacity rate. The essential difference is that in the present "heat exchanger" the transfer of heat is something to be avoided, not promoted. Indeed, the only function of the pipe is to deliver the stream at a temperature (T_{out}) that resembles as closely as possible the original temperature, T_h.

For this we must think of insulating the pipe, because it is clear that the more insulation we use the smaller the heat loss to the ambient, and the higher the outlet

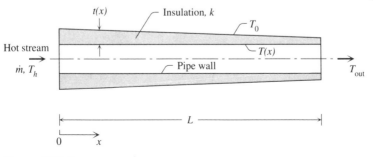

Figure PR9.4

temperature. The problem is tougher than this, because the amount of insulation material (its volume) is fixed. This constraint arises when the insulation material is expensive, or when the weight of the insulation can deflect the pipe and load the pipe supports beyond permissible levels, or when the pipe is to be used in an airborne application.

The design question that you are asked to consider is what is the best way of distributing over the pipe length L the fixed amount of insulation material? You may speculate from the start that the insulation should be relatively thicker near the entrance (hot end), because in that region the temperature difference between the stream and the ambient is the largest. Investigate the goodness of this idea by analyzing the geometry shown in the sketch. The insulation thickness varies linearly in the longitudinal direction, but the insulation volume is fixed and the taper (slope) parameter dt/dx is the only design variable. In other words, determine the optimal taper that minimizes the total heat transfer rate between the stream and the environment.

For simplicity assume that the pipe diameter is sufficiently large so that at every x the insulation is thin relative to the pipe radius. In other words, model the insulation layer as a plane wall. Assume also that the pipe wall is practically at the same temperature as the stream, $T(x)$, and that the temperature of the outer surface of the insulation is the same as T_0. This is to say that at every x the overall thermal resistance between the stream and the environment is due entirely to the insulation.

Note: the solution to the general problem in which all the simplifying assumptions made above are dropped is beyond the level of this course, and is reported in the Solutions Manual.

10

RADIATION

10.1 INTRODUCTION

In this chapter we consider the phenomenon of radiative heat transfer, which relative to conduction and convection represents the third distinct mechanism of heat exchange between two entities at different temperatures. The radiation mechanism was previewed in Section 1.5; it was noted then that unlike conduction and convection, radiation is the mechanism by which bodies can exchange heat *from a distance*, without making direct contact. A net heat transfer by radiation can take place even when the space between two surfaces is completely evacuated.

Thermal radiation is the stream of electromagnetic radiation emitted by a material entity (solid body, pool of liquid, cloud of reacting gaseous mixture) on account of its *finite* absolute temperature T. The temperature and the emitted thermal radiation are reflections of the degree of molecular agitation of the material.

The fundamental question with regard to radiation heat transfer is outlined in Fig. 10.1, in which the two bodies, (T_1) and (T_2), have arbitrary shapes: they were drawn spherically only for simplicity. The two bodies emit their respective streams of thermal radiation in all the directions to which they have access. In fact, every point (e.g., area element) of each body emits radiation in all the directions in which one can look from that point. Only a fraction of the total stream emitted by (T_1) is intercepted and possibly absorbed by body (T_2). This fraction depends not only on the shapes and sizes of the two bodies but also on their relative position, on the condition (e.g., smoothness, cleanliness) of their surfaces, and on the nature of the surroundings. Similarly, only a fraction of the radiation emitted by (T_2) is intercepted and possibly absorbed by (T_1).

With reference to just one of the bodies depicted in Fig. 10.1, the heat transfer problem reduces to calculating

1. The radiation heat transfer rate that leaves the body surface (e.g., the radiation directly emitted by the surface, plus the reflected portion of the radiation that strikes the surface); and
2. The radiation heat transfer rate that enters the surface (e.g., the absorbed portion of the radiation that strikes the surface).

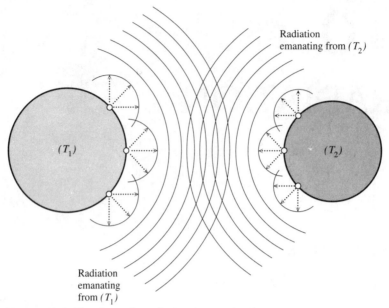

Figure 10.1 The thermal–radiation interaction between two entities at different temperatures.

The difference between quantities 1 and 2 represents the *net* heat transfer rate that leaves the body that is being analyzed. A similar accounting can be made of the heat transfer interaction experienced by the other surfaces (bodies) that take part in the configuration.

It is already apparent that radiation heat transfer calculations involve several new complications relative to the conduction and convection configurations examined until now. Chief among these is the optical aspect, that is, the manner in which one body "sees" its neighboring entities. For this reason, space geometry and the ability to see in three dimensions play an important role in the analyses we are about to develop. The surface condition and the manner in which it affects the emission and absorption of radiation represents another complication. In what follows, we begin with the simplest possible configuration of radiation heat exchange, and we continue with a case-by-case coverage of the complications that may alter the configuration.

10.2 BLACKBODY RADIATION

10.2.1 Definitions

The surface of a system (e.g., solid body) that participates in a radiation heat transfer process is classified according to its ability to absorb the radiation heat current that strikes it. Consider one unit area dA of such a surface, and let G represent the total radiation heat flux (W/m²) that arrives on dA from all the other surfaces that can see dA. This quantity is called *total irradiation* and will be defined in eq. (10.59). In general, three things can happen to the incident flux G:

1. A portion, αG, can be *absorbed* at the surface, that is, in the molecular layers situated immediately below the surface;
2. Another portion, ρG, can be *reflected* back toward the entire space that can be seen from dA; and possibly
3. A third portion, τG, can pass right through the body and exit through the other side.

Figure 10.2 (left side) shows that the conservation of energy in the coin-shaped control volume that encloses dA requires $\alpha G + \rho G + \tau G = G$; in other words,

$$\alpha + \rho + \tau = 1 \tag{10.1}$$

The nonnegative dimensionless numbers that appear on the left side are properties of the system that is being analyzed. They are known as the *total absorptivity* (α), *total reflectivity* (ρ), and *total transmissivity* (τ) of the material that resides under dA. These properties will be defined more rigorously in Section 10.4.2. Equation (10.1) implies that none of the numbers (α, ρ, τ) can be greater than 1.

Opaque is the system (or the region associated with dA) that is characterized by zero total transmissivity. For this class, eq. (10.1) reduces to

$$\alpha + \rho = 1 \quad \text{(opaque, } \tau = 0) \tag{10.2}$$

If, in addition, the reflectivity is zero, the surface is said to be *black*, and the system covered by the black surface is referred to as *blackbody*. This class is represented by

$$\alpha = 1 \quad \text{(black, } \rho = 0, \tau = 0) \tag{10.3}$$

This terminology is not meant to imply that the color of the surface is actually black. Many surfaces that are not colored black have α values that come close

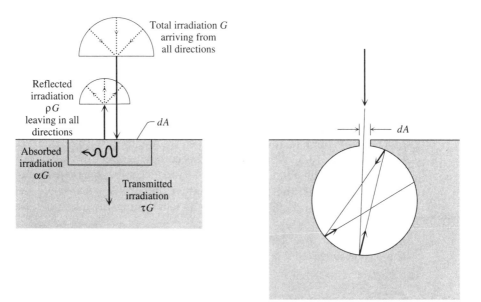

Figure 10.2 The definition of total absorptivity, reflectivity, and transmissivity (left) and the black surface behavior of the cavity aperture dA (right).

to 1 and, for engineering purposes, can be modeled as black. The "black surface" terminology is a useful reminder that all the incident radiation G disappears into the surface, so that none of it "shines back" at the sources that produced G. A black surface that illustrates this interpretation is shown on the right side of Fig. 10.2. The surface that appears black to an outside observer is the small opening dA — the aperture — through which the incident flux G enters the cavity that has been machined into the solid body. The incident flux bounces many times around the cavity, and during each bounce it is partially absorbed and reflected by the cavity wall.* When dA is small enough, there are so many bounces that the fraction of G that finally escapes (is "reflected") through the opening is negligible relative to the incident stream. The dA that the external observer sees is, therefore, a surface with $\alpha = 1$, namely, a black surface. The small-aperture cavity illustrated in Fig. 10.2 is the basic design by which a black surface of known temperature is constructed in the laboratory.

10.2.2   Temperature and Energy

Two alternative descriptions are available for the manner in which the radiation emitted by a surface *propagates* through the surrounding medium. One is the continuum description, in which the "shower" of energy G discussed in the preceding paragraphs is an electromagnetic wave, or a superposition of waves of many wavelengths. The alternative is the discrete-particle description provided by quantum theory. According to the latter, thermal radiation is a stream of particles (photons), each being characterized by zero rest mass and the energy quantum

$$e = h\nu \tag{10.4}$$

where $h = 6.6256 \times 10^{-34}$ J·s is Planck's constant. In either description, thermal radiation is characterized also by its frequency ν (s^{-1}) or range of frequencies. The frequency information can be conveyed also by means of the wavelength λ (m) or range of wavelengths that are present in a particular radiation stream. The relationship between frequency and wavelength is

$$\lambda = \frac{c}{\nu} \tag{10.5}$$

where c is the speed of propagation of the wave or of the photons. In an evacuated space that speed is the speed of light, $c = 2.998 \times 10^8$ m/s.

The one-to-one relationship between frequency and wavelength is also illustrated in Fig. 10.3, which shows the complete spectrum of electromagnetic radiation and the relatively narrow band occupied by thermal radiation. In wavelength terms, the thermal radiation domain is sandwiched between approximately 0.1 and 100 μm, where the length unit (the micron) is 1 μm = 10^{-6} m.

An even narrower band inside the thermal radiation domain is constituted by the *visible range*, that is, the thermal radiation that can be seen by the human

---

*Note that the cavity wall itself is not black: this is why it can partially absorb and reflect the radiation that impinges on it. Only the aperture dA is black, that is, "black" from the point of view of an external entity that radiates toward dA.

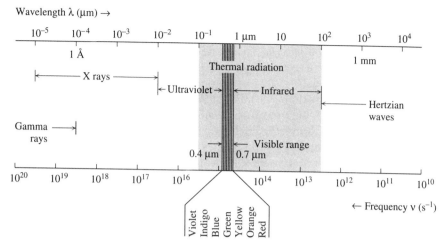

Figure 10.3 The wavelength and frequency domains of thermal radiation, and their position on the electromagnetic spectrum.

eye. The visible range is sandwiched roughly between the wavelengths of 0.4 and 0.7 μm. Proceeding in order of increasing wavelengths, the thermal radiation domain is therefore divided into three consecutive subdomains, namely, the ultraviolet range, the visible range, and the infrared range.

The visible range can be split and discussed further by recognizing the various wavelengths that pinpoint the principal colors distinguished by the human eye. No stream of thermal radiation is truly of a single color (i.e., of a single wavelength λ). The radiation characterized by an infinitesimally narrow band of wavelengths, $\lambda - (\lambda + d\lambda)$, is called *monochromatic* (literally, single-color) radiation of wavelength λ. This terminology is used over the entire thermal radiation domain, not just in the visible range.

An important and rarely acknowledged property of thermal radiation is the temperature. To develop an understanding for this property, consider the evacuated enclosure drawn in Fig. 10.4. The inner surface of the wall of this

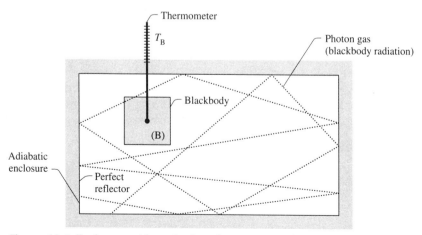

Figure 10.4 Enclosure with perfectly reflecting internal surfaces, and blackbody radiation of temperature T_B.

enclosure is a perfect reflector: all the radiation that impinges on it is reflected fully toward the enclosure. Seen from the outside, the enclosure is therefore surrounded by an adiabatic surface, because the net heat transfer rate across the wall is zero. The enclosure and all of its contents constitute an isolated thermodynamic system.

Next, assume that a certain blackbody (B) is introduced in the enclosure. This body emits radiation of all wavelengths and absorbs fully the radiation streams that strike it. In time, the enclosure fills with thermal radiation (a photon "gas") of all wavelengths; Fig. 10.4 shows the traces of only a few photons that travel through the enclosure. Simultaneously, the temperature of body (B)—a property that can be measured with a thermometer—approaches an equilibrium value T_B. At equilibrium, the body (B) can only have the same temperature as the photon gas that surrounds it, because the enclosure contents constitute an isolated system. The temperature of the photon gas—in this case, the temperature of blackbody radiation—is T_B.

The photons of the blackbody radiation trapped inside the enclosure cover all the frequencies (or wavelengths); however, some frequencies are represented by considerably more photons than other frequencies. The distribution of the black-body photon population over the frequency spectrum can be deduced on the basis of quantum-statistical thermodynamics (e.g., Refs. [1, 2]):

$$n_v = \frac{8\pi v^2 c^{-3}}{\exp(hv/kT) - 1} \tag{10.6}$$

where $k = 1.3805 \times 10^{-23}$ J/K is Boltzmann's constant. The number n_v represents the number of photons per unit volume and frequency interval; in other words, the units of n_v are photons/m$^3 \cdot$s^{-1}. The temperature T in the denominator of eq. (10.6) is the absolute (Kelvin) temperature of the particular blackbody radiation. In the preceding paragraph this temperature was labeled T_B.

Recalling that the energy of one photon is hv, eq. (10.4), the energy per unit volume and frequency interval u_v (J/m$^3 \cdot$s^{-1}) can be calculated by writing

$$u_v = n_v hv = \frac{8\pi hv^3 c^{-3}}{\exp(hv/kT) - 1} \tag{10.7}$$

This formula was first proposed in 1901 by Planck [3], which is why eq. (10.7) is commonly known as *Planck's radiation equation*. It is also the starting point in the thermodynamic treatment of thermal radiation as a photon gas system [4].

A more useful alternative to eq. (10.7) is the expression for the blackbody radiation energy per unit volume and *wavelength*, u_λ (J/m$^3 \cdot$m). This quantity can be obtained from eq. (10.7) by first noting that the frequency unit Δv and the wavelength unit $\Delta \lambda$ obey the relationship

$$|\Delta v| = \frac{c}{\lambda^2}|\Delta\lambda| \tag{10.8}$$

This relationship is obtained by differentiating eq. (10.5). The number of photons per unit volume and wavelength is, then,

$$u_\lambda = u_v \frac{\Delta v}{\Delta\lambda} = \frac{8\pi hc\lambda^{-5}}{\exp(hc/k\lambda T) - 1} \tag{10.9}$$

10.2.3 Intensity

These energy considerations allow us to finally calculate the *energy transport* that is associated with a certain stream of photons. Consider for this purpose an infinitesimal area dA that is exposed to equilibrium blackbody radiation of temperature T. For example, dA can be a small patch on the surface of body (B) in Fig. 10.4, or it can be the area bordered by a small wire loop suspended somewhere in the photon gas that fills the enclosure. This dA area has been magnified in Fig. 10.5 (left side). Drawn perpendicular to dA is the axis $dA-n$, which serves as centerline for the "pencil of rays" of infinitesimal solid angle $d\omega$.

The right side of Fig. 10.5 also shows the definition of the concept of *solid angle*,

$$d\omega = \frac{dA_n}{r^2} \tag{10.10}$$

where dA_n is an infinitesimal patch on the sphere of radius r. Note that when $d\omega$ is fixed, dA_n increases proportionally with r^2; this means that the solid angle is a measure of that fraction of space that can be viewed from the center of the sphere through the "window" represented by dA_n. The maximum solid angle is enjoyed by an observer who can look in all the directions (i.e., through all the points of the spherical surface $4\pi r^2$ drawn around him); therefore,

$$\omega_{\text{sphere}} = \frac{4\pi r^2}{r^2} = 4\pi \text{ sr} \tag{10.11}$$

The solid angle unit is the steradian (symbol sr); this particular word* serves as a reminder that the solid angle is the solid-body geometry analog of the concept of angle, whose unit is the radian (symbol rad) used in plane geometry.

Returning now to the original infinitesimal area dA shown on the left side of Fig. 10.5, we see that the product $u_\lambda c$ represents the energy flow rate per unit

---

*The first part of the word "steradian" is based on the Greek word *stereos* (hard, firm, solid).

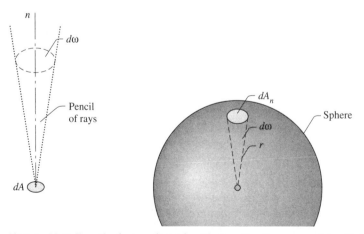

Figure 10.5 Pencil of rays aligned with the direction normal to dA (left), and the definition of infinitesimal solid angle (right).

time, wavelength, and area *normal* to a particular ray located inside the thin pencil. Dividing $u_\lambda c$ by 4π steradians, we obtain the *intensity of monochromatic blackbody radiation*, $I_{b,\lambda}$ (W/m$^3 \cdot$sr),

$$I_{b,\lambda} = \frac{u_\lambda c}{4\pi} = \frac{2hc^2\lambda^{-5}}{\exp(hc/k\lambda T) - 1} \tag{10.12}$$

in other words, the number of watts per unit wavelength (m), unit area normal to the ray (m^2), and unit solid angle (sr). The monochromatic energy current that flows through the pencil of rays and through the opening of area dA is therefore equal to $I_{b,\lambda}\, dA\, d\omega$. Worth noting is that the intensity of monochromatic blackbody radiation ($I_{b,\lambda}$) is a function of only wavelength and temperature; in the next section we examine this function by focusing on a quantity that is proportional to $I_{b,\lambda}$.

The *total intensity of blackbody radiation*, I_b (W/m$^2 \cdot$sr), is obtained by integrating the monochromatic intensity $I_{b,\lambda}$ over the entire radiation spectrum:

$$I_b(T) = \int_0^\infty I_{b,\lambda}(\lambda, T)d\lambda \tag{10.13}$$

It turns out that I_b is proportional to the fourth power of the absolute temperature of the particular ray of blackbody radiation; this proportionality will be derived in eq. (10.23). The total intensity I_b represents the energy conveyed by the ray (i.e., in a certain direction) per unit time, solid angle, and unit area *normal* to the direction of the ray.

10.2.4 Emissive Power

The actual direction of the photons that travel along the rays of the pencil of Fig. 10.5 was not discussed until now. The quantity $I_{b,\lambda}\, dA\, d\omega$ can represent either the monochromatic energy current that arrives through the pencil at dA, or the current that leaves dA through the pencil aligned with the normal direction $dA-n$. Of greater interest in heat transfer is the energy current that leaves dA in *all the directions* to which the dA observer has access. Those directions are all intercepted by the hemisphere that uses the plane of dA as base and dA itself as center. This new situation is illustrated in Fig. 10.6.

To calculate the per-unit-wavelength energy current "emitted" by dA in all the directions of the hemisphere, we recall the two angular coordinates (ϕ, θ) of the spherical system defined in Fig. 1.9. Consider the arbitrary direction represented by the radial line $dA-a$ in Fig. 10.6. The intensity of monochromatic blackbody radiation emitted along this ray is given by eq. (10.12). The solid angle of the pencil of rays centered around the direction $dA-a$ is

$$d\omega = \frac{dA_n}{r^2} = \frac{(r \sin\phi\, d\theta)(r\, d\phi)}{r^2} = \sin\phi\, d\theta\, d\phi \tag{10.14}$$

The area normal to the $dA-a$ direction is not dA but the *projected* area $dA \cos\phi$. The energy current per unit wavelength that leaves dA through the pencil of rays aligned with $dA-a$ is therefore $I_{b,\lambda}(\sin\phi\, d\theta\, d\phi)(dA \cos\phi)$. Integrating this quantity over all the directions intercepted by the hemisphere, we obtain the per-unit-wavelength energy current emitted by dA in all directions:

$$\int_{\theta=0}^{2\pi} \int_{\phi=0}^{\pi/2} I_{b,\lambda} \sin\phi \cos\phi\, d\phi\, d\theta\, dA = \pi I_{b,\lambda}\, dA \tag{10.15}$$

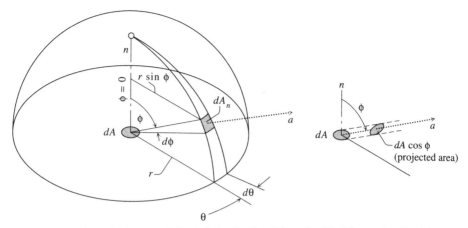

Figure 10.6 The calculation of the infinitesimal solid angle dA_n/r^2 associated with the direction dA–a (left), and the projection of dA on a plane normal to the direction dA–a (right).

The product $\pi I_{b,\lambda}$ represents the *monochromatic hemispherical emissive power* of the black surface to which dA belongs,

$$E_{b,\lambda} = \pi I_{b,\lambda} = \frac{C_1 \lambda^{-5}}{\exp(C_2/\lambda T) - 1} \tag{10.16}$$

for which the values of the constants C_1 and C_2 can be deduced from eq. (10.12):

$$\begin{aligned} C_1 &= 2\pi h c^2 = 3.742 \times 10^{-16} \text{ W} \cdot \text{m}^2 \\ C_2 &= \frac{hc}{k} = 1.439 \times 10^{-2} \text{ m} \cdot \text{K} \end{aligned} \tag{10.17}$$

The units of $E_{b,\lambda}$ are W/m·m², in other words, energy per unit time, wavelength, and surface area. The factor of π steradians appearing in the product $\pi I_{b,\lambda}$ is the result of performing the special integral listed in eq. (10.15). The π factor should not be confused with the hemispherical solid angle, which is equal to 2π steradians [cf. eq. (10.11)].

The main features of the $E_{b,\lambda}(\lambda, T)$ function are illustrated in Fig. 10.7. The monochromatic emissive power increases dramatically with the absolute temperature, as we'll discover in eq. (10.19). A maximum $E_{b,\lambda}$ value is registered at a characteristic wavelength for each temperature: solving $\partial E_{b,\lambda}/\partial\lambda = 0$ in conjunction with eq. (10.16), we obtain

$$\lambda T = 2.898 \times 10^{-3} \text{ m} \cdot \text{K} \tag{10.18}$$

This simple relation is known as *Wien's displacement law* [5]. It describes the locus of the $E_{b,\lambda}$ maxima, which on the logarithmic field of Fig. 10.7 is represented by a straight line. The wavelength of maximum $E_{b,\lambda}$ varies inversely with the absolute temperature. This means that as the temperature increases, the bulk of the energy emitted by a blackbody shifts to (is "displaced" toward) progressively shorter wavelengths. The maximum value of the monochromatic emissive power is obtained by substituting eq. (10.18) into eq. (10.16):

$$E_{b,\lambda, \text{ max}} = (12.87 \times 10^{-6} \text{ W/m}^3 \cdot \text{K}^5) T^5 \tag{10.19}$$

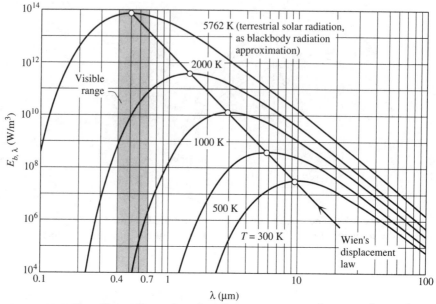

Figure 10.7 The effects of wavelength and temperature on the monochromatic hemispherical blackbody emissive power, and the meaning of Wien's displacement law.

It is possible now to replace all the $E_{b,\lambda}(\lambda,T)$ curves of Fig. 10.7 with a unique curve, by dividing the $E_{b,\lambda}$ ordinate by T^5 (or $E_{b,\lambda, \text{max}}$), and by multiplying the λ abscissa by T. The end result of this construction is the bell-shaped curve labeled $E_{b,\lambda}/E_{b,\lambda, \text{max}}$ on Fig. 10.8. The constant that appears on the right side of Wien's displacement law, eq. (10.18), is now an important entry on the λT abscissa of Fig. 10.8.

To summarize, the monochromatic emissive power $E_{b,\lambda}$ represents the heat flux that leaves dA along all the rays intersected by the hemisphere, per unit of wavelength interval. Of even greater interest in heat transfer engineering is the *total hemispherical emissive power* E_b, that is, the heat flux integrated over all the wavelengths of the radiation spectrum:

$$E_b = \int_0^\infty E_{b,\lambda}\,d\lambda \tag{10.20}$$

In view of eq. (10.16), the result of this integration is the *Stefan–Boltzmann law* [6, 7],

$$E_b = \sigma T^4 \tag{10.21}$$

where the constant $\sigma = 5.67 \times 10^{-8}$ W/m²·K⁴ is shorthand for the group

$$\sigma = \frac{C_1}{C_2^4}\int_0^\infty \frac{u^3\,du}{e^u - 1} = \frac{C_1}{C_2^4}\frac{\pi^4}{15} \tag{10.22}$$

Incidentally, eqs. (10.13), (10.16), and (10.20) to (10.22) can be used to prove that the total intensity of blackbody radiation I_b is also proportional to T^4:

$$I_b = \frac{\sigma}{\pi}T^4 = \frac{E_b}{\pi} \tag{10.23}$$

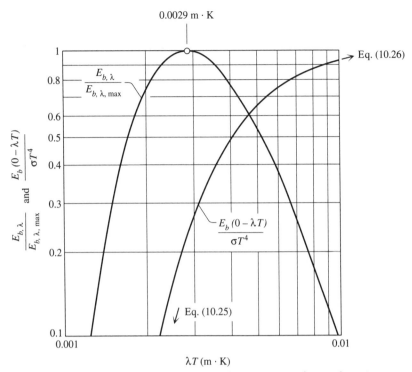

Figure 10.8 The dimensionless radiation function $E_b(0-\lambda T)/\sigma T^4$, and the single-curve substitute for the family of $E_{b,\lambda}(\lambda, T)$ curves shown in Fig. 10.7.

The total hemispherical emissive power σT^4 is equal to the area under the curve $E_{b,\lambda}(\lambda, T = \text{constant})$ when this curve is plotted in a Cartesian set of *linear* coordinates (Fig. 10.9a). The total number of watts per unit area emitted by the black surface in the *finite band* of wavelengths $\lambda_1 - \lambda_2$ can be calculated by first defining the *radiation function* [8]

$$E_b(0-\lambda_1 T) = \int_0^{\lambda_1} E_{b,\lambda}(\lambda, T)d\lambda \qquad (10.24)$$

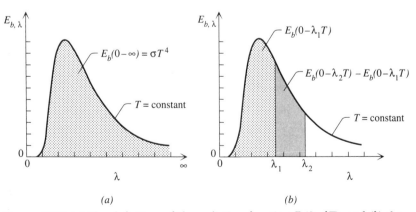

Figure 10.9 (a) The definition of the radiation function $E_b(0-\lambda T)$, and (b) the calculation of the total hemispherical emissive power associated with the finite-width band of wavelengths $\lambda_1 - \lambda_2$.

This quantity represents the area trapped under the $E_{b,\lambda}$ isotherm between $\lambda = 0$ and $\lambda = \lambda_1$ (Fig. 10.9b), which also means that $\sigma T^4 = E_b(0-\infty)$. The heat flux emitted hemispherically in the finite band $\lambda_1 - \lambda_2$ is therefore equal to the difference between the respective radiation functions:

$$\int_{\lambda_1}^{\lambda_2} E_{b,\lambda}(\lambda, T)d\lambda = E_b(0-\lambda_2 T) - E_b(0-\lambda_1 T) \tag{10.25}$$

The values of the radiation function $E_b(0-\lambda T)$ can be summarized in a surprisingly compact form, by noting first that the dimensionless ratio $E_b(0-\lambda T)/\sigma T^4$ depends only on the value of the group λT. Furthermore, the ratio $E_b(0-\lambda T)/\sigma T^4$ is greater than 0 and less than 1, because it represents the fraction of σT^4 that is emitted in the interval $0-\lambda$. That fraction is the same as the result of dividing the dotted area of Fig. 10.9b by the dotted area of Fig. 10.9a.

The values of the dimensionless radiation function $E_b(0-\lambda T)/\sigma T^4$ have been tabulated by Dunkle [8]. Table 10.1 shows a set of representative values; the intermediate-λT portion of the table accounts for the rising curve drawn in Fig. 10.8. In the small-λT limit, the dimensionless radiation function is approximated within less than 1 percent by the asymptote

$$\frac{E_b(0-\lambda T)}{\sigma T^4} \cong \frac{15}{\pi^4}(u^3 + 3u^2 + 6u + 6)e^{-u}$$

$$\text{where } u = \frac{C_2}{\lambda T} \quad \text{if } \lambda T < 0.0042 \text{ m} \cdot \text{K} \tag{10.26}$$

In the large-λT limit, the data of Table 10.1 approach within 1 percent the asymptote

$$\frac{E_b(0-\lambda T)}{\sigma T^4} \cong 1 - \left(\frac{0.00535 \text{ m} \cdot \text{K}}{\lambda T}\right)^3 \quad \text{if } \lambda T > 0.016 \text{ m} \cdot \text{K} \tag{10.27}$$

Table 10.1 Principal Values and the Asymptotic Behavior of the Dimensionless Radiation Function $E_b(0 - \lambda T)/\sigma T^4$

λT (m·K)	$\dfrac{E_b(0 - \lambda T)}{\sigma T^4}$	λT (m·K)	$\dfrac{E_b(0 - \lambda T)}{\sigma T^4}$
The small-λT limit: eq. (10.26)		0.0065	0.776
0.001	0.00213	0.007	0.808
0.002	0.0667	0.0075	0.834
0.0025	0.162	0.008	0.856
0.003	0.273	0.009	0.890
0.0035	0.383	0.010	0.914
0.004	0.481	0.011	0.932
0.0045	0.564	0.012	0.945
0.005	0.634	0.015	0.970
0.0055	0.691	0.020	0.986
0.006	0.738	The large-λT limit: eq. (10.27)	

Example 10.1

Intensity, Projected Area, and Solid Angle

The black surface $A_1 = 1$ cm² emits radiation with total intensity $I_{b,1} = 1400$ W/m²·sr. A portion of this radiation, $q_{1 \to 2}$(W), strikes the surface $A_2 = 1$ cm² shown in the figure. The relative position of A_1 and A_2 is fixed by the distance $r = 1$ m and the two angles (30° and 60°) indicated on the drawing. Calculate the heat current $q_{1 \to 2}$.

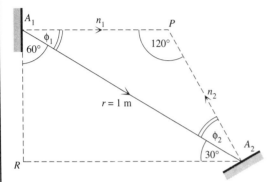

Figure E10.1

Solution. The angles ϕ_1 and ϕ_2 can be calculated by remembering the following lessons of plane geometry:

$$\phi_1 = 30° \tag{1}$$

$$\phi_1 + \phi_2 = 180° - 120° \tag{2}$$

Equation (1) refers to two angles formed in the "underarms" of the line $\overline{A_1 A_2}$ as it cuts the parallel lines $\overline{A_1 P}$ and $\overline{A_2 R}$. Equation (2) is the statement that in the triangle $\overline{A_1 P A_2}$ the angles add up to 180°. Combining eqs. (1) and (2), we conclude that

$$\phi_2 = 30° \tag{3}$$

The radiation of intensity $I_{b,1}$ is emitted by the surface A_1 in all the directions of the hemisphere that has A_1 as base. The heat current emitted by A_1 and intercepted by A_2 is

$$q_{1 \to 2} = I_{b,1} A_{n,1} \omega_{1-2} \tag{4}$$

in which $A_{n,1}$ is the projection of A_1 on the plane normal to the direction A_1–A_2:

$$A_{n,1} = A_1 \cos \phi_1 = \frac{3^{1/2}}{2} \, \text{cm}^2 \tag{5}$$

The solid angle ω_{1-2} is subtended by the projected area $A_{n,2}$,

$$\omega_{1-2} \cong \frac{A_{n,2}}{r^2} \tag{6}$$

where $A_{n,2}$ is the projection of A_2 on the plane perpendicular to the direction A_1–A_2:

$$A_{n,2} = A_2 \cos \phi_2 = \frac{3^{1/2}}{2} \, \text{cm}^2 \tag{7}$$

Note that the solid angle relationship (6) is approximately valid only when the target area A_2 is small relative to r^2. Numerically, eq. (6) yields

$$\omega_{1-2} = \frac{3^{1/2}}{2} 10^{-4} \text{sr} \tag{8}$$

so that the one-way heat current $q_{1 \to 2}$ amounts to

$$q_{1 \to 2} = \left(1400 \frac{W}{m^2 \cdot sr}\right)\left(\frac{3^{1/2}}{2} cm^2\right)\left(\frac{3^{1/2}}{2} 10^{-4} \ sr\right)$$

$$= 1.05 \times 10^{-5} W \tag{9}$$

Example 10.2

Radiation Function, and the Human Skin as a Selective Absorber

The human skin is "selective" when it comes to the absorption of the solar radiation that strikes it perpendicularly. The skin absorbs only 50 percent of the incident radiation with wavelengths between $\lambda_1 = 0.52$ μm and $\lambda_2 = 1.55 \mu$m. The radiation with wavelengths shorter than λ_1 and longer than λ_2 is fully absorbed. The solar surface may be modeled as black with temperature $T = 5800$ K. Calculate what fraction of the incident solar radiation is absorbed by the human skin.

Solution.   With an eye on the way in which Table 10.1 has been constructed, we calculate, in order,

$$\lambda_1 T = (0.52 \times 10^{-6} \ m)(5800 \ K) = 0.003 \ m \cdot K$$

$$\lambda_2 T = (1.55 \times 10^{-6} \ m)(5800 \ K) = 0.009 \ m \cdot K$$

Reading the table, we obtain

$$\frac{E_b(0 - \lambda_1 T)}{\sigma T^4} = 0.273$$

$$\frac{E_b(0 - \lambda_2 T)}{\sigma T^4} = 0.890$$

which means that the $\lambda_1 - \lambda_2$ band contains 61.7 percent of the total emissive power of the solar surface,

$$\frac{E_b(\lambda_1 T - \lambda_2 T)}{\sigma T^4} = 0.890 - 0.273 = 0.617$$

while the $\lambda_2 - \infty$ band contains 11 percent:

$$\frac{E_b(\lambda_2 T - \infty)}{\sigma T^4} = 1 - 0.89 = 0.11$$

The absorbed fraction of the incident solar radiation is due to 100 percent absorption below λ_1, 50 percent absorption between λ_1 and λ_2, and 100 percent absorption above λ_2:

$$1 \times 0.273 + 0.5 \times 0.617 + 1 \times 0.11 = 0.692$$

In conclusion, the human skin absorbs roughly 70 percent of the incident solar radiation. It is important to note that the group σT^4 is not the solar radiation heat flux that strikes the skin. The actual heat flux is considerably smaller than σT^4 because it depends on the relative size and position of the two surfaces (sun, skin). On this aspect we focus in the next section.

10.3   HEAT TRANSFER BETWEEN BLACK SURFACES

10.3.1   The Geometric View Factor

Consider now the problem of estimating the net heat transfer rate q_{1-2} (W) between the two isothermal black surfaces (A_1, T_1) and (A_2, T_2) shown in Fig. 10.10. Contrary to some of the simplifying features adopted in the construction of Fig. 10.10, the shapes and relative positions of the two surfaces are meant to be arbitrary. The q_{1-2} analysis that we are about to see consists of estimating, in order, the following:

1. The fraction of the radiation emitted by the area element dA_1 and intercepted (i.e., absorbed fully) by the area element dA_2;
2. The fraction of the radiation emitted by dA_2 and intercepted (absorbed) by dA_1;
3. The net heat transfer rate from dA_1 to dA_2, namely, the difference between the answers to parts (1) and (2); and, finally,
4. The net heat transfer rate from A_1 to A_2, that is, between the two isothermal and finite-size areas.

The thinking behind step (1) was presented already in the discussion of Fig. 10.6. If this time the segment r is the distance between the area elements dA_1 and dA_2, then the solid angle through which dA_2 is seen by an observer stationed at dA_1 is equal to $dA_2 \cos \phi_2/r^2$. Note that $dA_2 \cos \phi_2$ is the size of dA_2 after it has been projected onto a plane perpendicular to the dA_1–dA_2 line.

Traveling from dA_1 toward dA_2 (and toward the rest of the space) is blackbody radiation of total intensity $I_{b,1} = I_b(T_1)$. According to the definition (10.13), $I_{b,1}$ is the number of watts of T_1 radiation that are emitted in the r direction, per unit solid angle and per unit of area normal to the r direction. The size of the emitting area that is normal to the r direction is the "projected dA_1" area, namely, $dA_1 \cos \phi_1$. All this means that if we multiply $I_{b,1}$ by the projected emitting area $dA_1 \cos \phi_1$ and by the solid angle subtended by the target, $dA_2 \cos \phi_2/r^2$, we obtain the answer to step (1)

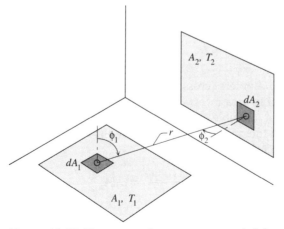

Figure 10.10 The geometric parameters needed for calculating the view factor integral (10.33).

$$q_{dA_1 \to dA_2} = I_{b,1} dA_1 \cos \phi_1 \frac{dA_2 \cos \phi_2}{r^2} \qquad (10.28)$$

The arrow used in the $dA_1 \to dA_2$ subscript is a reminder that $q_{dA_1 \to dA_2}$ represents a *one-way* transfer of energy per unit time, in this case, from dA_1 (emitter) to dA_2 (target). The argument of the preceding two paragraphs can be used one more time in step (2) of the analysis; however, the end result can be read in eq. (10.28) by simply switching the positions of the subscripts 1 and 2. The fraction of T_2 radiation emitted by dA_2 and intercepted (absorbed) by dA_1 is therefore (note again the one-way arrow in the $dA_2 \to dA_1$ subscript)

$$q_{dA_2 \to dA_1} = I_{b,2} dA_2 \cos \phi_2 \frac{dA_1 \cos \phi_1}{r^2} \qquad (10.29)$$

The third step consists of subtracting eq. (10.29) from eq. (10.28) to evaluate the *net* heat transfer rate that leaves dA_1 and arrives at dA_2:

$$q_{dA_1 - dA_2} = q_{dA_1 \to dA_2} - q_{dA_2 \to dA_1}$$

$$= (I_{b,1} - I_{b,2}) \frac{\cos \phi_1 \cos \phi_2}{r^2} dA_1 dA_2 \qquad (10.30)$$

This net heat transfer rate is distinguished from the previous one-way contributions ($q_{dA_1 \to dA_2}$ and $q_{dA_2 \to dA_1}$) by means of the new subscript $dA_1 - dA_2$: the first area in this subscript (dA_1) represents the origin of the net heat transfer rate, whereas the second area (dA_2) represents the destination. Equation (10.30) can also be written in terms of the total hemispherical emissive powers* associated with T_1 and T_2 [cf. eq. (10.23)]:

$$q_{dA_1 - dA_2} = \sigma(T_1^4 - T_2^4) \frac{\cos \phi_1 \cos \phi_2}{\pi r^2} dA_1 dA_2 \qquad (10.31)$$

The finite-size areas A_1 and A_2 communicate through a very large number of dA_1, dA_2 pairs of the kind analyzed until now. The net heat transfer rate from A_1 to A_2 is the sum of all the contributions of type (10.31) that are possible. This sum, q_{1-2}, can be obtained by integrating $q_{dA_1 - dA_2}$ over the two finite areas A_1 and A_2:

$$q_{1-2} = \sigma(T_1^4 - T_2^4) \int_{A_1} \int_{A_2} \frac{\cos \phi_1 \cos \phi_2}{\pi r^2} dA_1 dA_2 \qquad (10.32)$$

On the left side, the subscript 1–2 states that the net heat transfer rate q_{1-2} leaves the surface A_1 and enters (crosses) the surface A_2.

The units of the double-area integral on the right side of eq. (10.32) are the units of area (m²). It is permissible (and convenient) to replace this cumbersome integral expression with the product $A_1 F_{12}$, in which the dimensionless factor F_{12} is the *geometric view factor†* based on A_1:

$$F_{12} = \frac{1}{A_1} \int_{A_1} \int_{A_2} \frac{\cos \phi_1 \cos \phi_2}{\pi r^2} dA_1 dA_2 \qquad (10.33)$$

---

*Note the appearance of π in the denominator on the right side of eq. (10.31).

†This factor is also known as geometric configuration factor, view factor, radiation shape factor, and angle factor.

In the end, the q_{1-2} formula assumes the much simpler form

$$q_{1-2} = \sigma(T_1^4 - T_2^4)A_1 F_{12} \tag{10.34}$$

The geometric view factor is a purely geometric quantity, one that depends only on the sizes, orientations, and relative position of the two surfaces. The same definition holds when the surfaces are not black (Section 10.4.4). Here we derived the formula for F_{12} by considering two black surfaces because this model leads to the simplest analysis.

Alternatively, the double-area integral of eq. (10.32) can be replaced with the product $A_2 F_{21}$, in which F_{21} is the geometric view factor *based on A_2*:

$$F_{21} = \frac{1}{A_2} \int_{A_1} \int_{A_2} \frac{\cos\phi_1 \cos\phi_2}{\pi r^2} dA_1\, dA_2 \tag{10.35}$$

The net heat transfer rate from A_1 to A_2 is now given by the expression

$$q_{1-2} = \sigma(T_1^4 - T_2^4)A_2 F_{21} \tag{10.36}$$

Equations (10.34) and (10.36) show that the final calculation of q_{1-2} depends on being able to evaluate one geometric view factor (F_{12} or F_{21}). Before plunging into the blind manipulation of the double-area integral (10.33), it pays to review the physical meaning of the view factor F_{12}. For this we return to the expression (10.28) for the radiation heat transfer rate emitted by dA_1 and absorbed by dA_2. If we integrate this expression over both finite-size areas, we obtain the radiation heat transfer rate emitted by A_1 and absorbed by A_2:

$$q_{1\rightarrow2} = I_{b,1} \int_{A_1} \int_{A_2} \frac{\cos\phi_1 \cos\phi_2}{r^2} dA_1\, dA_2 = \cdots$$

$$= \sigma T_1^4 A_1 F_{12} \tag{10.37}$$

On the right side, the group $\sigma T_1^4 A_1$ is the same as $E_{b,1}A_1$, or the number of watts of blackbody radiation emitted by the surface A_1 *in all the directions* in which the points of A_1 can look. Only a portion of $E_{b,1}A_1$ is intercepted and absorbed by A_2 (because, in general, A_1 may be surrounded by more than just A_2): that portion* is $q_{1\rightarrow2}$ or [cf. eq. (10.37)] $E_{b,1}A_1 F_{12}$. In conclusion, the physical meaning of the view factor is this

$$F_{12} = \frac{q_{1\rightarrow2} \text{ (radiation leaving } A_1 \text{ and being intercepted by } A_2)}{E_{b,1}A_1 \text{ (radiation leaving } A_1 \text{ in all directions)}} \tag{10.38}$$

The "ratio" definition formulated above suggests that the value of a geometric view factor falls between 0 and 1. This important feature is evident on the ordinates of Figs. 10.11 and 10.12. These figures and Table 10.2 contain the view factors of some of the two-surface configurations that are encountered

---

*The *one-way* heat transfer rate $q_{1\rightarrow2}$ should not be confused with the *net* heat transfer rate q_{1-2} of eqs. (10.34) and (10.36). The relationship that unites these concepts is [see also the top line of eq. (10.30)]

$$q_{1-2} = q_{1\rightarrow2} - q_{2\rightarrow1}$$

where $q_{2\rightarrow1}$ is the portion of the T_2 temperature radiation emitted by A_2 and intercepted by A_1 ($q_{2\rightarrow1} = E_{b,2}A_2 F_{21}$).

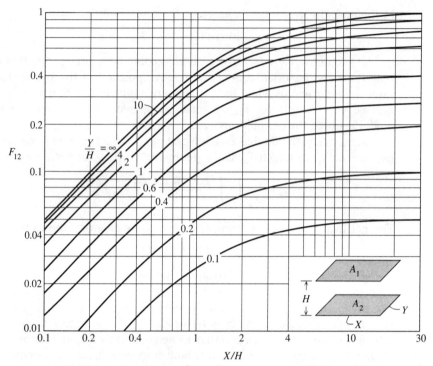

Figure 10.11 The geometric view factor between two parallel rectangles.

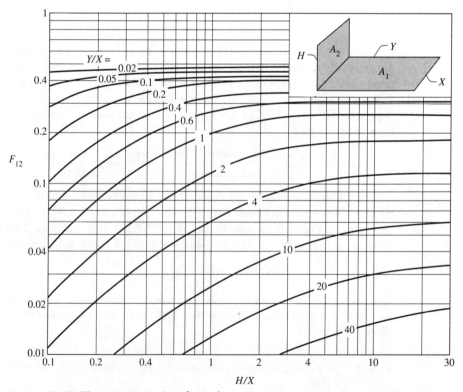

Figure 10.12 The geometric view factor between two perpendicular rectangles with a common edge.

Table 10.2 Minicatalog of Geometric View Factors (after Refs. [9,10])

Configuration	Geometric View Factor

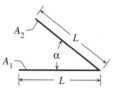

Two infinitely long plates of width L, joined along one of the long edges:

$$F_{12} = F_{21} = 1 - \sin\frac{\alpha}{2}$$

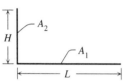

Two infinitely long plates of different widths (H, L), joined along one of the long edges and with a 90° angle between them:

$$F_{12} = \tfrac{1}{2}[1 + x - (1 + x^2)^{1/2}]$$

where $x = H/L$

Triangular cross section enclosure formed by three infinitely long plates of different widths (L_1, L_2, L_3):

$$F_{12} = \frac{L_1 + L_2 - L_3}{2L_1}$$

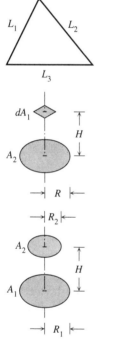

Disc and parallel infinitesimal area positioned on the disc centerline:

$$F_{12} = \frac{R^2}{H^2 + R^2}$$

Parallel discs positioned on the same centerline:

$$F_{12} = \frac{1}{2}\left\{X - \left[X^2 - 4\left(\frac{x_2}{x_1}\right)^2\right]^{1/2}\right\}$$

where $x_1 = \dfrac{R_1}{H}$, $x_2 = \dfrac{R_2}{H}$, and $X = 1 + \dfrac{1 + x_2^2}{x_1^2}$

Infinite cylinder parallel to an infinite plate of finite width $(L_1 - L_2)$:

$$F_{12} = \frac{R}{L_1 - L_2}\left(\tan^{-1}\frac{L_1}{H} - \tan^{-1}\frac{L_2}{H}\right)$$

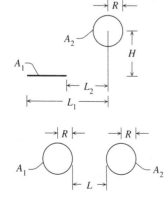

Two parallel and infinite cylinders:

$$F_{12} = F_{21} = \frac{1}{\pi}\left[\left(X^2 - 1\right)^{1/2} + \sin^{-1}\left(\frac{1}{X}\right) - X\right]$$

where $X = 1 + \dfrac{L}{2R}$

Table 10.2, (*continued*)

Configuration	Geometric View Factor
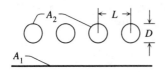	Row of equidistant infinite cylinders parallel to an infinite plate: $$F_{12} = 1 - (1 - x^2)^{1/2} + x \tan^{-1}\left(\frac{1 - x^2}{x^2}\right)^{1/2}$$ where $x = \dfrac{D}{L}$
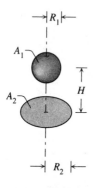	Sphere and disc positioned on the same centerline: $$F_{12} = \tfrac{1}{2}[1 - (1 + x^2)^{-1/2}]$$ where $x = \dfrac{R_2}{H}$

frequently in radiation heat transfer calculations, most notably the parallel coaxial discs. Several other configurations are covered in Howell's catalog [9] as well as in Siegel and Howell's treatise [10].

10.3.2   Relations Between View Factors

In principle, the calculation of each shape factor can be traced to the purely geometric definition (10.33): this procedure is illustrated in Example 10.3. In some instances, however, it is possible to deduce the F_{12} value based on a geometric "shortcut," that is, by manipulating the presumably known view factors of one or more related configurations. Essential to this simpler approach are the following relationships also known as view factor "algebra." These relationships are purely geometric, that is, independent of surface description (black vs. nonblack).

Reciprocity.   The first relation between two different view factors is obtained by comparing eqs. (10.34) and (10.36):

$$A_1 F_{12} = A_2 F_{21} \tag{10.38}$$

If we happen to know F_{21} from a chart, formula, or table, we can calculate F_{12} if we also know the areas involved in the current configuration (A_1, A_2). Equation (10.38) is known as the reciprocity rule, or the reciprocity property of the two view factors associated with the same pair of surfaces.

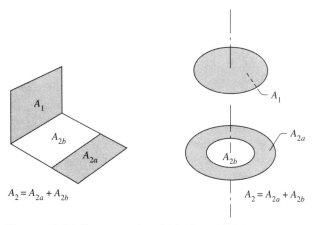

Figure 10.13 Two cases in which the additivity property can be used to calculate the view factor F_{12a}.

Additivity.   Let A_{2a} and A_{2b} be the two surfaces that, together, make up the A_2 surface discussed until now (Fig. 10.13). If F_{12} represents the fraction of the A_1 radiation that is intercepted by A_2, then F_{12} can only be the sum of F_{12a} and F_{12b}, because F_{12a} is the fraction absorbed by A_{2a}, and F_{12b} is the fraction absorbed by A_{2b}:

$$F_{12} = F_{12a} + F_{12b} \tag{10.39}$$

Of course, we can write as many terms on the right side of this equation as there are pieces in the A_2 mosaic, for example,

$$F_{12} = \sum_{i=1}^{n} F_{12_i} \tag{10.40}$$

when A_2 happens to be broken up into n pieces, $A_2 = A_{2_1} + A_{2_2} + \cdots + A_{2_n}$. Equation (10.40) expresses the additivity property, or the additivity rule of the view factors between one surface (A_1) and all the pieces of a neighboring surface (A_2).

This property can be used to great advantage in the calculation of the view factor F_{12a} in the two cases shown in Fig. 10.13. In the two-rectangle geometry on the left side, $F_{12a} = F_{12} - F_{12b}$, where both F_{12} and F_{12b} can be read off Fig. 10.12 (note that the A_1-A_{2b} and A_1-A_2 pairs touch along one edge).

On the right side of Fig. 10.13, the view factor between the small disc and the annulus is simply $F_{12a} = F_{12} - F_{12b}$. Since A_2 and A_{2b} are both discs, F_{12} and F_{12b} can be calculated with the parallel discs formula listed as the fifth entry in Table 10.2.

Enclosure.   Returning to the (A_1, A_2) pair of surfaces that started the study of geometric view factors, Fig. 10.10, we recall that not all the radiation emitted by A_1 is intercepted by A_2. The remainder is intercepted by the other surface (or surfaces) that surround A_1. The emitting surface (A_1) and all the surfaces that surround it $(A_2, A_3, \ldots, A_n)$ form the enclosure shown schematically in Fig. 10.14.

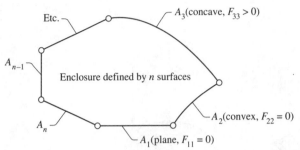

Figure 10.14 Enclosure formed by n surfaces, and the dependence of the view factor F_{ii} on the shape of each surface.

Think now of the radiation that is emitted by A_1 in all the directions in which it can look ($E_{b,1}A_1$): the portions intercepted by $A_2, A_3, \ldots, A_n$ are, respectively, $E_{b,1}A_1F_{12}, E_{b,1}A_1F_{13}, \ldots, E_{b,1}A_1F_{1n}$. And if the A_1 surface happens to be concave, some of the $E_{b,1}A_1$ current is intercepted by A_1 itself (that portion is $E_{b,1}A_1F_{11}$). The conservation of $E_{b,1}A_1$ inside the enclosure requires that

$$E_{b,1}A_1 = E_{b,1}A_1F_{11} + E_{b,1}A_1F_{12} + \cdots + E_{b,1}A_1F_{1n} \tag{10.41}$$

or, after dividing by $E_{b,1}A_1$,

$$1 = F_{11} + F_{12} + \cdots + F_{1n} \tag{10.42}$$

An equation of type (10.42) can be written for each of the surfaces that participates in the enclosure; therefore, the most we get out of the preceding energy conservation argument is a system of n equations:

$$1 = \sum_{j=1}^{n} F_{ij} \qquad (i = 1, 2, \ldots, n) \tag{10.43}$$

In this system, the subscript i indicates the surface that serves as emitter (i.e., the surface for which the equation is written), while the subscript j counts all the surfaces that make up the enclosure. Again, the enclosure relationships (10.43) hold true regardless of whether the surfaces are black or not. We return to system (10.43) when we analyze the net heat transfer through the surfaces of an enclosure (Section 10.4.5).

10.3.3   Two-Surface Enclosures

The simplest application of the preceding analysis is the problem of estimating the net heat transfer rate q_{1-2} between two black surfaces that form an enclosure. Three examples of two-surface enclosures are illustrated in Fig. 10.15. In each of these examples the enclosed space is bounded only by A_1 and A_2. The two cylinders and the two spheres *are not necessarily concentric.*

The formula for calculating q_{1-2} has already been derived and listed as eq. (10.34) or eq. (10.36). These relations apply to steady as well as unsteady (time-dependent) situations. The lower part of Fig. 10.15 shows an electrical circuit analog of the net heat transfer rate across the enclosure. This circuit is based on reading eq. (10.34) as a proportionality between the current q_{1-2} and the potential difference $E_{b,1} - E_{b,2}$, that is,

$$q_{1-2} = (E_{b,1} - E_{b,2})A_1F_{12} \tag{10.44}$$

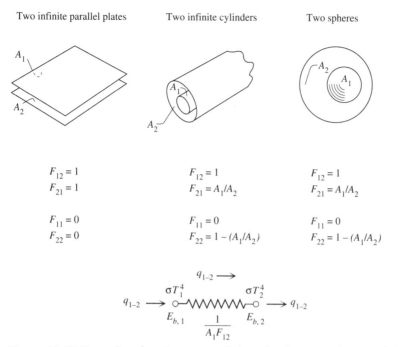

Figure 10.15 Examples of enclosures consisting of only two surfaces, and the corresponding thermal resistance diagram.

where $E_{b,1} = E_b(T_1) = \sigma T_1^4$ and $E_{b,2} = E_b(T_2) = \sigma T_2^4$. The product $A_1 F_{12}$ plays the role of *thermal conductance* for the net heat transfer rate between the potential nodes represented by $E_{b,1}$ and $E_{b,2}$. The inverse of $A_1 F_{12}$ (or $A_2 F_{21}$) is recognized as the *radiation thermal resistance*, R_r (m^{-2}):

$$R_r = \frac{1}{A_1 F_{12}} = \frac{1}{A_2 F_{21}} \tag{10.45}$$

One aspect that is brought to light by the resistance analog of the two-surface enclosure is that the net surface-to-surface heat transfer rate q_{1-2} must first enter A_1 through its back side, and then exit through the back side of A_2. If the physical sense of q_{1-2} is indeed the sense assumed in drawing Fig. 10.15, the steady state prevails only if an external device heats the A_1 surface at the rate q_{1-2}, and if another external device cools the A_2 surface at the same rate. Relative to the environment that surrounds the enclosure, the back sides of both A_1 and A_2 are *not* adiabatic.

Listed under each configuration in Fig. 10.15 are the respective geometric view factors. The F_{12} factors are equal to 1 because all the radiation emitted by the surface A_1 is intercepted by A_2.* The F_{21} factors are determined based on the reciprocity relationship $F_{21} = F_{12} A_1 / A_2$. The row of F_{21} values demonstrates that the infinite parallel-plate geometry is indeed the $A_1 / A_2 \to 1$ limit of the two-cylinder and two-sphere geometries.

Although not essential to the task of calculating q_{1-2}, the view factors F_{11} and F_{22} have also been listed to illustrate the observations written around the

---

*In the cylindrical and spherical geometries, A_1 is the *inner* (smaller) surface.

enclosure of Fig. 10.14. These particular factors are zero in the case of plane and convex surfaces, and finite in the case of concave surfaces. For example, the A_2 surface in the two-sphere configuration sees itself through the geometric view factor $F_{22} = 1 - (A_1/A_2)$.

Example 10.3

The View Factor Between Two Parallel Strips

Consider the problem of calculating the geometric view factor between the infinitesimal area dA_1 and an infinitely long strip, A_2, of small width Δy situated in a plane parallel to dA_1. The width Δy is small relative to the distance R between the two strips.

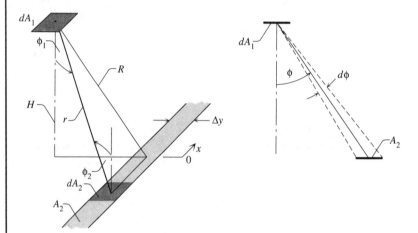

Figure E10.3a

Solution.   With reference to Fig. E10.3a and eq. (10.33), we note first that the angles ϕ_1 and ϕ_2 are equal and that $\cos \phi_1 = H/r$ and $r^2 = R^2 + x^2$. The segment of length R is the shortest distance between dA_1 and the A_2 strip; therefore, according to a theorem of space geometry, the R segment is perpendicular to the x direction. The double-area integral (10.33) becomes

$$F_{12} = \frac{1}{dA_1} \int_{dA_1} \int_{A_2} \frac{(H/r)^2}{\pi r^2} dA_1\, dA_2 = \frac{H^2 \Delta y}{\pi} \int_{-\infty}^{\infty} \frac{dx}{(R^2 + x^2)^2}$$

$$= \frac{H^2 \Delta y}{\pi} \left. \left| \frac{1}{2R^3} \tan^{-1} \frac{x}{R} + \frac{x}{2R^2(x^2 + R^2)} \right| \right|_{-\infty}^{\infty}$$

$$= \frac{H^2 \Delta y}{2R^3} \tag{1}$$

in which we have used $dA_2 = \Delta y\, dx$. Equation (1) can be rearranged using the terminology shown on the right side of the Fig. E10.3a, namely,

$$\frac{H}{R} = \cos \phi \qquad \frac{\Delta y \cos \phi}{R} = d\phi \tag{2}$$

That side of the figure shows the projection of both dA_1 and A_2 on the plane defined by R and the normal to dA_1. Combining eqs. (1) and (2), we obtain

$$F_{12} = \frac{\cos \phi}{2} d\phi = \tfrac{1}{2} d(\sin \phi) \tag{3}$$

Equation (3) is a powerful building block that can be used in the construction of the F_{12} values of considerably more complicated two-surface configurations. Worth noting is the fact that eq. (3) is more general than the case defined in Fig. E10.3a. The same view factor formula applies when (see, for example, Ref. [10]) dA_1 is not parallel to A_2, and/or dA_1 is itself a strip of infinitesimal width and *finite* length in the direction parallel to the A_2 strip. This general configuration is shown on the left side of Fig. 10.3b.

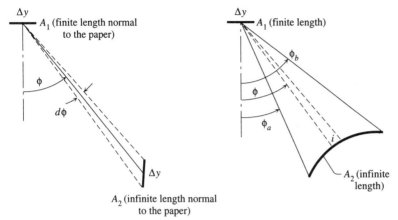

Figure E10.3b

The right side of Fig. E10.3b shows the even more general case in which the width of the infinitely long strip A_2 is finite. This new strip can be reconstructed by gluing edge-to-edge a large number of Δy-narrow strips of the type indicated by i. The additivity property (10.40) can be invoked to write that the view factor from A_1 to the finite-width strip A_2 is equal to the sum of all the view factors F_{1i} given by eq. (3):

$$F_{12} = \sum_i F_{1i} = \int_{\phi_a}^{\phi_b} \tfrac{1}{2} d(\sin \phi)$$

$$= \tfrac{1}{2}(\sin \phi_b - \sin \phi_a) \tag{4}$$

Example 10.4

Enclosure Formed by Two Black Surfaces

A much simplified model of a stack of vertical circuit boards is shown in the figure. The arrangement is long in the direction perpendicular to the plane of the figure. Consider the two-surface enclosure formed in the space between two consecutive boards, and assume that

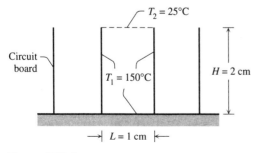

Figure E10.4

both surfaces (1 = solid walls, 2 = ambient) are black. Assume that radiation is the only significant mode of heat transfer. Calculate the net heat transfer rate from $T_1 = 150°C$ to $T_2 = 25°C$ through the top "surface" of the enclosure.

Solution. The solid walls represented by two boards and horizontal holder (T_1), and the ambient (T_2) form an enclosure in which the two surfaces are black. The net heat transfer rate from T_1 to T_2 through one enclosure of this kind is

$$q_{1-2} = A_1 F_{12}(\sigma T_1^4 - \sigma T_2^4)$$

The reciprocity property requires

$$A_1 F_{12} = A_2 F_{21}$$

in which $F_{21} = 1$ and $A_2 = LB$. The width B is the long dimension of the boards, in the direction perpendicular to the plane of the figure. The net heat transfer rate per enclosure and unit length is

$$\frac{1}{B} q_{1-2} = L\sigma(T_1^4 - T_2^4)$$

$$= (0.01 \text{ m})\left(5.67 \times 10^{-8} \frac{W}{m^2 \cdot K^4}\right)[(150 + 273.15)^4 - (25 + 273.15)^4]K^4$$

$$= 13.7 \text{ W/m}$$

An interesting aspect of this result is that it is independent of the board height H. Indeed, it is easy to obtain the same result when the boards are absent ($H = 0$ cm), that is, when the heat exchange is between two parallel and infinite black surfaces (the horizontal holder, 150°C, and the ambient, 25°C).

10.4 DIFFUSE-GRAY SURFACES

10.4.1 Emissivity

The reason we devoted so much space to developing the analysis of heat transfer between black surfaces is that, under certain conditions, the same type of analysis—the same approach (structure)—can be used in the case of real or "nonblack" surfaces.* For example, we shall learn that the geometric view factors of Section 10.3.1 apply unchanged to nonblack surfaces. The challenge is to identify the special conditions that make the preceding analysis extendable to nonblack surfaces. Those conditions are condensed in the so-called "diffuse-gray" surface model, to which we now turn.

The intensity of the radiation emitted by a real surface of temperature T is only a fraction of the corresponding intensity of blackbody radiation. For example, relative to the intensity of monochromatic blackbody radiation considered in the preceding sections, $I_{b,\lambda}(\lambda, T)$, a real surface emits monochromatic radiation of intensity $I_\lambda(\lambda, T, \phi, \theta)$. Figure 10.16 (left side) and the arguments of $I_\lambda(\lambda, T, \phi, \theta)$ stress that, in general, the intensity of the radiation emitted by a real surface depends on the direction (ϕ, θ) in which each particular ray is pointed.

The inequality between the real surface and black surface intensities, $I_\lambda \leq I_{b,\lambda}$, is expressed alternatively as

---

*Surfaces that do not absorb all the incoming radiation.

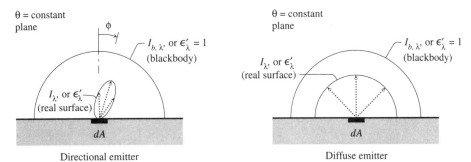

Figure 10.16 The emission characteristics of a directional emitter (left) and a diffuse emitter (right).

$$\epsilon'_\lambda(\lambda,T,\phi,\theta) = \frac{I_\lambda(\lambda,T,\phi,\theta)}{I_{b,\lambda}} \leq 1 \qquad (10.46)$$

where ϵ'_λ is the *directional monochromatic emissivity* of the real surface. The dimensionless number ϵ'_λ is a property of the emitting surface. The type of surface illustrated on the left side of Fig. 10.16, where ϵ'_λ and I_λ depend on the direction of the emitted ray, is recognized as a *directional nonblack* emitter. The prime notation (') indicates that ϵ'_λ is a "directional" property, that is, a number associated with a certain direction (ϕ,θ).

The right side of Fig. 10.16 shows a nonblack surface that emits uniformly in all the directions that stream out of the infinitesimal surface area dA. The intensity of the emitted monochromatic radiation is again smaller than in the blackbody limit, $I_{b,\lambda}$. This time, however, I_λ and the ratio $I_\lambda/I_{b,\lambda} = \epsilon'_\lambda$ are independent of the direction (ϕ,θ). The surface that has this feature is a *diffuse nonblack* emitter.

The comparison between the emission properties of real and black surfaces can be carried out also in terms of hemispherical quantities. Let $E_\lambda(\lambda,T)$ be the monochromatic emissive power of the real surface, that is, the number of watts emitted per unit of real surface area and wavelength, in all the directions intercepted (cut) by the hemisphere, in accordance with the integral (10.15):

$$E_\lambda(\lambda,T) = \int_{\theta=0}^{2\pi} \int_{\phi=0}^{\pi/2} I_\lambda(\lambda,T,\phi,\theta) \sin\phi \cos\phi \, d\phi \, d\theta \qquad (10.47)$$

The ratio between $E_\lambda(\lambda,T)$ and the blackbody limit of the same quantity—the blackbody monochromatic emissive power $E_{b,\lambda}$—is known as the *monochromatic hemispherical emissivity* of the real surface, and is labeled ϵ_λ:

$$\epsilon_\lambda(\lambda,T) = \frac{E_\lambda(\lambda,T)}{E_{b,\lambda}(\lambda,T)} \leq 1 \qquad (10.48)$$

Finally, an even more global measure of the difference between the emissions from a real surface and from a black surface of the same temperature is the *total hemispherical emissivity* ϵ:

$$\epsilon(T) = \frac{E(T)}{E_b(T)} \leq 1 \qquad (10.49)$$

The numerator in this definition is the actual emissive power (W/m²) of the real surface,

$$E(T) = \int_0^\infty E_\lambda(\lambda,T)d\lambda \qquad (10.50)$$

Combining eqs. (10.48), (10.49), and (10.50) we reach the conclusion that the total hemispherical emissivity ϵ is the $E_{b,\lambda}$-weighted average of the monochromatic hemispherical emissivity ϵ_λ:

$$\epsilon(T) = \frac{1}{\sigma T^4}\int_0^\infty E_\lambda(\lambda,T)d\lambda = \frac{1}{\sigma T^4}\int_0^\infty \epsilon_\lambda E_{b,\lambda}d\lambda \qquad (10.51)$$

Now examine Fig. 10.17. The left side of this figure shows the general outlook of the real-surface monochromatic hemispherical emissive power $E_\lambda(\lambda,T)$ next to the black surface-limiting curve $E_{b,\lambda}(\lambda,T)$. The right side of the figure tells the same story in terms of monochromatic hemispherical emissivities (the curves shown on the left side of Fig. 10.17 have been drawn to scale based on the ϵ_λ curves assumed on the right side).

In many instances, the measured $\epsilon_\lambda(\lambda,T)$ value of a real surface is a complicated function of wavelength. The simplest (first-cut) approximation of this function is the constant-ϵ_λ dashed line shown on the right side of Fig. 10.17, namely, the statement

$$\epsilon_\lambda (\lambda,T) \cong \epsilon_\lambda(T) \qquad \text{or} \qquad \epsilon_\lambda \neq \text{function}(\lambda) \qquad \text{(gray)} \qquad (10.52)$$

This approximation is appropriate in instances where the dominant values of the $\epsilon_\lambda(\lambda,T)$ function are located in that portion of the λ spectrum that is responsible for the bulk of the blackbody emissive power σT^4. The example illustrated in Fig. 10.17 is one such case: note the position of the $\lambda_1 - \lambda_2$ interval on the two abscissas of that figure. The $\epsilon_\lambda = 0.6$ value was chosen on the right side of Fig. 10.17 in such a way that it approximates the actual $\epsilon_\lambda(\lambda,T)$ curve over those wavelengths that contribute the most to the actual emissive power of the nonblack surface.

When the approximation (10.52) is justified, the "reduced" Planck curve $\epsilon_\lambda E_{b,\lambda}$ [in which $\epsilon_\lambda \neq \text{function}(\lambda)$] is an adequate approximation of the actual E_λ

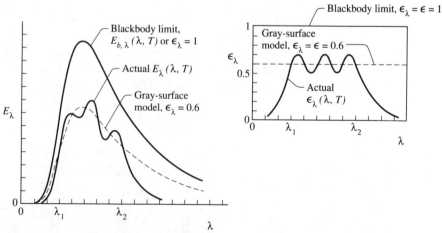

Figure 10.17 The gray-surface model: the monochromatic emissive power (left), and the corresponding monochromatic hemispherical emissivity (right).

curve of the real surface. On the left side of Fig. 10.17, the reduced Planck curve was drawn with a dashed line using the λ-independent reduction factor $\epsilon_\lambda = 0.6$.

A *gray surface* or *gray body* of temperature T is the surface whose monochromatic hemispherical emissivity is independent of wavelength (i.e., a constant if T is fixed), eq. (10.52). The weighted average indicated on the extreme right-hand side of eq. (10.51) shows further that the total hemispherical emissivity of a gray surface is equal to its monochromatic hemispherical emissivity,

$$\epsilon = \epsilon_\lambda \quad \text{(gray)} \tag{10.53}$$

In all the real-surface analyses presented in this chapter, it will be assumed that the surface is a *diffuse* emitter *and* that is can be modeled as *gray*. It will also be assumed that the surface is a diffuse absorber and a diffuse reflector, in accordance with the definitions that will be given in Figs. 10.18 and 10.19 and the accompanying discussion.

It turns out that the *diffuse-gray* model approximates fairly well the behavior of many surfaces encountered in heat transfer engineering, for example, copper, aluminum oxide, paints, and paper. Tables 10.3 and 10.4 show a compilation of total hemispherical emissivity values (ϵ) of many real surfaces that have been modeled as diffuse and gray. Clean, well-polished metallic surfaces are characterized by small ϵ values. Nonmetallic surfaces, on the other hand, have high emissivities: in fact, some of these come close to satisfying the blackbody model $\epsilon = 1$ (e.g., soot, smooth glass, frost). Metallic surfaces that, in time, become coated with oxides and other impurities also acquire considerably higher emissivity values.

In general, then, the total hemispherical emissivities listed in the tables are functions of the emitter temperature, type of material, and degree of surface smoothness and cleanliness. There are also difficulties that stem from the qualitative description of the surface condition in Tables 10.3 and 10.4, for example, "polished" versus "highly polished," or "oxidized" versus "black oxidized" (copper, Table 10.3). Furthermore, the temperature effect is itself a function of the type of material that makes up the surface. For example, the total hemispherical emissivity of metallic surfaces increases with the absolute temperature. At cryogenic temperatures (5 K–100 K), the ϵ value of a metallic surface is almost proportional to the temperature T. The ϵ values of nonmetallic surfaces can either increase or decrease as T increases.

10.4.2   Absorptivity and Reflectivity

A real opaque surface absorbs only a portion of the energy brought by the incoming (incident) radiation. The remainder—the portion that is not absorbed—is reflected back toward the surroundings. Let $I_\lambda(\lambda,T,\phi,\theta)$ represent the intensity of the general (nonblack) incident radiation that strikes the real-surface element from the direction (ϕ,θ). The relative size of the portion that is absorbed at the surface, $I_{a,\lambda}(\lambda,T,\phi,\theta)$, is indicated by the *directional monochromatic absorptivity* α_λ':

$$\alpha_\lambda'(\lambda,T,\phi,\theta) = \frac{I_{a,\lambda}(\lambda,T,\phi,\theta)}{I_\lambda(\lambda,T,\phi,\theta)} \tag{10.54}$$

Table 10.3 Metallic Surfaces: Representative Values of Total Hemispherical Emissivity[a]

Material		$\epsilon\ (T)$
Aluminum,	crude	0.07–0.08 (0°C–200°C)
	foil, bright	0.01 (−9°C), 0.04 (1°C), 0.087 (100°C)
	highly polished	0.04–0.05 (1°C)
	ordinarily rolled	0.035 (100°C), 0.05 (500°C)
	oxidized	0.11 (200°C), 0.19 (600°C)
	roughed with abrasives	0.044–0.066 (40°C)
	unoxidized	0.022 (25°C), 0.06 (500°C)
Bismuth,	unoxidized	0.048 (25°C), 0.061 (100°C)
Brass,	after rolling	0.06 (30°C)
	browned	0.5 (20°C–300°C)
	polished	0.03 (300°C)
Chromium,	polished	0.07 (150°C)
	unoxidized	0.08 (100°C)
Cobalt,	unoxidized	0.13 (500°C), 0.23 (1000°C)
Copper,	black oxidized	0.78 (40°C)
	highly polished	0.03 (1°C)
	liquid	0.15
	matte	0.22 (40°C)
	new, very bright	0.07 (40°C–100°C)
	oxidized	0.56 (40°C–200°C), 0.61 (260°C), 0.88 (540°C)
	polished	0.04 (40°C), 0.05 (260°C), 0.17 (1100°C), 0.26 (2800°C)
	rolled	0.64 (40°C)
Gold,	polished or electrolytically deposited	0.02 (40°C), 0.03 (1100°C)
Inconel,	sandblasted	0.79 (800°C), 0.91 (1150°C)
	stably oxidized	0.69 (300°C), 0.82 (1000°C)
	untreated	0.3 (40°C–260°C)
	rolled	0.69 (800°C), 0.88 (1150°C)
Inconel X		0.74–0.81 (100°C–440°C)
	stably oxidized	0.89 (300°C), 0.93 (1100°C)
Iron (see also Steel)		
	cast	0.21 (40°C)
	cast, freshly turned	0.44 (40°C), 0.7 (1100°C)
	galvanized	0.22–0.28 (0°C–200°C)
	molten	0.02–0.05 (1100°C)
	plate, rusted red	0.61 (40°C)
	pure polished	0.06 (40°C), 0.13 (540°C), 0.35 (2800°C)
	red iron oxide	0.96 (40°C), 0.67 (540°C), 0.59 (2800°C)
	rough ingot	0.95 (1100°C)
	smooth sheet	0.6 (1100°C)
	wrought, polished	0.28 (40°C–260°C)
Lead,	oxidized	0.28(0°C–200°C)
	unoxidized	0.05 (100°C)
Magnesium		0.13 (260°C), 0.18 (310°C)
Mercury		0.09 (0°C), 0.12 (100°C)
Molybdenum		0.071 (100°C), 0.13 (1000°C), 0.19 (1500°C)
	oxidized	0.78–0.81 (300°C–540°C)

Table 10.3, (*continued*)

	Material	$\epsilon\,(T)$
Monel,	oxidized	0.43 (20°C)
	polished	0.09 (20°C)
Nichrome,	rolled	0.36 (800°C), 0.8 (1150°C)
	sandblasted	0.81 (800°C), 0.87 (1150°C)
Nickel,	electrolytic	0.04 (40°C), 0.1 (540°C), 0.28 (2800°C)
	oxidized	0.31–0.39 (40°C), 0.67 (540°C)
	wire	0.1 (260°C), 0.19 (1100°C)
Platinum,	oxidized	0.07 (260°C), 0.11 (540°C)
	unoxidized	0.04 (25°C), 0.05 (100°C), 0.15 (1000°C)
Silver,	polished	0.01 (40°C), 0.02 (260°C), 0.03 (540°C)
Steel,	calorized	0.5–0.56 (40°C–540°C)
	cold rolled	0.08 (100°C)
	ground sheet	0.61 (1100°C)
	oxidized	0.79 (260°C–540°C)
	plate, rough	0.94–0.97 (40°C–540°C)
	polished	0.07 (40°C), 0.1 (260°C), 0.14 (540°C), 0.23 (1100°C), 0.37 (2800°C)
	rolled sheet	0.66 (40°C)
	type 347, oxidized	0.87–0.91 (300°C–1100°C)
	type AISI 303, oxidized	0.74–0.87 (300°C–1100°C)
	type 310, oxidized and	
	rolled	0.56 (800°C), 0.81 (1150°C)
	sandblasted	0.82 (800°C), 0.93 (1150°C)
Stellite		0.18 (20°C)
Tantalum		0.19 (1300°C), 0.3 (2500°C)
Tin,	unoxidized	0.04–0.05 (25°C–100°C)
Tungsten,	filament	0.03 (40°C), 0.11 (540°C), 0.39 (2800°C)
Zinc,	oxidized	0.11 (260°C)
	polished	0.02 (40°C), 0.03 (260°C)

*a*Data collected from Refs. [11, 12].

Table 10.4 Nonmetallic Surfaces: Representative Values of Total Hemispherical Emissivity*a*

Material	$\epsilon\,(T)$
Bricks	
chrome refractory	0.94 (540°C), 0.98 (1100°C)
fire clay	0.75 (1400°C)
light buff	0.8 (540°C), 0.53 (1100°C)
magnesite refractory	0.38 (1000°C)
sand lime red	0.59 (1400°C)
silica	0.84 (1400°C)
various refractories	0.71–0.88 (1100°C)
white refractory	0.89 (260°C), 0.68 (540°C)
Building materials	
asbestos, board	0.96 (40°C)
asphalt pavement	0.85–0.93 (40°C)

continued

Table 10.4, (*continued*)

Material	$\epsilon\,(T)$
clay	0.39 (20°C)
concrete, rough	0.94 (0°C–100°C)
granite	0.44 (40°C)
gravel	0.28 (40°C)
gypsum	0.9 (40°C)
marble, polished	0.93 (40°C)
mica	0.75 (40°C)
plaster	0.93 (40°C)
quartz	0.89 (40°C), 0.58 (540°C)
sand	0.76 (40°C)
sandstone	0.83 (40°C)
slate	0.67 (40°C–260°C)
Carbon	
baked	0.52–0.79 (1000°C–2400°C)
filament	0.95 (260°C)
graphitized	0.76–0.71 (100°C–500°C)
rough	0.77 (100°C–320°C)
soot (candle)	0.95 (120°C)
soot (coal)	0.95 (20°C)
unoxidized	0.8 (25°C–500°C)
Ceramics	
coatings	
alumina on inconel	0.65 (430°C), 0.45 (1100°C)
zirconia on inconel	0.62 (430°C), 0.45 (1100°C)
earthenware, glazed	0.9 (1°C)
matte	0.93 (1°C)
porcelain	0.92 (40°C)
refractory, black	0.94 (100°C)
light buff	0.92 (100°C)
white Al_2O_3	0.9 (100°C)
Cloth	
cotton	0.77 (20°C)
silk	0.78 (20°C)
Glass	
Convex D	0.8–0.76 (100°C–500°C)
fused quartz	0.75–0.8 (100°C–500°C)
Nonex	0.82–0.78 (100°C–500°C)
Pyrex	0.8–0.9 (40°C)
smooth	0.92–0.95 (0°C–200°C)
waterglass	0.96 (20°C)
Ice, smooth	0.92 (0°C)
Oxides	
Al_2O_3	0.35–0.54 (850°C–1300°C)
C_2O	0.27 (850°C–1300°C)
Cr_2O_3	0.73–0.95 (850°C–1300°C)
Fe_2O_3	0.57–0.78 (850°C–1300°C)
MgO	0.29–0.5 (850°C–1300°C)
NiO	0.52–0.86 (500°C–1200°C)
ZnO	0.3–0.65 (850°C–1300°C)

Table 10.4, (continued)

Material	$\epsilon (T)$
Paints	
aluminum	$0.27-0.7$ ($1°C-100°C$)
enamel, snow white	0.91 ($40°C$)
lacquer	$0.85-0.93$ ($40°C$)
lampblack	$0.94-0.97$ ($40°C$)
oil	$0.89-0.97$ ($0°C-200°C$)
white	$0.89-0.97$ ($40°C$)
Paper, white	0.95 ($40°C$), 0.82 ($540°C$)
Roofing materials	
aluminum surfaces	0.22 ($40°C$)
asbestos cement	0.65 ($1400°C$)
bituminous felt	0.89 ($1400°C-2800°C$)
enameled steel, white	0.65 ($1400°C$)
galvanized iron, dirty	0.90 ($1400°C-2800°C$)
new	0.42 ($1400°C$)
roofing sheet, brown	0.8 ($1400°C$)
green	0.87 ($1400°C$)
tiles, uncolored	0.63 ($1400°C-2800°C$)
brown	0.87 ($1400°C$)
black	0.94 ($1400°C$)
asbestos cement	0.66 ($1400°C-2800°C$)
weathered asphalt	0.88 ($1400°C-2800°C$)
Rubber	
hard, black, glossy	
surface	0.95 ($40°C$)
soft, gray	0.86 ($40°C$)
Snow	
fine	0.82 ($-10°C$)
frost	0.98 ($0°C$)
granular	0.89 ($-10°C$)
Soils	$0.92-0.96$ ($0°C-20°C$)
black loam	0.66 ($20°C$)
plowed field	0.38 ($20°C$)
Water	$0.92-0.96$ ($0°C-40°C$)
Wood	
beech	0.91 ($70°C$)
oak, planed	0.91 ($40°C$)
sawdust	0.75 ($40°C$)
spruce, sanded	0.82 ($100°C$)

[a]Data collected from Refs. [11, 12].

A surface is called a *directional absorber* when the directional monochromatic absorptivity α'_λ varies with the orientation of the incoming ray (Fig. 10.18, left side). When α'_λ exhibits the same value regardless of the direction (ϕ, θ), the surface is called a *diffuse absorber* (Fig. 10.18, right side). It is worth keeping in mind that the incident intensity I_λ that strikes dA generally depends on the direction from which it comes, because its magnitude is determined by the

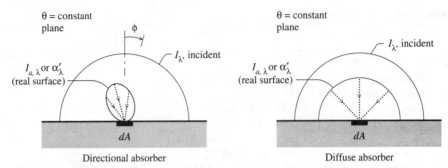

Figure 10.18 The absorption characteristics of a directional absorber (left) and a diffuse absorber (right).

various emitters that can look toward dA. In Fig. 10.18 the size of I_λ was drawn constant (independent of ϕ and θ) only to facilitate the comparison between it and its absorbed fraction $I_{a,\lambda}$, that is, the notion that $I_{a,\lambda}$ is generally smaller than the incident I_λ.

In a way that mimics the $(\epsilon'_\lambda, \epsilon_\lambda, \epsilon)$ string of emissivity definitions reviewed in the preceding section, we can define next the *monochromatic hemispherical absorptivity* α_λ:

$$\alpha_\lambda(\lambda,T) = \frac{G_{a,\lambda}(\lambda,T)}{G_\lambda(\lambda,T)} \tag{10.55}$$

The denominator G_λ (W/m²·m) is the *monochromatic irradiation*, or the number of watts that strike the unit area *from all the directions* and per unit wavelength. Note that G_λ is related to the intensity of the incident radiation [I_λ of eq. (10.54)] through the familiar double integral [see eq. (10.47)]

$$G_\lambda(\lambda,T) = \int_{\theta=0}^{2\pi} \int_{\phi=0}^{\pi/2} I_\lambda(\lambda,T,\phi,\theta) \sin\phi \cos\phi \, d\phi \, d\theta \tag{10.56}$$

The numerator $G_{a,\lambda}$ used in definition (10.55) is the absorbed portion of the monochromatic irradiation G_λ:

$$G_{a,\lambda}(\lambda,T) = \int_{\theta=0}^{2\pi} \int_{\phi=0}^{\pi/2} I_{a,\lambda}(\lambda,T,\phi,\theta) \sin\phi \cos\phi \, d\phi \, d\theta \tag{10.57}$$

Finally, the *total hemispherical absorptivity* is defined as the ratio

$$\alpha(T) = \frac{G_a(T)}{G(T)} \tag{10.58}$$

where $G(T)$ is the *total irradiation*, which was noted already in Fig. 10.2 (left side). The $G_a(T)$ numerator is the absorbed portion of the total irradiation. Both G and G_a have units of heat flux (W/m²).

The total irradiation $G(T)$ is obtained by integrating the monochromatic irradiation over the entire spectrum:

$$G(T) = \int_0^\infty G_\lambda(\lambda,T)d\lambda \tag{10.59}$$

Similarly, the absorbed portion G_a is the same as the λ integral of the absorbed monochromatic irradiation $G_{a,\lambda}$:

$$G_a(T) = \int_0^\infty G_{a,\lambda}(\lambda,T)d\lambda \qquad (10.60)$$

Combined, eqs. (10.55) and (10.58)–(10.60) demonstrate that the total hemispherical absorptivity α is the G_λ-weighted average of the monochromatic hemispherical absorptivity α_λ:

$$\alpha(T) = \frac{1}{G(T)} \int_0^\infty \alpha_\lambda(\lambda,T)G_\lambda(\lambda,T)d\lambda \qquad (10.61)$$

The difference between the total irradiation G and the absorbed portion G_a is the reflected portion G_r. We did this accounting once in eq. (10.2), which now can be restricted to an *opaque surface* and written in terms of the *total hemispherical reflectivity* $\rho = 1 - \alpha$:

$$G_r = G - G_a = (1 - \alpha)G = \rho G \qquad (10.62)$$

The reflected total-irradiation portion G_r is the hemispherical solid angle integral of all the reflected "offspring" generated by each incident ray. In most cases there are many reflected rays (directions) associated with a single incident ray. This feature is illustrated in Fig. 10.19. The surface is a *diffuse reflector* when the intensity of the reflected radiation is uniform in all directions, that is, independent of ϕ and θ. Conversely, when the reflected rays prefer one or a limited bundle of directions, the surface is called a *directional reflector*. The extreme directional reflector shown in the upper left corner of Fig. 10.19 — the *specular,** or mirrorlike, reflector — directs the reflected portion of each incident ray into a single direction.

To summarize, the *diffuse-gray model* we adopt now for the remainder of this chapter describes a surface that is simultaneously

1. Gray, eq. (10.52) and Fig. 10.17,
2. A diffuse emitter, Fig. 10.16 (right side),
3. A diffuse absorber, Fig. 10.18 (right side),
4. A diffuse reflector, Fig. 10.19 (right side), and
5. Opaque.

---

*The term "specular" has its origin in the Latin adjective *specularis* (like a mirror), which is related to the noun *speculum* (mirror, made of polished metal). The modern English word "speculate," which comes from the feminine Latin noun *specula* (a lookout, watchtower; or a little hope, ray of hope), is also related to these terms.

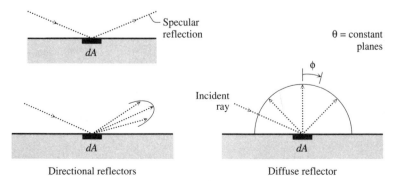

Directional reflectors   Diffuse reflector

Figure 10.19 The reflection characteristics of a directional reflector (left) and a diffuse reflector (right).

The total hemispherical emissivity of the diffuse-gray surface can be found in Tables 10.3 and 10.4 or in other handbooks [11, 12]. We now turn to the problem of estimating the total hemispherical absorptivity α, which is equivalent to the problem of determining the total hemispherical reflectivity $(\rho = 1 - \alpha)$.

10.4.3   Kirchhoff's Law

Consider first the inner face (Σ) of the larger surface of the enclosure shown in Fig. 10.20. This face is nonblack and isothermal, at the temperature T_Σ. In the discussion associated with Fig. 10.2 (right), we learned that a cavity bounded by a *nonblack* surface that is *isothermal* absorbs all the incident radiation that enters through a small orifice. Some of the radiation (photon gas) that fills the cavity of Fig. 10.2 (right) escapes through the orifice. Thermodynamically it can be argued that, in the eyes of an external observer, the radiation that streams out through the aperture can only be a function of the temperature of the cavity wall, that is, a blackbody radiation. This argument leads to the conclusion that a cavity surrounded by an *isothermal wall* fills with *blackbody radiation of the same temperature as the wall,* regardless of whether the wall is black or nonblack (good absorber or a good reflector). With respect to the enclosure in Fig. 10.20, this means that the cavity surrounded by the surface Σ fills with blackbody radiation of temperature T_Σ.

Any elementary area dA suspended inside this cavity is struck by blackbody radiation of monochromatic intensity $I_{b,\lambda}(\lambda,T_\Sigma)$. Now consider the solid body of temperature T_A which is delineated by the nonblack surface A. If this body is introduced in the cavity bounded by the surface Σ, then the entire surface A will absorb heat transfer at the rate (Ref. [4], p. 487)

$$q_{\text{absorbed by } A} = \int_{A=0}^{A} \int_{\omega=0}^{2\pi} \int_{\lambda=0}^{\infty} \alpha_\lambda'\,(\lambda,T_A,\phi,\theta) I_{b,\lambda}(\lambda,T_\Sigma)\, \cos \phi \; d\lambda \; d\omega \; dA$$

(10.63)

The second of these nested integrals accounts for the arrival of blackbody radiation from all the directions of the hemisphere ($0 \le \omega \le 2\pi$) based on the area element dA: this particular integral and its notation were spelled out in detail in eqs. (10.14) and (10.15). The factor α_λ' in the integrand is the directional monochromatic absorptivity of the surface A.

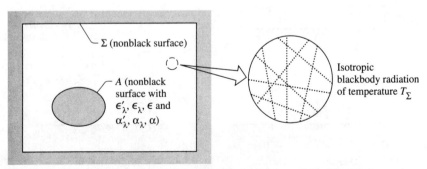

Figure 10.20 Two-surface enclosure for the derivation of Kirchhoff's law, $\alpha_\lambda' = \epsilon_\lambda'$.

Let $\epsilon'_\lambda(\lambda, T_A, \phi, \theta)$ be the directional monochromatic emissivity of the surface A. The total heat current (watts) emitted by the entire area A is the result of another triple integral:

$$q_{\text{emitted by } A} = \int_{A=0}^{A} \int_{\omega=0}^{2\pi} \int_{\lambda=0}^{\infty} \epsilon'_\lambda(\lambda, T_A, \phi, \theta) I_{b,\lambda}(\lambda, T_A) \cos \phi \, d\lambda \, d\omega \, dA \quad (10.64)$$

Worth noting in the integrand is the monochromatic intensity of blackbody radiation of temperature T_A, which is not the same as the $I_{b,\lambda}(\lambda, T_\Sigma)$ factor that appears in the integrand of eq. (10.63).

When the body (A) reaches *thermal equilibrium* with the enclosure wall (Σ), the temperature T_A becomes *steady* and equal to the outer wall temperature:

$$T_A = T_\Sigma \quad \text{(constant)} \quad (10.65)$$

The steadiness of T_A implies that the *net* heat transfer rate through the surface A is zero, meaning that the absorbed current, eq. (10.63), matches (and cancels) the effect of the emitted current, eq. (10.64). In addition, the thermal equilibrium condition $T_A = T_\Sigma$ guarantees that $I_{b,\lambda}(\lambda, T_A) = I_{b,\lambda}(\lambda, T_\Sigma)$.

All of these observations lead to the conclusion that, at equilibrium, the integrals of eqs. (10.63) and (10.64) are equal and that their equality requires

$$\alpha'_\lambda(\lambda, T_A, \phi, \theta) = \epsilon'_\lambda(\lambda, T_A, \phi, \theta) \quad (10.66)$$

This result is known as *Kirchhoff's law* [13]. It states that the α'_λ value of a nonblack surface is always equal to the ϵ'_λ value of the same surface *when the surface is in thermal equilibrium with the radiation that impinges on it*. In the problems that confront us in the field of heat transfer, surfaces are almost never in equilibrium with their surroundings.* In spite of the lack of thermodynamic equilibrium, eq. (10.66) is considered to be sufficiently accurate, and is an integral part of the foundation of radiation heat transfer theory.

From the point of view of a user of the diffuse-gray model, an important question is whether the total absorptivity α can be calculated based on Kirchhoff's law, that is, based solely on emissivity information. To answer this question, we note first that the hemispherical monochromatic absorptivity α_λ is a weighted average of the directional monochromatic absorptivity α'_λ:

$$\alpha_\lambda(\lambda, T) = \frac{1}{G_\lambda(\lambda, T)} \int_{\theta=0}^{2\pi} \int_{\phi=0}^{\pi/2} \alpha'_\lambda(\lambda, T, \phi, \theta) I_\lambda(\lambda, T, \phi, \theta) \sin \phi \cos \phi \, d\phi \, d\theta$$

$$(10.67)$$

This relationship follows from eqs. (10.54), (10.55), and (10.57). If the surface is a *diffuse absorber*, then α'_λ is not a function of the direction (ϕ, θ) (see Fig. 10.18, the right side); and, after using eq. (10.56), the relationship (10.67) reduces to

$$\alpha_\lambda(\lambda, T) = \alpha'_\lambda(\lambda, T) \quad \text{(diffuse absorber)} \quad (10.68)$$

Similarly, we find that the hemispherical monochromatic emissivity ϵ_λ is related to the directional monochromatic emissivity ϵ'_λ through the hemispherical solid-angle integral

---

*Recall that under the conditions of thermal equilibrium the fundamental heat transfer problem becomes trivial, eq. (1.14).

$$\epsilon_\lambda(\lambda,T) = \frac{1}{E_{b,\lambda}(\lambda,T)} \int_{\theta=0}^{2\pi} \int_{\phi=0}^{\pi/2} \epsilon'_\lambda(\lambda,T,\phi,\theta) I_{b,\lambda}(\lambda,T) \sin\phi \cos\phi \, d\phi \, d\theta \quad (10.69)$$

This is the result of combining eqs. (10.46) through (10.48). For the *diffuse emitter* illustrated in Fig. 10.16 (right side) the directional monochromatic emissivity is not a function of ϕ and θ. Therefore, pulling ϵ'_λ in front of the integral (10.69), and recognizing eqs. (10.15) and (10.16), we obtain

$$\epsilon_\lambda(\lambda,T) = \epsilon'_\lambda(\lambda,T) \quad \text{(diffuse emitter)} \quad (10.70)$$

In conclusion, for a surface that is both a diffuse absorber and a diffuse emitter, Kirchhoff's law states that

$$\alpha_\lambda(\lambda,T) = \epsilon_\lambda(\lambda,T) \quad \text{(diffuse absorber and emitter)} \quad (10.71)$$

This statement follows from eqs. (10.66), (10.68), and (10.70). If, in addition, the surface can be modeled as *gray*, ϵ_λ is independent of wavelength, and [cf. eq. (10.53)] $\epsilon_\lambda = \epsilon(T)$. From eq. (10.71) we conclude that α_λ is also λ independent and, more specifically, that

$$\alpha_\lambda = \epsilon(T) \quad (10.72)$$

This α_λ estimate can finally be substituted in the integrand on the right side of eq. (10.61) to conclude that

$$\alpha(T) = \epsilon(T) \quad \text{(diffuse-gray surface)} \quad (10.73)$$

Equation (10.73) represents Kirchhoff's law for a surface that obeys the diffuse-gray model. It states that the total hemispherical absorptivity α of a surface of temperature T is equal to the total hemispherical emissivity $\epsilon(T)$ of the same surface. To the user of the diffuse-gray model, eq. (10.73) means that α can be estimated based on the ϵ information assembled in Tables 10.3 and 10.4, *provided* that the incident radiation has the same temperature as the surface temperature T.

In applications in which the temperature of the incident radiation does not differ substantially from that of the target surface, the thermal equilibrium restriction can be overlooked so that $\alpha(T)$ can be estimated (approximately) by using the $\epsilon(T)$ value listed in Tables 10.3 and 10.4. This approximation may not be appropriate when the incident radiation and the target surface have vastly different temperatures. In the case of granular snow, for example, Table 10.4 shows that $\epsilon = 0.89$ when the snow temperature is $T = -10°C$. Kirchhoff's law assures us that $\alpha = 0.89$ when the snow temperature is $-10°C$ *and* the temperature of the incident radiation is not much different than $-10°C$. In those situations, the snow approaches the black surface model! On the other hand, when the incident radiation has a much higher temperature (e.g., solar radiation, $T \cong 5800$ K), the absorptivity of the snow is much lower, $\alpha \sim 0.4$, which is why the snow looks bright. The same observation holds true for the absorptivities exhibited by white paper.

To review the progress made in this section, the general form of Kirchhoff's law refers to directional monochromatic quantities, eq. (10.66). The equality between the total hemispherical quantities α and ϵ, eq. (10.73), was derived specifically for a diffuse-gray surface. It turns out that eq. (10.73) holds not only for diffuse-gray surfaces but also for several other special sets of circumstances. These can be found in the more advanced radiation heat transfer treatises (e.g., Ref. [10], p. 65).

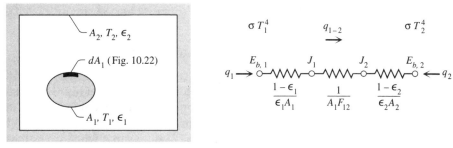

Figure 10.21 Enclosure defined by two diffuse-gray surfaces, and the thermal resistance analog of the net heat transfer rate from A_1 to A_2.

10.4.4 Two-Surface Enclosures

Consider now the problem of estimating the net heat transfer rate between two diffuse-gray surfaces that form an enclosure. The areas (A_1, A_2), temperatures (T_1, T_2) and total hemispherical emissivities (ϵ_1, ϵ_2) are specified (Fig. 10.21). Assume also that the smaller of the two surfaces, A_1, is not concave, that is, that $F_{11} = 0$. The object of the following analysis is the calculation of the net heat transfer rate q_{1-2}. This problem is related to the enclosure analyzed in Section 10.3.3, in which both surfaces were modeled as black. This time, both surfaces reflect some of the radiation that impinges on them.

Figure 10.22 shows an expanded view of the area element dA_1. Let G_1 (W/m^2) be the total irradiation that *arrives* at the dA_1 location. Proceeding in the opposite direction (i.e., *departing* from dA_1 toward the cavity) is the reflected portion $\rho_1 G_1$ *plus* the heat flux emitted by dA_1 itself, $\epsilon_1 E_{b,1}$. The total one-way heat flux that departs from dA_1 represents the surface *radiosity* and is labeled J_1 (W/m^2):

$$J_1 = \rho_1 G_1 + \epsilon_1 E_{b,1} \tag{10.74}$$

The difference between the heat flux that leaves, J_1, and the heat flux that arrives, G_1, is the net heat flux that leaves dA_1,

$$q_1'' = J_1 - G_1 \tag{10.75}$$

Eliminating G_1 between eqs. (10.74) and (10.75), and recognizing that for a diffuse-gray surface $\rho_1 = 1 - \alpha_1 = 1 - \epsilon_1$, we obtain, in order,

$$q_1'' = J_1 - \frac{J_1 - \epsilon_1 E_{b,1}}{\rho_1}$$

$$= \frac{\epsilon_1}{1 - \epsilon_1}(E_{b,1} - J_1) \tag{10.76}$$

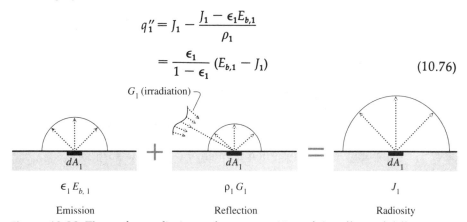

Figure 10.22 The surface radiosity as the superposition of the effects of diffuse emission and diffuse reflection.

The net heat current that leaves the entire A_1 area is simply $q_1 = q_1'' A_1$, or [14]

$$q_1 = \frac{\epsilon_1 A_1}{1 - \epsilon_1} (E_{b,1} - J_1) \qquad (10.77)$$

It pays to examine the structure of this result. The net heat current that leaves the A_1 surface, q_1, must be provided by an external agent (a heater); this current must be "pumped" through the A_1 surface, that is, from its back side to the side that faces the enclosure. Equation (10.77) shows that what drives the q_1 current through the A_1 surface is the potential difference $E_{b,1} - J_1$. What impedes the passage of q_1 through A_1 is the *internal resistance* $(1 - \epsilon_1)/\epsilon_1 A_1$, the generic form of which is

$$R_i = \frac{1 - \epsilon}{\epsilon A} \qquad (10.78)$$

These observations are summarized in the electrical resistance analog shown in Fig. 10.21. The diffuse-gray surface A_1 is represented by *two* nodes: by the radiosity node J_1 on the enclosure side of the surface A_1, and by the blackbody emissive power $E_{b,1} = E_b(T_1)$ on the back side of the same surface. An analogous internal resistance, $(1 - \epsilon_2)/\epsilon_2 A_2$, connects the J_2 and $E_{b,2}$ nodes that account for the surface A_2.

It remains to analyze what happens in the enclosure itself, that is, in the space confined between A_1 and A_2. These two surfaces make themselves seen through their respective radiosities. The surface A_1, for example, sends out the total heat current $J_1 A_1$ in all the directions to which A_1 has access. Furthermore, the radiosity heat flux is distributed uniformly over the hemispherical solid angle based on each area element dA_1. The diffuse character of J_1 is due to the assumed diffuse-gray model: note the diffuse emission and diffuse reflection that contribute to J_1 in Fig. 10.22.

The total heat current $J_1 A_1$ that originates from A_1 has all the features of the one-way heat current $E_{b,1} A_1$ discussed in Section 10.3.1. That discussion can be repeated word for word here, to conclude that the share of the one-way current $J_1 A_1$ that is intercepted by the surface A_2 is

$$q_{1 \to 2} = J_1 A_1 F_{12} \qquad (10.79)$$

This one-way radiation current is also equal to $J_1 A_2 F_{21}$ because of the reciprocity relation $A_1 F_{12} = A_2 F_{21}$.

In using the geometric view factors of Section 10.3.1 we are finally reaping the rewards promised by the diffuse-gray model. It is because of all the features of the diffuse-gray model that the view factor methodology can be applied to the present problem.

If we now position ourselves on the surface A_2 and look toward the enclosure, we see that out of the entire one-way radiation that leaves A_2 (namely, $J_2 A_2$) only a certain portion is intercepted by the surface A_1:

$$q_{2 \to 1} = J_2 A_2 F_{21} = J_2 A_1 F_{12} \qquad (10.80)$$

The *net* heat current that proceeds in the direction $A_1 \to A_2$ is therefore

$$q_{1-2} = q_{1 \to 2} - q_{2 \to 1} = A_1 F_{12}(J_1 - J_2) \qquad (10.81)$$

Comparing this last result with the resistance network started in Fig. 10.21, we conclude that the net heat current from node J_1 to node J_2 is impeded by the purely geometric resistance $1/A_1F_{12}$, which is also equal to $1/A_2F_{21}$. In Fig. 10.21, this new resistance has been inserted between the internal resistances that account for the two diffuse-gray surfaces.

The series of three resistances makes it easy to express the net current q_{1-2} in terms of the overall blackbody emissive power difference $E_{b,1} - E_{b,2}$ or $\sigma(T_1^4 - T_2^4)$:

$$q_{1-2} = \frac{\sigma(T_1^4 - T_2^4)}{\dfrac{1-\epsilon_1}{\epsilon_1 A_1} + \dfrac{1}{A_1 F_{12}} + \dfrac{1-\epsilon_2}{\epsilon_2 A_2}} \tag{10.82}$$

Note that the continuity of energy through the surface A_1 requires that $q_{1-2} = q_1$, where q_1 is the heat transfer rate supplied to the back side of A_1, eq. (10.77). If q_2 is the label of the heat transfer rate supplied by an external agent to the back side of the surface A_2, then the same energy continuity argument requires $q_{1-2} = -q_2$ and, in summary,

$$q_1 = q_{1-2} = -q_2 \tag{10.83}$$

Three important configurations that fall in the "two-surface enclosure" category were illustrated in Fig. 10.15. Substituting into the general solution (10.82) the $F_{12} = 1$ value listed under each of those configurations, it is easy to derive the following special relationships:

Two infinite parallel plates ($A_1 = A_2 = A$):

$$q_{1-2} = \frac{\sigma A(T_1^4 - T_2^4)}{\dfrac{1}{\epsilon_1} + \dfrac{1}{\epsilon_2} - 1} \tag{10.84}$$

The annular space between two infinite cylinders or between two spheres:

$$q_{1-2} = \frac{\sigma A_1(T_1^4 - T_2^4)}{\dfrac{1}{\epsilon_1} + \dfrac{A_1}{A_2}\left(\dfrac{1}{\epsilon_2} - 1\right)} \tag{10.85}$$

Extremely large surface (A_2) *surrounding a convex surface* ($A_1, F_{11} = 0$):

$$q_{1-2} = \sigma A_1 \epsilon_1(T_1^4 - T_2^4) \tag{10.86}$$

The two cylinders and two spheres referred to in eq. (10.85) do not have to be positioned concentrically: note the eccentric positions illustrated in Fig. 10.15.

Equations (10.84) and (10.85) are particularly useful in the analysis of thermal insulation systems in which the dominant mode of heat leakage is radiation. These systems employ "sandwiches" of two or more cavities of the kind defined by the surfaces A_1 and A_2 in the preceding analysis. In the sandwich, the cavities are separated by means of one or more opaque partitions called *radiation shields*. The thermal insulation effect of a single shield is illustrated in the following example.

Example 10.5

Radiation Shielding

Consider the heat transfer in the geometry defined in Fig. E10.5, in which a third surface, A_s, has been inserted in the space between the two surfaces of Fig. 10.21. The intermediate surface A_s "shields" A_2 from the radiosity emanating from A_1; at the same time, A_s shields A_1 from that fraction of the A_2 radiosity that would otherwise be intercepted by A_1. Determine the net heat transfer rate from A_1 to A_2.

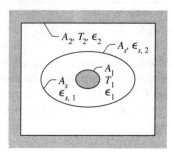

$$\sigma T_1^4 \qquad J_1 \qquad J_{s,1} \qquad \sigma T_s^4 \qquad J_{s,2} \qquad J_2 \qquad \sigma T_2^4$$

$q_{1-2} \rightarrow$ ⊶⟋⟍⟋⟍⊶⟋⟍⟋⟍⊶⟋⟍⟋⟍⊶⟋⟍⟋⟍⊶⟋⟍⟋⟍⊶⟋⟍⟋⟍⊶ $\rightarrow q_{1-2}$

or $\qquad \dfrac{1-\epsilon_1}{\epsilon_1 A_1} \quad \dfrac{1}{A_1 F_{1s}} \quad \dfrac{1-\epsilon_{s,1}}{\epsilon_{s,1} A_s} \quad \dfrac{1-\epsilon_{s,2}}{\epsilon_{s,2} A_s} \quad \dfrac{1}{A_s F_{s2}} \quad \dfrac{1-\epsilon_2}{\epsilon_2 A_2}$ or

q_1 $\qquad\qquad\qquad\qquad\qquad\qquad\qquad\qquad\qquad\qquad\qquad\qquad\qquad -q_2$

Figure E10.5

Solution.   This new configuration is clearly a sandwich of two of the two-surface enclosures analyzed in Fig. 10.21. The electrical network analog of the heat transfer path from A_1 all the way to A_2 consists of six resistances in series. Labeling with $\epsilon_{s,1}$ and $\epsilon_{s,2}$ the total hemispherical emissivities of the two sides of the radiation shield A_s, we can write immediately

$$q_{1-2} = \frac{\sigma(T_1^4 - T_2^4)}{\underbrace{\dfrac{1-\epsilon_1}{\epsilon_1 A_1} + \dfrac{1}{A_1 F_{1s}} + \dfrac{1-\epsilon_{s,1}}{\epsilon_{s,1} A_s}}_{\substack{\text{Total resistance} \\ \text{of the } A_1-A_s \text{ gap}}} + \underbrace{\dfrac{1-\epsilon_{s,2}}{\epsilon_{s,2} A_s} + \dfrac{1}{A_s F_{s2}} + \dfrac{1-\epsilon_2}{\epsilon_2 A_2}}_{\substack{\text{Total resistance} \\ \text{of the } A_s-A_2 \text{ gap}}}} \tag{1}$$

The net heat current q_{1-2} leaves A_1, penetrates the shield A_s, and arrives finally at the outermost surface A_2. Equation (1) should be compared with the no-shield limit, eq. (10.82), to observe the insulating effect of having inserted one shield in the space between A_1 and A_2: the denominator in eq. (1) contains six terms, whereas the denominator of eq. (10.82) contains only three.

This comparison acquires more meaning if we assume the special case where A_1, A_s, and A_2 are three infinite parallel plates of area A, and where all the emissivities have the same value, $\epsilon_1 = \epsilon_{s,1} = \epsilon_{s,2} = \epsilon_2 = \epsilon$. In this case eq. (1) reduces to

$$q_{1-2} = \frac{\sigma A(T_1^4 - T_2^4)}{2\left(\dfrac{2}{\epsilon} - 1\right)} \quad \text{(one shield)} \tag{2}$$

In the same special case the net heat transfer rate in the absence of the shield is [cf. eq. (10.82)]

$$q_{1-2} = \frac{\sigma A(T_1^4 - T_2^4)}{\dfrac{2}{\epsilon} - 1} \quad \text{(no shield)} \tag{3}$$

Equations (2) and (3) show that the shield reduces the net heat transfer rate to one half of the rate that prevails when the shield is absent.

Note that, unlike A_1 and A_2, the shield is not heated or cooled by an external entity. The shield is suspended between A_1 and A_2. The temperature T_s "floats" to that special level that guarantees the continuity of the net heat current through the shield node σT_s^4. In other words, the temperature T_s must be such that the heat current driven by the potential difference $\sigma T_1^4 - \sigma T_s^4$ is equal to the current driven by $\sigma T_s^4 - \sigma T_2^4$.

10.4.5   Enclosures with More Than Two Surfaces

As a generalization of the two-surface analysis of the preceding section, consider the enclosure of Fig. 10.23, in which all n surfaces are diffuse-gray. Equations (10.74) through (10.77) continue to apply to surface 1 in this new enclosure. What changes is the expression for the total irradiation current that strikes the surface A_1. In general, the observer seated on A_1 may see the radiosities of all n surfaces of the enclosure. For example, the irradiation current (watts) that emanates from the jth surface A_j and strikes the surface A_1 is $J_j A_j F_{j1}$. It follows that the irradiation current that impinges on A_1 is

$$A_1 G_1 = J_1 A_1 F_{11} + J_2 A_2 F_{21} + \cdots + J_n A_n F_{n1}$$

$$= \sum_{j=1}^{n} J_j A_j F_{j1}$$

$$= \sum_{j=1}^{n} J_j A_1 F_{1j} \tag{10.87}$$

Note that the last step in the above sequence is justified by the reciprocity relation $A_j F_{j1} = A_1 F_{1j}$.

From the point of view of surface A_1, the heat transfer problem is still the calculation of the net heat transfer rate q_1 that must be supplied to the back of A_1. This heat current can be evaluated using eq. (10.77) *provided* the radiosity J_1

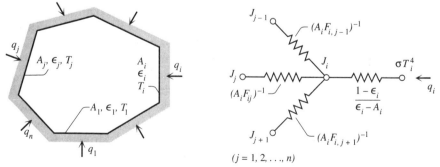

Figure 10.23 Enclosure formed by n diffuse-gray surfaces, and the resistance network portion associated with the surface A_i.

is known. The problem reduces then to the calculation of J_1, which, as we will soon discover, depends on the geometry and radiative properties of all the surfaces of the enclosure. The needed relation between J_1 and the rest of the enclosure parameters follows from substituting into the J_1 definition (10.74) the general expression that was just derived for the irradiation G_1, namely, eq. (10.87):

$$J_1 = (1 - \alpha_1) \sum_{j=1}^{n} J_j F_{1j} + \epsilon_1 \sigma T_1^4 \tag{10.88}$$

In starting with eq. (10.74), we have also used $\rho_1 = 1 - \alpha_1$ and $E_{b,1} = \sigma T_1^4$.

Equation (10.88) states that the radiosity of the surface A_1 depends on the properties of the A_1 surface $(\alpha_1, \epsilon_1, T_1)$, the radiosities of all the surfaces in the enclosure $(J_j; j = 1, 2, \ldots, n)$, and the respective view factors through which all these surfaces are visible from A_1. A system of n equations for the n radiosities J_j is obtained by writing an equation of type (10.88) for each surface:

$$J_i = (1 - \alpha_i) \sum_{j=1}^{n} J_j F_{ij} + \epsilon_i \sigma T_i^4 \qquad (i = 1, 2, \ldots, n) \tag{10.89}$$

If the geometry and all the surface properties are specified, then the system (10.89) pinpoints the values of the n radiosities; and, in the end, eq. (10.77) delivers the value of the heat current q_1. An equation of type (10.77) holds for every single surface that participates in the enclosure:

$$q_i = \frac{\epsilon_i A_i}{1 - \epsilon_i} (\sigma T_i^4 - J_i) \qquad (i = 1, 2, \ldots, n) \tag{10.90}$$

Therefore, the n radiosities delivered by system (10.89) can also be used to calculate the heat current q_i that must be supplied to the back of each surface. Equations (10.90) are restricted to gray surfaces, while equations (10.89) are not.

Note that, by convention, the heat transfer rate q_i has a positive value when it crosses the A_i surface and proceeds *toward the enclosure*. One important relationship among all the heat currents q_i follows from the steady-state first law of thermodynamics, applied to the thermodynamic system defined by the enclosure:

$$\sum_{i=1}^{n} q_i = 0 \tag{10.91}$$

The solution to the multisurface enclosure problem is summarized in eqs. (10.89) and (10.90). When all of the surface temperatures T_i are specified, the system (10.89) can be solved first to determine all the J_is, leaving the q_is to be calculated from eq. (10.90). When only some of the T_is and q_is are specified, the two systems (10.89) and (10.90) must be solved simultaneously. For example, when a surface A_k is insulated (adiabatic) with respect to the exterior of the enclosure, its net heat current is "specified," because $q_k = 0$.

The network analog [14] of this multisurface analysis is presented on the right side of Fig. 10.23. Each surface (A_i) is represented by two nodes, σT_i^4 and J_i, linked by an internal resistance of type $(1 - \epsilon_i)/\epsilon_i A_i$. On its enclosure side, the surface A_i is linked to each of the other surfaces through a resistance of type $1/A_i F_{ij}$. This feature can be illustrated also analytically by writing that the net heat current q_i deposited by A_i into the enclosure is the difference between what emanates from the surface (namely, $A_i J_i$) and what is intercepted by the same surface ($A_i G_i$):

$$q_i = A_i J_i - A_i G_i$$

$$= A_i J_i - \sum_{j=1}^{n} J_j A_i F_{ij} \tag{10.92}$$

$$= A_i J_i \sum_{j=1}^{n} F_{ij} - \sum_{j=1}^{n} J_j A_i F_{ij} \tag{10.93}$$

$$= \sum_{j=1}^{n} A_i F_{ij} (J_i - J_j) \tag{10.94}$$

The first step in this derivation, eq. (10.92), is based on writing an equation of type (10.87) for the surface A_i. The second step, eq. (10.93), consists of invoking the enclosure rule (10.43) for the surface A_i.

The end expression for q_i, eq. (10.94), shows that on the enclosure side of the surface A_i the current q_i breaks up into n minicurrents of type $A_i F_{ij}(J_i - J_j)$. Each minicurrent is driven by the respective radiosity difference between A_i and a neighboring surface A_j. The resistance overcome by each minicurrent is $1/A_i F_{ij}$. This resistance is shown in the network on the right side of Fig. 10.23.

A final observation concerns the geometric view factors F_{ij} of the enclosure. These can be arranged in the following matrix to show that the n-surface enclosure possesses n^2 view factors:

$$\begin{vmatrix} F_{11} & F_{12} \ldots F_{1n} \\ F_{21} & F_{22} \ldots F_{2n} \\ \cdot \\ \cdot \\ \cdot \\ F_{n1} & F_{n2} \ldots F_{nn} \end{vmatrix}$$

Not all of these numbers can be specified independently, in other words, only some of the F_{ij}s constitute true "degrees of freedom" in the design (sizing, shaping) of the enclosure. Reciprocity relations of type (10.38) relate the view factors situated under and to the left of the F_{11}–F_{nn} diagonal to their counterparts across the diagonal. There are $(n^2 - n)/2$ such relationships, because there are n view factors on the diagonal and $(n^2 - n)/2$ view factors on each side of the diagonal. Additionally, n enclosure relationships of type (10.43) can be written, one relationship for each surface. In conclusion, the number of independent geometric view factors is*

$$n^2 - \frac{1}{2}(n^2 - n) - n = \frac{n}{2}(n - 1) \tag{10.95}$$

For some of the simplest enclosures, which form the object of the problems proposed at the end of the chapter, this number is as follows:

---

*Note that the number $(n/2)(n - 1)$ is always an integer.

Number of Surfaces	2	3	4	5
Number of independent geometric view factors	1	3	6	10

This table shows that the number of independent view factors increases rapidly as the number of surfaces (isothermal patches) of the enclosure increases. The algebra associated with solving systems (10.89) and (10.90) becomes more challenging at the same time. Therefore, we have an incentive to model the heat transfer configuration (i.e., to define the enclosure) by means of the smallest number of isothermal surfaces possible. A modeling decision of this kind is illustrated in the next example.

Example 10.6

Enclosure with Three Surfaces, One of Which is Adiabatic ("Reradiating")

The attached figure shows the key features of an oven, namely, the source of heat (e.g., flame, electrical resistance), and the object that is to be heated (e.g., cut of meat, block of metal). The surfaces of both items can be modeled as diffuse-gray and can be characterized by A_1, ϵ_1, T_1 and A_2, ϵ_2, T_2, respectively

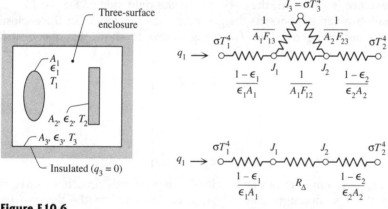

Figure E10.6

Surrounding the heater A_1 and the heat sink A_2 is a third surface that must be insulated (adiabatic) to channel all of the heating effect of A_1 into the heat sink A_2. In an industrial furnace, this third surface is built out of a high temperature-resistant (refractory) material. In real life, the temperature will vary along this surface, the highest temperatures being registered in the zone that is the closest to the heat source. The simplest model of the adiabatic surface is the isothermal surface (A_3, ϵ_3, T_3) identified in the figure. The temperature T_3 is nothing more than a first-cut (average) estimate of the temperatures reached by the real-life refractory surface. This isothermal and refractory surface model simplifies the analysis considerably, in fact, it makes a fully analytical solution possible.

Solution.   Of interest is the relationship between the heat-source and the heat-sink temperatures, and the net heat transfer rate between the two. This relationship can be read off the resistance network that corresponds to this problem. Starting with the upper version of the network, it is worth noting the three radiosity nodes (J_1, J_2, J_3), and that the internal resistance of the third surface, $(1 - \epsilon_3)/\epsilon_3 A_3$, is not shown. If this resistance had been drawn, say, above the node J_3, we would have discovered that the current through it is, in fact, zero because the surface A_3 is adiabatic ($q_3 = 0$). In conclusion, the potentials (nodes) σT_3^4 and J_3 coincide on the diagram, and this means that the internal resistance $(1 - \epsilon_3)/\epsilon_3 A_3$ does not play any role in the functioning of the network. In particular, the heat transfer rate from A_1 to A_2 does not depend on ϵ_3.

Examining the lower version of the resistance network, we can write immediately that

$$q_{1-2} = q_1 = \frac{\sigma(T_1^4 - T_2^4)}{\dfrac{1 - \epsilon_1}{\epsilon_1 A_1} + R_\Delta + \dfrac{1 - \epsilon_2}{\epsilon_2 A_2}} \tag{1}$$

The second term in the denominator, R_Δ, is shorthand for the resistance contributed by the triangular loop of the upper diagram,

$$R_\Delta = \frac{R_s(1/A_1 F_{12})}{R_s + (1/A_1 F_{12})} \tag{2}$$

where

$$R_s = \frac{1}{A_1 F_{13}} + \frac{1}{A_2 F_{23}} \tag{3}$$

A more accurate solution to the same problem can be obtained based on a model where the insulated surface A_3 is divided into two isothermal areas, A_{3a} and A_{3b}. This approach leads to a four-surface enclosure problem, for which an analytical solution of type (1)–(3) is not possible. Instead, the problem must be solved numerically by using eqs. (10.89)–(10.90), assuming that the known parameters of the problem have been specified numerically.

10.5 PARTICIPATING MEDIA

10.5.1 Volumetric Absorption

The preceding analyses of radiation between two or more surfaces were based on the assumption that the medium that separates the surfaces is perfectly transparent and does not participate in the heat transfer process. The nonparticipating medium model is appropriate when the space between the radiating surfaces is occupied by vacuum, or by a gas with monatomic or symmetrical diatomic molecule. Examples are the main components of air (N_2, O_2), H_2, Ne, Ar, and Xe.

The nonparticipating medium model is not appropriate when the medium absorbs and scatters the radiation that passes through it. Primary examples of such media are strongly polar gases the molecules of which exhibit asymmetry (e.g., H_2O, CO_2, NH_3, O_3, CO, SO_2, NO, hydrocarbons, and alcohols). Gases of this type are called *participating media*, because at each point in their volume they absorb and scatter the incident radiation while emitting their own.

In this section we consider the additional complication that is introduced in the previous model by a participating medium. The major difference is that, while until now we analyzed heat transfer processes that occurred strictly between *surfaces*, when the medium is a participating one, the analysis must account also for the *volumetric* absorption, transmission, and emission of radiation. This volumetric effect is usually associated with monochromatic radiation that falls within certain bands, or wavelength intervals. Radiation with a wavelength outside these bands will pass unattenuated through the medium: with respect only to this radiation the medium is nonparticipating.

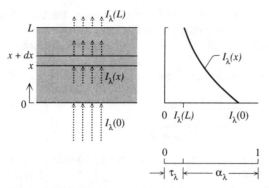

Figure 10.24 The attenuation of the monochromatic radiation that penetrates a volumetrically absorbing medium.

We develop an understanding for the radiative behavior of a participating medium by focusing on the one-dimensional layer of thickness L shown in Fig. 10.24. This layer is being penetrated by a beam of monochromatic radiation. The intensity of this beam, $I_\lambda(x)$, is attenuated locally by the volumetric absorption in the sublayer of thickness dx. Experiments have shown that the local attenuation dI_λ is proportional to the local intensity of the beam,

$$dI_\lambda = -\kappa_\lambda I_\lambda dx \qquad (10.96)$$

This proportionality serves as definition for the *monochromatic extinction coefficient* κ_λ, which has the units m^{-1}. The extinction coefficient is proportional to the concentration of absorbing molecules (i.e., the partial pressure of the absorbing gas) in the medium. Integrating eq. (10.96) across the layer, we obtain

$$I_\lambda(x) = I_\lambda(0)\exp(-\kappa_\lambda x) \qquad (10.97)$$

This exponential decay in beam intensity across the layer is known as *Beer's law*. In particular, when $x = L$, eq. (10.97) shows the fraction of the incident intensity that escapes through the other side of the layer:

$$\tau_\lambda = \frac{I_\lambda(L)}{I_\lambda(0)} = \exp(-\kappa_\lambda L) \qquad (10.98)$$

This fraction is less than 1, and it represents the monochromatic transmissivity of the layer. The remaining fraction is absorbed in the L-thick layer,

$$\alpha_\lambda = 1 - \tau_\lambda = 1 - \exp(-\kappa_\lambda L) \qquad (10.99)$$

as gases do not reflect the radiation that passes through them ($\alpha_\lambda + \tau_\lambda = 1$). The number α_λ is the monochromatic absorptivity of the L-thick medium.

Finally, if the gas temperature T_g is uniform and does not differ much from the temperature T_s of the surface that produced the beam $I_\lambda(0)$, we may invoke Kirchhoff's law:

$$\epsilon_\lambda = \alpha_\lambda = 1 - \exp(-\kappa_\lambda L) \qquad (10.100)$$

In this last equation ϵ_λ is the monochromatic emissivity of the layer.

Accurate analyses of gas absorption and emission must account for every band of wavelengths that contribute significantly to the process. In engineering calculations it is more convenient to work with *total* quantities, which are obtained by integrating eqs. (10.98) to (10.100) over the entire spectrum:

$$\tau_g = \frac{\displaystyle\int_0^\infty I_\lambda(L)\,d\lambda}{\displaystyle\int_0^\infty I_\lambda(0)\,d\lambda} \qquad (10.101)$$

$$\alpha_g = 1 - \tau_g \qquad (10.102)$$

$$\epsilon_g \cong \alpha_g \qquad \text{(if } T_g \cong T_s, \text{ i.e., if Kirchhoff's law applies)} \qquad (10.103)$$

These coefficients are, in order, the gas transmissivity, absorptivity, and emissivity.

As a special case, the *gray gas* is defined by $\alpha_g = \alpha_\lambda$, $\epsilon_g = \epsilon_\lambda$, and $\alpha_g = \epsilon_g$. The gray gas represents a medium in which the monochromatic coefficients (τ_λ, α_λ, ϵ_λ) are independent of wavelength, and where the gas emissivity is equal to the gas absorptivity regardless of the source of incident radiation.

10.5.2 Gas Emissivities and Absorptivities

The emission and absorption characteristics of participating gases are described quantitatively by the following procedure developed by Hottel [15]. Consider the hemispherical enclosure shown in Fig. 10.25, in which the gas mixture contains *only one participating component* (CO_2, for example) and one or more nonparticipating (transparent) species. The gas mixture is at the uniform temperature T_g and total pressure P, while the partial pressure of carbon dioxide is P_c. The radius of the hemispherical gas volume is L.

The gas mixture exchanges thermal radiation with the small area A_s situated in the center of the base.* The radiation heat current that is emitted by the gas and arrives at A_s is

$$q_{g \to s} = \epsilon_g A_s \sigma T_g^4 \qquad (10.104)$$

where ϵ_g is the emissivity of the participating gas. Note that $q_{g \to s}$ is a one-way radiation current, and that the surface A_s absorbs this current entirely only if A_s is black. Note further that $\epsilon_g \sigma T_g^4$ (with $\epsilon_g < 1$) represents only a fraction of the emissive power associated with a blackbody of temperature T_g.

Experimental measurements of the emitted radiation $q_{g \to s}$ have been used in conjunction with the ϵ_g definition (10.104) for the purpose of calculating and

---

*The subscript used for "wall" in this book is w. Here we make an exception by using s (for "surface") and reserve w for quantities associated with water vapor as a radiating component in a mixture.

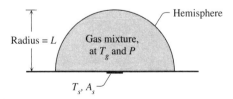

Figure 10.25 Hemispherical space filled with gas radiating to a black surface element.

recording the gas emissivity ϵ_g. For carbon dioxide the emissivity is labeled ϵ_c; it is a function of the mixture temperature, the CO_2 partial pressure times the radius of the gas volume, and the total pressure of the mixture:

$$\epsilon_c = f_c(T_g, P_c L, P) \tag{10.105}$$

The function f_c is reported graphically by the following combination of Figs. 10.26 and 10.27. When the mixture pressure is equal to 1 atm, the emissivity of carbon dioxide is simply the ϵ_c valued indicated on the ordinate of Fig. 10.26. If the total pressure of the mixture happens to be different from 1 atm, the ϵ_c reading provided by Fig. 10.26 must be multiplied by the correction factor C_c provided by Fig. 10.27. In general, then, the f_c function indicated in eq. (10.105) is reconstructed by multiplying the readings from the two charts:

$$\epsilon_{c,P \neq 1\,atm} = \epsilon_{c,P = 1\,atm} \times C_{c,P \neq 1\,atm} \tag{10.106}$$

Fig. 10.26   Fig. 10.27

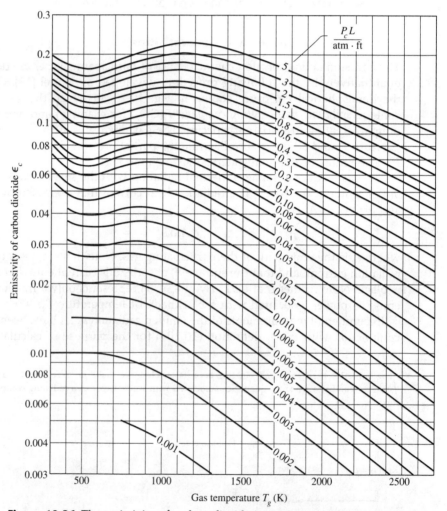

Figure 10.26 The emissivity of carbon dioxide in a mixture with nonparticipating gases at a mixture pressure of 1 atm. The length L is defined in Fig. 10.25 or in Table 10.5. (Hottel [15], with permission from McGraw-Hill Book Company.)

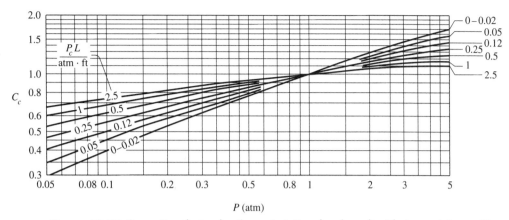

Figure 10.27 Correction factor for the emissivity of carbon dioxide in a mixture with nonparticipating gases at mixture pressures other than 1 atm. (Hottel [15], with permission from McGraw-Hill Book Company.)

The carbon dioxide charts show that the emissivity increases monotonically as the product P_cL increases (see Fig. 10.26, $P = 1$ atm). This trend is understandable because when the partial pressure and the gas volume increase, the number of CO_2 molecules that emit radiation toward the target A_s also increases. The same explanation holds true for the trend made visible in Fig. 10.27: ϵ_c also increases as the total pressure P increases.

Consider now the reverse of the interaction between the participating gas and the surface A_s in Fig. 10.25. Let $q_{s \to g}$ be the radiation heat flux that is emitted by A_s in all directions and reaches the gas (note that if A_s is black, $q_{s \to g} = A_s \sigma T_s^4$). Only a fraction of this one-way heat current is absorbed by the gas,

$$q_a = \alpha_g q_{s \to g} \qquad (10.107)$$

namely, the fraction represented by the gas absorptivity α_g. As noted previously in eq. (10.103), α_g is equal to ϵ_g when T_s and T_g are equal, or almost equal. When the surface and gas temperatures differ greatly, the gas absorptivity can be estimated based on the following empirical rule [15]. In the case of carbon dioxide as the lone participating gas in the mixture, the absorptivity α_c with respect to incident radiation of temperature T_s is

$$\alpha_c = f_c\left(T_s, P_cL\frac{T_s}{T_g}, P\right) \times \left(\frac{T_g}{T_s}\right)^{0.65} \qquad (10.108)$$

The f_c function must be compared with eq. (10.105) to see that the first factor on the right side of eq. (10.108) is obtained from Figs. 10.26 and 10.27, by replacing T_g with T_s and P_cL with $P_cL\,(T_s/T_g)$. In the ratio T_s/T_g the temperatures are absolute, that is, expressed in degrees Kelvin.

A similar procedure is available for calculating the total emissivity and absorptivity of water vapor as the lone radiating component in a gas mixture. The charts shown in Figs. 10.28 and 10.29 deliver the emissivity of water vapor as the function

$$\epsilon_w = f_w(T_g, P_wL, P + P_w) \qquad (10.109)$$

where P_w is the partial pressure of water vapor. Figure 10.28 alone provides the value for ϵ_w when $P + P_w = 1$ atm. For other values of $P + P_w$, the ϵ_w reading obtained from Fig. 10.28 must be multiplied by the correction factor C_w recommended by Fig. 10.29:

$$\epsilon_{w,P + P_w \neq 1 \text{ atm}} = \epsilon_{w,P + P_w = 1 \text{ atm}} \times C_{w,P + P_w \neq 1 \text{ atm}} \qquad (10.110)$$

$$\underbrace{\qquad}_{\text{Fig. 10.28}} \qquad \underbrace{\qquad}_{\text{Fig. 10.29}}$$

The water vapor absorptivity with respect to incident radiation produced by a surface of temperature T_s can be evaluated with the empirical formula [15]

$$\alpha_w = f_w \left(T_s, \ P_w L \frac{T_s}{T_g}, \ P + P_w \right) \left(\frac{T_g}{T_s} \right)^{0.45} \qquad (10.111)$$

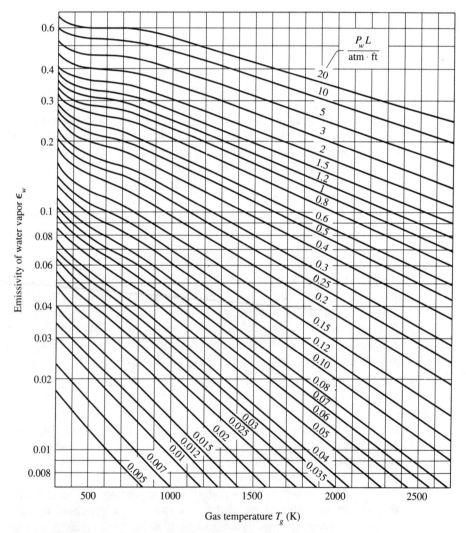

Figure 10.28 The emissivity of water vapor in a mixture with nonparticipating gases at a mixture pressure of 1 atm. The length L is defined in Fig. 10.25 or in Table 10.5. (Hottel [15], with permission from McGraw-Hill Book Company.)

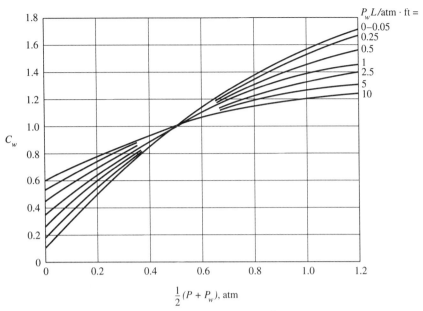

Figure 10.29 Correction factor for the emissivity of water vapor in a mixture with nonparticipating gases at mixture pressures other than 1 atm. (Hottel [15], with permission from McGraw-Hill Book Company.)

By comparing it with eq. (10.109), we note that the f_w factor on the right side of eq. (10.111) is obtained from Figs. 10.28 and 10.29 by replacing T_g with T_s and $P_w L$ with $P_w L(T_s/T_g)$.

The calculations described until now referred to cases in which only one participating component was present in the mixture. That component was carbon dioxide (Figs. 10.26 and 10.27) or water vapor (Figs. 10.28 and 10.29). When CO_2 and H_2O are present *together* in the mixture, the emitted radiation is a bit smaller than the radiation emitted by CO_2 alone plus the radiation emitted by H_2O alone. This effect is expected, because each gas obstructs somewhat the radiation emitted by the other. The overall gas emissivity ϵ_g is, therefore, smaller than the sum of the ϵ_c and ϵ_w values discussed above:

$$\epsilon_g = \epsilon_c \qquad\quad + \epsilon_w \qquad\quad - \Delta\epsilon \qquad\qquad (10.112)$$

Figs. 10.26–27   Figs. 10.28–29   Fig. 10.30

The $\Delta\epsilon$ correction is reported in Fig. 10.30; it is primarily a function of the amount of radiating molecules $L(P_c + P_w)$ and combination (proportions) of CO_2 and H_2O in the mixture. The $\Delta\epsilon$ correction is largest (a few percentage points) when the partial pressures of carbon dioxide and water vapor are comparable.

The emissivities of sulfur dioxide, carbon monoxide, and ammonia can be evaluated by consulting a set of similar charts compiled by Hottel [15].

The length L employed in the construction of Figs.10.26 through 10.30 refers strictly to the radius of the hemispherical gas volume (Fig. 10.25). The same charts and formulas can be used for calculating ϵ_g and α_g in other geometric combinations of gas volumes and radiating surfaces. Listed in Table 10.5 are

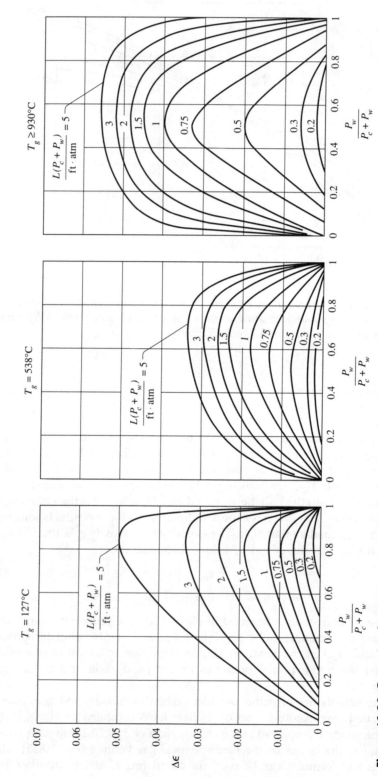

Figure 10.30 Correction for gas emissivity when carbon dioxide and water vapor are present simultaneously in a mixture with nonparticipating gases. (Hottel [15], with permission from McGraw-Hill Book Company.)

Table 10.5 The Equivalent Length L_e for Several Gas Volume Shapes[a]

Shape of Gas Volume	Actual Dimension	Equivalent Length
Sphere, radiation to internal surface	Diameter D	$L_e \cong 0.60D$
Infinite cylinder, radiation to entire internal surface	Diameter D	$L_e \cong 0.95D$
Circular cylinder with height $= D$, radiation to entire surface	Diameter D	$L_e \cong 0.6D$
Circular cylinder with height $= D$, radiation to spot in the center of base	Diameter D	$L_e \cong 0.77D$
Semi-infinite circular cylinder, radiation to entire base	Diameter D	$L_e \cong 0.65D$
Semi-infinite cylinder, radiation to spot in the center of base	Diameter D	$L_e \cong 0.9D$
Cube, radiation to one face	Side a	$L_e \cong 0.67a$
Space between two infinite parallel planes, radiation to both planes	Spacing s	$L_e \cong 1.8s$
Space between tubes in an infinite tube bundle with tube diameter $=$ clearance between two closest tube walls, radiation to a single tube, tube centers on		
Equilateral triangles	Tube diameter D	$L_e \cong 2.8D$
Squares	Tube diameter D	$L_e \cong 3.5D$
Arbitrary volume V surrounded by surface A, radiation to A	V, A	$L_e \cong 3.6\dfrac{V}{A}$

[a]L_e replaces L in Figs. 10.26–10.29 (after Hottel [15]).

several configurations to which this procedure can be applied by replacing L with the *equivalent length* L_e shown in the right column.

10.5.3 Gas Surrounded by Black Surface

As an application of the method described in the preceding section, consider the heat transfer between a radiating gas and an enclosure of surface area A_s and temperature T_s, which contains the gas. The gas is characterized by T_g, P, L_e, and the partial pressure(s) of the radiating component(s) in the gas mixture. This gas information can be used to determine ϵ_g and the α_g value that refers to radiation arriving from T_s. If the internal surface A_s is black, it absorbs all the radiation that is emitted by the gas:

$$q_{g \to s} = \epsilon_g \sigma T_g^4 A_s \qquad (10.113)$$

The radiation heat current emitted by A_s is $\sigma T_s^4 A_s$, while the portion that is absorbed by the entire gas volume is

$$q_{s \to g} = \alpha_g \sigma T_s^4 A_s \qquad (10.114)$$

In summary, the instantaneous *net* rate of heat transfer from the gas to its black enclosure is

$$q_{g-s} = q_{g \to s} - q_{s \to g}$$

$$= \sigma A_s (\epsilon_g T_g^4 - \alpha_g T_s^4) \tag{10.115}$$

The expression in round brackets shows that in this calculation the radiation emitted by the surface can be neglected if the gas temperature is sufficiently higher than the surface temperature.

10.5.4   Gray Medium Surrounded by Diffuse-Gray Surfaces

Consider now the two-surface enclosure shown in Fig. 10.31. Each surface is diffuse-gray and isothermal, $(A_1, T_1, \epsilon_1$ and $A_2, T_2, \epsilon_2)$, whereas the medium (e.g., gas) is isothermal and gray $(T_g, \epsilon_g = \alpha_g)$. Of interest is the net heat transfer between A_1 and A_2, and how the gas "interferes" with this process.

An observer standing on A_1 sees not only the medium but also a surface of different temperature, A_2. The latter is visible only to the extent that the medium is "transparent." The radiation that leaves A_1 is $J_1 A_1$. The portion that reaches A_2 is $J_1 A_1 F_{12} \tau_g$, where τ_g is the medium transmissivity $(\tau_g = 1 - \alpha_g)$. In the reverse direction, the portion of the radiation leaving A_2 and arriving through the medium at A_1 is $J_2 A_2 F_{21} \tau_g$, where $A_2 F_{21} = A_1 F_{12}$. The net heat transfer rate from A_1 to A_2 directly (i.e., by penetrating the medium) is therefore $A_1 F_{12} \tau_g (J_1 - J_2)$. In conclusion, the radiosity nodes J_1 and J_2 are linked by a direct resistance of size $(A_1 F_{12} \tau_g)^{-1}$ in the network constructed in Fig. 10.31.

The surfaces communicate also *indirectly*, by using the medium as an intermediary. Take the interaction between A_1 and the medium. The fraction of the radiation leaving A_1 and being absorbed volumetrically by the medium is $J_1 A_1 F_{1g} \alpha_g$, where $\alpha_g = \epsilon_g$. Traveling in the opposite direction is the medium emission that is intercepted by A_1, namely, $\epsilon_g \sigma T_g^4 A_1 F_{1g}$. The net heat current along this path is $A_1 F_{1g} \epsilon_g (J_1 - \sigma T_g^4)$. This shows that the J_1 node and the gas node σT_g^4 are linked by the resistance $(A_1 F_{1g} \epsilon_g)^{-1}$.

The radiation network is completed by a similar resistance between the gas volume (σT_g^4) and the second gray surface (J_2). Between each radiosity node and the respective blackbody emissive power node of the surface, we see the internal resistances defined in eq. (10.78). The node that represents the gas volume is a *floating* one when the gas is not heated by an independent mecha-

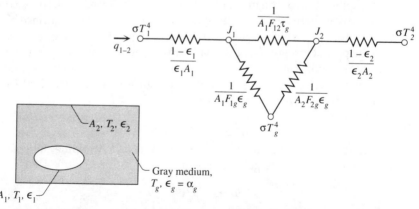

Figure 10.31 Gray medium enclosed by two diffuse-gray surfaces (left), and the radiation network for the case in which the temperature of the medium is floating (right).

nism (e.g., electrically or by chemical reactions). Under these circumstances only, the net rate of heat transfer from A_1 to A_2 can be written by reading the network:

$$q_{1-2} = \frac{\sigma(T_1^4 - T_2^4)}{\dfrac{1 - \epsilon_1}{\epsilon_1 A_1} + R_\Delta + \dfrac{1 - \epsilon_2}{\epsilon_2 A_2}} \qquad (10.116)$$

where R_Δ is the equivalent series resistance of the triangular loop of the network:

$$R_\Delta = \frac{(A_1 F_{12} \tau_g)^{-1} [(A_1 F_{1g} \epsilon_g)^{-1} + (A_2 F_{2g} \epsilon_g)^{-1}]}{(A_1 F_{12} \tau_g)^{-1} + (A_1 F_{1g} \epsilon_g)^{-1} + (A_2 F_{2g} \epsilon_g)^{-1}} \qquad (10.117)$$

The radiative heat transfer between surfaces and participating media is discussed in greater depth in the advanced treatments by Sparrow and Cess [16] and Siegel and Howell [10].

REFERENCES

1. W. M. Rohsenow and H. Y. Choi, *Heat, Mass and Momentum Transfer*, Prentice-Hall, Englewood Cliffs, NJ, 1961, pp. 364–370.

2. R. E. Sonntag and G. Van Wylen, *Introduction to Thermodynamics*, 2nd edition, Wiley, New York, 1982, pp. 653–658.

3. M. Planck, Distribution of energy in the spectrum, *Ann. Physik*, Vol. 4, No. 3, 1901, pp. 553–563.

4. A. Bejan, *Advanced Engineering Thermodynamics*, Wiley, New York, 1988, Chapter 9.

5. W. Wien, Temperatur und Entropie der Strahlung, *Ann. Phys.*, Ser. 2, Vol. 52, 1894, pp. 132–165.

6. J. Stefan, Über die Beziehung zwischen der Wärmestrahlung und der Temperatur, *Stiz. ber. Akad. Wiss. Wien*, Vol. 79, 1879, pp. 391–428.

7. L. Boltzmann, Ableitung des Stefan'schen Gesetzes, betreffend der Abhängigkeit der Wärmestrahlung von der Temperatur aus der electromagnetischen Lichtteorie, *Ann Phys. (Leipzig)*, Ser. 3, Vol. 22, 1884, pp. 291–294.

8. R. V. Dunkle, Thermal radiation tables and applications, *Trans. ASME*, Vol. 65, 1954, pp. 549–552.

9. J. R. Howell, *A Catalog of Radiation Configuration Factors*, McGraw-Hill, New York, 1982.

10. R. Siegel and J. R. Howell, *Thermal Radiation Heat Transfer*, McGraw-Hill, New York, 1972, pp. 783–791.

11. G. G. Gubareff, J. E. Janssen, and R. H. Torborg, *Thermal Radiation Properties Survey*, 2nd edition, Honeywell Research Center, Minneapolis, MN, 1960.

12. Y. S. Touloukian and C. Y. Ho, Eds., *Thermophysical Properties of Matter*, Vols. 7–9, Plenum, New York, 1972.

13. G. Kirchhoff, *Gesammelte Abhandlungen*, Johann Ambrosius Barth, Leipzig, 1882, p. 574.

14. A. K. Oppenheim, Radiation analysis by the network method, *Trans. ASME*, Vol. 78, 1956, pp. 725–735.

15. H. C. Hottel, Radiant-heat transmission, Chapter 4 in W. H. McAdams, *Heat Transmission*, McGraw-Hill, 3rd edition, New York, 1954.

16. E. M. Sparrow and R. D. Cess, *Radiation Heat Transfer*, Hemisphere, Washington, DC, 1978.

17. H. C. Hottel and A. F. Sarofim, *Radiative Heat Transfer*, McGraw-Hill, New York, 1967, pp. 31–39.

18. R. G. Watts, Climate change due to greenhouse gases: change, impacts, and responses, *Heat Transfer 1990*, G. Hetsroni, Ed., Hemisphere, Washington, DC, Vol. 1, 1990, pp. 419–434.

19. A. Bejan, *Solutions Manual for Advanced Engineering Thermodynamics*, Wiley, New York, 1988, Problem 7.6.

20. A. F. Mills, Private communication, January 18, 1990.

21. A. De Vos, Private communication, June 22, 1990.

22. G. Lauriat, A numerical study of a thermal insulation enclosure: influence of the radiative heat transfer, ASME HTD-Vol. 8, 1980, pp. 63–71.

23. A. Bejan, Surfaces covered with hair: optimal strand diameter and optimal porosity for minimum heat transfer, *Biomimetics*, Vol. 1, no. 1, 1992, pp. 23–38.

24. J. L. Lage, Fundamental aspects of convection in enclosed fluids, Ph.D. thesis, Duke University, Durham, NC, 1991.

PROBLEMS

Radiative Properties

10.1 Calculate the solid angle through which the sun can be seen from the earth. With reference to the attached figure, the dimensions of the earth–sun configuration are

$$d = 12{,}756 \text{ km, or } 7926 \text{ miles}$$

$$r = 1.447 \times 10^8 \text{ km, or } 9.302 \times 10^7 \text{ miles}$$

$$D = 1.392 \times 10^6 \text{ km, or } 8.65 \times 10^5 \text{ miles}$$

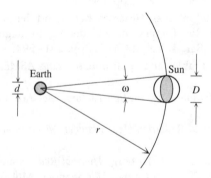

Figure P10.1

10.2 Determine analytically the wavelength at which the monochromatic hemispherical emissive power $E_{b,\lambda}$ is maximum, and verify that your result matches Wien's displacement law. Also determine the corresponding expression for the maximum monochromatic hemispherical emissive power $E_{b,\lambda,\text{max}}$.

10.3 Prove that the dimensionless radiation function $E_b(0-\lambda T)/\sigma T^4$ depends only on the value of the group λT.

10.4 Fresh snow absorbs 80 percent of the incident solar radiation with wavelengths shorter than $\lambda_1 = 0.4 \ \mu m$, 10 percent of the radiation between λ_1 and $\lambda_2 = 1 \ \mu m$, and 100 percent of the radiation with wavelengths longer than λ_2. The solar surface is approximated adequately by a black surface with the temperature $T = 5800$ K. Calculate the fraction of the total incident solar radiation that is absorbed by fresh snow.

10.5 The surface $A_1 = 1 \ cm^2$ shown in the figure is black and emits radiation with the total intensity $I_{b,1} = 3000 \ W/m^2 \cdot sr$. The target area is $A_2 = 2 \ cm^2$.

(a) Calculate the heat current $q_{1 \rightarrow 2}$ representing the portion of the radiation emitted by A_1 and intercepted by A_2.

(b) Repeat the preceding calculation by assuming that the surface A_2 is centered at the point P and is parallel to A_1. Comment on what the face-to-face alignment of A_1 and A_2 does to the value of q_{1-2}.

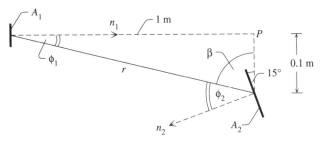

Figure P10.5

10.6 (a) In the small-λT limit, the denominator appearing on the right side of eq. (10.16) is approximately equal to exp $(C_2/\lambda T)$. Verify that in this limit the radiation function approaches the asymptote listed in eq. (10.26).

(b) In the large-λT limit, the denominator of the $E_{\lambda,b}$ expression (10.16) approaches $C_2/\lambda T$. Show that the corresponding asymptote of the radiation function is the expression listed in eq. (10.27).

View Factors

10.7 Determine analytically the view factor F_{12} between the infinitesimal area dA_1 and the parallel disc A_2 shown in the fourth entry to Table 10.2. Verify that in the $H/R \rightarrow 0$ limit the view factor F_{12} approaches 1, and discuss the physical meaning of this limiting configuration.

10.8 Consider the infinitely long enclosure with triangular cross section shown as the third configuration in Table 10.2. This enclosure has three surfaces. Invoke the reciprocity and enclosure relations that exist among the various view factors to derive (and prove the validity of) the F_{12} expression listed in the table.

10.9 On a clear day, a disc-shaped flat-plate solar collector is oriented so that its axis passes through the center of the sun. The sun diameter is 1.392×10^6 km, and the average radius of the earth's orbit is 1.447×10^8 km.

(a) Calculate the geometric view factor F_{21} in which surface 1 is the collector disc and surface 2 is the solar sphere.

(b) Calculate the same view factor by assuming that the solar surface is itself a flat disc of diameter 1.392×10^6 km that is parallel to the disc of the collector. Are the view factors of the coaxial disc–sphere and coaxial disc–disc configurations always equal? Why, then, are the numerical answers to parts (a) and (b) identical?

10.10 The last entry in Table 10.2 suggests that the view factor F_{12} from a sphere to a disc positioned on the same centerline does not depend on the sphere radius (R_1).

This result may seem surprising because view factors are geometry dependent, and R_1 is an important part of the sphere–disc geometry. Derive your own version of the F_{12} view factor from the sphere to the disc. (*Hint*: Consider the larger sphere that touches the rim of the disc, and use the view factor information listed for the two-sphere enclosure in Fig. 10.15.)

10.11 The figure shows two finite-width strips that are infinitely long in the direction normal to the paper. The widths of these strips are L_1 and L_2. Show that the geometric view factor from surface 1 to surface 2 is

$$F_{12} = \frac{a + b - (c + d)}{2L_1}$$

for which the dimensions a, b, c, and d have been defined directly on the figure. This method of calculating F_{12} is known as the *crossed-strings method* [17], in view of the diagonals a and b that cross in the space between L_1 and L_2.

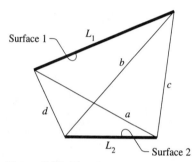

Figure P10.11

10.12 Consider the two infinitely long parallel plates of width X and spacing H shown in the figure. This configuration corresponds to the limit $Y/H \rightarrow \infty$ represented by the uppermost curve in Fig. 10.11. Rely on the answer to the preceding problem to determine a compact analytical expression for the view factor F_{12}.

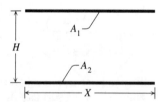

Figure P10.12

Diffuse-Gray Enclosures

10.13 The evacuated space between two infinite parallel plates maintained at different temperatures (T_H, T_L) contains n parallel radiation shields. All the surfaces —the inner surfaces of the end plates T_H and T_L and both sides of each of the radiation shields T_i $(i = 1, 2, \ldots, n)$—are diffuse-gray surfaces with the same total hemispherical emissivity ϵ.

(a) Determine the radiation heat transfer rate from T_H to T_L.

(b) Show that the heat transfer rate from T_H to T_L is $n + 1$ times smaller than what it would have been if all the shields were absent.

10.14 The figure shows a vertical cross section through the center of an enclosed barbecue system. The pile of charcoal $(T_1 = 800°C)$ is shaped as a disc of diameter

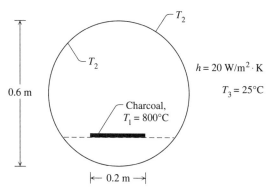

Figure P10.14

20 cm. The thin-walled metallic enclosure is shaped as a 0.6-m-diameter sphere. The enclosure wall can be modeled as isothermal (T_2). The heat transfer from the outside of the enclosure to the ambient ($T_3 = 25\,°C$) is by radiation and convection. The convection effect is characterized by the overall (forced and natural convection) heat transfer coefficient $h = 20 \text{ W/m}^2 \cdot \text{K}$. The charcoal, enclosure, and ambient can be modeled as black surfaces. Assume that the coarse grill that supports the charcoal is a transparent surface.

(a) Calculate the temperature of the enclosure wall, T_2.

(b) To what degree does radiation contribute to the total heat transfer rate from the outer surface of the enclosure, that is, from T_2 to T_3?

10.15 The figure shows the cross section through an enclosure that extends to infinity in the direction normal to the paper. The cross section is an isosceles triangle in which

$$A_2 = A_3 = 10A_1$$

Each of the three surfaces can be modeled as black and isothermal. The A_3 surface is insulated with respect to the surroundings, while the net heat current q_{1-2} enters the enclosure through the A_1 surface and exits through the A_2 surface.

(a) Calculate the numerical values of the view factors F_{12}, F_{13}, F_{21}, and F_{23}.

(b) Determine the net heat transfer rate q_{1-2}.

(c) Determine the "floating" temperature of the insulated surface, T_3, as a function of the temperatures of the differentially heated surfaces of the enclosure (T_1, T_2).

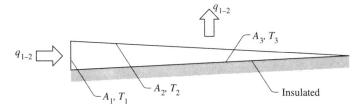

Figure P10.15

10.16 The reason why it is a good idea to keep turning the steak on the grill is that the heat flux on the bottom side of the steak is much larger than the heat flux on the top side. To see this, consider the enclosed barbecue shown in the figure. The coal pile is shaped as a disc of diameter 20 cm, with the upper area A_1 at $T_1 = 800\,°C$.

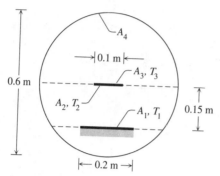

Figure P10.16

Assume that the underside of the coal pile is insulated by the ash mound collected under it. The enclosure area A_4 is a 0.6-m-diameter sphere with temperature $T_4 = 150°C$.

A fresh steak has just been put on the grill, $T_2 = T_3 = 25°C$. Its shape is approximated well by a disc with a diameter of 10 cm. It is positioned at 15 cm directly above the coal. All the surfaces inside the enclosure (A_1, A_2, A_3, A_4) can be modeled as black. Assume further that the coarse grills that hold the steak and the coal are transparent.

Calculate the average heat flux q_2'' that enters through the bottom of the steak (A_2) and the average heat flux q_3'' that enters through the top (A_3). Compare q_2'' with q_3''.

10.17   The ceramic wall shown in the figure holds an array of flat-plate electrical conductors whose normal operating temperature is $T_2 = 150°C$. The emissivity of the conductor surface is $\epsilon_2 = 0.7$. The ceramic wall is a refractory surface. The surroundings can be modeled as black with temperature $T_3 = 25°C$. The plates are long in the direction perpendicular to the figure. Analyze the rectangular three-surface enclosure (T_1, T_2, T_3) formed between two consecutive plates, and calculate

(a) The net heat transfer rate from T_2 to T_3 through the plane labeled A_3 in the figure, and

(b) The temperature of the ceramic wall, T_1.

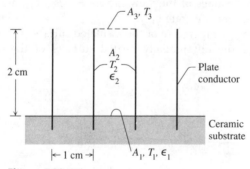

Figure P10.17

10.18   Energy is extracted from a geothermal reservoir by using the well shown in the figure. Cold water is forced to flow downward through the annular space of the coaxial cylinders, and it is heated by contact with the surrounding rock material (the temperature of this material increases almost linearly with depth). The hot water stream returns to the earth's surface through the inner pipe.

A critical component in the design of the well is the wall of the inner pipe. This

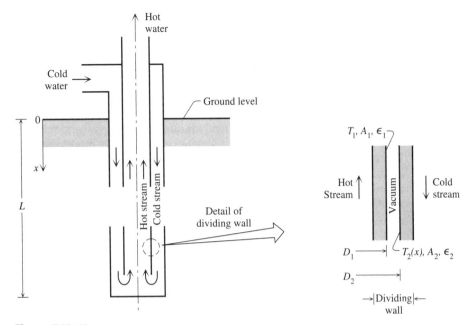

Figure P10.18

has to be a good insulator, to prevent the heat transfer that would take place in the radial direction, from the inner hot stream to the cold stream of the annular space.

A possible design is outlined on the right side of the figure. The dividing wall discussed until now will be made out of two concentric tubes, so that the narrow space created between them can be evacuated. This evacuated annular space has an inner surface of diameter $D_1 = 15$ cm and an outer surface of diameter $D_2 = 16$ cm. The total length (depth) of the well is $L = 1$ km.

Estimate the total heat transfer rate between the hot and cold water streams across the evacuated gap. Assume that this heat leakage is controlled (impeded) primarily by the radiation between the long cylindrical shells D_1 and D_2. The total hemispherical emissivities of these two surfaces are $\epsilon_1 \cong \epsilon_2 \cong 0.6$. The temperature of the inner surface (D_1) is controlled by the upflowing stream of pressurized hot water, $T_1 \cong 220°C$. The temperature of the outer surface of the evacuated gap is controlled by the downflowing cold stream and is given by the linear relation

$$T_2(x) = T_2(0) + \left(\frac{dT_2}{dx}\right)x$$

where $T_2(0)$ is the ground-level temperature, $T_2(0) \cong 20°C$, and $dT_2/dx \cong 200°C/km$.

10.19 The ice slab suspended vertically inside the cardboard box described in Project 7.1 is melted not only by natural convection but also by radiation. Model the ice slab and the cardboard surface that surrounds it as a two-surface enclosure in which the ice surface is much smaller than the cardboard surface. The ice slab has the height $H = 38.3$ cm, width $W = 25.5$ cm, thickness $L = 1$ cm, and temperature $T_w = 0°C$. The temperature of the cardboard surface is $T_\infty = 18°C$.

Calculate the net instantaneous radiation heat transfer rate from the cardboard enclosure to the ice surface. Calculate also the volumetric flowrate of meltwater that corresponds to this heating rate. Compare this last estimate with the meltwater flowrate due solely to laminar natural convection, the calculation of which is described in Project 7.1.

10.20   Hot ash from the boiler of a cogeneration plant is stored temporarily in a hopper located in a 5-m-deep pit in the floor of the plant. The hopper is long in the direction perpendicular to the plane of the figure. The distance from the side wall of the hopper to the vertical wall of the pit is 2.5 m. The temperature of the outer surface of the hopper is 300°C. This raises some questions concerning the safety of workers who may have to descend into the pit to unclog the flow of ash through the bottom end of the hopper.

A key safety parameter is the temperature of the pit wall and floor. Estimate this temperature by relying on the two dimensional enclosure model shown on the right side of the figure. The hopper surface is diffuse-gray, with $\epsilon_1 = 0.6$. The pit surface is adiabatic (reradiating). The ambient is represented by the black surface of temperature 40°C, which completes the enclosure. Assume that all the heat transfer is due to radiation.

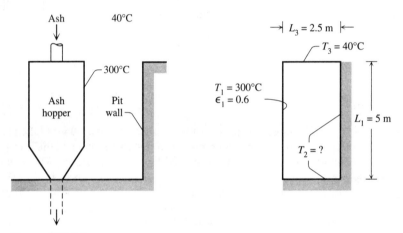

Figure P10.20

10.21   In the calculation of the pit wall temperature (T_2) in the preceding problem, you were advised to assume that all the heat transfer in the three-surface enclosure model is by radiation. The answer to that calculation turns out to be $T_2 = 232$°C.

Obtain a more realistic estimate for T_2 by taking into account the natural convection heat transfer from the pit wall to the air that fills the hopper–wall space. Assume that the heated air rises as a boundary layer along the vertical wall of the pit. Assume also that the air temperature midway between the hopper and the pit wall is 40°C, that is, equal to the ambient air that descends slowly into the 2.5-m-wide space (this ambient air replaces the heated air that rises along the pit wall). Finally, assume that the heat transfer coefficient calculated in this manner applies over the entire surface of the pit wall (5 m side and 2.5 m bottom).

10.22   Another aspect of the ash hopper analyzed in the preceding two problems is that it is designed not only to temporarily store the ash but also to cool it. Hot ash with a temperature of 900°C falls steadily into the hopper at the rate of 550 kg/h. The specific heat of the ash material is 1 kJ/kg·K. The freshly fallen ash is cooled to the new temperature T_1 as it rests on top of the ash pile.

(a) Estimate the steady-state temperature of the top of the ash pile, T_1, by analyzing the hopper model shown on the right side of the figure. The top of the ash pile is a diffuse-gray surface with $\epsilon_1 = 0.9$ and $A_1 = 11$ m². The wall of the upper portion of the hopper has the area $A_2 = 27$ m²; this is also modeled as diffuse-gray, with $\epsilon = 0.6$ on both sides. The 40°C environment is represented by

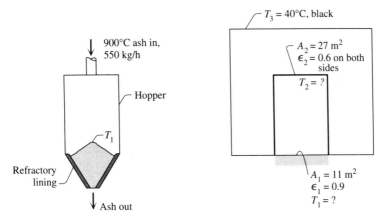

900°C ash in,
550 kg/h

Hopper

T_1

Refractory
lining

Ash out

$T_3 = 40°C$, black

$A_2 = 27$ m^2
$\epsilon_2 = 0.6$ on both
sides

$T_2 = ?$

$A_1 = 11$ m^2
$\epsilon_1 = 0.9$
$T_1 = ?$

Figure P10.22

the outermost surface, which is modeled as black. Assume that the heat transfer is due entirely to radiation.

(b) Calculate the total heat transfer rate released by the ash flow, that is, from T_1 to T_3 across the two-chamber enclosure model.

(c) Calculate the temperature of the hopper wall, T_2.

10.23 The intensity of the solar irradiation that strikes the earth is $I = 1360$ W/m^2. The earth's surface absorptivity for solar radiation is $\alpha = 0.7$. At the same time, the earth loses heat by radiation to the universe of temperature $T_\infty \cong 4$ K. The earth's emissivity is $\epsilon(T) = 0.95$, where T is the earth's average temperature (T is averaged annually and spatially).

(a) Assume that the atmosphere is absent, and calculate the earth's average temperature T.

(b) Compare the calculated T value with the actual average temperature of the earth (15°C). Comment on the difference and how that difference may be due to the role played by the atmosphere [18].

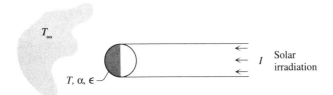

T_∞

T, α, ϵ

I Solar irradiation

Figure P10.23

Participating Media

10.24 A spherical vessel with a 1-m diameter and 200°C wall temperature contains a gas at a pressure of 2 atm and temperature of 1000°C. The gas composition on a molar basis is 50% N_2, 30% CO_2, and 20% O_2. The interior surface of the vessel is black. Calculate

(a) The gas emissivity,

(b) The gas absorptivity with respect to the radiation emitted by the surface, and

(c) The instantaneous net heat transfer rate from the gas to the wall of the container.

10.25 The equilibrium mixture of H_2, O_2, and H_2O at 1 atm and 3000 K has the following composition in terms of mole fractions (Ref. [4], pp. 358–359):

$$x_{H_2} = 0.137 \qquad x_{O_2} = 0.069 \qquad x_{H_2O} = 0.794$$

The mixture is placed instantly in a long cylinder with a 0.4-m internal diameter. The internal surface of the cylinder may be modeled as black. Estimate the instantaneous cooling rate that must be supplied to the external surface of the cylinder to maintain the wall temperature at 600 K.

10.26 The adiabatic burning of methane with theoretical air generates the ideal gas mixture of products $CO_2 + 2H_2O + 7.52N_2$, in which the numerical coefficients represent moles per one mole of CO_2. The mixture of products exits the combustion chamber at 1 atm and at the adiabatic flame temperature of 2320 K [19]. It passes through a cylindrical duct with a 0.2-m internal diameter and a wall temperature of 500 K. Its flowrate is high enough so that mixture temperature is fairly constant along the duct, this in spite of the radiation cooling effect provided by the duct wall. The latter may be modeled as black. Calculate the heat transfer rate removed from the products, per unit axial length. The emission from the duct wall may be neglected.

10.27 An industrial furnace that burns methane with 400 percent theoretical air exhausts a gas stream with the following composition: $CO_2 + 2H_2O + 6O_2 + 30.08N_2$. The numerical coefficients represent moles per one mole of CO_2. This mixture of products of combustion is at 700 K and 1 atm and occupies the space between two parallel walls. The distance between the walls is 0.5 m, and the area of one wall is 20 m². Both surfaces are at a temperature of 460 K and may be modeled as black. Calculate the net radiation heat transfer rate received by both surfaces. Note that the gas and wall temperatures are comparable; therefore, the radiation emitted by the walls should not be neglected.

10.28 The gap between two large parallel surfaces at different temperatures (T_1, T_2) is occupied by an isothermal gray gas $(T_g, \epsilon_g, \tau_g)$. Each surface has an area A and is diffuse-gray. Their respective emissivities are equal to the emissivity of the gas; their common value happens to be 0.5.

(a) Derive an expression for the net rate of heat transfer between the two surfaces.

(b) Compare your result with the heat transfer rate that occurs when the space between the plates is completely evacuated. Show in this way that the gas has a radiation shielding effect.

10.29 The 10-cm-wide space between the large parallel plates shown in the figure is filled with steam at atmospheric pressure. The steam temperature is uniform at a level that could be determined by analyzing the radiation heat transfer in the enclosure formed by the two plates (the method is outlined in the next problem). Model the steam as a gray gas, and determine its absorptivity α_g by assuming that T_g is halfway between T_1 and T_2. Evaluate the goodness of this model by comparing α_g with (1) the absorptivity of steam with respect to radiation arriving from T_1, and (2) the absorptivity of steam with respect to radiation arriving from T_2.

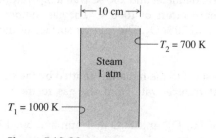

Figure P10.29

10.30 An isothermal layer of gray gas with the emissivity $\epsilon_g = 0.3$ is sandwiched between two large and parallel walls. The internal surfaces of these walls can be modeled as diffuse-gray, with temperatures and emissivities listed on the figure.
  (a) Calculate the net radiation heat flux from T_1 to T_2.
  (b) Determine the gas temperature T_g, and show in this way that T_g is closer to the higher of the two side-wall temperatures.

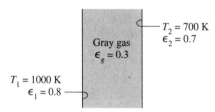

Figure P10.30

10.31 The heat transfer process that goes on in a gas furnace can be modeled as shown in the attached figure. The surface that is to be heated is cold and diffuse-gray (A_1, T_1, ϵ_1). The surrounding surface (A_2) is refractory, that is, adiabatic on its back side. The space between the two surfaces is the furnace volume, which is filled by the hot (participating) gas that acts as heat source. The gas is modeled as isothermal and gray (T_g, $\alpha_g = \epsilon_g$).
  Draw the radiation network for this furnace model, and derive an expression for the net instantaneous heat transfer rate from the gas to the cold surface, q_{g-1}.

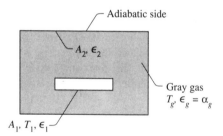

Figure P10.31

10.32 The space between the infinite parallel plates shown on the left side of the figure is filled with a gray gas, while the internal parallel surfaces are diffuse-gray. The net heat transfer originates from the gas, which is hot (T_g), and sinks into one of the surfaces, which is cold (T_1). The other surface is refractory, that is, insulated on its back side. The gas emissivity and the emissivity of the cold surface are equal: their common value is 0.5.

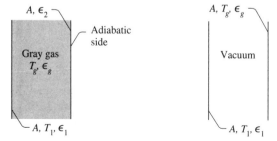

Figure P10.32

(a) Derive an expression for the net instantaneous heat transfer rate from the gas to the cold surface, q_{g-1},

(b) Consider the alternative shown on the right side of the figure, where the space has been evacuated and the right wall has assumed the temperature and emissivity of the gas. The transfer of heat is again from T_g to T_1, through vacuum this time. Derive an expression for the net heat transfer rate q_{g-1}, and compare it with the one derived in part (a). In this way you will be able to answer the question of when the heat transfer rate is higher, when the heat source fills the entire volume [gas, part (a)], or when the heat source is flattened into a sheet [part (b)].

PROJECTS

10.1 Extraterrestrial Power Plant Driven by Solar Radiation

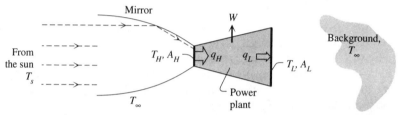

Figure PR10.1

(a) The figure shows the main components of an extraterrestrial heat engine powered by solar radiation and cooled by the cold background provided by the universe [4]. The collector A_H receives the net heat transfer q_H from the sun (T_S) via an ideally concentrating mirror system: the purpose of this system is to make sure that the only object that can be seen from A_H is the solar surface. In other words, there is no leakage of heat transfer from the collector A_H to the cold background (T_∞).

The radiator of area A_L rejects the net heat transfer rate q_L to the cold background. All the surfaces (sun, A_H, A_L, background) can be modeled as black. Economics places a limit on how much "surface" can be invested in this power plant, namely,

$$A_H + A_L = A$$

where A is fixed. The internal operation of the heat engine itself (the system between A_H and A_L) can be modeled as reversible.

As a designer of this power plant, you must find the best way of dividing (allocating) the area inventory A between A_H and A_L. Your objective is to maximize the instantaneous power output $\dot{W}$ subject to fixed A, T_S, and T_∞. In your analysis, exploit the fact that T_∞^4 is negligibly small compared with T_L^4. Show that the optimum design is characterized by

$$\frac{A_H}{A} = 0.35 \qquad \left(\frac{T_S}{T_H}\right)^4 = 1.538 \qquad \dot{W} = 0.0414\,\sigma A T_S^4$$

(b) Numerically, set $T_S = 5762$ K for the blackbody temperature of the solar surface, and comment on the feasibility of the optimum temperature levels that are recommended by your design for T_H and T_L. A more realistic model [20] for the hot end of the power plant is one where the mirror is not an ideal concentrator, that is, where the cold background can be seen from A_H. In this new model the solar surface (T_S), the collector (T_H) and the background (T_∞) form a three-surface enclosure in which all the surfaces are black. Let F_{HS} be the view factor associated with looking from A_H to the sun.

Consider again the $\dot{W}$ maximization problem described in part (a), and show that the only change in the solution listed above is the replacement of T_S^4 with $F_{HS}T_S^4$ [21]. Comment on the effect of the geometric factor F_{HS} (smaller than 1) on the optimum numerical values recommended for T_H and T_L.

10.2   Ash Hopper with Shield Cooled by Convection

The steady-flow hot ash hopper described in Problem 10.22 has to be redesigned because the temperature of its outer wall is too high. This high temperature poses a threat to the workers who must get close to unplug the flow of ash through the bottom end of the hopper.

It is proposed to surround the hopper wall A_2 with a metallic shield A_3. Fans will circulate air on the outside of the A_3 surface and through the space left between A_3 and A_2. The circulation is sufficiently intense so that the bulk temperature of the air stream does not rise significantly above its inlet temperature of $40°C$. Assume that each of the air-cooled surfaces is characterized by the same forced-convection heat transfer coefficient h_{FC}:

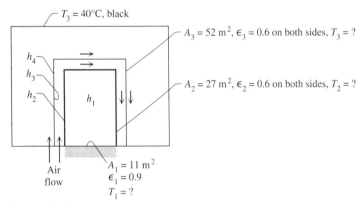

Figure PR10.2

$$h_2 = h_3 = h_4 = h_{FC}$$

Separate calculations of the natural convection effect of the gases trapped above the ash pile (between A_1 and A_2) indicated that the overall heat transfer coefficient between T_1 and T_2 is $h_1 = 5$ W/m²·K. This h_1 value is based on the area A_2 and remains fixed throughout the calculations that are proposed below.

As shown on the left side of the sketch that accompanies Problem 10.22, the high temperature (T_1) of the ash pile is maintained by a steady inflow of $900°C$ ash, with a flowrate of 550 kg/h. The specific heat of this ash material is 1 kJ/kg·K.

The shield temperature T_3 is considered safe when it does not exceed $70°C$. How high must the h_{FC} value be to meet this design objective? The h_{FC} value determined in this manner will be needed later for the design of the air-cooling system (fans, A_2–A_3 spacing, etc.).

10.3   Combined Radiation and Natural Convection in an Enclosure Heated from the Side

The heat transfer process in a room is a combination of radiation and convection (natural convection, if forced air circulation is absent). This project can be used to develop an understanding of the proportions in which radiation and natural con-

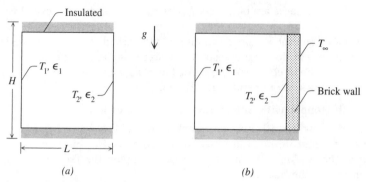

Figure PR10.3

vection participate in a room with square vertical cross section, $H = L = 3$ m. The room is long in the direction perpendicular to the $H \times L$ area.

(a) Consider first the room model shown on the left. The inner vertical surfaces are diffuse-gray with $\epsilon_1 = \epsilon_2 = 0.9$. They are also isothermal, and their temperatures are known, $T_1 = 30°C$ and $T_2 = 10°C$. The top and bottom surfaces are insulated. Calculate the total heat transfer rate from T_1 to T_2, and show that 70 percent of this is due to radiation.

(b) In the more realistic model sketched on the right [22], the cold wall is 30 cm thick and made out of common brick. This time the temperature of the outer surface of the cold wall is specified, $T_\infty = 10°C$. It is assumed that the temperature of the inner surface (T_2) is uniform over the height H. Calculate T_2, and the total heat transfer rate across the system (from T_1 to T_∞). By comparing these results with the results of part (a), show that the brick wall performs a significant insulation function.

10.4 Combined Radiation and Conduction: Optimal Density (Packing) of Fibrous Insulation

The heat transfer through a layer of fibrous insulation is a complex mix of fiber-to-fiber radiation and parallel conduction through the air trapped between the fibers. This is the case when the spaces between fibers are small enough so that the air motion is slow and the convection effect negligible.

An interesting characteristic of such layers is that their thermal insulation effect is the greatest when the packing (fiber density, or layer porosity) has a certain value. The porosity is defined as the ratio

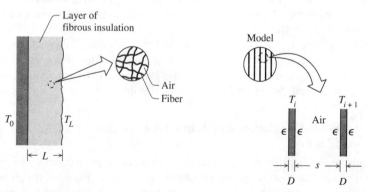

Figure PR10.4

$$\varphi = \frac{\text{air space}}{\text{total (air + fiber) space}}$$

Animal fur is a good example of a fibrous layer permeated by air. An even more intriguing characteristic of animal fur is that the observed porosity falls in the narrow range $\varphi \cong 0.95 - 0.99$, that is, in a way that seems to be independent of the animal body size or hair strand thickness.

To develop an understanding of the basis for the optimal fiber packing, consider the simple two-dimensional model shown in the figure [23]. The fibers are accounted for by a stack of n equidistant shields of thickness equal to the fiber thickness D. The layer thickness L and the two temperatures separated by the insulating layer (T_0, T_L) are given. The overall temperature difference is considerably smaller than the local absolute temperature, $T_0 - T_L \ll T_0$. Each surface is diffuse-gray, having the same total hemispherical emissivity ϵ. The air thermal conductivity is k.

(a) Derive an expression for the heat transfer rate across the layer, q, and show that q is minimized when the fiber-to-fiber spacing reaches the optimal value

$$s_{\text{opt}} = D\, B^{-1/2}, \quad \text{where } B = \frac{4\sigma T_0^3 D}{k\left(\dfrac{2}{\epsilon} - 1\right)}$$

(b) Consider the fiber thickness range $D \sim 10\ \mu m - 100\ \mu m$, which corresponds to the observed thickness of hair strands of animals of various body sizes (20 cm $-$ 3 m) [24]. Show that the optimal porosity that corresponds to s_{opt} agrees approximately with the observed porosities of animal fur.

10.5 Combined Radiation and Conduction: Critical Insulation Radius

A bare cylinder of length L, outer radius r_i, and wall temperature T_i is to be wrapped with a layer of insulation of conductivity k and outer radius r_o. The outer surface of the insulating layer is diffuse-gray, and its emissivity is ϵ. The wrapped cylinder is situated in an enclosure that has the wall temperature T_∞. The heat transfer from the cylinder wrapping to the enclosure wall is by radiation.

(a) Determine under what conditions the wrapping of the cylinder has an insulating effect. As in Section 2.4, determine the critical insulation radius by maximizing the heat transfer rate q from T_i to T_∞. Since the solution requires a numerical trial-and-error procedure, it is advisable to work with the dimensionless quantities

$$Q = \frac{q}{2\pi k L (T_i - T_\infty)} \qquad B = \frac{\epsilon \sigma r_i}{k} T_i^3$$

$$\rho = \frac{r_o}{r_i} \qquad \theta = \frac{T_o}{T_i}$$

where T_o is the temperature at the outer surface of the insulation. Assume that the enclosure temperature is low enough so that T_∞^4 is negligible when compared with T_o^4.

(b) Consider an application in which $r_i = 10$ cm, $T_i = 500$ K, $\epsilon = 0.2$, and $k = 0.02$ W/m·K. Will the wrapping of the cylinder have an insulating effect?

11

MASS TRANSFER PRINCIPLES

11.1 THE ANALOGY BETWEEN MASS TRANSFER AND HEAT TRANSFER

In this final chapter we turn our attention to mass transfer, or the transport of a certain chemical species through a medium in which the species acts as a component (contaminant). A general feature of the mass transfer problems that we shall consider is the fact that the chemical species of interest is distributed nonuniformly through the medium. The species nonuniformity will be described in terms of differences and gradients of species concentration, where the mass concentration ρ_i represents the number of kilograms of species i per unit volume.

Mass transfer processes abound in the world around us. The most familiar of these processes is the drying of the wet surface of a plane wall that is exposed to the wind. The air lamina that comes in contact with the surface of the water layer becomes saturated with water vapor. The latter is the species of interest, on the background of (i.e., as a component in) the ideal gas mixture represented by the humid air. In engineering thermodynmics terms, the relative humidity of the near-surface air becomes 100 percent, which means that the water vapor pressure of the humid air mixture equals the saturation pressure of water at the temperature of the wet surface. (Review the nomenclature of mixtures consisting of dry air and water vapor, e.g., Ref. [1] and the end of Appendix D.)

The mass concentration of water vapor in near-surface air, $\rho_{w,0}$, is generally different than its concentration in the free stream, $\rho_{w,\infty}$. The concentration difference $\rho_{w,0} - \rho_{w,\infty} > 0$ has the ability to drive more water vapor out of the surface water layer. This evaporation process will be enhanced in proportion with the longitudinal speed with which the free stream sweeps the wet surface, Fig. 11.1. The effect of the longitudinal air flow is to sharpen the normal concentration gradient $(\partial \rho_w / \partial y)_{y=0}$ precisely in the same way that the free stream U_∞ had the ability to sharpen the temperature gradient at the wall in forced-convection heat transfer problems (e.g., Fig. 5.7).

Even simpler mass transfer problems are found in those systems that do not "flow" along the mass transfer surface. If, as shown in Fig. 11.2, the space above the pool of liquid i is enclosed in a tall cylindrical column with impermeable lateral walls, and open at the top, the vapor i will diffuse toward the open

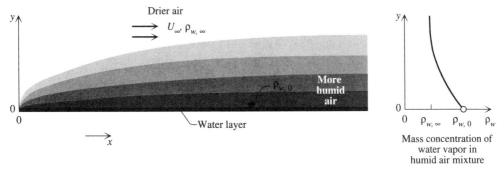

Figure 11.1 The forced-convection concentration boundary layer near a wet surface swept by air.

end through the stationary air. The entire column is isothermal. The vapor is driven in this direction by the gradient of the concentration of i in the i + air mixture, which is maintained from one end by the water pool and from the other end by contact with the rest of the atmosphere.

In drawing Fig. 11.2 we made no assumption about the orientation of the column with respect to gravity. For example, if gravity points downward in the figure, the diffusion process described in the preceding paragraph occurs only when the substance i is such that the density of the i + air mixture increases when the concentration of i increases. In other words, the mass transfer through the vertical column is by pure diffusion in one direction when the mixture is stably stratified, with the denser layers situated at lower levels.

It is worth mentioning that a mixture of water vapor and air (i.e., humid air) behaves in the opposite way, in the sense that humid air is less dense than dry air. Therefore, if i is water and g points downward in Fig. 11.2, and if the column is wide enough, humid air will rise by *natural convection* through the center of the column, while dry air takes its place by descending along the walls [2]. On the other hand, we can imagine that the column is rotated by 180°C (turned upside down), while its ceiling is wetted by a water film. In this case the humid air mixture is stably stratified, and the transport of i through the column is by one-dimensional diffusion, as in Fig. 11.2. This configuration is discussed further in connection with Fig. 11.4.

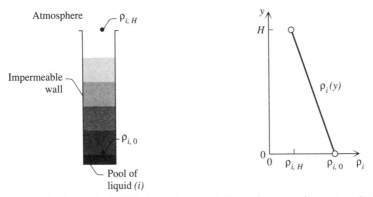

Figure 11.2 Vertical diffusion of vapor i through a one-dimensional air column above a pool of liquid i.

These introductory examples already illustrate the analogy that exists between mass transfer and heat transfer phenomena. The parallel flow drying of the wet surface in Fig. 11.1 is similar to the convective heat transfer effected by the boundary layer near a plane surface. The unidirectional diffusion of vapor i in Fig. 11.2 is in a way similar to the conduction of heat across a plane wall (e.g., Fig. 2.1, right).* It seems that the heat transfer role of temperature differences and gradients is played in the realm of mass transfer by concentration differences and gradients. Furthermore, instead of the heat flux that formed the subject of so many of our calculations in the heat transfer part of this course, in mass transfer we are concerned with the *mass flux*, or the mass flowrate of a certain species, per unit area.

The analogy between mass transfer and heat transfer (conduction and convection) is the reason we focus on mass transfer in this concluding chapter of the course. In this way, the study of mass transfer gives us one more opportunity to review the skeleton of the conduction and convection heat transfer material covered until now. In what follows, we place special emphasis on those features in which mass transfer processes differ from their heat transfer counterparts.

11.2   THE CONSERVATION OF CHEMICAL SPECIES

11.2.1   Species Velocity Versus Bulk Velocity

The conservation principle that governs the migration of a chemical species through a gaseous, liquid, or solid medium is the principle of mass conservation applied to the chemical species alone. In this section we derive the species conservation equation for the general case where the bulk of the medium moves relative to the boundaries. This general form of the species conservation equation will find direct application in problems of mass transfer by convection, as illustrated in Section 11.4.

The species conservation equation for a stationary medium can be written down as a special, much simpler case of the equation for convective mass transfer. This move consists of setting the bulk velocity components of the medium equal to zero. This simpler class of problems will be analyzed under the heading of pure diffusion mass transfer in Section 11.3.

Consider the flow of the chemical species i through the two-dimensional medium suggested by Fig. 11.3. The mass concentration of this species, $\rho_i(x,y,t)$, is another name for its partial density, that is, the number of kilograms of i per unit volume. Viewing the medium as a mixture of n species (components), the density of the medium, ρ, is clearly the sum of the individual species densities (mass concentrations):

$$\rho = \sum_{i=1}^{n} \rho_i \tag{11.1}$$

With reference to the $\Delta x \, \Delta y$ control volume of Fig. 11.3, the species conservation principle is the statement that the net flow of species i into the control volume must equal the rate of i accumulation inside the control volume:

---

*This analogy holds true only when the bulk motion of the i + air mixture in Fig. 11.2 is negligible. This bulk motion is caused by the migration of vapor i toward the open end.

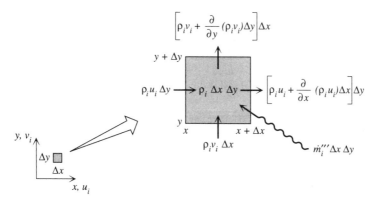

Figure 11.3 The conservation of the chemical species i in a two-dimensional control volume.

$$\frac{\partial \rho_i}{\partial t}\Delta x \Delta y = \rho_i u_i \Delta y - \left[\rho_i u_i + \frac{\partial}{\partial x}(\rho_i u_i)\Delta x + \cdots\right]\Delta y$$

$$+ \rho_i v_i \Delta x - \left[\rho_i v_i + \frac{\partial}{\partial y}(\rho_i v_i)\Delta y + \cdots\right]\Delta x + \dot{m}_i'' \Delta x \Delta y \qquad (11.2)$$

The velocity components u_i, v_i refer strictly to the motion of species i relative to the control volume. These velocity components are to be distinguished from the velocity components u, v that describe the bulk motion of the medium through the same control volume. We return in eqs. (11.5) to the distinction between u_i, v_i and u, v.

Worth noting is that the product $\rho_i u_i$ represents the mass flowrate of species i in the x direction, per unit area normal to the x direction. The term $\dot{m}_i'''$ on the right side of eq. (11.2) is the volumetric rate of the creation of species i. This term accounts for the effect of a chemical reaction in which the species i is, locally, a reaction product [3]. The species generation rate $\dot{m}_i'''$ is negative in cases where the species i is consumed during the reaction.

Dividing by $\Delta x \Delta y$ and invoking the limit* $(\Delta x, \Delta y) \to 0$, we can reduce the species conservation statement (11.2) to

$$\frac{\partial \rho_i}{\partial t} + \frac{\partial}{\partial x}(\rho_i u_i) + \frac{\partial}{\partial y}(\rho_i v_i) = \dot{m}_i''' \qquad (11.3)$$

Note that in the absence of chemical reactions ($\dot{m}_i''' = 0$) this equation has the same form as the mass conservation statement for the bulk flow of the medium, eq. (5.5). In fact, eq. (5.5) can be rederived based on the analysis of Fig. 11.3, by writing one equation of type (11.3) for each of the n species, adding the n equations term by term, and invoking eq. (11.1). Note also that the sum of all the $\dot{m}_i'''$ terms on the right-hand side is zero, as mass cannot be created even when chemical reactions are present:

---

*The Taylor expansions enclosed in square brackets in eq. (11.2) have been truncated to the first two terms already, based on the assumption that Δx and Δy are sufficiently small. The higher-order terms that were omitted in the writing of eq. (11.2) drop out in the limit $(\Delta x, \Delta y) \to 0$.

$$\frac{\partial}{\partial t}\left(\sum_{i=1}^{n}\rho_i\right) + \frac{\partial}{\partial x}\left(\sum_{i=1}^{n}\rho_i u_i\right) + \frac{\partial}{\partial y}\left(\sum_{i=1}^{n}\rho_i v_i\right) = 0 \qquad (11.4)$$

$$\frac{\partial \rho}{\partial t} + \frac{\partial}{\partial x}(\rho u) + \frac{\partial}{\partial y}(\rho v) = 0 \qquad (5.5)$$

This term-by-term comparison of eqs. (11.4) and (5.5) reveals a key concept in mass transfer, namely, the *bulk* or *mass-averaged velocity* components of the medium:

$$u = \frac{1}{\rho}\sum_{i=1}^{n}\rho_i u_i \qquad v = \frac{1}{\rho}\sum_{i=1}^{n}\rho_i v_i \qquad (11.5)$$

The u and v components have been used extensively in the convection part of this course, where they were viewed simply as the velocity components of the fluid medium. In mass transfer, u and v emerge as averages of the respective components of all the species in the mixture that constitutes the flowing medium.

That the bulk velocity component differs from the species velocity component can be seen by considering the experiment illustrated in Fig. 11.4. At time $t = 0$, a vertical column of dry air a is placed under a column of air saturated with water vapor v. In time, water vapor diffuses downward into the relatively dry (air + water vapor) mixture, while dry air diffuses upward into the relatively humid mixture. In every horizontal cross section, the vertical velocity of each species is *finite*, one positive ($v_a > 0$) and the other negative ($v_v < 0$). The bulk velocity of the mixture, however, is *zero*, because both ends of the column are sealed. With reference to the notation employed in the second of eqs. (11.5), we conclude that the species component v_i is, in general, different than the bulk component v. In other words, the species can move relative to the medium (mixture) as a whole.

11.2.2 Diffusion Mass Flux

The velocity difference $v_i - v$ is recognized as the *diffusion velocity* of species i in the y direction. The group $\rho_i(v_i - v)$ represents the per-unit-area flowrate of species i in the y direction and *relative* to the bulk motion of the medium. A

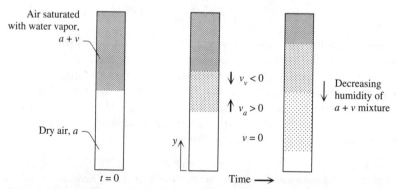

Figure 11.4 Column with both ends closed: downward diffusion of water vapor and upward diffusion of dry air in a stationary medium ($v = 0$).

shorter name for this group is the *diffusion mass flux* $j_{y,i}$, the units of which are kilograms of i per second and unit area. Combining the diffusion mass flux definitions

$$j_{x,i} = \rho_i(u_i - u) \tag{11.6}$$

$$j_{y,i} = \rho_i(v_i - v) \tag{11.7}$$

with the statement that the species i is conserved, eq. (11.3), we obtain

$$\frac{\partial \rho_i}{\partial t} + \frac{\partial}{\partial x}(\rho_i u) + \frac{\partial}{\partial y}(\rho_i v) = -\frac{\partial j_{x,i}}{\partial x} - \frac{\partial j_{y,i}}{\partial y} + \dot{m}_i'' \tag{11.8}$$

This equation can be modified further by expanding the spatial derivatives on the left side and recognizing that, for a constant-ρ mixture flow, we can invoke eq. (5.6):

$$\frac{\partial \rho_i}{\partial t} + u\frac{\partial \rho_i}{\partial x} + v\frac{\partial \rho_i}{\partial y} = -\frac{\partial j_{x,i}}{\partial x} - \frac{\partial j_{y,i}}{\partial y} + \dot{m}_i''' \tag{11.9}$$

11.2.3 Fick's Law

On the right side of eq. (11.9), we note the terms containing the diffusion mass flux components $j_{x,i}$ and $j_{y,i}$. Following Adolph Fick's hypothesis [4], we write that these fluxes are proportional to the corresponding concentration gradients in the same manner that the conduction heat flux is driven by the temperature gradient [review the Fourier law, eqs. (1.36)]. In a *binary mixture*, that is, in a medium that contains only two species, the species of interest ($i = 1$) and the rest of the mixture ($i = 2$), *Fick's law* of mass diffusion proclaims the proportionality between the mass flux and the corresponding mass concentration gradient:

$$j_{x,1} = -D_{12}\frac{\partial \rho_1}{\partial x} \tag{11.10}$$

$$j_{y,1} = -D_{12}\frac{\partial \rho_1}{\partial y} \tag{11.11}$$

The proportionality constant D_{12} is the *mass diffusivity** of species 1 into species 2. Examples of mass diffusivity values can be examined later in Tables 11.2 through 11.4. It is easy to verify that the units of D_{12} are m^2/s, which also happen to be the units of the thermal diffusivity α and the momentum diffusivity (kinematic viscosity) v. The diffusion flux is driven solely by the concentration gradient strictly in an isothermal and isobaric medium (see an introduction to Irreversible Thermodynamics, for example, in Ref. [1], pp. 706–712). Nevertheless, Fick's law, eqs. (11.10) and (11.11), is a good engineering approximation in many nonisothermal systems, where the concentration gradients are superimposed on temperature gradients.

Combining the species conservation argument with Fick's law, we reach the terminal point of this derivation, namely, the so-called *concentration equation*:

---

*This factor is also known as the *diffusion coefficient*, or the *diffusion constant*.

$$\frac{\partial \rho_i}{\partial t} + u\frac{\partial \rho_i}{\partial x} + v\frac{\partial \rho_i}{\partial y} = D\left(\frac{\partial^2 \rho_i}{\partial x^2} + \frac{\partial^2 \rho_i}{\partial y^2}\right) + \dot{m}''' \tag{11.12}$$

For simplicity in notation, the subscript of the mass diffusivity coefficient D has been omitted. The mass diffusivity D refers to the species of interest, as it diffuses through the rest of the medium. In the step from eq. (11.9) to (11.12), the D coefficient has been treated as a constant.

In summary, the concentration eq. (11.12) governs the distribution of the species of interest through the constant-properties medium. This distribution is described by the function $\rho_i(x,y,t)$. It is educational to compare the mass concentration equation with the energy equation for a two-dimensional medium with constant properties and negligible frictional heating [cf. eq. (5.20)]:

$$\frac{\partial T}{\partial t} + u\frac{\partial T}{\partial x} + v\frac{\partial T}{\partial y} = \alpha\left(\frac{\partial^2 T}{\partial x^2} + \frac{\partial^2 T}{\partial y^2}\right) + \frac{\dot{q}}{\rho c_P} \tag{11.13}$$

Equations (11.12) and (11.13) reveal again the analogy between mass transfer and heat transfer. In particular, the concentration field of a mass transfer problem without chemical reaction ($\dot{m}''' = 0$) is governed by an equation similar to that for the temperature distribution in a heat transfer problem without internal heating effect ($\dot{q} = 0$).

The corresponding forms of the mass concentration equation in three-dimensional Cartesian, cylindrical, and spherical coordinates are listed in Table 11.1. Worth keeping in mind is that in these equations ρ and D are being treated as constants, and that the volumetric chemical reaction effect is assumed absent. The mass concentration equation for a *stationary medium* can be read off Table 11.1 by setting all the bulk velocity components equal to zero.

11.2.4 Molar Concentration and Molar Flux

In some applications, especially for mass transfer through gaseous mixtures, it is customary to discuss the composition of the mixture in terms of *mole fractions* x_i,

Table 11.1 The Conservation of the Species with Mass Concentration ρ_i Through a Medium with Nearly Constant Properties (ρ, D) and Without Chemical Reaction

Cartesian coordinates (x,y,z: Table 5.1):

$$\frac{\partial \rho_i}{\partial t} + u\frac{\partial \rho_i}{\partial x} + v\frac{\partial \rho_i}{\partial y} + w\frac{\partial \rho_i}{\partial z} = D\left(\frac{\partial^2 \rho_i}{\partial x^2} + \frac{\partial^2 \rho_i}{\partial y^2} + \frac{\partial^2 \rho_i}{\partial z^2}\right)$$

Cylindrical coordinates (r,θ,z; Table 5.2):

$$\frac{\partial \rho_i}{\partial t} + v_r\frac{\partial \rho_i}{\partial r} + \frac{v_\theta}{r}\frac{\partial \rho_i}{\partial \theta} + v_z\frac{\partial \rho_i}{\partial z} = D\left[\frac{1}{r}\frac{\partial}{\partial r}\left(r\frac{\partial \rho_i}{\partial r}\right) + \frac{1}{r^2}\frac{\partial^2 \rho_i}{\partial \theta^2} + \frac{\partial^2 \rho_i}{\partial z^2}\right]$$

Spherical coordinates (r,ϕ,θ; Table 5.3):

$$\frac{\partial \rho_i}{\partial t} + v_r\frac{\partial \rho_i}{\partial r} + \frac{v_\phi}{r}\frac{\partial \rho_i}{\partial \phi} + \frac{v_\theta}{r \sin\phi}\frac{\partial \rho_i}{\partial \theta} =$$

$$= D\left[\frac{1}{r^2}\frac{\partial}{\partial r}\left(r^2\frac{\partial \rho_i}{\partial r}\right) + \frac{1}{r^2 \sin\phi}\frac{\partial}{\partial \phi}\left(\sin\phi\frac{\partial \rho_i}{\partial \phi}\right) + \frac{1}{r^2 \sin^2\phi}\frac{\partial^2 \rho_i}{\partial \theta^2}\right]$$

not mass concentrations ρ_i. The mole fraction of species i in a mixture sample containing N moles is defined as

$$x_i = \frac{N_i}{N} \tag{11.14}$$

where N_i is the number of moles of i in that sample. (Review the thermodynamics of nonreacting mixtures, e.g., Ref. [1], pp. 195–202.) Since the moles of all the mixture components add up to the total number of moles of the mixture, $N_1 + N_2 + \cdots N_n = N$, it is clear that all the mole fractions x_i add up to 1:

$$\sum_{i=1}^{n} x_i = 1 \tag{11.15}$$

An important observation is that under certain conditions that are discussed below, the mole fraction x_i is proportional to the mass concentration of the same species, ρ_i. To see this proportionality, consider the fact that

$$\rho_i = \frac{m_i}{V} \tag{11.16}$$

where V is the volume of the mixture sample and m_i is the mass of species i found in that sample. The number of moles of i is also proportional to m_i:

$$N_i = \frac{m_i}{M_i} \tag{11.17}$$

where we recognize M_i as the molar mass* of the species (the number of kilograms of i per kilomole of i). The special form taken by eq. (11.17) in the case of the entire mixture that occupies the sample is

$$N = \frac{m}{M} \tag{11.18}$$

In this last equation, m is the total mass of the mixture sample, and M is the molar mass of the mixture:

$$M = \sum_{i=1}^{n} x_i M_i \tag{11.19}$$

Combining eqs. (11.14), (11.17), and (11.18), and noting the definition of the mixture density $\rho = m/V$, it is easy to show that

$$x_i = \left(\frac{M}{\rho M_i} \right) \rho_i \tag{11.20}$$

In this relationship the molar mass of the mixture, M, depends on the local composition. When the changes in composition from one point to the next are sufficiently small, or when the M_i values of the individual components are not too dissimilar, the M value is practically constant through the mixture. Under these conditions, x_i and ρ_i are proportional, and all the species conservation equations of Table 11.1 can be rewritten in terms of mole fraction, by simply replacing the mass concentration of the species of interest (ρ_i) with the mole

---

*The molar mass M_i is also called *molecular weight* in some thermodynamics texts.

fraction of the same species (x_i). For example, the mole fraction distribution during time-dependent diffusion through a stationary medium in Cartesian coordinates is described by

$$\frac{\partial x_i}{\partial t} = D\left(\frac{\partial^2 x_i}{\partial x^2} + \frac{\partial^2 x_i}{\partial y^2} + \frac{\partial^2 x_i}{\partial z^2}\right) \tag{11.21}$$

The species conservation equation can be written also in terms of the *molar concentration* C_i, which represents the number of moles of i per unit volume in the mixture sample:

$$C_i = \frac{N_i}{V} \tag{11.22}$$

The proportionality between molar concentration and mole fraction is expressed by

$$C_i = \frac{\rho}{M} x_i \tag{11.23}$$

Therefore, the species conservation eq. (11.21) can be rewritten by replacing x_i with C_i. Similarly, in all the equations listed in Table 11.1, the mass concentration ρ_i can be replaced with the molar concentration C_i. It is important not to confuse ρ_i with C_i: the relationship between the two is

$$\rho_i = M_i C_i \tag{11.24}$$

Another alternative would be to describe the local composition in the mixture in terms of the *mass fraction*, which is defined as the dimensionless ratio ρ_i/ρ and labeled Φ_i. This alternative is not used in the present treatment.

Consistent with the molar restatement of the species conservation principle is the use of the *diffusion molar flux* $\hat{j}_i$, which represents the number of moles of i that flow per unit time and unit area, relative to the mixture as a whole. Instead of eq. (11.10), for example, Fick's law for the x direction proclaims the proportionality between the diffusion molar flux and the corresponding gradient of molar concentration,

$$\hat{j}_{x,i} = -D\frac{\partial C_i}{\partial x} \tag{11.25}$$

The mole fraction formulation presented above applies to any single-phase mixture, gaseous, liquid, or solid. In the special case of an *ideal gas mixture*, the mole fractions are also proportional to the respective *partial pressures* P_i of the ideal gas components:

$$x_i = \frac{P_i}{P} \tag{11.26}$$

If the mixture pressure P is uniform, the partial pressure P_i can be used as a variable in the species conservation equation, that is, by replacing ρ_i with P_i throughout each of the equations listed in Table 11.1

11.3 DIFFUSION THROUGH A STATIONARY MEDIUM

11.3.1 Steady Diffusion

The simplest class of mass transfer problems is that in which the mixture is stationary, and enough time has passed since the imposition of boundary conditions so that the concentration distribution of the diffusing species is time independent. These phenomena of *steady diffusion* are, analytically, analogous to the steady conduction processes studied in Chapters 2 and 3.

Consider the one-dimensional medium of thickness L shown on the left side of Fig. 11.5. The diffusion of a certain species across this layer is described by the steady, one-dimensional version of eq. (11.21),

$$\frac{d^2 x_i}{dy^2} = 0 \tag{11.27}$$

in which x_i is the mole fraction of that species. Alternative forms of this equation could be written by replacing x_i with ρ_i or C_i, as demonstrated in the preceding section. If we assume for the moment that the species concentrations (or mole fractions) are known at the two boundaries,

$$
\begin{aligned}
x_i &= x_0 \quad \text{at} \quad y = 0 \\
x_i &= x_L \quad \text{at} \quad y = L
\end{aligned}
\tag{11.28}
$$

it is a simple matter to integrate eq. (11.27) and to report that the mole fraction distribution is linear:

$$x_i = (x_L - x_0)\frac{y}{L} + x_0 \quad \text{(slab)} \tag{11.29}$$

This conclusion is equivalent to saying that the flux of species i is conserved (constant) as it progresses across the layer. This diffusion flux can be expressed either as a molar flux [cf., eq. (11.25)],

$$
\hat{j}_{y,i} = -D\frac{dC_i}{dy} = -D\frac{\rho}{M}\frac{dx_i}{dy}
$$

$$
= D\frac{\rho}{M}\frac{x_0 - x_L}{L}
$$

$$
= D\frac{C_0 - C_L}{L} \tag{11.30}
$$

or as a mass flux [cf. eq. (11.11)],

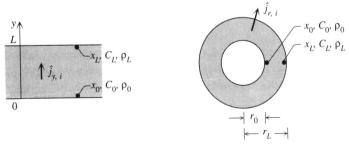

Figure 11.5 Steady unidirectional diffusion through a stationary medium: constant-thickness layer (left), and annulus formed between concentric cylinders or spheres (right).

$$j_{y,i} = -D\frac{d\rho_i}{dy} = -D\frac{\rho M_i}{M}\frac{dx_i}{dy}$$

$$= D\frac{\rho M_i}{M}\frac{x_0 - x_L}{L}$$

$$= D\frac{\rho_0 - \rho_L}{L} \tag{11.31}$$

In either instance, we learn that the "flowrate" of the species of interest across the layer is proportional to its mass diffusivity D and the mole fraction or concentration difference, and that is inversely proportional to the layer thickness. The molar flux and the mass flux are proportional, $j_{y,i} = M_i\hat{j}_{y,i}$, the role of proportionality constant being played by the molar mass of the species of interest, M_i. Note also that in the first step in the development of eq. (11.30) it has been assumed that the M value of the mixture is practically constant across the layer.

The similarities between, say, the last form of eq. (11.30) and the result for the heat flux across a plane wall, eq. (2.5), are worth contemplating. These similarities suggest that the solutions to other steady diffusion problems can be written by properly changing the notation in their counterparts in conduction heat transfer. For example, the mole fraction distribution across the annular layer formed between *two concentric cylindrical surfaces* (Fig. 11.5, right) has the same analytical form as the temperature distribution listed in eq. (2.31):

$$x_i = x_0 - (x_0 - x_L)\frac{\ln(r/r_0)}{\ln(r_L/r_0)} \qquad \text{(cylindrical shell)} \tag{11.32}$$

This time, we wrote r_0 for the inner radius and r_L for the outer radius, with the thought that L continues to represent the *length traveled* by the diffusing species. The molar flowrate from the inner surface to the outer surface of the annulus, per unit length in the direction normal to the plane of Fig. 11.5 (right), is

$$\dot{N}_i' = 2\pi r_0(\hat{j}_{r,i})_{r=r_0} = 2\pi r_0\left(-D\frac{dC_i}{dr}\right)_{r=r_0}$$

$$= \frac{2\pi D}{\ln(r_L/r_0)}(C_0 - C_L) \tag{11.33}$$

The corresponding mass flowrate per unit axial length is

$$\dot{m}_i' = 2\pi r_0(j_{r,i})_{r=r_0} = 2\pi r_0\left(-D\frac{d\rho_i}{dr}\right)_{r=r_0}$$

$$= \frac{2\pi D}{\ln(r_L/r_0)}(\rho_0 - \rho_L) \tag{11.34}$$

This quantity is proportional to the molar flowrate per unit length, $\dot{m}_i' = M_i\dot{N}_i'$.

Similarly, the solution for steady one-dimensional diffusion across the annular space formed by *two concentric spherical surfaces* (Fig. 11.5, right) is

$$x_i = x_L + \frac{x_0 - x_L}{r_0^{-1} - r_L^{-1}}\left(\frac{1}{r} - \frac{1}{r_L}\right) \qquad \text{(spherical shell)} \tag{11.35}$$

The total molar and mass flowrates of species i, from the spherical surface of radius r_0 to the outer surface of radius r_L, are, respectively,

$$\dot{N}_i = 4\pi r_0^2 (\hat{j}_{r,i})_{r=r_0} = 4\pi r_0^2 \left(-D\frac{dC_i}{dr} \right)_{r=r_0}$$

$$= \frac{4\pi D}{r_0^{-1} - r_L^{-1}}(C_0 - C_L) \qquad (11.36)$$

$$\dot{m}_i = 4\pi r_0^2 (j_{r,i})_{r=r_0} = 4\pi r_0^2 \left(-D\frac{d\rho_i}{dr} \right)_{r=r_0}$$

$$= \frac{4\pi D}{r_0^{-1} - r_L^{-1}}(\rho_0 - \rho_L) \qquad (11.37)$$

The conduction heat transfer analog of these results is listed in eq. (2.39). The proportionality between the total mass and molar flowrates is $\dot{m}_i = M_i \dot{N}_i$.

11.3.2 Mass Diffusivities

The solutions for steady unidirectional diffusion through a bulk-stationary medium, eqs. (11.29)–(11.37), show that to calculate the mass transfer rate we must first identify the proper values of two items:

(a) The mass diffusivity of the species of interest, D, and

(b) The species concentration at the two surfaces that define the mass transfer medium, x_0 (or C_0, ρ_0) and x_L (or C_L, ρ_L).

The second item — the specification of boundary conditions — will form the subject of the next section. Here, we review the contents of Tables 11.2 through 11.4, and the ways in which these mass diffusivity data can be extrapolated to temperatures and pressures that differ from those specified in the tables.

Consider first the case of a *binary gaseous mixture*, such as hydrogen and nitrogen at atmospheric pressure and room temperature. Let subscripts 1 and 2 represent the two components in the mixture. Table 11.2 shows the value of the mass diffusivity D, which is shorthand for the mass diffusivity of species 1 into 2 [namely, D_{12}, eq. (11.10)], as well as for the diffusivity of species 2 into species 1 (labeled D_{21}), in other words,

$$D \equiv D_{12} = D_{21} \qquad (11.38)$$

The equality of D_{12} and D_{21} can be demonstrated with reference to the left side of Fig. 11.5 where, *in addition* to the transversal diffusion of species 1 [see eq. (11.30)],

$$\hat{j}_{y,1} = -D_{12}\frac{\rho}{M}\frac{dx_1}{dy} \qquad (11.39)$$

we must also recognize the simultaneous diffusion of the second component of the binary mixture,

$$\hat{j}_{y,2} = -D_{21}\frac{\rho}{M}\frac{dx_2}{dy} \qquad (11.40)$$

However, since the diffusive medium remains stationary, the net flowrate (kmol/s, or kg/s) through any constant-y plane must be zero; for example,

$$\hat{j}_{y,1} + \hat{j}_{y,2} = 0 \qquad (11.41)$$

Table 11.2 Mass Diffusivities of Binary Gaseous Mixtures at Atmospheric Pressure[a]

Gaseous Mixture	D (m²/s)	T(K)
Air–acetone	1.09×10^{-5}	273
Air–ammonia	2.80×10^{-5}	298
Air–benzene	0.77×10^{-5}	273
Air–carbon dioxide	1.42×10^{-5}	276
	1.77×10^{-5}	317
Air–ethanol	1.45×10^{-5}	313
Air–helium	7.65×10^{-5}	317
Air–n-hexane	0.80×10^{-5}	294
Air–methanol	1.32×10^{-5}	273
Air–naphthalene	5.13×10^{-6}	273
Air–water vapor	2.60×10^{-5}	298
	2.88×10^{-5}	313
Ammonia–hydrogen	5.70×10^{-5}	263
	1.10×10^{-4}	358
Argon–carbon dioxide	1.33×10^{-5}	276
Argon–hydrogen	8.29×10^{-5}	295
Benzene–hydrogen	4.04×10^{-5}	311
Benzene–nitrogen	1.02×10^{-5}	311
Carbon dioxide–nitrogen	1.67×10^{-5}	298
Carbon dioxide–oxygen	1.53×10^{-5}	293
Carbon dioxide–water vapor	1.98×10^{-5}	307
Cyclohexane–nitrogen	0.73×10^{-5}	288
Helium–methane	6.76×10^{-5}	298
Hydrogen–nitrogen	7.84×10^{-5}	298
Hydrogen–water vapor	9.15×10^{-5}	307
Methane–water vapor	3.56×10^{-5}	352
Nitrogen–water vapor	3.59×10^{-5}	352
Oxygen–water vapor	3.52×10^{-5}	352

[a]Data collected from Refs. [5–7].

The binary mixture version of eq. (11.15) requires also that $x_1 + x_2 = 1$ in any constant-y plane; therefore, combining this last equation with eqs. (11.39)–(11.41) leads to the requirement that D_{12} must be equal to D_{21}. For this reason, the D values listed in Table 11.2 are known also as *mutual* diffusion coefficients.

The theoretical work of predicting the mass diffusivity in binary gaseous mixtures has been reviewed by Reid et al. [6] and Bird et al. [8]. Recommended by Reid et al. [6] is a semi-empirical correlation developed by Fuller et al. [9], according to whom D is proportional to the group $T^{1.75}/P$, where T and P are the mixture absolute temperature and pressure, respectively. This functional dependence can be used to extend the applicability of the mass diffusivity data of Table 11.2. If T_0 and P_0 are the mixture temperature and pressure specified in the table, then the diffusivity at different values of T and P can be evaluated by writing

$$\frac{D(T,P)}{D(T_0, P_0)} \cong \left(\frac{T}{T_0}\right)^{1.75} \frac{P_0}{P} \qquad (11.42)$$

The real-life effect of temperature on mass diffusivity deviates somewhat from this simple formula if the ratio T/T_0 is sizably greater than 1. Worth keeping in mind is that the mass diffusivity of a binary gaseous mixture does not depend on the concentration.

The evaluation of mass diffusivities is considerably more complicated for a *multicomponent gaseous mixture*, because the mixture composition plays an important role (see, for example, Ref. [7]). Nevertheless, Fick's law of diffusion and the equations of Table 11.1 continue to hold true, provided that D is an appropriately chosen (calculated or measured) coefficient. The method for making this selection is outlined in more advanced treatments such as Refs. [6, 8, 10–12].

The mass diffusivities of *liquid mixtures* are generally 10^4 to 10^5 times smaller than the diffusivities exhibited by gaseous mixtures. In *binary* liquid mixtures, the mass diffusivity is a function of composition (i.e., unlike in binary gaseous mixtures). The effect of mixture composition on D becomes negligible in the limit of "infinite dilution" when only small amounts of the diffusing species of interest (the solute) are mixed with a second species (the solvent).

Table 11.3 shows a collection of mass diffusivity data for dilute binary liquid mixtures in which the solvent is always liquid water. Many more data for other dilute binary liquid mixtures can be found in Refs. [5, 6]. In engineering terms, "dilute" means that the solute mole fraction does not exceed approximately 5 percent.

The D values of Table 11.3 refer strictly to the mixture temperature listed in the rightmost column. Let subscript 1 represent the solute and subscript 2 the solvent in the dilute binary liquid mixture. It has been shown [13] that the diffusion coefficient of interest (D_{12}, or D in Table 11.3) increases with the temperature as the group T/μ_2, where the viscosity of the *solvent* (μ_2) is, in general, a function of temperature (consult Appendix C). If T_0 is the absolute

Table 11.3 Mass Diffusivities of Gases and Organic Solutes at Low Concentrations in Water (Dilute Aqueous Solutions)

Solute	Solvent	D (m²/s)	T(K)
Acetone	Water	1.16×10^{-9}	293
Air	Water	2.5×10^{-9}	293
Aniline	Water	0.92×10^{-9}	293
Benzene	Water	1.02×10^{-9}	293
Carbon dioxide	Water	1.92×10^{-9}	298
Chlorine	Water	1.25×10^{-9}	298
Ethanol	Water	0.84×10^{-9}	298
Ethylene glycol	Water	1.04×10^{-9}	293
Glycerol	Water	0.72×10^{-9}	288
Hydrogen	Water	4.5×10^{-9}	298
Nitrogen	Water	2.6×10^{-9}	293
Oxygen	Water	2.1×10^{-9}	298
Propane	Water	0.97×10^{-9}	293
Urea	Water	1.2×10^{-9}	293
Vinyl chloride	Water	1.34×10^{-9}	298

[a]Data collected from Refs. [5, 6].

temperature that corresponds to the D value listed in Table 11.3, the temperature domain covered by that table can be extended by using the relationship

$$\frac{D(T)}{D(T_0)} \cong \frac{T}{T_0} \frac{\mu_2(T_0)}{\mu_2(T)} \qquad (11.43)$$

Diffusion processes through *solids* are characterized by mass diffusivities that are even smaller than those of liquids. These processes are considerably more complicated and diverse than what we have discussed until now, because the *structure* of the solid medium (e.g., pores, capillaries) can play a decisive role in how fast a certain species diffuses across the solid. The various mechanisms that contribute to what in an aggregate sense can be described as "diffusion" or "leakage" through a solid form the subject of specialized treatises of materials science [14, 15]. It has been found that the flowrate is described adequately by a relationship of type (11.10) or (11.25), in which the mass diffusion coefficient is determined from experiment. Table 11.4 lists the room-temperature diffusivities

Table 11.4 Mass Diffusivities of Gases and Interstitial Atoms Through Solid Materials[a]

Solid	D (cm^2/s) $\times 10^6$ ($T = 25°C$)					
	He	H$_2$	O$_2$	N$_2$	CO$_2$	CH$_4$
Natural rubber	21.6	10.2	1.58	1.1	1.1	0.89
Butyl rubber	5.93	1.52	0.081	0.045	0.058	
Silicone rubber (10% filler by weight, vulcanized)	53.4	67.1	17.0	13.2		
Polyethylene (density 0.964 g/cm^3)	3.07		0.17	0.093	0.124	0.057
Neoprene (polychloroprene)		4.31	0.43	0.29	0.27	
Polystyrene	10.4	4.36	0.11		0.058	

	D (cm^2/s)			
	He		H$_2$	
	20°C	500°C	20°C	500°C
Silica (Si O$_2$)	$(2.4-5.5) \times 10^{-10}$	$(0.17-1.4) \times 10^{-7}$	$(0.6-2.1) \times 10^{-8}$	
Pyrex	4.5×10^{-11}	2×10^{-8}		
Iron			2.6×10^{-9}	

Interstitial Atoms (Solute) in a Metal (Solvent), Forming a Very Dilute Alloy

$$D = D_0 \exp\left(-\frac{Q}{\bar{R}T}\right), \quad \bar{R} = 8.314 \frac{kJ}{kmol \cdot K}, \quad T = T(K)$$

Solvent	Solute	D_0 (cm^2/s)	Q (kJ/kmol)
Iron	C	0.02	84,200
	N	0.003	76,200
	H	0.001	13,400
Austenite	C	0.2	138,200
Nickel	H	0.0045	36,000

[a]Data collected from Refs. [14, 16].

of several fluids that diffuse through solids. Additional examples of gas diffusivities through polymers can be found in Refs. [15, 17].

11.3.3   Boundary Conditions

An important feature of the boundary conditions specified in the steady diffusion solutions of Section 11.3.1 is that each condition applies to the *inner side* of the boundary. The inner side is the side of the interface that faces the medium through which the species of interest diffuses. This feature has been stressed in the drawing of Fig. 11.5, where the boundary conditions (x_0, C_0, ρ_0) and (x_L, C_L, ρ_L) have been attached to the inner sides of the boundary, that is, in the *shaded domain*. It is an important feature because, unlike the temperature, the species concentration does not vary continuously across the interface. The exception, of course, is the case of the single-component system (pure substance), in which the mole fraction is equal to 1 on both sides of the interface.

Figure 11.6 illustrates four examples of interfaces that may confine a certain diffusive domain. In each example the diffusive domain of interest is indicated by the shaded area: the shaded side of the interface forms the subject of our mass diffusion analysis.

The first example (Fig. 11.6*a*) shows the type of boundary that has been alluded to in the discussion of Figs. 11.1 and 11.2, namely, the interface between an ideal gas mixture and the liquid phase of one of its components. For example, if 1 is a pure substance in the liquid phase (e.g., water), then 1 is also a component in the gaseous mixture above the interface (namely, water vapor in air). The proper boundary condition on the *gas side* of the interface states that the vapor pressure of the component of interest, P_1, is equal to its own saturation pressure at the temperature T of interface liquid:

$$P_1 = P_{1,\text{sat}}(T) \tag{11.44}$$

The second example (Fig. 11.6*b*) deals with the interface between a liquid medium and a gaseous mixture. The species of interest, species 1, diffuses

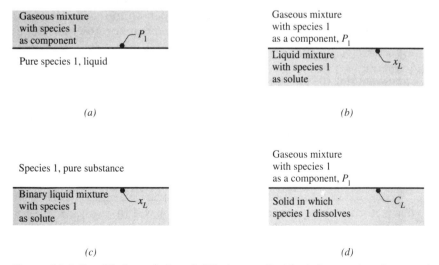

Figure 11.6 Possible boundaries of diffusive media (shaded spaces) and ways of specifying concentration boundary conditions.

through the liquid and is also present as a component in the gaseous mixture. The boundary condition of interest is the mole fraction of species 1 on the *liquid side* of the interface, namely, x_L (the same notation was used in Fig. 11.5, left). The boundary mole fraction x_L will be greater when larger quantities of species 1 are present in the mixture on the gaseous side of the interface, that is, when the partial pressure P_1 is higher. In the case of a dilute solution, where only small amounts of the solvent 1 are found in the liquid, x_L and P_1 are linked through the proportionality known as *Henry's law**:

$$x_L = \frac{P_1}{H} \qquad (11.45)$$

Table 11.5 shows a collection of values of Henry's constant H, the units of which are those of pressure. In addition to the dilute-solution requirement mentioned already, it should be noted that Henry's law applies only at low and moderate pressures, for example, when the partial pressure P_1 does not exceed 1 atm in the gaseous mixture [5]. At higher partial pressures, Henry's constant is a function of the partial pressure itself, and the use of a certain H value in eq. (11.45) is restricted to a limited range of partial pressures P_1.

The third boundary condition example (Fig. 11.6c) refers to a binary liquid mixture in which species 1 is the solute. The other side of the interface is occupied by pure 1, as a pure substance at the temperature and pressure of the liquid mixture. For example, this type of boundary condition occurs between a block of salt (NaCl), which would be situated on the upper side of the interface in Fig. 11.6c, and a pool of saline water (NaCl and H_2O). The concentration of NaCl on the *liquid side* of the interface can be determined by claiming thermodynamic equilibrium at the interface and consulting the solubility data available in the chemical engineering literature [5, 18–20]. For the above example, Ref. [5] indicates that the solubility of NaCl in H_2O is 35.7 at 0°C, which means that 35.7 grams of NaCl will coexist with 100 grams of H_2O in the solution. The mass fraction of NaCl on the liquid side of the interface is therefore 35.7/(35.7 + 100) = 0.263, if the interface temperature is indeed 0°C.

Finally, Fig. 11.6d shows the interface between a solid medium and a gaseous mixture. The species that diffuses through the solid (species 1) is also a component in the gaseous mixture. Its concentration on the solid side of the interface has been related empirically to its vapor pressure on the gas side, by writing

---

*William Henry (1774–1836) was the English chemist who first reported this relationship.

Table 11.5 Henry's Constant H for Several Gases in Water at Moderate Pressures[a]

$T(K)$			H (bar)			
	Air	N_2	O_2	H_2	CO_2	CO
290	6.2×10^4	7.6×10^4	3.8×10^4	6.7×10^4	1.3×10^3	5.1×10^4
300	7.4×10^4	8.9×10^4	4.5×10^4	7.2×10^4	1.7×10^3	6.0×10^4
320	9.2×10^4	1.1×10^5	5.7×10^4	7.6×10^4	2.7×10^3	7.4×10^4
340	1.04×10^5	1.24×10^5	6.5×10^4	7.6×10^4	3.7×10^3	8.4×10^4

[a]Data collected from Ref. [18].

$$C_L = SP_1 \tag{11.46}$$

in which S is the solubility coefficient of species 1 in the given solid. The S values of several gas–solid combinations are shown in Table 11.6. If species 1 is completely absent from the gas side, then $P_1 = 0$, and eq. (11.46) yields $C_L = 0$. The following example shows that the molar flux is proportional to the product $D \cdot S$.

Example 11.1

Steady Diffusion, Solid Between Two Parallel Planes

A thin rubber (neoprene) membrane separates a volume of high-pressure nitrogen gas (5 bar) from a volume of low-pressure nitrogen gas (1 bar). The thickness of the membrane is 0.5 mm, and the temperature of the entire system is 300 K. Calculate the molar and mass fluxes of the nitrogen that diffuses through the membrane.

Solution. Focusing first on the molar flux, we combine eqs. (11.30) and (11.46) and obtain

$$\hat{j}_{N_2} = D\frac{C_0 - C_L}{L} = DS\frac{P_0 - P_L}{L} \tag{1}$$

Table 11.4 shows that the room-temperature mass diffusivity of nitrogen through neoprene is $D \cong 0.29 \times 10^{-6}$ cm²/s. The solubility of nitrogen in neoprene is listed in Table 11.6, $S \cong 2.1 \times 10^{-3}$ kmol/m³·bar. The other data to be substituted in eq. (1) are $L = 0.5$ mm, $P_0 = 5$ bar, and $P_L = 1$ bar. The result is

$$\hat{j}_{N_2} = \left(0.29 \times 10^{-6}\frac{cm^2}{s}\right)\left(2.1 \times 10^{-3}\frac{kmol}{m^3 \cdot bar}\right)\left(\frac{(5-1)\ bar}{0.5\ mm}\right)$$

$$= 4.87 \times 10^{-10}\,kmol/m^2 \cdot s$$

The corresponding mass flux follows from eq. (11.31) in combination with eq. (11.24),

$$j_{N_2} = D\frac{\rho_0 - \rho_L}{L} = DM_{N_2}\frac{C_0 - C_L}{L}$$

$$= M_{N_2}\hat{j}_{N_2}$$

in which M_{N_2} is the molar mass of nitrogen listed in the table of Ideal Gas Constants and Specific Heats, at the end of Appendix D. Numerically, we obtain

$$j_{N_2} = \left(28.01\frac{kg}{kmol}\right)\left(4.87 \times 10^{-10}\frac{kmol}{m^2 \cdot s}\right)$$

$$= 1.36 \times 10^{-8}\ kg/m^2 \cdot s$$

Table 11.6 Solubilities of Several Gases in Solids[a]

Solid	Gas	T (K)	S (kmol/m³·bar)
Rubber (neoprene, vulcanized)	H_2	300	2.1×10^{-3}
	N_2	300	2.1×10^{-3}
Silica (SiO_2)	H_2	300–773	2.8×10^{-4}
	He	300–773	4.0×10^{-4}
Pyrex	He	300–773	3.4×10^{-4}
Nickel (Ni)	H_2	358	8.0×10^{-3}

[a]Data collected from Ref. [14].

11.3.4   Time-Dependent Diffusion

Mass diffusion problems in which the concentration field changes in time are handled using the same analytical methods as in problems of time-dependent heat conduction (Chapter 4). To see this correspondence, consider the process of unidirectional (vertical) diffusion into the semi-infinite medium shown in Fig. 11.7.

Initially, the concentration of the species of interest is uniform (C_{in}) through-out the semi-infinite medium. Beginning with the time $t = 0$, the concentration at the $y = 0$ boundary is maintained at a different level, C_0, by contact with another medium. The boundary concentration C_0 refers to the side of the $y = 0$ boundary that faces the medium through which the species of interest diffuses. When C_0 is greater than C_{in}, the species diffuses into the semi-infinite medium and forms a *concentration boundary layer* the thickness of which increases with time. This effect is illustrated on the right side of Fig. 11.7. On the other hand, when C_0 is smaller than C_{in}, the species diffuses out of the semi-infinite medium. In this case the concentration boundary layer is the region that becomes partially vacated by the diffusing species.

The mathematical statement of the time-dependent diffusion problem de-fined in Fig. 11.7 is as follows:

Concentration equation:
$$\frac{\partial^2 C}{\partial y^2} = \frac{1}{D}\frac{\partial C}{\partial t} \tag{11.47}$$

Initial condition:
$$C = C_{in} \quad \text{at} \quad t = 0 \tag{11.48}$$

Boundary conditions:
$$C = C_0 \quad \text{at} \quad y = 0 \tag{11.49}$$
$$C \to C_{in} \quad \text{as} \quad y \to \infty \tag{11.50}$$

This problem is line by line analogous to that listed in eqs. (4.22)–(4.25) for time-dependent conduction into a semi-infinite body with fixed surface temper-ature (Fig. 4.3). The solution to the present problem can be written down immediately by using the conduction solution (4.42), in which y replaces x, the mass diffusivity D replaces α, the boundary concentration C_0 replaces T_∞, and the initial concentration C_{in} replaces the initial temperature T_i:

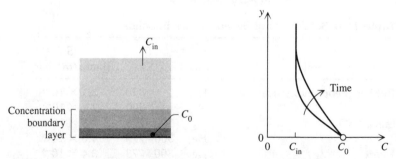

Figure 11.7 Concentration boundary layer in a semi-infinite medium with a different concentration imposed at the surface.

$$\frac{C - C_0}{C_{in} - C_0} = erf\left[\frac{y}{2(D \cdot t)^{1/2}}\right] \tag{11.51}$$

This solution can be used for calculating the concentration in the boundary layer, as a function of time and distance to the boundary of the medium, $C(y, t)$. The denominator of the error function argument in eq. (11.51) shows that the order of magnitude (the scale) of the thickness of the concentration boundary layer is $(D \cdot t)^{1/2}$. In conclusion, the mass diffusion effect triggered by a sudden change in surface concentration penetrates the medium to a distance proportional to the square root of both time and mass diffusivity.

The group $(D \cdot t)^{1/2}$ is a very important length scale in calculations of time-dependent diffusion. We learned in Chapter 4 that the semi-infinite medium is an adequate model for more complicated body shapes at short times, when the diffusion penetration distance is much shorter than the transversal dimension of the body. At longer times, when $(D \cdot t)^{1/2}$ is comparable with, or larger than, the transversal length scale (e.g., the radius of a spherical body, r_0), the concentration distribution through the body depends also on the transversal length scale.

The three mass diffusion solutions summarized in Fig. 11.8 show that at sufficiently long times the concentration history depends on body size and shape. Each of the bodies illustrated in this figure is initially in a state of uniform concentration, C_{in}. The concentration on the inner side of each surface

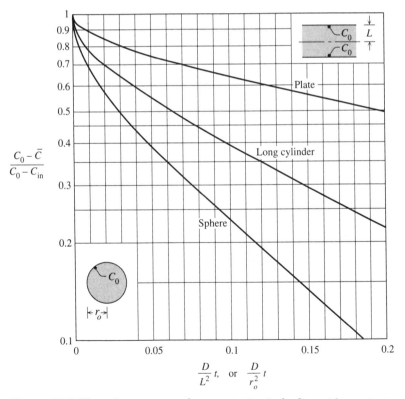

Figure 11.8 The volume-averaged concentration in bodies with constant concentration imposed at the surface.

is maintained at a different level C_0 (different from C_{in}) starting with time $t = 0$. The representative transversal length of the plate geometry is the half-thickness L; in the case of a long cylinder or a sphere, that length is the radius r_o. Plotted on the abscissa of Fig. 11.8 is the dimensionless time

$$\frac{D}{L^2}t \quad \text{or} \quad \frac{D}{r_o^2}t \tag{11.52}$$

This dimensionless group is the mass diffusion equivalent of the Fourier number Fo defined in eq. (4.63).

The ordinate of Fig. 11.8 refers to the concentration of the diffusing species *averaged* over the entire volume of the body, $\overline{C}$. The general downward trend of the three curves shows that the volume-averaged concentration $\overline{C}$ approaches the surface concentration level C_0 as the time increases. If we multiply the $\overline{C}$ value furnished by Fig. 11.8 with the volume of the respective body, we obtain the total amount (moles) of diffusing species found at a certain moment in the entire body. In this way, we can keep track of the amount of species absorbed (if $C_0 > C_{in}$) or released (if $C_0 < C_{in}$) by the body as a whole.

The domain covered by the graphic solutions displayed in Fig. 11.8 can be extended toward much longer times by noting the straight-line shape acquired in that limit by each of the three curves. When the abscissa parameter is greater than approximately 0.1, the curves are represented very closely by the asymptotes:

Slab:

$$\frac{C_0 - \overline{C}}{C_0 - C_{in}} \cong \frac{8}{\pi^2}\exp\left(-\frac{\pi^2}{4}\frac{D}{L^2}t\right) \tag{11.53}$$

Cylinder:

$$\frac{C_0 - \overline{C}}{C_0 - C_{in}} \cong \frac{4}{b_1^2}\exp\left(-b_1^2\frac{D}{r_o^2}t\right) \quad \text{with } b_1 = 2.405 \tag{11.54}$$

Sphere:

$$\frac{C_0 - \overline{C}}{C_0 - C_{in}} \cong \frac{6}{\pi^2}\exp\left(-\pi^2\frac{D}{r_o^2}t\right) \tag{11.55}$$

These exponential forms and the semilogarithmic grid employed in Fig. 11.8 are the reasons why the curves become straight as the time increases. Another worthwhile observation is that the b_1 constant used in eq. (11.54) is the value reached by the eigenvalue b_1 of Table 4.1 in the limit of infinite Biot number (i.e., in the limit of prescribed surface temperature). It follows that the solutions summarized in Fig. 11.8 are the mass diffusion counterparts of the heat conduction (Bi = ∞) solutions covered by Figs. 4.7 through 4.15.

Example 11.2

Time-Dependent Diffusion, Air in Water

A thin layer of pure water is placed in contact with air at atmospheric pressure and 20°C. Air begins to diffuse into the water. Of interest is the air mole fraction x registered at 1 mm away from the water–air interface. Calculate the time that must pass until x rises to half of the air mole fraction at the interface (x_0), that is, until $x = \frac{1}{2} x_0$. Calculate also the actual value of the air mole fraction x at that time.

Solution. If we neglect all convection effects in the thin water layer, the diffusion of air in water is described by eq. (11.51), in which the left side can be rewritten in terms of the air mole fraction in the dilute water solution:

$$\frac{x - x_0}{x_{in} - x_0} = \text{erf}\left[\frac{y}{2(D \cdot t)^{1/2}}\right] \tag{1}$$

The values to substitute on the left side are $x = \frac{1}{2} x_0$ and $x_{in} = 0$; therefore, eq. (1) reduces to

$$\frac{1}{2} = \text{erf}\left[\frac{y}{2(D \cdot t)^{1/2}}\right] \tag{2}$$

or, after consulting the inverse error function values listed in Appendix E,

$$\frac{y}{2(D \cdot t)^{1/2}} = 0.477 \tag{3}$$

The diffusivity of air in water at 20°C is $D = 2.5 \times 10^{-9}$ m²/s. The time t when the mole fraction x rises to the level $\frac{1}{2} x_0$ at the distance $y = 1$ mm away from the surface is, therefore,

$$t = \left(\frac{y}{2 \times 0.477}\right)^2 \frac{1}{D}$$

$$= \frac{(1 \text{ mm})^2}{0.91} \frac{\text{s}}{2.5 \times 10^{-9} \text{ m}^2} = 440 \text{ s} \cong 7.3 \text{ minutes}$$

To calculate the actual mole fraction registered at this time at the $y = 1$ mm location, we must first determine the mole fraction at the interface. Invoking Henry's law (11.45), in which $P = 1$ atm $= 1.013$ bar and $H = 6.2 \times 10^4$ bar (see Table 11.5), we obtain

$$x_0 = \frac{P}{H} = \frac{1.013 \text{ bar}}{6.2 \times 10^4 \text{ bar}} = 1.63 \times 10^{-5}$$

The mole fraction of interest is half this value, namely, $x = \frac{1}{2} x_0 = 8.2 \times 10^{-6}$.

11.4 CONVECTION

11.4.1 Forced Convection in Laminar Boundary Layer Flow

The analogy between mass transfer and heat transfer continues as we turn our attention to convection, that is, mass transfer problems in which the species of interest diffuses through a medium in motion (a flow). Consider first the laminar boundary layer flow alluded to in Fig. 11.1 and reproduced here in Fig. 11.9. A fluid mixture with uniform velocity U_∞ flows parallel to the surface $y = 0$, which is coated with (or made out of) a certain substance that is soluble in the stream. The concentration of this particular species *on the flow side* of the $y = 0$ wall is constant and equal to C_w. The concentration of the same species sufficiently far into the stream is equal to another constant, C_∞.

The flow of the mixture itself has the characteristics of the laminar boundary layer described in Section 5.3. From that discussion (in particular, Section 5.3.2) we know to expect the existence of a *concentration boundary layer* near the surface. The concentration boundary layer is the slender flow region across

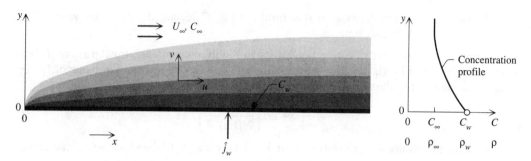

Figure 11.9 Mass transfer from a plane surface to a forced-convection laminar boundary layer flow.

which the species concentration varies smoothly from $C = C_w$ at the wall to $C = C_\infty$ in the mixture free stream.

The concentration profile (Fig. 11.9, right) has its steepest gradient right at the wall, $(\partial C/\partial y)_{y=0}$. Note further that since $u = 0$ in the blade (lamina) of flow situated immediately adjacent to the wall, the mixture in this flow blade is stationary (i.e., "stuck to the wall"). The migration of the species of interest across this very thin layer is by diffusion in what approaches a stationary medium; therefore, we invoke Fick's law (11.25) and relate the *local* surface molar flux $\hat{j}_w$ to the concentration gradient:

$$\hat{j}_w = -D\left(\frac{\partial C}{\partial y}\right)_{y=0} \tag{11.56}$$

The basic problem in an external flow mass convection configuration is to find the relationship between the species flux $\hat{j}_w$ and the "driving" concentration difference $C_w - C_\infty$. This problem is equivalent to determining the *local mass transfer coefficient for external flow,*

$$h_m = \frac{\hat{j}_w}{C_w - C_\infty} \tag{11.57}$$

which, in view of eq. (11.57), can also be written as

$$h_m = -\frac{D(\partial C/\partial y)_{y=0}}{C_w - C_\infty} \tag{11.58}$$

Recalling that D is the mass diffusivity of the species of interest relative to the mixture, and that the units of D are m^2/s, from eq. (11.58) we conclude that the mass transfer coefficient has the units m/s.

Equation (11.58) shows also that to calculate h_m, we must first determine the concentration distribution in the immediate vicinity of the wall. The C distribution can be obtained by solving the *boundary layer approximated** concentration equation

$$u\frac{\partial C}{\partial x} + v\frac{\partial C}{\partial y} = D\frac{\partial^2 C}{\partial y^2} \tag{11.59}$$

---

*Review the reasoning behind eq. (5.29), to see the origin of the simplification of the right sides of eqs. (5.32), (5.62), and (11.59).

subject to the boundary conditions identified in Fig. 11.9:

$$C = C_w \quad \text{at} \quad y = 0 \qquad (11.60)$$

$$C \rightarrow C_\infty \quad \text{as} \quad y \rightarrow \infty \qquad (11.61)$$

In this problem the mixture flow distribution (u, v) is known: its derivation formed the subject of Section 5.3.1. Furthermore, by comparing eqs. (11.59)–(11.61) with the boundary layer energy eq. (5.62) and the temperature boundary conditions of Fig. 5.9 (namely, $T = T_w$ at $y = 0$, and $T \rightarrow T_\infty$ as $y \rightarrow \infty$), we reach the important conclusion that the concentration boundary layer of Fig. 11.9 is *analogous* to the temperature boundary layer of Fig. 5.9. The only difference between these two problems is one of notation, in such a way that to each parameter from the left column corresponds another parameter in the right column*.

Temperature Boundary Layer (Fig. 5.9)		Concentration Boundary Layer (Fig. 11.9)	
T	$\rightarrow$	C	
T_w	$\rightarrow$	C_w	
T_∞	$\rightarrow$	C_∞	(11.62)
α	$\rightarrow$	D	
q_w''	$\rightarrow$	$\hat{j}_w$	
k	$\rightarrow$	D	

The most useful aspect of this analogy is that we can simply write down the known heat transfer solution and, by changing the notation in accordance with the preceding table, we can deduce the proper solution to the mass convection problem. For example, in eq. (5.79b) we learned that the solution for the local heat flux in a high Prandtl number flow near a flat plate is

$$\frac{q_w''}{T_w - T_\infty}\frac{x}{k} = 0.332 \left(\frac{\nu}{\alpha}\right)^{1/3}\left(\frac{U_\infty x}{\nu}\right)^{1/2} \qquad \left(\frac{\nu}{\alpha} \gtrsim 0.5\right) \qquad (5.79b)$$

The mass transfer equivalent of this result is, therefore,

$$\frac{\hat{j}_w}{C_w - C_\infty}\frac{x}{D} = 0.332 \left(\frac{\nu}{D}\right)^{1/3}\left(\frac{U_\infty x}{\nu}\right)^{1/2} \qquad \left(\frac{\nu}{D} \gtrsim 0.5\right) \qquad (11.63)$$

or, more succinctly,

$$Sh_x = 0.332 Sc^{1/3} Re_x^{1/2} \qquad (Sc \gtrsim 0.5) \qquad (11.64)$$

In this abbreviated form, we encounter for the first time the local *Sherwood number*[†]

---

*This table will be continued on pages 600 and 611.

[†]Thomas K. Sherwood (1903–1976) was of professor of chemical engineering at the Massachusetts Institute of Technology from 1930 until 1969. He is best known for his work on convection mass transfer with and without chemical reactions, drying, and for his impact on chemical engineering education.

$$Sh_x = \frac{\hat{j}_w x}{(C_w - C_\infty)D} = \frac{h_m x}{D} \tag{11.65}$$

and the *Schmidt number**

$$Sc = \frac{\nu}{D} \tag{11.66}$$

We also learn that the transformation table (11.62) can be extended by adding two new entries:

Heat Transfer		Mass Transfer	
Nu_x	$\rightarrow$	Sh_x	(11.62′)
Pr	$\rightarrow$	Sc	

In subsequent applications of the heat transfer–mass transfer analogy, it is more expedient to refer directly to this end of the transformation table, and to replace Nu_x with Sh_x, and Pr with Sc, in the correct Nusselt number formula. For example, the small-Sc companion of the mass transfer formula (11.64), can be obtained by first recognizing the Nu_x formula for the low-Pr range, eq. (5.79a),

$$Nu_x = 0.564 Pr^{1/2} Re_x^{1/2} \qquad (Pr \lesssim 0.5) \tag{5.79a}$$

and then applying the transformation $Nu_x \rightarrow Sh_x$ and $Pr \rightarrow Sc$:

$$Sh_x = 0.564 Sc^{1/2} Re_x^{1/2} \qquad (Sc \lesssim 0.5) \tag{11.67}$$

The local mass transfer coefficient h_m can be calculated with eqs. (11.64) and (11.67), in which $h_m = Sh_x D/x$. This coefficient can be averaged in the usual way over the total length L of the swept surface,

$$\overline{h}_m = \frac{1}{L} \int_0^L h_m \, dx \tag{11.68}$$

The overall Sherwood number based on this L-averaged mass transfer coefficient is defined as

$$\overline{Sh}_L = \frac{\overline{h}_m L}{D} \tag{11.69}$$

and, after using eqs. (11.64) and (11.67), can be calculated with the formulas

$$\overline{Sh}_L = 0.664 Sc^{1/3} Re_L^{1/2} \qquad (Sc \gtrsim 0.5) \tag{11.70}$$

$$\overline{Sh}_L = 1.128 Sc^{1/2} Re_L^{1/2} \qquad (Sc \lesssim 0.5) \tag{11.71}$$

In summary, the mass transfer coefficient is needed to calculate the rate at which the species leaves the surface swept by the flow. This relationship

---

*Ernst Schmidt (1892–1975) was professor of engineering thermodynamics at the Technical University of Munich (1919–1925 and 1952–1961), the Technical University of Danzig (1925–1937), and the University of Braunschweig (1937–1952). He has made contributions in many areas, for example, the analogy between heat and mass transfer, the optimization of fins, the visualization of convective flows, natural convection, the radiative properties of solids, dropwise condensation, and the international steam tables.

between h_m and the mass transfer rate can be expressed in molar terms, as in eq. (11.57), or in mass terms:

$$h_m = \frac{j_w}{\rho_w - \rho_\infty} \tag{11.72}$$

Relations similar to eqs. (11.57) and (11.72) exist between $\bar{h}_m$ and the L-averaged molar flux and mass flux. These relationships can be rewritten, respectively, as

$$\dot{N}' = \bar{h}_m L(C_w - C_\infty) \tag{11.73}$$

$$\dot{m}' = \bar{h}_m L(\rho_w - \rho_\infty) \tag{11.74}$$

for calculating the mole transfer rate $\dot{N}'$ (moles of species/s·m) and the mass transfer rate $\dot{m}'$ (kilograms of species/s·m). Note that $\dot{N}'$ and $\dot{m}'$ are both expressed per unit length in the direction perpendicular to Fig. 11.9. More general are the alternative relationships

$$\dot{N} = \bar{h}_m A(C_w - C_\infty) \tag{11.75}$$

$$\dot{m} = \bar{h}_m A(\rho_w - \rho_\infty) \tag{11.76}$$

where A is the area of the swept surface, $\dot{N}$ (moles of species/s) is the total mole transfer rate, and $\dot{m}$ (kilograms of species/s) is the total mass transfer rate.

11.4.2 The Impermeable Surface Model

The mass transfer–heat transfer analogy described in the preceding section rests on the important assumption that the transversal velocity of the flow (v in Fig. 11.9) is zero at the wall. Only then is the flow distribution [u, v, assumed known in eq. (11.59)] the same as in the velocity boundary layer studied in Section 5.3.1. In a mass transfer problem this assumption is by definition an *approximation*, because the transfer of mass through the surface amounts to a flow perpendicular to the surface. The $v = 0$ approximation is justified when the concentration of the species of interest is "low," namely, lower than a critical level.

To see this, consider the laminar boundary layer mass transfer configuration studied in relation to Fig. 11.9. The mass transfer conclusions (11.70)–(11.71) mean that the mass flux (kg/s·m²) through the $y = 0$ surface has the following order of magnitude:

$$j_w \sim (\rho_w - \rho_\infty)\frac{D}{x}\mathrm{Re}_x^{1/2}\mathrm{Sc}^n \tag{11.77}$$

The exponent n is an abbreviated notation for $\frac{1}{3}$ when $\mathrm{Sc} \gtrsim 0.5$ and for $\frac{1}{2}$ when $\mathrm{Sc} \lesssim 0.5$. The next challenge is to determine when this transversal mass flow (or surface "blowing" effect) is negligible.

One of the most basic features of the laminar boundary layer flow analyzed in Section 5.3.1 is that the transversal velocity component v is finite, or that the streamlines are not perfectly parallel to the $y = 0$ surface. According to eq. (5.37), the order of magnitude of the transversal velocity component is $v \sim U_\infty \mathrm{Re}_x^{-1/2}$. In conclusion, if we write ρ for the density of the flowing mixture, the *transversal* movement of mixture fluid (kg/s·m²) has the magnitude

$$\rho v \sim \rho U_\infty \mathrm{Re}_x^{-1/2} \tag{11.78}$$

The mass transfer through the surface can be neglected only when the transversal movement induced by it is small relative to the natural (always present) transversal movement of the boundary layer flow. Quantitatively, this occurs when

$$j_w < \rho v \qquad (11.79)$$

The two sides of this inequality are listed in eqs. (11.77) and (11.78); in the end, the condition (11.79) reduces to

$$\frac{\rho_w - \rho_\infty}{\rho} < Sc^{1-n} \qquad (11.80)$$

The impermeable surface assumption is therefore valid at "low" concentrations $(\rho_w - \rho_\infty)$ that satisfy the inequality (11.80). And since the exponent $1 - n$ is positive, the validity of this assumption improves as the Schmidt number increases. Table 11.7 lists the Sc values of several common substances that diffuse through air or water.

Table 11.7 The Schmidt Number, Lewis Number,[a] and Composition Buoyancy Coefficient at Low Concentration, 1 atm, and Approximately 25°C[b]

Main Fluid	Species at Low Concentrations	$Sc = \nu/D$	$Le = Sc/Pr$	$\rho\beta_c = -(\partial\rho/\partial\rho_i)_{T,P}$
Air (Pr = 0.7)	Ammonia	0.78	1.11	+1.07
	Carbon dioxide	0.94	1.34	−0.34
	Hydrogen	0.22	0.314	+13.4
	Oxygen	0.75	1.07	−0.094
	Water vapor	0.60	0.86	+0.61
	Benzene	1.76	2.51	−0.63
	Ether	1.66	2.37	−0.61
	Methanol	0.97	1.39	−0.095
	Ethyl alcohol	1.30	1.86	−0.37
	Ethylbenzene	2.01	2.87	−0.73
Water (Pr = 7)	Ammonia	445	63.57	−0.5
	Carbon dioxide	453	64.71	
	Hydrogen	152	21.71	
	Oxygen	356	50.86	
	Nitrogen	468	66.86	
	Chlorine	617	88.14	
	Sulfur dioxide	523	74.71	
	Calcium chloride	750	107.14	+0.8
	Sodium chloride	580	82.86	+0.7
	Methanol	556	79.43	−0.17
	Sucrose	1,700	242.86	

[a]The Lewis number is the ratio of the thermal diffusivity divided by the mass diffusivity, $Le = \alpha/D = Sc/Pr$. Its name is in honor of Warren K. Lewis, who was professor of chemical engineering at the Massachusetts Institute of Technology. Warren K. Lewis and Thomas K. Sherwood (p. 599) are credited with the development of the modern chemical engineering curriculum.

[b]After Gebhart and Pera [21].

Example 11.3

Laminar Boundary Layer, Humid Air Flow

The rain left a thin film of water on a roof tile. A 10 km/h wind sweeps the tile along its 10-cm exposed length. The atmospheric air and the tile surface have the temperature 25°C. The relative humidity of the atmospheric air is 40 percent.

(a) Calculate the average mass transfer coefficient between the tile surface and the humid air flow.

(b) Determine the mass transfer rate of the water that leaves the tile surface.

(c) Verify that the impermeable surface assumption is justified in this numerical case.

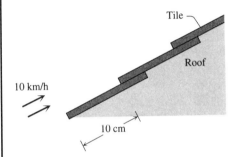

Figure E11.3

Solution. (a) The relevant properties of dry air at 25°C and atmospheric pressure are (Appendix D)

$$\rho_a = 1.185 \text{ kg/m}^3 \qquad \nu_a = 1.55 \times 10^{-5} \text{ m}^2/\text{s}$$

while for water vapor in air, from Tables 11.2 and 11.7 we collect

$$D = 2.88 \times 10^{-5} \text{ m}^2/\text{s} \qquad \text{Sc} = 0.6$$

The Reynolds number based on the swept length of $L = 0.1$ m shows that the boundary layer is laminar:

$$\text{Re}_L = \frac{U_\infty L}{\nu} \cong \frac{U_\infty L}{\nu_a} = 10 \frac{10^3 \text{ m}}{3600 \text{ s}} \frac{0.1 \text{ m}}{1.55 \times 10^{-5} \text{ m}^2/\text{s}}$$

$$= 17,921 \qquad \text{(laminar)}$$

This Re_L estimate is based on the assumption that the ν value of humid air (halfway between 100 percent humidity at the wall and 40 percent humidity in the free stream) is approximately equal to the kinematic viscosity of dry air, ν_a.

The L-averaged mass transfer coefficient follows from eq. (11.70) in two steps:

$$\overline{\text{Sh}}_L = 0.664(0.6)^{1/3}(17921)^{1/2} = 75$$

$$\overline{h}_m = \frac{D}{L} \overline{\text{Sh}}_L = \frac{2.88 \times 10^{-5} \text{ m}^2/\text{s}}{0.1 \text{ m}} 75$$

$$= 0.0216 \text{ m/s}$$

(b) The mass flowrate of the water that leaves the surface of the roof tile is given by

$$\dot{m}' = \overline{h}_m L(\rho_w - \rho_\infty)$$

To determine the water vapor density in the humid air mixture just above the surface, we review the engineering thermodynamics textbook, specifically, the section on psychrometry. We learn that the saturation pressure of water vapor at 25°C is

$$P_{sat}(25°C) = 3169 \text{ N/m}^2$$

and that the pressure of the air–water vapor mixture is 1 atm $= 1.0133 \times 10^5$ N/m² The mole fraction of water vapor at the surface is

$$x_w = \frac{P_{sat}(25°C)}{1 \text{ atm}} = \frac{3169}{1.0133 \times 10^5}$$

$$= 0.0313$$

The water vapor density ρ_w that corresponds to this mole fraction can be calculated by using eq. (11.24) and, later, eq. (11.23):

$$\rho_w = M_{H_2O}C_w$$

$$= M_{H_2O}\frac{\rho}{M}x_w \cong M_{H_2O}\frac{\rho_a}{M_a}x_w = 18.02\frac{\rho_a}{28.97}0.0313 \tag{1}$$

$$= 0.0195\rho_a$$

The mass density of saturated water vapor calculated above is $\rho_w = 0.0195\rho_a = 0.0195 \times 1.185$ kg/m³ $= 0.0231$ kg/m³ An alternative approach to the same ρ_w value consists of consulting a set of steam tables (saturated steam, the temperature table), and to invert the specific volume of saturated vapor at 25°C:

$$\rho_w = (43360 \text{ cm}^3/\text{g})^{-1} = 0.0231 \text{ kg/m}^3$$

Turning our attention to the calculation of ρ_∞, we recall that the relative humidity is defined as the ratio (see also the end of Appendix D)

$$\phi = \frac{P_v}{P_{sat}(T)}$$

In our case, $\phi_\infty = 0.4$ and $P_{sat}(25°C) = 3169$ N/m²; therefore, the partial pressure of water vapor outside the concentration boundary layer is

$$P_{v,\infty} = 0.4 \times 3169 \frac{N}{m^2} = 1268 \text{ N/m}^2 .$$

Beyond this point, the calculation of ρ_∞ follows the steps used earlier in the calculation of ρ_w:

$$x_\infty = \frac{1268}{1.0133 \times 10^5} = 0.0125$$

$$\rho_\infty \cong \frac{M_{H_2O}}{M_a}x_\infty \rho_a = \frac{18.02}{28.97}0.0125\rho_a \tag{2}$$

$$= 0.0078\rho_a$$

Now we have all the necessary information for calculating the water mass transfer rate:

$$\dot{m}' = \bar{h}_m L(\rho_w - \rho_\infty)$$

$$= \left(0.0216\frac{m}{s}\right)(0.1m)\left[(0.0195 - 0.0078) \, 1.185\frac{kg}{m^3}\right]$$

$$= 3 \times 10^{-5} \text{ kg/s·m}$$

In the calculation of ρ_w and ρ_∞, eqs. (1,2), we have assumed that the ratio ρ/M of the humid air mixture is approximately the same as the ratio for dry air, ρ_a/M_a. This approximation was noted also in the text, under eq. (11.20).

(c) The validity of the impermeable surface approximation is determined by substituting numerical values in the inequality (11.80):

$$\frac{\rho_w - \rho_\infty}{\rho_a} < Sc^{1-n}$$

$$\frac{0.0195\rho_a - 0.0078\rho_a}{\rho_a} < (0.6)^{1-1/3}$$

$$0.012 < 0.71$$

The last line shows that the criterion (11.80) is satisfied in an order-of-magnitude sense and that the impermeable surface approximation is justified.

11.4.3 Other External Forced-Convection Configurations

Analogous mass transfer rate formulas can be deduced for the other external forced-convection configurations treated in this text. For example, in the case of a *turbulent boundary layer flow over a flat surface*, we begin with the overall Nusselt number relationship developed in Chapter 5,

$$\overline{Nu}_L = 0.037Pr^{1/3}(Re_L^{4/5} - 23,550)$$
$$(Pr \gtrsim 0.5, \ 5 \times 10^5 < Re_L < 10^8) \quad (5.134)$$

and substitute $\overline{Sh}_L$ in place of $\overline{Nu}_L$ and Sc in place of Pr. In this way, we arrive at the formula for the overall Sherwood number, or the L-averaged mass transfer coefficient $\overline{h}_m$,

$$\overline{Sh}_L = \frac{\overline{h}_m L}{D} = 0.037Sc^{1/3} (Re_L^{4/5} - 23,550)$$
$$(Sc \gtrsim 0.5, \ 5 \times 10^5 < Re_L < 10^8) \quad (11.81)$$

An alternative expression for the high-Re_L range can be obtained by recognizing the Colburn relationship for the local heat transfer coefficient,

$$St_x = \tfrac{1}{2} C_{f,x} Pr^{-2/3} \qquad (Pr \gtrsim 0.5) \qquad (5.131)$$

and the corresponding local skin friction coefficient expression

$$\tfrac{1}{2} C_{f,x} = 0.0296 \left(\frac{U_\infty x}{\nu}\right)^{-1/5} \qquad (5.121)$$

Combined, eqs. (5.121) and (5.131) yield

$$St_x = 0.0296Pr^{-2/3}Re_x^{-1/5} \qquad (Pr \gtrsim 0.5) \qquad (11.82)$$

The next challenge is to find the mass transfer analog of the local Stanton number that appears on the left side of eq. (11.82), namely,

$$St_x = \frac{h_x}{\rho c_p U_\infty} = \frac{h_x \alpha}{k U_\infty} = \frac{q_w'' \alpha}{(T_w - T_\infty)k U_\infty} \qquad (11.83)$$

The tabulated transformation (11.62) through (11.62') shows that the mass transfer analog of St_x is obtained by replacing q_w'' with $\hat{j}_w$, $T_w - T_\infty$ with $C_w - C_\infty$, α with D, and k again with D:

$$\frac{q_w'' \alpha}{(T_w - T_\infty)k U_\infty} \rightarrow \frac{\hat{j}_w D}{(C_w - C_\infty)D U_\infty} \qquad (11.84)$$

What emerges on the right side of (11.84) is simply the ratio h_m/U_∞, which is called the *local mass transfer Stanton number*

$$St_m = \frac{h_m}{U_\infty} \tag{11.85}$$

In the end, the mass transfer analog of eq. (11.82) becomes

$$St_m = 0.0296Sc^{-2/3}Re_x^{-1/5} \qquad (Sc \gtrsim 0.5) \tag{11.86}$$

or, in terms of the L-averaged mass transfer coefficient,

$$\overline{St}_m = \frac{\overline{h}_m}{U_\infty} = 0.037Sc^{-2/3}Re_L^{-1/5} \qquad (Sc \gtrsim 0.5) \tag{11.87}$$

It is easy to verify that the same $\overline{h}_m$ result is approached by eq. (11.81) as Re_L exceeds 10^7.

For a single *cylinder in cross-flow*, the expression for the surface-averaged Sherwood number can be derived from the heat transfer correlation listed in Chapter 5:

$$\overline{Nu}_D = 0.3 + \frac{0.62Re_D^{1/2}Pr^{1/3}}{[1 + (0.4/Pr)^{2/3}]^{1/4}}\left[1 + \left(\frac{Re_D}{282,000}\right)^{5/8}\right]^{4/5} \tag{5.135}$$

In this expression, $\overline{Nu}_D$ is replaced by

$$\overline{Sh}_{D_o} = \frac{\overline{h}_m D_o}{D} \tag{11.88}$$

in which the outer diameter of the cylinder (D_o, Fig. 11.10) must not be confused with the mass diffusivity coefficient D. As usual, on the right side of eq. (5.135) the Reynolds number Re_D retains its position (it becomes labeled Re_{D_o}), and Pr is replaced by Sc:

$$\overline{Sh}_{D_o} = 0.3 + \frac{0.62Re_{D_o}^{1/2}Sc^{1/3}}{[1 + (0.4/Sc)^{2/3}]^{1/4}}\left[1 + \left(\frac{Re_{D_o}}{282,000}\right)^{5/8}\right]^{4/5}$$

$$(Re_{D_o}Sc > 0.2) \tag{11.89}$$

Similarly, the surface-averaged Sherwood number for mass transfer from a *sphere* of diameter D_o to a uniform flow (U_∞, C_∞, Fig. 11.10) can be deduced from eq. (5.139). The operation consists of replacing Pr with Sc, and $\overline{Nu}_D$ with the average Sherwood number defined in eq. (11.88):

$$\overline{Sh}_{D_o} = 2 + (0.4Re_{D_o}^{1/2} + 0.06\,Re_{D_o}^{2/3})Sc^{0.4}, \qquad (3.5 < Re_{D_o} < 7.6 \times 10^4) \tag{11.90}$$

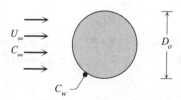

Figure 11.10 Uniform flow past a perpendicular cylinder or a sphere.

11.4.4 Internal Forced Convection

In configurations of internal forced convection, or duct flow, the local mass transfer coefficient is based on the difference between the species concentration of the exposed side of the wall and the bulk concentration of the stream,

$$h_m = \frac{\hat{j}_w}{C_w - C_b} = \frac{\hat{j}_w}{\rho_w - \rho_b} \tag{11.91}$$

In a way that parallels the definition of bulk temperature at the start of Chapter 6, the bulk concentration of the stream is defined by

$$C_b = \frac{1}{UA} \int_A uC \, dA \tag{11.92}$$

In this averaging process, u is the longitudinal velocity of the stream (Fig. 11.11), U is the cross section-averaged velocity, and A is the cross-sectional area.

The flow begins with an entrance region, followed downstream by a fully developed region. In the entrance region, the concentration profile changes shape from one longitudinal location to the next. In accordance with eq. (6.32) and the mass transfer–heat transfer analogy discussed in the preceding three subsections, the concentration entrance length X_C for *laminar flow* is approximately

$$\frac{X_C}{D_h} \cong 0.05 \mathrm{Re}_{D_h} \mathrm{Sc} \tag{11.93}$$

The Reynolds number is based on the mean velocity U and the hydraulic diameter D_h; in other words, $\mathrm{Re}_{D_h} = UD_h/\nu$. The X_C estimate provided by eq. (11.93) is valid for all Schmidt numbers.

In the case of laminar flow through a round tube with constant species concentration at the wall (C_w, Fig. 11.11), the local mass transfer coefficient h_m can be estimated with the help of Fig. 6.10. On the ordinate of that figure, Nu_D is replaced by the local Sherwood number based on the tube inner diameter D_i:

$$\mathrm{Sh}_{D_i} = \frac{h_m D_i}{D} \tag{11.94}$$

In this definition the tube inner diameter D_i should not be confused with the mass diffusivity D. The subscript i is meant as a reminder that the diameter D_i is *internal*. In other words, in this instance i does not refer to the species i mentioned in many parts of this chapter.

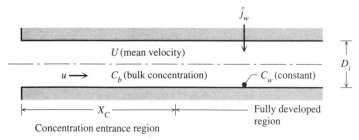

Figure 11.11 Flow through a round tube with constant species concentration at the wall.

The new dimensionless group for the abscissa of Fig. 6.10 is $(x/D_i)/(Re_{D_i}Sc)$, and the new parameter that distinguishes between the curves is Sc, instead of Pr. This modified version of Fig. 6.10 shows that in the fully developed region the mass transfer coefficient becomes independent of longitudinal position, approaching the value

$$\frac{h_m D_i}{D} = 3.66 \qquad (11.95)$$

In the *turbulent regime*, the concentration entrance length is about ten times the tube diameter, that is, approximately the same as the hydrodynamic entrance length X of eq. (6.65). The heat transfer correlations of Chapter 6 can be converted into their mass transfer equivalents through the substitutions $Nu_D \rightarrow Sh_{D_i}$, $Pr \rightarrow Sc$, and $Re_D \rightarrow Re_{D_i}$. For example, the heat transfer correlation (6.90) leads to the following estimate for the Sherwood number in fully developed turbulent flow:

$$Sh_{D_i} = 0.023 Re_{D_i}^{4/5} Sc^{1/3}, \qquad (Sc \gtrsim 0.5, \; 2 \times 10^4 < Re_{D_i} < 10^6) \qquad (11.96)$$

Example 11.4

Laminar Fully Developed Duct Flow, Asymmetric Boundary Conditions

The space between two parallel glass panes is 0.4 cm, and the length of this channel is 1.5 m. A very fine water mist covers one of the glass surfaces. The mist can be regarded as a thin film of water. The other surface does not have any water liquid on it. The temperature of the entire system is uniform and equal to 25°C.

Air is forced to flow through the channel for the purpose of drying the foggy surface. The cross section-averaged air velocity is 0.5 m/s. Calculate the mass transfer coefficient between the wet wall and the air stream.

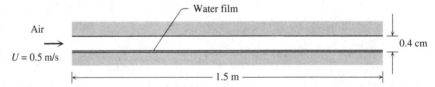

Figure E11.4

Solution.   First, we record the following properties of air and water vapor as a component in humid air:

$$\rho_a = 1.185 \text{ kg/m}^3 \qquad \nu_a = 1.55 \times 10^{-5} \text{ m}^2/\text{s} \qquad D = 2.88 \times 10^{-5} \text{ m}^2/\text{s}$$

Before we can answer any "convection" questions, we must determine the characteristics of the flow. The hydraulic diameter of the parallel-plate channel is twice the plate-to-plate spacing, $D_h = 2 \times 0.004 \text{ m} = 0.008 \text{ m}$. The Reynolds number is, therefore,

$$Re_{D_h} = \frac{U D_h}{\nu_a} = \left(0.5 \; \frac{\text{m}}{\text{s}}\right) \frac{0.008 \text{ m}}{1.55 \times 10^{-5} \text{ m}^2/\text{s}}$$

$$= 258 \qquad \text{(laminar)}$$

The entrance length of this laminar flow can be estimated based on eq. (6.4'):

$$X \cong 0.05 D_h Re_{D_h} = 0.05 \times 0.008 \text{ m} \times 258$$
$$\cong 0.1 \text{ m}$$

The entrance length is negligible relative to the total length of the channel. In conclusion, we can treat the flow as laminar and fully developed.

The mass transfer coefficient h_m appears in the D_h-based Sherwood number

$$Sh_{D_h} = \frac{h_m D_h}{D}$$

The Sherwood number Sh_{D_h} is analogous (and equal) to the D_h-based Nusselt number Nu_{D_h} whose numerical values are catalogued in Table 6.1. The next step is to select from Table 6.1 the proper Nu_{D_h}, that is, the value that corresponds to the boundary conditions of the present mass transfer problem. We note first that the wet wall (specifically the surface of the water film) acts as plane of constant species concentration, $C_w = $ constant. The opposite wall is impermeable; therefore, the mass transfer rate through it is equal to zero.

These asymmetric mass transfer boundary conditions correspond to the heat transfer configuration in which one wall is isothermal and the other is adiabatic. This configuration occupies the bottom line in Table 6.1; therefore,

$$Nu_{D_h} = 4.86$$

and, since $Nu_{D_h} = Sh_{D_h}$,

$$h_m = 4.86 \frac{D}{D_h} = 4.86 \frac{2.88 \times 10^{-5} \text{ m}^2/\text{s}}{0.008 \text{ m}}$$

$$= 0.0175 \text{ m/s}$$

11.4.5 Natural Convection

Local variations of species concentration can induce a buoyancy effect in the same way that temperature variations caused all the natural convection flows analyzed in Chapter 7. The basis for the concentration buoyancy effect is the fact that, in general, the density of the mixture varies with the concentration of the species of interest. In particular, the density of a binary (two-constituent) mixture ρ depends on the mixture temperature T, mixture pressure P, and mixture composition. The latter is described by the density of the species* of interest, ρ_i; therefore, at equilibrium the mixture density obeys a relation of the type

$$\rho = \rho(T, P, \rho_i) \tag{11.97}$$

When the temperature and pressure are uniform through the mixture and the ρ_i variations are sufficiently small, the equation of state (11.97) can be replaced with the linear relation

$$\rho \cong \rho_\infty + \left(\frac{\partial \rho}{\partial \rho_i}\right)_{T,P} (\rho_i - \rho_{i,\infty}) + \cdots$$

$$= \rho_\infty [1 - \beta_c (\rho_i - \rho_{i,\infty}) + \cdots] \tag{11.98}$$

In this expression ρ_∞ is the mixture density that corresponds to the "reference" composition represented by the species density $\rho_{i,\infty}$. The β_c factor is the *composition expansion coefficient* defined by

---

*Note that in this section it is necessary to use the subscript i for the species of interest. This subscript was omitted in Sections 11.4.1–11.4.4.

$$\beta_c = -\frac{1}{\rho}\left(\frac{\partial \rho}{\partial \rho_i}\right)_{T,P} \tag{11.99}$$

Note the analogy between eqs. (11.98) and (7.14) and between β_c and the thermal expansion coefficient β of thermal natural convection.

Representative numerical values of the product $\rho\beta_c$ (dimensionless) are listed in Table 11.7. A positive $\rho\beta_c$ value corresponds to a mixture that becomes lighter (less dense) as the presence of the species of interest intensifies (i.e., as ρ_i increases). An example of such a mixture is humid air. A negative $\rho\beta_c$ value, on the other hand, indicates a mixture that becomes denser as more of the species of interest is added to the mixture.

The mass transfer rates due to flows driven by the buoyancy effect of nonuniform composition can be deduced from the analogous heat transfer results presented in Chapter 7. To see how the mass transfer–heat transfer analogy extends to natural convection, consider the laminar boundary layer flow shown in Fig. 11.12. The momentum equation for this flow is obtained by combining eq. (7.10) with eq. (11.98):

$$u\frac{\partial v}{\partial x} + v\frac{\partial v}{\partial y} = \nu\frac{\partial^2 v}{\partial x^2} + g\beta_c\left(\rho_i - \rho_{i,\infty}\right) \tag{11.100}$$

This momentum equation is almost identical to that of the heat transfer-driven flow, eq. (7.17). The only change occurs in the last term (the buoyancy term), in which the group $\beta(T - T_\infty)$ is now replaced by $\beta_c(\rho_i - \rho_{i,\infty})$. Related to this observation is a comparison of Fig. 7.2 with Fig. 11.12, which shows that the boundary conditions of the two problems are such that the group $\beta(T_w - T_\infty)$ will be replaced by $\beta_c(\rho_{i,w} - \rho_{i,\infty})$.

The species conservation equation for the boundary layer of Fig. 11.12 reads

$$u\frac{\partial \rho_i}{\partial x} + v\frac{\partial \rho_i}{\partial y} = D\frac{\partial^2 \rho_i}{\partial x^2} \tag{11.101}$$

Compared with the corresponding energy equation for the thermally driven flow of Fig. 7.2, eq. (7.11), the species conservation equation shows that the role

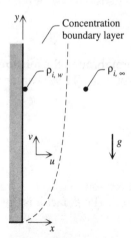

Figure 11.12 Natural convection boundary layer in an *isothermal* fluid in contact with a vertical surface with the same temperature and with constant species concentration.

of the thermal diffusivity α of Chapter 7 is now played by the mass diffusivity D. This last substitution occurred also in the forced-convection mass transfer problems treated earlier in this section.

In summary, because of the two substitutions noted above, the Rayleigh number of Chapter 7 is replaced by a new dimensionless group called *mass transfer Rayleigh number*, $Ra_{m,y}$:

$$Ra_y = \frac{g\beta(T_w - T_\infty)y^3}{v\alpha} \rightarrow \frac{g\beta_c(\rho_{i,w} - \rho_{i,\infty})y^3}{vD} = Ra_{m,y} \qquad (11.102)$$

The analogy between the results of Chapter 7 and those for mass transfer natural convection consists of these substitutions, in addition to the more general substitutions noted in the case of forced-convection mass transfer. Here is a natural-convection update of the tabulation begun in eqs. (11.62) through (11.62') on pages 599 and 600:

Heat Transfer		Mass Transfer
T	$\rightarrow$	ρ_i
$\beta(T_w - T_\infty)$	$\rightarrow$	$\beta_c(\rho_{i,w} - \rho_{i,\infty})$
α	$\rightarrow$	D
Ra_y	$\rightarrow$	$Ra_{m,y}$
k	$\rightarrow$	D
Pr	$\rightarrow$	Sc
Nu_y	$\rightarrow$	Sh_y

$$(11.62'')$$

In this way, the overall Sherwood number for the vertical surface of Fig. 11.12 can be deduced from the overall Nusselt number correlation (7.61):

$$\overline{Sh}_y = \left\{ 0.825 + \frac{0.387Ra_{m,y}^{1/6}}{[1 + (0.492/Sc)^{9/16}]^{8/27}} \right\}^2, \quad (10^{-1} < Ra_{m,y} < 10^{12}) \quad (11.103)$$

In accordance with eqs. (11.69) and (11.72), the overall Sherwood number $\overline{Sh}_y$ is a dimensionless way of expressing the height-averaged mass transfer coefficient,

$$\overline{Sh}_y = \frac{\overline{h}_m y}{D} = \frac{\overline{j}_w y}{(\rho_{i,w} - \rho_{i,\infty})D} = \frac{\hat{j}_w y}{(C_{i,w} - C_{i,\infty})D} \qquad (11.104)$$

The preceding discussion was based on the tacit assumption that the concentration of the species of interest is low enough so that the vertical surface of Fig. 11.12 can be treated as impermeable. The Sherwood number formulas for other flows driven by composition-induced buoyancy can be deduced from the results of Chapter 7, by making the substitutions (11.62''). More precise means for calculating the natural convection evaporation from a shallow water pan in contact with air have been developed by Sparrow et al. [22].

Vastly more complicated is the problem of mass transfer in a nonisothermal flow, which is being driven by a *combination of two buoyancy effects*, the effect of temperature variations, $g\beta(T - T_\infty)$, and the effect of nonuniform composition, $g\beta_c(\rho_i - \rho_{i,\infty})$. Results for this class of natural convection configurations can be found in the more specialized books on convection (e.g., Refs. [3, 23]).

REFERENCES

1. A. Bejan, *Advanced Engineering Thermodynamics*, Wiley, New York, 1988, pp. 224–232.
2. E. M. Sparrow and G. A. Nunez, Experiments on isothermal and non-isothermal evaporation from partially filled, open-topped vertical tubes, *Int. J. Heat Mass Transfer*, Vol. 31, 1988, pp. 1345–1355.
3. A. Bejan, *Convection Heat Transfer*, Wiley, New York, 1984, Chapter 9.
4. A. Fick, On liquid diffusion, *Philos. Mag.*, Vol. 4, No. 10, 1855, pp. 30–39.
5. H. R. Perry and C. H. Chilton, *Chemical Engineers' Handbook*, 5th edition, McGraw-Hill, New York, 1973.
6. R. C. Reid, J. M. Prausnitz, and T. K. Sherwood, *The Properties of Gases and Liquids*, 3rd edition, McGraw-Hill, New York, 1977.
7. T. R. Marrero and E. A. Mason, Gaseous diffusion coefficients, *J. Phys. Chem. Ref. Data*, Vol. 1, 1972, pp. 3–118.
8. R. B. Bird, W. E. Stewart, and E. N. Lightfoot, *Transport Phenomena*, Wiley, New York, 1960, Chapter 16.
9. E. N. Fuller, P. D. Schettler, and J. C. Giddins, A new method for prediction of binary gas-phase diffusion coefficients, *Industrial and Engineering Chemistry*, Vol. 58, No. 5, 1966, pp. 19–27.
10. J. O. Hirschfelder, C. F. Curtiss, and R. B. Bird, *Molecular Theory of Gases and Liquids*, Wiley, New York, 1954.
11. R. Krishna and G. L. Standart, Mass and energy transfer in multicomponent systems, *Chem. Eng. Commun.*, Vol. 3, 1979, pp. 201–275.
12. E. Obermeier and A. Schaber, A simple formula for multicomponent gaseous diffusion coefficients derived from mean free path theory, *Int. J. Heat Mass Transfer*, Vol. 20, 1977, pp. 1301–1306.
13. C. R. Wilke and P. Chang, Correlation of diffusion coefficients in dilute solutions, *AIChE J.*, Vol. 1, 1955, pp. 264–270.
14. R. M. Barrer, *Diffusion in and through Solids*, Macmillan, New York, 1941.
15. J. Crank and G. S. Park, *Diffusion in Polymers*, Academic Press, London, 1968.
16. P. G. Shewmon, *Transformations in Metals*, McGraw-Hill, New York, 1969, p. 53.
17. C. E. Rogers, *Engineering Design for Plastics*, Reinhold, New York, 1964.
18. J. H. Perry, Ed., *Chemical Engineers' Handbook*, McGraw-Hill, New York, 1950.
19. A. E. Markham and K. A. Kobe, The solubility of gases in liquids, *Chem. Rev.*, Vol. 28, 1941, pp. 519–588.
20. R. Battino and H. L. Clever, The solubility of gases in liquids, *Chem. Rev.*, Vol. 66, 1966, pp. 395–463.
21. B. Gebhart and L. Pera, The nature of vertical natural convection flows resulting from the combined buoyancy effects of thermal and mass diffusion, *Int. J. Heat Mass Transfer*, Vol. 14, 1971, pp. 2025–2050.
22. E. M. Sparrow, G. K. Kratz, and M. J. Schuerger, Evaporation of water from a horizontal surface by natural convection, *J. Heat Transfer*, Vol. 105, 1983, pp. 469–475.
23. B. Gebhart, Y. Jaluria, R. L. Mahajan, and B. Sammakia, *Buoyancy-Induced Flows and Transport*, Hemisphere, New York, 1988, Chapter 6.

PROBLEMS

Steady Diffusion

11.1 Derive eq. (11.19) and in this way prove that the mole fraction of a certain species in a mixture is proportional to its mass concentration.

11.2 A neoprene membrane 0.1 mm thin covers an orifice in the wall that separates a chamber filled with hydrogen at 2 bar and 300 K, from an adjacent chamber filled with hydrogen at 1 bar and 300 K. The orifice is small enough so that the membrane is practically plane.

  (a) Calculate the molar concentrations of hydrogen on the gas side and solid side of each of the neoprene–hydrogen interfaces.

  (b) Plot the steady-state distribution of molar concentration of hydrogen through and in the vicinity of the neoprene membrane.

  (c) Calculate the molar and mass fluxes of the hydrogen that diffuses through the membrane.

11.3 A straight tube made out of Pyrex contains helium gas at 2 bar pressure and room temperature. The outer and inner diameters of the tube are 4 mm and 3.5 mm, respectively. The tube is surrounded by vacuum. Calculate the steady-state molar flowrate of helium through the Pyrex wall, per unit length of tube.

11.4 Hydrogen is stored at high pressure in a stainless-steel bottle with a wall thickness of 1 mm. The solubility of hydrogen in stainless steel is such that the mass concentration of hydrogen at the internal surface of the wall is 0.01 g/cm^3. The diffusivity of hydrogen through stainless steel is approximately 10^{-5} cm^2/s. Calculate the rate of hydrogen leakage through the wall, and express this result in terms of standard cubic meters (H$_2$ at 1 atm and 25°C), per square meter and per second.

11.5 Assume that in the natural drying of a hazelnut in air, water vapor diffuses through the air trapped in the porous shell of the nut. Air at 20°C and 1 atm resides on both sides of the shell. On the inside, the air is saturated with water vapor (supplied by the "green" nut), while on the outside the air has a relative humidity of 10 percent. The hazelnut is approximated adequately by a sphere with an outside diameter of 1 cm. The shell has a nearly uniform thickness of 1 mm. Calculate the mass flowrate of the water that escapes through the shell.

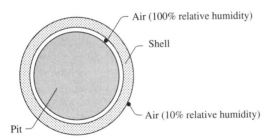

Figure P11.5

11.6 A spherical bead of silica (SiO$_2$) glass with a diameter of 5 mm is surrounded by room-temperature air. As a manufacturing defect, the bead contains a gaseous inclusion of helium at the pressure of 1 bar. The diameter of the spherical helium bubble is 2 mm. The glass shell and the helium bubble are positioned concentrically.

  (a) Calculate the molar flowrate of the helium that diffuses through the glass shell.

(b) How long will it take until one half of the original amount of trapped helium escapes?

Time-Dependent Diffusion

11.7   A shallow pan contains a layer of pure water 1 cm deep. The pan is exposed suddenly to the atmosphere, and air begins to diffuse into the water through the free surface. Assume that the water layer is shallow enough so that the water is motionless. In other words, neglect the effect of natural convection in the water and the surrounding air. Estimate the time needed for the entire volume of water to absorb half of the total amount of air of which it is capable of absorbing. The water temperature is 20°C and, initially, there is no air dissolved in the water.

11.8   A spherical ball of mild steel is being case hardened by contact with a carbon-rich packing at high temperature. The initial mass concentration of carbon in mild steel is 16 kg/m³, while the carbon concentration in steel at the surface of the sphere is 120 kg/m³. The sphere diameter is 1 cm, and the diffusivity of carbon in mild steel is $D \cong 6 \times 10^{-10}$ m²/s.

(a) Calculate the time when carbon reaches a concentration of 50 kg/m³ at a distance of 0.5 mm beneath the surface of the sphere.

(b) Calculate the time when the concentration of carbon averaged over the entire spherical volume reaches 100 kg/m³.

11.9   A vacuum is created suddenly on the outside of a silica bottle filled with helium at 20°C. The bottle wall is 2 mm thick. Estimate the order of magnitude of the time until the diffusion of helium through the wall becomes steady. How much shorter will this time be if the wall is made of polyethylene?

11.10   Steel rods with diameters of 2 cm are case hardened by storage in a carbon-rich environment. The initial concentration of carbon in the rods is 16 kg/m³. The carbon concentration imposed by the environment at the rod surface is 120 kg/m³. The diffusivity of carbon through mild steel is approximately 6×10^{-10} m²/s. Calculate the storage time that is needed to raise the volume-averaged concentration of carbon in the rods to 110 kg/m³.

11.11   A 1-mm-thick sheet of nickel contains atomic hydrogen, which is distributed uniformly through its volume. The hydrogen is removed by heating the nickel to 400°C and holding it in vacuum. This procedure is called vacuum degassing. Calculate the duration of this process if the average concentration of hydrogen is to decrease to 10 percent of its initial level.

11.12   A sheet of austenite is decarburized by heating to 900°C and exposure to vacuum. The initial concentration of carbon is uniform. How long will it be until the carbon concentration at 1 mm beneath the surface drops to half of its initial value?

11.13   An iron machine part is heated to 900°C and exposed to a carbon-rich environment. Carbon diffuses to a certain depth during a specified time interval. It is proposed to double the diffusion penetration depth associated with that time interval. Calculate the proper temperature to which the iron must be heated.

Forced Convection

11.14   A shallow baby pool is filled with water at 25°C. The pool surface is a 3 m × 3 m square. The wind blows at 30 km/h parallel to the water surface and parallel to one side of the square. The atmospheric air has a temperature of 25°C and a relative humidity of 30 percent. Assume that relative to this air flow the pool water surface behaves as a plane stationary wall.

(a) Calculate the instantaneous flowrate of the water that is being removed by the wind.

(b) How long does the wind have to blow for the pool level to drop by 2 mm?

11.15   A spray of room deodorant is released at approximately 5 m from an observer who smells the deodorant about 10 seconds later. Did the deodorant propagate through the room air by pure mass diffusion? Estimate the order of magnitude of the velocity of the air draft.

11.16   Consider the straight duct with fully developed mixture flow sketched in the attached figure. The mixture average velocity is U, the duct cross-sectional area is A_c, and the duct length is L. The cross section perimeter is $p = A/L$, where A is the total surface of the duct wall.

The bulk density of the species of interest is denoted by ρ (kg/m³). The ρ value at the duct inlet is known, ρ_{in}. Specified also is the ρ value on the duct wall, $\rho_w =$ constant. By following the steps of the equivalent heat transfer analysis presented in Section 6.4, show that the rate of mass transfer from the entire duct to the mixture stream is

$$\dot{m} = UA_c(\rho_w - \rho_{in})\left[1 - \exp\left(-\frac{h_m}{U}\frac{A}{A_c}\right)\right]$$

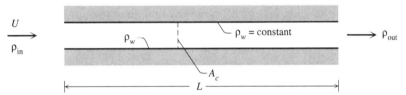

Figure P11.16

11.17   The double-pane window system shown in the figure is plagued by the deposition of liquid water (fogging) on the inner surfaces of the glass panes. On each surface the liquid is well approximated by a film with constant thickness $\delta = 10^{-5}$ m.

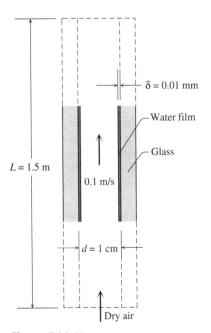

Figure P11.17

It is proposed to defog the window by blowing dry air (0 percent relative humidity) longitudinally through the parallel-plate channel. The mean velocity of the channel air stream is 0.1 m/s. The temperature in the entire system is uniform and equal to 25°C. The pressure is atmospheric. Neglecting natural convection effects, determine:

(a) The flow regime over most of the length L,

(b) The mass transfer coefficient between the liquid films and the channel flow,

(c) The rate at which the air stream removes water from the wetted surfaces (using the formula listed in the preceding problem statement), and

(d) The time needed for defogging completely the double-pane window system.

11.18   In the double-pane window described in Example 11.4 in the text, the air flow is laminar and fully developed, and the mass transfer coefficient is 0.0175 m/s. The channel is 1.5 m long, and the spacing is 0.4 cm. Air with 20 percent relative humidity enters with a mean speed of 0.5 m/s. The temperature is 25°C throughout the system. Calculate the rate at which water is being removed from the wet glass surface (note the relationship listed in Problem 11.16). Determine also the bulk relative humidity of the air stream that leaves the channel.

11.19   In the drying (aging) of the new hazelnut shown in the figure attached to Problem 11.5, the velocity of the ambient air is 1 m/s. The temperature is 20°C everywhere. The hazelnut can be modeled as a sphere with an outer diameter of 1 cm.

(a) Calculate the average mass transfer coefficient $\bar{h}$ between the shell and the ambient air.

(b) Compare the external mass transfer resistance $(1/\bar{h})$ with the pure diffusion resistance posed by the air trapped in the shell (assumed porous and fissured). The shell thickness is 1 mm. Is the external mass transfer resistance negligible?

11.20   A baseball pitcher throws a 70 miles per hour "spitball" in a game played in 25°C temperature and 20 percent relative humidity. The illegal act of preparing the spitball is approximated well by the even spreading of 2 cm³ of 25°C water over the ball. The distance traveled by the ball is 19.4 m, and the ball diameter is 7.4 cm.

Calculate and express in cm³ the amount of water that evaporates as the ball travels through the air. Compare this quantity with the original amount, and advise the umpire on whether he should check the ball after it is caught. In other words, is the "evidence" likely to disappear between the pitcher's mound and the catcher's mitt?

Natural Convection

11.21   (a) Calculate the $\rho\beta_c$ value of water vapor as a component in humid air at 1 atm and 25°C. Rely on eq. (11.99) and assume that the density of humid air varies (approximately) linearly as the relative humidity increases from 0 to 100 percent. In other words, calculate the density of humid air and the mass concentration of water vapor, for both ends of the relative humidity range. Compare the calculated $\rho\beta_c$ value with the value listed in Table 11.7.

(b) Show that the ρ/M ratio of humid air is practically constant as the relative humidity increases from 0 percent to 100 percent.

11.22   A thin film of water wets a vertical wall 1 m high and 1 m wide. The temperature of the wall and the surrounding air is 25°C. The surrounding air is still and has a relative humidity of 40 percent. Calculate the mass transfer Rayleigh number based on height, and the water mass transfer rate removed by the natural convection of humid air. Is the humid air rising or descending along the wall?

11.23   Consider the laminar natural convection boundary layer driven by mass transfer along a vertical wall of height H. The wall and the surrounding fluid mixture (density ρ) are at the same temperature, while the species mass concentration difference between the vertical surface and the mixture reservoir is $\rho_{i,w} - \rho_{i,\infty}$.

(a) Invoke the analogy between natural convection mass transfer and natural convection heat transfer, and obtain an order-of-magnitude expression for the horizontal (or "entrainment") velocity component of the mixture.

(b) Rely on the same analogy to estimate the order of magnitude of the mass flux of species i through the wall.

(c) Compare the horizontal mass fluxes estimated in parts (a) and (b), and show that the vertical surface may be modeled as an impermeable surface when

$$|\rho_{i,w} - \rho_{i,\infty}| << \rho$$

In other words, show that the analogy with the heat transfer phenomena of Chapter 7 holds true when the species of interest is present in small quantities in the mixture.

APPENDIX A
CONSTANTS AND CONVERSION FACTORS

CONSTANTS

Universal ideal gas constant

$$\bar{R} = 8.314 \text{ kJ/kmol} \cdot \text{K}$$
$$= 1.9872 \text{ cal/mol} \cdot \text{K}$$
$$= 1.9872 \text{ Btu/lbmol} \cdot \text{R}$$
$$= 1545.33 \text{ ft} \cdot \text{lbf/lbmol} \cdot \text{R}$$

Boltzmann's constant $\quad k = 1.38054 \times 10^{-23} \text{ J/K}$

Planck's constant $\quad h = 6.626 \times 10^{-34} \text{ J} \cdot \text{s}$

Speed of light in vacuum $\quad c = 2.998 \times 10^{8} \text{ m/s}$

Avogadro's number $\quad N = 6.022 \times 10^{23} \text{ molecules/mol}$

Stefan–Boltzmann constant

$$\sigma = 5.669 \times 10^{-8} \text{ W/m}^2 \cdot \text{K}^4$$
$$= 0.1714 \times 10^{-8} \text{ Btu/h} \cdot \text{ft}^2 \cdot \text{R}^4$$

Atmospheric pressure

$$P_{\text{atm}} = 0.101325 \text{ MPa}$$
$$= 1.01325 \text{ bar}$$
$$= 1.01325 \times 10^{5} \text{ N/m}^2$$

Ice point at 1 atm $\quad T_{\text{ice}} = 0°\text{C} = 273.15 \text{ K}$

Gravitational acceleration

$$g = 9.807 \text{ m/s}^2$$
$$= 32.17 \text{ ft/s}^2$$

Mole

1 mol = sample containing 6.022×10^{23} elementary entities (e.g., molecules); abbreviated also as 1 gmol, or

$$1 \text{ mol} = 10^{-3} \text{ kmol}$$
$$= 10^{-3} \text{ kgmol}$$
$$= \frac{1}{453.6} \text{ lbmol}$$

Natural logarithm

$$\ln x = 2.30258 \log_{10} x$$
$$\log_{10} x = 0.4343 \ln x$$

Important numbers

$$e = 2.71828$$
$$\pi = 3.14159$$
$$1° = 0.01745 \text{ radians}$$

CONVERSION FACTORS

Acceleration	$1 \text{ m/s}^2 = 4.252 \times 10^7 \text{ ft/h}^2$
Area	$1 \text{ in.}^2 = 6.452 \text{ cm}^2$
	$1 \text{ ft}^2 = 0.0929 \text{ m}^2$
	$1 \text{ yd}^2 = 0.8361 \text{ m}^2$
	$1 \text{ mile}^2 = 2.59 \text{ km}^2$
	$1 \text{ hectare} = (100 \text{ m})^2$
	$1 \text{ acre} = 4047 \text{ m}^2$
	$= 0.405 \text{ hectares}$
Density	$1 \text{ kg/m}^3 = 0.06243 \text{ lbm/ft}^3$
	$1 \text{ lbm/ft}^3 = 16.018 \text{ kg/m}^3$
Energy	$1 \text{ kJ} = 737.56 \text{ ft} \cdot \text{lbf}$
	$= 0.9478 \text{ Btu}$
	$= 3.725 \times 10^{-4} \text{ hp} \cdot \text{h}$
	$= 2.778 \times 10^{-4} \text{ kW} \cdot \text{h}$
	$1 \text{ Btu} = 1055 \text{ J}$
	$= 778.16 \text{ ft} \cdot \text{lbf}$
	$= 3412.14 \text{ kW} \cdot \text{h}$
	$= 2544.5 \text{ hp} \cdot \text{h}$
	$1 \text{ cal} = 4.187 \text{ J}$
	$1 \text{ erg} = 10^{-7} \text{ J}$
Force	$1 \text{ lbf} = 4.448 \text{ N}$
	$= 0.4536 \text{ kgf}$
	$1 \text{ dyne} = 10^{-5} \text{ N}$
Heat flux	$1 \text{ W/m}^2 = 0.317 \text{ Btu/h} \cdot \text{ft}^2$
	$1 \text{ Btu/h} \cdot \text{ft}^2 = 3.154 \text{ W/m}^2$
Heat transfer coefficient	$1 \text{ W/m}^2 \cdot \text{K} = 0.1761 \text{ Btu/h} \cdot \text{ft}^2 \cdot {}^\circ\text{F}$
	$= 0.8598 \text{ kcal/h} \cdot \text{m}^2 \cdot {}^\circ\text{C}$
	$1 \text{ Btu/h} \cdot \text{ft}^2 \cdot {}^\circ\text{F} = 5.6786 \text{ W/m}^2 \cdot \text{K}$
Heat transfer rate	$1 \text{ Btu/s} = 1055 \text{ W}$
	$1 \text{ Btu/h} = 0.2931 \text{ W}$
	$1 \text{ hp} = 745.7 \text{ W}$
	$1 \text{ ft} \cdot \text{lbf/s} = 1.3558 \text{ W}$
Kinematic viscosity (ν), thermal diffusivity (α), mass diffusivity (D)	$1 \text{ m}^2/\text{s} = 10^4 \text{ cm}^2/\text{s}$
	$= 10^4 \text{ stokes}$
	$= 3.875 \times 10^4 \text{ ft}^2/\text{h}$
	$= 10.764 \text{ ft}^2/\text{s}$
Latent heat, specific energy, specific enthalpy	$1 \text{ kJ/kg} = 0.4299 \text{ Btu/lbm}$
	$= 0.2388 \text{ cal/g}$
	$1 \text{ Btu/lbm} = 2.326 \text{ kJ/kg}$
Length	$1 \text{ in.} = 2.54 \text{ cm}$
	$1 \text{ ft} = 0.3048 \text{ m}$
	$1 \text{ yd} = 0.9144 \text{ m}$
	$1 \text{ mile} = 1.609 \text{ km}$
Mass	$1 \text{ lbm} = 0.4536 \text{ kg}$
	$1 \text{ kg} = 2.2046 \text{ lbm}$
	$= 1.1023 \times 10^{-3} \text{ U.S. ton}$
	$= 10^{-3} \text{ tonne}$

$$1 \text{ oz} = 28.35 \text{ g}$$

Mass transfer coefficient	$1 \text{ m/s} = 1.181 \times 10^4 \text{ ft/h}$
	$1 \text{ ft/h} = 8.467 \times 10^{-5} \text{ m/s}$
Power	$1 \text{ Btu/s} = 1055 \text{ W} = 1.055 \text{ kW}$
	$1 \text{ Btu/h} = 0.293 \text{ W}$
	$1 \text{ W} = 3.412 \text{ Btu/h}$
	$= 9.48 \times 10^{-4} \text{ Btu/s}$
	$1 \text{ hp} = 0.746 \text{ kW}$
	$= 0.707 \text{ Btu/s}$
Pressure, stress	$1 \text{ Pa} = 1 \text{ N/m}^2$
	$1 \text{ psi} = 6895 \text{ N/m}^2$
	$1 \text{ atm} = 14.69 \text{ psi}$
	$= 1.013 \times 10^5 \text{ N/m}^2$
	$1 \text{ bar} = 10^5 \text{ N/m}^2$
	$1 \text{ Torr} = 1 \text{ mm Hg}$
	$= 133.32 \text{ N/m}^2$
	$1 \text{ psi} = 27.68 \text{ in. H}_2\text{O}$
	$1 \text{ ft H}_2\text{O} = 0.4335 \text{ psi}$
Specific heat, specific entropy	$1 \text{ kJ/kg} \cdot \text{K} = 0.2388 \text{ Btu/lbm} \cdot {}^\circ\text{F}$
	$= 0.2389 \text{ cal/g} \cdot {}^\circ\text{C}$
	$1 \text{ Btu/lbm} \cdot {}^\circ\text{F} = 4.187 \text{ kJ/kg} \cdot \text{K}$
Speed	$1 \text{ mile/h} = 0.447 \text{ m/s}$
	$= 1.609 \text{ km/h}$
	$1 \text{ km/h} = 0.278 \text{ m/s}$
	$= 0.622 \text{ mile/h}$
	$1 \text{ m/s} = 3.6 \text{ km/h}$
	$= 2.237 \text{ miles/h}$
Temperature	$1 \text{ K} = 1{}^\circ\text{C}$
	$1 \text{ K} = (9/5){}^\circ\text{F}$
	$T(\text{K}) = T({}^\circ\text{C}) + 273.15$
	$T({}^\circ\text{C}) = (5/9) [T({}^\circ\text{F}) - 32]$
	$T({}^\circ\text{F}) = T(\text{R}) - 459.67$
Temperature difference	$\Delta T(\text{K}) = \Delta T({}^\circ\text{C})$
	$= (5/9) \Delta T({}^\circ\text{F}) = (5/9) \Delta T(\text{R})$
Thermal conductivity	$1 \text{ W/m} \cdot \text{K} = 0.5782 \text{ Btu/h} \cdot \text{ft} \cdot {}^\circ\text{F}$
	$= 0.01 \text{ W/cm} \cdot \text{K}$
	$= 2.39 \times 10^{-3} \text{ cal/cm} \cdot \text{s} \cdot {}^\circ\text{C}$
	$1 \text{ Btu/h} \cdot \text{ft} \cdot {}^\circ\text{F} = 1.7307 \text{ W/m} \cdot \text{K}$
Thermal resistance	$1 \text{ K/W} = 0.5275 {}^\circ\text{F/Btu} \cdot \text{h}$
	$1 {}^\circ\text{F/Btu} \cdot \text{h} = 1.896 \text{ K/W}$
Viscosity	$1 \text{ N} \cdot \text{s/m}^2 = 1 \text{ kg/s} \cdot \text{m}$
	$= 2419.1 \text{ lbm/ft} \cdot \text{h}$
	$= 5.802 \times 10^{-6} \text{ lbf} \cdot \text{h/ft}^2$
	$1 \text{ poise} = 1 \text{ g/s} \cdot \text{cm}$
Volume	$1 \text{ liter} = 10^{-3} \text{ m}^3 = 1 \text{ dm}^3$
	$1 \text{ in.}^3 = 16.39 \text{ cm}^3$
	$1 \text{ ft}^3 = 0.02832 \text{ m}^3$
	$1 \text{ yd}^3 = 0.7646 \text{ m}^3$
	$1 \text{ gal (U.S.)} = 3.785 \text{ liters}$

$$1 \text{ gal (imperial)} = 4.546 \text{ liters}$$
$$1 \text{ pint} = 0.5683 \text{ liters}$$

Volumetric heat
generation rate
$$1 \text{ W/m}^3 = 0.0966 \text{ Btu/h} \cdot \text{ft}^3$$
$$1 \text{ Btu/h} \cdot \text{ft}^3 = 10.35 \text{ W/m}^3$$

MULTIPLES USED IN SI UNITS

Multiplier	Prefix	Symbol
10^{12}	tera	T
10^{9}	giga	G
10^{6}	mega	M
10^{3}	kilo	k
10^{2}	hecto	h
10	deka	da
10^{-1}	deci	d
10^{-2}	centi	c
10^{-3}	milli	m
10^{-6}	micro	μ
10^{-9}	nano	n
10^{-12}	pico	p
10^{-15}	femto	f
10^{-18}	atto	a

DIMENSIONLESS GROUPS[a]

Biot number	$\text{Bi} = hL/k_s$
Boussinesq number	$\text{Bo} = g\beta \Delta T H^3/\alpha^2$
Eckert number	$\text{Ec} = U^2/c_p \Delta T$
Fourier number	$\text{Fo} = \alpha t/L^2$
Graetz number	$\text{Gz} = D^2 U/\alpha x = \text{Re}_D \text{Pr}\dfrac{D}{x}$
Grashof number	$\text{Gr} = g\beta \Delta T H^3/\nu^2$
Lewis number[b]	$\text{Le} = \alpha/D = \text{Sc}/\text{Pr}$
Mass transfer Stanton number	$\text{St}_m = h_m/U$
Mass transfer Rayleigh number[b]	$\text{Ra}_m = g\beta_c \Delta \rho_i H^3/\nu D$
Nusselt number	$\text{Nu} = hL/k_f$

---

[a]Subscripts $(\ )_s$ = solid, $(\ )_f$ = fluid.

[b]In this definition, D is the mass diffusivity (m^2/s).

Péclet number	$\mathrm{Pe} = UL/\alpha = \mathrm{Re} \cdot \mathrm{Pr}$
Prandtl number	$\mathrm{Pr} = v/\alpha = \mathrm{Sc}/\mathrm{Le}$
Pressure drop number	$\Pi = \Delta P \cdot L^2/\mu\alpha$
Rayleigh number	$\mathrm{Ra} = g\beta\,\Delta T H^3/\alpha v$
Rayleigh number based on heat flux	$\mathrm{Ra}^* = g\beta q'' H^4/\alpha v k$
Reynolds number	$\mathrm{Re} = UL/v$
Schmidt number[b]	$\mathrm{Sc} = v/D = \mathrm{Le} \cdot \mathrm{Pr}$
Sherwood number[b]	$\mathrm{Sh} = h_m L/D$
Stanton number	$\mathrm{St} = h/\rho c_p U = \mathrm{Nu}/\mathrm{Re} \cdot \mathrm{Pr}$
Stefan number	$\mathrm{Ste} = c_f \Delta T/h_{sf}$

APPENDIX B
PROPERTIES OF SOLIDS

NONMETALLIC SOLIDS[a]

Material	T (°C)	ρ (kg/m³)	c_P (kJ/kg·K)	k (W/m·K)	α (cm²/s)
Asbestos					
Cement board	20			0.6	
Felt (16 laminations per cm)	40			0.057	
Fiber	50	470	0.82	0.11	0.0029
Sheet	20			0.74	
	50			0.17	
Asphalt	20	2120	0.92	0.70	0.0036
Bakelite	20	1270	1.59	0.230	0.0011
Bark	25	340	1.26	0.074	0.0017
Beef (see Meat)					
Brick					
Carborundum	1400			11.1	
Cement	10	720		0.34	
Common	20	1800	0.84	0.38–0.52	0.0028–0.0034
Chrome	100			1.9	
Facing	20			1.3	
Firebrick	300	2000	0.96	0.1	0.00054
Magnesite (50% MgO)	20	2000		2.68	
Masonry	20	1700	0.84	0.66	0.0046
Silica (95% SiO$_2$)	20	1900		1.07	
Zircon (62% ZrO$_2$)	20	3600		2.44	
Brickwork, dried in air	20	1400–1800	0.84	0.58–0.81	0.0049–0.0054
Carbon					
Diamond (type IIb)	20	3250	0.51	1350	8.1
Graphite (firm, natural)	20	2000–2500	0.61	155	1.02–1.27
Carborundum (SiC)	100	1500	0.62	58	0.62
Cardboard	0–20	~790		~0.14	
Celluloid	20	1380	1.67	0.23	0.001

Cement (Portland, fresh, dry)	20	3100	0.75	0.3	0.0013
Chalk (CaCO₃)	20	2000–3000	0.74	2.2	0.01–0.015
Clay	20	1450	0.88	1.28	0.01
Fireclay	100	1700–2000	0.84	0.5–1.2	0.35–0.71
Sandy clay	20	1780		0.9	
Coal	20	1200–1500	1.26	0.26	0.0014–0.0017
Anthracite	900	1500		0.2	
Brown coal	900			0.1	
Bituminous in situ		1300		0.5–0.7	0.003–0.004
Dust	30	730	1.3	0.12	0.0013
Concrete, made with gravel, dry	20	2200	0.88	1.28	0.0066
Cinder	24			0.76	
Cork					
Board	20	150	1.88	0.042	0.0015
Expanded	20	120		0.036	
Cotton	30	81	1.15	0.059	0.0063
Earth					
Coarse-grained	20	2040	1.84	0.59	0.0016
Clayey (28% moisture)	20	1500		1.51	
Sandy (8% moisture)	20	1500		1.05	
Diatomaceous	20	466	0.88	0.126	0.0031
Fat	20	910	1.93	0.17	0.001
Felt, hair	−7	130–200		0.032–0.04	
	94	130–200		0.054–0.051	
Fiber insulating board	20	240		0.048	
Glass					
Borosilicate	30	2230		1.09	0.0036–0.0047
Fiber	20	220		0.035	0.0035
Lead	20	2890	0.68	0.7–0.93	0.0078
Mirror	20	2700	0.80	0.76	0.0087
Pyrex	60–100	2210	0.75	1.3	
Quartz	20	2210	0.73	1.4	0.0034
Window	20	2800	0.80	0.81	0.0028
Wool	0	200	0.66	0.037	

Continued

Material	T (°C)	ρ (kg/m³)	c_p (kJ/kg·K)	k (W/m·K)	α (cm²/s)
Granite	20	2750	0.89	2.9	0.012
Gypsum	20	1000	1.09	0.51	0.0047
Ice (see also p. 633)	0	917	2.04	2.25	0.012
Ivory	80			0.5	
Kapok	30			0.035	
Leather, dry	20	860	1.5	0.12–0.15	~0.001
Limestone (Indiana)	100	2300	0.9	1.1	~0.005
Linoleum	20	535		0.081	
Lunar surface dust, in high vacuum	250	1500 ± 300	~0.6	~0.0006	
Magnezia (85%)	38–204			0.067–0.08	
Marble	20	2600	0.81	2.8	0.013
Meat					
Beef					
chuck	25	1060			~0.0014
liver	43–66				0.0012
eye of loin, parallel to fiber	27			0.5	
ground	2–7			0.3	
lean	6			0.35	
	2–47			0.45	
Chicken					
muscle, perpendicular to fiber	5–27			0.41	
skin	5–27			0.03	
egg, white	33–38			0.55	
egg, whole	−8			0.46	
egg, albumen gel, freeze-dried	41			0.04	
egg, yolk	24–38			0.42	
Fish					
cod fillets	−19			1.17	
halibut	43–66	1080			0.0014
herring	−19			0.8	
salmon, perpendicular to fiber	−23			1.3	
	2			0.7	

	Temperature				
salmon, freeze-dried, parallel to fiber	−29			0.04	
Horse	25			0.41	
Lamb, lean	7–57			0.45	
Pork		1090			0.0014
ham, smoked	43–66			0.15	
fat	25			0.52	
lean, perpendicular to fiber	27–57			0.45	
lean, parallel to fiber	7–57			0.37	
pig skin	25				
Sausage					
23% fat	25			0.38	
15% fat	25			0.43	
Seal, blubber	(−13)–(−2)			0.21	
Turkey					
breast, perpendicular to fiber	−3			1.05	
	2			0.7	
breast, parallel to fiber	−8			1.4	
leg, perpendicular to fiber	2			0.7	
Whale					
blubber, perpendicular to fiber	18			0.21	
meat	−12			1.3	
Mica	20	2900	0.8	0.52	0.0061
Mortar	20	1900	1.2	0.93	0.0014
Paper	20	700	2.9	0.12	~0.001
Paraffin	30	870–925	0.8	0.24–0.27	0.0058
Plaster	20	1690	1.44	0.79	0.0011
Plexiglas (Acrylic glass)	20	1180		0.184	
Plums	−16		2.30	0.3	0.0017
Polyethylene	20	920		0.35	
Polystyrene	20	1050		0.157	
Polyurethane	20	1200	2.09	0.32	0.0013
Polyvinyl chloride (PVC)	20	1380	0.96	0.15	0.0011
Porcelain	95	2400	1.08	1.03	0.004

Continued

627

Material	T (°C)	ρ (kg/m³)	c_p (kJ/kg·K)	k (W/m·K)	α (cm²/s)
Quartz	20	2100–2500	0.78	1.40	~0.008
Rubber					
Foam	20	500	1.67	0.09	0.0011
Hard (ebonite)	20	1150	2.01	0.16	0.0006
Soft	20	1100	1.67	~0.2	~0.001
Synthetic	20	1150	1.97	0.23	0.001
Salt (rock salt)	0	2100–2500	0.92	7	0.03–0.036
Sand					
Dry	20			0.58	
Moist	20	1640		1.13	
Sandstone	20	2150–2300	0.71	1.6–2.1	0.01–0.013
Sawdust, dry	20	215		0.07	
Silica stone (85% SiC)	700	2720	1.05	1.56	0.055
Silica aerogel	0	140		0.024	
Silicon	20	2330	0.703	153	0.94
Silk (artificial)	35	100	1.33	0.049	0.0037
Slag	20	2500–3000	0.84	0.57	0.0023–0.0027
Slate					
Parallel to lamination	20	2700	0.75	2.9	0.014
Perpendicular to lamination	20	2700	0.75	1.83	0.009
Snow, firm	0	560	2.1	0.46	0.0039

Soil (see also Earth)					
Dry	15	1500	1.84	1	0.004
Wet	15	1930		2	
Strawberries, dry	−18			0.59	
Sugar (fine)	0	1600	1.25	0.58	0.0029
Sulfur	20	2070	0.72	0.27	0.0018
Teflon (polytetrafluoroethylene)	20	2200	1.04	0.23	0.001
Wood, perpendicular to grain					
Ash	15	740		0.15–0.3	
Balsa	15	100		0.05	
Cedar	15	480		0.11	
Mahogany	20	700		0.16	
Oak	20	600–800	2.4	0.17–0.25	~0.0012
Pine, Fir, Spruce	20	416–421	2.72	0.15	0.0012
Plywood	20	590		0.11	
Wool					
Sheep	20	100	1.72	0.036	0.0021
Mineral	50	200	0.92	0.042	0.0025
Slag	25	200	0.8	0.05	0.0031

[a]Constructed based on data compiled in Refs. [1–9].

METALLIC SOLIDS[a,b]

Metals, Alloys	Properties at 20°C (293 K)				Thermal conductivity k (W/m·K)[b]						
	ρ (kg/m³)	c_p (kJ/kg·K)	k (W/m·K)	α (cm²/s)	−100°C (173 K)	0°C (273 K)	100°C (373 K)	200°C (473 K)	400°C (673 K)	600°C (873 K)	1000°C (1273 K)
Aluminum											
Pure	2707	0.896	204	0.842	215	202	206	215	249		
Duralumin (94–96% Al, 3–5% Cu, trace Mg)	2787	0.883	164	0.667	126	159	182	194			
Silumin (87% Al, 13% Si)	2659	0.871	164	0.710	149	163	175	185			
Antimony	6690	0.208	17.4	0.125	19.2	17.7	16.3	16.0	17.2		
Beryllium	1850	1.750	167	0.516	126	160	191	215			
Bismuth, polycrystalline	9780	0.124	7.9	0.065	12.1	8.4	7.2	7.2			
Cadmium, polycrystalline	8650	0.231	92.8	0.464	97	93	92	91			
Cesium	1873	0.230	36	0.836							
Chromium	7190	0.453	90	0.276	120	95	88	85	77		
Cobalt (97.1% Co), polycrystalline	8900	0.389	70	0.202							
Copper											
Pure	8954	0.384	398	1.16	420	401	391	389	378	366	336
Commercial	8300	0.419	372	1.07							
Aluminum bronze (95% Cu, 5% Al)	8666	0.410	83	0.233							
Brass (70% Cu, 30% Zn)	8522	0.385	111	0.341	88		128	144	147		
Brass (60% Cu, 40% Zn)	8400	0.376	113	0.358							
Bronze (75% Cu, 25% Sn)	8666	0.343	26	0.086							
Bronze (85% Cu, 6% Sn, 9% Zn, 1% Pb)	8800	0.377	61.7	0.186							
Constantan (60% Cu, 40% Ni)	8922	0.410	22.7	0.061	21		22.2	26			
German silver (62% Cu, 15% Ni, 22% Zn)	8618	0.394	24.9	0.073	19.2		31	40	48		

Properties of metals (continued)

Material	ρ (kg/m³)	c_p	k	α							
Gold	19300	0.129	315	1.27	87	318	67	309	48	40	35
Iron											
Pure	7897	0.452	73	0.205		55	52	48	42	35	29
Cast (5% C)	7272	0.420	52	0.170		43	43	42	36	33	28
Carbon steel, 0.5% C	7833	0.465	54	0.148		36	36	36	33	31	28
1.0% C	7801	0.473	43	0.117		62	55	52	42	36	33
1.5% C	7753	0.486	36	0.097		40	38	36	33	29	29
Chrome steel, 1% Cr	7865	0.460	61	0.167		22	22	22	24	24	29
5% Cr	7833	0.460	40	0.111							
20% Cr	7689	0.460	22	0.064							
Chrome–Nickel steel,											
15% Cr, 10% Ni	7865	0.460	19	0.053		14	15.1	15.1	17	19	
20% Cr, 15% Ni	7833	0.460	15.1	0.042				15.1			
Invar (36% Ni)	8137	0.460	10.7	0.029							
Manganese steel, 1% Mn	7865	0.460	50	0.139			15	17	21	25	28
5% Mn	7849	0.460	22	0.064			15		20	23	
Nickel–Chrome steel											
80% Ni, 15% Cr	8522	0.460	17	0.045			16	18			
20% Ni, 15% Cr	7865	0.460	14	0.039							
Silicon steel, 1% Si	7769	0.460	42	0.116							
5% Si	7417	0.460	19	0.056							
Stainless steel, Type 304	7817	0.460	13.8	0.040	13						36
Type 347	7817	0.420	15	0.044							
Tungsten steel, 2% W	7961	0.444	62	0.176		62	59	54	48	45	
10% W	8314	0.419	48	0.139							
Wrought (0.5% CH)	7849	0.460	59	0.163	36.9	59	57	52	45	36	33
Lead	11340	0.130	34.8	0.236		35.1	33.4	31.6	23.3		
Lithium	530	3.391	61	0.340		61	61				
Magnesium											
Pure	1746	1.013	171	0.970	178	171	168	163			
6–8% Al, 1–2% Zn	1810	1.000	66	0.360		52	62	74			
electrolytic		1.000	114	0.640	93	111	125	130			
2% Mn	1778										

Continued

Metals, Alloys	Properties at 20°C (293 K)				Thermal conductivity k (W/m·K)[b]						
	ρ (kg/m³)	c_p (kJ/kg·K)	k (W/m·K)	α (cm²/s)	−100°C (173 K)	0°C (273 K)	100°C (373 K)	200°C (473 K)	400°C (673 K)	600°C (873 K)	1000°C (1273 K)
Manganese											
Pure	7300	0.486	7.8	0.022							
Manganin (84% Cu, 4% Ni, 12% Mn)	8400	0.406	21.9	0.064							
Molybdenum	10220	0.251	123	0.480	138	125	118	114	109	106	99
Monel 505 (at 60°C)	8360	0.544	19.7	0.043							
Nickel											
Pure	8906	0.445	91	0.230	114	94	83	74	64	69	78
Nichrome (24% Fe, 16% Cr)	8250	0.448	12.6	0.034							
90% Ni, 10% Cr	8666	0.444	17	0.044		17.1	18.9	20.9	24.6		
Niobium	8570	0.270	53	0.230							
Palladium	12020	0.247	75.5	0.254		75.5	75.5	75.5	75.5		
Platinum	21450	0.133	71.4	0.250	73	72	72	72	74	77	84
Potassium	860	0.741	103	1.62							
Rhenium	21100	0.137	48.1	0.166							
Rhodium	12450	0.248	150	0.486							
Rubidium	1530	0.348	58.2	1.09							
Silver, 99.99% Ag	10524	0.236	427	1.72	431	428	422	417	401	386	
99.90% Ag	10524	0.236	411	1.66	422	405		373	364		
Sodium	971	1.206	133	1.14							
Tantalum	16600	0.138	57.5	0.251	76	57.4					
Tin, polycrystalline	7304	0.220	67	0.417		68	63				
Titanium, polycrystalline	4540	0.523	22	0.093	26	22	21	20	19	21	22
Tungsten, polycrystalline	19300	0.134	179	0.692		182					
Uranium	18700	0.116	28	0.129	24	27	29	31	36	41	
Vanadium	6100	0.502	31.4	0.103		31.3					
Wood's metal (50% Bi, 25% Pb, 12.4% Cd, 12.5% Sn)	1056	0.147	12.8	0.825							
Zinc	7144	0.388	121	0.437	122	122	117	110	100		
Zirconium, polycrystalline	6570	0.272	22.8	0.128		23.2					

[a]Constructed based on data compiled in Refs. [1–7].

[b]The effect of temperature on thermal conductivity is illustrated further in Fig. 1.6.

ICE PROPERTIES[a]

T (0°C)	ρ^b (g/cm³)	h_{sf} (kJ/kg)	β^b (K⁻¹)
0	0.9164	333.4	
−5		308.5	
−10	0.9187	284.8	1.56×10^{-4}
−15		261.6	
−20	0.9203	241.4	1.38×10^{-4}
−40	0.9228		1.29×10^{-4}
−100	0.9273		1.24×10^{-4}
−200	0.9328		1.1×10^{-5}

[a]Constructed based on data compiled in Ref. [10].

[b]At atmospheric pressure.

ELECTRICAL RESISTIVITIES: $\rho_e \cong \rho_{e,0}[1 + \alpha_0(T - T_0)]$

Material	Reference Temperature T_0(°C)	Reference Resistivity $\rho_{e,0}(10^{-8} \text{ W} \cdot \text{m/A}^2)$	Temperature Coefficient α_0(°C⁻¹)
Aluminum	20	2.83	0.004
Brass	20	6.2	0.0015
Carbon			
Amorphous	20	3800–4100	—
Graphite	20	720–812	—
Copper			
Pure	20	1.68	0.004
Drawn	20	1.724	0.004
Gold		2.44	0.0034
Iron			
Pure	20	9.61	0.013
Cast	20	75–100	—
Wire	20	97.8	—
Lead	20	22	0.0039
Lithium	20	9.28	—
Magnesium	20	4.38	—
Manganese	20	144	—
Monel	20	43.5	0.0019
Mercury	20	96.8	0.0009
Molybdenum	20	5.34	—
Nickel	20	8.54	0.0041
Niobium	0	12.5	—
Palladium	20	10.54	—
Phosphorus, white	11	10^{17}	—
Platinum	20	10.72	0.003

Continued

Material	Reference Temperature $T_0(°C)$	Reference Resistivity $\rho_{e,0}(10^{-8}\ W\cdot m/A^2)$	Temperature Coefficient $\alpha_0(°C^{-1})$
Potassium	20	7.2	—
Silver	20	1.63	0.0038
Sodium	20	4.77	—
Steel			
Soft	20	15.9	0.0016
Transformer	20	11.1	—
Trolley wire	20	12.7	—
Tin	20	11.63	0.0042
Titanium	20	42	—
Tungsten	20	5.51	0.005
Vanadium	20	19.7	—
Ytterbium	25	25	—
Yttrium	25	59.6	—
Zinc	20	5.97	0.0037
Zirconium	20	42.1	—

POROUS MATERIALS[a]

Material	Porosity φ	Permeability K (cm²)	Contact Surface per Unit Volume (cm⁻¹)
Agar-agar		$2 \times 10^{-10}-4.4 \times 10^{-9}$	
Black slate powder	0.57–0.66	$4.9 \times 10^{-10}-1.2 \times 10^{-9}$	$7 \times 10^3-8.9 \times 10^3$
Brick	0.12–0.34	$4.8 \times 10^{-11}-2.2 \times 10^{-9}$	
Catalyst (Fischer–Tropsch, granules only)	0.45		5.6×10^5
Cigarette		1.1×10^{-5}	
Cigarette filters	0.17–0.49		
Coal	0.02–0.12		
Concrete (ordinary mixes)	0.02–0.07		
Concrete (bituminous)		$1 \times 10^{-9}-2.3 \times 10^{-7}$	
Copper powder (hot-compacted)	0.09–0.34	$3.3 \times 10^{-6}-1.5 \times 10^{-5}$	
Cork board		$2.4 \times 10^{-7}-5.1 \times 10^{-7}$	
Fiberglass	0.88–0.93		560–770
Granular crushed rock	0.45		
Hair (on mammals)	0.95–0.99		
Hair felt		$8.3 \times 10^{-6}-1.2 \times 10^{-5}$	
Leather	0.56–0.59	$9.5 \times 10^{-10}-1.2 \times 10^{-9}$	$1.2 \times 10^4-1.6 \times 10^4$
Limestone (dolomite)	0.04–0.10	$2 \times 10^{-11}-4.5 \times 10^{-10}$	
Sand	0.37–0.50	$2 \times 10^{-7}-1.8 \times 10^{-6}$	150–220
Sandstone ("oil sand")	0.08–0.38	$5 \times 10^{-12}-3 \times 10^{-8}$	
Silica grains	0.65		
Silica powder	0.37–0.49	$1.3 \times 10^{-10}-5.1 \times 10^{-10}$	$6.8 \times 10^3-8.9 \times 10^3$
Soil	0.43–0.54	$2.9 \times 10^{-9}-1.4 \times 10^{-7}$	
Spherical packings (well shaken)	0.36–0.43		
Wire crimps	0.68–0.76	$3.8 \times 10^{-5}-1 \times 10^{-4}$	29–40

[a]Nield and Bejan [11], based on data compiled by Scheidegger [12] and Bejan and Lage [13].

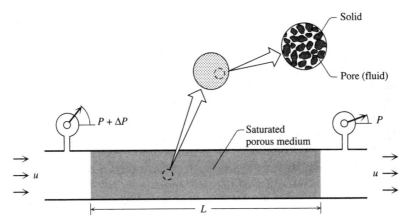

Figure Ap-B.1

Porosity:

$$\varphi = \frac{\text{void volume}}{\text{total volume of sample}}$$

Permeability K of porous medium saturated with one fluid:

$$u = \frac{K}{\mu} \frac{\Delta P}{L} \qquad \text{(Darcy's law)}$$

where ΔP = pressure drop across porous column of length L saturated with fluid

μ = fluid viscosity

u = fluid velocity averaged over the entire volume (solid + fluid) of the saturated porous medium sample.

Darcy's law is valid when the Reynolds number based on u and the pore length scale (D_p) is sufficiently small so that [11,12,14],

$$\frac{uD_p}{\nu} \lesssim 10$$

REFERENCES

1. E. R. G. Eckert and R. M. Drake, *Analysis of Heat and Mass Transfer*, McGraw-Hill, New York, 1972.
2. Y. S. Touloukian and C. Y. Ho, Eds., *Thermophysical Properties of Matter*, Plenum, New York, 1972.
3. K. Raznjevic, *Handbook of Thermodynamic Tables and Charts*, Hemisphere, Washington, DC, 1976.
4. U. Grigull and H. Sandner, *Heat Conduction*, translated by J. Kestin, Hemisphere, Washington, DC, 1984, Appendix E.
5. F. Kreith and W. Z. Black, *Basic Heat Transfer*, Harper and Row, New York, 1980, Appendix E.
6. J. H. Lienhard, *A Heat Transfer Textbook*, 2nd edition, Prentice-Hall, Englewood Cliffs, NJ, 1987, Appendix A.
7. L. C. Witte, P. S. Schmidt, and D. R. Brown, *Industrial Energy Management and Utilization*, Hemisphere, New York, 1988.

8. M. S. Qashou, R. I. Vachon, and Y. S. Touloukian, Thermal conductivity of foods, *ASHRAE Trans.*, Vol. 78, Pt. 1, 1972, pp. 165–183.

9. R. Dickerson, Jr., and R. B. Reed, Jr., Thermal diffusivity of meats, *ASHRAE Trans.*, Vol. 81, 1975, pp. 356–364.

10. P. V. Hobbs, *Ice Physics*, Oxford University Press, 1974, Chapter 5.

11. D. A. Nield and A. Bejan, *Convection in Porous Media*, Springer-Verlag, New York, 1992.

12. A. E. Scheidegger, *The Physics of Flow through Porous Media*, University of Toronto Press, Toronto, 1974.

13. A. Bejan and J. L. Lage, Heat transfer from a surface covered with hair, *Convective Heat and Mass Transfer in Porous Media*, S. Kakac, B. Kilkis, F. A. Kulacki, and F. Arinc, Eds., Kluwer Academic, Dordrecht, The Netherlands, 1991, pp. 823–845.

14. C. L. Tien and K. Vafai, Convective and radiative heat transfer in porous media, *Adv. Appl. Mech.*, Vol. 27, 1990, pp. 225–281.

WATER AT ATMOSPHERIC PRESSURE[a]

T (°C)	ρ (g/cm³)	c_P (kJ/kg·K)	c_v (kJ/kg·K)	h_{fg} (kJ/kg)	β (K⁻¹)
0	0.9999	4.217	4.215	2501	-0.6×10^{-4}
5	1	4.202	4.202	2489	$+0.1 \times 10^{-4}$
10	0.9997	4.192	4.187	2477	0.9×10^{-4}
15	0.9991	4.186	4.173	2465	1.5×10^{-4}
20	0.9982	4.182	4.158	2454	2.1×10^{-4}
25	0.9971	4.179	4.138	2442	2.6×10^{-4}
30	0.9957	4.178	4.118	2430	3.0×10^{-4}
35	0.9941	4.178	4.108	2418	3.4×10^{-4}
40	0.9923	4.178	4.088	2406	3.8×10^{-4}
50	0.9881	4.180	4.050	2382	4.5×10^{-4}
60	0.9832	4.184	4.004	2357	5.1×10^{-4}
70	0.9778	4.189	3.959	2333	5.7×10^{-4}
80	0.9718	4.196	3.906	2308	6.2×10^{-4}
90	0.9653	4.205	3.865	2283	6.7×10^{-4}
100[b]	0.9584	4.216	3.816	2257	7.1×10^{-4}

T (°C)	μ (g/cm·s)	v (cm²/s)	k (W/m·K)	α (cm²/s)	Pr	$\dfrac{g\beta}{\alpha v} = \dfrac{\mathrm{Ra}_H}{H^3 \Delta T}$ (K⁻¹·cm⁻³)
0	0.01787	0.01787	0.56	0.00133	13.44	-2.48×10^3
5	0.01514	0.01514	0.57	0.00136	11.13	$+0.47 \times 10^3$
10	0.01304	0.01304	0.58	0.00138	9.45	4.91×10^3
15	0.01137	0.01138	0.59	0.00140	8.13	9.24×10^3
20	0.01002	0.01004	0.59	0.00142	7.07	14.45×10^3
25	0.00891	0.00894	0.60	0.00144	6.21	19.81×10^3
30	0.00798	0.00802	0.61	0.00146	5.49	25.13×10^3

Continued

T (°C)	μ (g/cm·s)	v (cm²/s)	k (W/m·K)	α (cm²/s)	Pr.	$\dfrac{g\beta}{\alpha v}=\dfrac{Ra_H}{H^3\Delta T}$ (K^{-1}·cm^{-3})
35	0.00720	0.00725	0.62	0.00149	4.87	30.88×10^3
40	0.00654	0.00659	0.63	0.00152	4.34	37.21×10^3
50	0.00548	0.00554	0.64	0.00155	3.57	51.41×10^3
60	0.00467	0.00475	0.65	0.00158	3.01	66.66×10^3
70	0.00405	0.00414	0.66	0.00161	2.57	83.89×10^3
80	0.00355	0.00366	0.67	0.00164	2.23	101.3×10^3
90	0.00316	0.00327	0.67	0.00165	1.98	121.8×10^3
100	0.00283	0.00295	0.68	0.00166	1.78	142.2×10^3

[a]Data collected from Refs. [1–3].
[b]Saturated.

WATER AT SATURATION PRESSURE[a]

T (°C)	ρ (g/cm³)	c_P (kJ/kg·K)	μ (g/cm·s)	v (cm²/s)	k (W/m·K)	α (cm²/s)	Pr
0	0.9999	4.226	0.0179	0.0179	0.56	0.0013	13.7
10	0.9997	4.195	0.0130	0.0130	0.58	0.0014	9.5
20	0.9982	4.182	0.0099	0.0101	0.60	0.0014	7
40	0.9922	4.175	0.0066	0.0066	0.63	0.0015	4.3
60	0.9832	4.181	0.0047	0.0048	0.66	0.0016	3
80	0.9718	4.194	0.0035	0.0036	0.67	0.0017	2.25
100	0.9584	4.211	0.0028	0.0029	0.68	0.0017	1.75
150	0.9169	4.270	0.00185	0.0020	0.68	0.0017	1.17
200	0.8628	4.501	0.00139	0.0016	0.66	0.0017	0.95
250	0.7992	4.857	0.00110	0.00137	0.62	0.0016	0.86
300	0.7125	5.694	0.00092	0.00128	0.56	0.0013	0.98
340	0.6094	8.160	0.00077	0.00127	0.44	0.0009	1.45
370	0.4480	11.690	0.00057	0.00127	0.29	0.00058	2.18

[a]Data collected from Refs. [1, 2].

AMMONIA, SATURATED LIQUID[a]

T (°C)	ρ (g/cm³)	c_P (kJ/kg·K)	μ (kg/s·m)	v (cm²/s)	k (W/m·K)	α (cm²/s)	Pr
−50	0.704	4.46	3.06×10^{-4}	4.35×10^{-3}	0.547	1.74×10^{-3}	2.50
−25	0.673	4.49	2.58×10^{-4}	3.84×10^{-3}	0.548	1.81×10^{-3}	2.12
0	0.640	4.64	2.39×10^{-4}	3.73×10^{-3}	0.540	1.82×10^{-3}	2.05
25	0.604	4.84	2.14×10^{-4}	3.54×10^{-3}	0.514	1.76×10^{-3}	2.01
50	0.564	5.12	1.86×10^{-4}	3.30×10^{-3}	0.476	1.65×10^{-3}	2.00

[a]Constructed based on data compiled in Ref. [4].

CARBON DIOXIDE, SATURATED LIQUID[a]

T (K)	P (bar)	ρ (g/cm³)	c_P (kJ/kg·K)	μ (kg/s·m)	ν (cm²/s)	k (W/m·K)	α (cm²/s)	Pr
216.6	5.18	1.179	1.707	2.10×10^{-4}	1.78×10^{-3}	0.182	9.09×10^{-4}	1.96
220	6.00	1.167	1.761	1.86×10^{-4}	1.59×10^{-3}	0.178	8.26×10^{-4}	1.93
240	12.83	1.089	1.933	1.45×10^{-4}	1.33×10^{-3}	0.156	7.40×10^{-4}	1.80
260	24.19	1.000	2.125	1.14×10^{-4}	1.14×10^{-3}	0.128	6.03×10^{-4}	1.89
280	41.60	0.885	2.887	0.91×10^{-4}	1.03×10^{-3}	0.102	4.00×10^{-4}	2.57
300	67.10	0.680		0.71×10^{-4}	1.04×10^{-3}	0.081		
304.2	73.83	0.466		0.60×10^{-4}	1.29×10^{-3}	0.074		

[a]Constructed based on data compiled in Refs. [5, 6].

FUELS, LIQUIDS AT $P \cong 1$ ATM[a]

T (°C)	ρ (g/cm³)	c_P (kJ/kg·K)	μ (kg/s·m)	ν (cm²/s)	k (W/m·K)	α (cm²/s)	Pr
			Gasoline				
20	0.751	2.06	5.29×10^{-4}	7.04×10^{-3}	0.1164	7.52×10^{-4}	9.4
50	0.721	2.20	3.70×10^{-4}	5.13×10^{-3}	0.1105	6.97×10^{-4}	7.4
100	0.681	2.46	2.25×10^{-4}	3.30×10^{-3}	0.1005	6.00×10^{-4}	5.5
150	0.628	2.74	1.56×10^{-4}	2.48×10^{-3}	0.0919	5.34×10^{-4}	4.6
200	0.570	3.04	1.11×10^{-4}	1.95×10^{-3}	0.0800	4.62×10^{-4}	4.2
			Kerosene				
20	0.819	2.00	1.49×10^{-3}	1.82×10^{-2}	0.1161	7.09×10^{-4}	25.7
50	0.801	2.14	9.56×10^{-4}	1.19×10^{-2}	0.1114	6.50×10^{-4}	18.3
100	0.766	2.38	5.45×10^{-4}	7.11×10^{-3}	0.1042	5.72×10^{-4}	12.4
150	0.728	2.63	3.64×10^{-4}	5.00×10^{-3}	0.0965	5.04×10^{-4}	9.9
200	0.685	2.89	2.62×10^{-4}	3.82×10^{-3}	0.0891	4.50×10^{-4}	8.5
250	0.638	3.16	2.01×10^{-4}	3.15×10^{-3}	0.0816	4.05×10^{-4}	7.8

[a]Constructed based on data compiled in Ref. [7].

HELIUM, LIQUID AT $P = 1$ ATM[a]

T (K)	ρ (g/cm³)	c_P (kJ/kg·K)	μ (kg/s·m)	ν (cm²/s)	k (W/m·K)	α (cm²/s)	Pr
2.5	0.147	2.05	3.94×10^{-6}	2.68×10^{-4}	0.0167	5.58×10^{-4}	0.48
3	0.143	2.36	3.86×10^{-6}	2.69×10^{-4}	0.0182	5.38×10^{-4}	0.50
3.5	0.138	3.00	3.64×10^{-6}	2.64×10^{-4}	0.0191	4.62×10^{-4}	0.57
4	0.130	4.07	3.34×10^{-6}	2.57×10^{-4}	0.0196	3.71×10^{-4}	0.69
4.22[b]	0.125	4.98	3.17×10^{-6}	2.53×10^{-4}	0.0196	3.15×10^{-4}	0.80

[a]Data collected from Ref. [8].
[b]Saturated.

LITHIUM, SATURATED LIQUID[a]

T (K)	P (bar)	ρ (g/cm³)	c_P (kJ/kg·K)	μ (kg/s·m)	ν (cm²/s)	k (W/m·K)	α (cm²/s)	Pr
600	4.2×10^{-9}	0.503	4.23	4.26×10^{-4}	0.0085	47.6	0.223	0.038
800	9.6×10^{-6}	0.483	4.16	3.10×10^{-4}	0.0064	54.1	0.270	0.024
1000	9.6×10^{-4}	0.463	4.16	2.47×10^{-4}	0.0053	60.0	0.312	0.017
1200	0.0204	0.442	4.14	2.07×10^{-4}	0.0047	64.7	0.355	0.013
1400	0.1794	0.422	4.19	1.80×10^{-4}	0.0043	68.0	0.384	0.011

[a]Constructed based on data compiled in Refs. [5, 6].

NITROGEN, LIQUID AT $P = 1$ ATM[a]

T (K)	ρ (g/cm³)	c_P (kJ/kg·K)	μ (kg/s·m)	ν (cm²/s)	k (W/m·K)	α (cm²/s)	Pr
65	0.861	1.988	2.77×10^{-5}	3.21×10^{-3}	0.161	9.39×10^{-4}	3.42
70	0.840	2.042	2.12×10^{-5}	2.53×10^{-3}	0.151	8.77×10^{-4}	2.88
75	0.819	2.059	1.77×10^{-5}	2.17×10^{-3}	0.141	8.36×10^{-4}	2.59
77.3[b]	0.809	2.065	1.64×10^{-5}	2.03×10^{-3}	0.136	8.15×10^{-4}	2.49

[a]Interpolated from data in Ref. [9].
[b]Saturated.

POTASSIUM, SATURATED LIQUID[a]

T (K)	P (bar)	ρ (g/cm³)	c_P (kJ/kg·K)	μ (kg/s·m)	ν (cm²/s)	k (W/m·K)	α (cm²/s)	Pr
400	1.84×10^{-7}	0.814	0.805	4.13×10^{-4}	0.0051	52.0	0.794	0.0064
600	9.26×10^{-4}	0.767	0.771	2.38×10^{-4}	0.0031	43.9	0.742	0.0042
800	0.0612	0.720	0.761	1.71×10^{-4}	0.0024	37.1	0.677	0.0035
1000	0.7322	0.672	0.792	1.35×10^{-4}	0.0020	31.3	0.589	0.0034
1200	3.963	0.623	0.846	1.14×10^{-4}	0.0018	26.3	0.499	0.0037
1400	12.44	0.574	0.899	0.98×10^{-4}	0.0017	21.5	0.416	0.0041

[a]Constructed based on data compiled in Refs. [5, 6].

MERCURY, SATURATED LIQUID[a]

T (K)	P (bar)	ρ (g/cm³)	c_p (kJ/kg·K)	μ (kg/s·m)	ν (cm²/s)	k (W/m·K)	α (cm²/s)	Pr	β (K⁻¹)
260	6.9×10^{-8}	13.63	0.141	1.79×10^{-3}	1.31×10^{-3}	8.00	0.042	0.0316	1.8×10^{-4}
300	3.1×10^{-6}	13.53	0.139	1.52×10^{-3}	1.12×10^{-3}	8.54	0.045	0.0248	1.8×10^{-4}
340	5.5×10^{-5}	13.43	0.138	1.34×10^{-3}	1.00×10^{-3}	9.06	0.049	0.0205	1.8×10^{-4}
400	1.4×10^{-3}	13.29	0.137	1.17×10^{-3}	8.83×10^{-4}	9.80	0.054	0.0163	1.8×10^{-4}
500	0.053	13.05	0.135	1.01×10^{-3}	7.72×10^{-4}	10.93	0.062	0.0125	1.8×10^{-4}
600	0.578	12.81	0.136	9.10×10^{-4}	7.10×10^{-4}	11.94	0.071	0.0100	1.9×10^{-4}
800	11.18	12.32	0.140	8.08×10^{-4}	6.56×10^{-4}	13.57	0.079	0.0083	1.9×10^{-4}
1000	65.74	11.79	0.149	7.54×10^{-4}	6.40×10^{-4}	14.69	0.084	0.0076	1.9×10^{-4}

[a]Constructed based on data compiled in Refs. [5, 6].

REFRIGERANT 12 (FREON 12, CCl$_2$F$_2$), SATURATED LIQUID[a]

P (bar)	T (K)	ρ (g/cm^3)	c_P (kJ/kg·K)	μ (kg/s·m)	ν (cm^2/s)	k (W/m·K)	α (cm^2/s)	Pr
0.2	211.1	1.579	0.865	5.28×10^{-4}	3.34×10^{-3}	0.101	7.40×10^{-4}	4.52
0.4	223.5	1.554	0.876	4.48×10^{-4}	2.88×10^{-3}	0.097	7.12×10^{-4}	4.05
0.6	231.7	1.522	0.884	4.08×10^{-4}	2.68×10^{-3}	0.094	6.98×10^{-4}	3.84
1.0	243.0	1.488	0.894	3.59×10^{-4}	2.41×10^{-3}	0.089	6.68×10^{-4}	3.61
2.0	260.6	1.435	0.914	2.95×10^{-4}	2.06×10^{-3}	0.083	6.33×10^{-4}	3.25
3.0	272.3	1.392	0.930	2.62×10^{-4}	1.88×10^{-3}	0.079	6.11×10^{-4}	3.08
6.0	295.2	1.321	0.969	2.13×10^{-4}	1.61×10^{-3}	0.070	5.47×10^{-4}	2.95
10.0	314.9	1.247	1.023	1.88×10^{-4}	1.51×10^{-3}	0.063	4.94×10^{-4}	3.05
20.0	346.3	1.099	1.234	1.49×10^{-4}	1.36×10^{-3}	0.053	3.91×10^{-4}	3.47
30.0	367.2	0.955	1.520	1.16×10^{-4}	1.21×10^{-3}	0.042	2.89×10^{-4}	4.20

[a]Constructed based on data compiled in Refs. [5, 6].

REFRIGERANT 22 (FREON 22, CHClF$_2$), SATURATED LIQUID[a]

T (K)	P (bar)	ρ (g/cm^3)	c_P (kJ/kg·K)	μ (kg/s·m)	ν (cm^2/s)	k (W/m·K)	α (cm^2/s)	Pr
180	0.037	1.545	1.058	6.47×10^{-5}	4.19×10^{-4}	0.146	8.93×10^{-4}	0.47
200	0.166	1.497	1.065	4.81×10^{-5}	3.21×10^{-4}	0.136	8.53×10^{-4}	0.38
220	0.547	1.446	1.080	3.78×10^{-5}	2.61×10^{-4}	0.126	8.07×10^{-4}	0.32
240	1.435	1.390	1.105	3.09×10^{-5}	2.22×10^{-4}	0.117	7.62×10^{-4}	0.29
260	3.177	1.329	1.143	2.60×10^{-5}	1.96×10^{-4}	0.107	7.04×10^{-4}	0.28
280	6.192	1.262	1.193	2.25×10^{-5}	1.78×10^{-4}	0.097	6.44×10^{-4}	0.28
300	10.96	1.187	1.257	1.98×10^{-5}	1.67×10^{-4}	0.087	5.83×10^{-4}	0.29
320	18.02	1.099	1.372	1.76×10^{-5}	1.60×10^{-4}	0.077	5.11×10^{-4}	0.31
340	28.03	0.990	1.573	1.51×10^{-5}	1.53×10^{-4}	0.067	4.30×10^{-4}	0.36

[a]Constructed based on data compiled in Ref. [6].

SODIUM, SATURATED LIQUID[a]

T (K)	P (bar)	ρ (g/cm^3)	c_P (kJ/kg·K)	μ (kg/s·m)	ν (cm^2/s)	k (W/m·K)	α (cm^2/s)	Pr
500	7.64×10^{-7}	0.898	1.330	4.24×10^{-4}	0.0047	80.0	0.67	0.0070
600	5.05×10^{-5}	0.873	1.299	3.28×10^{-4}	0.0038	75.4	0.66	0.0057
700	9.78×10^{-4}	0.850	1.278	2.69×10^{-4}	0.0032	70.7	0.65	0.0049
800	0.00904	0.826	1.264	2.30×10^{-4}	0.0028	65.9	0.63	0.0044
900	0.0501	0.802	1.258	2.02×10^{-4}	0.0025	61.4	0.61	0.0041
1000	0.1955	0.776	1.259	1.81×10^{-4}	0.0023	56.7	0.58	0.0040
1200	1.482	0.729	1.281	1.51×10^{-4}	0.0021	54.5	0.58	0.0036
1400	6.203	0.681	1.330	1.32×10^{-4}	0.0019	52.2	0.58	0.0034
1600	17.98	0.633	1.406	1.18×10^{-4}	0.0019	49.9	0.56	0.0033

[a]Constructed based on data compiled in Refs. [5, 6].

UNUSED ENGINE OIL[a]

T (K)	ρ (g/cm³)	c_P (kJ/kg·K)	μ (kg/s·m)	ν (cm²/s)	k (W/m·K)	α (cm²/s)	Pr	β (K⁻¹)
260	0.908	1.76	12.23	135	0.149	9.32×10^{-4}	144500	7×10^{-4}
280	0.896	1.83	2.17	24.2	0.146	8.90×10^{-4}	27200	7×10^{-4}
300	0.884	1.91	0.486	5.50	0.144	8.53×10^{-4}	6450	7×10^{-4}
320	0.872	1.99	0.141	1.62	0.141	8.13×10^{-4}	1990	7×10^{-4}
340	0.860	2.08	0.053	0.62	0.139	7.77×10^{-4}	795	7×10^{-4}
360	0.848	2.16	0.025	0.30	0.137	7.48×10^{-4}	395	7×10^{-4}
380	0.836	2.25	0.014	0.17	0.136	7.23×10^{-4}	230	7×10^{-4}
400	0.824	2.34	0.009	0.11	0.134	6.95×10^{-4}	155	7×10^{-4}

[a]Constructed based on data compiled in Refs. [5, 6].

CRITICAL POINT DATA[a]

Liquid	Critical Temperature K	Critical Temperature °C	Critical Pressure MPa	Critical Pressure atm	Critical Specific Volume (cm³/g)
Air	133.2	−140	3.77	37.2	2.9
Alcohol (methyl)	513.2	240	7.98	78.7	3.7
Alcohol (ethyl)	516.5	243.3	6.39	63.1	3.6
Ammonia	405.4	132.2	11.3	111.6	4.25
Argon	150.9	−122.2	4.86	48	1.88
Butane	425.9	152.8	3.65	36	4.4
Carbon dioxide	304.3	31.1	7.4	73	2.2
Carbon monoxide	134.3	−138.9	3.54	35	3.2
Carbon tetrachloride	555.9	282.8	4.56	45	1.81
Chlorine	417	143.9	7.72	76.14	1.75
Ethane	305.4	32.2	4.94	48.8	4.75
Ethylene	282.6	9.4	5.85	57.7	4.6
Helium	5.2	−268	0.228	2.25	14.4
Hexane	508.2	235	2.99	29.5	4.25
Hydrogen	33.2	−240	1.30	12.79	32.3
Methane	190.9	−82.2	4.64	45.8	6.2
Methyl chloride	416.5	143.3	6.67	65.8	2.7
Neon	44.2	−288.9	2.7	26.6	2.1
Nitric oxide	179.3	−93.9	6.58	65	1.94
Nitrogen	125.9	−147.2	3.39	33.5	3.25
Octane	569.3	296.1	2.5	24.63	4.25
Oxygen	154.3	−118.9	5.03	49.7	2.3
Propane	368.7	95.6	4.36	43	4.4
Sulfur dioxide	430.4	157.2	7.87	77.7	1.94
Water	647	373.9	22.1	218.2	3.1

[a]Based on a compilation from Ref. [10].

REFERENCES

1. A. Bejan, *Convection Heat Transfer*, Wiley, New York, 1984, pp. 462–465.
2. K. Raznjevic, *Handbook of Thermodynamic Tables and Charts*, Hemisphere, Washington, DC, 1976.
3. G. K. Batchelor, *An Introduction to Fluid Dynamics*, Cambridge University Press, Cambridge, England, 1967.
4. E. R. G. Eckert and R. M. Drake, *Analysis of Heat and Mass Transfer*, McGraw-Hill, New York, 1972.
5. D. W. Green and J. O. Maloney, Eds., *Perry's Chemical Engineers' Handbook*, 6th edition, McGraw-Hill, New York, 1984, pp. 3-1–3-263.
6. P. E. Liley, Thermophysical properties, in S. Kakac, R. K. Shah, and W. Aung, Eds., *Handbook of Single-Phase Convective Heat Transfer*, Wiley, New York, 1987, Chapter 22.
7. N. B. Vargaftik, *Tables on the Thermophysical Properties of Liquids and Gases*, 2nd edition, Hemisphere, Washington, DC, 1975.
8. R. D. McCarty, Thermophysical properties of Helium-4 from 2 to 1500K with pressures to 1000 atmospheres, NBS TN 631, Washington, DC, November 1972.
9. R. T. Jacobsen, R. B. Stewart, R. D. McCarty, and H. J. M. Hanley, Thermophysical properties of nitrogen from the fusion line to 3500 R (1944 K) for pressures to 150,000 psia (10342 $\times$ 10^5 N/m^2), NBS TN 648, Washington, DC, December 1973.
10. A. Bejan, *Advanced Engineering Thermodynamics*, Wiley, New York, 1988.

APPENDIX D

PROPERTIES OF GASES

AMMONIA, GAS AT $P = 1$ ATM[a]

T (°C)	ρ (kg/m^3)	c_P $(kJ/kg \cdot K)$	μ $(kg/s \cdot m)$	ν (cm^2/s)	k $(W/m \cdot K)$	α (cm^2/s)	Pr
0	0.793	2.18	9.35×10^{-6}	0.118	0.0220	0.131	0.90
50	0.649	2.18	1.10×10^{-5}	0.170	0.0270	0.192	0.88
100	0.559	2.24	1.29×10^{-5}	0.230	0.0327	0.262	0.87
150	0.493	2.32	1.47×10^{-5}	0.297	0.0391	0.343	0.87
200	0.441	2.40	1.65×10^{-5}	0.374	0.0467	0.442	0.84

[a]Constructed based on data compiled in Ref. [4].

CARBON DIOXIDE, GAS AT $P = 1$ BAR[a]

T (K)	ρ (kg/m^3)	c_P $(kJ/kg \cdot K)$	μ $(kg/s \cdot m)$	ν (cm^2/s)	k $(W/m \cdot K)$	α (cm^2/s)	Pr
300	1.773	0.852	1.51×10^{-5}	0.085	0.0166	0.109	0.78
350	1.516	0.898	1.75×10^{-5}	0.115	0.0204	0.150	0.77
400	1.326	0.941	1.98×10^{-5}	0.149	0.0243	0.195	0.77
500	1.059	1.014	2.42×10^{-5}	0.229	0.0325	0.303	0.76
600	0.883	1.075	2.81×10^{-5}	0.318	0.0407	0.429	0.74
700	0.751	1.126	3.17×10^{-5}	0.422	0.0481	0.569	0.74
800	0.661	1.168	3.50×10^{-5}	0.530	0.0551	0.714	0.74
900	0.588	1.205	3.81×10^{-5}	0.648	0.0618	0.873	0.74
1000	0.529	1.234	4.10×10^{-5}	0.775	0.0682	1.043	0.74

[a]Constructed based on data compiled in Refs. [5, 6].

DRY AIR AT ATMOSPHERIC PRESSURE[a]

T (°C)	ρ (kg/m³)	c_P (kJ/kg·K)	μ (kg/s·m)	ν (cm²/s)	k (W/m·K)	α (cm²/s)	Pr	$\dfrac{g\beta}{\alpha\nu} = \dfrac{\mathrm{Ra}_H}{H^3\,\Delta T}$ (cm⁻³·K⁻¹)
−180	3.72	1.035	6.50×10^{-6}	0.0175	0.0076	0.019	0.92	3.2×10^4
−100	2.04	1.010	1.16×10^{-5}	0.057	0.016	0.076	0.75	1.3×10^3
−50	1.582	1.006	1.45×10^{-5}	0.092	0.020	0.130	0.72	367
0	1.293	1.006	1.71×10^{-5}	0.132	0.024	0.184	0.72	148
10	1.247	1.006	1.76×10^{-5}	0.141	0.025	0.196	0.72	125
20	1.205	1.006	1.81×10^{-5}	0.150	0.025	0.208	0.72	107
30	1.165	1.006	1.86×10^{-5}	0.160	0.026	0.223	0.72	90.7
60	1.060	1.008	2.00×10^{-5}	0.188	0.028	0.274	0.70	57.1
100	0.946	1.011	2.18×10^{-5}	0.230	0.032	0.328	0.70	34.8
200	0.746	1.025	2.58×10^{-5}	0.346	0.039	0.519	0.68	9.53
300	0.616	1.045	2.95×10^{-5}	0.481	0.045	0.717	0.68	4.96
500	0.456	1.093	3.58×10^{-5}	0.785	0.056	1.140	0.70	1.42
1000	0.277	1.185	4.82×10^{-5}	1.745	0.076	2.424	0.72	0.18

[a]Data collected from Refs. [1–3].

HELIUM, GAS AT $P = 1$ ATM[a]

T (K)	ρ (kg/m^3)	c_P (kJ/kg·K)	μ (kg/s·m)	ν (cm^2/s)	k (W/m·K)	α (cm^2/s)	Pr
4.22	16.9	9.78	1.25×10^{-6}	7.39×10^{-4}	0.011	6.43×10^{-4}	1.15
7	7.53	5.71	1.76×10^{-6}	2.34×10^{-3}	0.014	3.21×10^{-3}	0.73
10	5.02	5.41	2.26×10^{-6}	4.49×10^{-3}	0.018	6.42×10^{-3}	0.70
20	2.44	5.25	3.58×10^{-6}	0.0147	0.027	0.0209	0.70
30	1.62	5.22	4.63×10^{-6}	0.0286	0.034	0.0403	0.71
60	0.811	5.20	7.12×10^{-6}	0.088	0.053	0.125	0.70
100	0.487	5.20	9.78×10^{-6}	0.201	0.074	0.291	0.69
200	0.244	5.19	1.51×10^{-5}	0.622	0.118	0.932	0.67
300	0.162	5.19	1.99×10^{-5}	1.22	0.155	1.83	0.67
600	0.0818	5.19	3.22×10^{-5}	3.96	0.251	5.94	0.67
1000	0.0487	5.19	4.63×10^{-5}	9.46	0.360	14.2	0.67

[a]Data collected from Ref. [7].

n-HYDROGEN, GAS AT $P = 1$ ATM[a]

T (K)	ρ (kg/m^3)	c_P (kJ/kg·K)	μ (kg/s·m)	ν (cm^2/s)	k (W/m·K)	α (cm^2/s)	Pr
250	0.0982	14.04	7.9×10^{-6}	0.804	0.162	1.17	0.69
300	0.0818	14.31	8.9×10^{-6}	1.09	0.187	1.59	0.69
350	0.0702	14.43	9.9×10^{-6}	1.41	0.210	2.06	0.69
400	0.0614	14.48	1.09×10^{-5}	1.78	0.230	2.60	0.68
500	0.0491	14.51	1.27×10^{-5}	2.59	0.269	3.78	0.68
600	0.0408	14.55	1.43×10^{-5}	3.50	0.305	5.12	0.68
700	0.0351	14.60	1.59×10^{-5}	4.53	0.340	6.62	0.68

[a]Constructed based on the data compiled in Refs. [5, 6].

NITROGEN, GAS AT $P = 1$ ATM[a]

T (K)	ρ (kg/m^3)	c_P (kJ/kg·K)	μ (kg/s·m)	ν (cm^2/s)	k (W/m·K)	α (cm^2/s)	Pr
77.33	4.612	1.123	5.39×10^{-6}	0.0117	0.0076	0.0147	0.80
100	3.483	1.073	6.83×10^{-6}	0.0197	0.0097	0.0261	0.76
200	1.711	1.044	1.29×10^{-5}	0.0754	0.0185	0.103	0.73
300	1.138	1.041	1.78×10^{-5}	0.156	0.0259	0.218	0.72
400	0.854	1.045	2.20×10^{-5}	0.258	0.0324	0.363	0.71
500	0.683	1.056	2.58×10^{-5}	0.378	0.0386	0.535	0.71
600	0.569	1.075	2.91×10^{-5}	0.511	0.0442	0.722	0.71
700	0.488	1.098	3.21×10^{-5}	0.658	0.0496	0.925	0.71

[a]Data collected from Ref. [8].

OXYGEN, GAS AT P = ATM[a]

T (K)	ρ (kg/m³)	c_p (kJ/kg·K)	μ (kg/s·m)	ν (cm²/s)	k (W/m·K)	α (cm²/s)	Pr
250	1.562	0.915	1.79×10^{-5}	0.115	0.0226	0.158	0.73
300	1.301	0.920	2.07×10^{-5}	0.159	0.0266	0.222	0.72
350	1.021	0.929	2.34×10^{-5}	0.229	0.0305	0.321	0.71
400	0.976	0.942	2.58×10^{-5}	0.264	0.0343	0.372	0.71
500	0.780	0.972	3.03×10^{-5}	0.388	0.0416	0.549	0.71
600	0.650	1.003	3.44×10^{-5}	0.529	0.0487	0.748	0.71
700	0.557	1.031	3.81×10^{-5}	0.684	0.0554	0.963	0.71

[a]Constructed based on the data compiled in Refs. [5, 6].

REFRIGERANT 12 (FREON 12, CCl_2F_2), GAS AT P = 1 BAR[a]

T (K)	ρ (kg/m³)	c_p (kJ/kg·K)	μ (kg/s·m)	ν (cm²/s)	k (W/m·K)	α (cm²/s)	Pr
300	4.941	0.614	1.26×10^{-5}	0.026	0.0097	0.032	0.80
350	4.203	0.654	1.46×10^{-5}	0.035	0.0124	0.045	0.77
400	3.663	0.684	1.62×10^{-5}	0.044	0.0151	0.061	0.73
450	3.248	0.711	1.75×10^{-5}	0.054	0.0179	0.077	0.70
500	2.918	0.739	1.90×10^{-5}	0.065	0.0208	0.097	0.67

[a]Constructed based on the data compiled in Refs. [5, 6].

REFRIGERANT 22 (FREON 22, $CHClF_2$), GAS AT P = 1 ATM[a]

T (K)	ρ (kg/m³)	c_p (kJ/kg·K)	μ (kg/s·m)	ν (cm²/s)	k (W/m·K)	α (cm²/s)	Pr
250	4.320	0.587	1.09×10^{-5}	0.025	0.008	0.032	0.80
300	3.569	0.647	1.30×10^{-5}	0.036	0.011	0.048	0.77
350	3.040	0.704	1.51×10^{-5}	0.050	0.014	0.065	0.76
400	2.650	0.757	1.71×10^{-5}	0.065	0.017	0.085	0.76
450	2.352	0.806	1.90×10^{-5}	0.081	0.020	0.105	0.77
500	2.117	0.848	2.09×10^{-5}	0.099	0.023	0.128	0.77

[a]Constructed based on the data compiled in Refs. [5, 6].

STEAM AT $P = 1$ BAR[a]

T (K)	ρ (kg/m³)	c_P (kJ/kg·K)	μ (kg/s·m)	ν (cm²/s)	k (W/m·K)	α (cm²/s)	Pr
373.15	0.596	2.029	1.20×10^{-5}	0.201	0.0248	0.205	0.98
400	0.547	1.996	1.32×10^{-5}	0.241	0.0268	0.246	0.98
450	0.485	1.981	1.52×10^{-5}	0.313	0.0311	0.324	0.97
500	0.435	1.983	1.73×10^{-5}	0.398	0.0358	0.415	0.96
600	0.362	2.024	2.15×10^{-5}	0.594	0.0464	0.633	0.94
700	0.310	2.085	2.57×10^{-5}	0.829	0.0581	0.899	0.92
800	0.271	2.151	2.98×10^{-5}	1.10	0.0710	1.22	0.90
900	0.241	2.219	3.39×10^{-5}	1.41	0.0843	1.58	0.89
1000	0.217	2.286	3.78×10^{-5}	1.74	0.0981	1.98	0.88
1200	0.181	2.43	4.48×10^{-5}	2.48	0.130	2.96	0.84
1400	0.155	2.58	5.06×10^{-5}	3.27	0.160	4.00	0.82
1600	0.135	2.73	5.65×10^{-5}	4.19	0.210	5.69	0.74
1800	0.120	3.02	6.19×10^{-5}	5.16	0.330	9.10	0.57
2000	0.108	3.79	6.70×10^{-5}	6.20	0.570	13.94	0.45

[a]Constructed based on data compiled in Refs. [5, 6].

IDEAL GAS CONSTANTS AND SPECIFIC HEATS[a,b]

Gas		M (kg/kmol)	R (kJ/kg·K)	c_P (kJ/kg·K)	c_v (kJ/kg·K)
Air, dry	—	28.97	0.287	1.005	0.718
Argon	Ar	39.944	0.208	0.525	0.317
Carbon dioxide	CO_2	44.01	0.189	0.846	0.657
Carbon monoxide	CO	28.01	0.297	1.040	0.744
Helium	He	4.003	2.077	5.23	3.15
Hydrogen	H_2	2.016	4.124	14.31	10.18
Methane	CH_4	16.04	0.518	2.23	1.69
Nitrogen	N_2	28.016	0.297	1.039	0.743
Oxygen	O_2	32.000	0.260	0.918	0.658
Water vapor	H_2O	18.016	0.461	1.87	1.40

[a]The c_P and c_v values correspond to the temperature 300 K. This ideal gas model is valid at low and moderate pressures ($P \lesssim 1$ atm).
[b]After Ref. [9].

HUMID AIR AS AN IDEAL GAS MIXTURE OF DRY AIR AND WATER VAPOR

Mole fraction of water vapor:

$$x_v = \frac{P_v}{P} \qquad P_v = \text{partial pressure of water vapor}$$

$$P = \text{mixture pressure}$$

Relative humidity:

$$\phi = \frac{P_v}{P_{sat}(T)} \qquad P_{sat} = \text{pressure of saturated water vapor at mixture temperature } T$$

$$\phi = \frac{x_v \text{ in the actual mixture, at } T \text{ and } P}{x_v \text{ in the saturated mixture, at the same } T \text{ and } P}$$

Specific humidity, or humidity ratio:

$$\omega = \frac{m_v \text{ (kg water vapor in the mixture)}}{m_a \text{ (kg dry air in the mixture)}}$$

$$\omega = \frac{0.622}{\dfrac{P}{\phi P_{sat}(T)} - 1}$$

Relationships between relative humidity and specific humidity:

$$\phi = \frac{\omega}{\omega + 0.662} \frac{P}{P_{sat}(T)}$$

$$\phi = \frac{\omega P_a}{0.662\, P_{sat}(T)}, \qquad P_a = P - P_v$$

ADDITIONAL PHYSICAL QUANTITIES INCLUDED IN THE TEXT

REFERENCES

1. A. Bejan, *Convection Heat Transfer*, Wiley, New York, 1984, p.465.

2. K. Raznjevic, *Handbook of Thermodynamic Tables and Charts*, Hemisphere, Washington, DC, 1976.

3. G. K. Batchelor, *An Introduction to Fluid Dynamics*, Cambridge University Press, Cambridge, England, 1967.

4. E. R. G. Eckert and R. M. Drake, *Analysis of Heat and Mass Transfer*, McGraw-Hill, New York, 1972.

5. D. W. Green and J. O. Maloney, Eds., *Perry's Chemical Engineers' Handbook*, 6th edition, McGraw-Hill, New York, 1984, pp. (3-1)–(3-263).

6. P. E. Liley, Thermophysical properties, in S. Kakac, R. K. Shah, and W. Aung, Eds., *Handbook of Single-Phase Convective Heat Transfer*, Wiley, New York, 1987, Chapter 22.

7. R. D. McCarty, Thermophysical properties of Helium-4 from 2 to 1500 K with pressures to 1000 atmospheres, NBS TN 631, Washington, DC, November 1972.

8. R. T. Jacobsen, R. B. Stewart, R. D. McCarty, and H. J. M. Hanley, Thermophysical properties of nitrogen from the fusion line to 3500 R (1944 K) for pressures to 150,000 psia (10342×10^5 N/m^2), NBS TN 648, Washington, DC, December 1973.

9. A. Bejan, *Advanced Engineering Thermodynamics*, Wiley, New York, 1988.

APPENDIX E
MATHEMATICAL FORMULAS

AREAS AND VOLUMES OF SIMPLE BODIES

Right circular cylinder

Lateral area $= 2\pi RH$
Total area $\quad= 2\pi R(H + R)$
Volume $\quad\quad= \pi R^2 H$

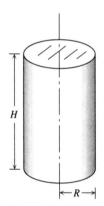

Figure Ap-E.1

Truncated right circular cylinder

Lateral area $= \pi R(H_1 + H_2)$

Total area $\quad= \pi R\left\{ H_1 + H_2 + R + \left[R^2 + \frac{1}{4}(H_2 - H_1)^2 \right]^{1/2} \right\}$

Volume $\quad\quad= \dfrac{\pi}{2} R^2 (H_1 + H_2)$

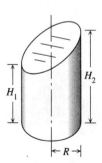

Figure Ap-E.2

Right circular cone

Lateral area $= \pi R(R^2 + H^2)^{1/2}$
Total area $= \pi R[R + (R^2 + H^2)^{1/2}]$
Volume $= \dfrac{\pi}{3} R^2 H$

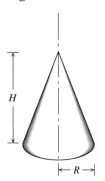

Figure Ap-E.3

Frustum of right circular cone

Lateral area $= \pi[H^2 + (R_1 - R_2)^2]^{1/2}(R_1 + R_2)$
Volume $= \dfrac{\pi}{3} H (R_1^2 + R_2^2 + R_1 R_2)$

Figure Ap-E.4

Sphere of radius R

Area $= 4\pi R^2 = \pi D^2$
Volume $= \dfrac{4}{3} \pi R^3$

Spherical sector

Total area $= \pi R(L + 2H)$
Volume $= \dfrac{2}{3} \pi R^2 H$

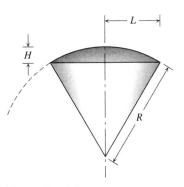

Figure Ap-E.5

Spherical segment

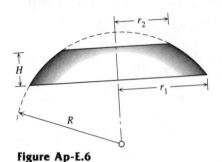

Lateral (spherical) area	$= 2\pi RH$
Total area	$= 2\pi RH + \pi \, (r_1^2 + r_2^2)$
Volume	$= \dfrac{\pi}{6} H \, (3 \, r_1^2 + 3 \, r_2^2 + H^2)$

Figure Ap-E.6

ERROR FUNCTION

Definition and properties:

$$\text{erf}(x) = \frac{2}{\pi^{1/2}} \int_0^x e^{-m^2} dm$$

$$\text{erf}(-x) = -\text{erf}(x)$$

$$\text{erfc}(x) = 1 - \text{erf}(x) \qquad \text{(complimentary error function)}$$

$$\frac{d}{dx} [\text{erf}(x)]_{x=0} = \frac{2}{\pi^{1/2}} = 1.12838$$

The usual error function table: finding $\text{erf}(x)$ when x is specified:

x	$\text{erf}(x)$	x	$\text{erf}(x)$
0	0	0.9	0.79691
0.01	0.01128	1	0.8427
0.1	0.11246	1.2	0.91031
0.2	0.2227	1.4	0.95229
0.3	0.32863	1.6	0.97635
0.4	0.42839	1.8	0.98909
0.5	0.5205	2	0.99532
0.6	0.60386	2.5	0.99959
0.7	0.6778	3	0.99998
0.8	0.74210	∞	1

The inverted table: finding x when $\text{erf}(x)$ is specified:

$\text{erf}(x)$	x
0	0
0.1	0.08886
0.2	0.17914
0.25	0.22531
0.3	0.27246
0.4	0.37081
0.5	0.47694
0.6	0.59512
0.7	0.73287
0.75	0.81342
0.8	0.90619
0.9	1.16309
1	∞

Closed-form approximate expressions for $\text{erf}(x)$ and $\text{erfc}(x)$, accurate within 0.42 percent [1]:

$$\text{erf}(x) \cong 1 - A \exp[-B(x + C)^2]$$

$$\text{erfc}(x) \cong A \exp[-B(x + C)^2]$$

where A, B, and C are three constants,

$$A = 1.5577 \qquad B = 0.7182 \qquad C = 0.7856$$

EXPONENTIAL INTEGRAL FUNCTION

x	$\int_x^\infty \dfrac{e^{-u}}{u}\,du$	x	$\int_x^\infty \dfrac{e^{-u}}{u}\,du$
0.00	$+\infty$		
0.01	4.0379	1.60	0.08631
0.02	3.3547	1.70	0.07465
0.04	2.6813	1.80	0.06471
0.07	2.1508	1.90	0.05620
0.10	1.8229	2.00	0.04890
0.15	1.4645	2.20	0.03719
0.20	1.2227	2.40	0.02844
0.30	0.9057	2.60	0.02185
0.40	0.7024	2.80	0.01686
0.50	0.5598	3.00	0.01305
0.60	0.4544	3.20	0.01013
0.70	0.3738	3.40	0.00789
0.80	0.3106	3.60	0.00616
0.90	0.2602	3.80	0.00482
1.00	0.2194	4.00	0.00378
1.10	0.18599	4.20	0.00297
1.20	0.15841	4.40	0.00234
1.30	0.13545	4.60	0.00184
1.40	0.11622	4.80	0.00145
1.50	0.10002	5.00	0.00115

LEIBNIZ'S FORMULA FOR DIFFERENTIATING AN INTEGRAL

$$\frac{d}{dx}\left[\int_{a(x)}^{b(x)} F(x,m)\,dm \right] = \int_{a(x)}^{b(x)} \frac{\partial F(x,m)}{\partial x}\,dm + F(x,b)\frac{db}{dx} - F(x,a)\frac{da}{dx}$$

HYPERBOLIC FUNCTIONS

Definitions:

$$\sinh u = \frac{1}{2}(e^u - e^{-u}) \qquad \cosh u = \frac{1}{2}(e^u + e^{-u})$$

$$\tanh u = \frac{\sinh u}{\cosh u} = \frac{e^u - e^{-u}}{e^u + e^{-u}}$$

$$e^u = \cosh u + \sinh u \qquad e^{-u} = \cosh u - \sinh u$$

Derivatives:

$$\frac{d}{dx}(\sinh u) = (\cosh u)\frac{du}{dx} \qquad \frac{d}{dx}(\cosh u) = (\sinh u)\frac{du}{dx}$$

$$\frac{d}{dx}(\tanh u) = \frac{1}{\cosh^2 u}\frac{du}{dx}$$

Identities:

$$\sinh(u+v) = \sinh u \cosh v + \cosh u \sinh v$$
$$\sinh(u-v) = \sinh u \cosh v - \cosh u \sinh v$$
$$\cosh(u+v) = \cosh u \cosh v + \sinh u \sinh v$$
$$\cosh(u-v) = \cosh u \cosh v - \sinh u \sinh v$$
$$\cosh^2 u - \sinh^2 u = 1$$
$$\sinh 2u = 2\sinh u \cosh u$$
$$\cosh 2u = \cosh^2 u + \sinh^2 u$$
$$\sinh iu = i\sin u \qquad \cosh iu = \cos u$$
$$\sinh u = -i\sin iu \qquad \cosh u = \cos iu$$

Inverse hyperbolic functions:

$$\sinh^{-1} u = \ln[u + (u^2+1)^{1/2}]$$
$$\cosh^{-1} u = \ln[u \pm (u^2-1)^{1/2}] \qquad (u \geq 1)$$
$$\tanh^{-1} u = \frac{1}{2}\ln\left(\frac{1+u}{1-u}\right) \qquad (u^2 < 1)$$

x	$\sinh x$	$\cosh x$	$\tanh x$
00.0	0.0000	1.0000	0.0000
0.1	0.1002	1.0050	0.0997
0.2	0.2013	1.0201	0.1974
0.3	0.3045	1.0453	0.2913
0.4	0.4108	1.0811	0.3800
0.5	0.5211	1.1276	0.4621
0.6	0.6367	1.1855	0.5370
0.7	0.7586	1.2552	0.6044
0.8	0.8881	1.3374	0.6640
0.9	1.0265	1.4331	0.7163
1.0	1.1752	1.5431	0.7616
1.1	1.3356	1.6685	0.8005
1.2	1.5095	1.8107	0.8337
1.3	1.6984	1.9709	0.8617
1.4	1.9043	2.1509	0.8854
1.5	2.1293	2.3524	0.9052
1.6	2.3756	2.5775	0.9217
1.7	2.6456	2.8283	0.9354
1.8	2.9422	3.1075	0.9468
1.9	3.2682	3.4177	0.9562

x	sinh x	cosh x	tanh x
2.0	3.6269	3.7622	0.9640
2.1	4.0219	4.1443	0.9705
2.2	4.4571	4.5679	0.9757
2.3	4.9370	5.0372	0.9801
2.4	5.4662	5.5569	0.9837
2.5	6.0502	6.1323	0.9866
2.6	6.6947	6.7690	0.9890
2.7	7.4063	7.4735	0.9910
2.8	8.1919	8.2527	0.9926
2.9	9.0596	9.1146	0.9940
3.0	10.018	10.068	0.9951
3.5	16.543	16.573	0.9982
4.0	27.290	27.308	0.9993
4.5	45.003	45.014	0.9998
5.0	74.203	74.210	0.9999
6.0	201.71	201.72	1.0000
7.0	548.32	548.32	1.0000
8.0	1490.5	1490.5	1.0000
9.0	4051.5	4051.5	1.0000
10.0	11013	11013	1.0000

BESSEL FUNCTIONS

x	$J_0(x)$	$J_1(x)$	x	$J_0(x)$	$J_1(x)$
0.0	1.0000	0.0000	1.5	0.5118	0.5579
0.1	0.9975	0.0499	1.6	0.4554	0.5699
0.2	0.9900	0.0995	1.7	0.3980	0.5778
0.3	0.9776	0.1483	1.8	0.3400	0.5815
0.4	0.9604	0.1960	1.9	0.2818	0.5812
0.5	0.9385	0.2423	2.0	0.2239	0.5767
0.6	0.9120	0.2867	2.1	0.1666	0.5683
0.7	0.8812	0.3290	2.2	0.1104	0.5560
0.8	0.8463	0.3688	2.3	0.0555	0.5399
0.9	0.8075	0.4059	2.4	0.0025	0.5202
1.0	0.7652	0.4400			
1.1	0.7196	0.4709			
1.2	0.6711	0.4983			
1.3	0.6201	0.5220			
1.4	0.5669	0.5419			

REFERENCE

1. P. R. Greene, A useful approximation to the error function: applications to mass, momentum and energy transport in shear layers, *J. Fluids Engineering*, Vol. 111, 1989, pp. 224–226.

APPENDIX F

LOCAL REYNOLDS NUMBER TRANSITION CRITERION

One of the most critical aspects of fluids engineering calculations is knowing the regime in which a specified flow is most likely to exist. This aspect is particularly important in convection heat and mass transfer calculations, as the laminar flow formulas predict transfer rates that differ significantly from the rates predicted by turbulent flow correlations.

Over the 100 years that passed since Osborne Reynolds' demonstration of the transition to turbulence in pipe flow, we have accumulated a long listing of empirical observations of transition. The most frequently used observations of this kind are summarized in the first column of Table F.1.

The laminar–turbulent transition of each flow is marked by a critical range of values of a characteristic dimensionless group of the flow. In laminar boundary layer flow over a flat plate, for example, the Reynolds number based on the distance x to the leading edge, $U_\infty x/\nu$, is as low as 2×10^4 when the free stream is highly disturbed, and as high as 10^6 and even higher when the free stream is without disturbances. A single, representative value was adopted in eq. (5.85) only for the sake of simplicity and convenience.

Continuing the reading of Table F.1, we see that the natural convection flow along a vertical wall undergoes transition at a Rayleigh number based on local altitude y and wall–fluid temperature difference of order 10^9. This particular observation refers to common fluids such as air and water, that is, to fluids with Prandtl numbers of order 1.

The transition to turbulence in a round jet that discharges freely into a reservoir is characterized by a considerably smaller "critical" number (roughly 30) when the latter is the Reynolds number based on nozzle diameter and cross section-averaged velocity.

In summary, the traditional record of transition observations consists of a collection of special "critical" numbers the order of which can be as low as 30 and as high as 10^{12}, depending on the particular flow configuration. The objective of this appendix is to draw attention to a new point of view that has crystallized during the last ten years. According to this view, the various critical numbers are manifestations of a unique transition criterion, which is suggested by the numbers assembled in the second column of Table F.1. These new

Table F.1 Traditional Critical Numbers for Transition in Several Key Flows, and the Corresponding Local Reynolds Number Scale (After Ref. [1])

Flow	Traditional Critical Number	Local Reynolds Number
Boundary layer flow over flat plate	$Re_x \sim 2 \times 10^4$–10^6	$Re_l \sim 94$–660
Natural convection boundary layer along vertical wall with uniform temperature (Pr ~ 1)	$Ra_y \sim 10^9$	$Re_l \sim 178$
Natural convection boundary layer along vertical wall with constant heat flux (Pr ~ 1)	$Ra_y^* \sim 4 \times 10^{12}$	$Re_l \sim 330$
Round jet	$Re_{nozzle} \sim 30$	$Re_l \gtrsim 30$
Wake behind long cylinder in cross-flow	$Re \sim 40$	$Re_l \gtrsim 40$
Pipe flow	$Re \sim 2000$	$Re_l \sim 500$
Film condensation on a vertical wall	$Re \sim 450$	$Re_l \sim 450$

numbers represent the so-called *local Reynolds number* of the laminar–turbulent transition region of the flow. The local Reynolds number Re_l is defined as

$$Re_l = \frac{VD}{\nu} \tag{F.1}$$

in which V is the local *longitudinal* velocity scale of the flow and D is the local *transversal* length scale.

In a relatively narrow order-of-magnitude band, we see that all the traditional critical numbers from the first column of Table F.1 correspond to local Reynolds numbers of order 10^2:

$$Re_l \sim 10^2 \quad \text{(at transition)} \tag{F.2}$$

For example, the critical local Reynolds number of laminar boundary layer flow over a flat plate is nearly the same as that of the buoyancy-driven jet along a heated vertical wall. It can be argued that the Re_l range for transition in round jet flow is actually higher than the listed nozzle Reynolds number, because the laminar jet expands rapidly outside the nozzle (i.e., the D scale of the jet is larger than the nozzle diameter). The same observation applies to the transition in the wake behind a long cylinder, where the listed Reynolds number is based on diameter of the cylinder, not on the transversal length scale of the wake.

On the high side of the transition criterion $Re_l \sim 10^2$, we note that the transition in pipe flow occurs at diameter-based Reynolds numbers of order 2000. The actual thickness of the centerline flow "fiber" that exhibits the sinuous motion is considerably smaller than the pipe diameter, therefore the *local* Reynolds number is correspondingly smaller than 2000. This is why a smaller value ($Ra_l \sim 500$) is listed in the second column of the table.

Figure F.1 reviews the local Reynolds numbers that correspond to the transitions considered in Table F.1. One remarkable aspect of this figure is that it condenses the transition observations to a relatively narrow band of values

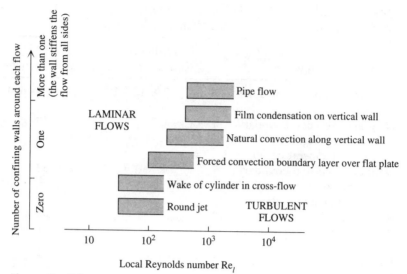

Figure Ap-F.1

centered around $Re_l \sim 10^2$. (Compare this narrow band with the $10 - 10^{12}$ range covered by the traditional critical numbers.)

The other interesting aspect of Fig. F.1 is that it unveils the flow-straightening effect that solid walls have on transition. Flows without solid walls (jets, wakes, plumes) exhibit Re_l values that are on the low side of 10^2. Flows stiffened by one solid wall have somewhat higher local Reynolds numbers at transition. The pipe flow is straightened by solid surfaces from all sides, and, consequently, its transition Re_l value is on the high side of 10^2.

The 100 years of fluid mechanics research on laminar–turbulent transition have also produced theoretical explanations of the phenomenon. The method is that of hydrodynamic stability theory [2], in which each flow of the first column of Table F.1 has been subjected analytically to a whole range of postulated "disturbances." The result of the stability analysis is an understanding of how (to what degree) a certain flow becomes unstable under the influence of a certain disturbance. The answer of whether the flow is not likely to remain laminar is noncommittal, as it depends on the characteristics of the postulated disturbance.

The empirical Re_l criterion, recommended by the second column of Table F.1, has invited a new look at the mechanism of transition. The new theoretical advances that have been made during the past ten years have been reviewed in Refs. [1, 3] and are illustrated here by means of the following argument.

Consider a two-dimensional wall jet of thickness D and longitudinal velocity V. If the jet flow is treated as inviscid, the stability analysis [3] indicates that the flow becomes unstable to sinusoidal disturbances of wavelength λ if

$$\lambda > 1.883D \quad \text{(inviscid, unstable)} \tag{F.3}$$

If the above inequality holds, the jet will wrinkle, and its elbows will grow and roll up into eddies in a time of order

$$t_{\text{eddy}} \sim \frac{\lambda}{V/2} \tag{F.4}$$

where the $V/2$ denominator represents the approximate speed of the λ wave. (As the "interface" between the jet and the motionless ambient, the wave

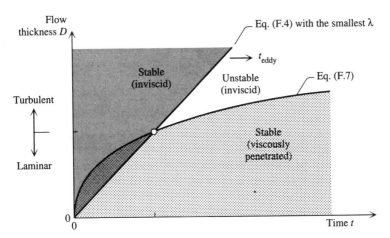

Figure Ap-F.2

acquires an intermediate speed between V and zero.) Combined with the $\lambda >$ 1.883D inequality, the roll-up time scale (t_{eddy}) occupies the wedge area situated to the right of the solid shadow in Fig. F.2. The same area would represent the domain of instability when the jet is inviscid.

This leads to the next question, which is, how long will it take for a flow to be penetrated transversally by the effect of viscous diffusion? The answer is provided by the classical solution for the flow parallel to an infinite solid wall that, starting with the time $t = 0$, moves at constant speed V parallel to itself and an originally motionless fluid. Figure F.3 shows the velocity distribution in the viscously entrained fluid [4]:

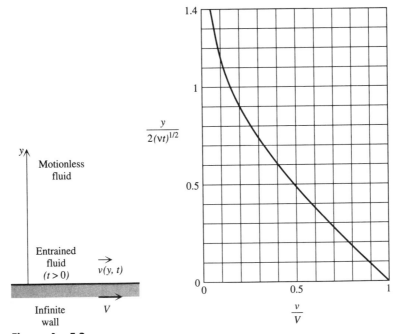

Figure Ap-F.3

$$\frac{v}{V} = \text{erfc} \left[\frac{y}{2(vt)^{1/2}} \right]$$ (F.5)

The "knee" of this curve is located where the erfc argument is of order 1; therefore, we can write

$$\frac{y}{2(vt_v)^{1/2}} \sim 1$$ (F.6)

where t_v is the time of viscous penetration across the transversal distance y. An inviscid jet of thickness D is being penetrated by viscous diffusion from both sides; therefore, we set $y \sim D/2$ in eq. (F.6), and the time of viscous penetration becomes

$$t_v \sim \frac{D^2}{16v}$$ (F.7)

This time scale serves as boundary for the dotted area shown in Fig. F.2, which represents $t > t_v$, that is, jets that have enough time to become penetrated by viscous diffusion and remain laminar. Figure F.2 shows that the domain of unstable inviscid flows is limited to a sharp-pointed area the tip of which is located at the intersection $t_{\text{eddy}} \sim t_v$, in which t_{eddy} is based on the shortest unstable wavelength $(1.833D)$. Using the above formulas, it can be shown that the intersection of the two time curves corresponds to a local Reynolds number of order 10^2, more exactly,

$$\frac{VD}{v} \sim 59$$ (F.8)

and that the laminar regime is indeed represented in an order-of-magnitude sense by $\text{Re}_l < 10^2$.

The earliest theoretical derivation of the $\text{Re}_l \sim 10^2$ criterion was reported in 1981, without any use of hydrodynamic stability arguments [5, 6]. Instead of the $\lambda > 1.833D$ inequality that started the preceding analysis, the original derivation relied on the natural meandering (buckling) wavelength $\lambda = 1.814D$. It is easy to see that in this case the $t_{\text{eddy}} \sim t_v$ time intersection leads to the same Re_l criterion for transition.

Regardless of whether the $\text{Re}_l \sim 10^2$ mark of transition is accepted empirically or theoretically, the local Reynolds number criterion is unambiguous with respect to the wavelength λ that prevails at transition. This wavelength is roughly twice the thickness of the flow, meaning, unique. In addition, the buckling theory alluded to in the preceding paragraph anticipates analytically the observed fact that the original deformation of the flow is a *sinusoid*. These attributes distinguish fundamentally the Re_l criterion from the classical results of hydrodynamic stability analyses in which the shape of the initial deformation of the flow (the "disturbance") is arbitrary, and it must be assumed.

The local Reynolds number criterion and the theory behind it have several important consequences that can only be enumerated in these concluding paragraphs. One consequence is that the Reynolds number of the smallest eddy in the ensuing flow field is of order 10^2. This discovery contradicts the established view in modern fluid mechanics that holds that the smallest eddy Reynolds number is of order 1 (see, for example, Tennekes and Lumley [7] and Bradshaw [8]).

To appreciate the truth in this last observation, examine once more the flow past the cylinder photographed in Fig. 5.19. Had an eddy been shed from that cylinder, the Reynolds number of that eddy would have been of order 1. Of course, the photograph, and many like it, show that no eddies are present. The first eddies are shed only when the Reynolds number based on cylinder diameter (i.e., the eddy Reynolds number) is of order 10^2 or higher.

While on the topic of the cylinder in cross-flow, it is worth noting that the λ and t_{eddy} scales identified above can be used to predict the vortex shedding frequency [5, 6]

$$f_v \sim \frac{1}{t_{\text{eddy}}} \sim 0.27 \frac{U_\infty}{D} \tag{F.9}$$

This agrees very well (in an order-of-magnitude sense, certainly) with the empirical correlation (5.137). Note that in eq. (F.4) we used the free-stream velocity U_∞ as longitudinal velocity scale V.

A similar back-of-the-envelope argument has been used to predict the pulsating frequency of the flames of large pool fires [9],

$$f_v \sim \left(\frac{2.3 \text{ m/s}^2}{D}\right)^{1/2} \tag{F.10}$$

where D is the length scale (width, diameter) of the pool (base) shape. Until 1991 the correlation (F.10) was one of the unanswered questions in fluid mechanics [10], having been discovered and rediscovered empirically based on fire observations involving a wide variety of fuels and the fire pool shapes. In addition to deriving eq. (F.10) analytically, the scale analysis [9] showed that the fire will pulsate only when its base D is greater than approximately 2 cm.

Another consequence of the Re_l criterion is that in turbulent flow near a wall it predicts the existence of a viscous sublayer. Furthermore, the thickness of this sublayer must be of order 10 in the classical y^+ coordinate [3]. This y^+ thickness can be verified by intersecting the two u^+ expressions listed on the right side of eq. (5.116).

The same theoretical starting point has been used to derive for the first time analytically Colburn's analogy between heat transfer and friction in turbulent flow near a wall,

$$St_x Pr^{2/3} = \tfrac{1}{2} C_{f,x} \tag{F.11}$$

where St_x and $C_{f,x}$ are the local Stanton number and skin friction coefficient, respectively. An integral part of that analysis [3] is the proof that Colburn's analogy is valid only for fluids with Prandtl numbers of order 1 or greater.

The length and time scales of the unstable flow and the process of eddy formation can be used to anticipate the growth (mixing) of turbulent jets, shear layers, and plumes. The time-averaged version of these mixing regions turns out to be one that increases linearly in the downstream direction. The kinematic analysis based on the eddy length and time scales, however, reveals also the instantaneous structure of the mixing region, that is, the main eddy sizes and their distribution over the mixing region (see Chapter 8 of Ref. [3]).

Finally, the predicted sinusoidal shape and the proportionality between wavelength and flow thickness during the first stages of inviscid-stream deformation provide a theoretical explanation for the many "meandering" flows that surround us. Examples of such flows are the river meanders, the waving of flags, the fall of paper ribbons, the deformation of fast liquid jets shot through the air, and the sinuous structure of all turbulent plumes. The same scales are the basis for the geometric similarity that exists between the laminar sections of straight boundary layer-type flows. This geometric similarity was noted recently [11]. The local Reynolds number criterion was used by Shenoy [12] to predict the transition in vertical boundary layer natural convection in non-newtonian fluids.

REFERENCES

1. A. Bejan, Buckling flows: a new frontier in fluid mechanics, *Annual Review of Numerical Fluid Mechanics and Heat Transfer*, Vol. 1, T. C. Chawla, Ed., Hemisphere, Washington, D.C., 1987, pp. 262–304.

2. P. G. Drazin and W. H. Reid, *Hydrodynamic Stability*, Cambridge University Press, Cambridge, England, 1981.

3. A. Bejan, *Convection Heat Transfer*, Wiley, New York, 1984, pp. 214–219.

4. H. Schlichting, *Boundary Layer Theory*, translated by J. Kestin, 4th edition, McGraw-Hill, New York, 1960, pp. 72–73.

5. A. Bejan, On the buckling property of inviscid jets and the origin of turbulence, *Lett. Heat Mass Transfer*, Vol. 8, 1981, pp. 187–194.

6. A. Bejan, *Entropy Generation Through Heat and Fluid Flow*, Wiley, 1982, pp. 64–97.

7. H. Tennekes and J. L. Lumley, *A First Course in Turbulence*, MIT Press, Cambridge, MA 1972, p. 20.

8. P. Bradshaw, Ed., *Turbulence*, 2nd edition, *Topics in Applied Physics*, Vol. 12, Springer-Verlag, Berlin, 1978, p. 22.

9. A. Bejan, Predicting the pool fire vortex shedding frequency, *J. Heat Transfer*, Vol. 113, 1991, pp. 261–263.

10. P. J. Pagni, Pool vortex shedding frequencies, in L. M. Trefethen and R. L. Panton, Some unanswered questions in fluid mechanics, ASME Paper No. 89-WA/FE-5.

11. R. A. Gore, C. T. Crowe, and A. Bejan, The geometric similarity of the laminar sections of boundary layer-type flows, *Int. Comm. Heat Mass Transfer*, Vol. 17, 1990, pp. 465–475.

12. A. V. Shenoy, Criterion for transition to turbulence during natural convection heat transfer from a flat vertical plate to a power-law fluid, *Int. Comm. Heat Mass Transfer*, Vol. 18, 1991, pp. 385–396.

AUTHOR INDEX

665

SUBJECT INDEX